Metric Spaces

$\mathbb{E}^n$	Euclidean n-space 50
$\|x - y\|$	Euclidean distance 42
$\|x\|$	Euclidean norm 49
$\mathbb{B}^n$	open unit ball in $\mathbb{E}^n$ 51
$\mathbb{D}^n$	closed unit ball in $\mathbb{E}^n$ 51
$\mathbb{S}^n$	unit n-sphere 51, 57
D1-D3	metric axioms 49
(X, d)	metric space 55
$B(x, r)$	open ball in (X, d) 55
$D(x, r)$	closed ball (X, d) 55
$diam(A)$	diameter of subset 69
$\mathcal{C}([0, 1])$	space of cont. fns. 55
$d_\infty(f, g)$	uniform metric 53
δ	discrete metric 53
h	Hausdorff metric 97
$(\mathfrak{H}(X), h)$	hyperspace 97
$\lambda(A)$	Lebesgue measure in $\mathbb{E}^n$ 83
$\mu^p(A)$	Hausdorff measure 84
$h\text{-}dim(A)$	Hausdorff dimension 84

Topology

T1-T3	topology axioms 116
$(X, \mathcal{T})$	topological space 116
C1-C3	closed set properties 118
$\text{INT}(A) = \mathring{A}$	set interior 61, 127
$\text{CLS}(A) = \overline{A}$	set closure 61, 127
$\text{BDY}(A)$	set boundary 61, 127
X/R	quotient modulo R 150
$q_R : X \to X/R$	quotient map 149
X/A	quotient modulo $A \subseteq X$ 150
$q_A : X \to X/A$	quotient map 150
$\mathcal{S}_X$	subspace topology 118
$\mathcal{T} \times \mathcal{R}$	product topology 118
$\dot{\bigcup}_\alpha X_\alpha$	topological disj. union 156
$\prod_\alpha X_\alpha$	topological product 159
$\mathbb{R}^n$	real n-space 49
$\mathbb{R}_{(]}$	right interval space 153
$\mathbb{P}^2$	projective plane 169, 363
$\mathbb{P}^n$	projective n-space 426
$\mathbb{P}^\infty$	infinite projective space 465
P_n	pseudo-projective plane 431
$T_0\text{-}T_5$	separation properties 123
$\mathcal{P}(X)$	discrete topology 117
$\mathcal{T}(p)$	particular pt. topology 118
$\mathcal{D}(p)$	excluded pt. topology 118
$\mathcal{CC}$	cocomplement topology 117
$\mathcal{FC}$	cofinite topology 117
X_∞	one-point compactification 216
$\beta(X)$	Stone-Cech compactification 245
CX	cone over X 309
C_f, W_H	mapping cone, wedge 327
$p : \mathbb{S}^3 \to \mathbb{S}^2$	Hopf's map 313

An Introduction to Topology and Homotopy

THE PRINDLE, WEBER & SCHMIDT SERIES IN MATHEMATICS

Althoen and Bumcrot, *Introduction to Discrete Mathematics*
Boye, Kavanaugh, and Williams, *Elementary Algebra*
Boye, Kavanaugh, and Williams, *Intermediate Algebra*
Burden and Faires, *Numerical Analysis, Fourth Edition*
Cass and O'Connor, *Fundamentals with Elements of Algebra*
Cullen, *Linear Algebra and Differential Equations, Second Edition*
Dick and Patton, *Calculus, Volume I*
Dick and Patton, *Calculus, Volume II*
Dick and Patton, *Technology in Calculus: A Sourcebook of Activities*
Eves, *In Mathematical Circles*
Eves, *Mathematical Circles Adieu*
Eves, *Mathematical Circles Squared*
Eves, *Return to Mathematical Circles*
Fletcher, Hoyle, and Patty, *Foundations of Discrete Mathematics*
Fletcher and Patty, *Foundations of Higher Mathematics, Second Edition*
Gantner and Gantner, *Trigonometry*
Geltner and Peterson, *Geometry for College Students, Second Edition*
Gilbert and Gilbert, *Elements of Modern Algebra, Third Edition*
Gobran, *Beginning Algebra, Fifth Edition*
Gobran, *Intermediate Algebra, Fourth Edition*
Gordon, *Calculus and the Computer*
Hall, *Algebra for College Students*
Hall, *Beginning Algebra*
Hall, *College Algebra with Applications, Third Edition*
Hall, *Intermediate Algebra*
Hartfiel and Hobbs, *Elementary Linear Algebra*
Humi and Miller, *Boundary-Value Problems and Partial Differential Equations*
Kaufmann, *Algebra for College Students, Fourth Edition*
Kaufmann, *Algebra with Trigonometry for College Students, Third Edition*
Kaufmann, *College Algebra, Second Edition*
Kaufmann, *College Algebra and Trigonometry, Second Edition*
Kaufmann, *Elementary Algebra for College Students, Fourth Edition*
Kaufmann, *Intermediate Algebra for College Students, Fourth Edition*
Kaufmann, *Precalculus, Second Edition*
Kaufmann, *Trigonometry*
Kennedy and Green, *Prealgebra for College Students*
Laufer, *Discrete Mathematics and Applied Modern Algebra*
Nicholson, *Elementary Linear Algebra with Applications, Second Edition*
Pence, *Calculus Activities for Graphic Calculators*
Pence, *Calculus Activities for the TI-81 Graphic Calculator*
Plybon, *An Introduction to Applied Numerical Analysis*
Powers, *Elementary Differential Equations*
Powers, *Elementary Differential Equations with Boundary-Value Problems*
Proga, *Arithmetic and Algebra, Third Edition*
Proga, *Basic Mathematics, Third Edition*

Rice and Strange, *Plane Trigonometry, Sixth Edition*
Schelin and Bange, *Mathematical Analysis for Business and Economics, Second Edition*
Strnad, *Introductory Algebra*
Swokowski, *Algebra and Trigonometry with Analytic Geometry, Seventh Edition*
Swokowski, *Calculus, Fifth Edition*
Swokowski, *Calculus, Fifth Edition (Late Trigonometry Version)*
Swokowski, *Calculus of a Single Variable*
Swokowski, *Fundamentals of College Algebra, Seventh Edition*
Swokowski, *Fundamentals of Algebra and Trigonometry, Seventh Edition*
Swokowski, *Fundamentals of Trigonometry, Seventh Edition*
Swokowski, *Precalculus: Functions and Graphs, Sixth Edition*
Tan, *Applied Calculus, Second Edition*
Tan, *Applied Finite Mathematics, Third Edition*
Tan, *Calculus for the Managerial, Life, and Social Sciences, Second Edition*
Tan, *College Mathematics, Second Edition*
Trim, *Applied Partial Differential Equations*
Venit and Bishop, *Elementary Linear Algebra, Third Edition*
Venit and Bishop, *Elementary Linear Algebra, Alternate Second Edition*
Wiggins, *Problem Solver for Finite Mathematics and Calculus*
Willard, *Calculus and Its Applications, Second Edition*
Wood and Capell, *Arithmetic*
Wood and Capell, *Intermediate Algebra*
Wood, Capell, and Hall, *Developmental Mathematics, Fourth Edition*
Zill, *A First Course in Differential Equations with Applications, Fourth Edition*
Zill and Cullen, *Advanced Engineering Mathematics*
Zill, *Calculus, Third Edition*
Zill, *Differential Equations with Boundary-Value Problems, Second Edition*

THE PRINDLE, WEBER & SCHMIDT SERIES IN ADVANCED MATHEMATICS

Brabenec, *Introduction to Real Analysis*
Ehrlich, *Fundamental Concepts of Abstract Algebra*
Eves, *Foundations and Fundamental Concepts of Mathematics, Third Edition*
Keisler, *Elementary Calculus: An Infinitesimal Approach, Second Edition*
Kirkwood, *An Introduction to Real Analysis*
Ruckle, *Modern Analysis: Measure Theory and Functional Analysis with Applications*
Sieradski, *An Introduction to Topology and Homotopy*

To Karen, Matthew, and Damian

An Introduction to Topology and Homotopy

ALLAN J. SIERADSKI
UNIVERSITY OF OREGON

PWS-KENT Publishing Company
Boston

PWS-KENT
Publishing Company

20 Park Plaza
Boston, Massachusetts 02116

Copyright © 1992 by PWS-KENT Publishing Company.

All rights reserved. No part of this book may be reproduced, stored in a retrieval system, or transcribed, in any form or by any means—electronic, mechanical, photocopying, recording, or otherwise—without the prior written permission of PWS-KENT Publishing Company.

PWS-KENT Publishing Company is a division of Wadsworth, Inc.

Printed in the United States of America.

Library of Congress Cataloging-in-Publication Data
Sieradski, Allan J.
   An introduction to topology and homotopy/Allan J. Sieradski.
     p.   cm.
   Includes bibliographical references (p.   ) and index.
   ISBN 0-534-92960-5
   1. Topology. 2. Homotopy theory. I. Title.
QA611.S48 1992   91-32672
514--dc20   CIP

Sponsoring Editor Steve Quigley
Production Editor Helen Walden
Manufacturing Coordinator Lisa Flanagan
Text and Art Preparation Allan J. Sieradski
Cover Design Designworks
Printing and Binding Maple-Vail Book Mfr. Group

 This text is printed on recycled, acid-free paper

92 93 94 95 96 — 10 9 8 7 6 5 4 3 2 1

PREFACE

In this text, my aim is to provide an introduction to topology and homotopy for students at the senior and first-year graduate level. It is intended to be an enjoyable excursion into these topics for these students, rather than an encyclopedic treatment suitable as a reference manual for experts. Most topics are developed slowly in their historic manner, in order that a newcomer not be overwhelmed by the ultimate achievements of several generations of mathematicians. A careful reader will see how topologists have gradually refined and extended the work of their predecessors and how most good ideas reach beyond what their originators envisioned.

The first half of the text treats the topology of complete metric spaces, including their hyperspaces of sequentially compact subspaces (an arena for spaces of fractional Hausdorff dimension). It also presents general topology through the characterizations of normality due to Urysohn and Tietze and the metrization theorems of Urysohn, Nagata, and Smirnov. The second half of the text develops the homotopy category and the fundamental group functor. It emphasizes CW complexes and the portion of their homotopy theory detectable by the fundamental group functor. It examines the role of the fundamental group in the classification of surfaces, in knot theory, in the Co-Galois classification of covering spaces, and in the qualitative theory of foliations and vector fields on surfaces.

The text presumes the mathematical maturity required in a rigorous upper division mathematics course. Most students will have had such an experience with an analysis course, but this text assumes no results from real or complex analysis except the existence of the real numbers. To encourage the development of topological intuition, the text is amply illustrated. Examples, too numerous to be completely covered in two semesters of lectures, make the text suitable for independent study and allow an instructor the freedom to select topics to be emphasized in class.

I use this material in a year-long undergraduate/graduate course. It provides a strong foundation for a subsequent algebraic topology course devoted to the higher homotopy groups, homology, and cohomology. For a one-semester course of general topology, one can use Chapters **1**-**8**.

For a one-semester course for students familiar with metric topology, one can begin with Section 3.4 on the hyperspace of a complete metric space and Section 3.5 on Peano's curve as a refresher. Then proceed to Chapter **4** (Topological Spaces), Chapter **6** (Connectedness), Chapter **7** (Compactness), Chapter **9** (Fundamental Group), Chapter **13** (Surfaces), and possibly Chapter **14** (Covering Spaces). Chapter **13** uses fundamental group calculations developed in Chapters **11** and **12**.

A word about notation. Exercises, examples, and results (e.g., theorems, propositions, lemmas) are items that are numbered independently. For example, Exercise **3**.2.1 is the first exercise in Section 2 of Chapter **3**. When the chapter or section number is missing, the reference is to the numbered item in the current chapter or section. Section **3**.2 is Section 2 of Chapter **3**. To minimize citation chases, the key results are referred to by name as well as number, and proofs of major results are located as near as possible to the supporting material. The symbol □ is used to indicate the end of a proof. Exercises to which the text refers the reader for further details are starred.

I would like to acknowledge the help, encouragement, and contributions of the following people. Over the past several years, my colleague Richard Koch has generously shared his time and computer expertise, patiently answering my questions and solving my technical problems. Chris Lane made a critical reading of the penultimate version of this text and provided numerous improvements in its presentation and accuracy. Users of early versions of this material, including students in my recent topology classes and my colleague Micheal Dyer, encouraged me to continue its development and significantly influenced its final form. At PWS-KENT Publishing Company, Stephen Quigley, senior mathematics editor, guided this text to market and Helen Walden edited it to meet their standards. The software designers at Adobe Systems Inc. created the graphics design package Illustrator® that I used to produce the figures in this text, and those at Altsys Corporation created the font editor Fontographer® that I used for specialized symbols and characters. Finally, my wife Karen and sons Matthew and Damian accepted deferred projects and cancelled outings to support the completion of this text. I thank all of them.

Allan J. Sieradski
Eugene, Oregon

CONTENTS

PART ONE: TOPOLOGY 1

Chapter 1 **Preliminaries** 3
    1 The Topological Viewpoint 3
    2 Set Theory 8
    3 The Real Line 16
    4 Functions 21
    5 Relations 28
       Exercises, Sections 1-5 36

Chapter 2 **Metric Topology** 42
    1 Topology of the Euclidean Line 42
    2 Topology of Euclidean Space 49
    3 Metric Topology 54
    4 Continuous Functions 63
    5 Completeness 68
       Exercises, Sections 1-5 75

Chapter 3 **Metric Space Properties** 85
    1 Connected Spaces 85
    2 Sequentially Compact Metric Spaces 89
    3 Hyperspace of Compact Subspaces 96
    4 Topological Equivalence 101
    5 Peano's Curve 105
       Exercises, Sections 1-5 109

Chapter 4 **Topological Spaces** 116
    1 Topologies 116
    2 Topological Separation 122
    3 Interior, Closure, and Boundary 127
    4 Continuous Functions 131
       Exercises, Sections 1-4 136

Chapter 5	**Construction of Spaces**		**143**
	1 Subspaces and Identification Spaces	143	
	2 Topological Bases	151	
	3 Topological Unions and Products	156	
	4 Weak Topologies and Adjunction Spaces	163	
	Exercises, Sections 1-4	168	
Chapter 6	**Connectedness**		**175**
	1 Connected Spaces	175	
	2 Components	180	
	3 Path-Connectedness	186	
	Exercises, Sections 1-3	192	
Chapter 7	**Compactness**		**197**
	1 Compact Spaces	197	
	2 Topological Aspects of Compactness	201	
	3 Compactness in Metric Spaces	207	
	4 Locally Compact Spaces	211	
	5 A Characterization of Arcs	218	
	Exercises, Sections 1-5	226	
Chapter 8	**Separation and Covering Properties**		**233**
	1 Normal Spaces	233	
	2 Completely Regular Spaces	241	
	3 Metrization Theorems	247	
	4 Paracompact Spaces	254	
	Exercises, Sections 1-5	259	

PART TWO: HOMOTOPY 263

Chapter 9	**Fundamental Group**		**265**
	1 The Fundamental Groupoid	265	
	2 Path-Homotopy in Spheres	272	
	3 The Fundamental Group	280	
	4 Poincaré Index Formula	285	
	Exercises, Sections 1-4	292	
Chapter 10	**Homotopy**		**296**
	1 Categories and Functors	296	
	2 Basic Homotopy Concepts	301	
	3 Homotopy Category	305	
	4 Fibrations and Cofibrations	312	
	5 Homotopy Equivalences	317	
	Exercises, Sections 1-5	323	

Chapter 11	**Group Theory**		**328**
	1 Overview	328	
	2 Groups and Homomorphisms	331	
	3 Products and Sums of Groups	335	
	4 Quotient Groups	340	
	Exercises, Sections 2-4	347	
Chapter 12	**Calculations of π_1**		**350**
	1 The Seifert-Van Kampen Theorem	350	
	2 Polygonal Complexes	357	
	3 Knot Theory	364	
	Exercises, Sections 1-3	372	
Chapter 13	**Surfaces**		**374**
	1 Polygonal Surfaces	374	
	2 Classification of Surfaces	379	
	3 Euler Characteristic	385	
	4 Poincaré's Index Theorem	390	
	Exercises, Sections 1-4	394	
Chapter 14	**Covering Spaces**		**399**
	1 Fundamentals	399	
	2 Liftings of Paths and Homotopies	403	
	3 General Liftings	409	
	4 Fundamental Theorem	413	
	5 Intermediate Covering Spaces	418	
	6 Group Actions and Geometries	425	
	Exercises, Sections 1-6	430	
Chapter 15	**CW Complexes**		**435**
	1 Cell Complexes	435	
	2 Fundamental Group of a CW Complex	444	
	3 Covering Spaces of CW Complexes	454	
	Exercises, Sections 1-3	462	
Bibliography			**466**
Index			**468**

PART ONE : TOPOLOGY

PRELIMINARIES

1 The Topological Viewpoint

Geometry uses the notions of distance, length, area, and volume to express certain properties of objects in space. There are, however, features of objects that are not based upon rigid measurements.

For example, when a spherical balloon is inflated, it may lose the geometric property that all its points are equidistant from a central point in space. But all its deformed versions, whether ellipsoid-like or whatever, retain the feature that they divide space into a bounded region and an unbounded region, and that it is impossible to travel within space from one region to the other region without passing through the deformed sphere. In addition, if the balloon is punctured by the removal of a single point, it ceases to divide space into two regions.

Stretching and shrinking objects are examples of *continuous functions*. Puncturing and tearing objects are examples of *discontinuous functions*. These distinctions can be made only when the objects under consideration are sets of points in which it is known which subsets are near which points. The fundamental requirement for continuity is that the function respect all the nearness relationships between subsets and points of the object.

The features of the object that are preserved by continuous functions with continuous inverses are called *topological properties*. There are enough such features to warrant their own study in the discipline of topology.

TOPOLOGICAL CONCEPTS

Here is a rough version of the basic concepts of topology.

A *topological space* X is a set, each of whose points x is enclosed in a family of subsets of X, called its *neighborhoods*.

A subset A of a topological space X is *arbitrarily near* a point x of X if every neighborhood of x contains at least one member of the subset A.

A *topological* or *continuous* function $f : X \rightarrow Y$ between two topological spaces is required to respect the nearness relationships in this way: whenever a subset A of X is arbitrarily near a point x of X, the image set $f(A)$ in Y must be arbitrarily near the image point $f(x)$ in Y.

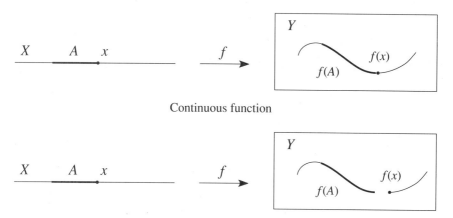

Continuous function

Discontinuous function

Two spaces X and Y are *topologically equivalent* or *homeomorphic* if there exist continuous inverse functions $X \leftrightarrow Y$, called *homeomorphisms*.

A TOPOLOGICAL SCRAPBOOK

The following illustrations of the topological viewpoint are offered to whet your appetite for this study. We will eventually return to them with further explanation, once we master the appropriate terminology and tools.

The nearness relationships in the circular, elliptical, and square regions in the plane are preserved when rubber versions of them are stretched and shrunk, and they can be deformed one into the other in this manner. Thus, the *circular, elliptical, and square discs*, below, are homeomorphic spaces.

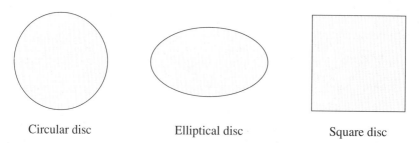

Circular disc Elliptical disc Square disc

A *sphere with two holes*, a *cylinder*, an *annulus*, and a *disc with one hole* are homeomorphic. A sphere with two holes is just an inflated version of a cylinder, which flattens into an annulus (a disc with one hole).

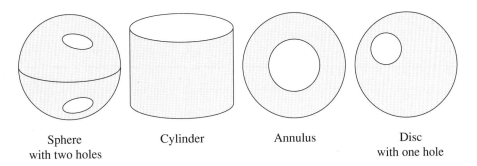

| Sphere with two holes | Cylinder | Annulus | Disc with one hole |

There is no homeomorphism of the *sphere* (balloon surface) with the *torus* (inner tube surface). They are topologically inequivalent spaces.

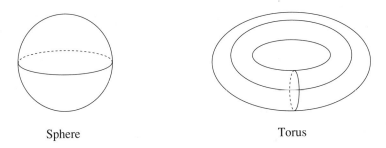

Sphere   Torus

Cutting alters the nearness relationships in a topological space, and converts the space into an inequivalent one. If the cut edges are glued together in a manner that reproduces the nearness relationships, the original topological space is recovered, although it may be *embedded* (placed) differently in the larger arena where the cutting and gluing took place.

For example, a topological *circle* (not a disc) can be cut into a topologically inequivalent curved arc (with one endpoint), and then glued together in space in the form of a *trefoil knot*:

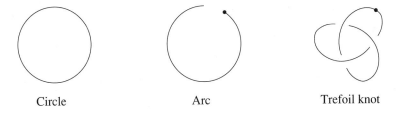

Circle   Arc   Trefoil knot

The topological circle and the trefoil knot are homeomorphic topological spaces because the correspondence of their points preserves the nearness relationships. However, there is no self-homeomorphism of Euclidean space that carries the unknotted circle to the trefoil knot. So the trefoil knot is a different *embedding* (placement) of the circle into Euclidean space.

The cylinder and the singly-twisted *Möbius strip* are not homeomorphic.

The border of the cylinder consists of two disjoint circles, while the border of the Möbius strip is a single circle. Trace them for yourself.

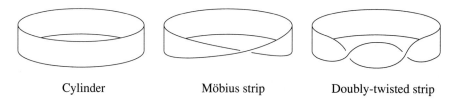

Cylinder   Möbius strip   Doubly-twisted strip

But the cylinder can be cut-and-glued into the topologically equivalent *doubly-twisted strip*, above right. The doubly-twisted strip is simply a different embedding of the cylinder into Euclidean space.

The *pin-cushion space*, below left, is obtained by pinching the north and south pole of a spherical balloon surface into a single point. The *sausage-link space*, below right, is obtained by pinching a circular wire band on a torus surface into a single point. They appear to be quite different:

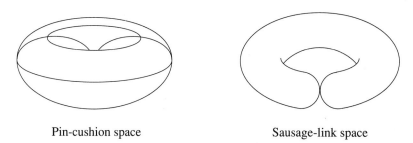

Pin-cushion space   Sausage-link space

But they are homeomorphic topological spaces as there is a correspondence of their points that preserves all nearness relationships. To visualize this correspondence, cut the pin-cushion space apart to recover a spherical balloon. When the poles are identified along a path outside the balloon, rather than along a diameter of the sphere, the result is the sausage-link space. There is no homeomorphism of Euclidean space that molds the pin-cushion space into the sausage-link space; the sausage-link space is a different embedding of the pin-cushion space into Euclidean space.

The wire handcuff and the wire eight, depicted below, are topologically inequivalent because certain points have exceptional neighborhoods.

Wire handcuff   Wire eight

There are two points of the wire handcuff that have *T*-shaped neighborhoods; in the wire eight, no point has a *T*-shaped neighborhood and one point has some *X*-shaped neighborhoods. Curiously, the thickened versions of the wire handcuff and wire eight are homeomorphic because each one can be continuously molded into a solid ball with two handles:

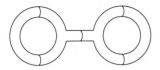 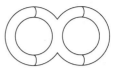

Thickened handcuff   Thickened eight

A *punctured torus*, below left, can be manipulated in space to become a disc with a pair of untwisted, but overlapping, bands glued to its edge:

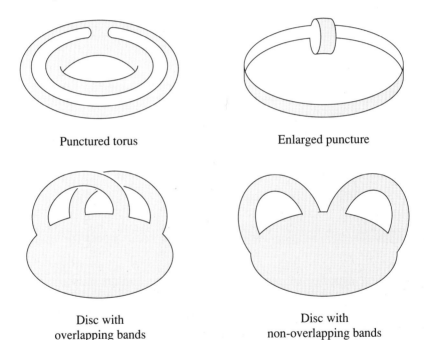

Punctured torus   Enlarged puncture

Disc with   Disc with
overlapping bands   non-overlapping bands

The disc with two untwisted, non-overlapping bands glued to its edge, above right, is homeomorphic to a sphere with three holes. They are topologically distinct from the punctured torus whose border is a single circle.

Thickened versions of the punctured torus surface and the sphere with three holes are homeomorphic because each one can be continuously molded into a solid ball with two handles, like the thickenings of the wire handcuff and the wire eight. This makes the topologically inequivalent punctured torus, triply punctured sphere, wire handcuff, and wire eight equivalent in a weaker sense; they are *homotopy equivalent*.

2 Set Theory

Topological spaces are sets with extra structure. Working with topological spaces requires a basic facility with the underlying sets. This section presents the minimum prerequisite of set theory for this text.

Elementary set theory results from careful application of the logic of propositions. So we begin with some logical conventions, as well as the specialized notation used throughout the text.

LOGIC

The building blocks required for logical demonstrations are **propositions**, by which we mean complete sentences that are either true or false, but not both. By use of the modifier *not* and the connectives *and* and *or*, simple propositions can be combined to form compound propositions, as follows.

Let $\mathcal{P}$ and $\mathcal{R}$ be any propositions. The proposition "*not $\mathcal{P}$*," called the **negation** of $\mathcal{P}$, is denoted by $\sim \mathcal{P}$. The proposition "$\mathcal{P}$ *and* $\mathcal{R}$" is denoted by $\mathcal{P} \wedge \mathcal{R}$ and is called the **conjunction** of $\mathcal{P}$ and $\mathcal{R}$. The proposition "$\mathcal{P}$ *or* $\mathcal{R}$" is denoted by $\mathcal{P} \vee \mathcal{R}$ and is called the **disjunction** of $\mathcal{P}$ and $\mathcal{R}$.

To avoid ambiguity, we give these constructions the constant interpretation provided by the following tables that express their truth values as functions of the truth values of $\mathcal{P}$ and $\mathcal{R}$:

$\mathcal{P}$	$\sim\mathcal{P}$
T	F
F	T

$\mathcal{P}$	$\mathcal{R}$	$\mathcal{P} \wedge \mathcal{R}$
T	T	T
T	F	F
F	T	F
F	F	F

$\mathcal{P}$	$\mathcal{R}$	$\mathcal{P} \vee \mathcal{R}$
T	T	T
T	F	T
F	T	T
F	F	F

Any new proposition formed by successive applications of these constructions is called a **compound proposition**, whose **constituents** are the original propositions. The truth value of any composite proposition depends upon only the truth value of its constituent propositions.

Different compound propositions can have the same meaning; i.e., their truth values can coincide for all possible truth values of their constituents.

Example 1 "*It is false that both $\mathcal{P}$ and $\mathcal{R}$ are true*" means the same as "*either $\mathcal{P}$ is false or $\mathcal{R}$ is false*." For $\sim(\mathcal{P} \wedge \mathcal{R})$ and $(\sim\mathcal{P}) \vee (\sim\mathcal{R})$ have the same truth values:

$\mathcal{P}$	$\mathcal{R}$	$\mathcal{P} \wedge \mathcal{R}$	$\sim(\mathcal{P} \wedge \mathcal{R})$	$\sim\mathcal{P}$	$\sim\mathcal{R}$	$(\sim\mathcal{P}) \vee (\sim\mathcal{R})$
T	T	T	F	F	F	F
T	F	F	T	F	T	T
F	T	F	T	T	F	T
F	F	F	T	T	T	T

There is this terminology for propositions with the same meaning: two propositions $\mathcal{P}$ and $\mathcal{R}$ are called **logically equivalent**, denoted by $\mathcal{P} \Leftrightarrow \mathcal{R}$, if they have the same truth value for all truth values of their constituents.

The truth table in Example 1 establishes a logical equivalence which is one of De Morgan's laws: $\sim (\mathcal{P} \wedge \mathcal{R}) \Leftrightarrow (\sim \mathcal{P}) \vee (\sim \mathcal{R})$.

Similarly, each of the following logical equivalences is easily established by a direct construction of an appropriate truth table (with 2, 4, or 8 rows).

2.1 Laws of Logic *Conjunctions, disjunctions, and negations of propositions satisfy the following properties.*

(a) **Idempotent laws:**
$\mathcal{P} \wedge \mathcal{P} \Leftrightarrow \mathcal{P}$
$\mathcal{P} \vee \mathcal{P} \Leftrightarrow \mathcal{P}$

(b) **Associative laws:**
$\mathcal{P} \wedge (\mathcal{R} \wedge \mathcal{S}) \Leftrightarrow (\mathcal{P} \wedge \mathcal{R}) \wedge \mathcal{S}$
$\mathcal{P} \vee (\mathcal{R} \vee \mathcal{S}) \Leftrightarrow (\mathcal{P} \vee \mathcal{R}) \vee \mathcal{S}$

(c) **Commutative laws:**
$\mathcal{P} \wedge \mathcal{R} \Leftrightarrow \mathcal{R} \wedge \mathcal{P}$
$\mathcal{P} \vee \mathcal{R} \Leftrightarrow \mathcal{R} \vee \mathcal{P}$

(d) **Distributive laws:**
$\mathcal{P} \wedge (\mathcal{R} \vee \mathcal{S}) \Leftrightarrow (\mathcal{P} \wedge \mathcal{R}) \vee (\mathcal{P} \wedge \mathcal{S})$
$\mathcal{P} \vee (\mathcal{R} \wedge \mathcal{S}) \Leftrightarrow (\mathcal{P} \vee \mathcal{R}) \wedge (\mathcal{P} \vee \mathcal{S})$

(e) **Identity laws:**
$\mathcal{P} \wedge T \Leftrightarrow \mathcal{P}$ $\mathcal{P} \vee F \Leftrightarrow \mathcal{P}$
$\mathcal{P} \wedge F \Leftrightarrow F$ $\mathcal{P} \vee T \Leftrightarrow T$

(f) **De Morgan's laws:**
$\sim (\mathcal{P} \wedge \mathcal{R}) \Leftrightarrow (\sim \mathcal{P}) \vee (\sim \mathcal{R})$
$\sim (\mathcal{P} \vee \mathcal{R}) \Leftrightarrow (\sim \mathcal{P}) \wedge (\sim \mathcal{R})$

(g) **Negation laws:**
$\mathcal{P} \wedge (\sim \mathcal{P}) \Leftrightarrow F$ $\mathcal{P} \vee (\sim \mathcal{P}) \Leftrightarrow T$ $\sim (\sim \mathcal{P}) \Leftrightarrow \mathcal{P}$

A weaker relation than equivalence is implication. A proposition $\mathcal{P}$ **logically implies** a proposition $\mathcal{R}$, written $\mathcal{P} \Rightarrow \mathcal{R}$, provided that the **conclusion** $\mathcal{R}$ is true if the **hypothesis** $\mathcal{P}$ is true. The implication $\mathcal{P} \Rightarrow \mathcal{R}$ is often stated "$\mathcal{P}$ is a **sufficient condition** for $\mathcal{R}$" or "$\mathcal{R}$ is a **necessary condition** for $\mathcal{P}$."

Logical arguments that deduce valid conclusions from valid hypotheses are based on the standard rules of inference below. They are easily verified.

2.2 Rules of Inference

(a) **Tautology rule:** $\mathcal{P} \Rightarrow \mathcal{P}$.

(b) **Equivalence rule:** $\mathcal{P} \Leftrightarrow \mathcal{R}$ if and only if $\mathcal{P} \Rightarrow \mathcal{R}$ and $\mathcal{R} \Rightarrow \mathcal{P}$.

(c) **Detachment rule:** If $\mathcal{P}$ is true and $\mathcal{P} \Rightarrow \mathcal{R}$, then $\mathcal{R}$ is true.

(d) **Syllogism rule:** If $\mathcal{P} \Rightarrow \mathcal{R}$ and $\mathcal{R} \Rightarrow \mathcal{S}$, then $\mathcal{P} \Rightarrow \mathcal{S}$.

(e) **Contrapositive rule:** $\mathcal{P} \Rightarrow \mathcal{R}$ if and only if $\sim \mathcal{R} \Rightarrow \sim \mathcal{P}$.

LOGICAL QUANTIFIERS

Let $\mathcal{P}(x)$ denote an expression that involves a variable x and that becomes a proposition upon substitution of suitable objects for the variable x.

The proposition $\mathcal{P}(x)$ can be qualified by use of either the **universal quantifier** $\forall$ (*for all*) or the **existential quantifier** $\exists$ (*there exists*), as follows.

Let $(\exists x, \mathcal{P}(x))$ mean that "*there exists an object x such that $\mathcal{P}(x)$ is true.*"
Let $(\forall x, \mathcal{P}(x))$ mean that "*for all objects x, $\mathcal{P}(x)$ is true.*"

There are these generalized De Morgan laws in logic:

$$\sim((\exists x, \mathcal{P}(x))) \Leftrightarrow (\forall x, \sim \mathcal{P}(x)) \qquad \sim((\forall x, \mathcal{P}(x))) \Leftrightarrow (\exists x, \sim \mathcal{P}(x))$$

The first holds because to say that "*there does not exist an object such that $\mathcal{P}(x)$ is true*" means that "*for all objects x, $\mathcal{P}(x)$ is false.*" Similarly, the second holds because to say that "*it is false that, for all objects x, $\mathcal{P}(x)$ is true,*" is to say that "*there exists an object x such that $\mathcal{P}(x)$ is false.*"

SETS AND MEMBERSHIP

We do not rigorously develop an axiomatic theory of sets. Rather, we are content to work informally with the following ideas. A **set** is a collection of objects, called its **members**, **points**, or **elements**. Sets are denoted by uppercase letters A, B, C, etc.; their members by lowercase letters a, b, c, etc.

The proposition, *The object x is a member of the set S*, is symbolized by $x \in S$; its negation, $\sim (x \in S)$, is abbreviated by $x \notin S$. For a logical development of set theory, it is necessary that $x \in S$ be a *proposition*, i.e., a sentence that is either true or false for each specific object x and set S. This prevents some collections from being sets (e.g., see Theorem 5.10).

Two sets A and B are called **equal**, written $A = B$, provided that they have the same members. Thus, $A = B$ means that $\forall x, (x \in A) \Leftrightarrow (x \in B)$.

We say that A is a **subset** of B or that B **contains** A, written $A \subseteq B$, if each member of A is a member of B. So $A \subseteq B$ means that $\forall x, (x \in A) \Rightarrow (x \in B)$.

A subset $A \subseteq B$ is a **proper subset**, written $A \subset B$, if $A \neq B$. Two sets A and B are called **disjoint** if they have no members in common.

Example 2 There are the proper subsets $\mathbb{P} \subset \mathbb{N} \subset \mathbb{Z} \subset \mathbb{Q} \subset \mathbb{R}$ of positive integers, non-negative integers, integers, and rational numbers in the set $\mathbb{R}$ of real numbers.

The Rules of Inference (2.2) and the definition of the subset relation give the following rules of containment.

2.3 Rules of Containment

(a) **Tautology rule:** $\qquad A \subseteq A.$

(b) **Equivalence rule:** $\qquad A = B$ if and only if $A \subseteq B$ and $B \subseteq A$.

(c) **Detachment rule:** $\qquad$ If $x \in A$ and $A \subseteq B$, then $x \in B$.

(d) **Syllogism rule:** $\qquad$ If $A \subseteq B$ and $B \subseteq C$, then $A \subseteq C$.

(e) **Contrapositive rule:** $\qquad A \subseteq B \subseteq X$ if and only if $B^c \subseteq A^c \subseteq X$.

SET CONSTRUCTIONS

Some sets are easily specified by a complete listing of their members. Traditional notation for such a set has the listing of its members surrounded by the *set braces* { and }.

Example 3 $\mathbb{P} = \{1, 2, \ldots, k, \ldots\}$ (*positive integers*)
$\mathbb{N} = \{0, 1, 2, \ldots, k, \ldots\}$ (*non-negative integers*)
$\mathbb{Z} = \{\ldots, -k, \ldots, -2, -1, 0, 1, 2, \ldots, k, \ldots\}$ (*integers*)

A set { } that has no members is called the *empty set*; it is unique and is denoted by $\emptyset$.

When a listing of the members of a set is impractical, a description via a characterizing property may be feasible. Suppose that $\mathcal{P}(x)$ is a proposition that involves an unspecified object x; that is, $\mathcal{P}(x)$ is a sentence that is either true or false for each substitution of a specific object for x. Then the collection of all objects $x \in X$ from some set X for which $\mathcal{P}(x)$ is true is also a *set*, usually denoted by $\{x \in X : \mathcal{P}(x)\}$.

Example 4   Let $\leq$ and $<$ denote the order relation and strict order relation on the set $\mathbb{R}$ of real numbers. Associated with each pair of real numbers r and s are these four types of *finite intervals*, each with *length* $0 \leq s - r$ when $r \leq s$ and empty otherwise:

Closed: $[r, s] = \{x \in \mathbb{R} : r \leq x \leq s\}$ *Open-Closed:* $(r, s] = \{x \in \mathbb{R} : r < x \leq s\}$
Open: $(r, s) = \{x \in \mathbb{R} : r < x < s\}$ *Closed-Open:* $[r, s) = \{x \in \mathbb{R} : r \leq x < s\}$

And there are these four types of *infinite intervals* or *rays*:

Closed: $[r, \infty) = \{x \in \mathbb{R} : r \leq x\}$ *Closed:* $(-\infty, s] = \{x \in \mathbb{R} : x \leq s\}$
Open: $(r, \infty) = \{x \in \mathbb{R} : r < x\}$ *Open:* $(-\infty, s) = \{x \in \mathbb{R} : x < s\}$

The following constructions of *complement*, *union*, and *intersection* are three basic ways to utilize sets to form other sets. Let $A, B \subseteq X$ be sets.

The (*relative*) *complement of A in B* is the set, denoted by $B - A$, of all members of X that are members of B but not members of A:

$$B - A = \{x \in X : (x \in B) \wedge (x \notin A)\}.$$

When only subsets $A \subseteq X$ are considered, then X is called the *universe* and $X - A$ is called the (*absolute*) *complement* of A and is denoted by A^c.

The *union* of A and B is the set, denoted by $A \cup B$, of all members of X that are members of either A, or B, or both A and B:

$$A \cup B = \{x \in X : (x \in A) \vee (x \in B)\}.$$

The *intersection* of A and B is the set, denoted by $A \cap B$, of all members of X that are members of both A and B:

$$A \cap B = \{x \in X : (x \in A) \wedge (x \in B)\}.$$

2.4 Summary *The definitions are captured by these statements:*
(a) $A = B$ means $(x \in A) \Leftrightarrow (x \in B)$. (b) $x \in A \cup B \Leftrightarrow (x \in A) \vee (x \in B)$.
(c) $A \subseteq B$ means $(x \in A) \Rightarrow (x \in B)$. (d) $x \in A \cap B \Leftrightarrow (x \in A) \wedge (x \in B)$.
(e) $x \in B - A \Leftrightarrow (x \in B) \wedge (x \notin A)$. (f) $x \in A^c \Leftrightarrow (x \in X) \wedge (x \notin A)$.

Example 5 For two intervals $A, B \subseteq \mathbb{R}$, the intersection $A \cap B$ is always an interval (possibly empty) and the union $A \cup B$ is an interval in $\mathbb{R}$ if and only if $A \cap B \neq \emptyset$. The complement of a nonempty interval is the union of two disjoint rays.

The following laws arise from corresponding logical laws for disjunction, conjunction, and negation. We prove one and leave the rest as Exercise 1.

2.5 Laws of Set Theory *Unions, intersections, and complements of subsets of a set X satisfy the following properties.*

(a) **Idempotent laws:**
$A \cap A = A$
$A \cup A = A$

(b) **Associative laws:**
$A \cap (B \cap C) = (A \cap B) \cap C$
$A \cup (B \cup C) = (A \cup B) \cup C$

(c) **Commutative laws:**
$A \cap B = B \cap A$
$A \cup B = B \cup A$

(d) **Distributive laws:**
$A \cap (B \cup C) = (A \cap B) \cup (A \cap C)$
$A \cup (B \cap C) = (A \cup B) \cap (A \cup C)$

(e) **Identity laws:**
$A \cap X = A \quad A \cup \emptyset = A$
$A \cap \emptyset = \emptyset \quad A \cup X = X$

(f) **De Morgan's laws:**
$X - (A \cap B) = (X - A) \cup (X - B)$
$X - (A \cup B) = (X - A) \cap (X - B)$

(g) **Complementation laws:**
$A \cap (X - A) = \emptyset \qquad A \cup (X - A) = X \qquad X - (X - A) = A$

Proof: We verify the first distributive law, $A \cap (B \cup C) = (A \cap B) \cup (A \cap C)$. By 2.4(a), this law follows from these logical equivalences:

$x \in A \cap (B \cup C)$
$\Leftrightarrow (x \in A) \wedge (x \in B \cup C)$, by the definition of intersection $\cap$,
$\Leftrightarrow (x \in A) \wedge ((x \in B) \vee (x \in C))$, by the definition of union $\cup$,
$\Leftrightarrow ((x \in A) \wedge (x \in B)) \vee ((x \in A) \wedge (x \in C))$, by the distributive law 2.1(d),
$\Leftrightarrow (x \in (A \cap B)) \vee (x \in A \cap C))$, by the definition of intersection $\cap$,
$\Leftrightarrow x \in (A \cap B) \cup (A \cap C)$, by the definition of union $\cup$. □

THE POWER SET

The collection $\{A : A \subseteq S\}$ of all subsets A of a set S is a set, denoted by $\mathcal{P}(S)$ and called the **power set** of S. By this definition, $A \in \mathcal{P}(S) \Leftrightarrow A \subseteq S$. Hence, $W \subseteq \mathcal{P}(S)$ if and only if $A \in W \Rightarrow A \subseteq S$.

The relationship between membership in some power set and containment in some other set can be a terrible source of confusion. Here are several examples to guide you through some of the possibilities:

Example 6 Since $\emptyset = \{\ \} \subseteq S$, then $\emptyset \in \mathcal{P}(S)$; hence, $\{\emptyset\} \subseteq \mathcal{P}(S)$. So $\{\emptyset\}$ is a nonempty subset of $\mathcal{P}(S)$ whose single member is $\emptyset \in \mathcal{P}(S)$.

Example 7 As $S \subseteq S$, then $S \in \mathcal{P}(S)$; hence, $\{S\} \subseteq \mathcal{P}(S)$ is a subset whose single member is S. For example, if $S = \{a, b\}$, then $\{S\} = \{\{a, b\}\} \neq S$.
  When $S \neq \emptyset$, the subset $\{\emptyset, S\} \subseteq \mathcal{P}(S)$ does not equal the singleton $\{S\}$; $\{\emptyset, S\}$ is a doubleton subset with distinct members $\emptyset$ and S.

Example 8 If $x \in S$, then $\{x\} \subseteq S$ and $\{x\} \in \mathcal{P}(S)$. Thus we have $\{\{x\}\} \subseteq \mathcal{P}(S)$ and also $\{\{x\}\} \in \mathcal{P}(\mathcal{P}(S))$.

INDEXED FAMILIES OF SETS

Let $\mathcal{A}$ be a nonempty set. Suppose that, for each $\alpha \in \mathcal{A}$, there is a set A_α. The system $\mathcal{F} = \{A_\alpha : \alpha \in \mathcal{A}\}$ is called a ***family of sets indexed by*** $\mathcal{A}$.

The ***union*** of a family $\mathcal{F} = \{A_\alpha \subseteq X : \alpha \in \mathcal{A}\}$ of subsets of a set X is the set of all members $x \in X$ that are members of A_α *for some* $\alpha \in \mathcal{A}$:

$$\cup_{\alpha \in \mathcal{A}} A_\alpha = \{x \in X : \exists\, \alpha \in \mathcal{A}, x \in A_\alpha\}.$$

The ***intersection*** of a family $\mathcal{F} = \{A_\alpha \subseteq X : \alpha \in \mathcal{A}\}$ of subsets of a set X is the set of all members $x \in X$ that are members of A_α *for all* $\alpha \in \mathcal{A}$:

$$\cap_{\alpha \in \mathcal{A}} A_\alpha = \{x \in X : \forall\, \alpha \in \mathcal{A}, x \in A_\alpha\}.$$

For a family $\mathcal{F} = \{A_\alpha : \alpha \in \mathcal{A}\}$, the union may be denoted by $\cup_\alpha A_\alpha$, $\cup \{A_\alpha : \alpha \in \mathcal{A}\}$, or $\cup \mathcal{F}$; the intersection by $\cap_\alpha A_\alpha$, $\cap \{A_\alpha : \alpha \in \mathcal{A}\}$, or $\cap \mathcal{F}$.

Example 9 The family of finite open intervals $\{(-1 + 1/k, +1 - 1/k) : k \geq 1\}$ indexed by the positive integers has union $\cup_k (-1 + 1/k, +1 - 1/k) = (-1, +1)$.
  The family $\{(-1/k, +1/k) : k \geq 1\}$ has intersection $\cap_k (-1/k, +1/k) = \{0\}$.

2.6 Laws of Set Theory *Unions and intersections of indexed families* $\{A_\alpha : \alpha \in \mathcal{A} = \cup_\gamma \mathcal{A}_\gamma\}$ *and* $\{B_\beta : \beta \in \mathcal{B}\} = \{B_\delta : \delta \in \mathcal{D}\}$ *of subsets of* X *satisfy:*

(a) **Commutative laws:** $\cup_\beta B_\beta = \cup_\delta B_\delta \qquad \cap_\beta B_\beta = \cap_\delta B_\delta$

(b) **Associative laws:** $\cup \{A_\alpha : \alpha \in \mathcal{A}\} = \cup_\gamma \{\cup \{A_\alpha : \alpha \in \mathcal{A}_\gamma\}\}$
  $\cap \{A_\alpha : \alpha \in \mathcal{A}\} = \cap_\gamma \{\cap \{A_\alpha : \alpha \in \mathcal{A}_\gamma\}\}$

(c) **Distributive laws:** $(\cup_\alpha A_\alpha) \cap (\cup_\beta B_\beta) = \cup_{(\alpha, \beta)} (A_\alpha \cap B_\beta)$
  $(\cap_\alpha A_\alpha) \cup (\cap_\beta B_\beta) = \cap_{(\alpha, \beta)} (A_\alpha \cup B_\beta)$

(d) **De Morgan's laws:** $X - (\cup_\alpha A_\alpha) = \cap_\alpha (X - A_\alpha)$
  $X - (\cap_\alpha A_\alpha) = \cup_\alpha (X - A_\alpha)$

PRODUCT OF SETS

The ***ordered pair*** of two objects x and y is the set $<x, y> = \{\{x\}, \{x, y\}\}$. By this definition, the ordered pair $<x, y>$ is determined by x and y; and the order, *x first* and *y second*, is important unless $x = y$. In fact, this definition yields the following characterizing property of ordered pairs (Exercise 14):

$$<x, y> = <u, v> \Leftrightarrow x = u \text{ and } y = v.$$

For any two sets X and Y, the ***product set*** $X \times Y$, with ***factors*** X and Y, is the set of all ordered pairs $<x, y>$, where $x \in X$ and $y \in Y$:

$$X \times Y = \{<x, y> : x \in X \text{ and } y \in Y\}.$$

For each $x \in X$, the product set $\{x\} \times Y$ is a copy of Y in $X \times Y$; for each $y \in Y$, the product set $X \times \{y\}$ is a copy of X in $X \times Y$. Each of the subsets $\{x\} \times Y$ and $X \times \{y\}$ is called a ***slice*** of the product set $X \times Y$.

The product set $X \times Y$ is the union of either family of slices

$$\{\{x\} \times Y : x \in X\} \quad \text{or} \quad \{X \times \{y\} : y \in Y\}.$$

So a picture for $X \times Y$ arises from pictures for the factor sets X and Y:

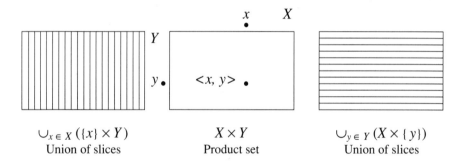

$\bigcup_{x \in X} (\{x\} \times Y)$ $X \times Y$ $\bigcup_{y \in Y} (X \times \{y\})$
Union of slices   Product set   Union of slices

Example 10   The product $\mathbb{R} \times \mathbb{R}$ of two copies of the real line $\mathbb{R}$ is the real plane $\mathbb{R}^2$; it is a union $\bigcup_{x \in X} (\{x\} \times \mathbb{R})$ of vertical lines and a union $\bigcup_{y \in Y} (\mathbb{R} \times \{y\})$ of horizontal lines, as in the previous display.

Example 11   The product $\mathbb{R}^2 \times \mathbb{R}$ of the real plane $\mathbb{R}^2$ and line $\mathbb{R}$ is real space $\mathbb{R}^3$:

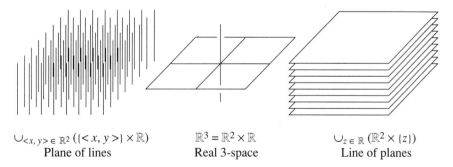

$\bigcup_{<x, y> \in \mathbb{R}^2} (\{<x, y>\} \times \mathbb{R})$ $\mathbb{R}^3 = \mathbb{R}^2 \times \mathbb{R}$ $\bigcup_{z \in \mathbb{R}} (\mathbb{R}^2 \times \{z\})$
Plane of lines   Real 3-space   Line of planes

SECTION 2   SET THEORY   15

Example 12 The product $\mathbb{S}^1 \times \mathbb{I}$ of the circle

$$\mathbb{S}^1 = \{<x, y> \in \mathbb{R}^2 : x^2 + y^2 = 1\}$$

and the interval $\mathbb{I} = [0, 1] \subseteq \mathbb{R}$ is a cylinder in $\mathbb{R}^3 = \mathbb{R}^2 \times \mathbb{R}$:

$\cup_{<x, y> \in \mathbb{S}^1} (\{<x, y>\} \times \mathbb{I})$ $\mathbb{S}^1 \times \mathbb{I}$ $\cup_{z \in \mathbb{I}} (\mathbb{S}^1 \times \{z\})$
Circle of intervals   Cylinder   Interval of circles

Example 13 The product $\mathbb{S}^1 \times \mathbb{S}^1$ of two copies of the circle $\mathbb{S}^1$ can be pictured as a torus (inner tube surface):

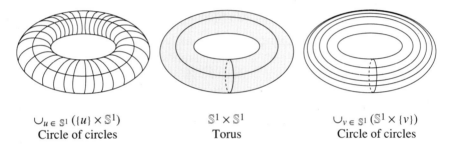

$\cup_{u \in \mathbb{S}^1} (\{u\} \times \mathbb{S}^1)$ $\mathbb{S}^1 \times \mathbb{S}^1$ $\cup_{v \in \mathbb{S}^1} (\mathbb{S}^1 \times \{v\})$
Circle of circles   Torus   Circle of circles

2.7 Product Properties *Let $A, C \subseteq X$ and $B, D \subseteq Y$. Then in $X \times Y$, we have the following relations:*
(a) $A \times (B \cap D) = (A \times B) \cap (A \times D)$,
(b) $A \times (B \cup D) = (A \times B) \cup (A \times D)$,
(c) $A \times (Y - B) = (A \times Y) - (A \times B)$,
(d) $(A \times B) \cap (C \times D) = (A \cap C) \times (B \cap D)$,
(e) $(A \times B) \cup (C \times D) \subseteq (A \cup C) \times (B \cup D)$, and
(f) $(X \times Y) - (A \times B) = (X \times (Y - B)) \cup ((X - A) \times Y)$.

Proof: The following pictures of $X \times Y$ illustrate the last three product properties. Exercise 15 asks for proofs of all six of these claims.   □

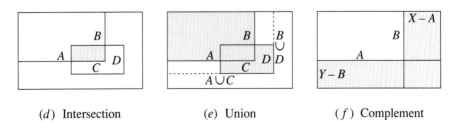

(d) Intersection   (e) Union   (f) Complement

The product of sets X_i $(1 \leq i \leq n)$ is defined inductively to be the set
$$X_1 \times \ldots \times X_n = (X_1 \times \ldots \times X_{n-1}) \times X_n$$
of all ***n-tuples***
$$(x_1, \ldots, x_{n-1}, x_n) = <(x_1, \ldots, x_{n-1}), x_n>$$
with ***coordinates*** $x_i \in X_i$ $(1 \leq i \leq n)$. When $X_i = X$ $(1 \leq i \leq n)$, the product set $X_1 \times \ldots \times X_n$ is written as X^n. For example, $\mathbb{R}^n$ is ***real n-space*** $\mathbb{R} \times \ldots \times \mathbb{R}$.

More general product sets are presented in Section **5**.3.

3 The Real Line

The set $\mathbb{R}$ of real numbers, given its order relation $\leq$, is called the ***real line***; it is a precise mathematical construction for the intuitive geometric line.

COMPLETENESS OF THE REAL NUMBERS

The real line $\mathbb{R}$ has a certain richness or completeness property not shared by the set $\mathbb{Q} = \{p/q : p \in \mathbb{Z}, q \in \mathbb{P}\}$ of rational numbers. This *completeness* of $\mathbb{R}$ involves bounds of sets of real numbers and is described below.

A subset $S \subset \mathbb{R}$ is ***bounded above*** by a real number r if $s \leq r$ for all $s \in S$. Such a real number r is called an ***upper bound*** of the subset $S \subset \mathbb{R}$.

A subset $S \subseteq \mathbb{R}$ is ***bounded below*** by a real number t if $t \leq s$ for all $s \in S$. Then t is called a ***lower bound*** of the subset $S \subset \mathbb{R}$.

The set $\mathbb{Z} \subset \mathbb{R}$ of integers is neither bounded above nor bounded below. Because its members are a unit distance apart, the unbounded set $\mathbb{Z}$ enters every open interval of length > 1. Similarly, the set $\{p/q : p \in \mathbb{Z}\}$ of equally spaced integral multiples of a positive fraction $1/q > 0$ ($q \in \mathbb{P}$) is unbounded and enters every open interval of length $> 1/q$. For any open interval (a, b) of positive length $b - a > 0$, there is some $q \in \mathbb{P}$ satisfying the inequality $q > 1/(b - a)$, since $\mathbb{P}$ is not bounded above. Then the length of (a, b) exceeds $1/q$; hence, $a < p/q < b$, for some $p \in \mathbb{Z}$ and $q \in \mathbb{P}$. This proves:

3.1 Rational Density Property
Between any two distinct real numbers there is a rational number.

The crucial feature distinguishing $\mathbb{R}$ from its subset $\mathbb{Q}$ of rational numbers is the following property, which we accept from real analysis.

3.2 Order Completeness Property
(a)   Any nonempty subset $S \subseteq \mathbb{R}$ that is bounded above has a unique upper bound that is less than or equal to every upper bound of S. This upper bound is called the ***least upper bound*** of S and is denoted by lub S.

(b) Any nonempty subset $S \subseteq \mathbb{R}$ that is bounded below has a unique lower bound that is greater than or equal to every lower bound of S. This lowerbound is called the **greatest lower bound** of S and is denoted by glb S.

If $lub\ S \in S$, then it is called the **maximum** of S and is denoted by max S; if $glb\ S \in S$, then it is called the **minimum** of S and is denoted by min S.

In this text, we accept the completeness property as an axiom about the real numbers. But you should understand that the real numbers are constructed from the set $\mathbb{Q}$ of rational numbers with precisely the completeness property in mind. The completeness property gives the real line an amazing topological elasticity, as we shall show in Section **3**.4.

NESTED INTERVAL PROPERTY

Real numbers satisfying the order relations

$$r_1 \leq r_2 \leq \ldots \leq r_k \leq \ldots \leq t_k \leq \ldots \leq t_2 \leq t_1$$

define closed intervals satisfying these containment relations:

$$[r_1, t_1] \supseteq [r_2, t_2] \supseteq [r_3, t_3] \supseteq \ldots \supseteq [r_k, t_k] \supseteq \ldots .$$

Such a collection of intervals is called a sequence of **nested** closed intervals.

Georg Cantor (1845-1918) used the completeness of the real line to deduce that the intersection of any sequence of nested closed intervals is a closed interval $[r, t]$, though possibly a singleton $\{r = t\}$.

3.3 Nested Intervals Theorem *The intersection $\cap_k [r_k, t_k]$ of a sequence of nested closed intervals is a nonempty closed interval $[r, t]$. If the lengths*

$$t_1 - r_1 \geq t_2 - r_2 \geq \ldots \geq t_k - r_k \geq \ldots$$

become smaller than each positive real number, then $\cap_k [r_k, t_k] = \{r = t\}$.

Proof: By the nested feature of the intervals, the set $\{r_k : 1 \leq k\}$ is bounded above by each t_k. So $r = lub\{r_k : 1 \leq k\}$ exists and $r \leq t_k$ for all $k \geq 1$. Then the set $\{t_k : 1 \leq k\}$ is bounded below by r; hence, $t = glb\{t_k : 1 \leq k\}$ exists and $r \leq t$. Then $[r, t] \subseteq \cap_k [r_k, t_k]$, by the relations just derived:

$$r_1 \leq r_2 \leq \ldots \leq r_k \leq \ldots \leq r \leq t \leq \ldots \leq t_k \leq \ldots \leq t_2 \leq t_1.$$

Any real number $s \in \cap_k [r_k, t_k]$ is an upper bound for $\{r_k : 1 \leq k\}$ and a lower bound for $\{t_k : 1 \leq k\}$. Thus, by the definitions of *lub* and *glb*, we have

$$r = lub\{r_k : 1 \leq k\} \leq s \leq glb\{t_k : 1 \leq k\} = t.$$

This shows that $\cap_k [r_k, t_k] \subseteq [r, t]$. Thus, $[r, t] = \cap_k [r_k, t_k]$.

If the differences $t_k - r_k$ become smaller than all positive real numbers, then their non-negative lower bound $t - r \geq 0$ must be 0, i.e., $r = t$. □

The Nested Intervals Theorem is probably the clearest expression of the completeness of the real numbers. No matter how devious your selection of the nested closed intervals in a sequence, there is some real number common to all the intervals in the sequence. The story can be quite different when one considers a sequence of nested open intervals (see Exercise 10).

REPRESENTING REAL NUMBERS

Using the Nested Intervals Theorem and any integer $B \geq 2$, it is possible to create a *base B*, or *B-adic*, system of representing each real number by a sequence of digits from the set $\{0, 1, \ldots, B-1\}$. The standard decimal system has base ten; the binary system has base two; the ternary system has base three. Several startling topological consequences of the completeness of the real numbers are expressed in later chapters using the different systems with base *two*, *three*, *nine*, and *ten*.

Consider any integer $B \geq 2$. The **B-adic system** for representing real numbers in the closed unit interval $\mathbb{I} = [0, 1]$ is based upon the Nested Intervals Theorem and special intervals that arise when one inductively subdivides each interval at hand into B equal subintervals, beginning with the unit interval $\mathbb{I}$. The subdivision points and intervals that arise are called **B-adic fractions** and **B-adic intervals**.

The first subdivision of $\mathbb{I}$ produces the 1st stage B-adic fractions:

$$\frac{0}{B} < \frac{1}{B} < \ldots < \frac{b}{B} < \frac{b+1}{B} < \ldots < \frac{B-1}{B} < \frac{B}{B}.$$

They bound the 1st stage B-adic intervals that we index this way:

$$B[b] = \left[\frac{b}{B}, \frac{b+1}{B}\right] \qquad (b \in \{0, \ldots, B-1\}).$$

The subdivision of the 1st stage B-adic interval $B[b_1]$ into B equal subintervals introduces the 2nd stage B-adic fractions:

$$\frac{b_1 0}{B^2} < \frac{b_1 1}{B^2} < \ldots < \frac{b_1 b_2}{B^2} < \frac{b_1 b_2 + 1}{B^2} < \ldots < \frac{b_1 + 1}{B}.$$

They bound the 2nd stage B-adic intervals that we index using a 2nd digit b_2:

$$B[b_1 b_2] = \left[\frac{b_1 b_2}{B^2}, \frac{b_1 b_2 + 1}{B^2}\right] \qquad (b_2 \in \{0, \ldots, B-1\}).$$

In its turn, each $(k-1)$st stage B-adic interval $B[b_1 b_2 \ldots b_{k-1}]$ is subdivided into B equal subintervals, each of which has the form

$$B[b_1 b_2 \ldots b_k] = \left[\frac{b_1 b_2 \ldots b_k}{B^k}, \frac{b_1 b_2 \ldots b_k + 1}{B^k}\right]$$

for a value of the kth digit $b_k \in \{0, \ldots, B-1\}$.

Any sequence of digits $b_1, b_2, \ldots, b_k, \ldots$ from $\{0, \ldots, B-1\}$ determines a sequence of nested B-adic intervals

$$[0, 1] = \mathbb{I} \supset B[b_1] \supset B[b_1 b_2] \supset \ldots \supset B[b_1 b_2 \ldots b_k] \supset \ldots ,$$

whose intersection $\cap_k B[b_1 b_2 \ldots b_k]$ is a unique real number $0 \le r \le 1$, by the Nested Intervals Theorem (3.3). The expression $(.b_1 b_2 \ldots b_k \ldots)_B$ is called a ***B*-adic expansion** of the real number r.

Each endpoint B-adic fraction is given by

$$\frac{b_1 b_2 \ldots b_k}{B^k} = \frac{b_1}{B} + \frac{b_2}{B^2} + \ldots + \frac{b_k}{B^k}.$$

So we can view $r = (.b_1 b_2 \ldots b_k \ldots)_B$ as the sum of the infinite series

$$\frac{b_1}{B} + \frac{b_2}{B^2} + \ldots + \frac{b_k}{B^k} + \ldots .$$

But the series notation presumes more familiarity with limits than we wish to make at this point and it also obscures the useful technique of basing coordinate systems on sequences of nested subsets in more general spaces. The latter technique is applied in Section **3**.4 to describe Peano's original space-filling curve and in Section **7**.5 to characterize topological arcs.

Example 1   The long division $1/11 = .\overline{0909}\ldots$, in base ten, (where the overlined digits repeat indefinitely) is another way of saying that the real number $1/11$ is the intersection point of the nested decimal intervals

$$\left[\frac{0}{10}, \frac{1}{10}\right] \supset \left[\frac{09}{100}, \frac{10}{100}\right] \supset \left[\frac{090}{1000}, \frac{091}{1000}\right] \supset \left[\frac{0909}{10000}, \frac{0910}{10000}\right] \supset \ldots .$$

So $1/11$ is the unique real number that appears alternately in the first and last tenth of each decimal subinterval of $[0, 1]$ that contains it.

Each real number $0 \le r \le 1$ has a B-adic expansion, because, at each stage of the subdivision process, r belongs to at least one of the B equal subintervals of each B-adic interval containing r. In this way, r is located as the intersection point of at least one sequence of nested B-adic intervals

$$[0, 1] = \mathbb{I} \supset B[b_1] \supset B[b_1 b_2] \supset \ldots \supset B[b_1 b_2 \ldots b_k] \supset \ldots .$$

The digits $b_1, b_2, \ldots, b_k, \ldots$ from $\{0, \ldots, B-1\}$ that name the nested intervals provide the B-adic expansion $(.b_1 b_2 \ldots b_k \ldots)_B$ of the real number r.

A B-adic fraction b that appears as a subdivision point at some k^{th} stage in the B-adic subdivision process can be viewed as a member of the kth stage subinterval to the left of it and then as a member of the last subinterval of every subsequent subdivision, or as a member of the kth stage subinterval to the right of it and then as a member of the first subinterval of every subsequent subdivision. In this way, the B-adic fraction d acquires two distinct B-adic expansions, one that terminates in the digit *zero*, the other in $B-1$.

A real number r that is not a B-adic fraction belongs to a unique B-adic subinterval at each stage of the subdivision process. Thus r lies in a unique sequence of nested B-adic intervals and so it has a unique B-adic expansion.

Thus, we have established the following:

3.4 B-adic System

(a) Each B-adic expression $(.b_1 b_2 \ldots b_k \ldots)_B$ determines a unique real number $r = \cap_k B[b_1 b_2 \ldots b_k]$ in $\mathbb{I} = [0, 1]$.
(b) Each real number in $\mathbb{I}$ has some B-adic expansion.
(c) Each B-adic fraction in $\mathbb{I}$ has exactly two B-adic expansions, one terminating in zero, the other in $B - 1$; every other real number in $\mathbb{I}$ has a unique B-adic expansion.

Example 2 A decimal interval $D[d_1 d_2 \ldots d_k]$ or expansion $(.d_1 d_2 \ldots d_k \ldots)_{10}$ is called **even** provided that each of its digits, $d_i \in \{0, \ldots, 9\}$, is even. A real number $0 \leq r \leq 1$ has an even decimal expansion if and only if r is the intersection point of a sequence of even nested decimal intervals:

$$[0, 1] = \mathbb{I} \supset D[d_1] \supset D[d_1 d_2] \supset \ldots \supset D[d_1 d_2 \ldots d_k] \supset \ldots .$$

Neither decimal expansion of the right hand endpoint of an *even* decimal interval is *even*. So r is actually the intersection point $\cap_k D[d_1 d_2 \ldots d_k)$ of the corresponding left-closed and right-open, even, decimal intervals.

Let $E \subset [0, 1)$ denote the set of all real numbers $0 \leq r < 1$ with an even decimal expansion. If $E_0 = [0, 1)$ and E_k $(k \geq 1)$ is the union of all k^{th} stage, left-closed and right-open, even, decimal subintervals, then $E = \cap_k E_k$.

Notice that

$$E_k = DAT(E_{k-1}), \qquad (k \geq 1)$$

where DAT is the procedure that subdivides each left-closed and right-open interval (in a disjoint union of such intervals) into ten equal left-closed right-open subintervals and deletes the alternate 2nd, 4th, 6th, 8th, and 10th tenth. Therefore, $E = \cap_k E_k$ can also be viewed as the subset of points of $[0, 1)$ that survive iteration of the *deleted alternate tenths* procedure DAT.

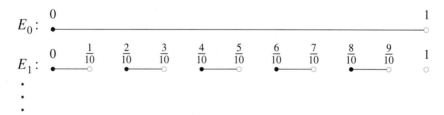

Example 2: Deleted alternate tenths procedure

Example 3 Cantor's Deleted Middle Third Set. A ternary interval $T[t_1 t_2 \ldots t_k]$ or ternary expansion $(.t_1 t_2 \ldots t_k \ldots)_3$ is called **even** provided that each of its digits is even: $t_i \in \{0, 2\}$ for all i. A real number $0 \leq r \leq 1$ has an even ternary expansion if and only if r is the intersection of a sequence of nested even ternary intervals:

$$[0, 1] = \mathbb{I} \supset T[t_1] \supset T[t_1 t_2] \supset \ldots \supset T[t_1 t_2 \ldots t_k] \supset \ldots .$$

In such a sequence, each interval is either the first or last third of the previous ternary interval. So each endpoint of an *even* ternary interval has an *even* ternary expansion.

Cantor's set $C \subset \mathbb{I}$ is the set of all real numbers $0 \leq r \leq 1$ that have an even ternary expansion. If we let $C_0 = \mathbb{I}$ and let C_k ($k \geq 1$) be the union of all k^{th} stage, even, closed, ternary intervals, then we have $C = \cap_k C_k$.

Notice that

$$C_k = DMT(C_{k-1}), \qquad (k \geq 1)$$

where *DMT* denotes the procedure that deletes the open middle third from each closed interval in any union of disjoint closed intervals. Thus, Cantor's set $C = \cap_k C_k$ can also be viewed as the subset of points of $\mathbb{I}$ that survive iteration of the *deleted middle third* procedure *DMT*, as in the following diagram:

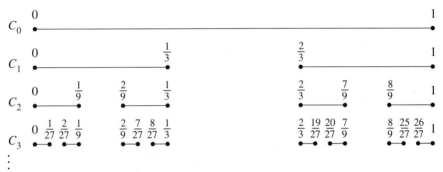

Example 3: Cantor's deleted middle third procedure

4 Functions

In most mathematical disciplines, the objects of interest are sets with extra structure. They are compared via correspondences of their elements that respect the extra structure. *Functions* between the underlying sets provide the foundation for these comparisons.

FUNCTIONS

Let X and Y be sets. A *function* from X to Y, denoted by $f : X \to Y$, is a rule that assigns to each member $x \in X$ some definite member $y \in Y$, denoted by $f(x)$. The set X is called the **domain** of the function f and may be denoted by $domain(f)$. The set Y is called the **codomain** of the function f and may be denoted by $codomain(f)$. The member $f(x) = y$ of the codomain set Y is called the *image* of $x \in X$ under the function f.

More formally, a *function* f is a subset $f \subseteq X \times Y$ such that, for each $x \in X$, there is a unique $y \in Y$, denoted by $y = f(x)$, for which $<x, y> \in f$.

Two functions $f, g : X \to Y$ with the same domain X and the same codomain Y are *equal*, written $f = g$, provided that they make the same assignment to each member of the domain set: $f(x) = g(x)$ in Y for every $x \in X$. According to this notion of equality, equal functions must agree on all three items: *domain set*, *codomain set*, and *assignment rule*.

A function $f : X \to Y$ is called **one-to-one** or **injective** provided that distinct members of the domain X have distinct images in the codomain Y:

$$\forall\, x, x' \in X : f(x) = f(x') \implies x = x'.$$

A function $f : X \to Y$ is called **onto** or **surjective** provided that each member of the codomain Y is the image of a member of the domain X:

$$\forall\, y \in Y, \exists\, x \in X \text{ such that } f(x) = y.$$

A function $f : X \to Y$ that is injective and surjective is called **bijective**.

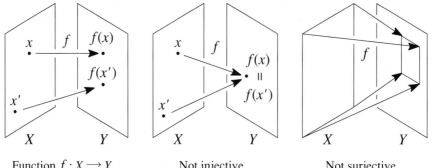

Function $f : X \to Y$ Not injective   Not surjective

Example 1   For any set X, the *identity function* $1_X : X \to X$ is defined by $1_X(x) = x$ for all $x \in X$; it is bijective. For any subset $A \subseteq X$, the *inclusion function* $i_A : A \to X$ is defined by $i_A(a) = a$ for all $a \in A$. The inclusion i_A is injective; it is surjective only when $A = X$.

Example 2   For any product set $X \times Y$, there are *projection functions*

$$p_X : X \times Y \to X \quad \text{and} \quad p_Y : X \times Y \to Y$$

defined by $p_X(<x, y>) = x$ and $p_Y(<x, y>) = y$ for all $<x, y> \in X \times Y$.

The projections p_X and p_Y are always surjections; the projection p_X or p_Y is an injection only if Y or X is a singleton, respectively.

Example 3   Two functions $f : X \to Y$ and $g : W \to Z$ determine a *product function*

$$f \times g : X \times W \to Y \times Z,$$

defined by $f \times g(<x, w>) = <f(x), g(w)> \in Y \times Z$, for all $<x, w> \in X \times W$. When both f and g are injective or surjective, then $f \times g$ is also.

Example 4   Any function $a : \mathbb{P} \to X$ from the set $\mathbb{P}$ of positive integers to a set X is called a *sequence* in X. If images $a(k)$ of the integers $k \in \mathbb{P}$, called the *terms* of the sequence, are denoted by a_k, then the sequence a is usually denoted by (a_k). For each integer $1 \le m$, the set of terms $A_m = \{a_k : m \le k\}$ is called a *tail* of the sequence (a_k).

When $k : \mathbb{P} \to \mathbb{P}$ is a strictly increasing sequence (i.e., $k(i) < k(i+1)$), the sequence $b : \mathbb{P} \to X$, $b(i) = a(k(i))$, is called a *subsequence* of a and is usually denoted by (a_{k_i}).

IMAGE AND PRE-IMAGE

Let $f : X \to Y$ be any function. For a subset $A \subseteq X$, the subset

$$f(A) = \{ f(a) \in Y : a \in A \}$$

of the images in Y of the members of A is called the *image* of $A \subseteq X$ under f. The image $f(X)$ of the domain set X is called the *image* of the function f.

For a subset $B \subseteq Y$, the subset

$$f^{-1}(B) = \{x \in X : f(x) \in B\}$$

of all members of X whose image under f lies in B is the *pre-image* of B.

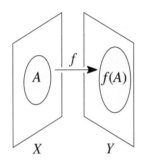
Image of $A \subseteq X$

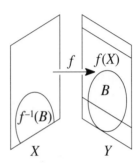
Pre-Image of $B \subseteq Y$

4.1 Summary *The definitions are expressed by these implications:*
(a) $x \in A \Rightarrow f(x) \in f(A)$.
(b) $y \in f(A) \Rightarrow \exists\, x \in A : y = f(x)$.
(c) $x \in f^{-1}(B) \Rightarrow f(x) \in B$.
(d) $f(x) \in B \Rightarrow x \in f^{-1}(B)$.

4.2 Image and Pre-image Properties *Let $f : X \to Y$ be any function. For all subsets $A, A_\alpha, C \subseteq X$ and $B, B_\beta, D \subseteq Y$ we have:*
(1a) $A \subseteq f^{-1}(f(A))$.
(1b) $f(f^{-1}(B)) \subseteq B$.
(2a) $f(\cup_\alpha A_\alpha) = \cup_\alpha f(A_\alpha)$.
(2b) $f^{-1}(\cup_\beta B_\beta) = \cup_\beta f^{-1}(B_\beta)$.
(3a) $f(\cap_\alpha A_\alpha) \subseteq \cap_\alpha f(A_\alpha)$.
(3b) $f^{-1}(\cap_\beta B_\beta) = \cap_\beta f^{-1}(B_\beta)$.
(4a) $f(C - A) \supseteq f(C) - f(A)$.
(4b) $f^{-1}(D - B) = f^{-1}(D) - f^{-1}(B)$.

Proof: Statements (1b - 4b) are left as exercises (Exercise 2).
(1a) If $x \in A$, then $f(x) \in f(A)$. This implies that $x \in f^{-1}(f(A))$.
(2a) If $y \in f(\cup_\alpha A_\alpha)$, then we have $y = f(x)$ for some $x \in \cup_\alpha A_\alpha$. Hence we have $y \in f(A_\alpha)$ for some $\alpha \in \mathcal{A}$, so that $y \in \cup_\alpha f(A_\alpha)$. The converse implications hold as well; thus, $f(\cup_\alpha A_\alpha) = \cup_\alpha f(A_\alpha)$.
(3a) If $y \in f(\cap_\alpha A_\alpha)$, then we have $y = f(x)$ for some $x \in \cap_\alpha A_\alpha$. Hence we have $y \in f(A_\alpha)$ for all $\alpha \in \mathcal{A}$, so that $y \in \cap_\alpha f(A_\alpha)$. The converse implications fail in general; thus, we have only $f(\cap_\alpha A_\alpha) \subseteq \cap_\alpha f(A_\alpha)$.
(4a) If $y \in f(C) - f(A)$, then $y = f(x)$ for some $x \in C$, but for no $x \in A$. Hence we have $y \in f(C - A)$. This proves that $f(C) - f(A) \subseteq f(C - A)$. □

COMPOSITE AND INVERSE FUNCTIONS

Functions $f : X \to Y$ and $g : Y \to Z$, with $codomain(f) = domain(g)$, combine to form a function $h : X \to Z$ by the rule $h(x) = g(f(x))$ for all $x \in X$. The function h is called the ***composite*** of f and g, and it is denoted $h = g \circ f$.

Whenever the triple composite functions $(g \circ f) \circ e$ and $g \circ (f \circ e)$ are defined, they are equal and may be written as $g \circ f \circ e$. Also, for any function $f : X \to Y$, we have $1_Y \circ f = f = f \circ 1_X$.

Functions $f : X \to Y$ and $g : Y \to X$ with opposite domain and codomain sets are ***inverse functions*** if their composites are the identity functions:

$$1_X = g \circ f : X \to Y \to X \qquad \text{and} \qquad 1_Y = f \circ g : Y \to X \to Y.$$

If a function f has an inverse, it is unique and may be denoted by f^{-1}. For, if $g, g' : Y \to X$ both qualify as an inverse for $f : X \to Y$, we have:

$$g = g \circ 1_Y = g \circ f \circ g' = 1_X \circ g' = g'.$$

Example 5 Inverse linear functions $f, g : \mathbb{R} \to \mathbb{R}$ are defined by $f(x) = ax + b$ and $g(y) = 1/a\,(y - b)$, where $a, b \in \mathbb{R}$ and $a \neq 0$.

Example 6 The *reciprocal* functions $f : (0, 1] \leftrightarrow [1, \infty) : g$, defined by $f(x) = 1/x$ and $g(y) = 1/y$, are inverse functions.

4.3 Inversion Theorem
(a) (Inversion Rule) *If $f : X \to Y$ and $h : Y \to Z$ have inverses, then their composite $h \circ f : X \to Y \to Z$ has the inverse $(h \circ f)^{-1} = f^{-1} \circ h^{-1} : Z \to Y \to X$.*
(b) (Existence) *A function has an inverse if and only if it is bijective.*

Proof: (a) The functions $f^{-1} \circ h^{-1}$ and $h \circ f$ are inverses, as

$$f^{-1} \circ h^{-1} \circ h \circ f = f^{-1} \circ 1_Y \circ f = 1_X \qquad \text{and} \qquad h \circ f \circ f^{-1} \circ h^{-1} = h \circ 1_Y \circ h^{-1} = 1_Z.$$

(b) ($\Rightarrow$) When the relation $g \circ f = 1_X$ holds, we have the implications

$$f(x) = f(x') \Rightarrow g(f(x)) = g(f(x')) \Rightarrow x = x',$$

for all $x, x' \in X$. Thus, f is injective. When the relation $f \circ g = 1_Y$ holds, then $y \in Y$ is the image of $x = g(y)$: $f(x) = f(g(y)) = y$. Thus, f is surjective.

(b) ($\Leftarrow$) Since f is surjective, we can choose a member $g(y) \in f^{-1}(\{y\})$ for each $y \in Y$. These choices define a function $g : Y \to X$ with $f(g(y)) = y$ for all $y \in Y$, so that $f \circ g = 1_Y$. So for all $x \in X$, we have $f(g(f(x))) = f(x)$, hence, $g(f(x)) = x$, as f is injective. This proves $g \circ f = 1_X$. Thus $g = f^{-1}$. $\square$

SET EQUIVALENCE

Sets X and Y are ***equivalent***, written $X \sim Y$, if inverse functions $X \leftrightarrow Y$ exist.

By the Inversion Theorem, two sets X and Y are equivalent when there's a bijective function $f : X \to Y$, for then an inverse function $g : Y \to X$ exists.

By the Inversion Rule, $X \sim Y$ and $Y \sim Z$ imply $X \sim Z$, for any sets X, Y, and Z.

The concept of equivalent sets was proposed by Bernhard Bolzano (1781-1848) and was developed by Georg Cantor (1845-1918).

Example 7 *The interval $(-1, +1)$ is equivalent to $\mathbb{R}$.* Inverse functions
$$f : (-1, +1) \hookrightarrow \mathbb{R} : g$$
are given by
$$f(y) = \frac{y}{1 - |y|} \quad \text{and} \quad g(x) = \frac{x}{1 + |x|}.$$

They can be viewed in $\mathbb{R} \times \mathbb{R}$ as interchanging the vertical intervals $\{0\} \times [0, 1)$ and $\{0\} \times (-1, 0]$ with the horizontal rays $[0, \infty) \times \{0\}$ and $(-\infty, 0] \times \{0\}$ along lines of sight from the points $<-1, +1>$ and $<+1, -1>$, as in the following figure:

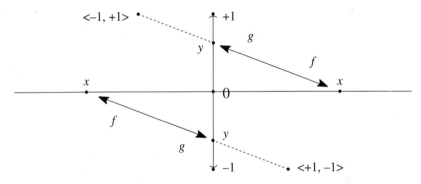

Example 7: $f : (-1, +1) \hookrightarrow \mathbb{R} : g$

Example 8 *The intervals $(0, 1]$ and $(0, 1)$ are equivalent.* There is a bijection,
$$(0, 1] \to (0, 1),$$
that pairs each number in $(0, 1]$ from the list
$$\ldots < \frac{1}{2^k} < \ldots < \frac{1}{16} < \frac{1}{8} < \frac{1}{4} < \frac{1}{2} < 1$$
with its predecessor viewed in $(0, 1)$, and fixes all points in the complementary open intervals between consecutive members of that list:

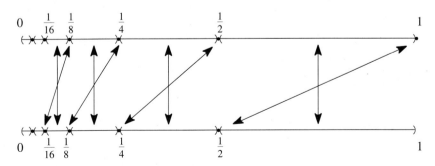

Example 8: Bijection $(0, 1] \to (0, 1)$

Example 9 For any set X, the power set $\mathcal{P}(X)$ is equivalent to the set 2^X of all functions $X \to \{0, 1\}$ into the two-point set $2 = \{0, 1\}$. Each subset $A \subseteq X$ determines a *characteristic function* $\chi_A : X \to \{0, 1\}$, defined by $\chi_A(x) = 1 \Leftrightarrow x \in A$. Conversely, each function $f : X \to \{0, 1\}$ has a *supporting subset*, defined as $f^{-1}(\{1\}) \subseteq X$. These correspondences define inverse functions $\mathcal{P}(X) \to 2^X$ and $2^X \to \mathcal{P}(X)$.

Example 10 The product set $\mathbb{P}^2 = \mathbb{P} \times \mathbb{P}$ is equivalent to $\mathbb{P}$: the function $g : \mathbb{P}^2 \to \mathbb{P}$, $g(m, n) = 2^m(2n - 1)$, is a bijection by the uniqueness of prime factorizations in $\mathbb{P}$.

Cantor showed that a surprising number of objects, notably $\mathbb{R}$ and $\mathbb{R} \times \mathbb{R}$, that are different from a geometric viewpoint are actually equivalent as sets. Such examples are most easily exposed using a fact that Cantor conjectured, but could not show: two sets X and Y are equivalent if X *fits into* Y and Y *fits into* X, i.e., if there exist injections $X \to Y$ and $Y \to X$. This conjecture was established in the 1890's by Schröder and Cantor's student Bernstein.

4.4 Bernstein-Schröder Theorem *Let X and Y be sets. If there exist injections $X \to Y$ and $Y \to X$, then X and Y are equivalent sets.*

Proof: Let $f : X \to Y$ and $g : Y \to X$ be injections. Each point $x \in g(Y) \subseteq X$ has a unique pre-image $y \in Y$ under g; no $x \in X - g(Y)$ has one. Each point $y \in f(X) \subseteq Y$ has a unique pre-image $x \in X$ under f; no $y \in Y - f(X)$ has one.

Consider the two *composition telescopes*:

$$\cdots \xrightarrow{g} (X) \xrightarrow{f} (Y) \xrightarrow{g} (X) \xrightarrow{f} (Y) \xrightarrow{g} (X) \xrightarrow{f} (Y) \xrightarrow{g} (X)$$

and

$$\cdots \xrightarrow{f} (Y) \xrightarrow{g} (X) \xrightarrow{f} (Y) \xrightarrow{g} (X) \xrightarrow{f} (Y) \xrightarrow{g} (X) \xrightarrow{f} (Y)$$

Peering through these telescopes leftward from X and Y, we see that each $x \in X$ and each $y \in Y$ has a unique maximal sequence of pre-images under alternating copies of the injections f and g, which we call their *ancestry*:

$$\cdots \to y_{2k+1} \to x_{2k} \to \cdots \to x_5 \to y_4 \to x_3 \to y_2 \to x_1 = x$$

and

$$\cdots \to x_{2k+1} \to y_{2k} \to \cdots \to y_5 \to x_4 \to y_3 \to x_2 \to y_1 = y.$$

The ancestry of each $x \in X$ and $y \in Y$ either has finite even length $2k \geq 2$, finite odd length $2k + 1 \geq 1$, or infinite length. These properties give disjoint sets X_{even}, X_{odd}, and X_∞ with union X, and disjoint sets Y_{even}, Y_{odd}, and Y_∞ with union Y. And the injections f and g restrict to three injections

$$f : X_{odd} \to Y_{even} \qquad g : Y_{odd} \to X_{even} \qquad f : X_\infty \to Y_\infty$$

that are surjective. So they determine a bijection $X \to Y$. $\square$

4.5 Corollary *A set X is equivalent to a subset $A \subseteq X$ if there exists an injection $X \to A$.*

Example 11 *The real line $\mathbb{R}$ is equivalent to any subset A that contains a nonempty open interval.* We may assume that $(-1, +1) \subseteq A \subseteq \mathbb{R}$ by Example 5, and there is an injection $\mathbb{R} \to (-1, +1) \subseteq A$ by Example 7. Then it follows from Corollary 4.5 that $\mathbb{R} \sim A$. Thus, *all intervals with positive length in $\mathbb{R}$ are equivalent.*

More generally, *$\mathbb{R}^n$ is equivalent to any subset A that contains a nonempty open cube.* By Examples 3 and 5, we may assume that $(-1, +1)^n \subseteq A \subseteq \mathbb{R}^n$; by Examples 3 and 7, there is an injection $\mathbb{R}^n \to (-1, +1)^n \subseteq A$. Thus, $\mathbb{R}^n \sim A$ by Corollary 4.5.

Example 12 *The real line $\mathbb{R}$, the power set $\mathcal{P}(\mathbb{P})$, and the set $2^\mathbb{P}$ of binary sequences are equivalent.* According to the ternary version of the B-adic System (3.4), the function $2^\mathbb{P} \to \mathbb{R}$, defined by $(a_k) \to (.a_1 a_2 \ldots a_k \ldots)_3$, is injective. Also, according to the binary version of the B-adic System (3.4), each real number $0 < r < 1$ has a unique *binary* expansion $(.a_1 a_2 \ldots a_k \ldots)_2$ that doesn't terminate in ones and the assignment $r \to (a_k) \in 2^\mathbb{P}$ defines an injection $\mathbb{R} \sim (0, 1) \to 2^\mathbb{P}$. Thus, $\mathbb{R}$ is equivalent to $2^\mathbb{P}$ and $\mathcal{P}(\mathbb{P})$, by the Bernstein-Schröder Theorem (4.4) and Example 9.

Example 13 *The real plane $\mathbb{R}^2 = \mathbb{R} \times \mathbb{R}$ is equivalent to $\mathbb{R}$.* First of all, there is a bijection $2^\mathbb{P} \times 2^\mathbb{P} \to 2^\mathbb{P}$ defined by Cantor's *shuffle*:

$$< (a_1, a_2, \ldots, a_k, \ldots), (b_1, b_2, \ldots, b_k, \ldots) > \; \to \; (a_1, b_1, a_2, b_2, \ldots, a_k, b_k, \ldots).$$

Then it follows that $\mathbb{R} \times \mathbb{R} \sim \mathbb{R}$ by use of the equivalence $\mathbb{R} \sim 2^\mathbb{P}$ of Example 12.

SIZE OF SETS

A set X is called *finite* if it is equivalent to an initial portion $I_m = \{1, \ldots, m\}$ of the set $\mathbb{P}$ of positive integers.

A set X is called *denumerable* if it is equivalent to $\mathbb{P}$. An explicit bijection $\mathbb{P} \to X$ is called an *enumeration* of X. So a denumerable set X is one whose members can be listed (enumerated) in a sequence: $x_1, \ldots, x_k, \ldots$.

A set is called *countable* if it is finite or denumerable; otherwise, it is called *uncountable*.

Example 14 *Any subset of a finite set is finite*, because any subset of an initial portion of $\mathbb{P}$ is equivalent to an initial portion of $\mathbb{P}$.

The product of a finite family of finite sets is finite: the function

$$I_m \times I_n \to I_{m+n}, \text{ defined by } < i, k > \; \to (k-1)m + i,$$

is a bijection; hence, $I_m \times \ldots \times I_r \sim I_{m + \ldots + r}$, by finite induction.

The union of a finite family of disjoint finite sets is finite: there is an injection

$$\cup \{I_{m_k} : k \in I_n\} \to I_m \times I_n$$

given by $i \in I_{m_k} \to \; < i, k > \; \in I_m \times I_n$, where $m = max\{m_1, \ldots, m_n\}$.

Example 15 *Any subset of a denumerable set is countable*, because the complementary subset can be removed from the enumeration of the denumerable set.

The product of a finite family of denumerable sets is denumerable: $\mathbb{P}^2 \sim \mathbb{P}$, by

Example 10; hence, $\mathbb{P}^n$ ($n \geq 1$) is denumerable by finite induction.

The union of a denumerable family of denumerable sets is denumerable: there is a bijection $\mathbb{P}^2 \to \cup\{\mathbb{P}_k = \mathbb{P} : k \in \mathbb{P}\}$, defined by $<k, m> \to m \in \mathbb{P}_k$.

The sets $\mathbb{Z}$ and $\mathbb{Q}$ of integers and rational numbers are denumerable. $\mathbb{Z}$ is enumerated by the listing: $0, 1, -1, 2, -2, \ldots, k, -k, \ldots$. The function $\mathbb{Q} \to \mathbb{Z}^2$, defined by $k/m \to <k, m>$ for relatively prime $k \in \mathbb{Z}$ and $m \in \mathbb{P}$, is an injection.

Example 16 *The real line $\mathbb{R}$, the power set $\mathcal{P}(\mathbb{P})$, and the set $2^{\mathbb{P}}$ of binary sequences are uncountable.* Any enumeration $a(1), a(2), \ldots, a(k), \ldots$ of sequences $a \in 2^{\mathbb{P}}$ is incomplete: Terms $b_1, \ldots, b_k, \ldots$ selected from $\{0, 1\}$ so that

$$b_1 \neq a(1)_1 \text{ (the 1st term of } a(1)\text{)}, \ldots, b_k \neq a(k)_k \text{ (the } k\text{th term of } a(k)\text{)}, \ldots,$$

determine a sequence $b \in 2^{\mathbb{P}}$, which is not among those enumerated. Thus, the equivalent sets $\mathbb{R}$, $2^{\mathbb{P}}$, and $\mathcal{P}(\mathbb{P})$ in Example 12 are not denumerable.

This diagonal construction technique is due to G. Cantor. Cantor's Power Theorem (4.6), below, also proves that $2^{\mathbb{P}}$ is uncountable.

Is there a largest set? Cantor answered this question in the negative using the power set $\mathcal{P}(X)$ of each set X.

4.6 Cantor's Power Theorem *For any set X, there is an injection $X \to \mathcal{P}(X)$ but no surjection $X \to \mathcal{P}(X)$. So $\mathcal{P}(X)$ has a larger size than X.*

Proof: The assignment $x \to \{x\} \in \mathcal{P}(X)$ defines an injection $f : X \to \mathcal{P}(X)$. There can be no surjection $X \to \mathcal{P}(X)$ by the following reasoning. For any function $g : X \to \mathcal{P}(X)$, we have $g(x) \in \mathcal{P}(X)$ or $g(x) \subseteq X$, for all $x \in X$.

The subset,

$$A = \{x \in X : x \notin g(x) \subseteq X\},$$

of all members of X that do not belong to their image $g(x) \subseteq X$ cannot be a member of the image of $g : X \to \mathcal{P}(X)$. For, if $g(x^*) = A$ for some $x^* \in X$, then both possibilities $x^* \in A \subseteq X$ and $x^* \notin A \subseteq X$ lead to contradictions:

$$x^* \in A = g(x^*) \Rightarrow x^* \text{ is not a member of } A \Rightarrow x^* \notin A$$

and

$$x^* \notin A = g(x^*) \Rightarrow x^* \text{ is a member of } A \Rightarrow x^* \in A. \qquad \square$$

5 Relations

A subset $R \subseteq X \times X$ is called a ***binary relation*** on X. Two elements $x, y \in X$ are ***related*** by R, written xRy, if $<x, y> \in R$.

Every two elements $x, y \in X$ are related by the ***indiscriminant relation*** $R = X \times X$. Two elements $x, y \in X$ are related by the ***diagonal*** or ***identity relation*** $\Delta(X) = \{<x, x> \in X \times X : x \in X\}$ if and only if $x = y$.

EQUIVALENCE RELATIONS

A binary relation $R \subseteq X \times X$ on X is an *equivalence relation* if it has the following three features of the identity relation.

- $\forall x \in X, xRx.$ *(reflexive)*
- $\forall x, y \in X, xRy \Rightarrow yRx.$ *(symmetric)*
- $\forall x, y, z \in X, (xRy) \wedge (yRz) \Rightarrow xRz.$ *(transitive)*

The *equivalence class* of $x \in X$ under the equivalence relation R is the set of all elements of X that are related by R to x:

$$[x]_R = \{ y \in X : xRy \}.$$

The classes $[x]_R$ form a set X/R, called the *quotient set of X modulo R*. The *quotient function* $q_R : X \to X/R$ is defined by $q_R(x) = [x]_R$ for all $x \in X$.

Example 1 For any equivalence relation R, the quotient function $q_R : X \to X/R$ is surjective. But q_R is injective only if R equals the diagonal relation $\Delta(X)$.

Example 2 Two integers $k, m \in \mathbb{Z}$ are *congruent modulo n*, written $k \equiv m \pmod{n}$, if their difference $k - m$ is a multiple of n, i.e., $k - m = nj$ for some $j \in \mathbb{Z}$.

This is an equivalence relation on $\mathbb{Z}$ with n equivalence classes, $[k]_n$ $(0 \leq k < n)$. The equivalence classes have no member in common and their union is $\mathbb{Z}$. The quotient set $\mathbb{Z}_n = \{[0]_n, \ldots, [n-1]_n\}$ is called the set of *integers modulo n*. For example, $\mathbb{Z}_3$ is the set of integers modulo 3; it consists of these three congruence classes:

$$[0]_3 = \{3k : k \in \mathbb{Z}\} \quad [1]_3 = \{1 + 3k : k \in \mathbb{Z}\} \quad [2]_3 = \{2 + 3k : k \in \mathbb{Z}\}.$$

Example 3 Real numbers $x, y \in \mathbb{R}$ are *congruent modulo $\mathbb{Z}$*, written $x \equiv_\mathbb{Z} y$, if and only if their difference is an integer: $y - x = k \in \mathbb{Z}$. This is an equivalence relation on $\mathbb{R}$ for which each equivalence class, $[x]_\mathbb{Z} = \{x + k : k \in \mathbb{Z}\}$, is a translated copy $x + \mathbb{Z}$ of $\mathbb{Z}$ for a unique $0 \leq x < 1$. No two equivalence classes have a member in common and their union is $\mathbb{R}$. The quotient set $\mathbb{R}/\equiv_\mathbb{Z}$ can be viewed as a circular set obtained from $0 \leq x \leq 1$ by identifying just 0 and 1, since numbers $0 \leq x < y \leq 1$ have the same equivalence class, $x + \mathbb{Z} = y + \mathbb{Z}$, if and only if $x = 0$ and $y = 1$.

The reflexive, symmetric, and transitive features of an equivalence relation provide its equivalence classes with these two properties: subsets in an indexed family $\mathcal{P} = \{A_\alpha \subseteq X : \alpha \in \mathcal{A}\}$ are *pairwise disjoint* if $A_\alpha \cap A_\beta \neq \emptyset$ implies $\alpha = \beta$; they *cover* X if $X = \cup_\alpha A_\alpha$.

5.1 Equivalence Class Properties *Let R be an equivalence relation on a set X. The equivalence classes of R are pairwise disjoint and cover X.*

Proof: If $w \in [x]_R \cap [z]_R$, then xRw and zRw; so, xRz by the symmetry and transitivity of R. The same properties imply that $xRy \Leftrightarrow zRy$, or equivalently, $y \in [x]_R \Leftrightarrow y \in [z]_R$. Thus, $[x]_R \cap [z]_R \neq \emptyset \Rightarrow [x]_R = [z]_R$.

By the reflexive property of R, we have $x \in [x]_R$ for all $x \in X$; hence, $X = \cup \{[x]_R \subseteq X : x \in X\}$. □

The equivalence class properties suggest the following definition. A family of subsets $\mathcal{P} = \{A_\alpha \subseteq X : \alpha \in \mathcal{A}\}$ is a ***partition*** of X if the sets in $\mathcal{P}$ are *pairwise disjoint* ($A_\alpha \cap A_\beta \neq \varnothing \Rightarrow \alpha = \beta$) and they *cover* X ($\cup_\alpha A_\alpha = X$).

Thus, the equivalence classes of an equivalence relation constitute a partition of the underlying set. The converse also holds (Exercise 3):

5.2 Relation by Partition *Given a partition $\mathcal{P}$ of a set X, there is an equivalence relation on X whose equivalence classes are the members of $\mathcal{P}$.*

PARTIAL, TOTAL, WELL ORDERS

A binary relation $\leq$ on a set X is a ***partial order*** if it has the three properties:

- $\forall\, x \in X,\, x \leq x.$ (***reflexive***)
- $\forall\, x, y \in X,\, (x \leq y) \wedge (y \leq x) \Rightarrow x = y.$ (***antisymmetric***)
- $\forall\, x, y, z \in X,\, (x \leq y) \wedge (y \leq z) \Rightarrow x \leq z.$ (***transitive***)

A set X, equipped with a partial order $\leq$, is called a ***partially ordered set***.

In a partially ordered set $(X, \leq)$, there are these specialized objects:

(*a*) An element $m \in X$ is called a ***maximal element*** if each $x \in X$ related to m satisfies $x \leq m$, equivalently, whenever $m \leq x$ and $x \in X$, then $m = x$.

(*b*) An element $b \in X$ is called an ***upper bound*** for a subset $A \subseteq X$ if the relation $a \leq b$ holds for all $a \in A$.

(*c*) A subset $A \subseteq X$ is called a ***chain*** if all the elements of A are related, that is, for all $x, y \in A$, either $x \leq y$ or $y \leq x$.

A partially ordered set $(X, \leq)$ that is a chain is called ***totally ordered***.

A partially ordered set $(X, \leq)$ is ***well-ordered*** if each nonempty subset $S \subseteq X$ has a first element; that is, for each $S \neq \varnothing$, there exists a member $s \in S$ satisfying $s \leq x$ for all $x \in S$.

A well-ordered set is totally ordered since each two element subset $\{x, y\} \subseteq X$ has a first element. Any subset $S \subseteq X$ of a totally (well-) ordered set $(X, \leq)$ gives a totally (well-) ordered set $(S, \leq)$.

Example 4 Given any set X, the power set $\mathcal{P}(X)$ is partially ordered by the inclusion relation $A \subseteq B$. But $(\mathcal{P}(X), \subseteq)$ is neither well-ordered nor totally ordered if X has more than one element. The set X is the only maximal element of $(\mathcal{P}(X), \subseteq)$.

Example 5 The set of real numbers $\mathbb{R}$, and each of its subsets, is totally ordered by the usual order relation $\leq$. $(\mathbb{N}, \leq)$ is well-ordered, but $(\mathbb{Z}, \leq)$ and $(\mathbb{R}, \leq)$ are not.

Example 6 A maximal element of a partially ordered set need not be unique. Let $A = \{a, b, c\}$ have the partial order relation $\leq\, = \{<a, b>, <a, c>\}$. Then $(A, \leq)$ is a non-totally ordered set with two maximal elements, b and c.

Example 7   Given partial order relations $\leq_A$ and $\leq_B$ on sets A and B, we define the *lexicographic order relation* $\leq$ on the product set $A \times B$, as follows:

$$<a_1, b_1> \leq <a_2, b_2> \Leftrightarrow (a_1 \leq_A a_2, \text{ or } a_1 = a_2 \text{ and } b_1 \leq_B b_2).$$

Let the interval $\mathbb{I} = [0, 1]$ have its usual order. Then the lexicographic order relation $\leq$ on the product $\mathbb{I} \times \mathbb{I}$ makes it a chain with maximal element $<1, 1>$. But $(\mathbb{I} \times \mathbb{I}, \leq)$ is not well-ordered.

Example 8   More generally, for any partially ordered set $(A, \leq)$, the *lexicographic order relation* $\leq$ on the set $A^{\mathbb{P}} = \{a : \mathbb{P} \to A\}$ of sequences $a = (a_k)$ in A is defined as follows:

$$a \leq b \Leftrightarrow a_k \leq b_k \text{ for the smallest integer } k \text{ such that } a_k \neq b_k.$$

Let $(X, \leq_X)$ and $(Y, \leq_Y)$ be partially ordered sets. An *order-preserving* function $f : (X, \leq_X) \to (Y, \leq_Y)$ is a function $f : X \to Y$ such that $f(x) \leq_Y f(z)$ whenever $x \leq_X z$. The partially ordered sets $(X, \leq_X)$ and $(Y, \leq_Y)$ have the same *order type* if there is an order-preserving bijection between them.

Example 9   Let $\mathbb{N}$ have its usual order $\leq$. Then the product $\mathbb{N} \times \mathbb{N}$, with the lexicographic order relation $\leq$, is well ordered, as is $\mathbb{N}$. In $(\mathbb{N}, \leq)$, but not in $(\mathbb{N} \times \mathbb{N}, \leq)$, every member has a finite number of predecessors. So there is no order-preserving bijection $(\mathbb{N} \times \mathbb{N}, \leq) \to (\mathbb{N}, \leq)$; $\mathbb{N} \times \mathbb{N}$ and $\mathbb{N}$ are set equivalent but $(\mathbb{N} \times \mathbb{N}, \leq)$ and $(\mathbb{N}, \leq)$ have different order type.

Example 10   When B is any integer $B \geq 2$ and A is the set of digits $\{0, 1, \dots B-1\}$, the sequence set $A^{\mathbb{P}}$ may be viewed as the set of *B-adic expansions*. Each finite sequence $(b_1, \dots, b_k, 0, \dots) \in A^{\mathbb{P}}$ defines a B-adic fraction

$$\frac{b_1 b_2 \dots b_k}{B^k} = \frac{b_1}{B} + \frac{b_2}{B^2} + \dots + \frac{b_k}{B^k}.$$

An arbitrary sequence $b \in A^{\mathbb{P}}$ defines a sequence of nested B-adic intervals

$$[0, 1] = \mathbb{I} \supset B[b_1] \supset B[b_1 b_2] \supset \dots \supset B[b_1 b_2 \dots b_k] \supset \dots,$$

each of which has B-adic fraction endpoints:

$$B[b_1 b_2 \dots b_k] = \left[\frac{b_1 b_2 \dots b_k}{B^k}, \frac{b_1 b_2 \dots b_k + 1}{B^k} \right] \qquad (k \geq 1).$$

The function $A^{\mathbb{P}} \to \mathbb{I}$ sending the expansion $b \in A^{\mathbb{P}}$ to the real number

$$(.b_1 b_2 \dots b_k \dots)_B = \bigcap_k B[b_1 b_2 \dots b_k \dots] \in \mathbb{I}$$

is the unique order-preserving function, from the lexicographic order on $A^{\mathbb{P}}$ to the usual order on $\mathbb{R}$, that corresponds each finite sequence with its B-adic fraction.

The familar induction principle for $\mathbb{N}$ extends to each well-ordered set $(X, \leq)$. Let 0 denote the first element of X. For any $x \in X$, let

$$[0, x) = \{z \in X : z < x\} \quad \text{and} \quad [0, x] = \{z \in X : z \leq x\},$$

where $z < x$ means $z \leq x$ and $z \neq x$.

5.3 Transfinite Induction *Let $(X, \leq)$ be a well-ordered set, and let $A \subseteq X$. Suppose that $[0, x) \subseteq A \Rightarrow x \in A$ for each $x \in X$. Then $A = X$.*

Proof: Because $[0, 0) = \emptyset \subseteq A$, we have $0 \in A$. Secondly, by hypothesis, the set $X - A$ can have no first member and, hence, it must be empty. □

SET THEORY ASSUMPTIONS

Given a nonempty finite collection $\{S_1, \ldots, S_n\}$ of nonempty sets, an element may be chosen from each one of the sets in the collection. This means there exists a function $s : \{1, \ldots, n\} \to S_1 \cup \ldots \cup S_n$ such that $s(i) \in S_i$ for every $1 \leq i \leq n$. This property has the following generalization.

5.4 Axiom of Choice *Let $\{S_\alpha : \alpha \in \mathcal{A}\}$ be an indexed collection of nonempty sets. Then there exists a set S consisting of exactly one element from each S_α; equivalently, there exists a function $s : \mathcal{A} \to \cup_\alpha S_\alpha$ such that $s(\alpha) \in S_\alpha$ for every $\alpha \in \mathcal{A}$. Such a function is called a **choice function**.*

Choice functions appear frequently in mathematical work. In 1938, K. Gödel showed that if the standard axiomatic theory of sets is consistent then it remains so when the axiom of choice is included. In 1963, P. Cohen proved that the axiom of choice is not derivable from the other axioms in the standard axiomatic treatment; it is an independent axiom.

We accept the axiom of choice, as well as the following two alternatives.

5.5 Zorn's Lemma *Let $(X, \leq)$ be a partially ordered set. If each chain in X has an upper bound then X has at least one maximal element m.*

5.6 Well-Ordering Principle *Every set X has a partial order relation $\leq$ that makes $(X, \leq)$ well-ordered.*

The axiom of choice, Zorn's lemma, and the well-ordering principle are variations of the same theme. They are logically equivalent forms of the same assumption that we make about set theory at several points in subsequent chapters of this text (in particular, Sections **7**.2 and **8**.3). We do not completely establish this equivalence, but are content to show the following two implications.

5.7 Theorem *The well-ordering principle implies the axiom of choice.*

Proof: Let $\{S_\alpha : \alpha \in \mathcal{A}\}$ be any nonempty indexed family of nonempty sets. Well-order the set $X = \cup_\alpha S_\alpha$ and define $s(\alpha)$ to be the first element of the subset $S_\alpha \subseteq X$. This defines a function $s : \mathcal{A} \to \cup_\alpha S_\alpha$ such that $s(\alpha) \in S_\alpha$ for every $\alpha \in \mathcal{A}$. This is a choice function for the family $\{S_\alpha : \alpha \in \mathcal{A}\}$. □

5.8 Theorem   *Zorn's lemma implies the axiom of choice.*

Proof: Let $\{S_\alpha : \alpha \in \mathcal{A}\}$ be any nonempty indexed family of nonempty sets. Consider the set F of all choice functions $b : \mathcal{B} \to \cup_\alpha S_\alpha$ (i.e., $b(\alpha) \in S_\alpha$ for all $\alpha \in \mathcal{B}$) defined on any subset $\mathcal{B} \subseteq \mathcal{A}$. Order F by

$$b \leq c \iff c : \mathcal{C} \to \cup_\alpha S_\alpha \text{ extends } b : \mathcal{B} \to \cup_\alpha S_\alpha.$$

This simply means that $\mathcal{B} \subseteq \mathcal{C} \subseteq \mathcal{A}$ and $c(\alpha) = b(\alpha)$ for all $\alpha \in \mathcal{B}$.

Each chain of functions in $(F, \leq)$ has an upper bound, namely, the function that those in the chain define on the union of their domains. By Zorn's Lemma, $(F, \leq)$ has a maximal element $s : \mathcal{S} \to \cup_\alpha S_\alpha$. It must be that $\mathcal{S} = \mathcal{A}$, in which case $s : \mathcal{A} \to \cup_\alpha S_\alpha$ is the desired choice function. Otherwise, we would have $\mathcal{S} \subset \mathcal{A}$ and there would exist an index $\beta \in \mathcal{A} - \mathcal{S}$ and an element $s_\beta \in S_\beta$, hence, a choice function $t : \mathcal{S} \cup \{\beta\} \to \cup_\alpha S_\alpha$, which extends s by the assignment $t(\beta) = s_\beta$. This would contradict the fact that s is maximal. □

ORDINALS

For the purpose of large-scale examples of topological spaces, we consider the special well-ordered sets called *ordinals*. This topic is useful, but not particularly crucial for the treatment of topology in this text.

It is possible for a proper subset of a set X to also be a member of X. For example, the set $X = \{a, b, \{a, b\}\}$ has one subset $\{a, b\}$ that is a member of X. Set membership isn't allowed to get too wild though.

5.9 Membership Lemma   *No nonempty set is a member of itself. More generally, there is no circle of set memberships*

$$A_1 \in A_2 \in \ldots \in A_n = A_0 \in A_1 \qquad (n \geq 1).$$

Proof: In the traditional axiomatic development of set theory, the *Axiom of Foundation* postulates in each nonempty set $X \neq \emptyset$ a member x such that $x \cap X = \emptyset$. In short, $X \neq \emptyset$ has *atoms* x. This implies that no set $A \neq \emptyset$ can be a member of itself, as follows. Were $A \in A$, then $X = \{A\}$ would be a set with no atom: its only member A would satisfy $A \cap X = \{A\}$. Similarly, a circle of set memberships, as above, would yield a set $X = \{A_1, A_2, \ldots, A_n\}$ with no atom: each member A_i would have a member in common with X, namely, the predecessor, A_{i-1}, of A_i in the circle. □

We now consider special properties of a set α whose members are sets.

A set α can have the property that *for any two distinct members of α, one of them is a member of the other*: $\forall\ \beta \neq \gamma \in \alpha, (\beta \in \gamma) \vee (\gamma \in \beta)$. The two member set $\alpha = \{\{b\}, \{a, \{b\}\}\}$ is an example.

A set α can also have the property that *the members of each member of α*

are members of α: $(\gamma \in \beta) \wedge (\beta \in \alpha) \Rightarrow (\gamma \in \alpha)$. An example is the three member set $\alpha = \{\emptyset, \{\emptyset\}, \{\emptyset, \{\emptyset\}\}\}$.

An **ordinal** is a set α with both of these properties:

(α1) *each member of α is a set whose members are members of α, and*

(α2) *given distinct members of α, one of them is a member of the other.*

Notice that, by Lemma 5.9, there is a trichotomy law in any ordinal α. For each pair $\beta \in \alpha$ and $\gamma \in \alpha$, exactly one of the following relations holds:

$$\beta \in \gamma, \qquad \beta = \gamma, \quad \text{or} \quad \gamma \in \beta.$$

5.10 Ordinal Theorem
(a) *Any member $\beta \in \alpha$ of an ordinal α is an ordinal.*
(b) *If α is an ordinal, so is the union $\alpha \cup \{\alpha\}$.*
(c) *If α and β are ordinals, then $\alpha \subset \beta \Leftrightarrow \alpha \in \beta$.*
(d) *If α and β are ordinals, then either $\alpha \subseteq \beta$ or $\beta \subseteq \alpha$.*
(e) *The collection $\mathbb{O}$ of all ordinals is not a set.*

Proof: (a) Suppose that $\delta \in \gamma \in \beta \in \alpha$. Then $\gamma \in \alpha$, and $\delta \in \alpha$ by two applications of (α1). Because $\beta \in \alpha$ and $\delta \in \alpha$, then $\delta \in \beta$ or $\beta \in \delta$ by (α2). But the second option would give an impermissible circle of memberships: $\delta \in \gamma \in \beta \in \delta$. Thus, the first option, $\delta \in \beta$, holds. This proves (β1).

Suppose $\gamma \in \beta$ and $\delta \in \beta$. Then by (α1), both $\gamma \in \alpha$ and $\delta \in \alpha$; hence, either $\gamma \in \delta$ or $\delta \in \gamma$, by (α2). This proves (β2).

(b) If α satisfies (α1)-(α2), then $\sigma = \alpha \cup \{\alpha\}$ satisfies (σ1)-(σ2).

(c) If $\alpha \subset \beta$, let $\gamma \in \beta$ be the first member of the nonempty complement $\beta - \alpha \subset \beta$. Then $\gamma \subseteq \alpha$: for all $\delta \in \gamma$, we have $\delta \in \beta$ by (β1); hence, $\delta \in \alpha$ by the minimality of $\gamma \in \beta - \alpha$. Also $\alpha \subseteq \delta$: for all $\delta \in \alpha \subset \beta$, we have $\delta \in \gamma$ by the trichotomy in β, as the options $\gamma \in \delta$ and $\gamma = \delta$ would imply $\gamma \in \alpha$ by (α1). This proves $\alpha = \gamma$, hence, $\alpha \in \beta$.

(d) Inspection shows that $\alpha \cap \beta = \{\gamma : (\gamma \in \alpha) \wedge (\gamma \in \beta)\}$ is an ordinal. Were $\alpha \cap \beta \neq \alpha$ and $\alpha \cap \beta \neq \beta$, then, by (c), we would have $\alpha \cap \beta \in \alpha$ and $\alpha \cap \beta \in \beta$, hence, $\alpha \cap \beta \in \alpha \cap \beta$, contrary to Lemma 5.9. Hence, $\alpha \cap \beta$ is either α or β; in other words, either $\alpha \subseteq \beta$ or $\beta \subseteq \alpha$.

(e) Were $\mathbb{O}$ a set, we would have ($\mathbb{O}$1) by (a) and ($\mathbb{O}$2) by (d); hence, $\mathbb{O}$ would be an ordinal. Then we would have $\mathbb{O} \in \mathbb{O}$, contrary to the Membership Lemma 5.9. $\square$

We call any collection that may not be a set a *class*. By 5.10(c, d), there is a trichotomy law for the class $\mathbb{O}$ of all ordinals: exactly one of the relations $\beta \in \gamma$, $\beta = \gamma$, and $\gamma \in \beta$ holds for any two ordinals β and γ.

We order $\mathbb{O}$ and any ordinal α via the **membership relation**:

$$\beta < \gamma \Leftrightarrow \beta \in \gamma \qquad \text{and} \qquad \beta \leq \gamma \Leftrightarrow (\beta = \gamma) \vee (\beta \in \gamma).$$

5.11 Theorem *Each ordinal and set of ordinals is well-ordered by $\leq$.*

Proof: By 5.10(*a*), an ordinal is a set of ordinals. So it suffices to consider a set X of ordinals. The relation $\leq$ is *reflexive*, by the trichotomy law; it is *anti-symmetric* $((\gamma \leq \beta) \wedge (\beta \leq \gamma) \Rightarrow \gamma = \beta)$ for there can be no membership circle, $\beta \in \gamma$ and $\gamma \in \beta$; and it is *transitive* $((\delta \leq \gamma) \wedge (\gamma \leq \beta) \Rightarrow \delta \leq \beta)$, since $(\delta \in \gamma) \wedge (\gamma \in \beta) \Rightarrow \delta \in \beta$, by ($\beta$1).

$(X, \leq)$ is totally ordered; by 5.10(*c*, *d*), any two elements are related.

Finally, $(X, \leq)$ is well-ordered. Given any nonempty set $A \subseteq X$, the axiom of foundation provides a $\beta \in A$ such that $\beta \cap A = \emptyset$. Then for all $\gamma \in A$, we have $\beta \leq \gamma$, because the option $\gamma \in \beta$ would make $\gamma \in \beta \cap A = \emptyset$. □

By Theorems 5.10 and 5.11, each ordinal α is the well-ordered set $[\emptyset, \alpha)$ of all ordinals $\beta < \alpha$. So by a transfinite construction, each well-ordered set $(X, \leq)$ has the same order type as an ordinal (Exercise 14).

For any ordinal α, the ordinal $\alpha \cup \{\alpha\}$ is denoted by $\alpha + 1$. The ordinal $\alpha + 1$ is called the *immediate successor* of α, and α is called the *immediate predecessor* of $\alpha + 1$, because (*i*) $\alpha < \alpha \cup \{\alpha\}$, (*ii*) $\alpha \cup \{\alpha\} \leq \beta$ for any ordinal $\alpha \leq \beta$, and (*iii*) $\beta \leq \alpha$ for any ordinal $\beta \leq \alpha \cup \{\alpha\}$ (Exercise 12).

Example 11 *Finite ordinals.* The first ordinal is the empty set, $\emptyset$, and is denoted by 0. The next ordinal is $\{\emptyset\}$ and is denoted by 1. The next ordinal is $\{\emptyset, \{\emptyset\}\} = \{0, 1\}$ and is denoted by 2. The next ordinal, $\{\emptyset, \{\emptyset\}, \{\emptyset, \{\emptyset\}\}\} = \{0, 1, 2\}$, is denoted by 3. When n is a finite ordinal, the next ordinal is $n + 1 = \{0, 1, 2, \ldots, n\}$.

Example 12 *The first infinite ordinal.* Another axiom of set theory, the *Axiom of Infinity*, postulates the existence of a set such that $\emptyset \in A$ and, whenever $a \in A$, we have $a \cup \{a\} \subseteq A$. The intersection of all the subsets of such a set A that have the same properties as A is the ordinal:

$$\omega = \{\emptyset, \{\emptyset\}, \{\emptyset, \{\emptyset\}\}, \{\emptyset, \{\emptyset\}, \{\emptyset, \{\emptyset\}\}\}, \ldots \}.$$

This ordinal ω may be taken as a definition of the set $\mathbb{N}$ of natural numbers.

Example 13 *Other countable ordinals.* There are ordinals

$$0 < 1 < 2 < \ldots < \omega < \omega + 1 < \omega + 2 < \ldots$$

and the set of these ordinals is an ordinal denoted by $\omega + \omega = \omega 2$. Continuing in this fashion we obtain a sequence of ordinals:

$$0 < 1 < 2 < \ldots < \omega < \omega + 1 < \omega + 2 < \ldots < \omega 2 <$$
$$\omega 2 + 1 < \ldots < \omega 3 < \omega 3 + 1 < \ldots < \omega n < \omega n + 1 < \ldots < \omega^2 <$$
$$\omega^2 + 1 < \ldots < \omega^2 + \omega < \ldots < \omega^2 + \omega 2 < \ldots < \omega^2 + \omega n < \ldots .$$

This sequence consists of two sequences of the order type of ω^2 and, hence, it is denoted by $\omega^2 2$. Subsequent ordinals include those of the form

$$\omega^k n_k + \omega^{k-1} n_{k-1} + \ldots + \omega^2 n_2 + \omega n_1 + n_0,$$

where k is finite and each n_i is finite. These are all countable ordinals.

The first ordinal and each ordinal that has an immediate predecessor is called a ***non-limit ordinal***. Other ordinals are called ***limit ordinals***.

5.12 Theorem *Let Ω be the class of all countable ordinals.*
(a) *The class Ω is a set and hence an ordinal; it is the first uncountable ordinal.*
(b) *Any countable set in Ω has an upper bound in Ω.*

Proof: (a) We accept the Well-Ordering Principle (5.6) and the existence of an ordinal with the same order type as a given well-ordered set (Exercise 14). Then the existence of uncountable sets implies the existence of uncountable ordinals. Such an uncountable ordinal Φ either consists entirely of countable ordinals and, hence, equals Ω, or the set of uncountable ordinals contained in Φ has Ω as its first member.

(b) Let $A \subseteq \Omega$ be any set. Each $\alpha \in A \subseteq \Omega$ is a countable ordinal, hence, a countable set $\alpha \subset \Omega$, by 5.10(c). The union $\beta = \cup \{\alpha \subset \Omega : \alpha \in A\}$ is a set that is an ordinal; and $\alpha \leq \beta$ for all $\alpha \in A$, by 5.10(c) (see Exercise 16). When A is countable, β the countable union of countable sets. So this upper bound β of A belongs to Ω. □

Exercises

SECTION 1

1. Devise your own homeomorphisms between circular, elliptical, and square discs.

2. Describe a homeomorphism between the pin-cushion space and the sausage-link space by drawing corresponding longitudinal and meridional curves on them.

3. Show how to mold a punctured torus into a disc with two overlapping bands. Deduce that a topologist can turn an inner tube inside out through its valve stem.

SECTION 2

1* Prove these laws of logic and set theory:
   (a) Idempotent laws 2.1(a) and 2.5(a)
   (b) Associative laws 2.1(b) and 2.5(b)
   (c) Distributive law 2.1(d) and 2.5(d)
   (d) De Morgan's laws 2.1(f) and 2.5(f)
   (e) Identity laws 2.1(e) and 2.5(e)
   (f) Complementation laws 2.1(g) and 2.5(g)

2. Use the laws of set theory to simplify the following notation:
   (a) $[A^c \cap (B \cup A^c)^c]^c$
   (b) $(A \cap B) \cup (A^c \cap B) \cup (A^c \cap B^c)$

3 Prove that for subsets $A, B \subseteq X$: $A \subseteq B \Leftrightarrow A \cup B = B \Leftrightarrow A \cap B = A \Leftrightarrow B^c \subseteq A^c$.

4 Call $\cup$ and $\cap$ *dual operations*, A and A^c *dual subsets*, and $\subseteq$ and $\supseteq$ *dual relations*.
  (a) Check that the stated laws of set theory appear in dual pairs, in the sense that the replacement of each entry in any law by its dual produces the dual law.
  (b) What does this duality imply about theorems deduced from the stated laws in set theory?

5 Consider an indexed family of subsets $\{A_\alpha \subseteq X : \alpha \in \mathcal{A}\}$. Verify:
  (a) $A_\gamma \subseteq \cup_\alpha A_\alpha, \forall \gamma \in \mathcal{A}$.
  (b) If $A_\gamma \subseteq C \subseteq X, \forall \gamma \in \mathcal{A}$, then $\cup_\alpha A_\alpha \subseteq C$.

6 Consider an indexed family of subsets $\{A_\alpha \subseteq X : \alpha \in \mathcal{A}\}$. Verify:
  (a) $\cap_\alpha A_\alpha \subseteq A_\gamma$ for all $\gamma \in \mathcal{A}$.
  (b) If $C \subseteq A_\gamma$ for all $\gamma \in \mathcal{A}$, then $C \subseteq \cap_\alpha A_\alpha$.

7 Consider indexed families of subsets $\{A_\alpha \subseteq X : \alpha \in \mathcal{A}\}$ and $\{B_\beta \subseteq X : \beta \in \mathcal{B}\}$. Establish the distributive laws for indexed unions and intersections:
  (a) $(\cup_\alpha A_\alpha) \cap (\cup_\beta B_\beta) = \cup_{(\alpha,\beta)} (A_\alpha \cap B_\beta)$.
  (b) $(\cap_\alpha A_\alpha) \cup (\cap_\beta B_\beta) = \cap_{(\alpha,\beta)} (A_\alpha \cup B_\beta)$.

8 Consider an indexed family of subsets $\{A_\alpha \subseteq X : \alpha \in \mathcal{A}\}$. Establish De Morgan's laws for indexed unions and intersections:
  (a) $X - (\cup_\alpha A_\alpha) = \cap_\alpha (X - A_\alpha)$.
  (b) $X - (\cap_\alpha A_\alpha) = \cup_\alpha (X - A_\alpha)$.

9 Consider indexed families of subsets $\{A_\alpha \subseteq X : \alpha \in \mathcal{A}\}$ and $\{B_\beta \subseteq Y : \beta \in \mathcal{B}\}$. Establish the distributive laws for unions and intersections in the product set $X \times Y$:
  (a) $(\cup_\alpha A_\alpha) \times (\cup_\beta B_\beta) = \cup_{(\alpha,\beta)} (A_\alpha \times B_\beta)$.
  (b) $(\cap_\alpha A_\alpha) \times (\cap_\beta B_\beta) = \cap_{(\alpha,\beta)} (A_\alpha \times B_\beta)$.

10 Let $S = \{a, b\}$. Form the power sets $\mathcal{P}(S)$ and $\mathcal{P}(\mathcal{P}(S))$.

11 Find all membership and containment relationships among the following sets:
$$\{a, b, c\}, \{a, b, \{a, b\}\}, \{a, b, c, \{a, b\}\}, \{\{a\}, \{b\}, \{c\}, \{a, b\}\}, \mathcal{P}(\{a, b, c\}).$$

12 For any finite set S, count the number of members of the iterated power sets:
$$\mathcal{P}(S), \ \mathcal{P}(\mathcal{P}(S)), \ \mathcal{P}(\mathcal{P}(\mathcal{P}(S))), \text{ etc.}$$

13 Characterize the sets X with the property: $X \subseteq \{X\}$.

14* Let $x, y, u,$ and v be any objects, not necessarily distinct. Prove that
$$\{x, \{x, y\}\} = \{u, \{u, v\}\}$$
if and only if both $x = u$ and $y = v$.

15* Let $A, C \subseteq X$ and $B, D \subseteq Y$ and form $X \times Y$. Prove that $\times$ distributes over $\cup$ and $\cap$:
  (a) $A \times (B \cap D) = (A \times B) \cap (A \times D)$.
  (b) $A \times (B \cup D) = (A \times B) \cup (A \times D)$.
  (c) $A \times (Y - B) = (A \times Y) - (A \times B)$.
  Prove the more general product properties:
  (d) $(A \times B) \cap (C \times D) = (A \cap C) \times (B \cap D)$.
  (e) $(A \times B) \cup (C \times D) \cup (A \times D) \cup (C \times B) = (A \cup C) \times (B \cup D)$.
  (f) $(X \times Y) - (A \times B) = (X \times (Y - B)) \cup ((X - A) \times Y)$.

16 Let A and B be subsets of X. Prove:
   (a) $A \cap B = \emptyset \Leftrightarrow A \subseteq X - B \Leftrightarrow B \subseteq X - A$.
   (b) $A \cup B = X \Leftrightarrow X - A \subseteq B \Leftrightarrow X - B \subseteq A$.

17 Describe the product sets:
   (a) $(\mathbb{Z} \times \mathbb{R}) \cup (\mathbb{R} \times \mathbb{Z}) \subseteq \mathbb{R} \times \mathbb{R}$,
   (b) $\mathbb{S}^1 \times \mathbb{S}^1 \times [0, 1]$, where $\mathbb{S}^1 = \{<x, y> \in \mathbb{R}^2 : x^2 + y^2 = 1\}$,
   (c) $\mathbb{S}^2 \times [0, 1]$, where $\mathbb{S}^2 = \{<x, y, z> \in \mathbb{R}^3 : x^2 + y^2 + z^2 = 1\}$, and
   (d) $W \times \mathbb{S}^1$, where W is a wire eight, as in Section 1.1.

18 Characterize the pairs of sets X and Y for which $X \times Y = Y \times X$.

19 Characterize the functions $f : X \to Y$ for which $F : \mathcal{P}(X) \to \mathcal{P}(Y), F(A) = f(A)$, is
   (a) injective, (b) surjective, or (c) bijective.

20 Characterize the functions $g : X \to Y$ for which $G : \mathcal{P}(Y) \to \mathcal{P}(X), G(B) = g^{-1}(A)$, is
   (a) injective, (b) surjective, or (c) bijective.

SECTION 3

1 Prove that the sequence $\left(\frac{1}{k^2}\right)$ is decreasing and that the greatest lower bound of the set of values of this sequence is 0.

2 Prove that the sequence $\left(\frac{k^3}{k^3+1}\right)$ is increasing and that the least upper bound of the set of values of this sequence is 1.

3 (a) Construct a bounded increasing sequence of rational numbers whose least upper bound is not a rational number.
   (b) Construct a bounded decreasing sequence of rational numbers whose greatest lower bound is not a rational number.

4 Find the least upper bound of the set of values of the increasing sequence:
$$\frac{1}{2}, \quad \frac{1}{2} + \frac{1}{6}, \quad \frac{1}{2} + \frac{1}{6} + \frac{1}{12}, \ldots, \quad \frac{1}{2} + \frac{1}{6} + \frac{1}{12} + \ldots + \frac{1}{k(k+1)}, \ldots.$$
Hint: First show that the kth term equals
$$\left(1 - \frac{1}{2}\right) + \left(\frac{1}{2} - \frac{1}{3}\right) + \left(\frac{1}{3} - \frac{1}{4}\right) + \ldots + \left(\frac{1}{k} - \frac{1}{k+1}\right) = \left(1 - \frac{1}{k+1}\right).$$

5 Find the least upper bound of the set of values of the increasing sequence:
$$\frac{1}{2}, \quad \frac{1}{2} + \frac{1}{4}, \quad \frac{1}{2} + \frac{1}{4} + \frac{1}{8}, \ldots, \quad \frac{1}{2} + \frac{1}{4} + \frac{1}{8} + \ldots + \frac{1}{2^k}, \ldots.$$

6 Describe the binary (base two) system of representing real numbers $0 \le r \le 1$.

7 Determine the binary (base two) and decimal expansion of the real number that is found alternately in the left, then right, binary subinterval when the unit interval $\mathbb{I}$ is repeatedly bifurcated:
$$[0, 1] \supset \left[0, \frac{1}{2}\right] \supset \left[\frac{1}{4}, \frac{1}{2}\right] \supset \left[\frac{1}{4}, \frac{3}{8}\right] \supset \left[\frac{5}{16}, \frac{3}{8}\right] \supset \ldots.$$
(Hint: The numbers are rational because the discovery procedures are periodic. Convert their binary expansions into fractional form and then convert to base ten.)

SECTIONS 1-5   EXERCISES

8   Determine the ternary and decimal expansions of the real number that is found alternately in the right, then left, ternary subinterval when the unit interval $\mathbb{I}$ is repeatedly trifurcated:

$$[0, 1] \supset \left[\tfrac{2}{3}, 1\right] \supset \left[\tfrac{2}{3}, \tfrac{7}{9}\right] \supset \left[\tfrac{20}{27}, \tfrac{7}{9}\right] \supset \left[\tfrac{20}{27}, \tfrac{61}{81}\right] \supset \cdots .$$

9   Use the Nested Intervals Theorem to prove:
(a)   Any sequence of nested rectangles with sides parallel to the axes of the plane has a nonempty intersection.
(b)   Any sequence of nested rectangular solids with sides parallel to the axes of space has a nonempty intersection.

10*   Find examples to show that the intersection of a nested family of open intervals may be empty, a single real number, or any type of interval.

11   (a)   Prove this density feature of the decimal fractions in $\mathbb{I}$: Between any two distinct real numbers $0 \leq r < s \leq 1$, there exists some decimal fraction.
(b)   Prove this density feature of the ternary fractions in $\mathbb{I}$: Between any two distinct real numbers $0 \leq r < s \leq 1$, there exists some ternary fraction.

12   (a)   Prove that both ternary fraction endpoints of each even ternary interval in $\mathbb{I}$ belong to Cantor's set C.
(b)   Prove that the ternary fractions in C do not exhaust C.

13   Describe the set of real numbers $0 \leq r \leq 1$ that have an even base four expansion.

SECTION 4

1   Show that equality can fail in the image properties 4.2.(3a-4a).

2*   Give proofs of the image and pre-image properties 4.2.(1b-4 b):
(1b) $f(f^{-1}(B)) \subseteq B$.
(2b) $f^{-1}(\cup_\beta B_\beta) = \cup_\beta f^{-1}(B_\beta)$.
(3b) $f^{-1}(\cap_\beta B_\beta) = \cap_\beta f^{-1}(B_\beta)$.
(4b) $f^{-1}(D - B) = f^{-1}(D) - f^{-1}(B)$.

3   Prove: For any function $f: X \to Y$ and subset $B \subseteq Y$, we have $f(f^{-1}(B)) = B \cap f(X)$.

4   Prove that a function $f : X \to Y$ is injective if and only if $f(A \cap B) = f(A) \cap f(B)$ for all subsets $A, B \subseteq X$.

5   Prove or disprove: For any sets X and Y, we have $X \sim Y \Leftrightarrow \mathcal{P}(X) \sim \mathcal{P}(Y)$.

6   Consider two functions $f : X \to Y$ and $g : Y \to X$ that satisfy the relation $g \circ f = 1_X$. Prove that f is injective and g is surjective.

7   Prove the following by induction:
(a)   A function $f: \{1, \ldots, n\} \to \{1, \ldots, n\}$ is injective $\Leftrightarrow f$ is surjective.
(b)   There exists a bijection $\{1, \ldots, m\} \to \{1, \ldots, k\} \Leftrightarrow m = n$.

The **graph** of a function $f : X \to Y$ is the relation $\Gamma(f) = \{<x, y> \in X \times Y : f(x) = y\}$.

8   Prove:
(a)   A relation $R \subseteq X \times Y$ is the graph $\Gamma(f)$ of a function $f : X \to Y \Leftrightarrow$ each slice $\{x\} \times Y$ meets R in exactly one point.

(b) f is surjective $\Leftrightarrow$ each slice $X \times \{y\}$ meets $\Gamma(f)$ in at least one point.
(c) f is injective $\Leftrightarrow$ each slice $X \times \{y\}$ meets $\Gamma(f)$ in at most one point.

9 Prove that two functions $f : \mathbb{R} \hookrightarrow \mathbb{R} : g$ are inverses if and only if their graphs $\Gamma(f)$ and $\Gamma(g)$ are mirror images across the line $y = x$ in the Euclidean plane $\mathbb{R} \times \mathbb{R}$ (i.e., $<x, y> \in \Gamma(f) \Leftrightarrow <y, x> \in \Gamma(g)$).

10 Find a non-identity polynomial function $\mathbb{R} \to \mathbb{R}$ that is its own inverse.

11 For all sets X, Y, and Z, construct inverse functions

$$(X \times Y) \times Z \hookrightarrow X \times (Y \times Z) \quad \text{and} \quad X \times Y \hookrightarrow Y \times X.$$

12 Let $f : X \to W$ and $g : Y \to W$ be functions with the same codomain but disjoint domains. Prove that there is a unique function $h : X \cup Y \to W$ such that $h \circ i_X = f$ and $h \circ i_Y = g$, where $i_X : X \to X \cup Y$ and $i_Y : Y \to X \cup Y$ are inclusion functions.

13 Let $f : W \to X$ and $g : W \to Y$ be functions with the same domain. Prove that there exists a unique function $h : W \to X \times Y$ such that $p_X \circ h = f$ and $p_Y \circ h = g$, where $p_X : X \times Y \to X$ and $p_Y : X \times Y \to Y$ are projection functions.

14 Let $f : X \to W$ and $g : Y \to Z$ be functions.
(a) Prove there exists a unique function $f \times g : X \times Y \to W \times Z$ such that

$$p_W \circ (f \times g) = f \circ p_X \quad \text{and} \quad p_Z \circ (f \times g) = g \circ p_Y.$$

(b) Does $(f \times g)(A \times B) = f(A) \times g(B)$ for all subsets $A \subseteq X$ and $B \subseteq Y$?
(c) Does $(f \times g)^{-1}(C \times D) = f^{-1}(C) \times g^{-1}(D)$ for all subsets $C \subseteq W$ and $D \subseteq Z$?
(d) Prove that if the functions f and g have inverses, then so does $f \times g$.

15 Let $f : X \to Y$ be a function. Prove:
(a) f is injective $\Leftrightarrow$ $f(X - A) \subseteq Y - f(A)$ for all $A \subseteq X$.
(b) f is surjective $\Leftrightarrow$ $f(X - A) \supseteq Y - f(A)$ for all $A \subseteq X$.

16 Prove that if X and Y are finite (countable) sets, then so are $X \cap Y$ and $Y - X$.

17 Prove that $\mathbb{R}^n \sim \mathbb{R}^m$ for all $n \geq 1$ and $m \geq 1$.

18 Prove that the set of ternary fractions in Cantor's set C (i.e., its set of *endpoints*) is countable, but that C is uncountable.

19 Let $2\mathbb{N} \subseteq \mathbb{N}$ be the subset of positive even integers. Use the injection $f : \mathbb{N} \to 2\mathbb{N}$, $f(k) = 4k$, and the injection $g : 2\mathbb{N} \to \mathbb{N}$, $g(2k) = 2k$, in the proof of the Bernstein-Schröder Theorem to produce a bijection $\mathbb{N} \to 2\mathbb{N}$.

SECTION 5

1 Two integers $k, m \in \mathbb{Z}$ are *congruent modulo* $n \in \mathbb{Z}$, written $k \equiv m \pmod{n}$, if there exists $j \in \mathbb{Z}$ such that $k - m = nj$. Prove that this congruence relation is an equivalence relation and determine its equivalence classes.

2 (a) Prove that the relation R on $\mathbb{R} \times \mathbb{R}$, defined by

$$<x, y> R <u, v> \Leftrightarrow ((x = v) \wedge (y = u)) \vee ((x = u) \wedge (y = v)),$$

is an equivalence relation.
(b) Determine the equivalence classes $[<x, y>]_R$ of $\mathbb{R} \times \mathbb{R}$ modulo R and describe the quotient set $\mathbb{R} \times \mathbb{R}/R$.

3* Given a partition $\mathcal{P}$ of a set X, define an equivalence relation on X whose equivalence classes are the members of $\mathcal{P}$.

4   Prove that a relation R on a set X is an equivalence relation if and only if there is a partition $\mathcal{P}$ of a set X such that $R = \cup \{A \times A : A \in \mathcal{P}\}$.

5   Show how any function $f : X \to Y$ has a factorization,

$$f = j \circ h \circ q_R : X \to X_R \to f(X) \subseteq Y,$$

as the composite of a quotient function q_R, a bijection h, and an inclusion j.

6   Let $\{A_\alpha : \alpha \in \mathcal{A}\}$ and $\{B_\beta : \beta \in \mathcal{B}\}$ be two partitions of the same set X. Prove that $\{A_\alpha \cap B_\beta : <\alpha, \beta> \in \mathcal{A} \times \mathcal{B}\}$ is a partition of X.

7   Let $\{A_\alpha : \alpha \in \mathcal{A}\}$ and $\{B_\beta : \beta \in \mathcal{B}\}$ be partitions of sets X and Y, respectively. Prove that $\{A_\alpha \times B_\beta : <\alpha, \beta> \in \mathcal{A} \times \mathcal{B}\}$ is a partition of $X \times Y$.

8   In the set $\mathbb{M}$ of integers $m > 1$, let $m \leq n$ be the relation that n divides m. Prove:
    (a) The relation $\leq$ is a partial ordering in which every chain has an upper bound.
    (b) The maximal elements of $(\mathbb{P}, \leq)$ are the prime integers.

9   Prove that the B-adic expansion function $A^{\mathbb{P}} \to \mathbb{I}$ (Example 10) is surjective, two-to-one on the finite sequences, and one-to-one on their complement.

10 Prove: In a well-ordered set,
    (a) each element that has a successsor has an immediate successor, and
    (b) an element that has a predecessor need not have an immediate predecessor.

11 Let the set $\mathbb{P}$ of positive integers and the interval $(0, 1]$ have their usual orders.
    (a) Show that the product set $\mathbb{P} \times (0, 1]$, with the lexicographic order, is order-equivalent to the ordered set $\mathbb{R}_+$ of positive real numbers. Define a set bijection

$$\mathbb{P} \times (0, 1] \to \mathbb{R}_+$$

that respects the order relations. Is $\mathbb{P} \times (0, 1]$ totally ordered, well-ordered?
    (b) Show that in the product set $(0, 1] \times \mathbb{P}$, with the lexicographic order, each element has an immediate successor, but only some have an immediate predecessor. Is $(0, 1] \times \mathbb{P}$ totally ordered, well-ordered?

12* (a) Prove that for any ordinal α, we have $\alpha \cup \{\alpha\} \leq \beta$ for any ordinal $\alpha \leq \beta$.
    (b) Prove that for any ordinal α, we have $\beta \leq \alpha$ for any ordinal $\beta < \alpha \cup \{\alpha\}$.

13 Prove that any non-increasing sequence of ordinals is eventually constant.

14* Use Transfinite Induction (5.3) to prove that each well-ordered set $(X, \leq)$ is order isomorphic to an ordinal.

15 (a) Identify the ordinal that has the same order type as the product set $\mathbb{N} \times \mathbb{N}$ with the lexicographic order (Example 7).
    (b) Identify the ordinal that has the same order type as the function set $\mathbb{N}^{\mathbb{P}}$ with the lexicographic order (Example 8).

16* Prove that any set X of ordinals has a first upper bound. Verify:
    (a) Each $\alpha \in X$ is a set and the union $\beta = \cup \{\alpha : \alpha \in X\}$ is an ordinal;
    (b) $\alpha \leq \beta$, for all $\alpha \in X$; and
    (c) $\beta \leq \gamma$, whenever γ is an ordinal such that $\alpha \leq \gamma$ for all $\alpha \in X$.

2

METRIC TOPOLOGY

Topology investigates sets of points, called topological spaces, in which one can formulate *nearness relationships* between subsets and points. These nearness relationships express how the points hang together to give the space its own character. It is easiest to formulate these ideas in a metric space, where it is possible to measure distance between points. But the topological considerations are really independent of any specific distance-measuring function, or even the ability to measure distance. What is crucial to express the nearness relationships is a *system of neighborhoods* for each point in the space. Let's run through these ideas first for the real line, then for higher dimensional Euclidean spaces and general metric spaces.

1   Topology of the Euclidean Line

Let $\leq$ denote the usual ordering on the set $\mathbb{R}$ of real numbers. We view $(\mathbb{R}, \leq)$ as the real line and each real number $x \in \mathbb{R}$ as a point in the real line.

For $x, y \in \mathbb{R}$, let $d(x, y)$ be the Euclidean distance between them, namely, the absolute value $|x - y|$ of their difference. When $\mathbb{R}$ is equipped with the Euclidean distance d, it is called the **Euclidean line** and is denoted by $\mathbb{E}^1$.

NEIGHBORHOODS IN THE EUCLIDEAN LINE

For $x \in \mathbb{E}^1$ and $r > 0$, the open interval $(x - r, x + r)$ is the set,

$$\{y \in \mathbb{E}^1 : d(x, y) < r\},$$

of all real numbers of distance less than r from x. So $(x - r, x + r)$ is called the **open interval of radius r centered on** x:

A subset $N \subseteq \mathbb{E}^1$ is called a ***neighborhood*** of the real number $x \in \mathbb{E}^1$ if N contains an open interval of positive radius centered on x.

Example 1 *An open interval is a neighborhood of each of its own members.* If $x \in (a, b)$, then each radius $0 < r < min\{d(a, x), d(x, b)\}$ determines an open interval $(x - r, x + r)$ centered on x that is contained in the given open interval (a, b).

Example 2 *A set containing a neighborhood of x is itself a neighborhood of x.* For instance, the closed interval $[0, 1]$ is neither a neighborhood of 0 nor 1, but is a neighborhood of each point of the open interval $(0, 1) \subset [0, 1]$.

ACCUMULATION POINTS

The collection of all neighborhoods of a point $x \in \mathbb{E}^1$ can be used to capture topological features of the Euclidean line near that point, as follows.

A subset $A \subseteq \mathbb{E}^1$ ***accumulates*** at $x \in \mathbb{E}^1$ if each neighborhood of x contains at least one point of $A - \{x\}$. Then x is called an ***accumulation point*** of A and each neighborhood of x contains infinitely many points of A (Exercise 5). On the other hand, when some neighborhood of x contains no members of A, except possibly x itself, then x is called a ***non-accumulation point*** of A.

Since each neighborhood of a point contains some open interval centered on that point, the accumulation terminology can be symbolized as follows.

The subset $A \subseteq \mathbb{E}^1$ accumulates at x provided that

$$\forall\, r > 0,\ (x - r, x + r) \cap (A - \{x\}) \neq \emptyset,$$

equivalently,

$$\forall\, r > 0,\ \exists\, a \in A : 0 < d(a, x) < r.$$

The subset $A \subseteq \mathbb{E}^1$ does not accumulate at x provided that

$$\exists\, r > 0 : (x - r, x + r) \cap (A - \{x\}) = \emptyset,$$

equivalently,

$$\exists\, r > 0 : \forall\, a \in A - \{x\},\ d(a, x) \geq r.$$

(This illustrates the logical De Morgan rules described in Section **1.2**.)

A set A can accumulate at some of its own points $a \in A$ and other points of $\mathbb{E}^1$ as well. The following examples expose some of the possibilities.

Example 3 The open interval (a, b) accumulates at each $a \leq x \leq b$, because each open interval $(x - r, x + r)$, with radius $r > 0$, contains the nonempty open interval $(a, b) \cap (x - r, x + r) = (m, M)$ between $m = max\ \{a, x - r\}$ and $M = min\ \{x + r, b\}$:

But (a, b) does not accumulate at any point $x < a$ nor at any point $b < y$, because an interval $(x - r, x + r)$, with radius $0 < r < a - x$, and an interval $(y - s, y + s)$, with radius $0 < s < y - b$, contain no points of (a, b):

```
    x - r   x   x + r       a                        b   y - s   y   y + s
─────(─────•─────)──────────(────────────────────────)─────(─────•─────)─────
```

Example 4 The arguments of Example 3 also establish that the closed interval $B = [a, b]$ accumulates at precisely its own points.

Example 5 The set $A = \{1, 1/2, \ldots, 1/k, \ldots\}$ of reciprocals of the natural numbers accumulates at $x = 0$: each interval $(0 - r, 0 + r)$ with radius $r > 0$ contains the tail $\{1/k, 1/(k + 1), \ldots\}$ of A whenever $k > 1/r$.

The set A does not accumulate at any other point: Any $x \neq 0$ has some member of $A - \{x\}$ nearest to it; hence, a suitably small interval centered on $x \neq 0$ contains no member of $A - \{x\}$.

DERIVED SETS

For a subset $A \subseteq \mathbb{E}^1$, let A' denote the set of all the accumulation points in $\mathbb{E}^1$ of A. The subset A' is called the ***derived set*** of A.

Example 6 *The derived set of any finite interval with endpoints $c < d$ in $\mathbb{E}^1$ is the closed interval $[c, d]$, with the endpoints included.* So the four types of intervals in $\mathbb{E}^1$ differ in their relationship with their accumulation points: each closed interval contains all its accumulation points, while the other types of intervals lack one or two of their accumulation points.

Example 7 *The set $\mathbb{Q}$ of rational numbers has derived set $\mathbb{Q}' = \mathbb{E}^1$.* Each real number is an accumulation point in $\mathbb{E}^1$ of $\mathbb{Q}$, since each interval of positive length contains rational numbers. This omnipresence of $\mathbb{Q}$ in $\mathbb{E}^1$ is due to the Rational Density Property (**1.3.1**).

Example 8 *The set D of all decimal fractions in the unit interval $\mathbb{I} = [0, 1]$ has derived set $D' = \mathbb{I}$.* The decimal fractions $d_1 d_2 \ldots d_k / 10^k$ in $\mathbb{I}$ constitute the set D of division points of the decimal system. Each real number $0 \leq x \leq 1$ is an accumulation point of D. For, if $.x_1 x_2 \ldots x_k \ldots$ is the expansion of x, then any interval $(x - r, x + r)$ with radius $r > 1/10^k > 0$ contains this k^{th} stage decimal interval containing x:

$$D[x_1 x_2 \ldots x_k] = \left[\frac{x_1 x_2 \ldots x_k}{10^k}, \frac{x_1 x_2 \ldots x_k + 1}{10^k} \right].$$

Therefore, the interval $(x - r, x + r)$ also contains the endpoint decimal fractions of $D[x_1 x_2 \ldots x_k]$ and all decimal fractions in between:

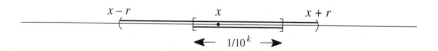

By definition, a point $x \in \mathbb{E}^1$ is a non-accumulation point of $A \subseteq \mathbb{E}^1$ if some neighborhood of x contains no point of $A - \{x\}$. So a member of A that is a non-accumulation point of $A \subseteq \mathbb{E}^1$ is one that has a neighborhood containing no other member of A; we call it an *isolated* member of A.

Example 9 *The set $\mathbb{Z}$ of integers has no accumulation points in $\mathbb{E}^1$.* Each integer $n \in \mathbb{Z}$ is isolated in its neighborhood $(n - 1, n + 1)$; each non-integer $x \notin \mathbb{Z}$ has an interval neighborhood $(n, n + 1)$ disjoint from $\mathbb{Z}$.

Example 10 A bounded set $S \subseteq \mathbb{E}^1$ accumulates at its least upper bound *lub S* and at its greatest lower bound *lub S*, unless they are isolated members of S.

LIMITS OF SEQUENCES

There are connections between the concept of accumulation points of sets of real numbers and the standard notion in calculus of the limit of a convergent sequence of real numbers. Let's review the basic definitions and then explore these connections.

Let $x \in \mathbb{E}^1$. A sequence (a_k) ***converges*** in $\mathbb{E}^1$ to x, or has ***limit*** x, if each neighborhood of x contains some tail $A_m = \{a_k : m \leq k\}$ of the sequence (a_k).

This means that, for each real number $r > 0$, there is a positive integer $m \geq 1$ such that all the terms of the sequence (a_k) with index $k \geq m$ belong to the interval neighborhood $(x - r, x + r)$ of x with radius r. In symbols,

$$\forall\, r > 0,\, \exists\, m \geq 1 : \forall\, k \geq m,\, |x - a_k| < r.$$

When (a_k) converges in $\mathbb{E}^1$ to x, we write $a_k \longrightarrow x$ in $\mathbb{E}^1$.

Roughly speaking, a convergent sequence (a_k) *eventually enters and remains in* each neighborhood of the limit point x. Typically, the smaller the radius r of the interval neighborhood $(x - r, x + r)$ of x, the larger the integers m that determine tails $A_m = \{a_k : m \leq k\}$ contained in that neighborhood.

A sequence that fails to converge to a limit is said to ***diverge***.

Example 11 *The alternating sequence $((-1)^k) = (-1, +1, -1, \ldots)$ diverges.* Each real number x has at least one of the three intervals $(-\infty, 0)$, $(-\frac{1}{2}, +\frac{1}{2})$, and $(0, +\infty)$ as a neighborhood. None of these intervals contains a tail of the sequence $((-1)^k)$, so no real number x qualifies as a limit of the sequence $((-1)^k)$.

Example 12 *The alternating harmonic sequence $((-1)^k/k) = (-1, \frac{1}{2}, \ldots)$ has limit $x = 0$.* The interval neighborhood $(-r, +r)$ of $x = 0$ contains the tail $\{(-1)^k/k : m \leq k\}$ of the sequence, whenever $0 < 1/r < m$.

1.1 Proposition *A convergent sequence in $\mathbb{E}^1$ has a unique limit.*

Proof: The reason is that, for distinct real numbers x and y, any radius $0 < r < \frac{1}{2} d(x, y)$ determines disjoint open interval neighborhoods centered on

x and y. Since disjoint intervals cannot both contain tails of the sequence, the distinct points x and y cannot both be limit points. □

Since each neighborhood of the limit of a convergent sequence contains all but finitely many members of the sequence, each convergent sequence in $\mathbb{E}^1$ is bounded in the following senses.

A sequence (a_k) is **bounded above** or **bounded below** if its set of values $\{a_k : 1 \leq k\}$ has an upper bound or a lower bound in $(\mathbb{R}, \leq)$, respectively.

While boundedness is a necessary condition for the convergence of a sequence in $\mathbb{E}^1$, it is not a sufficient one. This is illustrated by the bounded divergent sequence $((-1)^k)$ of Example 11; its terms alternate between its lower bound -1 and its upper bound $+1$. However, the Order Completeness Property (**1.3.2**) of $(\mathbb{R}, \leq)$ guarantees the convergence of any bounded sequence satisfying either of the following two *monotonicity conditions*.

A sequence (a_k) is **increasing** or **decreasing** if the relations

$$a_1 \leq a_2 \leq \ldots \leq a_k \leq a_{k+1} \leq \ldots \quad \text{or} \quad \ldots \leq a_{k+1} \leq a_k \leq \ldots \leq a_2 \leq a_1$$

hold for all integers $k \geq 1$.

1.2 Monotonic Limits Theorem
(a) *A bounded increasing sequence (a_k) has $x = lub\{a_k : 1 \leq k\}$ as a limit.*
(b) *A bounded decreasing sequence (a_k) has $x = glb\{a_k : 1 \leq k\}$ as a limit.*

Proof: When the sequence (a_k) is bounded above and increasing, consider the least upper bound, $x = lub\{a_k : 1 \leq k\}$. For any $r > 0$, the strictly smaller number $x - r < x$ is not an upper bound of the set of values $\{a_k : 1 \leq k\}$. So, for some integer $m \geq 1$, we have

$$x - r < a_m \leq a_{m+1} \leq \ldots \leq x.$$

When the sequence (a_k) is bounded below and decreasing, consider the greatest lower bound, $x = glb\{a_k : 1 \leq k\}$. For each $r > 0$, the strictly larger number $x < x + r$ is not a lower bound of the set of values $\{a_k : 1 \leq k\}$. So, for some integer $m \geq 1$, we have

$$x \leq \ldots \leq a_{m+1} \leq a_m < x + r.$$

In each case, the interval neighborhood $(x - r, x + r)$ contains a tail of the sequence (a_k). Hence, the sequence has limit x. □

We write $a_k \nearrow x$ and $a_k \searrow x$ in the cases (a) and (b) of monotonic limits.

Example 13 The increasing sequence

$$\frac{1}{3}, \frac{1}{3} + \frac{1}{9} = \frac{4}{9}, \ldots, \frac{1}{3} + \frac{1}{9} + \frac{1}{27} + \ldots + \frac{1}{3^k} = \frac{3^k - 1}{3^k(3-1)}, \ldots$$

is bounded above by $\frac{1}{2}$ since $2(3^k - 1) < 3^k (3 - 1)$ for all k. The difference between $\frac{1}{2}$ and the kth term of the sequence equals

$$\frac{1}{2} - \frac{3^k - 1}{3^k(3-1)} = \frac{1}{2(3^k)},$$

which becomes smaller than each real number $r > 0$ as the index k increases. Therefore, $\frac{1}{2}$ is the least upper bound and the limit of this sequence.

By definition, each neighborhood of the limit of a convergent sequence in $\mathbb{E}^1$ contains a tail of the sequence. So the terms of the convergent sequence not only get closer and closer to the limit, they must get closer and closer to one another. The latter property, which is formulated more precisely below, was observed by Augustin-Louis Cauchy (1789-1857) to be equivalent to the convergence property for sequences of real numbers.

A sequence (a_k) is a **Cauchy sequence** in $\mathbb{E}^1$ if, for each real number $r > 0$, the sequence has some tail, all of whose members are within Euclidean distance r of one another. This means that, for each $r > 0$, there exists an integer $m \geq 1$ such that $d(a_k, a_n) < r$ for all indices $k, n \geq m$.

1.3 Convergence Characterization *A sequence is a convergent sequence in $\mathbb{E}^1$ if and only if it is Cauchy sequence in $\mathbb{E}^1$.*

Proof: Let the sequence (a_k) converge in $\mathbb{E}^1$ to x. For each $r > 0$, there is some tail $A_m = \{a_k : m \leq k\}$ contained within the open interval neighborhood $(x - \frac{1}{2}r, x + \frac{1}{2}r)$ of x with diameter r. Then all the members of A_m are within Euclidean distance r of one another. So (a_k) is a Cauchy sequence in $\mathbb{E}^1$.

Conversely, suppose that (a_k) is a Cauchy sequence in $\mathbb{E}^1$. Then the set of values of (a_k) and each tail $A_m = \{a_k : m \leq k\}$ of (a_k) are bounded above and below (Exercise 21). Notice also that the tails of (a_k) are nested sets:

$$\ldots \subseteq A_{m+1} \subseteq A_m \subseteq \ldots \subseteq A_2 \subseteq A_1.$$

So their least upper bounds $b_m = lub\ A_m$, $1 \leq m$, form a decreasing sequence

$$\ldots \leq b_{m+1} \leq b_m \leq \ldots \leq b_2 \leq b_1,$$

bounded below by each lower bound of (a_k). Then $x = glb\ \{b_m : 1 \leq m\}$ exists and $b_m \searrow x$ by the Monotonic Limits Theorem (1.2).

To complete the proof, we show that $a_k \to x$. Let $r > 0$. Because $b_m \searrow x$ and (a_k) is a Cauchy sequence, there exists some integer $m \geq 1$ such that $x \leq b_m < x + \frac{1}{2}r$ and the members of the tail $A_m = \{a_k : m \leq k\}$ are within distance $\frac{1}{2}r$ of one another. But $b_m = lub\ A_m$, so there exists an integer $n \geq m$ such that $b_m - \frac{1}{2}r < a_n \leq b_m$. So all the members $a_k \in A_m$ satisfy the relations

$$x - r \leq b_m - r < a_k \leq b_m + \frac{1}{2}r < x + r.$$

So the interval $(x - r, x + r)$ contains the tail A_m of the sequence (a_k). This proves that $a_k \to x$. $\square$

LIMITS VERSUS ACCUMULATION POINTS

The following examples illustrate the distinctions between the limit of a convergent sequence and the accumulation points of its set of values.

Example 14 The sequence $((-1)^k(1 - 1/k))$ diverges; its set of values

$$\{\tfrac{1}{2}, -\tfrac{2}{3}, \tfrac{3}{4}, -\tfrac{4}{5}, \tfrac{5}{6}, \ldots\}$$

has two accumulation points, namely, $+1$ and -1.

Example 15 A sequence (a_k) that is eventually constant, say with terms $a_k = x$ for all $k \geq m$, has limit x. But its set of values is the finite set $\{a_1, \ldots, a_m\}$, which has no accumulation point in $\mathbb{E}^1$ (Exercise 2).

The connection between limit points and accumulation points is closer for convergent sequences that are not eventually constant.

1.4 Limit-Accumulation Properties in $\mathbb{E}^1$

(a) *The limit of a convergent sequence (a_k) in $\mathbb{E}^1$ that is not eventually constant is the unique accumulation point of its set of values $\{a_k : 1 \leq k\}$.*
(b) *Each accumulation point of a subset $B \subseteq \mathbb{E}^1$ is the limit of some convergent sequence (b_k) of distinct members of B.*

Proof: (a) Each interval $(x - r, x + r)$ centered on the limit point x contains a tail of the sequence (a_k). Because (a_k) is not eventually constant, the tail involves more than one member of the sequence. Thus, x is an accumulation point of the set of values $A = \{a_k : 1 \leq k\}$.

The intervals centered on distinct points $x \neq y$ with a radius $r < \tfrac{1}{2}d(x, y)$ are disjoint. Since $(x - r, x + r)$ contains some tail A_m of the sequence (a_k), then $(y - r, y + r)$ can contain at most members of the complementary initial portion $\{a_1, a_2, \ldots, a_{m-1}\}$ of the sequence. Then the interval centered on y with smaller radius, chosen less than $\tfrac{1}{2}d(x, y)$ and less than any of the nonzero distances $d(y, a_j) \neq 0$ ($1 \leq j \leq m - 1$), contains at most one member of the sequence, and then only if $y = a_j$ for some $j = 1, 2, \ldots, m - 1$. Thus, $y \neq x$ is not an accumulation point of the set of values, $A = \{a_k : 1 \leq k\}$, in $\mathbb{E}^1$.

(b) When x is an accumulation point of B, each interval $(x - r, x + r)$ centered on x contains some point of B distinct from x. This makes it possible to construct a sequence (b_k) of distinct points in B such that the distance of each b_k to x is less than half the distance of its predecessor, b_{k-1}, to x. Then this sequence (b_k) of distinct members of B has limit x, because each neighborhood of x contains a tail of (b_k) (Exercise 22). □

1.5 Corollary *A point x is an accumulation point of $A \subseteq \mathbb{E}^1$ if and only if each neighborhood of x contains infinitely many members of A.*

2 Topology of Euclidean Space

This section investigates neighborhoods and the nearness relationships in Euclidean spaces $\mathbb{E}^n$ of higher dimensions $n \geq 1$.

EUCLIDEAN SPACE

For each *dimension* $n \geq 1$, **real n-space** $\mathbb{R}^n$ is the set, $\mathbb{R} \times \ldots \times \mathbb{R}$, of *n*-tuples $x = (x_1, \ldots, x_n)$ of real numbers. For example, $\mathbb{R}^1$ is the real line $\mathbb{R}$, $\mathbb{R}^2$ is the real plane, and $\mathbb{R}^3$ is real space, etc.

Real *n*-space $\mathbb{R}^n$ is a real vector space. Two *n*-tuples $x = (x_1, \ldots, x_n)$ and $y = (y_1, \ldots, y_n)$ have a **vector sum:**

$$x + y = (x_1 + y_1, \ldots, x_n + y_n) \in \mathbb{R}^n.$$

A scalar $t \in \mathbb{R}$ and *n*-tuple $x = (x_1, \ldots, x_n)$ determine a **scalar product:**

$$t\,x = (t\,x_1, \ldots, t\,x_n) \in \mathbb{R}^n.$$

The **line segment** between any two points $x, y \in \mathbb{R}^n$ is the set

$$[x, y] = \{(1 - t)\,x + t\,y \in \mathbb{R}^n : 0 \leq t \leq 1\}.$$

Real *n*-space $\mathbb{R}^n$ also has a metric or distance function. The **Euclidean distance** between two points $x, y \in \mathbb{R}^n$ is given by

$$d(x, y) = \sqrt{|x_1 - y_1|^2 + \ldots + |x_n - y_n|^2}\ .$$

Euclidean distance $d(x, O)$ of a point $x \in \mathbb{R}^n$ to the origin $O = (0, \ldots, 0)$ is abbreviated by $\|x\|$ and is called the **Euclidean norm** of x. Notice that the Euclidean distance can be expressed using the norm as $d(x, y) = \|x - y\|$.

There are three crucial properties of this distance function (Exercise 1):

(**D1**) $0 \leq d(x, y)$ *for all* $x, y \in \mathbb{R}^n$, *with equality if and only if* $x = y$.

(**D2**) $d(x, y) = d(y, x)$ *for all* $x, y \in \mathbb{R}^n$, *and*

(**D3**) $d(x, z) \leq d(x, y) + d(y, z)$ *for all* $x, y, z \in \mathbb{R}^n$.

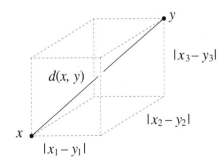
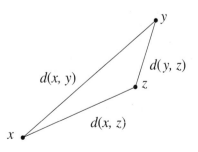

Euclidean distance   Triangle inequality (**D3**)

NEIGHBORHOODS IN EUCLIDEAN SPACE

Real n-space $\mathbb{R}^n$, when equipped with the Euclidean distance d, is called **Euclidean n-space** and is denoted by $\mathbb{E}^n$.

Consider a point $x \in \mathbb{E}^n$ and real number $r > 0$. The **open n-ball** about x with radius r is the set of points $y \in \mathbb{E}^n$ of distance strictly less than r from x:

$$B(x, r) = \{ y \in \mathbb{E}^n : d(x, y) < r \}.$$

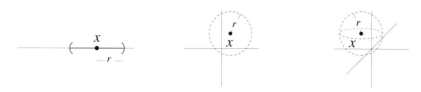

The **closed n-ball** about x with radius r is the set of points $y \in \mathbb{E}^n$ of distance less than or equal to r from x:

$$D(x, r) = \{ y \in \mathbb{E}^n : d(x, y) \leq r \}.$$

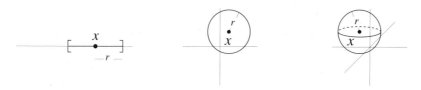

The complement, $S(x, r) = D(x, r) - B(x, r)$, is called the **sphere** about x with radius r; it is the set of points $y \in \mathbb{E}^n$ of distance r from x:

$$S(x, r) = \{ y \in \mathbb{E}^n : d(x, y) = r \}.$$

A subset $N \subseteq \mathbb{E}^n$ is called a **neighborhood** in $\mathbb{E}^n$ of a point $x \in \mathbb{E}^n$ if N contains some open n-ball $B(x, r)$ about x with positive radius $r > 0$:

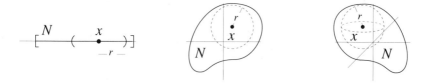

2.1 Neighborhood Property   *In Euclidean n-space $\mathbb{E}^n$,*
(a)   an open n-ball is a neighborhood of each of its points, and
(b)   the complement of a closed n-ball is a neighborhood of each of its points.

Proof: (*a*) If $y \in B(z, s)$, then $d(y, z) < s$. For any radius $0 < r < s - d(z, y)$, we have $B(y, r) \subseteq B(z, s)$, by these calculations:

$$x \in B(y, r) \implies d(x, y) < r < s - d(y, z)$$
$$\implies d(x, z) \leq d(x, y) + d(y, z) < s \quad \text{(by (D3))}$$
$$\implies x \in B(z, s).$$

Thus, the open n-ball $B(z, s)$ is a neighborhood of each of its points.

(b) If $y \notin D(z, s)$, then $d(z, y) > s$. For any radius $0 < r < d(z, y) - s$, we have $B(y, r) \subseteq X - D(z, s)$, by these calculations:

$$x \in B(y, r) \implies d(y, x) < r < d(z, y) - s$$
$$\implies d(z, x) \geq d(z, y) - d(y, x) > s \quad \text{(by (D3))}$$
$$\implies x \notin D(z, s).$$

Thus the complement $X - D(z, s)$ is a neighborhood of each of its points. □

ACCUMULATION POINTS

Neighborhoods in $\mathbb{E}^n$ determine nearness relationships in $\mathbb{E}^n$, as follows.

A subset $A \subseteq \mathbb{E}^n$ *accumulates* at the point $x \in \mathbb{E}^n$ if each neighborhood of x contains at least one point of A different from x:

$$\forall r > 0, \exists a \in A - \{x\} : d(a, x) < r.$$

Then x is called an *accumulation point* of A and each of its neighborhoods contains infinitely many points of A (Exercise 15).

The set of all accumulation points in $\mathbb{E}^n$ of a subset $A \subseteq \mathbb{E}^n$ is denoted by A' and is called the *derived set* of A.

Example 1 Consider the half-plane H of points below a horizontal line L in the Euclidean plane $\mathbb{E}^2$. The accumulation points of H include the points of H, as well as all the points of L, because each open 2-ball $B(x, r)$ centered about a point $x \in L \cup H$ contains points of $H - \{x\}$.

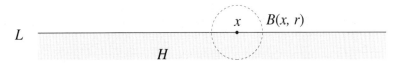

Example 1: Half-space $H \subset \mathbb{E}^2$

Points $x \in \mathbb{E}^n$ with norm $\|x\| \leq 1$, $= 1$, or < 1 form the *closed unit n-ball* $\mathbb{D}^n$, the *unit (n − 1)-sphere* $\mathbb{S}^{n-1}$, or the *open unit n-ball* $\mathbb{B}^n$, respectively.

For example, the case $n = 2$ involves the closed unit disc $\mathbb{D}^2$, the boundary circle $\mathbb{S}^1$, and the open unit disc $\mathbb{B}^2$ in the Euclidean plane $\mathbb{E}^2$.

Beware: $\mathbb{S}^2$, $\mathbb{S}^3$, etc. are not products of copies of $\mathbb{S}^1$.

Example 2 *The closed n-ball $\mathbb{D}^n = \mathbb{S}^{n-1} \cup \mathbb{B}^n$ is the derived set of $\mathbb{B}^n$ and $\mathbb{D}^n$.* Any open n-ball $B(x, r)$ about $x \in \mathbb{D}^n - \{\mathbf{0}\}$, with radius $0 < r < \|x\|$, contains the segment

$$\{t x : 1 - (r/\|x\|) < t < 1\}$$

in $\mathbb{B}^n$ on the radial line through x; and any open n-ball $B(\mathbf{O}, r)$ about $\mathbf{O} \in \mathbb{D}^n$, with radius $0 < r < 1$, lies in $\mathbb{B}^n$. So $\mathbb{D}^n \subseteq (\mathbb{B}^n)'$. Any open n-ball $B(x, r)$ about $x \notin \mathbb{D}^n$ with radius $0 < r < \|x\| - 1$ is disjoint from $\mathbb{B}^n$ by (**D3**). So $(\mathbb{B}^n)' \subseteq \mathbb{D}^n$.

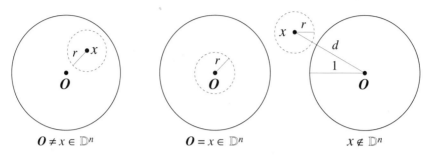

Example 2: $(\mathbb{B}^n)' = \mathbb{D}^n = (\mathbb{D}^n)'$

A point $x \in \mathbb{E}^n$ that has a neighborhood containing no point of the subset $A \subseteq \mathbb{E}^n$ different from x (i.e., $\exists\, r > 0$, $\forall a \in A - \{x\}$, $d(a, x) \geq r$) is called a **non-accumulation point** of A. A member of A that is a non-accumulation point of A is **isolated**, in the sense that it has a neighborhood in $\mathbb{E}^n$ containing no other member of A. The sets in the previous examples do not have any isolated members.

Example 3 A finite set $F \subseteq \mathbb{E}^n$ has no accumulation point in $\mathbb{E}^n$. If $x \in \mathbb{E}^n$, then any radius $0 < r < min\{d(x, y) \neq 0 : y \in F - \{x\}\}$ gives an open n-ball $B(x, r)$ disjoint from $F - \{x\}$. So F consists of isolated points that accumulate at no point outside F.

Example 4 *Asymptotic approach to a subset needn't imply accumulation.* The hyperbola $H = \{< x, y > \in \mathbb{E}^2 : xy = 1\}$ (below left) asymptotically approaches the coordinate axes $A = \{< x, y > \in \mathbb{E}^2 : xy = 0\}$. But H accumulates only at its own points, as any point $< x, y > \notin H$ is the center of some 2-ball disjoint from H. Thus, H is its own derived set: $H' = H$. Determine the derived set A'.

Example 4: Hyperbola and axes   Example 5: Polar spirals and unit circle

Example 5 *Asymptotic approach to a point implies accumulation.* The polar coordinate equations $r = 1 - 1/\theta$ and $r = 1 + 1/\theta$ ($\pi \leq \theta$) define an expanding spiral Σ_- in $\mathbb{E}^2$ and a contracting polar spiral Σ_+ in $\mathbb{E}^2$, respectively. Both spirals Σ_- and Σ_+ (above right) asymptotically approach, hence accumulate at, each point of the circle $\mathbb{S}^1 \subseteq \mathbb{E}^2$. In fact, $\Sigma_\pm' = \Sigma_\pm \cup \mathbb{S}^1$. However, $\mathbb{S}^1$ accumulates only at its own points.

OPEN-CLOSED SETS

The open intervals and the closed intervals in $\mathbb{E}^1$ have important topological properties that are captured by the following general definitions.

A subset $U \subseteq \mathbb{E}^n$ is **open in** $\mathbb{E}^n$ if, for each point $x \in U$, there is an open n-ball $B(x, r)$ of positive radius contained within U. So a subset is open in $\mathbb{E}^n$ if and only if it is a neighborhood in $\mathbb{E}^n$ of each of its points. A subset $F \subseteq \mathbb{E}^n$ is **closed in** $\mathbb{E}^n$ if F contains all its accumulation points in $\mathbb{E}^n$: $F' \subseteq F$.

Example 6 By the Neighborhood Property (2.1), each open n-ball $B(x, r)$ and the complement $X - D(x, r)$ of each closed n-ball $D(x, r)$ are *open in* $\mathbb{E}^n$. So each closed n-ball $D(x, r)$ is *closed in* $\mathbb{E}^n$: Each point outside $D(x, r)$ has $X - D(x, r)$ as a neighborhood disjoint from $D(x, r)$ and so cannot be an accumulation point of $D(x, r)$.

Example 7 The hyperbola $H = \{<x, y> \in \mathbb{E}^2 : xy = 1\}$ and the coordinate axes $A = \{<x, y> \in \mathbb{E}^2 : xy = 0\}$ in Example 4 are closed in $\mathbb{E}^2$.

Example 8 Any finite subset $F \subseteq \mathbb{E}^n$ is closed in $\mathbb{E}^n$; $F' = \emptyset$ by Example 3.

2.2 Proposition
(a) A subset is open in $\mathbb{E}^n$ if and only if its complement is closed in $\mathbb{E}^n$.
(b) Any union of sets that are open in $\mathbb{E}^n$ is open in $\mathbb{E}^n$.
(c) Any intersection of sets that are closed in $\mathbb{E}^n$ is closed in $\mathbb{E}^n$.

Proof: (a) Let $U = \mathbb{E}^n - F$ and $F = \mathbb{E}^n - U$ be complementary subsets of $\mathbb{E}^n$. Then each point $x \in U$ has a neighborhood contained within U if and only if F contains each point $x \in \mathbb{E}^n$ whose every neighborhood intersects F. In other words, U is open in $\mathbb{E}^n$ if and only if F is closed in $\mathbb{E}^n$.

(b) Let $U = \cup_\alpha U_\alpha$, where U_α is open in $\mathbb{E}^n$ for each $\alpha \in \mathcal{A}$. If $x \in U$, then $x \in U_\alpha$ for some $\alpha \in \mathcal{A}$. Hence, $B(x, r) \subseteq U_\alpha \subseteq U$ for some $r > 0$, because U_α is open in $\mathbb{E}^n$. Thus, U is open in $\mathbb{E}^n$.

(c) Let $F = \cap_\alpha F_\alpha$, where F_α is closed in $\mathbb{E}^n$ for each $\alpha \in \mathcal{A}$. By (b), $F^c = \cup_\alpha (F_\alpha)^c$ is open in $\mathbb{E}^n$. Hence, by (a), F is closed in $\mathbb{E}^n$. □

Example 9 Some subsets of $\mathbb{E}^n$ are neither open in $\mathbb{E}^n$ nor closed in $\mathbb{E}^n$. The cube $[0, 1)^n \subseteq \mathbb{E}^n$ is neither open nor closed in $\mathbb{E}^n$.

Example 10 An intersection of sets which are open in $\mathbb{E}^n$ need not be open in $\mathbb{E}^n$. The intersection of the open intervals $U_k = (-1/k, +1/k)$, indexed by the integers $k \geq 1$, is the singleton $\{0\}$. But $\{0\}$ is not open in $\mathbb{E}^1$; it contains no neighborhood of 0.

Example 11 A union of sets which are closed in $\mathbb{E}^n$ need not be closed in $\mathbb{E}^n$. The union of the closed intervals $F_k = [1/k, +1]$, indexed by the positive integers $k \geq 1$, is the interval $(0, 1]$. But $(0, 1]$ is not closed in $\mathbb{E}^1$, because it fails to contain its accumulation point 0.

2.3 Proposition
(a) Any intersection of a finite number of open sets in $\mathbb{E}^n$ is open in $\mathbb{E}^n$.
(b) Any union of a finite number of closed sets in $\mathbb{E}^n$ is closed in $\mathbb{E}^n$.

Proof: (a) Let $U = \cap_k U_k$, where U_k is open in $\mathbb{E}^n$ for each $k = 1, \ldots, m$. If $x \in U$, then $x \in U_k$ for each k. Hence, $B(x, r_k) \subseteq U_k$ for some $r_k > 0$, because U_k is open in $\mathbb{E}^n$. Thus,

$$B(x, r) \subseteq \cap_k B(x, r_k) \subseteq \cap_k U_k = U,$$

where $r = min\{r_1, \ldots, r_m\} > 0$. This proves that U is open in $\mathbb{E}^n$.

(b) This follows from (a), by use of a De Morgan Law as in the proof of (c) of Proposition 2.2. □

PERFECT SETS

A subset $A \subseteq \mathbb{E}^n$ is called **perfect** if it accumulates at precisely each of its own points. In short, a perfect subset A is one for which $A = A'$.

For example, each closed interval and each union of finitely many closed intervals is a perfect set in $\mathbb{E}^1$.

Example 12 *Cantor's Deleted Middle Third set is perfect.* Cantor's set C (Example 1.3.3) is the set of all real numbers $0 \le t \le 1$ that have an *even* ternary expansion $(.t_1 t_2 \ldots t_k \ldots)_3$, i.e., an expansion with digits $t_k \in \{0, 2\}$. For an alternative description of C, let $C_0 = [0, 1]$ and, for all $k \ge 1$, let C_k be the union of the 2^k disjoint, k^{th} stage, even ternary intervals obtained by deleting the open middle third from each of the closed intervals that constitute C_{k-1}. Then $C = \cap_k C_k$.

The set $T \cap C$ of all ternary fractions in C accumulates at each $t \in C$. If $t \in C$ has an even ternary expansion $(.t_1 t_2 \ldots t_k \ldots)_3$ that doesn't terminate in zeros, then

$$(.t_1 0 \ldots)_3, \; (.t_1 t_2 0 \ldots)_3, \; \ldots, \; (.t_1 t_2 \ldots t_k 0 \ldots)_3, \ldots$$

is an increasing (not eventually constant) sequence of ternary fractions in C that has limit (and therefore accumulates at) t. If $t \in C$ has an even ternary expansion $(.t_1 t_2 \ldots t_k \ldots)_3$ that terminates in zeros, then it is the limit of a strictly decreasing sequence of ternary fractions in C (Exercise 11). This proves that $C \subseteq C'$.

To prove the converse containment, observe that an accumulation point of C is an accumulation point of the larger closed set C_k, hence, a member of C_k, for each $k \ge 0$. This proves that $C' \subseteq \cap_k C_k = C$. So $C = C'$ and C is perfect.

3 Metric Topology

The Euclidean spaces themselves are so homogeneous and predictable that our topological investigation gets more interesting when we restrict our attention to various subsets of them, or, more generally, expand our focus to consider abstract sets that admit a distance function. The latter approach is developed in this section.

METRIC SPACES AND NEIGHBORHOODS

Let X be any subset of some Euclidean space. The Euclidean distance, when calculated between just points of X, gives a function $d : X \times X \to \mathbb{R}$ with the three properties (**D1-D3**) recorded in Section 2. This distance function is all we need to introduce balls, neighborhoods, and the nearness relation in X, without reference to the Euclidean space that contains it. This suggests the following concept of a metric space.

Let X be any set. Suppose that for each pair of points $x, y \in X$, it is possible to assign a real number $d(x, y)$, called the *distance* from x to y. Let the function $d : X \times X \to \mathbb{R}$ have these properties:

(**D1**) $0 \leq d(x, y)$ for all $x, y \in X$, with equality if and only if $x = y$,
(**D2**) $d(x, y) = d(y, x)$ for all $x, y \in X$, and
(**D3**) $d(x, z) \leq d(x, y) + d(y, z)$ for all $x, y, z \in X$.

Then d is called a *metric* on X, and the pair (X, d) is called a *metric space*. For example, *Euclidean n-space* $(\mathbb{E}^n, d)$ is a metric space.

For each point $x \in X$ and each real number $r > 0$, the set,

$$B(x, r) = \{ y \in X : d(x, y) < r \},$$

of points X of distance strictly less than r from the specific point x is called the *open ball* about x with *radius* r. At times, $B(x, r)$ may be called the open d-ball or X-ball, and may be denoted by $B_d(x, r)$ or $B_X(x, r)$, accordingly.

For each point $x \in X$ and each real number $r > 0$, the set,

$$D(x, r) = \{ y \in X : d(x, y) \leq r \},$$

of points X of distance less than or equal to r from the specific point x is called the *closed ball* about x, with *radius* r. A closed ball $D(x, r)$ in (X, d) may be denoted by $D_d(x, r)$ or $D_X(x, r)$.

A subset $N \subseteq X$ is called a *neighborhood* in X of the point $x \in X$ if N contains some open ball $B_d(x, r)$ about x with radius $r > 0$. The neighborhood relationship between $N \subseteq X$ and $x \in X$ is abbreviated by $N(x)$.

Example 1 For any set X, the *discrete metric* δ is given by

$$\delta(x, y) = 1, \text{ if } x \neq y, \quad \text{and} \quad \delta(x, y) = 0, \text{ if } x = y.$$

The pair (X, δ) is called a *discrete metric space*. In (X, δ), $B_\delta(x, r) = \{x\}$ if $r < 1$, and $B_\delta(x, r) = X$ if $1 \leq r$. Thus, any subset $N \subseteq X$ is a neighborhood of each of its points.

Example 2 Let $\mathcal{C}([0, 1])$ be the set of continuous functions $f : [0, 1] \to \mathbb{E}^1$. Then

$$d_\infty(f, g) = lub\{|f(t) - g(t)| : 0 \leq t \leq 1\}$$

is a metric on $\mathcal{C}([0, 1])$ (Exercise 2), called the *uniform metric*. The open ball $B(f, r)$ in $\mathcal{C}([0, 1])$ consists of all continuous functions g whose graph remains within a *tube* of radius r about the graph of f, as indicated in the following figures:

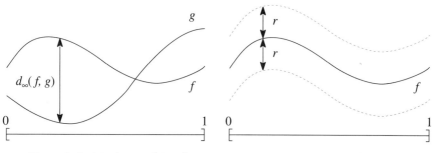

Example 2: Metric on $\mathcal{C}([0, 1])$ Example 2: Tube $B(f, r)$

For each subset $X \subseteq Y$ of a metric space (Y, d), there is the restricted metric $d : X \times X \to \mathbb{R}$ on X obtained by evaluating d on pairs of members of X. Then (X, d) is a metric space, called a ***metric subspace*** of (Y, d).

In the subspace (X, d), the open ball $B_X(x, r)$ is the intersection $B_Y(x, r) \cap X$ of X with the open Y-ball $B_Y(x, r)$ about x.

Unless otherwise specified, a subset X of real n-space $\mathbb{R}^n$ is always viewed as a metric subspace of Euclidean n-space $(\mathbb{E}^n, d)$.

Example 3   View $X = [0, 1) \cup (1, 2] \subseteq \mathbb{R}$ as a metric subspace of the Euclidean line $(\mathbb{E}^1, d)$. Then, for suitably small radii $r > 0$, the open X-balls about the points $x \in X$,

$$B_X(x, r) = (x - r, x + r) \cap X,$$

have the form $B_X(0, r) = [0, r)$, $B_X(2, r) = (2 - r, 2]$, and $B_X(x, r) = (x - r, x + r)$ for $x \neq 0, 2$. It follows that the subsets $[0, 1)$ and $(1, 2]$ are *neighborhoods in X* of precisely their own points.

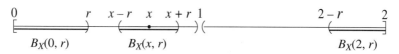

Example 3: $X = [0, 1) \cup (1, 2] \subseteq \mathbb{E}^1$

EQUIVALENT METRICS

Two metrics d and m on a space X are called ***numerically equivalent*** if there exist positive constants $c, k > 0$ such that, for all $<x, y> \in X \times X$, we have

$$c\, d(x, y) < m(x, y) \quad \text{and} \quad k\, m(x, y) < d(x, y).$$

Two metrics d and m on a space X are called ***topologically equivalent*** if they generate the same neighborhood relationships in X. This simply means that each open d-ball contains a concentric open m-ball, and vice versa.

These are equivalence relations on the set of metrics on X. Numerically equivalent metrics d and m are topologically equivalent (but not conversely). Specifically, there are these containment relationships (Exercise 3):

$$B_m(x, c\, r) \subseteq B_d(x, r) \quad \text{and} \quad B_d(x, k\, r) \subseteq B_m(x, r).$$

Example 4 Three useful alternatives to the Euclidean metric d on real n-space $\mathbb{R}^n$ are the *max metric* m, the *taxicab metric* τ, and the *radar screen metric* b:

$$m(x, y) = max\{ |x_k - y_k| : 1 \leq k \leq n\},$$
$$\tau(x, y) = \Sigma\{ |x_k - y_k| : 1 \leq k \leq n\}, \text{ and}$$
$$b(x, y) = min\{1, d(x, y)\}.$$

In $\mathbb{R}^2$, the three metrics d, m, and τ give these differently shaped open balls:

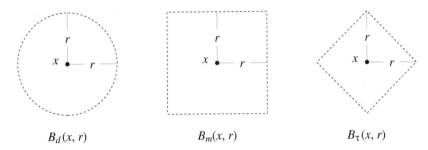

$B_d(x, r)$ $B_m(x, r)$ $B_\tau(x, r)$

The Euclidean, the max, and the taxicab metrics d, m, and τ on $\mathbb{R}^n$ are numerically equivalent; the following relationships hold for all $x, y \in \mathbb{R}^n$ (Exercise 4(b)):

$$m(x, y) \leq d(x, y) \leq \sqrt{n}\, m(x, y)) \quad \text{and} \quad m(x, y) \leq \tau(x, y) \leq n\, m(x, y).$$

The radar screen metric b is topologically equivalent to d, but it is not numerically equivalent to d (Exercise 4(d)).

ACCUMULATION POINTS AND DERIVED SETS

Let (X, d) be a metric space. As in $\mathbb{E}^n$, we have the accumulation concept.

A subset $A \subseteq X$ *accumulates* at a point $x \in X$ if each neighborhood of x contains at least one point of $A - \{x\}$ (hence, infinitely many points of A):

$$\forall\, r > 0,\, \exists\, a \in A - \{x\} : d(a, x) < r.$$

Then x is called an *accumulation point in X* of the subset A; otherwise, x is a *non-accumulation point in X* of A. The *derived set of A in X* is the subset of X, denoted by A', consisting of all of the accumulation points of A in X.

In a metric subspace (X, d) of a metric space (Y, d), the open X-ball $B_X(x, r)$ is the intersection $B_Y(x, r) \cap X$ of X with the open Y-ball $B_Y(x, r)$ about x. Therefore, for a subset $A \subseteq X$, the derived set of A in (X, d) is the intersection with X of the derived set of A in (Y, d).

Example 5 Consider the metric subspace $X = [0, 1) \cup (1, 2] \subseteq \mathbb{E}^1$ of Example 3. The subsets $A = [0, 1)$ and $B = (1, 2]$ equal their derived sets in X.

Example 6 Consider the unit 2-sphere $\mathbb{S}^2$ as a metric subspace of Euclidean space $\mathbb{E}^3$. An open $\mathbb{S}^2$-ball about a point $x \in \mathbb{S}^2$ is the intersection $B_{\mathbb{S}^2}(x, r) = B(x, r) \cap \mathbb{S}^2$ of $\mathbb{S}^2$ with an open ball $B(x, r)$ in $\mathbb{E}^3$ centered on $x \in \mathbb{S}^2$. When $r > 2$, $B_{\mathbb{S}^2}(x, r)$ is the sphere $\mathbb{S}^2$; when $r < 2$, it is a proper curved open disc in $\mathbb{S}^2$ centered on x.

Thus the northern and southern hemispheres $\mathbb{H}_+$ and $\mathbb{H}_-$ are neighborhoods in $\mathbb{S}^2$ of each of their points off the equator. Also, they equal their derived sets in $\mathbb{S}^2$.

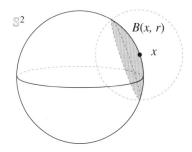
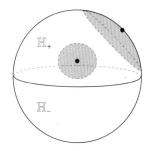

Example 6: $B_{\mathbb{S}^2}(x, r) = B(x, r) \cap \mathbb{S}^2$ Example 6: Neighborhood $\mathbb{H}_+$

LIMIT POINTS VERSUS ACCUMULATION POINTS

A sequence (a_k) **converges** in (X, d) to $x \in X$, or has **limit** x, if each neighborhood of x contains a tail of the sequence (a_k). This means that, for each real number $r > 0$, there exists a positive integer $1 \leq m$ such that

$$A_m = \{a_k : m \leq k\} \subseteq B_d(x, r).$$

A sequence that fails to converge in (X, d) is said to **diverge** in (X, d).

When (a_k) converges in (X, d) to x, we write "$a_k \to x$ in (X, d)." Notice that $a_k \to x$ in (X, d) if and only if $d(a_k, x) \to 0$ in $\mathbb{E}^1$ (Exercise 12).

This is consistent with the terminology in Section 1 for the Euclidean line $(\mathbb{E}^1, d)$. Trivial modifications of two of the results there yield:

3.1 Proposition *A limit of a convergent sequence in (X, d) is unique.*

3.2 Limit-Accumulation Properties *Let (X, d) be a metric space. Then:*
(a) *The limit of a convergent sequence (a_k) in (X, d) that is not eventually constant is the unique accumulation point of the set of values $\{a_k : 1 \leq k\}$.*
(b) *Each accumulation point of a subset $B \subseteq X$ is the limit of some convergent sequence (b_k) of distinct members of B.*

OPEN-CLOSED SUBSETS

Let (X, d) be a metric space. The open subsets $U \subseteq X$ and closed subsets $F \subseteq X$ are defined just as in Euclidean space $\mathbb{E}^n$.

A subset $U \subseteq X$ is **open in** X if, for each $x \in U$, there is some open ball $B_d(x, r)$ of positive radius $r > 0$ centered on x and contained within U.

For $U \subseteq X$ to be open in X, it is equivalent to require (Exercise 13):

- U is a neighborhood in X of each of its points;
- U contains no accumulation point of its complement $X - U$; or
- every convergent sequence outside U has its limit point outside U.

A subset $F \subseteq X$ is **closed in** X if F contains all of its accumulation points in X, i.e., $F' \subseteq F$.

For $F \subseteq X$ to be closed in X, it is equivalent to require (Exercise 14):

- *F contains any $x \in X$, all of whose neighborhoods intersect $F - \{x\}$;*
- *each point of $X - F$ has a neighborhood disjoint from F; or*
- *every convergent sequence contained in F has its limit point in F.*

When competing metrics are present for X, we may call a subset of X ***d-open*** or ***d-closed*** to clarify the metric d under consideration.

Example 7 *In a metric space (X, d), any open ball $B(z, s)$ is open in X and any closed ball $D(z, s)$ is closed in X.* The proof of the Neighborhood Property in $\mathbb{E}^n$ (2.1) suffices to establish these facts.

Example 8 By Example 3, the intervals $[0, 1)$ and $(1, 2]$ are open in the metric subspace $X = [0, 1) \cup (1, 2] \subseteq \mathbb{E}^1$. By Example 5, both $[0, 1)$ and $(1, 2]$ contain all of their own accumulation points in X, so they are also closed in X.

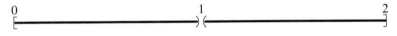

Example 8: Closed-open subsets in $X = [0, 1) \cup (1, 2]$

Example 8 suggests that a metric space with nonempty, complementary, open subsets is *disconnected*. This leads to the property of connectedness, which is studied in Section **3.1** and in Chapter **6**.

Example 9 Numerically equivalent and topologically equivalent metrics d and m determine the same neighborhoods, open sets, and closed sets in X. The reason is that each open d-ball contains a concentric open m-ball, and vice versa.

3.3 Theorem *Let (X, d) be a metric space. Then:*
(a) A subset is open in X if and only if its complement is closed in X.
(b) Any union of sets that are open in X is open in X.
(c) Any intersection of sets that are closed in X is closed in X.
(d) Any intersection of a finite number of open sets in X is open in X.
(e) Any union of a finite number of closed sets in X is closed in X.

Proof: The arguments used in Section 2 for $(\mathbb{E}^n, d)$ are valid here. □

By definition, an open set is the union of the open balls that it contains; conversely, by Theorem 3.3(b), any union of open balls is an open set. Thus, *a subset is open in a metric space if and only if it is the union of open balls*.

The open subsets and the closed subsets of a metric subspace are closely related to the open subsets and the closed subsets of the larger metric space.

3.4 Theorem *Let (X, d) be a metric subspace of (Y, d). Then:*
(a) *$V \subseteq X$ is open in $(X, d) \Leftrightarrow V = X \cap U$, where $U \subseteq Y$ is open in (Y, d).*
(b) *$G \subseteq X$ is closed in $(X, d) \Leftrightarrow G = X \cap F$, where $F \subseteq Y$ is closed in (Y, d).*

Proof: (a) Any open subset V of (X, d) is a union of X-balls; it has the form
$$V = \cup_\alpha B_X(x_\alpha, r_\alpha) = \cup_\alpha (X \cap B_Y(x_\alpha, r_\alpha)) = X \cap U,$$
for an open subset $U = \cup_\alpha B_Y(x_\alpha, r_\alpha)$ of (Y, d). Conversely, any set of the form $V = X \cap U$ is open in (X, d); it contains an X-ball $B_X(x, r) = X \cap B_Y(x, r)$ about its point $x \in V$, since the open subset U contains a Y-ball $B_Y(x, r)$.

(b) If $F \subseteq Y$, then $G = X \cap F \Leftrightarrow X - G = X \cap (Y - F)$. So, by 3.3-4(a),

G is closed in X $\Leftrightarrow$ $X - G$ is open in X
$\quad\quad\quad\quad\quad\quad\quad \Leftrightarrow X - G = X \cap (Y - F)$
$\quad\quad\quad\quad\quad\quad\quad \Leftrightarrow G = X \cap F$, for some closed set F in Y. $\quad\square$

Example 10 The *harmonic hayrick* $Y \subseteq \mathbb{E}^2$ (below left) is the union of the vertical lines $x = c$ in the plane that have x-intercept $c \in \{1, 1/2, \ldots, 1/k, \ldots, 0\}$.

Each vertical line $x = 1/k$ in Y, save $x = 0$, is an open set in Y because each equals the intersection of Y with an open vertical strip in $\mathbb{E}^2$.

Each vertical line, $x = 1/k$ in Y is a closed set in Y, as each one is closed in $\mathbb{E}^2$.

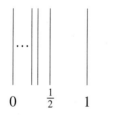

Example 10: Hayrick Example 11: Doubled topologist's sine curve

Example 11 The *doubled topologist's sine curve*,
$$Z = \{<x, \sin 1/x> \in \mathbb{E}^2 : 0 < |x| \leq 1/\pi\},$$
is the metric subspace of $\mathbb{E}^2$ depicted above, right. The two halves of Z are open and closed in Z, although they are neither open nor closed in $\mathbb{E}^2$.

3.5 Theorem *Every metric space (X, d) has these two properties:*
(a) *(Hausdorff property) Distinct points x and y have disjoint neighborhoods in X.*
(b) *(Normality) Disjoint closed sets A and B have disjoint neighborhoods in X, i.e., there exist disjoint open sets such that $A \subseteq U$ and $B \subseteq V$.*

Proof: (a) When $x \neq y$, $d(x, y) > 0$. Then any radius $0 < r < \frac{1}{2}d(x, y)$ determines open balls $B(x, r)$ and $B(y, r)$ that are disjoint by (**D3**).

(b) Since A and B are disjoint and closed, there exists for each $a \in A$ an open ball $B(a, r_a) \subseteq X - B$ and for each $b \in B$ an open ball $B(b, r_b) \subseteq X - A$. Then $U = \cup \{B(a, \frac{1}{2}r_a) : a \in A\}$ and $V = \cup \{B(b, \frac{1}{2}r_b) : b \in B\}$ are open sets containing A and B, respectively. The sets U and V are disjoint. Otherwise, there would exist $x \in B(a, \frac{1}{2}r_a) \cap B(b, \frac{1}{2}r_b)$ for some $a \in A$ and $b \in B$. Then, by the triangle inequality (**D3**), we would have

$$d(a, b) \leq d(a, x) + d(x, b) < \tfrac{1}{2} r_a + \tfrac{1}{2} r_b \leq \max \{r_a, r_b\}.$$

So either $a \in B(b, r_b)$ or $b \in B(a, r_a)$, contrary to the choices of r_a and r_b. □

CLOSURE, INTERIOR, AND BOUNDARY

The following theorem gives standard methods to enlarge any subset of a space to form a closed subset and to shrink any subset of a space to form an open subset.

3.6 Theorem *Consider any subset A of a metric space (X, d).*
(a) *The union $A \cup A'$ is closed in X and it contains A.*
(b) *The complement $A - (X - A)'$ is open in X and it is contained in A.*

Proof: (a) Suppose that $x \notin A \cup A'$. Then some X-ball $B_X(x, r)$ is disjoint from $A - \{x\} = A$. Since $B_X(x, r)$ is a neighborhood of each of its points, then $B_X(x, r)$ consists of non-accumulation points of A. So $B_X(x, r) \subseteq X - A \cup A'$. This proves that $X - A \cup A'$ is a neighborhood of each of its points; hence, it is open in X. Thus $A \cup A'$ is closed in X by Theorem 3.4(a).

(b) By (a), the set $(X - A) \cup (X - A)'$ is closed in X, and hence its complement $X - [(X - A) \cup (X - A)']$ is open in X. By De Morgan's Law (**1.2.5**(*f*)), the latter equals $A \cap [X - (X - A)'] = A - (X - A)'$. □

For any subset A of a metric space (X, d), the union $A \cup A'$ is called the *closure* of A in X and is denoted by $\text{CLS}_X(A)$, or by just $\text{CLS}(A)$ or $\overline{A}$.

The relative complement $A - (X - A)'$ is called the *interior* of A in X and is denoted by $\text{INT}_X(A)$, or by just $\text{INT}(A)$ or $\mathring{A}$ when X is implicit.

Finally, $\text{BDY}(A) = \text{CLS}(A) - \text{INT}(A)$ is called the *boundary* of A in X.

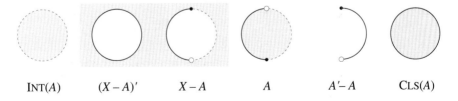

INT(A)   ($X - A$)' $X - A$ A $A' - A$ CLS(A)

Example 12 The closure, $\text{CLS}(Z)$, in $\mathbb{E}^2$ of the doubled topologist's sine curve Z (Example 11) is the union of Z with the vertical line segment $\{0\} \times [-1, +1]$ separating its two halves. The interior, $\text{INT}(Z)$, in $\mathbb{E}^2$ is empty. So $\text{BDY}(Z) = \text{CLS}(Z)$.

3.7 Theorem   *Let U and F be subsets of a metric space (X, d).*
(a)   *The subset F is closed in X if and only if $F = \text{CLS}(F)$.*
(b)   *The subset U is open in X if and only if $U = \text{INT}(U)$.*

Proof: (a) By definition, F is closed if and only if $F' \subseteq F$, or equivalently, $F = F \cup F' = \text{CLS}(F)$.

(b) By definition, U is open if and only if U is a neighborhood of each point $x \in U$. Equivalently, no point $x \in U$ is an accumulation point of the set $X - U$. So U is open if and only if U equals $\text{INT}(U) = U - (X - U)'$.   □

3.8 Theorem   *Let U and F be subsets of a metric space (X, d).*
(a)   *If $A \subseteq F$ and F is closed in X, then $\text{CLS}(A) \subseteq F$.*
(b)   *If $U \subseteq A$ and U is open in X, then $U \subseteq \text{INT}(A)$.*

Proof: (a) If $A \subseteq F$ and F is closed in X, then $A' \subseteq F' \subseteq F$. Hence, the closure of A, $\text{CLS}(A) = A \cup A'$, is contained in F.

(b) If $U \subseteq A$ and U is open in X, then no point of U is an accumulation point of $X - A$. Hence, U is contained in $\text{INT}(A) = A - (X - A)'$.   □

NEARNESS

Let (X, d) be a metric space, $x \in X$ a point, and $A \subseteq X$ a subset.

The **distance** from $x \in X$ to $A \subseteq X$ is defined to be the *greatest lower bound* of the distances from x to points of A:

$$d(x, A) = glb\{d(x, a) : a \in A\}.$$

We say that the subset $A \subseteq X$ is **arbitrarily near** the point $x \in X$ if each neighborhood of x contains at least one point of A:

$$\forall r > 0, \exists\, a \in A : d(x, a) < r.$$

Arbitrary nearness is equivalent to the distance condition $d(x, A) = 0$.

Arbitrary nearness is a weakening of the definition of accumulation: a subset A is arbitrarily near its own points and its accumulation points, namely, the points of its closure, $\text{CLS}(A) = A \cup A'$.

By these definitions, we have:

$$d(x, A) = 0 \Leftrightarrow A \text{ is arbitrarily near } x \Leftrightarrow x \in \text{CLS}(A).$$

There are two extreme possibilities for a subset $A \subseteq X$. First of all, $A \subseteq X$ may be arbitrarily near each point of X, as in this next definition.

A subset $A \subseteq X$ of the metric space (X, d) is called **dense in X** provided

$$\text{CLS}(A) = X.$$

There are the dense subsets $\mathbb{Q} \subseteq \mathbb{E}^1$ and $D \subseteq \mathbb{I}$ (Examples 1.7 and 1.8).
The second extreme possiblity involves a subset $A \subseteq X$, such as $\mathbb{Z} \subseteq \mathbb{E}^1$,

that is arbitrarily near so few points that its closure contains no open sets.

A subset $A \subseteq X$ of the metric space (X, d) is called **nowhere dense in X** if

$$\text{INT}(\text{CLS}(A)) = \emptyset.$$

The nearness relationships among subsets and points determine how the space hangs together to give it its own topological character.

Example 13 The nearness relationships in the metric subspaces

$$\mathbb{M} = \{0, 1, 1/2, \ldots, 1/k, \ldots\} \quad \text{and} \quad \mathbb{N} = \{0, 1, 2, \ldots, k, \ldots\}$$

of $\mathbb{E}^1$ are quite different. Any infinite subset $A \subseteq \mathbb{M}$ is arbitrarily near 0; but each subset $B \subseteq \mathbb{N}$ is arbitrarily near only its own members $x \in B$. So $\mathbb{M}$ has the proper dense subset $\mathbb{M} - \{0\}$, but $\mathbb{N}$ has none. Thus, no bijection $\mathbb{M} \leftrightarrow \mathbb{N}$ respects all their nearness relationships; we shall call M and N *topologically inequivalent* spaces.

Example 14 If you grasp the poles of the 2-sphere $\mathbb{S}^2$ and pull them apart without tearing $\mathbb{S}^2$, as you can with two points on a balloon, then you change the geometric shape of the sphere and its curved neighborhoods. However, you don't change the nearness relationships between points and subsets. So stretching carries the round sphere into what we shall call a *topologically equivalent* or *homeomorphic* space.

Functions from one metric space to another that respect the nearness relationships are called *continuous*; they are the subject of the next section.

4 Continuous Functions

The main task of topology is to determine which equivalent sets underlie topologically equivalent spaces. This section develops the appropriate concept of *topological* or *continuous* functions between metric spaces.

The open-closed interval $(0, 1]$ and the open interval $(0, 1)$ are set equivalent. The set bijection $(0, 1] \leftrightarrow (0, 1)$ constructed in Example **1**.4.8 tears the open-closed interval apart and reassembles the pieces to form the open interval $(0, 1)$. This procedure does not respect the nearness relationships in these metric spaces, and so it has no place in any topological investigation of these intervals. It is simply not *continuous* in the sense defined below.

CONTINUITY

Let (X, d_X) and (Y, d_Y) be any pair of metric spaces.

A function $f : X \to Y$ is **continuous** provided that, for each subset $A \subseteq X$ and each point $x \in X$, $x \in \text{CLS}_X(A)$ implies that $f(x) \in \text{CLS}_Y(f(A))$.

This means that whenever a subset $A \subseteq X$ is arbitrarily near a point $x \in X$, the image set $f(A) \subseteq Y$ is arbitrarily near the image point $f(x)$. So a continuous function is one that preserves all nearness relationships between points and subsets of the domain space X: whenever every neighborhood of a point $x \in X$ contains some point of a subset $A \subseteq X$, then every neighborhood of the image point $f(x) \in Y$ contains some point of the image set $f(A) \subseteq Y$.

Example 1 *Cutting is not continuous.* The function $k : [1, 3] \to \mathbb{E}^1$ that carries each point $1 \le x \le 2$ to $k(x) = x + 3$ and each point $2 < x \le 3$ to $k(x) = x + 4$ cuts the interval $[1, 3]$ into two halves at the point $x = 2$. The function k is *not continuous*: the subset $A = (2, 3]$ of $[1, 3]$ is arbitrarily near the point $x = 2$, yet the image set $k(A) = (6, 7]$ is not arbitrarily near the image point $k(x) = k(2) = 5$ in $\mathbb{E}^1$.

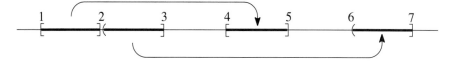

Example 1: A discontinuous function that cuts

Example 2 *Infinite stretching is continuous.* The function $g : (0, 1] \to [1, \infty)$, $g(x) = 1/x$, continuously stretches the finite interval $(0, 1]$ onto the ray $[1, \infty)$. Continuity holds because the open interval about $g(x)$ with radius $0 < s < 1 + 1/x$ contains the stretched image of the open interval about x with radius $r = x^2 s/(1 + xs)$. If $A \subseteq (0, 1]$ is arbitrarily near $x \in (0, 1]$, the interval about x contains some point $a \in A$. Hence, the interval about $g(x)$ contains the image point $g(a) \in g(A)$. This proves that the image set $g(A) \subseteq [1, \infty)$ is arbitrarily near the image point $g(x) = 1/x$.

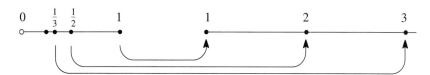

Example 2: A continuous infinite stretch

Example 3 There is no way to extend the function $s : (0, 1] \to [1, \infty)$ in Example 2 to become a continuous function $s : [0, 1] \to \mathbb{E}^1$. All the subsets $(0, 1/n] \subset [0, 1]$, where $n \ge 1$, are arbitrarily near 0, but there is no real number $s(0) \in \mathbb{E}^1$ to which all their image sets $s((0, 1/n]) = [n, \infty)$ are arbitrarily near.

A continuous function $f : X \to Y$ is allowed to carry a subset $A \subseteq X$ and point $x \in X$ that are not arbitrarily near in X into a subset $f(A) \subseteq Y$ and point $f(x) \in Y$ that are arbitrarily near in Y. *In other words, a continuous function can introduce new nearness relationships.*

Example 4 *Pinching is continuous.* The continuous function $p : \mathbb{D}^2 \to \mathbb{S}^2$ (opposite left) pinches the boundary circle $\mathbb{S}^1$ of the closed disc $\mathbb{D}^2$ into the north pole of the sphere $\mathbb{S}^2$, while carrying the center $O \in \mathbb{D}^2$ to the south pole and concentric circles in $\mathbb{D}^2$ to longitudinal circles on $\mathbb{S}^2$. This exemplifies the quotient map construction presented in Section **5**.1.

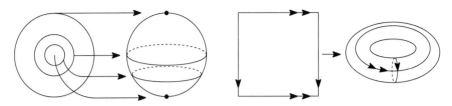

Example 4: A function that pinches   Example 5: A function that glues

Example 5 *Gluing is continuous.* It is possible to identify (i.e., glue together) two edges of a rectangle X to form a cylinder, and then to identify the two circular ends of the cylinder to form a torus (inner tube surface) Y. These instructions describe the continuous function $g : X \to Y$ (above right) which wraps the rectangle around the torus. This is an example of an *identification map*, as defined in Section **5**.1.

Try to invent a continuous function that glues the semicircles of the closed unit disc $\mathbb{D}^2 \subset \mathbb{E}^2$ together to produce the sphere $\mathbb{S}^2$.

CHARACTERIZATIONS OF CONTINUITY

By definition, a function $f : X \to Y$ between metric spaces is continuous provided that, whenever a subset $A \subseteq X$ is arbitrarily near a point $x \in X$, the image subset $f(A) \subseteq Y$ is arbitrarily near the image point $f(x)$. This is equivalent to the traditional definition of continuity given in calculus and has many other reformulations.

4.1   Continuity Characterizations   *The following are equivalent conditions for a function $f : X \to Y$ between metric spaces (X, d_X) and (Y, d_Y):*
(a)   *The function f is continuous.*
(b)   *For each subset $A \subseteq X$, we have $f(\mathrm{CLS}_X(A)) \subseteq \mathrm{CLS}_Y(f(A))$.*
(c)   *For each subset $B \subseteq Y$, we have $\mathrm{CLS}_X(f^{-1}(B)) \subseteq f^{-1}(\mathrm{CLS}_Y(B))$.*
(d)   *If $F \subseteq Y$ is closed in Y, then its pre-image $f^{-1}(F) \subseteq X$ is closed in X.*
(e)   *If $V \subseteq Y$ is open in Y, then its pre-image $f^{-1}(V) \subseteq X$ is open in X.*
(f)   *For $x \in X$ and $s > 0$, there exists an $r > 0$ such that*

$$d_X(x, z) < r \implies d_Y(f(x), f(z)) < s,$$

equivalently,

$$f(B_X(x, r)) \subseteq B_Y(f(x), s).$$

(g)   *If (a_k) converges to $x \in X$, then $(f(a_k))$ converges to $f(x) \in Y$.*

Proof: $(a) \Leftrightarrow (b)$. The continuity property is simply the condition that f carries the closure of A into the closure of $f(A)$, for each subset $A \subseteq X$.

$(b) \Rightarrow (c)$. Given $B \subseteq Y$, we apply (b) to the pre-image $A = f^{-1}(B)$ and we use the containment relation $f(f^{-1}(B)) \subseteq B$ to obtain

$$f(\mathrm{CLS}_X(f^{-1}(B))) \subseteq \mathrm{CLS}_Y(f(f^{-1}(B))) \subseteq \mathrm{CLS}_Y(B).$$

Then we have $\mathrm{CLS}_X(f^{-1}(B)) \subseteq f^{-1}(\mathrm{CLS}_Y(B))$, by the definition of pre-image.

$(c) \Rightarrow (b)$. Given $A \subseteq X$, we apply (c) to the image $B = f(A) \subseteq Y$ and we use the containment relation $A \subseteq f^{-1}(f(A))$ to obtain

$$\mathrm{CLS}_X(A) \subseteq \mathrm{CLS}_X(f^{-1}(f(A))) \subseteq f^{-1}(\mathrm{CLS}_Y(f(A))).$$

Then we have $f(\mathrm{CLS}_X(A)) \subseteq \mathrm{CLS}_Y(f(A))$, by the definition of pre-image.

$(c) \Rightarrow (d)$. If F is closed in Y, then $F = \mathrm{CLS}_Y(F)$. Then by (c), we have

$$\mathrm{CLS}_X(f^{-1}(F)) \subseteq f^{-1}(\mathrm{CLS}_Y(F)) = f^{-1}(F);$$

hence, $f^{-1}(F)$ equals its closure in X.

$(d) \Rightarrow (c)$. For $B \subseteq Y$, $\mathrm{CLS}_Y(B)$ is closed and contains B, so $f^{-1}(\mathrm{CLS}_Y(B))$ is closed and contains $f^{-1}(B)$, by (d). Thus, by Theorem 3.8, we have

$$\mathrm{CLS}_X(f^{-1}(B)) \subseteq f^{-1}(\mathrm{CLS}_Y(B)).$$

$(d) \Leftrightarrow (e)$. The pre-image process respects complements, and open sets have closed complements, and vice versa.

$(e) \Rightarrow (f)$. For $x \in X$ and $s > 0$, $B_Y(f(x), s)$ is open in Y and contains $f(x)$. So $f^{-1}(B_Y(f(x), s))$ is open in X by (e) and contains x, thus, some $B_X(x, r)$.

$(f) \Rightarrow (e)$. If V is open in Y and if $x \in f^{-1}(V)$, then $f(x) \in V$, so that $B_Y(f(x), s) \subseteq V$ for some $s > 0$. By (f), there exists an $r > 0$ such that

$$B_X(x, r) \subseteq f^{-1}(B_Y(f(x), s)) \subseteq f^{-1}(V).$$

This proves that $f^{-1}(V)$ is open in X.

$(f) \Rightarrow (g)$. If $a_k \to x$, then each X-ball $B_X(x, r)$ contains a tail A_m of (a_k). By (f), each Y-ball $B_Y(f(x), s)$ contains the image of some X-ball $B_X(x, r)$; hence, it contains a tail $f(A_m)$ of the sequence $(f(a_k))$. So $f(a_k) \to f(x)$.

$(\sim f) \Rightarrow (\sim g)$. Suppose that, for some $x \in X$ and $s > 0$, the relation

$$f(B_X(x, r)) \subseteq B_Y(f(x), s)$$

fails for all $r > 0$. Then, in particular, there exists for each $k \geq 1$ a member $a_k \in B_X(x, 1/k)$ with $f(a_k) \notin B_Y(f(x), s)$. Then the sequence (a_k) converges to $x \in X$, but $(f(a_k))$ does not converge to $f(x) \in Y$. □

Example 6 *Polynomials are continuous.* A function of the form

$$p(x) = c_0 + c_1 x + c_2 x^2 + \ldots + c_d x^d,$$

with coefficients $c_j \in \mathbb{E}^1$ ($0 \leq j \leq d$) and $c_d \neq 0$, is called a **polynomial of degree d**. Because $a_k \to x \Rightarrow p(a_k) \to p(x)$, the polynomial $p : \mathbb{E}^1 \to \mathbb{E}^1$ is continuous by (g).

Example 7 For any scalar $k \geq 0$, the **dilation** $h : \mathbb{E}^n \to \mathbb{E}^n$, given by scalar multiplication $h(x) = k x$, rescales Euclidean distance: $d(h(a), h(x)) = k\, d(a, x)$. Thus,

$$d(a_k, x) \to 0 \text{ in } \mathbb{E}^1 \Rightarrow d(h(a_k), h(x)) \to 0 \text{ in } \mathbb{E}^1.$$

Equivalently, h has the continuity property (g):

$$a_k \to x \text{ in } \mathbb{E}^n \Rightarrow h(a_k) \to h(x) \text{ in } \mathbb{E}^n.$$

Example 8 *Mapping the dust onto the firmament.* Cantor's space C consists precisely of points of $\mathbb{I} = [0, 1]$ that have an even ternary expansion (Example 1.3.3). Halving the digits in the even ternary expansion of any $t = (.t_1 t_2 \ldots t_k \ldots)_3 \in C$ forms binary digits $b_k = \frac{1}{2} t_k \in \{0, 1\}$ for some point $b = (.b_1 b_2 \ldots b_k \ldots)_2 \in \mathbb{I}$.

We define a function $\kappa : C \to \mathbb{I}$ by $\kappa(t) = b$. The ternary fraction endpoints,

$$1/3 = (.0\, 2\, \overline{2} \ldots)_3 \quad \text{and} \quad 2/3 = (.2\, 0\, \overline{0} \ldots)_3,$$

for the first open interval $(1/3, 2/3)$ deleted in the construction of C are both sent by κ to the binary fraction $(.0\, 1\, \overline{1} \ldots)_2 = \frac{1}{2} = (.1\, 0\, \overline{0} \ldots)_2$. The situation is similar for each interval deleted in the construction of C, as indicated in the figure below.

Doubling the binary digits of any point $b = (.b_1 b_2 \ldots b_k \ldots)_2 \in \mathbb{I}$ gives ternary digits for a point $t = (.t_1 t_2 \ldots t_k \ldots)_3 \in C$ with image $\kappa(t) = b$. So τ pinches the *dust-like* space C into a complete copy of the *firm* interval $\mathbb{I}$.

The Euclidean distance between points $t, t' \in C$ and the Euclidean distance between their images $\kappa(t), \kappa(t') \in \mathbb{I}$ are related this way:

$$d_C(t, t') < 1/3^k$$
$\Leftrightarrow \quad t, t' \in C$ lie in the same even k^{th} stage ternary interval
$\Leftrightarrow \quad$ the even ternary expansions of $t, t' \in C$ coincide through the kth digit
$\Rightarrow \quad$ some binary expansions of $\kappa(t), \kappa(t') \in \mathbb{I}$ coincide through the kth digit
$\Leftrightarrow \quad \kappa(t), \kappa(t') \in \mathbb{I}$ lie in the same k^{th} stage binary interval
$\Rightarrow \quad d_\mathbb{I}(\kappa(t), \kappa(t')) < 1/2^k.$

This proves that $\kappa : C \to \mathbb{I}$ is continuous, by (f), as follows. Given $s > 0$, we can find an integer k such that $1/2^k < s$ and then select a real number $0 < r < 1/3^k$. Then $d_C(t, t') < r$ implies $d_\mathbb{I}(\kappa(t), \kappa(t')) < s$, by the implications displayed above.

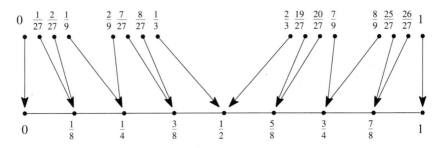

Example 8: Continuous surjection $\kappa : C \to \mathbb{I}$

Example 9 For any metric space (X, d) and point $p \in X$, the distance function $d_p : X \to \mathbb{E}^1$, defined by $d_p(x) = d(p, x)$, is continuous: We have

$$| d_p(a) - d_p(x) | \leq d(a, x), \qquad\qquad (\forall\, x \in X)$$

by the triangle inequality in (X, d). Thus, if $a_k \to x$ in X, then $d_p(a_k) \to d_p(x)$ in $\mathbb{E}^1$.

Example 10 We say that a function $g: X \to Y$ *preserves the metrics* d_X and d_Y if

$$d_Y(g(x), g(z)) = d_X(x, z) \qquad\qquad (\forall\, x, z \in X).$$

Then g is continuous by (f). Any subspace inclusion preserves the metrics and so is continuous. Any bijection $g : X \to Y$ that preserves the metrics is called an **isometry**.

COMPOSITION AND CONTINUITY

The composite $g \circ f : X \to Z$ of two functions $f : X \to Y$ and $g : Y \to Z$ is defined by $g \circ f(x) = g(f(x))$ for all $x \in X$. The image and pre-image under $g \circ f$ of subsets $A \subseteq X$ and $C \subseteq Z$ can be calculated in stages:

$$g \circ f(A) = g(f(A)) \quad \text{and} \quad (g \circ f)^{-1}(C) = f^{-1}(g^{-1}(C)).$$

4.2 Composition Theorem *If $f : X \to Y$ and $g : Y \to Z$ are continuous functions between metric spaces, then so is their composite $g \circ f : X \to Z$.*

Proof: If $A \subseteq X$ is arbitrarily near $x \in X$, then $f(A) \subseteq Y$ is arbitrarily near $f(x) \in Y$, by the continuity of f. Hence $g(f(A)) \subseteq Z$ is arbitrarily near $g(f(x)) \in Z$, by the continuity of g. This proves that $g \circ f$ is continuous.

Alternatively, if V is open in Z, then $g^{-1}(V)$ is open in Y, by the continuity of g. Hence, $f^{-1}(g^{-1}(V)) = (g \circ f)^{-1}(V)$ is open in X, by the continuity of f. Thus, $g \circ f$ is continuous, by the continuity characterization 4.1(e). □

When $i : X \to Y$ is an inclusion and $g : Y \to Z$ is any function, the composite function $g|_X = g \circ i : X \to Z$ is called the **restriction** of g to X.

4.3 Corollary *The restriction of a continuous function is continuous.*

Proof: When X is a metric subspace of Y, the inclusion i preserves the metrics and so is continuous. Then the restriction $g|_X = g \circ i : X \to Z$ of a continuous function $g : Y \to Z$ is continuous by the Composition Theorem (4.2). □

5 Completeness

The Convergence Characterization (1.3) expresses the *order completeness* of the real line $(\mathbb{R}, \leq)$ in terms of the Euclidean metric d this way: a sequence converges in $(\mathbb{E}^1, d)$ if and only if it is a Cauchy sequence in $(\mathbb{E}^1, d)$. This equivalence between the convergence and the Cauchy property is not available in all metric spaces, and is called *metric completeness*.

This section investigates the completeness of general metric spaces.

CAUCHY SEQUENCES

Let (X, d) be a metric space.

A sequence (a_k) in X is a **Cauchy sequence** in (X, d) if, for each real number $r > 0$, the sequence has some tail $\{a_k : m \leq k\}$, all of whose members are within distance r of one another. In detail, for each $r > 0$, there exists a positive integer $m \geq 1$ such that $d(a_k, a_n) < r$ for all indices $k, n \geq m$.

To aid the discussion of Cauchy sequences, we introduce these concepts:

A subset $A \subseteq X$ is ***d-bounded*** if the set of distances between points of A,

$$\{d(x, y) \in \mathbb{R} : x, y \in A\},$$

is a bounded set of real numbers. It is equivalent to require that A be contained in one of the open balls $B_d(x, r)$ in (X, d). It is possible for X itself to equal one of its open balls $B_d(x, r)$ of finite radius and so be d-bounded.

The ***diameter*** of a d-bounded subset $A \subseteq X$, denoted by $diam(A)$, is defined to be the *least upper bound* of the distances between points of A:

$$diam(A) = lub\{d(x, y) : x, y \in A\}.$$

5.1 Proposition Let (a_k) be a sequence in a metric space (X, d).
(a) If (a_k) converges in (X, d), then it is a Cauchy sequence in (X, d).
(b) If (a_k) is a Cauchy sequence in (X, d), then $\{a_k : 1 \leq k\}$ is d-bounded.

Proof: (a) Let $a_k \to x$ in (X, d). For each $r > 0$, the ball $B(x, \frac{1}{2}r)$ contains some tail $A_m = \{a_k : m \leq k\}$ of the sequence (a_k). Thus, by the triangle inequality (**D**3), all the members of A_m are within distance r of one another.

(b) Let (a_k) be a Cauchy sequence in (X, d). Then there is some positive integer $m \geq 1$ such that $d(a_k, a_n) < 1$ for all indices $k, n \geq m$. Then

$$1 + max\{d(a_k, a_n) : 1 \leq k \leq m, 1 \leq n \leq m\}$$

is an upper bound for the set $\{d(a_k, a_n) \in \mathbb{R} : 1 \leq k, n\}$. □

Neither implication (a) nor implication (b) in (5.1) has a valid converse:

Example 1 The harmonic sequence $(1/k)$ is a Cauchy sequence in the Euclidean interval $((0, 1), d)$. However, $(1/k)$ does not converge in $((0, 1), d)$. JSS

Example 2 The image, $\{-1, +1\}$, of the alternating sequence $((-1)^k)$ is bounded in the Euclidean metric d on $\mathbb{R}$. However, $((-1)^k)$ is not a Cauchy sequence in $\mathbb{E}^1 = (\mathbb{R}, d)$, since the Euclidean distance between its consecutive entries is always 2.

Using 5.1(b), the Cauchy sequence definition becomes the following.

5.2 Cauchy Characterization A sequence (a_k) is a Cauchy sequence in a metric space (X, d) if and only if its tails $A_m = \{a_k : m \leq k\}$ are d-bounded and their diameters, $diam(A_m)$, have limit 0.

COMPLETENESS PROPERTY

Some metric spaces, but not all, have a limit point for each Cauchy sequence that they contain. The following definition formalizes this property.

A metric space (X, d) and its metric d are called ***complete*** if each Cauchy sequence in (X, d) converges in (X, d).

Example 3 *The completeness property depends critically upon the metric d.* The Euclidean metric d on $\mathbb{R}$ is complete by the Convergence Characterization (1.3), but

$$g(x, y) = \left| \frac{x}{1+|x|} - \frac{y}{1+|y|} \right| \qquad (\forall\, x, y \in \mathbb{R})$$

is a *topologically equivalent* non-complete metric on $\mathbb{R}$ (Exercise 6). Also, the Euclidean metric on the set of irrationals, $\mathbb{R} - \mathbb{Q}$, is not complete, but there is a *topologically equivalent* complete metric on $\mathbb{R} - \mathbb{Q}$ given by $h(x, y) = 1/n$, where the nth digits of the decimal expansions of x and y are the first to disagree (Exercise 10).

Example 4 *A subspace of a complete metric space need not be complete.* Any sequence of rational numbers that converges in $(\mathbb{E}^1, d)$ to an irrational number is a Cauchy sequence in $(\mathbb{Q}, d)$ that fails to converge in $(\mathbb{Q}, d)$. So $(\mathbb{Q}, d)$ is not complete.

The following two results expand our catalog of metric spaces that have the completeness property.

5.3 Theorem *Numerically equivalent metric spaces (X, d) and (X, m) have the same convergent sequences and the same Cauchy sequences. Thus, one is a complete metric space if and only if the other is one.*

Proof: When d and m are numerically equivalent metrics, there are some positive constants $c, k > 0$ such that the order relations

$$c\, d(x, y) \leq m(x, y) \qquad \text{and} \qquad k\, m(x, y) \leq d(x, y)$$

hold for all $x, y \in X$.

These relations show that, for any sequence (a_k) in X and point $x \in X$, $d(a_k, x) \to 0$ if and only if $m(a_k, x) \to 0$. In other words, the sequence (a_k) converges to x in (X, d) if and only if it does so in (X, m). So these metric spaces have the same convergent sequences.

Similarly, the relations show that the tails $A_m = \{a_k : m \leq k\}$ of a sequence (a_k) in X are bounded and have diameters with limit 0 in the d metric if and only if that is the case in the m metric. So, by the Cauchy Characterization 5.2, these metric spaces have the same Cauchy sequences. □

Consider metric spaces $(X_1, d_1), \ldots, (X_n, d_n)$. The metrics $d_1, \ldots, d_n$ determine the *max metric* (see Exercise 3.10)

$$m(x, y) = max\{d_1(x_1, y_1), \ldots, d_n(x_n, y_n)\} \qquad (\forall\, x, y \in X)$$

on the set

$$X = X_1 \times \ldots \times X_n$$

of all n-tuples $x = (x_1, \ldots, x_n)$ with coordinates $x_1 \in X_1, \ldots, x_n \in X_n$.

5.4 Theorem *Let $(X_1, d_1), \ldots, (X_n, d_n)$ be complete metric spaces. Then the max metric $m = max\{d_1, \ldots, d_n\}$ on $X = X_1 \times \ldots \times X_n$ is complete.*

Proof: By definition of the max metric, the order relations

$$d_i(x_i, y_i) \leq m(x, y) \qquad (\forall\, x, y \in X)$$

hold for each index $1 \leq i \leq n$. Thus, when $(a(k))$ is a Cauchy sequence in the product space (X, m), the coordinate sequence $(a_i(k))$ is a Cauchy sequence in the coordinate space (X_i, d_i) for each $1 \leq i \leq n$.

For each index $1 \leq i \leq n$, the metric space (X_i, d_i) is complete and so $(a_i(k))$ converges there to some point x_i. So as k increases, $d_i(a_i(k), x_i) \to 0$.

Let $x = (x_1, \ldots, x_n)$. Then, as k increases,

$$m(a(k), x) = max\{d_1(a_1(k), x_1), \ldots, d_n(a_n(k), x_n)\} \to 0$$

and so the sequence $(a(k))$ converges in (X, m) to x. □

Example 5 *Euclidean n-space* $(\mathbb{E}^n, d)$ *is complete.* By Theorem 5.4, the max metric m determined by n copies of the Euclidean line $(\mathbb{E}^1, d)$ is a complete metric on real n-space $\mathbb{R}^n = \mathbb{R} \times \ldots \times \mathbb{R}$. Then the numerically equivalent Euclidean metric on $\mathbb{R}^n$ (see Example 3.4) is complete by Theorem 5.3.

Completeness in a metric space requires the existence of limits for all Cauchy sequences. Cantor's Nested Intervals Theorem (1.3.3) is one concrete expression of the completeness of the Euclidean line. The following generalization of Cantor's theorem to arbitrary complete metric spaces is due to Felix Hausdorff (1868-1942).

5.5 Nested Set Theorem *Let (X, d) be a complete metric space and let*

$$F_1 \supseteq F_2 \supseteq \ldots \supseteq F_m \supseteq F_{m+1} \supseteq \ldots$$

be a sequence of nonempty, nested, closed, d-bounded subsets whose diameters have limit 0. Then the intersection $\cap_m F_m$ is a single point of X.

Proof: First, we show $\cap_m F_m \neq \emptyset$. We select a point $a_k \in F_k$ for each $k \geq 1$ to produce a sequence (a_k) in (X, d). By the nested feature of the sequence $\{F_m\}$, each tail $A_m = \{a_k : m \leq k\}$ of (a_k) is contained in the corresponding closed set F_m. Thus, $\{A_m\}$ also consists of d-bounded sets whose diameters have limit 0. This proves that (a_k) is a Cauchy sequence in (X, d), by 5.2.

Since (X, d) is complete, (a_k) converges to some $x \in X$. For each $m \geq 1$, the subsequence (a_{m+k}), which is contained in F_m, also converges to x. As F_m is a closed set, it contains the limit x (Exercise 3.14). Thus $x \in \cap_m F_m$.

Second, we show that all points $x, y \in \cap_m F_m$ are equal. For each $m \geq 1$, we have $x, y \in F_m$, hence, $d(x, y) \leq diam\ F_m$. Because $diam\ F_m \to 0$, we conclude that $d(x, y) = 0$. Hence, $x = y$ by the metric axiom (**D**1). □

In Euclidean spaces and certain other complete metric spaces, the hypothesis that the diameters have limit 0 is not required to guarantee a nonempty intersection. See Section 3.2 and Cantor's Theorem (7.1.3).

Another consequence of completeness of a metric space (X, d) involves dense subsets. As in Section 3, we say that a set $A \subseteq X$ is *dense* in (X, d) or *d-dense* if CLS $(A) = X$. This just means that each open ball in X contains at least one point of A. The next result was established in 1899 by R. Baire.

5.6 Baire's Category Theorem *In a complete metric space (X, d), the intersection of any countable family of open dense sets in X is dense in X.*

Proof: Consider a countable family $\{A_m\}$ of open dense sets in (X, d). We must show that each open ball $B(x, r)$ intersects $\cap_m A_m$. Because A_1 is an open dense set, $B(x, r) \cap A_1$ is nonempty and open, hence, contains some closed ball $D(a_1, r_1)$ such that $0 < r_1 < \frac{1}{2} r$. Similarly, $B(a_1, r_1) \cap A_2$ contains some closed ball $D(a_2, r_2)$, $0 < r_2 < \frac{1}{2} r_1$. By induction, $B(a_{m-1}, r_{m-1}) \cap A_m$ contains some closed ball $D(a_m, r_m)$ with radius $0 < r_m < \frac{1}{2} r_{m-1}$, for all $m \geq 1$.

This construction produces a sequence of nested, closed, d-balls

$$D(a_1, r_1) \supseteq D(a_2, r_2) \supseteq \ldots \supseteq D(a_{m-1}, r_{m-1}) \supseteq D(a_m, r_m) \supseteq \ldots$$

that are contained in $B(x, r)$ and have diameters that satisfy the inequalities

$$\text{diam } D(a_m, r_m) \leq 2 r_m \leq (\tfrac{1}{2})^{m-1} r,$$

hence, have limit zero. By this construction,

$$B(x, r) \cap (\cap_m A_m) \supseteq \cap_m D(a_m, r_m);$$

by the Nested Set Theorem (5.5), $\cap_m D(a_m, r_m) \neq \emptyset$. Thus $B(x, r)$ intersects $\cap_m A_m$. This proves that $\cap_m A_m$ is a dense subset of X. $\square$

Example 6 The complement $\mathbb{E}^1 - \{r\}$ of each rational number $r \in \mathbb{Q}$ is a dense open set in the Euclidean line $\mathbb{E}^1$. The countable intersection $\cap \{\mathbb{E}^1 - \{r\} : r \in \mathbb{Q}\}$ is the dense set $\mathbb{E}^1 - \mathbb{Q}$ of irrational numbers. This illustrates Baire's Theorem (5.6).

The intersection of arbitrary dense subsets need not be dense, as the disjoint dense sets $\mathbb{Q}$ and $\mathbb{E}^1 - \mathbb{Q}$ in $\mathbb{E}^1$ illustrate.

There are numerous applications of Baire's theorem to analysis. For example, S. Banach used it to prove that the set of continuous functions $f : [0, 1] \to \mathbb{E}^1$ having a derivative at no point in $[0, 1)$ is dense in the metric space $(\mathcal{C}([0, 1]), d_\infty)$ (Example 3.2) of all continuous functions on $[0, 1]$.

Just as the rational numbers are a dense metric subspace of the complete Euclidean line, *any metric space (X, d) is isometric to a dense metric subspace of a complete metric space (Y, d)*. The latter is called a **completion** of (X, d). In 1914, F. Hausdorff gave a general completion technique modeled on Cauchy's vintage 1821 construction of the real numbers as equivalence classes of Cauchy sequences of rational numbers. Exercise 16 outlines the construction and asks for the details.

CONTRACTIONS

A function $f : X \to X$ on a metric space (X, d) is a ***contraction*** if there is a constant $0 \leq c < 1$ such that
$$d(f(x), f(y)) \leq c\, d(x, y) \qquad (\forall\, x, y \in X).$$
Any such number c is called a (***contractivity***) ***scalar*** for f.

Example 7 Given a point $p \in \mathbb{E}^n$ and scalar $r \geq 0$, the function
$$g_p : \mathbb{E}^n \to \mathbb{E}^n,\ g_p(x) = r\,x + (1-r)\,p,$$
is called a ***similarity of*** $\mathbb{E}^n$ ***centered on p with ratio r***. It rescales Euclidean distance:
$$d(g_p(x), g_p(y)) = r\, d(x, y) \qquad (\forall\, x, y \in \mathbb{E}^n).$$
So g_p is a contraction of Euclidean n-space $\mathbb{E}^n$ if and only if it has ratio $0 \leq r < 1$.

5.7 Proposition *Any contraction of a metric space is continuous.*

Proof: Let $f : X \to X$ be a contraction of (X, d). Then we have
$$d(a_k, x) \to 0 \text{ in } \mathbb{E}^1 \implies d(f(a_k), f(x)) \to 0 \text{ in } \mathbb{E}^1.$$
Equivalently, f has the continuity property:
$$a_k \to x \text{ in } (X, d) \implies f(a_k) \to f(x) \text{ in } (X, d). \qquad \square$$

Apparently, any contraction f of Euclidean space $\mathbb{E}^n$ moves points towards a point $p \in \mathbb{E}^n$ left fixed: $f(p) = p$. For example, a similarity $g_p : \mathbb{E}^n \to \mathbb{E}^n$ with ratio $0 \leq r < 1$ draws all points toward its center $p \in \mathbb{E}^n$.

This illustrates a principle discovered by Stefan Banach (1892-1945). We say that $x \in X$ is a ***fixed point*** of a function $f : X \to X$ if $f(x) = x$.

5.8 Banach's Contraction Mapping Principle
Let $f : X \to X$ be a contraction of a complete metric space (X, d). Then:
(a) The function f has a unique fixed point $x \in X$.
(b) For each $z \in X$, the sequence
$$z,\ f(z),\ f(f(z)),\ f(f(f(z))),\ \ldots$$
of iterated images of z under f converges to the unique fixed point x.

Proof: If x and y are both fixed points of f, then
$$d(x, y) = d(f(x), f(y)) \leq c\, d(x, y).$$
Since $0 \leq c < 1$, then $d(x, y) = 0$ and $x = y$. So f has at most one fixed point.

Given $z \in X$, we construct the sequence $(f^k(z))$ of iterated images of z under f: $f^1(z) = f(z)$ and $f^{k+1}(z) = f(f^k(z))$ for all $k \geq 1$. By the contraction property of f, successive steps of this sequence $(f^k(z))$ shrink by the factor

$0 \le c < 1$. Therefore, for each $k \ge 1$, we have
$$d(f^k(z), f^{k+1}(z)) \le c\, d(f^{k-1}(z), f^k(z)) \le \ldots \le c^k\, d(z, f(z)).$$
Therefore, when $n \ge k \ge 1$, we have
$$\begin{aligned}d(f^k(z), f^n(z)) &\le d(f^k(z), f^{k+1}(z)) + \ldots + d(f^{n-1}(z), f^n(z))\\ &\le (c^k + \ldots + c^{n-1})\, d(z, f(z))\\ &= \frac{c^k - c^n}{1-c}\, d(z, f(z)) \le \frac{c^k}{1-c}\, d(z, f(z)).\end{aligned}$$

Since $c^k \to 0$ ($c < 1$), this proves that $(f^k(z))$ is a Cauchy sequence in (X, d).

Since (X, d) is complete, $(f^k(z))$ converges to some $x \in X$. Because the contraction f is continuous, $(f(f^k(z)))$ must converge to $f(x)$. But $(f(f^k(z)))$ is the subsequence $(f^{k+1}(z))$ of $(f^k(z))$ and, therefore, also must converge to x. By the uniqueness of limits, we have that $f(x) = x$. □

By Exercise 4, every closed subspace of a complete metric space is complete. Therefore, the Contraction Mapping Principle (5.8) applies not only to all Euclidean spaces $\mathbb{E}^n$ but to all closed subspaces of $\mathbb{E}^n$, as well.

Example 8 By the Contraction Mapping Principle, any contraction of the closed unit Euclidean ball $\mathbb{D}^n$ has a fixed point. According to a more general theorem of L. E. J. Brouwer, any continuous function $\mathbb{D}^n \to \mathbb{D}^n$ has a fixed point. Both of these results hold in each dimension $n \ge 1$. But we will only establish the *Brouwer Fixed Point Theorem* in dimension $n = 1$ (Section **3.1**) and $n = 2$ (Section **9.2**).

Example 9 The contraction property of a real valued function $f : [a, b] \to [a, b]$ is the *Lipschitz condition*: $|f(x) - f(z)| \le c|x - z|$ for all $x, z \in [a, b]$, where $0 \le c < 1$.

When f is a differentiable function on $[a, b]$ and $|f'(x)| \le c < 1$ for all $x \in [a, b]$, the Mean Value Theorem in differential calculus implies that f satisfies the Lipschitz condition and, hence, is a contraction. A sequence $(z_k = f^k(z))$ converging to the fixed point $x \in [a, b]$ is *monotonic* if $0 < f'(x) < 1$ on $[a, b]$ (as in the *web diagram*, below left), and *oscillatory* if $-1 < f'(x) < 0$ on $[a, b]$ (below right).

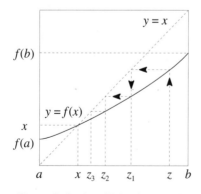

Example 9: $0 < f'(x) < 1$ on $[a, b]$

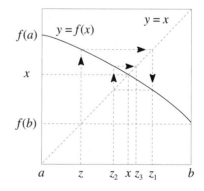

Example 9: $-1 < f'(x) < 0$ on $[a, b]$

Applications of the principle in analysis include Picard's Theorem on the existence of solutions to a differential equation $y' = y$, where $y = f(x, y)$ satisfies a Lipschitz condition with respect to y. The complete metric space involved is $(\mathcal{C}([a, b]), d_\infty)$ (see Exercise 8). Other applications involve approximations to linear equations (Exercise 13).

We apply the principle in Section 3.3 to contractions of the *hyperspace* of sequentially compact subspaces of a complete metric space. Often, the *fixed points* in the hyperspace are compact subspaces with unusual metric features, in particular, with *fractional Hausdorff dimension*. This Hausdorff dimension is a metric, not topological, concept whose discussion we confine to certain exercises in Sections 5, 3.3, and 3.5.

Exercises

SECTION 1

1. Determine the derived sets in $\mathbb{E}^1$ of the intervals $(0, 1)$, $(0, 1]$, $[0, 1)$, and $[0, 1]$.

2*. Prove that a finite subset $A \subseteq \mathbb{E}^1$ has no accumulation points in $\mathbb{E}^1$.

3. (a) Construct an infinite subset $A \subseteq \mathbb{E}^1$ that has no accumulation points in $\mathbb{E}^1$.
   (b) Find an infinite subset $A \subseteq \mathbb{E}^1$ that has just one accumulation point in $\mathbb{E}^1$.
   (c) Find an infinite subset $A \subseteq \mathbb{E}^1$ that equals its set of accumulation points in $\mathbb{E}^1$.

4. (a) Prove that if $F \subseteq \mathbb{E}^1$ is a finite set, then $\mathbb{E}^1 - F$ is a neighborhood of each of its own points.
   (b) Show that (a) fails for some infinite sets $F \subseteq \mathbb{E}^1$.

5*. Prove that if $x \in \mathbb{E}^1$ is an accumulation point of $A \subseteq \mathbb{E}^1$, then each neighborhood of x contains infinitely many members of A.

6. Determine accumulation points of these subsets of the unit interval $[0, 1]$:
   (a) the set of all binary fractions $0 \le b/2^k \le 1$, and
   (b) the set of all ternary fractions $0 \le t/3^k \le 1$.

7. Show that the members of $A = \{0, 1/2, 2/3, \ldots, 1 - 1/k, \ldots\}$ are isolated in $\mathbb{E}^1$. Find all the accumulation points of A.

8. Find a pair of disjoint subsets of $\mathbb{E}^1$ that have exactly the same derived set in $\mathbb{E}^1$.

A subset A is called *dense* in $\mathbb{E}^1$ if $A \cup A' = \mathbb{E}^1$.

9. (a) Prove that no finite set is dense in $\mathbb{E}^1$.
   (b) Find dense subsets A and B of $\mathbb{E}^1$ that are complementary subsets ($A \cup B = \mathbb{E}^1$ and $A \cap B = \emptyset$).

10. Prove that the sequence (k) of positive integers diverges in $\mathbb{E}^1$.

11. Prove that if $A \subseteq B$ in $\mathbb{R}$, then $A' \subseteq B'$ in $\mathbb{E}^1$.

12. Prove that the following sequences are monotonic and bounded. Determine the

greatest lower bound, least upper bound, and limit of each one.

(a) $\left(\frac{1}{k^2}\right)$ (b) $\left(\frac{k^3}{k^3+1}\right)$

13   Construct a bounded monotonic sequence of rational numbers whose limit is not a rational number.

14   Find the least upper bound and limit of the increasing sequence

$$\frac{1}{3}, \quad \frac{1}{3}+\frac{1}{9}, \quad \frac{1}{3}+\frac{1}{9}+\frac{1}{27}, \quad \ldots, \quad \frac{1}{3}+\frac{1}{9}+\frac{1}{27}+\ldots+\frac{1}{3^k}, \quad \ldots.$$

15   Prove that two open intervals centered on distinct points $x \neq y$ in $\mathbb{R}$ with a positive radius $r \leq \frac{1}{2} d(x, y)$ are disjoint.

16   (a)   Construct a sequence in $\mathbb{E}^1$ that has no limit point, but whose set of values accumulates at each positive integer.
     (b)   Construct a sequence in $\mathbb{E}^1$ that has a limit point, but whose set of values has no accumulation point.
     (c)   Construct a sequence in $\mathbb{E}^1$ that has no limit point, but whose set of values accumulates at each real number.

17   Construct a countable set $A \subseteq \mathbb{E}^1$ of irrational numbers whose accumulation points in $\mathbb{E}^1$ include all the real numbers.

18   (a)   Prove that Cantor's set C equals its derived set C' in $\mathbb{E}^1$. (In Section 2, a set $A \subseteq \mathbb{E}^1$ such that $A = A'$ is called *perfect*.)
     (b)   Prove that Cantor's set C contains no nonempty open intervals. (In Section 3, a set $A \subseteq \mathbb{E}^1$ that contains no nonempty open intervals is called *nowhere dense*.)

19   Use the base five notational system for real numbers and a variation of Cantor's construction to produce a perfect, nowhere dense, subset of $\mathbb{E}^1$. (See Exercise 18.)

20   (a)   Find an infinite subset $A \subseteq (0, 1)$ that has no accumulation point in $(0, 1)$.
     (b)   Is there an infinite subset $A \subseteq [0, 1]$ that has no accumulation point in $[0, 1]$?

21*   Prove that each Cauchy (a_k) sequence in $\mathbb{E}^1$ is bounded above and below.

22*   Let (b_k) be a sequence in $\mathbb{E}^1$ such that the distance of each term b_k to x is less than half the distance of its predecessor b_{k-1} to x. Prove that (b_k) has limit point x.

23   In $\mathbb{E}^1$, construct a convergent sequence and a divergent sequence that have the same set of values.

24   Given a Cauchy sequence (a_k) in $\mathbb{E}^1$, consider the greatest lower bounds $c_m = glb \, A_m$ of the tails $A_m = \{a_k : m \leq k\}$ $(1 \leq m)$ of (a_k). Prove that (a_k) converges to the limit of the bounded increasing sequence (c_m).

SECTION 2

1*   (a)   Derive the **Cauchy-Schwarz inequality** $|x \cdot y| \leq \|x\| \, \|y\|$, for $x, y \in \mathbb{R}^n$, from the inequality $0 \leq \sum_{i,k} (x_i y_k - x_k y_i)^2$. (The dot product $x \cdot y$ equals $\sum_i x_i y_i$.)
     (b)   Derive the **Minkowski inequality**, $\|x + y\| \leq \|x\| + \|y\|$ ($x, y \in \mathbb{R}^n$), from the Cauchy-Schwarz inequality.
     (c)   Establish the properties (**D1–D3**) for Euclidean distance $d(x, y) = \|x - y\|$.

2   (a) Prove that the intersection of any two open n-balls is an open set in $\mathbb{E}^n$.
    (b) Prove that $\mathbb{E}^n$ and the empty subset $\emptyset$ are both open and closed in $\mathbb{E}^n$.
    (c) Prove that the $(n-1)$-sphere $\mathbb{S}^{n-1}$ is closed, but not open, in $\mathbb{E}^n$.

3   Let $0 < r < 1$ be the limit of the increasing sequence $(a_k) \subset [0, 1]$, as well as the limit of the decreasing sequence $(b_k) \subset [0, 1]$.
    (a) Find the accumulation points of the subset $A = \cup_k [a_k, b_k] \times \{1/k\}$ of the Euclidean plane $\mathbb{E}^2$ that do not belong to A.
    (b) Find the accumulation points of the subset $B = \cup_k ([0, a_k] \cup [b_k, 1]) \times \{1/k\}$ of the Euclidean plane $\mathbb{E}^2$ that do not belong to B.

4   In the open unit 2-ball $\mathbb{B}^2$, construct disjoint subsets, A and B, whose derived sets contain the boundary 1-sphere $\mathbb{S}^1$.

5   Consider arbitrary subsets $A, B \subseteq \mathbb{E}^n$.
    (a) Prove that the derived set of the union $A \cup B$ equals the union of the derived sets of A and B: $(A \cup B)' = A' \cup B'$.
    (b) Decide whether the derived set of the intersection $A \cap B$ must equal the intersection of the derived sets of A and B.

6   Prove that a bounded closed set S in $\mathbb{E}^1$ contains both its *glb* and its *lub*.

7   Prove that $\mathbb{E}^n - \mathbb{D}^n$ and $\mathbb{E}^n - \mathbb{B}^n$ have the same accumulation points in $\mathbb{E}^n$.

8   (a) Prove that a subset $U \subseteq \mathbb{E}^n$ is open in $\mathbb{E}^n$ if and only if U contains none of the accumulation points of its complement $\mathbb{E}^n - U$.
    (b) Prove that a subset $F \subseteq \mathbb{E}^n$ is closed in $\mathbb{E}^n$ if and only if F contains each point $x \in \mathbb{E}^n$ whose every neighborhood intersects F.

9   Prove that each open set in $\mathbb{E}^n$ is the union of countably many open n-balls.

10 (a) Prove that the unions $\Sigma_- \cup \mathbb{S}^1$ and $\Sigma_+ \cup \mathbb{S}^1$ are closed in $\mathbb{E}^2$, where Σ_- and Σ_+ are the polar spirals in Example 5.
    (b) Prove that for any subset $A \subseteq \mathbb{E}^n$ the union $A \cup A'$ is closed in $\mathbb{E}^n$.

11* Prove that if a point $t \in C$ of Cantor's set (Example 12) has an even ternary expansion $(.t_1 t_2 \ldots t_k \ldots)_3$ that terminates in zeros, then it is the limit of a strictly decreasing sequence of ternary fractions in C.
    (Hint: 0 is the limit of the sequence $(.02\bar{2}\ldots)_3$, $(.002\bar{2}\ldots)_3$, $(.0002\bar{2}\ldots)_3$, $\ldots$.)

12 Prove that the set $C \times C \subset \mathbb{E}^1 \times \mathbb{E}^1$ of ordered pairs of members of Cantor's set $C \subseteq \mathbb{E}^1$ is a perfect subset of $\mathbb{E}^2 = \mathbb{E}^1 \times \mathbb{E}^1$.

13 Let $z \in \mathbb{E}^n$ be any point. Let (a_k) be any sequence of points such that the distance of each a_k to z is less than half the distance of its predecessor a_{k-1} to z. Prove that $A = \{a_k : 1 \le k\}$ is an infinite set whose only accumulation point in $\mathbb{E}^n$ is z.

14 (a) Determine whether the set E (Example 1.3.2) of real numbers in $\mathbb{I}$ that have an even decimal expansion is a perfect subset of $\mathbb{E}^1$.
    (a) Determine whether the set of real numbers in $\mathbb{I}$ that have an even base four expansion is a perfect subset of $\mathbb{E}^1$.

15* Prove that a point $x \in \mathbb{E}^n$ is an accumulation point of a subset $A \subseteq \mathbb{E}^n$ if and only if each neighborhood of x contains infinitely many points of A.

SECTION 3

1. Decide whether any of the following formulas defines a metric on $\mathbb{R}$:
   (a) $d(x, y) = |x - y|^2$
   (b) $d(x, y) = |x^2 - y^2|$
   (c) $d(x, y) = |x - y|^3$
   (d) $d(x, y) = |x^3 - y^3|$

2* Let $\mathcal{C}([0, 1])$ be the set of real-valued continuous functions $f : [0, 1] \to \mathbb{E}^1$. Prove that the function $d_\infty(f, g) = lub\{|f(t) - g(t)| : 0 \le t \le 1\}$ is a metric on $\mathcal{C}([0, 1])$.

3* (a) Let d and m be numerically equivalent metrics on X, with constants $c, k > 0$ such that $c\, d(x, y) < m(x, y)$ and $k\, m(x, y) < d(x, y)$, for all $<x, y> \in X \times X$. Prove:

$$B_m(x, cr) \subseteq B_d(x, r) \quad \text{and} \quad B_d(x, kr) \subseteq B_m(x, r).$$

   (b) Prove that numerically equivalent metrics are topologically equivalent.
   (c) Prove that topologically equivalent metrics on a set X determine the same open sets and the same closed sets.

4* (a) Verify the metric properties (**D1-D3**) for the max metric m, the taxicab metric τ, and the radar screen metric b on $\mathbb{R}^n$ (Example 4).
   (b) Establish the numerical equivalence of the Euclidean metric d, the max metric m, and the taxicab metric τ on $\mathbb{R}^n$ with these relationships:

$$m(x, y) \le d(x, y) \le \sqrt{n}\, m(x, y) \quad \text{and} \quad m(x, y) \le \tau(x, y) \le n\, m(x, y).$$

   (c) Draw open balls in $\mathbb{R}^3$ determined by the four metrics d, m, τ, and b.
   (d) Prove that the radar screen metric b is topologically equivalent, but not numerically equivalent, to the Euclidean metric d on $\mathbb{R}^n$.

5. Determine whether the following sets are open or closed in $\mathbb{E}^2$, in $\mathbb{D}^2$, or in $\mathbb{B}^2$:
   (a) $\{<x, y> \in \mathbb{B}^2 : \tfrac{1}{2} \le x^2 + y^2 < 1\}$
   (b) $\{<x, y> \in \mathbb{B}^2 : x^2 < y\}$

6. Find the closure and interior of the following subsets and their complements in the Euclidean line $\mathbb{E}^1$:
   (a) the set $\mathbb{Z}$ of integers,
   (b) the set $\mathbb{Q}$ of rational numbers, and
   (c) Cantor's set C.
   Are any of these sets dense in $\mathbb{E}^1$?

7. Find the closure and interior of the following subsets:
   (a) $\{<x, y> \in \mathbb{E}^2 : x^2 \le y\}$ in the Euclidean plane $\mathbb{E}^2$,
   (b) $\{<x, y, z> \in \mathbb{E}^3 : z = 0\}$ in the Euclidean space $\mathbb{E}^3$, and
   (c) $\{<x, y> \in \mathbb{E}^2 : y = 1/x, x \ne 0\}$ in the Euclidean plane $\mathbb{E}^2$.

8. (a) Prove that $A \subseteq X$ is dense in (X, d) if and only if A intersects all open sets in X.
   (b) Construct a countable dense subset in Euclidean n-space $\mathbb{E}^n$.
   (c) Find a metric space (X, d) and a point $x \in X$ whose complement $X - \{x\}$ is not dense in X.
   (d) Find a proper subset that is both open and dense in the Euclidean line $\mathbb{E}^1$.

9. Let (X, d) be a metric space. Prove:
   (a) $\cup \{B(x, s)) : 0 < s < r\} = B(x, r)$;
   (b) $\cup \{\text{CLS}(B(x, s)) : 0 < s < r\} = B(x, r)$; and
   (c) $B(x, r) \subseteq \text{INT}(\text{CLS}(B(x, r)))$, but they are not equal, in general.

SECTIONS 1-5   EXERCISES   79

10* Let (X, d_X) and (Y, d_Y) be metric spaces and let $X \times Y$ be the product set. Prove:
   (a) $m(<x, y>, <w, z>) = max\{d_X(x, w), d_Y(y, z)\}$ is a metric on $X \times Y$.
   (b) $W \subseteq X \times Y$ is m-open if and only if W is the union of sets of the form $U \times V$, where $U \subseteq X$ is d_X-open and $V \subseteq Y$ is d_Y-open.

11 For any subsets $A \subseteq B \subseteq X$ in a metric space (X, d), prove:
   (a) $X - \text{INT}(A) = \text{CLS}(X - A)$.   (b) $X - \text{CLS}(A) = \text{INT}(X - A)$.
   (c) $\text{CLS}(A) \subseteq \text{CLS}(B)$.   (d) $\text{INT}(A) \subseteq \text{INT}(B)$.

12* Prove the following for any sequence (a_k) in a metric space (X, d):
   (a) $a_k \to x$ in $(X, d) \Leftrightarrow d(a_k, x) \to 0$ in $\mathbb{E}^1$.
   (b) If δ is a discrete metric, $a_k \to x$ in $(X, \delta) \Leftrightarrow (a_k)$ is eventually constant at x.

13* Prove the equivalence of these properties of a subset U of a metric space (X, d):
   (a) U is open in X.
   (b) U contains no accumulation point of its complement $X - U$.
   (c) Any convergent sequence outside U has its limit point outside U.

14* Prove the equivalence of these properties of a subset F of a metric space (X, d):
   (a) F is closed in X.
   (b) F contains any point $x \in X$, all of whose neighborhoods intersect $F - \{x\}$.
   (c) any convergent sequence contained in F has its limit point in F.

15 In a metric space (X, d), the distance from a point $x \in X$ to a subset $A \subseteq X$ is defined by $d(x, A) = glb\{d(x, a) : a \in A\}$. Prove:
   (a) $d(x, A - \{x\}) = 0 \Leftrightarrow x \in A'$.
   (b) $d(x, A) = 0 \Leftrightarrow x \in \text{CLS}(A)$.
   (c) $d(x, X - A) > 0 \Leftrightarrow x \in \text{INT}(A)$.
   (d) $F \subseteq X$ is closed in $(X, d) \Leftrightarrow [d(x, F) = 0 \Rightarrow x \in F]$.

16 **Hilbert space** $\mathbb{H}$ consists of all sequences $x = (x_k)$ in $\mathbb{R}$ for which $\Sigma_{k=1}^{\infty} x_k^2$ is a convergent infinite series (i.e., has a convergent sequence of partial sums). Prove:
$$d(x, y) = \sqrt{\Sigma_{k=1}^{\infty} |x_k - y_k|^2}$$
is a metric on $\mathbb{H}$.

17 Let $\mathbb{Z}$ be the set of integers. Let $p > 0$ be a prime integer. Define $d_p : \mathbb{Z} \times \mathbb{Z} \to \mathbb{R}$ by $d_p(m, n) = 0$ when $m = n$; otherwise, $d_p(m, n) = 1/p^r$, where p^r is the largest non-negative power of p that divides $m - n$.
   (a) Prove that $(\mathbb{Z}, d_p)$ is a metric space.
   (b) Describe the open balls $B(0, 1)$ and $B(0, 1/p^s)$ $(s \geq 1)$ in $(\mathbb{Z}, d_p)$.

18 Invent a metric on the punctured plane $\mathbb{R}^2 - \{O\}$ that is appropriate for a traveler who walks only along rays emanating from the origin O or along arcs of circles centered on the origin O. Prove that your expression is a metric on $\mathbb{R}^2 - \{O\}$.

In a metric space (X, d), a subset $A \subseteq X$ is **d-bounded** if $\{d(x, y) : x, y \in A\}$ is a bounded set of real numbers. The **diameter** of a d-bounded subset $A \subseteq X$, denoted by $diam(A)$, is defined to be the least upper bound $lub\{d(x, y) : x, y \in A\}$.

19 Prove: (a_k) is a Cauchy sequence in $\mathbb{E}^1 \Leftrightarrow diam(A_m) \to 0$, where $A_m = \{a_k : m \leq k\}$.

20 In any metric space, characterize the Cauchy sequences that have only a finite set of distinct values.

A **pseudo-metric** on a set X is a function $d : X \times X \to \mathbb{R}$ satisfying the metric properties (**D3**) and (**D2**), and this weakened version of (**D1**):

(**D*1**) $0 \leq d(x, y)$ for all $x, y \in X$.

21* Let d be a pseudo-metric on X. Prove:
   (a) All the definitions in Section 3 apply to the pseudo-metric space (X, d) and all the results of Section 3 follow except Theorem 3.5.
   (b) $R = \{<x, y> \in X \times X : d(x, y) = 0\}$ is an equivalence relation on X.
   (c) There is a unique metric d^* on the quotient set $X^* = X/R$ such that $d^*(x^*, y^*) = d(x, y)$ for all $<x, y> \in X \times X$, where $x^* = [x]_R$ and $y^* = [y]_R$.

22 Let (X, d) be a metric space. Prove that $\min(1, d(x, y))$ is a metric on X that determines the same open sets, closed sets, and neighborhoods as the original metric d.

SECTION 4

1 Prove: The identity $1_X : X \to X$ is a continuous function for any metric space (X, d).

2 For any set X, the **constant** function $c_y : X \to Y$ at $y \in Y$ is defined by $c_y(x) = y$ for all $x \in X$. Prove that $c_y : X \to Y$ is a continuous function of metric spaces X and Y.

3 For any subset $A \subseteq X$, the **characteristic** function $\chi_A : X \to \{0, 1\}$ is defined by the rule $\chi_A(x) = 1$, if $x \in A$, and $\chi_A(x) = 0$, if $x \notin A$. Every function $f : X \to \{0, 1\}$ equals the characteristic function χ_A for its **support**, $A = \{x \in X : f(x) = 1\}$. Prove:
   (a) If $A = [0, 1] \subseteq [0, 1] \cup [2, 3] = X$, then $\chi_A : X \to \{0, 1\}$ is continuous.
   (b) If $A = [0, 1] \subseteq [0, 3] = X$, then $\chi_A : X \to \{0, 1\}$ is not continuous.
   (c) $\chi_A : X \to \{0, 1\}$ is continuous if and only if A is open and closed in (X, d).

4 Let $f, g : X \to \mathbb{E}^1$ be continuous real-valued functions on a metric space X. Prove:
   (a) Each linear combination $\alpha f + \beta g$ $(\alpha, \beta \in \mathbb{R})$ is contiunous.
   (b) The product function $f \cdot g$ is continuous.
   (c) If $f(x) \neq 0$ for all $x \in X$, then the reciprocal function $1/f$ is continuous.

5 Establish the continuity with respect to the Euclidean metrics of these functions:
   (a) $g : \mathbb{R}^2 \to \mathbb{R}$, $g(<x_1, x_2>) = x_1 + x_2$, and
   (b) $h : \mathbb{R}^2 \to \mathbb{R}$, $h(<x_1, x_2>) = x_1 x_2$.

6 Find a continuous function of the unit interval $\mathbb{I} = [0, 1] \subset \mathbb{E}^1$ onto a cross $\times \subseteq \mathbb{E}^2$.

7 Construct a continuous surjection $(0, 1) \to [0, 1]$. What about $[0, 1] \to (0, 1)$?

8 Prove: Continuity means $d_X(x, A) = 0 \Rightarrow d_Y(f(x), f(A)) = 0$, for all $A \subseteq X, x \in X$.

9 Prove that the identity functions $1_X : (X, d) \to (X, m)$ and $1_X : (X, m) \to (X, d)$ are continuous when d and m are numerically equivalent metrics on X.

10 Prove that if (X, d) is a metric space and (Z, δ) is a discrete metric space, then any function $f : (Z, \delta) \to (X, d)$ is continuous.

11 (a) Construct a continuous function that fails to have this property: Whenever a subset $A \subseteq X$ accumulates at a point $x \in X$, the image subset $f(A) \subseteq Y$ accumulates at the image point $f(x) \in Y$.
   (b) Prove that an injective function $f : X \to Y$ between metric spaces is continuous if and only if it has the property in (a).

12 Prove that a function $f : X \to Y$ between metric spaces is continuous if and only if, for each neighborhood V in Y of an image point $f(x)$, the pre-image $f^{-1}(V)$ is a neighborhood in X of x.

13 Sketch a picture of a continuous pinching function $p : X \to Y$, in these cases:
(a) X is the torus and Y is the 2-sphere,
(b) X is the 2-sphere and Y is the pincushion space (Section **1.1**), and
(c) X is the 2-sphere and Y is a pair of tangent 2-spheres.

14 Define a non-constant continuous function $f: \mathbb{I} \to \mathbb{I}$ such that $f(C) = \{0\}$, where C is Cantor's set (Example **1.3.3**).

15 Make use of different characterizations of continuity to reprove the continuity of compositions of continuous functions.

A function $f : X \to Y$ between metric spaces (X, d_X) and (Y, d_Y) is ***continuous at*** $x \in X$ if, for each $s > 0$, there exists $r > 0$ such that $f(B_X(x, r)) \subseteq B_Y(f(x), s)$.

16 (a) Prove that $f : X \to Y$ is continuous if and only if it is continuous at each $x \in X$.
(b) Prove that the function $f : \mathbb{E}^1 \to \mathbb{E}^1$, given by $f(x) = x$ for irrational x and $f(x) = 0$ for rational x, is continuous only at $x = 0$.

17 A function $f : X \to X$ on a metric space (X, d) is a ***contraction*** if there is a constant $0 \le c < 1$ such that $d(f(x), f(y)) \le c\, d(x, y)$ for all $x, y \in X$. Prove that any contraction is a continuous function.

18 Let $f : X \to Y$ be a continuous function between metric spaces. Prove that for each subset $A \subseteq X$, the restriction $f|_A : A \to f(A)$ is a continuous function between the metric subspaces $A \subseteq X$ and $f(A) \subseteq Y$.

19 Let $X \times Y$ have the max metric m determined by metric spaces (X, d_X) and (Y, d_Y) (Exercise 3.10). Prove that the metric m has these properties:
(a) The projection functions $p_X : X \times Y \to X$ and $p_Y : X \times Y \to Y$ are continuous.
(b) A function $f : W \to X \times Y$ is continuous if and only if the composite functions $p_X \circ f : W \to X$ and $p_Y \circ f : W \to Y$ are continuous.

20 Let $X \times Y$ have the max metric m determined by metric spaces (X, d_X) and (Y, d_Y).
(a) Prove that if a function $f : (X, d_X) \to (Y, d_Y)$ is continuous, then the graph $\Gamma(f) = \{<x, y> \in X \times Y : y = f(x)\}$ is a closed subset of $(X \times Y, m)$.
(b) Find a discontinuous function $f : \mathbb{E}^1 \to \mathbb{E}^1$ with closed graph in $\mathbb{E}^2$.

21* For any metric space (X, d) and subset $A \subseteq X$, the distance function $d_A : X \to \mathbb{E}^1$ is defined by $d_A(x) = d(x, A)$ (see Section **2.3**). Prove that d_A is continuous. (Hint: First show that $|d_A(x) - d_A(z)| \le d(x, z)$ for all $x, z \in X$.)

Let (X, d) be a metric space. A function $g : X \to \mathbb{E}^1$ is **upper semicontinuous** if for each $r \in \mathbb{E}^1$, the pre-image $g^{-1}((-\infty, r))$ is open in X; it is **lower semicontinuous** if for each $r \in \mathbb{E}^1$, the pre-image $g^{-1}((r, \infty))$ is open in X.

22 Let (X, d) be a metric space.
(a) Prove that a function $g : X \to \mathbb{E}^1$ is continuous if and only if it is both upper and lower semicontinuous.
(b) Let $\{g_\alpha : \alpha \in \mathscr{A}\}$ be a family of continuous functions $g_\alpha : X \to [m, M] \subseteq \mathbb{E}^1$. Prove that the function $M(x) = lub\, \{g_\alpha(x) : \alpha \in \mathscr{A}\}$ is lower semicontinuous and the the function $m(x) = glb\, \{g_\alpha(x) : \alpha \in \mathscr{A}\}$ is upper semicontinuous.

A ***pseudo-metric*** on a set X is a function $d : X \times X \to \mathbb{R}$ satisfying the metric properties (**D3**) and (**D2**), and this weakened version of (**D1**):

(**D*1**) $0 \leq d(x, y)$ for all $x, y \in X$.

23 A pseudo-metric d on X determines an equivalence relation R on X by $x R y \Leftrightarrow d(x, y) = 0$. Prove that the quotient function $q_R : X \to X/R$ is a continuous surjection that carries open (closed) sets to open (respectively, closed) sets.

SECTION 5

1 (a) Prove that in Euclidean n-space $(\mathbb{E}^n, d)$, any open ball $B_d(x, r)$ has the same diameter as its closure $\text{CLS}(B_d(x, r))$ and its boundary $\text{BDY}(B_d(x, r))$.
(b) Let δ be the discrete metric on $\mathbb{R}^n$ (Example 3.1). Compare the diameters of an open ball $B_\delta(x, r)$, its closure $\text{CLS}(B_\delta(x, r))$, and its boundary $\text{BDY}(B_\delta(x, r))$.

2* Prove that a metric space (X, d) is d-bounded if and only if there is some open ball $B_d(x, r)$ equal to X.

3* Let (X, d) be a metric space and let (a_k) be a Cauchy sequence in (X, d). Prove:
(a) Each subsequence of (a_k) is a Cauchy sequence in (X, d).
(b) If some subsequence of (a_k) converges in (X, d), then (a_k) converges in (X, d).

4* Prove that a metric subspace (X, d) of a complete metric space (Y, d) is complete if and only if X is closed in Y.

5 Let (X, d) be a metric space. Prove the following statements for the function
$$b(x, y) = min\{1, d(x, y)\} \qquad (x, y \in X).$$
(a) b is a metric on X.
(b) $1_X : (X, d) \to (X, b)$ and $1_X : (X, b) \to (X, d)$ are continuous.
(c) X is b-bounded.
(d) (X, d) is complete if and only if (X, b) is complete.

6* Prove the following statements for the function $g : \mathbb{R} \times \mathbb{R} \to \mathbb{R}$ defined by
$$g(x, y) = \left| \frac{x}{1 + |x|} - \frac{y}{1 + |y|} \right| \qquad (x, y \in \mathbb{R}).$$
(a) g is a metric on $\mathbb{R}$ that is topologically equivalent to the Euclidean metric d.
(b) $\mathbb{R}$ is g-bounded.
(c) $(\mathbb{R}, g)$ is not a complete metric space.
(d) Find an isometry $g : ((0, 1), d) \to (\mathbb{R}, g)$ from the Euclidean interval $(0, 1)$ to the metric space $(\mathbb{R}, g)$.

7 For any set X, let $\mathcal{B}(X, \mathbb{E}^1)$ be the set of all functions $f : X \to \mathbb{E}^1$ with bounded image in the Euclidean line. Prove the following statements for the function
$$d_\infty(f, g) = lub \, \{|f(x) - g(x)| : x \in X\} \qquad (\forall \, f, g \in \mathcal{B}(X, \mathbb{E}^1)).$$
(a) d_∞ is a metric on $\mathcal{B}(X, \mathbb{E}^1)$.
(b) $(\mathcal{B}(X, \mathbb{E}^1), d_\infty)$ is a complete metric space.

8* For any metric space (X, d), let $\mathcal{C}(X, \mathbb{E}^1) \subseteq \mathcal{B}(X, \mathbb{E}^1)$ be the subset of continuous d-bounded functions $f : X \to \mathbb{E}^1$. Prove:
(a) $\mathcal{C}(X, \mathbb{E}^1)$ is closed in $(\mathcal{B}(X, \mathbb{E}^1), d_\infty)$.

(b) $\mathcal{E}(X, \mathbb{E}^1)$ is a complete metric subspace of $(\mathcal{B}(X, \mathbb{E}^1), d_\infty)$.

9   Prove that Hilbert space (Exercise 3.16) is a complete metric space.

10* For irrational numbers x and y, let $h(x, y) = 1/n$, where the nth digits of the decimal expansions of x and y are the first to disagree. Prove that h is a metric on $\mathbb{R} - \mathbb{Q}$ that is complete and is topologically equivalent to the Euclidean metric on $\mathbb{R} - \mathbb{Q}$.

11 Prove this converse of the Nested Set Theorem (5.5): A metric space (X, d) is complete if each decreasing sequence $\{F_m\}$ of nonempty, nested, closed, d-bounded subsets whose diameters have limit 0 has nonempty intersection $\cap_m F_m$.

12 (a) Prove that in any metric space (X, d), the containment relation
$$\text{CLS } B(x, r) \subseteq \{z \in X : d(x, z) \le r\}$$
holds for all points $x \in X$ and radii $r > 0$.
(b) Find a metric space in which equality always holds in (a).
(c) Find a metric space in which equality always fails to hold in (a).

13* (a) Consider an affine function $T : \mathbb{E}^n \to \mathbb{E}^n$, defined by
$$T(x) = A x + b,$$
where A is the $n \times n$ matrix (a_{ij}) and $b \in \mathbb{E}^n$. Prove T is a contraction of Euclidean n-space $\mathbb{E}^n$ when the sum of the squares of the entries of A, $\sum_{i,j} a_{ij}^2$, is less than 1.
(b) Suppose that M is an $n \times n$ matrix such that $A = (I - cM)$ satisfies the condition in (a) for some scalar c. Prove that a solution of the linear system $M x + m = 0$ in $\mathbb{E}^n$, where $m \in \mathbb{E}^n$, arises as the fixed point of some contractive affine function.

14 Prove the converse of the Nested Set Theorem (5.5): A metric space (X, d) is complete if each sequence of nested nonempty, closed, d-bounded sets whose diameters have limit 0 has nonempty intersection.

15 Let (X, d) be a complete metric space. Prove:
(a) Any countable family of closed sets that cover X has some member with nonempty interior.
(b) The union of any countable family of nowhere dense sets has empty interior.

A *completion* of a metric space (X, d) is an isometry $f : X \to f(X)$ onto a d-dense metric subspace $f(X)$ of a complete metric space (Y^*, d^*).

16* Construct **Hausdorff's completion** of any metric space (X, d):
(a) Let Y be the set of Cauchy sequences $a = (a_k)$ in the metric space (X, d). Prove that $\Delta(a, b) = \lim d(a_k, b_k)$ defines a pseudo-metric (Exercise 3.21) on Y.
(b) Prove that the associated metric space (Y^*, Δ^*) (Exercise 3.21) is complete.
(c) Let $f : X \to Y$ assign to each $x \in X$ the equivalence class $[(x_k)]$ of the sequence (x_k) with the constant value $x_k = x$. Prove that $f : X \to f(X)$ is an isometry onto a dense metric subspace $f(X)$ of the complete metric space (Y^*, Δ^*).

17 Show that a nonempty, countable, complete metric space (X, d) has an isolated point, i.e., a point x such that $\{x\}$ is open in X.

Lebesgue Measure in Euclidean space.
For $a < b$ in $\mathbb{E}^n$, the nonempty rectangle
$$R(a, b) = \{x \in \mathbb{E}^n : a_i < x_i < b_i\}$$

has **volume** $vol(R(a, b)) = \Pi_i (b_i - a_i)$. The **Lebesgue measure** of a subset $A \subseteq \mathbb{E}^n$ is
$$\lambda(A) = glb \{ \Sigma_{k=1}^{\infty} vol(R_k) : A \subseteq \cup \{R_k : 1 \leq k\}\}.$$

18 Prove:
(a) The Lebesgue measure of a subset $A \subseteq \mathbb{E}^n$ equals
$$\mu(A) = lub \{\mu_\delta(A) : 0 < \delta\}, \text{ where}$$
$$\mu_\delta(A) = glb \{ \Sigma_{k=1}^{\infty} vol(R_i) : A \subseteq \cup \{R_k : 1 \leq k\}, diam(R_k) \leq \delta\}.$$
(b) Lebesgue measure λ satisfies the properties of a measure:
 (i) $A \subseteq B \Rightarrow \lambda(A) \leq \lambda(B)$.
 (ii) $\lambda(\cup_i A_i) \leq \Sigma_i \lambda(A_i)$ for countable families $\{A_i\}$.
(c) The measure μ is a metric measure:
$\mu(A \cup B) = \mu(A) + \mu(B)$ if $0 < d(A, B) = glb\{d(a, b) : a \in A, b \in B\}$.

Hausdorff Measure in a metric space.
Let A be any subset of a metric space (X, d) and consider any diameter $\delta > 0$. A **countable δ-cover** of A is a countable family $\{S_k : 1 \leq k\}$ of subsets of X such that
$$A \subseteq \cup \{S_k : 1 \leq k\} \text{ and } diam(S_k) \leq \delta \text{ for all } k \geq 1.$$

For an arbitrary real number $0 \leq p$, the **Hausdorff p-dimensional measure** of the subset A of a metric space (X, d) is
$$\mu^p(A) = lub \{\mu^p_\delta(A) : 0 < \delta\}, \text{ where}$$
$$\mu^p_\delta(A) = glb \{ \Sigma_{k=1}^{\infty} (diam(S_k))^p : \{S_k : 1 \leq k\} \text{ is a countable } \delta\text{-cover of } A\}.$$

19 Prove:
(a) $\mu^0(A) = 0$ if $A = \emptyset$, $\mu^0(A) = 0$ if $A = \{x_1, \ldots, x_n\}$, and $\mu^0(A) = \infty$ otherwise (assuming the conventions: $(diam(S))^0 = 0$ if $S = \emptyset$ and $(diam(S))^0 = 1$ if $S \neq \emptyset$).
(b) For $0 \leq p < q$, we have $\mu^q(A) \leq \mu^p(A)$, and $\mu^p(A) < \infty$ implies that $\mu^q(A) = 0$.
(c) In the Euclidean line $\mathbb{E}^1$, Lebesgue measure and Hausdorff 1-measure coincide.
(d) In Euclidean n-space $\mathbb{E}^n$, Lebesgue measure and Hausdorff n-measure differ by a positive constant for (i) rectangles, (ii) open subsets, and (iii) arbitrary subsets.

The **Hausdorff dimension** of a subset A of a metric space (X, d) is
$$h\text{-}dim(A) = lub\{p : 0 < \mu^p(A)\}.$$

20* Prove:
(a) The closed unit interval $\mathbb{I}$ in the Euclidean line $\mathbb{E}^1$ has Hausdorff dimension 1; its p-measure is $\mu^p(\mathbb{I}) = 0, 1,$ or ∞ accordingly as $p >, =,$ or < 1.
(b) Cantor's set C in the Euclidean line $\mathbb{E}^1$ has Hausdorff dimension $log\,2/log\,3$; its p-measure is $\mu^p(C) = 0, 1,$ or ∞ accordingly as $2/3^p < 2, 2/3^p = 1$, or $2/3^p > 1$.
(c) The circle $\mathbb{S}^1$ in the Euclidean plane $\mathbb{E}^2$ has Hausdorff dimension 1; its p-measure is $\mu^p(\mathbb{S}^1) = 0, 2\pi,$ or ∞ accordingly as $p > 1, p = 1$, or $p < 1$.

3

METRIC SPACE PROPERTIES

This chapter introduces the purely topological properties of connectedness and compactness and employs them to distinguish among various metric spaces. These are the simplest examples of the many topological invariants that constitute the core of point-set topology. Connectedness and compactness of topological spaces are studied more fully in Chapters **6** and **7**.

1 Connected Spaces

The metric subspaces $X = [0, 1) \cup (1, 2]$ and $Y = [0, 2]$ of the Euclidean line $\mathbb{E}^1$ are distinguished by this feature: X is the union of its two disjoint, nonempty, open subsets $[0, 1)$ and $(1, 2]$, while Y is not the union of two disjoint, nonempty, open subsets of itself. (The latter is verified by Theorem 1.3.)

A metric space X is called ***disconnected*** if it is the union of two disjoint, nonempty, open subsets E and F; such a decomposition $X = E \cup F$ is called a ***separation*** of X. A space X that has no separation is called ***connected***.

Since the complement of an open set is closed, and vice versa, the connectedness of a space X can be expressed in these alternate ways:

- X is not the union of two disjoint, nonempty, closed sets.

- X and $\emptyset$ are the only simultaneously open and closed subsets in X.

Before verifying the connectedness of any specific metric space, we consider two theorems involving the property of connectedness. They show that connectedness is inherited by continuous images and certain unions.

1.1 Theorem *The continuous image of a connected space is connected.*

Proof: Let $f : X \to Y$ be a continuous surjection between metric spaces. Then any separation $E \cup F$ of Y determines a separation $f^{-1}(E) \cup f^{-1}(F)$ of

X, because the pre-images of the disjoint, nonempty, open subsets E and F with union Y are disjoint, nonempty, open subsets with union X. Thus, the disconnectedness of Y implies the disconnectedness of X; or equivalently, the connectedness of X implies the connectedness of its continuous image Y. □

1.2 Theorem   *A space is connected if it is the union of connected subspaces that have a point in common.*

Proof: Let $X = \cup_\alpha X_\alpha$ and $x \in \cap_\alpha X_\alpha$, where $\{X_\alpha : \alpha \in \mathcal{A}\}$ is a family of connected subspaces of X. We show that any open and closed subset $E \subseteq X$ equals $\emptyset$ or X. For each $\alpha \in \mathcal{A}$, the subset $E \cap X_\alpha$ is open and closed in the connected subspace X_α; hence, $E \cap X_\alpha$ is either $\emptyset$ or X_α.

If $x \in E$, then $x \in E \cap X_\alpha$, hence, $\emptyset \neq E \cap X_\alpha = X_\alpha$ for each $\alpha \in \mathcal{A}$. Then

$$E = E \cap (\cup_\alpha X_\alpha) = \cup_\alpha (E \cap X_\alpha) = \cup_\alpha X_\alpha = X.$$

If $x \notin E$, then $x \notin E \cap X_\alpha$, hence, $X_\alpha \neq E \cap X_\alpha = \emptyset$ for each $\alpha \in \mathcal{A}$. Then

$$E = E \cap (\cup_\alpha X_\alpha) = \cup_\alpha (E \cap X_\alpha) = \emptyset.$$

□

CONNECTEDNESS IN THE EUCLIDEAN LINE

The next theorem characterizes the connected subspaces of the Euclidean line $\mathbb{E}^1$. The proof utilizes the order completeness of $(\mathbb{R}, \leq)$ (**1.3.2**).

1.3 Interval Theorem   *The following are equivalent for $C \subseteq \mathbb{E}^1$:*
(a) *C is a connected subspace of the Euclidean line $\mathbb{E}^1$.*
(b) *C has the **interval property**: if $a, b \in C$ and $a < x < b$, then $x \in C$.*
(c) *$C = (g, l), [g, l), (g, l],$ or $[g, l]$, for some $-\infty \leq g \leq l \leq +\infty$.*

Proof: $(a) \Rightarrow (b)$. If C is connected and x lies between points $a, b \in C$, then $x \in C$. Otherwise, the two sets $C \cap (-\infty, x)$ and $C \cap (x, +\infty)$ would constitute a separation of C.

$(b) \Rightarrow (c)$. Let $g = glb\ C$ and $l = lub\ C$, when these real numbers exist; otherwise, let $g = -\infty$ and $l = +\infty$. Then

$$C = (g, l),\ [g, l),\ (g, l],\ \text{or}\ [g, l],$$

by the interval property (b) and the definitions of the *glb* and the *lub*.

$(c) \Rightarrow (b)$. This is immediate from the definition of the intervals.

$(b) \Rightarrow (a)$. Suppose that $C = A \cup B$, where A and B are nonempty and disjoint. Select members $a \in A$ and $b \in B$, and label so that $a < b$. By (b), $C = A \cup B$ contains the interval $[a, b]$. Then $\alpha = lub(A \cap [a, b])$ belongs to $[a, b] \subseteq A \cup B$, and $[a, \alpha)$ has members of A arbitrarily near its right-hand endpoint α, while $(\alpha, b] \subseteq B$. Thus, $\alpha \in A \cup B = C$ belongs to the closure of both A and B in the subspace $C \subseteq \mathbb{E}^1$. This proves that the disjoint sets A and B are not both closed in C. Hence, C has no separation. □

SECTION 1　　CONNECTED SPACES　87

CONSEQUENCES OF CONNECTEDNESS OF INTERVALS

Several consequences of the Interval Theorem follow. They often appear in calculus courses without full explanation or reference to connectedness.

1.4　Intermediate Value Theorem　*If $f : [a, b] \to \mathbb{E}^1$ is continuous and if $c \in \mathbb{E}^1$ lies between $f(a)$ and $f(b)$, then $c = f(x)$ for some $a \leq x \leq b$.*

Proof: Since the continuous image $f([a, b])$ of the connected space $[a, b]$ is connected, it must be an interval according to the Interval Theorem (1.3). So $f([a, b])$ contains each point c between its two members $f(a)$ and $f(b)$. □

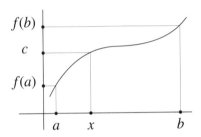

Intermediate Value Theorem (1.4)

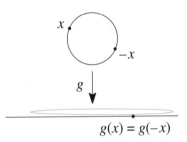

Borsuk-Ulam Theorem (1.5)

The following corollary of the Intermediate Value Theorem has this physical interpretation: *for any continuous mountain range on a sphere, each great circle contains a pair of antipodal points at the same elevation.*

1.5　Borsuk-Ulam Theorem　*Any continuous real-valued function $g : \mathbb{S}^1 \to \mathbb{E}^1$ on the 1-sphere $\mathbb{S}^1 \subset \mathbb{E}^2$ takes the same value on some pair of antipodal points $x = <x_1, x_2>$ and $-x = <-x_1, -x_2>$ in $\mathbb{S}^1$.*

Proof: The continuous function $f : [0, \pi] \to \mathbb{E}^1$, defined by

$$f(t) = g(<\cos t, \sin t>) - g(<-\cos t, -\sin t>),$$

calculates the difference between the values of g at each point of the northern hemisphere and the antipodal point in the southern hemisphere. Thus, f takes on opposite values at the endpoints 0 and π:

$$f(0) = g(<1, 0>) - g(<-1, 0>) = -(g(<-1, 0>) - g(<1, 0>)) = -f(\pi).$$

If the value $f(0) = -f(\pi)$ is 0, then g has the same value on the antipodal points $<1, 0>$ and $<-1, 0>$.

Otherwise, these opposite values, $f(0)$ and $f(\pi)$, are separated by the value $c = 0$. Then the Intermediate Value Theorem provides some $0 < t < \pi$ such that $0 = f(t)$. This means that g takes on the same value on the antipodal points $<\cos t, \sin t>$ and $<-\cos t, -\sin t>$. □

Here is an application of the Borsuk-Ulam Theorem to the problem of portioning a pair of pancakes on a griddle (Exercise 8):

1.6 Pancake Theorem *Any two connected open subsets in $\mathbb{E}^2$ are simultaneously bisected by some line in $\mathbb{E}^2$.*

By Banach's Contraction Mapping Principle (2.5.8), any contraction of a closed interval must leave at least one point of the interval fixed. The next application of the Intermediate Value Theorem says that, in fact, any attempt to move each point of a closed interval to a new location in that interval in a continuous manner is bound to fail.

Example 1 The continuous function $s : \mathbb{I} \to \mathbb{I}$, $s(x) = x^2$, moves each point of the closed unit interval $\mathbb{I} = [0, 1]$ to a new location in $\mathbb{I}$, except its endpoints 0 and 1.

If we follow s by a trade of the endpoints of $\mathbb{I}$, we obtain the continuous function $h : \mathbb{I} \to \mathbb{I}$, $h(x) = 1 - x^2$, which moves each point of the interval $\mathbb{I}$, except $x = \frac{1}{2}(\sqrt{5} - 1)$.

These functions are not contractions of the Euclidean interval $\mathbb{I}$; the existence of their fixed points is not implied by Banach's Contraction Mapping Principle (2.5.8).

As in Section **2.5**, we say that a point $x \in \mathbb{I}$ is a ***fixed point*** of the function $f : \mathbb{I} \to \mathbb{I}$ provided that $f(x) = x$.

Each fixed point x of f gives a point $< x, f(x) > = < x, x > \in \mathbb{I} \times \mathbb{I}$ in the graph of f that lies on the diagonal line $y = x$ of the unit square $\mathbb{I}^2 = \mathbb{I} \times \mathbb{I}$:

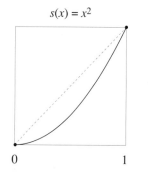
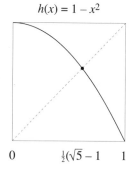
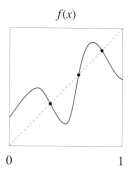

1.7 Fixed Point Theorem
Any continuous function $f : \mathbb{I} \to \mathbb{I}$ has at least one fixed point.

Proof: If either $f(0) = 0$ or $f(1) = 1$, we're done.

Otherwise, $0 < f(0)$ and $f(1) < 1$ since f takes values in $\mathbb{I} = [0, 1]$. Then the continuous function $h : \mathbb{I} \to \mathbb{E}^1$, given by $h(x) = f(x) - x$, has values

$$h(1) = f(1) - 1 < 0 < f(0) - 0 = h(0).$$

By the Intermediate Value Theorem (1.4), there exists a point $0 < x < 1$ at which h has the intermediate value $0 = h(x) = f(x) - x$. Thus $f(x) = x$. □

Connected Subspaces of $\mathbb{E}^n$

Since $\mathbb{E}^1$ satisfies the interval property, it is a connected space by the Interval Theorem (1.3). Each line in $\mathbb{E}^n$ is the continuous image of $\mathbb{E}^1$; hence, it is a connected subspace of $\mathbb{E}^n$ by Theorem 1.1. Since $\mathbb{E}^n$ is the union of all of its lines through the origin O, then $\mathbb{E}^n$ is a connected space by Theorem 1.2.

Similar applications of Theorems 1.1-1.3 show that the following types of subsets of $\mathbb{E}^n$ determine connected subspaces:

A subset $C \subseteq \mathbb{E}^n$ is *convex* if it contains the line segment $(1 - t)x + ty$ $(0 \le t \le 1)$ for each pair of points $x, y \in C$.

A subset $S \subseteq \mathbb{E}^n$ is *star-like* with respect to a point $p \in S$ if it contains the line segment $(1 - t)x + tp$ $(0 \le t \le 1)$ for each point $x \in S$.

Further applications of Theorem 1.2, involving convex subspaces and star-like subspaces, produce a wide variety of other connected subspaces of $\mathbb{E}^n$. Such examples show that, unlike the situation in $\mathbb{E}^1$, there is no concise characterization of the connected subspaces of $\mathbb{E}^n$.

The Fixed Point Theorem is valid for all continuous functions $f: \mathbb{I}^n \to \mathbb{I}^n$ and the Borsuk-Ulam Theorem is valid for all flattenings $g: \mathbb{S}^n \to \mathbb{E}^n$ in each dimension $n \ge 1$. Their proofs in higher dimensions require higher order connectedness. For dimension $n = 2$, see Section **9.2**.

2 Sequentially Compact Metric Spaces

The completeness property for a metric space requires that every sequence that has a chance of converging, i.e., every Cauchy sequence, does in fact converge. It is quite different to require that every sequence have a convergent subsequence. This new demand is not met in the Euclidean line; in $\mathbb{E}^1$ some sequences have all subsequences tending to infinity. But sequences in a bounded metric space seem to always have a convergent subsequence. This section studies this new property of metric spaces, called *sequential compactness*, and the equivalent *Bolzano-Weierstrass property*.

Bolzano-Weierstrass Property

An important topological difference between closed and non-closed intervals involves the existence of accumulation points of infinite subsets. The infinite subset $A = \{1/2, \ldots, 1/n, \ldots\}$ of the open interval $(0, 1)$ has an accumulation point 0 in the Euclidean line $\mathbb{E}^1$, but none in $(0, 1)$. However, any infinite subset of the closed interval $\mathbb{I} = [0, 1]$ has an accumulation point in $\mathbb{I}$. This fact relates to the completeness of $\mathbb{E}^1$ and is developed next.

2.1 Accumulation Theorem *Any infinite subset A of a closed interval $[s, t]$ of the Euclidean line $\mathbb{E}^1$ has an accumulation point in $[s, t]$.*

Proof: We use the binary procedure of subdividing intervals into two equal halves, beginning with the closed interval $[s, t]$. The interval $B_0 = [s, t]$ contains the infinite set A, so one of the two halves of B_0, say B_1, contains infinitely many members of A. By induction, there exist closed subintervals,

$$B_0 \supset B_1 \supset \ldots \supset B_{k-1} \supset B_k \supset \ldots ,$$

such that B_k is half its predecessor B_{k-1} and $A \cap B_k$ is infinite, for each $k \geq 1$.

By the Nested Intervals Theorem (**1.3.3**), the intersection $\cap_k B_k$ is a single real number $x \in [s, t]$. This real number x is an accumulation point of A. For any radius $r > 0$, there exists an integer $k \geq 1$ such that $2^k > (t - s)/r$. Then r exceeds the width, $(t - s)/2^k$, of the interval B_k that contains x:

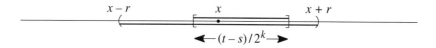

Therefore, the open interval $(x - r, x + r)$, centered on x with radius r, contains B_k, which in turn contains infinitely many members of A. □

The Accumulation Theorem is a variation of the following classical result in real analysis attributed to Bernhard Bolzano (1781-1848) and Karl Weierstrass (1815-1897). Theorems 2.1 and 2.2 are equivalent because any bounded subset of the Euclidean line $\mathbb{E}^1$ is contained in some closed interval.

2.2 Bolzano-Weierstrass Theorem *Any bounded infinite subset of the Euclidean line $\mathbb{E}^1$ has an accumulation point in $\mathbb{E}^1$.*

The topological feature of closed intervals stated in the Accumulation Theorem is not shared by all metric spaces.

Hereafter, we say that a metric space (X, d) has the ***Bolzano-Weierstrass property*** if every infinite subset of X has an accumulation point in (X, d).

Example 1 The Euclidean line $\mathbb{E}^1$ is complete, but does not have the Bolzano-Weierstrass property. The set $\mathbb{N}$ of natural numbers has no accumulation point in $\mathbb{E}^1$.

The Bolzano-Weierstrass property is expressed with sequences this way:

2.3 Theorem *The following are equivalent for a metric space (X, d):*
(a) *Every infinite subset in X has an accumulation point in X.*
(b) *Every sequence in X has a subsequence that converges in X.*

Proof: $(a) \Rightarrow (b)$. Consider any given sequence (a_k) in X. When the image set $A = \{a_k : 1 \leq k\}$ of the sequence (a_k) is finite, there is a constant, hence, convergent subsequence (a_{k_i}). When the image set A in X is infinite, it has an accumulation point $x \in X$, by (a). Then each neighborhood of x intersects each tail $A_m = \{a_k : m \leq k\}$ of the sequence. So we may select a subsequence (a_{k_i}) of (a_k) such that $a_{k_i} \in B(x, 1/i)$ for all $i \geq 1$. Then $a_{k_i} \to x$ in X.

$(b) \Rightarrow (a)$. Let A be any infinite subset of X. Then there is an infinite sequence (a_k) in A with distinct terms. By (b), this sequence has a convergent subsequence (a_{k_i}). By the Limit-Accumulation Property (**2**.3.2(a)), the limit x of the subsequence (a_{k_i}) is an accumulation point in (X, d) of the infinite subset A containing the sequence (a_k). $\square$

SEQUENTIAL COMPACTNESS

We give the equivalent property (b) in Theorem 2.3 its traditional name:

A metric space (X, d) is called *sequentially compact* if every sequence in X has a subsequence that converges in X.

Example 2 A nonempty interval in $(\mathbb{R}, \leq)$ is a sequentially compact subspace of the Euclidean line $\mathbb{E}^1$ if and only if it is a closed interval of finite length.

To characterize the sequentially compact subspaces of Euclidean n-space, we generalize the Nested Intervals Theorem (**1**.3.3) to boxes in $\mathbb{E}^n$.

The *box* $B \subset \mathbb{E}^n$ with *sides* $[s_i, t_i] \subset \mathbb{E}^1$ $(1 \leq i \leq n)$ is the set,

$$[s_1, t_1] \times \ldots \times [s_n, t_n] \subset \mathbb{E}^1 \times \ldots \times \mathbb{E}^1,$$

of points $x \in \mathbb{E}^n$ with coordinates $s_i \leq x_i \leq t_i$ for all $1 \leq i \leq n$. If the n sides have equal length L, the box B is called a *cube*. It has diagonal length $\sqrt{n}L$.

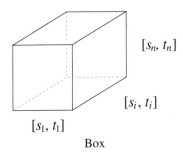
Box

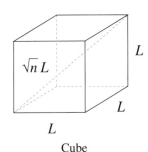
Cube

2.4 Nested Boxes Theorem
The intersection $\cap_k B_k$ of a sequence of nested boxes in $\mathbb{E}^n$,

$$B_1 \supseteq B_2 \supseteq \ldots \supseteq B_k \supseteq \ldots,$$

is a nonempty box B. If the lengths of all sides of the nested boxes become smaller than each positive real number, then B is a singleton in $\mathbb{E}^n$.

Proof: The completeness of Euclidean n-space $\mathbb{E}^n$ and Hausdorff's Nested Set Theorem (**2**.5.5) give the second half of this theorem. More directly, both halves follow from set theory and the Nested Intervals Theorem (**1.3.3**): the intersection of a sequence of nested boxes B_k ($k \geq 1$) in $\mathbb{E}^n$ with sides $[s_{ik}, t_{ik}]$ ($1 \leq i \leq n$) is the box B with sides $[s_i, t_i] = \cap_k [s_{ik}, t_{ik}]$ ($1 \leq i \leq n$). □

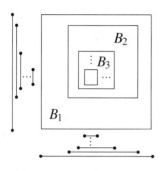
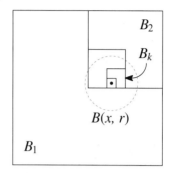

Nested Boxes Theorem (2.4)   Compact Cubes Theorem (2.5)

2.5 Compact Cubes Theorem   *Each cube is a sequentially compact metric subspace of Euclidean n-space $\mathbb{E}^n$.*

Proof: By Theorem 2.3, it is sufficient to establish the Bolzano-Weierstrass property for any cube. For this, we mimic the proof of the Accumulation Theorem (2.1). Let A be any infinite subset of a cube $B \subset \mathbb{E}^n$, with side length L. The procedure of subdividing intervals into two equal halves can be applied to each of the n sides of a cube in $\mathbb{E}^n$ to produce 2^n subcubes. We iterate this subdivision procedure, beginning with the cube $B_0 = B$. We select at each stage a subcube that contains infinitely many members of A. There results a sequence of nested subcubes, each having side length half that of its predecessor and containing infinitely many members of A:

$$B_0 \supset B_1 \supset \ldots \supset B_{k-1} \supset B_k \supset \ldots .$$

By the Nested Boxes Theorem (2.4), the intersection $\cap_k B_k$ is a single point $x \in B \subset \mathbb{R}^n$. This point $x \in B$ is an accumulation point of A. For any $r > 0$, there exists an integer $k > 0$ such that $2^k > \sqrt{n}L/r$. Then r exceeds the diameter, $\sqrt{n}L/2^k$, of the cube B_k that contains x. Thus, the Euclidean ball $B(x, r)$ centered on x contains B_k, hence, infinitely many members of A. □

CLOSED SETS AND COMPACTNESS

The unit sphere $\mathbb{S}^{n-1}$ and the closed unit ball $\mathbb{D}^n$ are the subspaces consisting of all points $x \in \mathbb{E}^n$ of norm $\|x\| = 1$ and $\|x\| \leq 1$, respectively. $\mathbb{S}^{n-1}$ and $\mathbb{D}^n$ are sequentially compact; their infinite subsets have accumulation points.

According to the following theorem, one way to establish the sequential compactness of the sphere $\mathbb{S}^{n-1}$ and the ball $\mathbb{D}^n$ is to observe that each of them is a *closed* subset of the sequentially compact n-cube $[-1, +1]^n$.

2.6 Theorem *Any closed subspace X of a sequentially compact metric space Y is sequentially compact.*

Proof: Consider any infinite subset $A \subseteq X \subseteq Y$. Because Y is sequentially compact, A has an accumulation point y in Y. Because X is closed in Y, the accumulation point y of the subset $A \subseteq X$ must belong to X. This proves that X is a sequentially compact subspace by Theorem 2.3. $\square$

> **Example 3** *Cantor's space C is sequentially compact.* In Example 1.3.3, C is described as the countable intersection $\cap_k C_k$ of sets, each of which is a finite union of disjoint closed subintervals of $\mathbb{I} = [0, 1]$. Thus, C is a closed subspace of the sequentially compact interval $\mathbb{I}$, and so it is sequentially compact by Theorem 2.6.

2.7 Theorem *Any sequentially compact subspace X of a metric space Y is closed.*

Proof: To prove that X is closed in Y, it suffices to show that any sequence (a_k) in X that converges in Y has its limit y in X (Exercise 2.3.14). Every subsequence (a_{k_i}) of the convergent sequence (a_k) also converges to y. And by the sequential compactness, some subsequence (a_{k_i}) converges in X. So the limit point y in Y belongs to X, by the uniqueness of limits. $\square$

The Compact Cubes Theorem (2.5) and Theorems 2.6 and 2.7 give the following characterization of sequential compactness of metric subspaces of Euclidean space. A variation of this result for the Euclidean line was first proved by Emile Borel (1871-1956) in 1894.

2.8 Heine-Borel Theorem *A metric subspace of Euclidean n-space is sequentially compact if and only if it is closed in $\mathbb{E}^n$ and bounded in the Euclidean metric of $\mathbb{E}^n$.*

Proof: A d-bounded subset X of Euclidean n-space $\mathbb{E}^n$ is contained in some cube $[-k, +k]^n$. When the subspace X is also closed in $\mathbb{E}^n$, it is sequentially compact by Theorem 2.6 and the Compact Cubes Theorem (2.5).

Conversely, when X is a sequentially compact subspace of $\mathbb{E}^n$, it is closed in $\mathbb{E}^n$ by Theorem 2.7. Also X is d-bounded. Otherwise, there would exist points $a_k \in X$ with $d(a_k, \boldsymbol{O}) > k$ for all indices $k \geq 1$, and then the sequence (a_k) would have no convergent subsequence. $\square$

Warning: A closed and bounded subset of a general metric space (X, m) need not be sequentially compact, even if $X = \mathbb{R}$ and m is topologically equivalent to the Euclidean metric d (i.e., determines the same open sets). There exists such a metric m on $\mathbb{R}$ for which $\mathbb{R}$ is m-bounded (Exercise 2).

TOTAL BOUNDEDNESS AND COMPACTNESS

Recall that a metric space (X, d) is called d-bounded whenever its points are a bounded d-distance apart. Then X equals some open d-ball $B(x, r)$ of finite radius $r > 0$ (Exercise **2.5.2**). Here is a stronger form of boundedness:

A metric space (X, d) is **totally bounded** if, for each real number $r > 0$, there is a finite collection of open d-balls of radius r that have union X:

$$X = \cup \{B(x_i, r) : 1 \leq i \leq k\}.$$

Total boundedness guarantees compactness of any complete metric space.

2.9 Compactness Characterization *A metric space is sequentially compact if and only if it is complete and totally bounded.*

Proof: First, we suppose that (X, d) is a sequentially compact metric space and show that it is complete and totally bounded.

By hypothesis, any sequence (a_k) in (X, d) has a convergent subsequence (a_{k_i}). When (a_k) is a Cauchy sequence, it must also converge in (X, d) by Exercise **2.5.3**. This proves that (X, d) is complete.

The metric space (X, d) is also totally bounded. Given any real number $r > 0$, we select any point $a_1 \in X$. Then we inductively choose points

$$a_{k+1} \notin \cup \{B(a_i, r) : 1 \leq i \leq k\} \qquad (1 \leq k).$$

Eventually, some a_{k+1} fails to exist and $X = \cup \{B(a_i, r) : 1 \leq i \leq k\}$. Otherwise, an infinite sequence (a_k) would result. Then, by hypothesis, (a_k) would have a convergent, hence, Cauchy subsequence (a_{k_i}). This is impossible; by choice, $d(a_k, a_n) \geq r$ for all indices $k > n \geq 1$.

We now suppose that (X, d) is complete and totally bounded. We verify its sequential compactness via the Bolzano-Weierstrass property in (2.3).

Consider any infinite subset $A \subseteq X$. For each $k \geq 1$, the totally bounded space X is covered by a finite collection of open balls of radius $(\frac{1}{2})^k$. Since A is infinite, one of the balls $B_1 = B(x_1, \frac{1}{2})$ of radius $\frac{1}{2}$ contains infinitely many members of A. Then $A \cap B_1$ is infinite. We inductively choose, for each $k \geq 1$, an open ball $B_k = B(x_k, (\frac{1}{2})^k)$ such that $A \cap B_1 \cap \ldots \cap B_k$ is infinite.

Then (x_k) is a Cauchy sequence in (X, d); indeed, the inequality

$$d(x_k, x_n) \leq (\tfrac{1}{2})^k + (\tfrac{1}{2})^n$$

holds by the triangle inequality and the nonempty intersection $B_k \cap B_n \neq \varnothing$. Because (X, d) is complete, the sequence (x_k) converges to a point $x \in X$.

The limit x is an accumulation point of A. Given $r > 0$, the ball $B(x, \frac{1}{2}r)$ contains some tail $\{x_k : m \leq k\}$ of the sequence (x_k). Whenever both $m \leq k$ and $(\frac{1}{2})^k < \frac{1}{2}r$, the larger ball $B(x, r)$ contains $B_k = B(x_k, (\frac{1}{2})^k)$, which in turn contains the infinite set $A \cap B_k$.

This proves the sequential compactness of X, by Theorem 2.3. □

CONTINUITY AND COMPACTNESS

The figure, below left, shows that it is easy to construct a continuous surjection $(0, 1) \to [0, 1]$ of the open interval $(0, 1)$ onto the closed interval $[0, 1]$. But there exists no continuous surjection $[0, 1] \to (0, 1)$. According to the Compact Image Theorem that follows, one explanation is that the closed interval $[0, 1]$ is sequentially compact, while the open interval $(0, 1)$ is not.

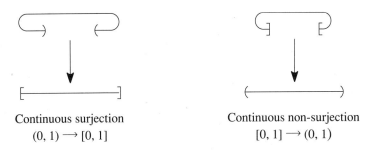

Continuous surjection
$(0, 1) \to [0, 1]$

Continuous non-surjection
$[0, 1] \to (0, 1)$

2.10 Compact Image Theorem *The continuous image of a sequentially compact metric space is sequentially compact.*

Proof: Let $f : X \to Y$ be a continuous function from a sequentially compact metric space X onto a metric space Y. We show that each sequence (b_k) in Y has a convergent subsequence. Since $f(X) = Y$, there corresponds a sequence (a_k) in X, with $f(a_k) = b_k$ for all $k \geq 1$. Since X is sequentially compact, (a_k) has a convergent subsequence (a_{k_i}), with limit $x \in X$. Since f is continuous, then $(b_{k_i} = f(a_{k_i}))$ is a convergent subsequence of (b_k), with limit $f(x) \in Y$. This proves that Y is sequentially compact. □

> **Example 4** By the Compact Image Theorem (2.10) and Theorems 1.1 and 1.3, any continuous image in the Euclidean line $\mathbb{E}^1$ of a connected, sequentially compact, metric space must be a closed interval $[r, s]$.

2.11 Closed Function Theorem *A continuous function between metric spaces, with sequentially compact domain, carries closed sets to closed sets.*

Proof: Let $f : X \to Y$ be a continuous function from a sequentially compact metric space X to a metric space Y. If $F \subseteq X$ is a closed subset, then F is a sequentially compact metric subspace of X, by Theorem 2.6. Then, by Theorem 2.10, the image of the continuous function $f : F \to f(F)$ is a sequentially compact metric subspace $f(F)$ of Y. Hence, $f(F) \subseteq Y$ is a closed subset of Y, by Theorem 2.7. □

> **Example 5** The squaring function $s : [-2, +2] \to \mathbb{E}^1$, $s(x) = x^2$, carries closed sets to closed sets. But it doesn't follow that s carries open sets to open sets; indeed, s carries the open interval $(-1, +1)$ to the non-open interval $[0, 1)$.

2.12 Extreme Value Theorem

Let X be a nonempty, sequentially compact, metric space. Each continuous real-valued function $f : X \to \mathbb{E}^1$ attains both a minimum value m and a maximum value M on X: there exist $x_m, x_M \in X$ such that, for all $x \in X$, we have

$$m = f(x_m) \leq f(x) \leq f(x_M) = M.$$

Proof: The image $f(X)$ is a sequentially compact subspace of the Euclidean line $\mathbb{E}^1$, by Theorem 2.10; hence, $f(X)$ is closed in $\mathbb{E}^1$ and bounded in the Euclidean metric d, by Theorem 2.8. Because $f(X)$ is d-bounded, the real numbers $m = glb\ f(X)$ and $M = lub\ f(X)$ exist, by the order completeness (**1.3.2**) of $(\mathbb{R}, \leq)$. Because $f(X)$ is closed, these numbers m and M belong to $f(X)$ and so they are the *minimum* and the *maximum* of the image set $f(X)$.

In other words, there exist members $x_m \in X$ and $x_M \in X$ with values $m = f(x_m)$ and $M = f(x_M)$ such that, for all $x \in X$, we have

$$m = f(x_m) \leq f(x) \leq f(x_M) = M. \qquad \square$$

Example 6 Any polynomial function $p : \mathbb{E}^1 \to \mathbb{E}^1$ of degree $d \geq 1$ fails to have either a minimum value, or a maximum value, or both, depending upon the parity of its degree d and the sign of the coefficient of x^d. But, according to the Extreme Value Theorem (2.12), the restriction of p to any finite closed interval $[a, b]$ always has both a minimum value m and a maximum value M.

Example 7 Consider any point $p \in X$ in a metric space (X, d). The distance function $d_p : X \to \mathbb{E}^1$, $d_p(x) = d(p, x)$, is continuous (Example **2**.4.9). Therefore, it has a minimum value on each sequentially compact subspace $A \subseteq X$. This means that the greatest lower bound of the distances $\{d(p, a) : a \in A\}$, which is denoted by $d(p, A)$ and called *the distance from p to the subspace A* (Section **2**.3), is achieved by some member $a_p \in A$ *nearest* to p:

$$d(p, A) = min\ \{d(p, a) : a \in A\} = d(p, a_p).$$

3 Hyperspace of Compact Subspaces

Let (X, d) be a complete metric space and let $\mathfrak{H}(X)$ denote the set whose members $A \in \mathfrak{H}(X)$ are the sequentially compact subspaces $A \subseteq X$. We seek a metric h for $\mathfrak{H}(X)$, i.e., an assignment of a distance $h(A, B) \geq 0$ for every pair $A, B \in \mathfrak{H}(X)$ such that the metric axioms (**D1-D3**) hold.

The distance calculation

$$d(A, B) = glb\ \{d(a, b) : a \in A, b \in B\}$$

measures how close the subspaces lie in the metric space X. But $d(A, B)$ is not a metric $\mathfrak{H}(X)$, because $d(A, B) = 0$ if and only if $A \cap B \neq \emptyset$ (Exercise 1). That is, $d(A, B) = 0$ does not imply that $A = B$ as is required by (**D1**).

HAUSDORFF METRIC

By Example 2.7, the function $d_A : X \to \mathbb{E}^1$ recording distance in X to A is

$$d_A(x) = d(x, A) = min\ \{d(x, a) : a \in A\}.$$

The function d_A is continuous (Exercise **2.4.21**). Thus, by the Extreme Value Theorem (2.12), d_A has a *maximum* value on B. This is the maximum distance $d(b, a_b)$ from a member $b \in B$ to a member $a_b \in A$ *nearest* to it.

We denote the maximum value of d_A on B by

$$\begin{aligned} d(B \to A) &= max\ \{\ d(b, A) : b \in B\} \\ &= max\ \{\ min\ \{d(b, a) : a \in A\} : b \in B\}. \end{aligned}$$

Similarly, $d(A \to B)$ is defined to be the maximum value of d_B on A. This is the maximum distance $d(a, b_a)$ from an $a \in A$ to a $b_a \in B$ *nearest* to it.

Example 1 In $\mathbb{E}^2$, the unit semi-circle and the line segment

$$A = \{<x, y> \in \mathbb{S}^1 : 0 \leq y\] \quad \text{and} \quad B = \{<x, 2> \in \mathbb{E}^2 : -2 \leq x \leq +2\}$$

give the distances $d(B \to A) = 2\sqrt{2} - 1$ and $d(A \to B) = 2$.

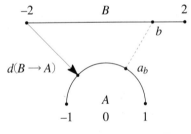
Example 1: $d(B \to A) = 2\sqrt{2} - 1$

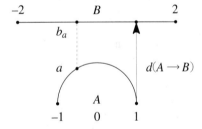
Example 1: $d(A \to B) = 2$

Notice that, for a pair of members $A, B \in \mathfrak{H}(X)$ (i.e., sequentially compact subspaces $A, B \subseteq X$), the distances $d(B \to A)$ and $d(B \to A)$ need not agree.

Using both calculations $d(B \to A)$ and $d(B \to A)$, Felix Hausdorff (1868-1942) devised the following metric for $\mathfrak{H}(X)$.

The ***Hausdorff distance*** for a pair of members $A, B \in \mathfrak{H}(X)$ is defined by

$$h(A, B) = max\ \{d(B \to A), d(A \to B)\},$$

namely, *the maximum distance from a member of one of the two sets to a member of the other set nearest to it.*

The Hausdorff distance h is a complete metric on $\mathfrak{H}(X)$, when d is a complete metric on X (Exercises 3 and 4). (See Barnsley, *Fractals Everywhere*.)

The complete metric space $(\mathfrak{H}(X), h)$ is called the ***hyperspace of compact subspaces*** of the complete metric space (X, d).

CONTRACTIONS OF $\mathfrak{H}(X)$

Consider any contraction $f : X \to X$ of the complete metric space (X, d). Since f is continuous, the image $f(A) \subseteq X$ of a sequentially compact subspace $A \subseteq X$ is another one according to the Compact Image Theorem (2.10). Therefore, the assignment $A \to f(A)$ defines a function $f : \mathfrak{H}(X) \to \mathfrak{H}(X)$.

3.1 Lemma *If $f : X \to X$ is a contraction of (X, d) with scalar c, then*
$$f : \mathfrak{H}(X) \to \mathfrak{H}(X)$$
is a contraction of $(\mathfrak{H}(X), h)$ with scalar c.

Proof: For $A, B \in \mathfrak{H}(X)$, we have

$$d(f(B) \to f(A)) = max \{min \{d(f(b), f(a)) : a \in A\} : b \in B\}$$
$$\leq max \{min \{c\, d(b, a) : a \in A\} : b \in B\}$$
$$= c\, d(B \to A).$$

Similarly, $d(f(A) \to f(B)) \leq c\, d(A \to B)$. Hence, we have

$$h(f(B), f(A)) = max \{d(f(B) \to f(A)),\, d(f(A) \to f(B))\}$$
$$\leq max \{c\, d(B \to A),\, c\, d(A \to B)\}$$
$$= c\, h(A, B). \qquad \square$$

3.2 Hyperspace Contraction Theorem *Let $(\mathfrak{H}(X), h)$ be the hyperspace of the complete metric space (X, d); let $G_i : \mathfrak{H}(X) \to \mathfrak{H}(X)$ ($1 \leq i \leq n$) be contractions of $(\mathfrak{H}(X), h)$ with scalars c_i ($1 \leq i \leq n$). Then the function*

$$\Gamma : \mathfrak{H}(X) \to \mathfrak{H}(X),\ \Gamma(A) = \cup\, \{G_i(A) \subseteq X : 1 \leq i \leq n\},$$

is a contraction of $(\mathfrak{H}(X), h)$ with scalar $c = max\, \{c_i : 1 \leq i \leq n\}$.

Proof: The Hausdorff metric has this property (Exercise 5):

$$h(A \cup B, C \cup D) \leq max\, \{h(A, C), h(B, D)\}.$$

This proves the theorem when $n = 2$; induction completes the argument. $\qquad \square$

By Banach's Contraction Mapping Principle (**2.5.8**), each such contraction $\Gamma : \mathfrak{H}(X) \to \mathfrak{H}(X)$ has a *unique fixed point P*. This a sequentially compact subspace $P \subseteq X$ such that $\Gamma(P) = P$. Moreover, the unique fixed point P is the limit in $(\mathfrak{H}(X), h)$ of the sequence $(\Gamma^k(Z))$ obtained by iterated application of Γ, beginning with any sequentially compact subspace $Z \subseteq X$ as a *catalyst*. Because each of these sequences of sequentially compact subspaces of X converges to the subspace P, it is called the ***attractor*** of $\Gamma : \mathfrak{H}(X) \to \mathfrak{H}(X)$.

Example 2 The two linear functions $g_1, g_2 : \mathbb{E}^1 \to \mathbb{E}^1$ given by

$$g_1(x) = \tfrac{1}{2}x \quad \text{and} \quad g_2(x) = \tfrac{1}{2}x + \tfrac{1}{2}$$

determine a contraction $\Gamma : \mathfrak{H}(\mathbb{E}^1) \to \mathfrak{H}(\mathbb{E}^1)$ by the assignment $\Gamma(A) = g_1(A) \cup g_2(A)$. The contraction Γ has $\mathbb{I} \in \mathfrak{H}(\mathbb{E}^1)$ as attractor, since Γ fixes $\mathbb{I}$:

$$\Gamma(\mathbb{I}) = [0, \tfrac{1}{2}] \cup [\tfrac{1}{2}, 1] = \mathbb{I}.$$

The catalyst $Z = \mathbb{I}$ gives the constant sequence $(\Gamma^k(\mathbb{I}) = \mathbb{I})$ in $(\mathfrak{H}(\mathbb{E}^1), h)$.
The two-point catalyst $Z = \{0, 1\}$ gives the sequence $(B_k = \Gamma^k(\{0, 1\}))$ of the finite sets B_k of all k^{th} stage binary fractions $0 \leq m/2^k \leq 1$.
By Banach's Contraction Mapping Principle (**2.5.8**), both sequences $(\mathbb{I} = \Gamma^k(\mathbb{I}))$ and $(B_k = \Gamma^k(\{0, 1\}))$ converge in $(\mathfrak{H}(\mathbb{E}^1), h)$ to $\mathbb{I}$.

Example 3 The two linear functions $g_1, g_2 : \mathbb{E}^1 \to \mathbb{E}^1$ given by

$$g_1(x) = \tfrac{1}{3}x \quad \text{and} \quad g_2(x) = \tfrac{1}{3}x + \tfrac{2}{3}$$

determine a contraction $\Gamma : \mathfrak{H}(\mathbb{E}^1) \to \mathfrak{H}(\mathbb{E}^1)$ by the assignment $\Gamma(A) = g_1(A) \cup g_2(A)$. Its attractor is Cantor's set $C \in \mathfrak{H}(\mathbb{E}^1)$, as Γ fixes C:

$$\Gamma(C) = g_1(C) \cup g_2(C) = C.$$

The catalyst $Z = \mathbb{I}$ gives the sequence $(C_k = \Gamma^k(\mathbb{I}))$, in which each C_k is the union of all even, closed, k^{th} stage ternary intervals in $\mathbb{I}$:

$$C_0 = [0, 1],$$

$$C_1 = \left[0, \tfrac{1}{3}\right] \cup \left[\tfrac{2}{3}, 1\right],$$

$$C_2 = \left[0, \tfrac{1}{9}\right] \cup \left[\tfrac{2}{9}, \tfrac{1}{3}\right] \cup \left[\tfrac{2}{3}, \tfrac{7}{9}\right] \cup \left[\tfrac{8}{9}, 1\right], \ldots .$$

The single point catalyst $Z = \{\tfrac{1}{2}\}$ gives a sequence $(M_k = \Gamma^k(\{\tfrac{1}{2}\}))$ of finite sets; for $k \geq 1$, M_k is the set of the midpoints of all complementary intervals to C_{k+1} in $\mathbb{I}$.
Both sequences $(C_k = \Gamma^k(\mathbb{I}))$ and $(M_k = \Gamma^k(\{\tfrac{1}{2}\}))$ have limit C in $(\mathfrak{H}(\mathbb{E}^1), h)$.

Example 3: $C_k \to C$ \qquad\qquad Example 3 $M_k \to C$

These examples show that the attractor of a contraction

$$\Gamma : \mathfrak{H}(X) \to \mathfrak{H}(X), \ \Gamma(A) = \cup \{G_i(A) : 1 \leq i \leq n\},$$

is not just the union of the attractors of the individual contractions $\{G_i\}$.

SEEDS FOR ATTRACTORS

We can use any sequentially compact subspace $K \subseteq X$ as a *seed* to influence the shape of an attractor under design. The constant function

$$G : \mathfrak{H}(X) \to \mathfrak{H}(X), \quad G(A) = K \in \mathfrak{H}(X),$$

is *a contraction with scalar $c = 0$*. When G is chosen as the first constituent of the contraction $\Gamma : \mathfrak{H}(X) \to \mathfrak{H}(X)$ of Theorem 3.2, then Γ is evaluated by

$$\Gamma(A) = K \cup G_2(A) \cup \ldots \cup G_n(A).$$

When the contractions G_i of $(\mathfrak{H}(X), h)$ arise from contractions g_i of (X, d), they satisfy this monotonicity condition: $A \subseteq B \Rightarrow G(A) \subseteq G(B)$. Then the sequence formed using K as the catalyst is nested and limits on the attractor:

$$K \subseteq \Gamma(K) \subseteq \Gamma^2(K) \subseteq \ldots \subseteq \Gamma^k(K) \subseteq \ldots.$$

Example 4 When the top two sides of a triangle based on [1/3, 2/3] are used to seed the contraction in Example 3, the attractor that results has this appearance:

Example 4

Example 5 With the unit radius circle centered on $<3, 0>$ taken as the seed K, the similarities $g_1, g_2 : \mathbb{E}^2 \to \mathbb{E}^2$ (centered on $\boldsymbol{O} = <0, 0>$ and $p = <4, 0>$, respectively)

$$g_1(x) = \tfrac{1}{2} x \qquad \text{and} \qquad g_2(x) = \tfrac{1}{2} x + \tfrac{1}{2} p$$

give the contraction $\Gamma : \mathfrak{H}(\mathbb{E}^2) \to \mathfrak{H}(\mathbb{E}^2)$, $\Gamma(A) = K \cup g_1(A) \cup g_2(A)$, with this attractor:

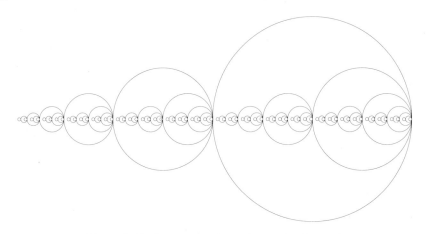

Example 5: An attractor formed on a circular seed

4 Topological Equivalence

Two metric spaces should be considered topologically identical only if there exists a bijection of their underlying sets that respects all their topological features. But how can we express this requirement using continuity? The following pair of examples suggests the answer.

Example 1 *Some continuous bijections have a continuous inverse.* The continuous bijection $f: \mathbb{D}^3 \to \mathbb{E}$ (below left) deforms the closed 3-ball $\mathbb{D}^3$ into a solid ellipsoid $\mathbb{E}$; f has a continuous inverse $g: \mathbb{E} \to \mathbb{D}^3$ that rounds $\mathbb{E}$ into $\mathbb{D}^3$.

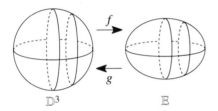
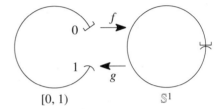

Example 1: Continuous inverse   Example 2: Discontinuous inverse

Example 2 *Some continuous bijections have a discontinuous inverse.* The continuous function $f: [0, 1) \to \mathbb{S}^1$, $f(t) = <\cos 2\pi t, \sin 2\pi t>$, (above right) wraps the interval $[0, 1) \subset \mathbb{E}^1$ around the circle $\mathbb{S}^1 \subset \mathbb{E}^2$. Because f is a bijection, it has an inverse $g: \mathbb{S}^1 \to [0, 1)$. But this function g cuts the circle apart, and therefore it is not continuous.

TOPOLOGICAL EQUIVALENCE

The previous examples suggest the following definition.
    Two metric spaces X and Y are **homeomorphic** or **topologically equivalent** if there exist inverse functions $f: X \leftrightarrow Y : g$, both of which are continuous. The homeomorphism relation is denoted by $X \cong Y$.
    Continuous inverse functions f and g are called **homeomorphisms**; their continuity requirements can be summarized this way:

$$A = g(B) \text{ is arbitrarily near } x = g(y) \text{ in } X$$
$$\Leftrightarrow \quad f(A) = B \text{ is arbitrarily near } f(x) = y \text{ in } Y.$$

Example 3 The boundary of a square disc □ and the boundary of an octagonal disc ○ are homeomorphic subspaces of the Euclidean plane. When they are placed concentrically in the plane, the radial lines through their common center define inverse homeomorphisms □ ↔ ○. Thus, *plain topology* is blind to corners. Detection of corners involves *differential topology*, which we don't consider in this text.

Example 4 The interval $(0, 1]$ and ray $[1, \infty)$ are homeomorphic metric subspaces of $\mathbb{E}^1$. Inverse homeomorphisms $f: (0, 1] \leftrightarrow [1, \infty) : g$ are given by the reciprocal functions $f(x) = 1/x$ and $g(y) = 1/y$. Topology ignores geometric measurements.

Example 5 Consider the tangent circle metric subspaces $X, Y \subset \mathbb{E}^2$, below. In $\mathbb{E}^3$, one can convert X into Y by a flip of the smaller circle. One can visualize this homeomorphism $X \cong Y$ within $\mathbb{E}^2$ as the continuous composite of the discontinuous cut-and-glue processes that are depicted below. Such processes that re-establish the original nearness relationships make it possible to visualize homeomorphisms that might not otherwise be apparent. In this way, one can visualize within $\mathbb{E}^3$ a homeomorphism between the pinched torus and the pincushion space of Section **1**.1. Try it.

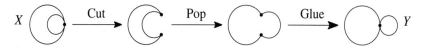

Example 6 *Numerically equivalent metrics give homeomorphic metric spaces.* As numerically equivalent metrics d and m on a set X determine the same open sets in X, the identity functions $1_X : (X, d) \hookrightarrow (X, m) : 1_X$ are inverse homeomorphisms.

Example 7: Sierpinski's dust C^2

Example 7 *Cantor's space is homeomorphic to Sierpinski's dust.* Cantor's space $C \subset \mathbb{I}$ (Example **1**.3.3) consists of the points of the closed unit interval $\mathbb{I}$ that survive an iteration of a deleted middle third procedure. *Sierpinski's dust* (above) is the product set $C^2 \subset \mathbb{I}^2$ of all ordered pairs from C; it consists of the points of the closed unit square $\mathbb{I}^2$ that survive an iteration of a deleted central cross procedure.

There is a bijection $f: C \to C^2$ defined by

$$f(t) = <f_1(t), f_2(t)> = <(.t_1 t_3 \ldots t_{2k-1} \ldots)_3, (.t_2 t_4 \ldots t_{2k} \ldots)_3>,$$

where $(.t_1 t_2 t_3 t_4 \ldots t_{2k-1} t_{2k} \ldots)_3$ is the unique even ternary expansion of t.

The Euclidean distance between points $t, t' \in C$ and the max distance between their images $f(t), f(t') \in C^2$ are related this way:

$$d(t, t') < 1/3^{2k} \Leftrightarrow \quad t, t' \in C \text{ lie in the same } (2k)^{\text{th}} \text{ stage ternary interval}$$
$$\Leftrightarrow \quad t_i = t_i' \text{ for } 1 \le i \le 2k$$
$$\Leftrightarrow \quad f_1(t), f_1(t') \in C \text{ lie in the same } k^{\text{th}} \text{ stage ternary interval and}$$
$$\quad f_2(t), f_2(t') \in C \text{ lie in the same } k^{\text{th}} \text{ stage ternary interval}$$
$$\Leftrightarrow \quad \text{max distance } m(f(t), f(t')) < 1/3^k.$$

Thus, f and its inverse are continuous, by the continuity characterization **2.4.1**(f). These dust-like spaces C and C^2 are homeomorphic, even though they are distinguishable metric subspaces of $\mathbb{E}^2$, with Hausdorff dimensions

$$\log 2 / \log 3 \quad \text{and} \quad \log 4 / \log 3,$$

respectively (see Exercise **2.5.20**).

DISTINGUISHING SPACES

A property of metric spaces is a ***topological property***, or ***topological invariant***, provided that, whenever a metric space has the property, so does every metric space homeomorphic to it.

4.1 Topological Invariance *Connectedness and sequential compactness are topological properties of metric spaces, but completeness is not.*

Proof: In fact, any continuous image of a connected or sequentially compact metric space is connected or sequentially compact (Theorems 1.1 and 2.10). So these properties are topological invariants of metric spaces.

The inverse functions $f: (-1, +1) \hookrightarrow \mathbb{R} : g$ in Example **1**.4.7 are homeomorphisms in the Euclidean metric d. These homeomorphic spaces $(\mathbb{E}^1, d)$ and $((-1, +1), d)$ show that completeness of a metric space (X, d) is not a topological property, but rather a property of the metric. Of course, by the Compactness Characterization (2.9), the two metric properties of completeness and total boundedness, taken together, are equivalent to the topological property of sequential compactness. □

This invariance provides a basic tool for distinguishing among certain metric spaces: *two metric spaces are not homeomorphic if one is connected or sequentially compact and the other is not.*

Example 8 The intervals [0, 1] and [0, 1) are set equivalent, but they are not homeomorphic subspaces of the Euclidean line $\mathbb{E}^1$. Being intervals, they are both connected. But [0, 1] is sequentially compact and [0, 1) is not.

Example 9 Example 2 gives a continuous bijection $f: [0, 1) \to \mathbb{S}^1$ that wraps the interval around the 1-sphere $\mathbb{S}^1$. But [0, 1) and $\mathbb{S}^1$ are not homeomorphic subspaces of Euclidean space. The sphere $\mathbb{S}^1$ is a closed subspace of the sequentially compact square $[-1, +1]^2$; by Theorem 2.6, it is sequentially compact. The interval [0, 1) is a non-closed subspace of $\mathbb{E}^1$; by Theorem 2.7, it is not sequentially compact.

Example 10 Let X be a sequentially compact metric space and let $x \in X$ be any accumulation point of X. The *punctured subspace* $X - \{x\} \subset X$ cannot be homeomorphic to X. In fact, because the punctured subspace $X - \{x\}$ has the accumulation point x in X, it is a non-closed, hence, non-sequentially compact subspace, by Theorem 2.7.

For example, Cantor's *dust-like* space C is not homeomorphic to any punctured version $C - \{x\}$ of itself, for C is perfect (each $x \in C$ is an accumulation point of C).

Example 11 Let $S \subset \mathbb{I}^2$ denote the zigzag subset (below left) consisting of the line segments joining consecutive entries of the sequence

$$<1, 1>, \; <0, \tfrac{1}{2}>, \; <1, \tfrac{1}{3}>, \; <0, \tfrac{1}{4}>, \; \ldots, \; <0, \tfrac{1}{2k}>, \; <1, \tfrac{1}{2k+1}>, \; \ldots,$$

together with the origin $O = <0, 0>$ as a bottom point.

The metric subspace $S \subset \mathbb{E}^2$ is called the *saw* space. The saw space S is not closed in $\mathbb{E}^2$, since it does not contain its accumulation points $<t, 0>$ $(0 < t \le 1)$. So S is a non-sequentially compact metric subspace of $\mathbb{E}^2$ according to Theorem 2.7.

The projection function $f: S \to \mathbb{I}$ given by $f(<x, y>) = y$ flattens the zigzag against the vertical copy of the closed unit interval $\mathbb{I} = [0, 1]$. The function f is bijective and continuous; but f cannot be a homeomorphism of the non-compact saw space S and the sequentially compact interval $\mathbb{I}$.

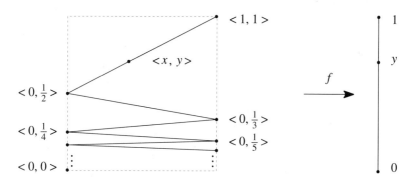

Example 11: Continuous flattening of the saw space

4.2 Homeomorphism Theorem

Let X be a sequentially compact metric space and let Y be any metric space. Any continuous bijection $f : X \to Y$ is a homeomorphism.

Proof: Let $g : Y \to X$ be the inverse of the function $f : X \to Y$. If $F \subseteq X$ is closed in X, then its pre-image, $g^{-1}(F) = f(F) \subseteq Y$, is closed in Y by the Closed Function Theorem (2.11). This proves that g is continuous. □

Example 12 Let $T \subset \mathbb{I}^2$ denote the tapered zigzag subset (on p. 105) consisting of the line segments joining consecutive entries of the sequence

$$<1, 1>, \; <0, \tfrac{1}{2}>, \; <\tfrac{1}{3}, \tfrac{1}{3}>, \; <0, \tfrac{1}{4}>, \; \ldots, \; <0, \tfrac{1}{2k}>, \; <\tfrac{1}{2k+1}, \tfrac{1}{2k+1}>, \ldots,$$

together with the origin $O = <0, 0>$ as a bottom point.

The metric subspace $T \subset \mathbb{I}^2$ is called the *tapered saw* space. Unlike the saw space in Example 11, the tapered saw space T contains all its accumulation points in $\mathbb{I}^2$. So it is a closed, hence, sequentially compact metric subspace of $\mathbb{I}^2$, by Theorem 2.6.

The projection function $f : T \to \mathbb{I}$, $f(<x, y>) = y$, is a continuous bijection of the sequentially compact tapered saw space T onto the unit interval $\mathbb{I}$. Hence, f is a homeomorphism by Theorem 4.2.

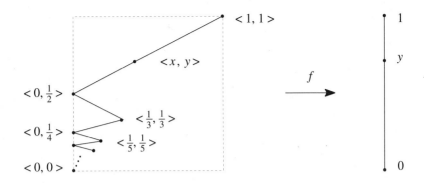

Example 12: Homeomorphic flattening of the tapered saw space

5 Peano's Curve

In the last half of the nineteenth century, mathematicians searched for a rigorous topological definition of dimension. They wanted one that would agree with the intuitive view that the Euclidean line $\mathbb{E}^1$, Euclidean plane $\mathbb{E}^2$, and Euclidean space $\mathbb{E}^3$ are distinguishable spaces of dimensions 1, 2, and 3.

The set theoretic work of G. Cantor in the 1870's complicated the problem. He exhibited a set bijection $\mathbb{R} \to \mathbb{R}^2$ (Example **1**.4.13). The Euclidean spaces $\mathbb{E}^1$ and $\mathbb{E}^2$ of apparently different dimensions are not distinguished by a count of points; the plane is not richer in points than the line. What survived Cantor's work was the hope that the dimension of a space is the number of *continuous* independent real variables required to define it.

In 1890, G. Peano (1858-1932) produced a startling example of a continuous function $\mathbb{I} \to \mathbb{I}^2$ that stretches the unit interval onto the entire unit square in the Euclidean plane. A similar example gave a continuous function $\mathbb{I} \to \mathbb{I}^3$ of the unit interval onto the entire unit cube in Euclidean space. For the moment, it seemed that the basic mathematical concepts of curve, surface, and solid had lost their distinction, so that a mathematically rigorous theory of dimension was not possible. The well-known French mathematician Henri Poincaré (1854-1912) was not amused. "How was it possible that intuition could so deceive us?" he asked.

BASE 9 REPRESENTATION

It is easiest to express Peano's square-filling curve by using the base 9 version of the *B*-adic system (**1**.3.4) of representing the real numbers in the unit interval $\mathbb{I} = [0, 1]$. This system involves iterating the following subdivision process, beginning with the unit interval $\mathbb{I}$, directed from 0 to 1.

- Each directed interval $I[\beta]$ is divided into 9 equal directed subintervals and these are labeled from 0 to 8, in order:

$I[\beta]$

$I[\beta 0]$ $I[\beta 1]$ $I[\beta 2]$ $I[\beta 3]$ $I[\beta 4]$ $I[\beta 5]$ $I[\beta 6]$ $I[\beta 7]$ $I[\beta 8]$

Each subinterval that arises in this iterated subdivision of $\mathbb{I}$ is called a **base 9 interval**. Given a real number $0 \leq r \leq 1$, it is possible to select at each stage of the subdivision process one of the new subintervals that contains r. In this way, r is located, by the Nested Intervals Theorem (**1.3.3**), as the intersection point of some sequence of nested base 9 intervals:

$$\mathbb{I} \supset I[b_1] \supset I[b_1 b_2] \supset \ldots \supset I[b_1 b_2 \ldots b_k] \supset \ldots .$$

The sequence of digits $b_1, b_2, \ldots, b_k, \ldots$ from $\{0\text{--}8\}$ that label these intervals determines the **base 9 expansion** $(.b_1 b_2 \ldots b_k \ldots)_9$ of this real number r.

There is a similar base 9 system of representing points of the unit square $\mathbb{I}^2 = [0, 1]^2$. It is developed by repeatedly applying the following **Peano subdivision process** for directed squares, beginning with $\mathbb{I}^2$, directed from its lower left hand corner $<0, 0>$ to its upper right hand corner $<1, 1>$:

- Each directed square $S(\beta)$ is divided into 9 equal sub-squares, called **Peano squares**, which are directed and labeled as indicated below.

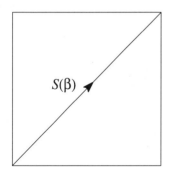 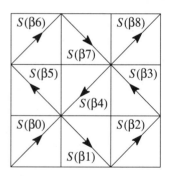

Then each expression $.b_1 b_2 \ldots b_k \ldots$, with digits $b_k \in \{0\text{--}8\}$, determines a sequence of nested Peano squares:

$$\mathbb{I}^2 \supset S(b_1) \supset S(b_1 b_2) \supset \ldots \supset S(b_1 b_2 \ldots b_k) \supset \ldots .$$

By the Nested Boxes Theorem (**2.4**), the intersection of this sequence of nested squares is a unique point $z \in \mathbb{I}^2$. The expression $.b_1 b_2 \ldots b_k \ldots$, with digits $b_k \in \{0\text{--}8\}$, is called a **Peano expansion** of this point z.

Conversely, given a point $z \in \mathbb{I}^2$, it is possible to select at each stage of the subdivision process one of the Peano squares that contains z. So z belongs to at least one sequence of nested Peano squares. This proves that each point has a Peano expansion $z = .b_1 b_2 \ldots b_k \ldots$. Peano's amazing representation of each point of the square $\mathbb{I}^2$ by a single expression is distinct from the usual cartesian coordinate representation of each point of $\mathbb{I}^2$ by an ordered pair $<x, y>$ of real numbers.

PEANO'S SQUARE FILLING CURVE

Peano's curve $P : \mathbb{I} \to \mathbb{I}^2$ sends the point $r \in \mathbb{I}$ with base 9 expansion $(.b_1 b_2 \ldots b_k \ldots)_9$ to the point $P(r) \in \mathbb{I}^2$ with Peano expansion $.b_1 b_2 \ldots b_k \ldots$. Thus, the intersection point of each sequence of nested base 9 intervals:

$$\mathbb{I} \supset I[b_1] \supset I[b_1 b_2] \supset \ldots \supset I[b_1 b_2 \ldots b_k] \supset \ldots$$

is sent to the intersection point of the sequence of nested Peano squares:

$$\mathbb{I}^2 \supset S(b_1) \supset S(b_1 b_2) \supset \ldots \supset S(b_1 b_2 \ldots b_k) \supset \ldots .$$

Peano's curve is well-defined. A real number r that has different base 9 expansions is an endpoint shared by consecutive base 9 intervals; one expansion of r terminates in *eights* and another in *zeros*. The corresponding Peano expansions that terminate in *eights* and *zeros* describe the same point $P(r)$ of $\mathbb{I}^2$, a corner-point shared by consecutive Peano squares.

Peano's curve maps the unit interval onto the unit square. Any point z of the square $\mathbb{I}^2$ belongs to at least one sequence of nested Peano squares. So z is the image of the intersection point in $\mathbb{I}$ of the corresponding sequence of nested base 9 intervals.

Peano's curve is continuous. By construction, Peano's curve P maps each base 9 interval $I[b_1 b_2 \ldots b_k]$ into the corresponding Peano square $S(b_1 b_2 \ldots b_k)$, with just the endpoints of the interval sent to the initial and terminal corner of the square. If $A \subseteq \mathbb{I}$ is arbitrarily near x, then A intersects any open base 9 interval that contains x and also any pair of consecutive base 9 intervals that share x as an endpoint. Therefore, the image $P(A) \subseteq \mathbb{I}^2$ intersects every Peano square that contains $P(x)$ as some non-initial and non-terminal point, and also every pair of consecutive Peano squares that share $P(x)$ as a corner point. It follows that $P(A)$ enters every neighborhood of $P(x)$, so that this image set $P(A)$ is arbitrarily near the image point $P(x)$. This establishes the continuity of the Peano curve P.

This completes the argument. We have proved the following.

5.1 Peano's Curve *There exists a continuous surjection $P : \mathbb{I} \to \mathbb{I}^2$ of the closed unit interval onto the closed unit square.*

PEANO'S CURVE AS A RUBBER BAND

The base 9 fraction $b_1 b_2 \ldots b_k / 9^k = (b_1 b_2 \ldots b_k 0 \ldots)_9$ is the initial endpoint of the base 9 interval $I[b_1 b_2 \ldots b_k]$; it is mapped by P to the initial corner of the Peano square $S(b_1 b_2 \ldots b_k)$. Thus, Peano's curve can be viewed as a limit of the following sequence of functions $P_k : \mathbb{I} \to \mathbb{I}^2$ (Exercise 10):

- P_1 treats $\mathbb{I}$ as a rubber band, stretching it along the diagonal of $\mathbb{I}^2$, thumbtacking its initial and terminal endpoints 0 and 1 at the initial and terminal corners < 0, 0 > and < 1, 1 >.

- P_2 retains the previous two thumbtacks, breaks $\mathbb{I}$ into the nine base 9 subintervals $I[0]$-$I[8]$ and stretches them across the diagonals of the squares $S(0)$-$S(8)$, thumbtacking the base nine fractions 1/9, 2/9, ... , 8/9 at the initial corners of the Peano squares $S(1)$-$S(8)$.

- P_3 retains the previous thumbtacks and repeats the previous base 9 subdivision process on each of the diagonal images of P_2, thumbtacking down the images of the next stage base 9 fractions at the initial corners of the next stage Peano squares, and so on.

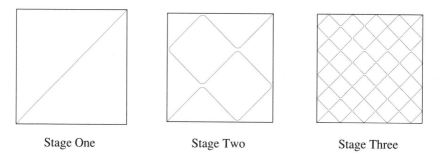

Stage One   Stage Two   Stage Three

DIMENSION THEORY SURVIVES

Were Peano's continuous surjection $P : \mathbb{I} \to \mathbb{I}^2$ also injective, it would be a homeomorphism according to the Homeomorphism Theorem (4.2). The interval $\mathbb{I}$ and the square $\mathbb{I}^2$ would be topologically indistinguishable. Of course, P is not one-to-one. Instead, it is two-to-one on the edges shared by two Peano squares and is four-to-one at the inside corners shared by four Peano squares. Because of this, Peano's work does not eliminate the possibility of a rigorous mathematical theory of dimension. It simply illustrates the need for a more sophisticated definition of dimension than the number of continuous independent real variables required to describe a space.

In 1913, L. J. E. Brouwer (1881-1966) devised a topological definition of dimension. His paper went unnoticed for several years. In 1922, P. Urysohn (1898-1924) and K. Menger (1902-) independently recreated Brouwer's work and organized it into a precise dimension theory. The definition of dimension is inductive: The empty set has dimension -1. The dimension of a space is the least integer n for which every point has arbitrarily small neighborhoods whose boundaries have dimension less than n.

All is well. Euclidean space $\mathbb{E}^n$ and cube $\mathbb{I}^n$ have topological dimension n. Euclidean spaces $\mathbb{E}^n$ and $\mathbb{E}^m$, and cubes $\mathbb{I}^n$ and $\mathbb{I}^m$, are homeomorphic if and only if $m = n$. We leave this work to more advanced topological texts.

Exercise 11 shows how to use connectedness to prove $\mathbb{I} \not\cong \mathbb{I}^2$.

Exercises

SECTION 1

1. Prove that the closed unit disc $\mathbb{D}^n$ and its boundary $(n-1)$–sphere $\mathbb{S}^{n-1}$ are connected subspaces of $\mathbb{E}^n$.

2. Prove that convex and star-like subspaces of $\mathbb{E}^n$ are connected.

3. Prove that the only connected subspaces of Cantor's space C are singletons.

4. Prove that a metric space (X, d) is disconnected if there are subsets $A, B \subseteq X$ with union $A \cup B = X$ and $d(A, B) = glb\{d(a, b) : a \in A, b \in B\} > 0$. Show that this sufficient condition is not necessary for the disconnectedness of a metric space.

5. (a) Prove that a metric space (X, d) is disconnected if and only if there exists a continuous function $f : X \to \mathbb{E}^1$ with a two-point image $f(X) = \{r, s\}$.
   (b) Prove that a metric space (X, d) is connected if and only if every continuous function $f : X \to \mathbb{E}^1$ has the *intermediate value property*: For each real number c between any two values $f(x_1), f(x_2) \in f(X)$, there exists an $x \in X$ such that $f(x) = c$.

6. Prove that a connected metric space (X, d) is either a singleton or is uncountable.

7. Show that there is no fixed point theorem for continuous functions of the open interval $(0, 1)$ nor for continuous functions of the open-closed interval $(0, 1]$.

8.* Prove the Pancake Theorem (1.6), as follows:
   (a) Rescale to make both subsets A and B lie in the open unit disc $\mathbb{B}^2$.
   (b) For each of the two subsets $C = A, B$ and for each $x \in \mathbb{S}^1$, let $t_C(x)$ be the scalar $-1 < t < +1$ such that the line l_C passing through $t x$ and perpendicular to the ray from O through x bisects the set C.
   (c) Prove that the function $g : \mathbb{S}^1 \to \mathbb{E}^1$ defined by $g(x) = t_A(x) - t_B(x)$ is continuous and satisfies $g(-x) = -g(x)$ for all $x \in \mathbb{S}^1$.
   (d) Apply the Borsuk-Ulam Theorem (1.5) to $g : \mathbb{S}^1 \to \mathbb{E}^1$.

9. Prove that the complement of a countable subspace of $\mathbb{E}^1$ is connected.

SECTION 2

1. Find two non-compact metric subspaces X and Y of a metric space Z whose union $X \cup Y$ is a sequentially compact metric subspace of Z.

2.* Let $b(x, y) = min(1, d(x, y))$ for $x, y \in \mathbb{R}$, where d is the Euclidean metric. Prove:
   (a) b is a metric on $\mathbb{R}$.
   (b) Each subset $A \subseteq \mathbb{R}$ has bounded b-$diam(A) = lub\{b(x, y) : x, y \in A\}$.
   (c) The metrics d and b give the same neighborhoods and open sets in $\mathbb{R}$.
   (d) The Heine-Borel Theorem (2.8) fails in $(\mathbb{R}, b)$.

3. Find the smallest sequentially compact subspace of the Euclidean plane $\mathbb{E}^2$ that contains the two polar spiral curves Σ_+ and Σ_- of Example **2.2.5**.

4. (a) Prove that the intersection of finitely many sequentially compact subspaces of a metric space is sequentially compact.

(b) Prove that the union of finitely many sequentially compact metric subspaces of any metric space is sequentially compact.

5 Prove: If X is a metric subspace of the Euclidean line $\mathbb{E}^1$ and if every continuous function $X \to \mathbb{E}^1$ attains both a minimum and maximum, then X is sequentially compact.

6 Prove: A metric space (X, d) is complete if there is a real number $r > 0$ such that the closure of each open ball $B(x, r)$ is a sequentially compact metric subspace of X.

7 Prove that, under a continuous function between metric spaces, the image of each sequentially compact subspace of the domain space is a sequentially compact subspace of the range space.

8 Prove that any continuous bijection $g : \mathbb{I} \to X$ of the closed unit interval $\mathbb{I}$ onto a metric space X has a continuous inverse.

9 Prove that the restriction $\mathrm{Sin} : [-\tfrac{1}{2}\pi, +\tfrac{1}{2}\pi] \to [-1, +1]$ of the trigonometric function $\sin : \mathbb{E}^1 \to \mathbb{E}^1$ has a continuous inverse $\mathrm{Arcsin} : [-1, +1] \to [-\tfrac{1}{2}\pi, +\tfrac{1}{2}\pi]$.

10 Prove that a denumerable, sequentially compact, metric space must have infinitely many isolated points.

11 Prove that a subspace in a complete metric space is totally bounded if and only if its closure is a compact subspace.

SECTION 3

1* Let (X, d) be a complete metric space and let $A, B \subseteq X$ be any subsets.
(a) Prove that the distance
$$d(A, B) = \mathrm{glb}\, \{d(a, b) : a \in A, b \in B\}$$
is achieved by some pair $a \in A$ and $b \in B$, so that
$$d(A, B) = \min\, \{d(a, b) : a \in A, b \in B\}.$$
(b) Prove that $d(A, B) = 0$ if and only if $A \cap B \neq \emptyset$.
(c) Does $d : \mathscr{P}(X) \times \mathscr{P}(X) \to \mathbb{R}$ satisfy any of the metric axioms (**D1**)-(**D3**)?

2 For $A \in \mathfrak{H}(X)$ and $r > 0$, let $A(r) = \{x \in X : d(x, A) \leq r\}$. Prove that for $A, B \in \mathfrak{H}(X)$, we have $h(A, B) \leq r \Leftrightarrow A \subseteq B(r) \Leftrightarrow B \subseteq A(r)$.

3* Prove that the Hausdorff distance h is a metric on the space $\mathfrak{H}(X)$ of sequentially compact subspaces of any complete metric space (X, d).

4* Let h be the Hausdorff metric on the space $\mathfrak{H}(X)$ of sequentially compact subspaces of a complete metric space (X, d). Let $\{A_k\}$ be a Cauchy sequence in $(\mathfrak{H}(X), h)$. Consider the set A of limits of Cauchy sequences (a_k) in X that have terms $a_k \in A_{n_k}$ for all $k \geq 1$, where (n_k) is any subsequence of (n).
Prove:
(a) A is a nonempty, closed, hence, complete metric subspace of (X, d).
(b) A is a totally bounded, hence, compact metric space of (X, d).
(c) $A_k \to A$ in $(\mathfrak{H}(X), h)$.

5* Let h be the Hausdorff metric on the hyperspace $\mathfrak{H}(X)$, as in Exercise 3. Prove that for all $A, B, C, D \in \mathfrak{H}(X)$, we have

$$h(A \cup B, C \cup D) \leq \max\{h(A, C), h(B, D)\}.$$

6. Prove: If (X, d) is a compact metric space, then $(\mathfrak{H}(X), h)$ is a compact metric space.

7. Sierpinski's dust is the product $C^2 \subset \mathbb{I}^2$, where $C \subset \mathbb{I}$ is Cantor's space.
   (a) Find four contractions (similarities) of the Euclidean plane $\mathbb{E}^2$ that determine a contraction Γ of $(\mathfrak{H}(\mathbb{E}^2), h)$ whose attractor is Sierpinski's dust $C^2 \subset \mathbb{E}^2$.
   (b) Construct a convergent sequence of nested compact subspaces of $\mathbb{I}^2$ whose limit in $(\mathfrak{H}(\mathbb{E}^2), h)$ is C^2.
   (c) Construct a convergent sequence of finite subsets of $\mathbb{I}^2$ disjoint from C^2 whose limit in $(\mathfrak{H}(\mathbb{E}^2), h)$ is C^2.

8. Let $(\mathfrak{H}(X), h)$ be the metric space of sequentially compact subspaces of a complete metric space (X, d).
   (a) Prove that a decreasing sequence of nested sequentially compact subspaces,

   $$A_1 \supseteq \ldots \supseteq A_k \supseteq A_{k+1} \supseteq \ldots,$$

   converges in $(\mathfrak{H}(X), h)$. Must the limit equal the intersection $\cap_k A_k$?
   (b) Does a strictly increasing sequence of nested sequentially compact subspaces,

   $$A_1 \subset \ldots \subset A_k \subset A_{k+1} \subset \ldots,$$

   necessarily converge in $(\mathfrak{H}(X), h)$? Does the limit ever equal the union $\cup_k A_k$?

9. Consider the contraction $\Gamma : \mathfrak{H}(\mathbb{E}^1) \to \mathfrak{H}(\mathbb{E}^1)$, $\Gamma(A) = g_1(A) \cup g_2(A)$, constructed from the contractions $g_1, g_2 : \mathbb{E}^1 \to \mathbb{E}^1$ given by

   $$g_1(x) = \tfrac{1}{4}x \quad \text{and} \quad g_2(x) = \tfrac{1}{2}x + \tfrac{1}{2}.$$

   Determine the attractor, P, of Γ and characterize the base four expansions of the points $x \in P \subset \mathbb{E}^1$.

10. In each case, determine a contraction of the hyperspace $(\mathfrak{H}(\mathbb{E}^2), h)$ whose attractor is the sequentially compact subspace of the Euclidean plane $\mathbb{E}^2$ depicted below:

    (a)   (b)

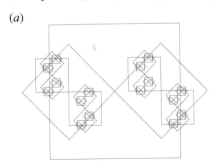
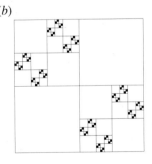

11. Let (X, d) be a sequentially compact metric space.
    (a) Prove that any increasing sequence of nested sequentially compact subspaces,

    $$A_1 \subseteq \ldots \subseteq A_k \subseteq A_{k+1} \subseteq \ldots,$$

    is a Cauchy sequence in the complete metric space $(\mathfrak{H}(X), h)$, hence, has a limit.
    (b) Let $\Gamma : \mathfrak{H}(X) \to \mathfrak{H}(X)$ be a contraction of $(\mathfrak{H}(X), h)$. Consider $K \in \mathfrak{H}(X)$ such that $K \subseteq \Gamma(A)$ for all $A \in \mathfrak{H}(X)$. Use $P = \cup_k \Gamma^k(K)$ to express the attractor of Γ.

SECTION 4

1. In each case, construct homeomorphisms among the given Euclidean metric spaces:
   (a) a semi-circle (with its two endpoints) and a closed interval,
   (b) circular, elliptical, and square discs,
   (c) the sphere with two holes, a cylinder, and a disc with one hole, and
   (d) a punctured 2-sphere and the plane.

2. Construct homeomorphisms among the following wire-like metric subspaces of $\mathbb{E}^2$:

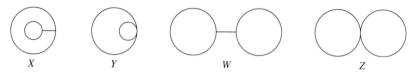

3. Classify up to homeomorphism the nonempty intervals and rays in $\mathbb{E}^1$.

4. Decide which of the letters of a sans-serif version of the roman alphabet

   ABCDEFGHIJKLMNOPQRSTUVWXYZ

   are topologically equivalent. Interpret each letter as a finite set of line segments.

5. Take four strips of paper and paste each one end-to-end into a loop, making each have one more (half) twist than the previous one. Use cut-and-glue techniques to determine which of these four twisted loops are topologically equivalent.

6. Prove that a topologist can't distinguish between a doughnut and a coffee cup: Show how to continuously mold a solid clay torus into the shape of a cup.

7. The following wire-like metric subspaces of Euclidean 3-space $\mathbb{E}^3$ are obtained from the first one by reconnecting the three vertical strands according to the six possible permutations of their ends. Classify these spaces up to homeomorphism using the cut-and-glue techniques:

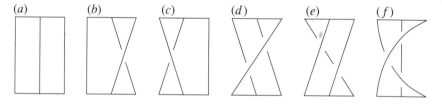

8. Find a continuous surjection of a non-compact metric space onto a sequentially compact metric space.

9. Let X be the union of the origin and the polar coordinate spiral $r = 1/\theta$ ($\pi \leq \theta$) that winds around the origin. Decide whether X is a sequentially compact subspace of $\mathbb{E}^2$. Is X homeomorphic to the closed interval $\mathbb{I} = [0, 1]$?

10. Prove that any continuous bijection $f : [a, b] \to [c, d]$ is a homeomorphism.

11* The subset $NC = ([0, \infty) \times \{1\}) \cup (\mathbb{N} \times \mathbb{I}) \subset \mathbb{E}^2$ is called the **natural comb**. The subset $HC = (\mathbb{I} \times \{1\}) \cup (\mathbb{M} \times \mathbb{I}) \subset \mathbb{E}^2$, where $\mathbb{M} = \{0, 1, 1/2, \ldots, 1/k, \ldots\}$, is called the

harmonic comb. Decide whether any of the three metric subspaces NC, HC, and $HC - (\{0\} \times \mathbb{I})$ of $\mathbb{E}^2$ are homeomorphic.

Exercise 11: Natural comb   Exercise 11: Harmonic comb

12* Consider the ***doubled harmonic comb***, the ***harmonic maze***, and the ***harmonic filter***, displayed below. Decide whether any two of them are homeomorphic metric subspaces of $\mathbb{E}^2$. State fully any intuitive claims on which you base your decisions.

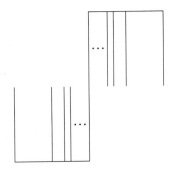 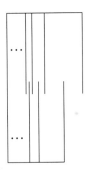 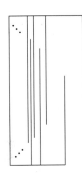

Doubled harmonic comb   Harmonic maze   Harmonic filter

13 Find a metric space X and an accumulation point x of X for which there is a homeomorphism $X \cong X - \{x\}$.

14* Determine the closure in $\mathbb{E}^2$ of the ***topologist's sine curve***
$$X = \{<x, \sin 1/x> \in \mathbb{E}^2 : 0 < x \leq 1/\pi\}.$$
Prove that X is homeomorphic to a metric subspace of $\mathbb{E}^1$, but that $\mathrm{CLS}(X)$ is not.

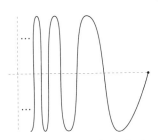 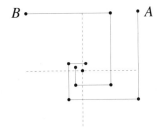

Exercise 14: Topologist's sine curve   Exercise 15: Cubical spirals

15 Let A denote the ***cubical spiral*** comprised of the straight line segments in the Euclidean plane $\mathbb{E}^2$ connecting the consecutive points in the following sequence:

$$<1,1>, <1,-\tfrac{1}{2}>, <-\tfrac{1}{4},-\tfrac{1}{2}>, <-\tfrac{1}{4},\tfrac{1}{8}>, <\tfrac{1}{16},\tfrac{1}{8}>, \ldots.$$

Let B denote the cubical spiral obtained by rotating A about the origin $\mathbf{0} = <0,0>$ by 90°. Prove that the metric subspace $A \cup \{\mathbf{0}\} \cup B \subset \mathbb{E}^2$ is homeomorphic to the closed unit interval $\mathbb{I} \subset \mathbb{E}^1$.

16 Let $C_r \subset \mathbb{E}^2$ be the circle of radius $r > 0$, centered at $<r, 0>$. Are the unions

$$X = \cup \{C_k : k \geq 1\} \qquad \text{and} \qquad Y = \cup \{C_{1/k} : k \geq 1\}$$

homeomorphic metric subspaces of $\mathbb{E}^2$? Explain.

17 Prove directly that the continuous flattening $f : S \to \mathbb{I}$ of the saw space S (Example 11) given by $f(<x, y>) = y$ does not have a continuous inverse: Find a subset $A \subset S$ that is not arbitrarily near the point $<0, 0>$, but whose image $f(A) \subset \mathbb{I}$ is arbitrarily near the image point $f(<0, 0>) = 0$.

SECTION 5

1 Peano's square-filling curve is simply the system of representing each point of the square by some base 9 expression. Catalog all the points of the square whose representation is unique, as well as those that have multiple representations (duplicate or quadruplicate). What is the size of each set in your catalog?

2 Consider the set $E \subset \mathbb{I}^2$ of points in the closed unit square that have an ***even*** Peano expansion. (See Exercise **6**.3.8.)
(*a*) Find five contractions (similarities) of the Euclidean plane $\mathbb{E}^2$ that determine a contraction Γ of $(\mathfrak{H}(\mathbb{E}^2), h)$ whose attractor is the set $E \subset \mathbb{I}^2$.
(*b*) Construct a convergent sequence of nested sequentially compact subspaces of $\mathbb{I}^2$ whose limit in $(\mathfrak{H}(\mathbb{E}^2), h)$ is E.
(*c*) Construct a convergent sequence of finite subsets of $\mathbb{I}^2$ disjoint from C^2 whose limit in $(\mathfrak{H}(\mathbb{E}^2), h)$ is E.
(*d*) Determine the Hausdorff p-measure and Hausdorff dimension of the metric subspace $E \subset \mathbb{E}^2$. (See Exercises **2**.5.19 and **2**.5.20.)

3 Develop a base 4 system of representing points in an equilateral triangle T. Use Hausdorff's Nested Set Theorem (**2**.5.5) and the following subdivision and labeling process for directed triangles:

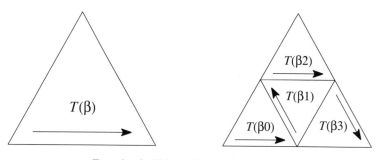

Exercise 3: Triangular subdivision

4 Construct a triangle-filling curve, using base 4 representations of points of the triangle as in Exercise 3; and describe your curve as a limit of stretched rubber bands.

5 *Sierpinski's triangle* $\mathcal{S}$ (below left) is the set of points of the coordinatized triangle $\triangle$ in Exercises 3 and 4 that have a base 4 expansion in which no 1 occurs.
   (*a*) Construct a contraction of the space $(\mathfrak{H}(\mathbb{E}^2), h)$ of compact subspaces of $\triangle$ whose attractor is Sierpinski's triangle $\mathcal{S}$.
   (*b*) Construct a convergent sequence of nested compact subspaces of $\triangle$ whose limit in $(\mathfrak{H}(\mathbb{E}^2), h)$ is $\mathcal{S}$.
   (*c*) Construct a convergent sequence of finite subsets of $\triangle$ disjoint from $\mathcal{S}$ whose limit in $(\mathfrak{H}(\mathbb{E}^2), h)$ is $\mathcal{S}$.
   (*d*) Determine the Hausdorff p-measure and Hausdorff dimension of Sierpinski's triangle as a metric subspace $\mathcal{S} \subset \mathbb{E}^2$. (See Exercises **2.5.19** and **2.5.20**.)

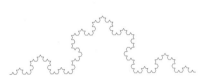

Exercise 5: Sierpinski's Triangle   Exercise 6: Koch's Curve

6 *Koch's curve* K (above right) is obtained from $\mathbb{I}$ by repeatedly replacing the middle third of any line segment by the other two sides of an equilateral triangle based on that middle third.
   (*a*) Find a contraction Γ of $(\mathfrak{H}(\mathbb{E}^2), h)$ whose attractor is Koch's curve K.
   (*b*) Construct a convergent sequence of nested compact subspaces of $\mathbb{E}^2$ whose limit in $(\mathfrak{H}(\mathbb{E}^2), h)$ is K.
   (*c*) Construct a convergent sequence of finite subsets of $\mathbb{E}^2$ disjoint from K whose limit in $(\mathfrak{H}(\mathbb{E}^2), h)$ is K.
   (*d*) Determine the Hausdorff p-measure and Hausdorff dimension of the metric subspace $K \subset \mathbb{E}^2$. (See Exercises **2.5.19** and **2.5.20**.)

7 Sierpinski's dust is the product $C^2 \subset \mathbb{I}^2$, where $C \subset \mathbb{I}$ is Cantor's space. Characterize the points of Sierpinski's dust in terms of their Peano expansions.

8 Adapt Peano's ideas to define a continuous space-filling curve $\mathbb{I} \rightarrow \mathbb{I}^3$.

9 Invent your own space-filling curve that fills an annular region of the plane.

10* Express Peano's curve as a limit in the metric space $\mathcal{C}(\mathbb{I}, \mathbb{I}^2)$ of continuous functions $f: \mathbb{I} \rightarrow \mathbb{I}^2$, with max metric $d_\infty(f, g) = lub \{d(f(t), g(t)) : 0 \leq t \leq 1\}$.

11 Prove the following statements and deduce that there is no homeomorphism $\mathbb{I} \cong \mathbb{I}^2$:
   (*a*) Any homeomorphism $f: X \rightarrow Y$ of metric spaces restricts to a homeomorphism $X - \{x\} \rightarrow Y - \{f(x)\}$ of point complements, for each $x \in X$.
   (*b*) Some points $t \in \mathbb{I}$ have a disconnected complement $\mathbb{I} - \{t\}$, but every point $<r, t> \in \mathbb{I}^2$ has a connected complement $\mathbb{I}^2 - \{<r, t>\}$.

4

TOPOLOGICAL SPACES

In this chapter, we abstract from the metric setting just what is needed to discuss continuity, without reference to a metric. Continuity can be expressed in terms of either set closure, set interior, neighborhoods of points, open sets, or closed sets. These concepts are so intertwined that any one of them can be used to formulate the remaining ones. So it is just a matter of choice which of them is used as the primitive notion in a definition of an abstract topological space. Each of these concepts had their promoters among early topologists, but the use of open sets has become the standard.

1   Topologies

The properties of the collection of open sets in a metric space (Theorem 2.3.3) suggest the following definitions. Let X be any set.

A *topology* on X is a set $\mathcal{T}$ of subsets $U \subseteq X$ satisfying these properties:

(T1)   Both $\emptyset$ and X are members of $\mathcal{T}$.

(T2)   Any union of members of $\mathcal{T}$ is a member of $\mathcal{T}$.

(T3)   Any intersection of finitely many members of $\mathcal{T}$ is a member of $\mathcal{T}$.

A pair $(X, \mathcal{T})$, where $\mathcal{T}$ is a topology on X, is called a *topological space*. When it is clear which topology $\mathcal{T}$ on X is involved, we shall abbreviate the topological space $(X, \mathcal{T})$ by X.

EXAMPLES AND DEFINITIONS

The following catalog offers various specialized topological spaces. Some of them are unusual and incompatible with intuition developed in metric topology. Because of this, they serve as useful test cases in upcoming investigations of abstract topological concepts.

Example 1 Let (X, d) be a metric space. The collection
$$\mathcal{T}_d = \{U \subseteq X : U \text{ is } d\text{-open}\}$$
is a topology on X by Proposition 2.3.3. $\mathcal{T}_d$ is called the **metric topology** induced by d. A topology $\mathcal{T}$ is **metrizable** if it is the topology $\mathcal{T}_d$ induced by some metric d on X.

We may rephrase a definition of Section 2.3 this way. Two metrics d and m are **topologically equivalent** if they induce the same topology: $\mathcal{T}_d = \mathcal{T}_m$.

Hereafter, we give $\mathbb{R}^n$ the topology induced by the Euclidean and equivalent metrics, and we call $\mathbb{R}^n$ (**topological**) **real n-space**.

Example 2 For any set X, the collection $\{\emptyset, X\}$ is a topology on X, called the **indiscrete topology**. If X has distinct points, this topology is not metrizable. For distinct points $x \neq y$ cannot be separated by disjoint open sets in X as can be achieved in a metric space using open balls of radius $r = \frac{1}{2}d(x, y)$ (Theorem 2.3.5).

Example 3 For any set X, the power set $\mathcal{P}(X) = \{U \subseteq X\}$ of all subsets of X is a topology on X, called the **discrete topology**. The discrete metric δ assigns the distance $\delta(x, y) = 1$ whenever $x \neq y$; it gives $B(x, r) = \{x\}$ if $r < 1$. This makes each subset of X a δ-open set. Thus, the discrete metric topology $\mathcal{T}_\delta$ is the discrete topology $\mathcal{P}(X)$.

Any metric topology on a finite set is discrete because each ball of radius less than the minimum of the distances between the finitely many points is a singleton.

Example 4 There are exactly four topologies on the doubleton set $X = \{0, 1\}$:

$\{\emptyset, \{0, 1\}\}$ $\{\emptyset, \{0\}, \{0, 1\}\}$ $\{\emptyset, \{1\}, \{0, 1\}\}$ $\{\emptyset, \{0\}, \{1\}, \{0, 1\}\}$

The middle two topologies are called **Sierpinski** topologies; they are not metrizable.

Example 5 Let $X = \{a, b, c\}$. The family $\{\emptyset, \{a, b\}, \{a, c\}, \{a, b, c\}\}$ is not a topology on X; it fails to contain the intersection $\{a, b\} \cap \{a, c\} = \{a\}$.

Example 6 For any set X, the collection
$$\mathcal{FC} = \{\emptyset\} \cup \{U \subseteq X : X - U \text{ is finite}\}$$
is a topology on X, called the **finite complement topology**. The empty subset $\emptyset$ is included in $\mathcal{FC}$ as a special case, since its complement, X, may not be finite.

A subset $U \subseteq X$ with finite complement, $X - U$, is called **co-finite**, and $\mathcal{FC}$ is often called the **co-finite topology**. When X is a finite set, the topology $\mathcal{FC}$ is discrete.

Example 7 For any set X, the collection
$$\mathcal{CC} = \{\emptyset\} \cup \{U \subseteq X : X - U \text{ is countable}\}$$
is a topology on X, called the **countable complement (co-countable) topology**. When X is a countable set, the topology $\mathcal{CC}$ is discrete (Example 3).

Example 8 If X is any set and p is any member of X, then
$$\mathcal{D}(p) = \{X\} \cup \{U \subseteq X : p \notin U\}$$
is a topology on X, called the **excluded point topology**.

Example 9 If X is any set and p is any member of X, then

$$\mathcal{T}(p) = \{\emptyset\} \cup \{U \subseteq X : p \in U\}$$

is a topology on X, called the ***particular point topology***.

Example 10 If X is any subset of a topological space $(Y, \mathcal{S})$, then

$$\mathcal{S}_X = \{U = V \cap X \subseteq X : V \in \mathcal{S}\}$$

is a topology on X (Exercise 2), called the ***subspace topology*** or ***relative topology*** induced by $\mathcal{S}$. The space $(X, \mathcal{S}_X)$ is called a ***topological subspace*** of $(Y, \mathcal{S})$.
The subspace topology is discussed more fully in Section **5.1**.

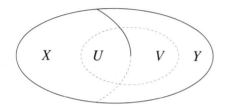
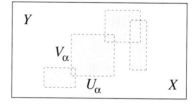

Example 10: Subspace topology   Example 11: Product topology

Example 11 Let $(X, \mathcal{T})$ and $(Y, \mathcal{R})$ be topological spaces. The collection of all unions of products $U \times V \subseteq X \times Y$ of members $U \in \mathcal{T}$ and $V \in \mathcal{R}$,

$$\mathcal{T} \times \mathcal{R} = \{W = \cup_\alpha (U_\alpha \times V_\alpha) \subseteq X \times Y : U_\alpha \in \mathcal{T}, V_\alpha \in \mathcal{R}\},$$

is a topology on $X \times Y$ (Exercise 3), called the ***product topology***. The topological space $(X \times Y, \mathcal{T} \times \mathcal{R})$ is called the ***topological product*** of $(X, \mathcal{T})$ and $(Y, \mathcal{R})$.
The product topology is examined more thoroughly in Section **5.3**.

OPEN SETS, CLOSED SETS, AND NEIGHBORHOODS

Let $(X, \mathcal{T})$ be a topological space. The members $U \in \mathcal{T}$ are called the ***open sets*** in $(X, \mathcal{T})$; their complements $X - U$ are called the ***closed sets*** in $(X, \mathcal{T})$.

A subset $N \subseteq X$ is called a ***neighborhood*** of x in X if there exists an open set $U \subseteq X$ such that $x \in U \subseteq N$.

We write $N(x)$ to abbreviate the neighborhood relationship. Notice that a subset $U \subseteq X$ is open if and only if it is a neighborhood of each point $x \in U$.

As in the metric setting, the properties (**T1**)-(**T3**) of the open sets in X imply dual properties for their complements, the closed sets:

(**C1**) *Both the empty set $\emptyset$ and X are closed sets in X.*

(**C2**) *Any intersection of closed sets in X is closed in X.*

(**C3**) *Any union of finitely many closed sets in X is closed in X.*

Example 12 The open sets, closed sets, and neighborhoods in the metric space (X, d) coincide with those subsets in the induced topological space $(X, \mathcal{T}_d)$.

Example 13 In the indiscrete topology on X (Example 2), the closed subsets are $\emptyset$ and X; in the discrete topology (Example 3), every subset of X is closed.

Example 14 In the finite complement topology $\mathcal{FC}$ (Example 6), the closed sets are X and its finite subsets. In the countable complement topology $\mathcal{CC}$ (Example 7), the closed sets are X and its countable subsets.

Example 15 In the particular point topology $\mathcal{P}(p)$ (Example 9), the closed sets are X and the subsets that exclude the point p. In the excluded point topology $\mathcal{D}(p)$ (Example 8), the closed sets are $\emptyset$ and the subsets that include p.

The open sets, closed sets, and neighborhoods in a subspace are related to those in the topological space by intersection with the subspace, as follows.

1.1 Theorem *Let $(X, \mathcal{S}_X)$ be a topological subspace of $(Y, \mathcal{S})$. Then,*
(a) *$U \subseteq X$ is open in X if and only if $U = V \cap X$ for some open set V in Y;*
(b) *$F \subseteq X$ is closed in X if and only if $F = G \cap X$ for some closed set G in Y;*
(c) *M is a neighborhood of x in X if and only if $M = N \cap X$ for some neighborhood N of x in Y.*

Proof: (a) This is the definition in Example 10 of the subspace topology.
  (b) If F is closed in X, then $X - F$ is open in X, so that $X - F = V \cap X$ for some open set V in Y. Then $F = X - (V \cap X) = (Y - V) \cap X = G \cap X$ for the closed set $G = Y - V$ in Y. Part (c) is left as an exercise. □

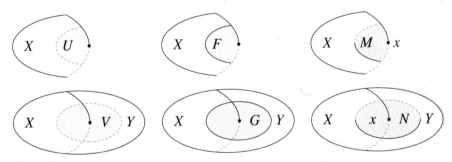

Relatively open $U \subseteq Y$ Relatively closed $F \subseteq Y$ Relative neighborhood $M \subseteq Y$

An open set in the subspace $(X, \mathcal{S}_X)$ is called **relatively open** in Y; a closed set in the subspace $(X, \mathcal{S}_X)$ is called **relatively closed** in Y; and a neighborhood in the subspace $(X, \mathcal{S}_X)$ is called a **relative neighborhood** in Y.

1.2 Theorem *In the topological product $(X \times Y, \mathcal{S} \times \mathcal{R})$,*
(a) *a subset W is open in $X \times Y$ if and only if, for each point $<x, y> \in W$, there exist $U \in \mathcal{S}$ and $V \in \mathcal{R}$ such that $<x, y> \in U \times V \subseteq W$, and*
(b) *a subset F is closed in $X \times Y$ if and only if, for each point $<x, y> \notin F$, there exist $U \in \mathcal{S}$ and $V \in \mathcal{R}$ such that $<x, y> \in U \times V \subseteq (X \times Y) - F$.*

Proof: (*a*) This expresses the definition (Example 11) that the open sets in $X \times Y$ are the unions of products $U \times V$ of open sets $U \in \mathcal{T}$ and $V \in \mathcal{R}$ of the two factor spaces X and Y.

(*b*) Both conditions say that the complement $(X \times Y) - F$ is open. □

CONVERGENCE

In any metric space (X, d), the closed sets and the open sets can be detected using convergent sequences in X. A set $F \subseteq X$ is closed if and only if F contains the limit of each convergent sequence of points in F; a set $U \subseteq X$ is open if and only if U contains the limit of no convergent sequence of points in $X - U$.

Convergence has the expected meaning in a topological space X.

A *sequence* in X is a function $a : \mathbb{N} \to X$; $a(k) = a_k$ is called the kth *term* of a; and, for each $m \geq 1$, the set $A_m = \{a_k : m \leq k\}$ is called a *tail* of a.

A sequence $a : \mathbb{N} \to X$ *converges* in X to a *limit* $x \in X$ if, for every neighborhood $U(x)$, there is some tail $A_m = \{a_k : m \leq k\}$ of a contained in U.

Unlike the metric situation, convergent sequences are inadequate to determine the open and closed subsets in all topological spaces. Here are two examples that indicate this *inadequacy of sequences* in topology.

> **Example 16** Let X be an uncountable set with the countable complement topology $\mathcal{CC}$ (Example 7). Every sequence in X has a countable set of values, whose complement is thus open in X. A convergent sequence in X must eventually enter and remain (i.e., must have a tail) in every neighborhood of its limit; so, its limit cannot belong to the (open) complement of its countable set of values. The limit must be a term of the sequence. Thus, every subset of X contains the limit of each of its convergent sequences, even a non-closed subset such as a proper uncountable subset $F \subset X$.

> **Example 17** The first uncountable ordinal Ω (Theorem **1.**5.12) determines the ordinal interval $[0, \Omega] = \Omega + 1$ that we give the *order topology* consisting of all rays $[0, \alpha)$ and $(\beta, \Omega]$ in $[0, \Omega]$, their finite intersections, and unions of such finite intersections. (The order topology exists for any totally ordered set; see Example **5.**2.5.)
>
> Because any ray $(\beta, \Omega]$ containing Ω contains a countable ordinal (e.g., $\beta + 1$), the singleton $\{\Omega\}$ is not open in the order topology. Thus, $[0, \Omega] - \{\Omega\} = [0, \Omega)$ is not closed in the order topology. Because each countable subset $A \subset [0, \Omega)$ has an upper bound $\beta \in [0, \Omega)$ by Theorem **1.**5.12, each sequence in $[0, \Omega)$ fails to enter some neighborhood $(\beta, \Omega]$ of Ω. So $[0, \Omega)$ contains the limit of each of its convergent sequences, even though it is not closed in the order topology on $[0, \Omega]$.

This inadequacy of convergent sequences led E. H. Moore and H. L. Smith in 1922 to generalize the concept of convergent sequences to that of *convergent nets*, which are adequate for all topological purposes. Their idea involves the replacement of the ordered set $(\mathbb{N}, \leq)$ by arbitrary directed sets.

A *directed set* $(\mathcal{D}, \leq)$ is a set $\mathcal{D}$ with a binary relation $\leq$ that has the three features that follow:

- $\forall\, d \in \mathcal{D}$: $d \leq d$. *(reflexive)*
- $\forall\, a, b, c \in \mathcal{D}$: $(a \leq b) \wedge (b \leq c) \Rightarrow a \leq c$. *(transitive)*
- $\forall\, a, b \in \mathcal{D}, \exists\, d \in \mathcal{D}$: $(a \leq d) \wedge (b \leq d)$. *(directed)*

A ***net*** in a space X is a function $\lambda : \mathcal{D} \to X$ of some directed set $(\mathcal{D}, \leq)$. For each $d \in \mathcal{D}$, the set $\Lambda_d = \{\lambda(c) : d \leq c \in \mathcal{D}\}$ is called a ***tail*** of λ.

A net $\lambda : \mathcal{D} \to X$ ***converges*** to a ***limit*** $x \in X$ if, for every neighborhood $U(x)$ in X, there is some tail $\Lambda_d = \{\lambda(c) : d \leq c\}$ of λ that is contained in U.

When $(\mathcal{D}, \leq) = (\mathbb{N}, \leq)$, a convergent net is simply a convergent sequence. So nets generalize sequences. Unlike the metric situation, the limit of a net or sequence in a topological space need not be unique.

Example 18 Let X be a space and let $x \in X$. The set $\mathcal{U}(x)$ of open neighborhoods U of x in X, ordered by reverse containment $\supseteq$, is a directed set $(\mathcal{U}(x), \supseteq)$. Any choice of points $\lambda(U) \in U$, for all $U \in \mathcal{U}(x)$, defines a net $\lambda : \mathcal{U}(x) \to X$ that converges to x. This net λ also converges to any point $z \in X$, all of whose neighborhoods contain x.

All basic topological concepts can be expressed through convergent nets.

1.3 Theorem *Let X be a topological space.*
(a) A subset F is closed in a space X if and only if F contains the limit of each convergent net $\lambda : \mathcal{D} \to X$ contained in F.
(b) A subset V is open in a space X if and only if V contains no limit of a convergent net $\lambda : \mathcal{D} \to X$ contained in $X - V$.

Proof: We establish the second of the two equivalent statements (a) and (b).

An open set V is a neighborhood of each $x \in V$. Since a net $\lambda : \mathcal{D} \to X$ that is contained in $X - V$ has no tail in V, it cannot have limit $x \in V$.

Conversely, let $X - V$ contain each limit of each convergent net contained in $X - V$. We show that V contains an open neighborhood of each $x \in V$ and, hence, is open by (**T**3). Were there to exist a point $\lambda(U) \in U - V$ for each member $U \in \mathcal{U}(x)$ of the directed set of open neighborhoods of x (Example 18), the net $\lambda : \mathcal{U}(x) \to X$ would be contained in $X - V$, but would converge to $x \in V$. By hypothesis, this doesn't happen. So some $U \in \mathcal{U}(x)$ has empty complement $U - V = \emptyset$, and therefore V contains the neighborhood U of x. $\square$

Since the tails $\{\Lambda_d : d \in \mathcal{D}\}$ of a net λ determine its convergence, their features suggest the following generalization of a net in a space X.

A ***filterbase*** in X is a set $\mathcal{A} = \{A\}$ of nonempty subsets of X such that

- $\forall\, A, B \in \mathcal{A}, \exists\, C \in \mathcal{A} : A \cap B \supseteq C$. *(directed)*

A filterbase $\mathcal{A}$ in a topological space X ***converges*** to a ***limit*** $x \in X$ if, for every neighborhood $U(x)$, there is some member $A \in \mathcal{A}$ contained in U.

Example 19 The neighborhood set $\mathcal{U}(x)$ is a filterbase in X that converges to x.

Example 20 Riemann integrability of a function $f : [0, 1] \to \mathbb{R}$ is most easily expressed using the convergence of a filterbase in $\mathbb{R}$. For each *partition* of $\mathbb{I} = [0, 1]$,

$$P : 0 = t_0 \leq t_1 \leq \ldots \leq t_n = 1,$$

and choices of points $\xi_i \in [t_{i-1}, t_i]$, there is the Riemann sum $\Sigma_i f(\xi_i)(t_i - t_{i-1}) \in \mathbb{R}$. Let Σ_P denote the set of all Riemann sums for all partitions that contain P. The set $\mathcal{R} = \{\Sigma_P\}$ is a filterbase in $\mathbb{R}$, and one says that f is *Riemann integrable* on $[0, 1]$ if and only if $\mathcal{R}$ converges in $\mathbb{R}$.

But it is really a matter of convenience whether to use nets or filterbases in a particular situation. The tails of any net determine a filterbase that converges to a point if and only if the net does so, and conversely, any filterbase determines a net with the same convergence features:

1.4 Theorem *There is a correspondence between nets and filterbases in a space such that one converges to a point if and only if the other does.*

Proof: We show that any filterbase $\mathcal{A}$ in X is the set of tails of some net λ in X. Form the set $\mathcal{D}$ of ordered pairs $<a, A>$, where $a \in A \in \mathcal{A}$; order $\mathcal{D}$ by $<a, A> \leq <b, B> \Leftrightarrow A \supseteq B$. Then $(\mathcal{D}, \leq)$ is a directed set, since $\mathcal{A}$ is a filterbase. The net $\lambda : \mathcal{D} \to X$, given by $\lambda(<a, A>) = a$, has the sets $\Lambda_{<a, A>} = A$ as tails. Therefore, $\mathcal{A}$ converges to x if and only if λ converges to x. $\square$

2 Topological Separation

The richness of the neighborhood system defined by a topology is reflected in the ability to separate various types of disjoint subsets of the space by disjoint neighborhoods. Six different levels of separation capability occur frequently enough to deserve their own names. These *separation properties* or *axioms* (T_0, T_1, T_2, T_3, T_4, and T_5) were introduced as either defining axioms or convenient properties by several pioneering topologists of the early decades of the twentieth century. Kolmogorov, Fréchet, and Hausdorff. Felix Hausdorff (1868-1942) used the T_2 property as part of his axiomatic development of abstract topological spaces, *Grundzüge der Mengenlehre*, published in 1914. The separation properties were cataloged as "trennungsaxiomen" by H. Tietze in 1923. They are defined below.

In Section 1, subset $N \subseteq X$ of a topological space $(X, \mathcal{T})$ is called a neighborhood of x in X if there exists an open set $U \subseteq X$ such that $x \in U \subseteq N$.

Similarly, a subset $N \subseteq X$ is called a **neighborhood** of a subset $A \subseteq X$ if there exists an open set $U \subseteq X$ such that $A \subseteq U \subseteq N$. We write $N(A)$ to abbreviate the neighborhood relationship.

Subsets $A, B \subseteq X$ are called **separated in X** if the complement of each one is a neighborhood of the other; e.g., disjoint closed subsets are separated.

SEPARATION PROPERTIES

A topological space $(X, \mathcal{T})$ is called

- a T_0 space if, for each pair of distinct points $x, y \in X$, there is either a neighborhood $N(x)$ such that $y \notin N(x)$ or one $M(y)$ such that $x \notin M(y)$;
- a T_1 space if, for each pair of distinct points $x, y \in X$, there exist neighborhoods $N(x)$ and $M(y)$ such that $y \notin N(x)$ and $x \notin M(y)$;
- a T_2 or **Hausdorff** space if, for each pair of distinct points $x, y \in X$, there exist disjoint neighborhoods $N(x)$ and $M(y)$;

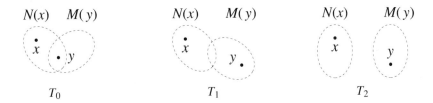

- a T_3 space if, for each closed set $B \subseteq X$ and each point $x \notin B$, there exist disjoint neighborhoods $N(x)$ and $M(B)$;
- a T_4 space if, for each pair of disjoint closed sets $A, B \subseteq X$, there exist disjoint neighborhoods $N(A)$ and $M(B)$; and
- a T_5 space if, for each pair of separated sets $A, B \subseteq X$, there exist disjoint neighborhoods $N(A)$ and $M(B)$.

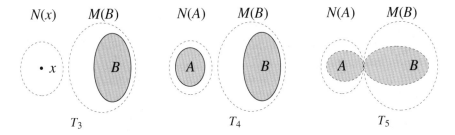

Example 1 Any metric topology $\mathcal{T}_d$ is T_2 and T_4, by Theorem **2.3.5**. For example, $B_d(x, r)$ and $B_d(y, r)$ are disjoint neighborhoods of $x \neq y$ whenever $0 < r \leq \frac{1}{2} d(x, y)$.

Example 2 In a T_1 space all points and, hence, all finite subsets are closed. Thus, a finite T_1 space is discrete; all subsets are simultaneously open and closed.

BASIC RELATIONSHIPS

Several combinations of the separation properties have specialized names.

A space that is both T_3 and T_0 is called a **regular** space. A space that is both T_4 and T_1 is called a **normal** space. A space that is both T_5 and T_1 is called a *completely normal* space. Regularity, normality, and another property, called complete regularity, are investigated more thoroughly in Chapter **8**. By these definitions, we have the following theorem.

2.1 Theorem   *The separation properties have these relationships.*
(*a*)   *A space is T_0 if and only if for each pair of distinct points there exists a closed set containing one but not the other.*
(*b*)   *A space is T_1 if and only if each of its points is a closed set.*
(*c*)   *Property T_5 implies T_4, T_2 implies T_1, and T_1 implies T_0.*
(*d*)   *A normal space is regular, and a regular space is Hausdorff.*

Proof: (*a*) A space is T_0 if and only if, for each pair of distinct points, there exists an open set containing one but not the other, or equivalently, a closed set containing one but not the other.

(*b*) A space X is T_1 if and only if the complement of each point is a neighborhood of every other point, hence, open.

(*c*) Since disjoint closed sets are separated in X, T_5 implies T_4. Properties T_2, T_1, and T_0 are successively weaker conditions on point neighborhoods.

(*d*) By (*b*), properties T_4 and T_1 imply T_3. By (*a*), T_3 and T_0 imply T_2. □

2.2 Theorem   *The separation properties T_0, T_1, T_2, T_3, and T_5 are hereditary: each subspace X of a T_i space Y is T_i ($i = 0, 1, 2, 3,$ or 5).*

Proof: First of all, neighborhoods or closed subsets in the subspace X arise as intersections with X of neighborhoods or closed subsets in the space Y.

T_0, T_1, and T_2 are hereditary. Neighborhoods $N(x)$ and $N(y)$ in Y of distinct points $x, y \in X$ give neighborhoods $N(x) \cap X$ and $N(y) \cap X$ in X with the same separation features.

T_3 is hereditary. Any closed set A in X disjoint from a point $x \in X$ is the intersection $F \cap X$ for some closed set F in Y that is necessarily disjoint from x. Then disjoint neighborhoods $N(F)$ and $N(x)$ in Y yield disjoint neighborhoods $N(F) \cap X$ and $N(x) \cap X$ in X for A and x.

T_5 is hereditary. Any separated sets A and B in X are separated in Y. Then disjoint neighborhoods $N(A)$ and $N(B)$ in Y yield disjoint neighborhoods $N(A) \cap X$ and $N(B) \cap X$ in X for A and B. □

DISTINGUISHING EXAMPLES

The following topological examples chart distinctions among the separation axioms.

Example 3   A T_0-space. The topology

$$\{\emptyset, \{a, b\}, \{b\}, \{b, c\}, \{a, b, c\}\}$$

on $X = \{a, b, c\}$ is T_0. As there is no neighborhood of the closed set $\{a\}$ disjoint from the point b, the space X is neither T_3, T_2, nor T_1. As there are no disjoint neighborhoods for the closed sets $\{a\}$ and $\{c\}$, X is neither T_4 nor T_5.

Example 4   A T_0-T_1-*space*. The finite complement topology $\mathcal{FC}$ (Example 1.6) on an infinite set X is the smallest topology on X for which points are closed; hence, this is a minimal T_1 topology. Nonempty open sets $U, V \in \mathcal{FC}$ cannot have an empty intersection, because the complement $(U \cap V)^c = U^c \cup V^c$ is a union of two finite sets and so cannot equal the infinite set X. Hence, the space X is neither T_5, T_4, T_3, nor T_2.

Example 5   A T_0-T_1-T_2-*space*. Give $X = \{p\} \cup ((0, 1) \times (0, 1)) \cup \{q\}$ the Euclidean topology on the open unit square $(0, 1) \times (0, 1)$, augmented by sets of the form

$$\{p\} \cup ((0, \tfrac{1}{2}) \times (0, r)) \quad \text{and} \quad ((\tfrac{1}{2}, 1) \times (0, r)) \cup \{q\} \quad (0 < r < 1).$$

Then X is T_2, T_1, and T_0. But X is neither T_5, T_4, nor T_3. The point q and the closed subset $A = \{p\} \cup ((0, \tfrac{1}{2}] \times (0, 1))$ do not have disjoint neighborhoods.

Example 6   A T_0-T_1-T_2-T_3-*space*. The tangent-disc topology (Exercise 1.8) on the closed upper half plane $\mathbb{R} \times [0, \infty) \subset \mathbb{R}^2$ is the smallest topology containing the real subspace topology on $\mathbb{R} \times [0, \infty)$ and all *tangent discs* $\{x\} \cup B$, where B is an open Euclidean disc in $\mathbb{R} \times (0, \infty)$ tangent to the axis $\mathbb{R} \times \{0\}$ at x. Since the standard topology on $\mathbb{R} \times [0, \infty)$ is T_2, the larger tangent-disc topology is T_2, T_1, and T_0.

The tangent-disc topology is T_3 (Exercise 9), because the open upper half plane $\mathbb{R} \times (0, \infty)$ acquires the real subspace topology and the axis $\mathbb{R} \times \{0\}$ acquires a discrete subspace topology.

The tangent-disc topology is neither T_4 nor T_5. Since $\mathbb{R} \times \{0\}$ is closed and acquires the discrete topology, any subset of $\mathbb{R} \times \{0\}$ is closed in the tangent-disc topology. The closed subsets A and B consisting of the rational and irrational numbers in $\mathbb{R} \times \{0\}$ do not have disjoint neighborhoods in the tangent-disc space.

Example 7   A T_3-*space*. Let X be any space that is T_3, but not T_4. Consider its topological product $Y = X \times \{0, 1\}$ (Example 1.11) with the indiscrete 2-point space $\{0, 1\}$. The nonempty open (closed) subsets of Y are the products $A \times \{0, 1\}$, where A is open (respectively, closed) in X.

Since X is not T_4, neither is Y: disjoint closed subsets $A, B \subseteq X$ without disjoint neighborhoods in X form sets $A \times \{0, 1\}$ and $B \times \{0, 1\}$ with the same property in Y.

Since X is T_3, so also is Y: any closed subset $A \times \{0, 1\}$ of Y and any point $<x, i> \notin A \times \{0, 1\}$ have disjoint neighborhoods in Y, because the closed subset $A \subset X$ and the point $x \notin A$ have disjoint neighborhoods in X.

The space $Y = X \times \{0, 1\}$ is neither T_2, T_1, nor T_0: each neighborhood of either $<x, 0>$ or $<x, 1>$ contains the other point.

Example 8   A *nothing-space*. Let X be an infinite set with the finite complement topology; and let $\{0, 1\}$ be indiscrete 2-point space. Consider the topological product $Y = X \times \{0, 1\}$ (Example 1.11). The nonempty open (closed) subsets of Y are the products $A \times \{0, 1\}$, where $A \subseteq X$ is co-finite (respectively, nonempty finite).

So Y is neither T_2, T_1, nor T_0, since it contains the non-T_0 subspaces $\{x\} \times \{0, 1\}$. Also Y is neither T_5, T_4, nor T_3, since no two nonempty open subsets of Y are disjoint.

Example 9   A T_4-*space*. Given any space $(X, \mathcal{T})$ and point $p \notin X$, let the union

$Y = X \cup \{p\}$ have the topology $\mathcal{T} \cup \{Y\}$. In other words, the proper open subsets of Y are the open subsets of X; and the nonempty closed subsets of Y have the form $F \cup \{p\}$, where F is closed in X.

The space Y is T_4, since there exist no nonempty disjoint closed subsets in Y.

Since the separation properties T_i ($i \neq 4$) are hereditary, then Y fails to be T_0, T_1, T_2, T_3, and T_5 when the subspace X is the space in Example 8.

Example 10 *A T_0-T_4-T_5-space.* The two-point space $\{0, 1\}$ with the Sierpinski topology $\{\{0\}, \{0, 1\}\}$ is T_0, T_4 and T_5, but is neither T_1, T_2, nor T_3.

Example 11 *A T_4-T_5-space.* Topologize the open interval $X = (0, 1)$ so that its nonempty, proper, open subsets are the intervals $(1/n, 1)$, where $n \geq 2$. Since every neighborhood of a point in X contains the interval $(\frac{1}{2}, 1)$, X is neither T_2, T_1, nor T_0. Since any neighborhood of a subset A contains every interval $(a, 1)$, $a \in A$, there are no nonempty, separated subsets of X (Exercise 10). So X is vacuously T_5 and T_4. But X is not T_3, since every neighborhood of the closed set $(0, \frac{1}{2}] = (0, 1) - (\frac{1}{2}, 1)$ contains $3/4$.

The previous examples show:

2.3 Theorem *The separation properties have these relationships:*
(a) A T_5 space is T_4, a T_2 space is T_1, and a T_1 space is T_0. These are the only valid implications of the form $T_i \Rightarrow T_j$.
(b) A normal ($T_4 + T_1$) space is T_3 and a regular ($T_3 + T_0$) space is Hausdorff (T_2). But neither a ($T_4 + T_0$) space nor a ($T_5 + T_0$) space need be T_3.
(c) A subspace of a T_4 space need not be T_4.

Example 12 *A non-metrizable T_2 space.* An example of a non-metrizable T_2 space is any uncountable set with the **Fort topology** with respect to a point $p \in X$:

$$\mathcal{F} = \{U \subseteq X : X - U \text{ is finite or } X - U \text{ includes } p\}.$$

The collection $\mathcal{F}$ is the union of the finite complement topology $\mathcal{FC}$ (Example 1.6) and the excluded point topology $\mathcal{D}(p)$ (Example 1.8). Although, in general, the union of two topologies need not be a topology, $\mathcal{F}$ is a topology (Exercise 1.1(e)).

The space $(X, \mathcal{F})$ is Hausdorff. For $x \neq y$, either p is a third point and then $\{x\}$ and $\{y\}$ are disjoint open neighborhoods of x and y, or p is not a third point, say $x = p \neq y$, and then $X - \{y\}$ and $\{y\}$ are disjoint neighborhoods of x and y, respectively.

In any metric space (X, d), each point x can be expressed as the countable intersection of open sets: $\{x\} = \bigcap_k B(x, 1/k)$. When X is uncountable, $(X, \mathcal{F})$ fails to have this property of metrizable spaces. If $\{U_k\}$ is any countable family of open sets that contain p, then $X - (\bigcap_k U_k) = \bigcup_k (X - U_k)$ is the countable union of finite sets and hence is countable. It follows that $\bigcap_k U_k$ cannot equal $\{p\}$ when X is uncountable.

So the Fort topology on an uncountable set is Hausdorff but not metrizable.

The following feature of the separation axioms is left as an exercise.

2.4 Theorem *When $i = 0, 1,$ or 2, a topological product of two spaces is a T_i space if and only if both of the factors are T_i spaces.*

3 Interior, Closure, and Boundary

Consider the interval $A = (0, 1]$ in the real line $\mathbb{R}$. In the metric terminology of Section **2**.3, the open interval $(0, 1)$ is the *interior* of A, the closed interval $[0, 1]$ is the *closure* of A, and the endpoint set $\{0, 1\}$ is the *boundary* of A. The points of $\mathbb{R}$ that belong to the interior $(0, 1)$, the closure $[0, 1]$, or the boundary $\{0, 1\}$ are distinguished by the relationships between their neighborhoods and A; the interior points have neighborhoods contained within A, the closure points have no neighborhoods contained in A^c, and the boundary points have no neighborhoods contained in A or in A^c.

INT, CLS, AND BDY

Similarly, in a topological space $(X, \mathcal{T})$ each subset $A \subseteq X$ has an interior, closure, and boundary, defined as follows:

The ***interior*** of A in $(X, \mathcal{T})$ is the set, denoted by $\text{INT}(A)$ or $\mathring{A}$, of all points $x \in X$ such that some neighborhood $N(x)$ is contained within A.

The ***closure*** of A in $(X, \mathcal{T})$ is the set, denoted by $\text{CLS}(A)$ or $\overline{A}$, of all points $x \in X$ such that each neighborhood $N(x)$ intersects A.

The ***boundary*** of A in $(X, \mathcal{T})$ is the set, denoted by $\text{BDY}(A)$ or $\mathring{A}$, of all points $x \in X$ such that each neighborhood $N(x)$ intersects A and A^c. Thus,

- $x \in \text{INT}(A) \Leftrightarrow \exists\, N(x),\, N(x) \subseteq A$.
- $x \in \text{CLS}(A) \Leftrightarrow \forall\, N(x),\, N(x) \cap A \neq \emptyset$.
- $x \in \text{BDY}(A) \Leftrightarrow \forall\, N(x),\, N(x) \cap A \neq \emptyset \neq N(x) \cap A^c$.

So $x \in \text{CLS}(A)$ if and only if there exists a point $\lambda(U) \in U \cap A$ for each $U \in \mathcal{U}(x)$ (Example 1.19), hence, a net $\lambda : \mathcal{U}(x) \to A \subseteq X$ converging to x.

The definitions also imply the following relationships:

$$\text{INT}(A) \subseteq A \subseteq \text{CLS}(A). \qquad \text{BDY}(A) = \text{CLS}(A) - \text{INT}(A).$$

$$\text{BDY}(A) = \text{CLS}(A) \cap \text{CLS}(A^c) = \text{BDY}(A^c).$$

3.1 Characterization of INT(A) *In any topological space $(X, \mathcal{T})$, $\text{INT}(A)$ is the largest open subset of X contained in the subset $A \subseteq X$:*
(a) *if $U \subseteq A$ and U is open in X, then $U \subseteq \text{INT}(A)$;*
(b) *$\text{INT}(A) \subseteq A$ and $\text{INT}(A)$ is open in X.*
Therefore, $\text{INT}(A)$ equals the union of all open subsets of X contained in A.

Proof: (a) An open set is a neighborhood of each of its points, and $\text{INT}(A)$ contains each point that has a neighborhood contained in A. Thus, when U is open, $U \subseteq A$ implies that $U \subseteq \text{INT}(A)$.

(b) Each $x \in \text{INT}(A)$ has an open neighborhood $U \subseteq A$, which by (a) is contained in $\text{INT}(A)$. Thus, $\text{INT}(A)$ is open in X. □

3.2 Characterization of CLS(A) In any topological space $(X, \mathcal{T})$, CLS(A) is the smallest closed subset of X that contains the subset $A \subseteq X$:
(a) if $A \subseteq F$ and F is closed in X, then CLS(A) $\subseteq F$;
(b) $A \subseteq$ CLS(A) and CLS(A) is closed in X.
Therefore, CLS(A) equals the intersection of all closed sets that contain A.

Proof: (a) An open set is a neighborhood of each of its points, and CLS(A) consists of all points each of whose neighborhoods intersects A. So, when U is open and disjoint from A, then U is disjoint from CLS(A). Equivalently by complementation, when F is closed and contains A, then F contains CLS(A).
(b) Any $x \notin$ CLS(A) has an open neighborhood U disjoint from A and CLS(A) by the proof of (a). Thus, (CLS(A))^c is open and CLS(A) is closed. □

Example 1 In any topological space $(X, \mathcal{T})$, we have:
$$\text{CLS}(X) = X = \text{INT}(X) \qquad \emptyset = \text{CLS}(\emptyset) = \text{INT}(\emptyset) = \text{BDY}(\emptyset) \qquad \text{BDY}(X) = \emptyset.$$

Example 2 In the indiscrete topology (Example 1.2), a subset $A \subseteq X$ has:
$$\text{INT}(A) = \begin{cases} \emptyset, & \text{if } A \neq X \\ X, & \text{if } A = X \end{cases} \quad \text{CLS}(A) = \begin{cases} \emptyset, & \text{if } A = \emptyset \\ X, & \text{if } A \neq \emptyset \end{cases} \quad \text{BDY}(A) = \begin{cases} \emptyset, & \text{if } A = \emptyset \\ X, & \text{if } A \neq \emptyset, X \\ \emptyset, & \text{if } A = X. \end{cases}$$

Example 3 In the discrete topology (Example 1.3), a subset $A \subseteq X$ has:
$$\text{INT}(A) = A \qquad \text{CLS}(A) = A \qquad \text{BDY}(A) = \emptyset.$$

Example 4 In a metric topology $\mathcal{T}_d$ (Example 1.1), a subset $A \subseteq X$ has:
$$\begin{aligned}
\text{INT}(A) &= \{x \in X : \exists\, r > 0,\, B(x, r) \subseteq A\} \\
&= \{x \in X : d(x, A^c) > 0\} \\
\text{CLS}(A) &= \{x \in X : \forall\, r > 0,\, B(x, r) \cap A \neq \emptyset\} \\
&= \{x \in X : d(x, A) = 0\} \\
\text{BDY}(A) &= \{x \in X : \forall\, r > 0,\, B(x, r) \cap A \neq \emptyset \neq B(x, r) \cap A^c\} \\
&= \{x \in X : d(x, A) = 0 = d(x, A^c)\}.
\end{aligned}$$

Example 5 In the particular point topology $\mathcal{T}(p)$ (Example 1.9), the nonempty open sets are those sets that include p, and the proper closed sets are those sets that exclude p. Therefore, a subset $A \subseteq X$ has:
$$\text{INT}(A) = \begin{cases} A, & \text{if } p \in A \\ \emptyset, & \text{if } p \notin A \end{cases} \quad \text{CLS}(A) = \begin{cases} X, & \text{if } p \in A \\ A, & \text{if } p \notin A \end{cases} \quad \text{BDY}(A) = \begin{cases} A^c, & \text{if } p \in A \\ A, & \text{if } p \notin A. \end{cases}$$

Example 6 In the finite complement topology $\mathcal{FC}$ on an infinite set X (Example 1.6), the nonempty open sets are those sets with finite complement, and the proper closed sets are the finite sets. Therefore, a subset $A \subseteq X$ has:
$$\text{INT}(A) = \begin{cases} \emptyset, & \text{if } A^c \text{ infinite} \\ A, & \text{if } A^c \text{ finite} \end{cases} \quad \text{CLS}(A) = \begin{cases} A, & \text{if } A \text{ finite} \\ X, & \text{if } A \text{ infinite} \end{cases} \quad \text{BDY}(A) = \begin{cases} A, & \text{if } A \text{ finite} \\ A^c, & \text{if } A^c \text{ finite} \\ X, & \text{otherwise.} \end{cases}$$

COMPLEMENTS OF CLOSURES AND INTERIORS

By definition, $x \in X$ belongs to the interior of A in $(X, \mathcal{T})$ if and only if there exists an open neighborhood $U(x) \subseteq A$. Also, $x \in X$ belongs to the closure of A in $(X, \mathcal{T})$ if and only if each open neighborhood $U(x)$ intersects A. The complements of $\text{INT}(A)$ and $\text{CLS}(A)$ in X can be formulated as follows:

3.3 Theorem *The complement of the interior or closure of a set is the closure or interior of the complement of the set:*

$$(\text{INT}(A))^c = \text{CLS}(A^c) \quad \text{and} \quad (\text{CLS}(A))^c = \text{INT}(A^c).$$

Proof: First, $x \in \text{INT}(A)$ if and only if there is a neighborhood $U(x) \subseteq A$. So $x \in (\text{INT}(A))^c$ if and only if every neighborhood $U(x)$ satisfies $U(x) \cap A^c \neq \emptyset$, equivalently, $x \in \text{CLS}(A^c)$. This proves that $(\text{INT}(A))^c = \text{CLS}(A^c)$.

The second claim, $(\text{CLS}(A))^c = \text{INT}(A^c)$, follows by complementation. $\square$

DENSE AND NOWHERE DENSE SUBSETS

The sets $\mathbb{Z}$ and $\mathbb{Q}$ of integers and rationals are both countable subsets of the set $\mathbb{R}$ of real numbers that are so spread out that neither contains an open subset of the real line $\mathbb{R}$. But $\mathbb{Z}$ and $\mathbb{Q}$ are quite differently situated in $\mathbb{R}$, as is revealed by their closure in the real line $\mathbb{R}$. The closure $\text{CLS}(\mathbb{Z}) = \mathbb{Z}$ contains no open set, while $\text{CLS}(\mathbb{Q}) = \mathbb{R}$.

We adopt the following vocabulary (introduced in Section **2**.3) for such disparate types of subsets $A, D \subseteq X$ in a general topological space X.

$D \subseteq X$ is called *dense* in X if these equivalent conditions hold:

- D intersects each nonempty open subset $\emptyset \neq U \subseteq X$.
- $\text{CLS}(D) = X$.
- If $D \subseteq F$ and F closed in X, then $F = X$.
- $\text{INT}(D^c) = (\text{CLS}(D))^c = \emptyset$.

$A \subseteq X$ is called *nowhere dense* in X if these equivalent conditions hold:

- $\text{CLS}(A)$ contains no nonempty open subset $U \subseteq X$.
- $\text{INT}(\text{CLS}(A)) = \emptyset$.
- $\text{CLS}((\text{CLS}(A))^c) = (\text{INT}(\text{CLS}(A)))^c = X$.
- $(\text{CLS}(A))^c$ is dense in X.

3.4 Theorem *Let D and A be subsets of any topological space X.*
(a) *D cannot be both dense and nowhere dense if X is nonempty.*
(b) *A is a nowhere dense closed set in X if and only if its complement, A^c, is a dense open set in X.*

Proof: (a) If D is both dense and nowhere dense in X, then $\text{CLS}(D) = X$ and $\text{INT}(\text{CLS}(D)) = \emptyset$. Hence,

$$X = \text{INT}(X) = \text{INT}(\text{CLS}(D)) = \emptyset.$$

(b) The subset A is nowhere dense in X if and only if $(\text{CLS}(A))^c = \text{INT}(A^c)$ is dense in X. It follows that A is a nowhere dense closed set in X if and only if its complement A^c is a dense open set in X. □

Example 7 *A subset and its complement can both be dense in a topological space.* e.g., the sets $\mathbb{Q}$ and $\mathbb{Q}^c$ of rationals and irrationals are both dense in the real line $\mathbb{R}$.

Example 8 *A subset and its complement cannot both be nowhere dense in a nonempty topological space X.* If A is nowhere dense, then $\text{INT}(\text{CLS}(A)) = \emptyset$, hence, $\text{INT}(A) = \emptyset$. If, in addition, A^c is nowhere dense, then

$$X = [\text{CLS}(\emptyset)]^c = [\text{CLS}(\text{INT}(A))]^c = \text{INT}(\text{CLS}(A^c)) = \emptyset.$$

Example 9 Give X the particular point topology $\mathcal{T}(p)$ (Example 1.9). Any subset $A \subseteq X$ containing p has $\text{CLS}(A) = X$; hence, it is dense in X. So $\{p\}$ is dense in X!

Any subset $A \subset X$ not containing p has $\text{INT}(\text{CLS}(A)) = \text{INT}(A) = \emptyset$; hence, it is nowhere dense in X. For example, $X - \{p\}$ is nowhere dense in X.

A space is called *separable* if it contains a countable dense subset.

Example 10 Real n-space $\mathbb{R}^n$ is separable, with countable dense set $\mathbb{Q}^n$.

Example 11 In an infinite set X with the finite complement topology $\mathcal{FC}$ (Example 1.6), any infinite subset $A \subseteq X$ is dense. Hence, X is separable.

Example 12 The ordinal spaces $[0, \Omega]$ and $[0, \Omega)$, given their order topology (Example 1.17), are not separable. Any countable subset $D \subset [0, \Omega)$ has an upper bound $\beta \in [0, \Omega)$, by Theorem **1.**5.12; hence, there is always a non-empty open interval (β, Ω) in the complement of any countable subset.

The *long line* L results when $[0, \Omega)$ is enlarged by the insertion of a copy of the interval $(0, 1)$ between each countable ordinal α and its immediate successor $\alpha + 1$. The order topology (Example **5.**2.5) on this totally ordered space L is not separable.

Any countable union of nowhere dense sets is called a *first category set in X*. Any set not of the first category is called a *second category set in X*.

3.5 Theorem *In a complete metric space, the complement of a first category set is a second category set.*

Proof: Else, X would be a countable union, $\cup_k \text{CLS}(A_k)$, of nowhere dense closed sets. Then $X^c = \emptyset$ would be the countable intersection $\cap_k \text{INT}(A_k^c)$ of dense open sets, by Theorem 3.4(b), contrary to Baire's Theorem (**2.**5.6). □

Example 13 $\mathbb{Q}$ is a first category set in $\mathbb{R}$; so, $\mathbb{R} - \mathbb{Q}$ is a second category set in $\mathbb{R}$.

Example 14 Cantor's set C is a first category set in $\mathbb{I}$ and a second category set in C!

4   Continuous Functions

CONTINUITY

Let $f : X \to Y$ be a function between topological spaces X and Y.

The function f is **continuous at** $x \in X$ if for each neighborhood N of $f(x)$ in Y, there is a neighborhood M of x in X such that $f(M) \subseteq N$. It is equivalent to require that the pre-image $f^{-1}(N)$ of each neighborhood N of $f(x)$ in Y is a neighborhood of x in X.

Example 1   No function $f : \mathbb{R} \to \mathbb{R}$ such that $f(1/k) = (-1)^k$ for all $k \geq 1$ is continuous at 0. At least one of the three intervals $(-\infty, 0)$, $(-\frac{1}{2}, +\frac{1}{2})$, or $(0, \infty)$ is a neighborhood of $f(0) \in \mathbb{R}$. But none of the pre-images, $f^{-1}((-\infty, 0))$, $f^{-1}((-\frac{1}{2}, +\frac{1}{2}))$, and $f^{-1}((0, \infty))$, is a neighborhood of 0, because these sets fail to contain the sequences $(1/2k)$, $(1/k)$, and $(1/(2k+1))$, respectively, each of which has limit 0.

4.1   Continuity Characterizations   *The following are equivalent conditions for a function $f : X \to Y$ between topological spaces X and Y:*
- (a)   *The function f is continuous at each point $x \in X$.*
- (b)   *If $V \subseteq Y$ is open in Y, then its pre-image $f^{-1}(V) \subseteq X$ is open in X.*
- (c)   *If $F \subseteq Y$ is closed in Y, then its pre-image $f^{-1}(F) \subseteq X$ is closed in X.*
- (d)   *For each subset $A \subseteq X$, we have $f(\operatorname{CLS}_X(A)) \subseteq \operatorname{CLS}_Y(f(A))$.*
- (e)   *For each subset $B \subseteq Y$, we have $\operatorname{CLS}_X(f^{-1}(B)) \subseteq f^{-1}(\operatorname{CLS}_Y(B))$.*
- (f)   *Whenever a filterbase $\mathcal{F}$ in X converges to x, the filterbase $f(\mathcal{F})$ in Y converges to $f(x)$.*

Proof: (a) $\Rightarrow$ (b). Let V be open in Y and let $x \in f^{-1}(V) \subseteq X$. Then V is a neighborhood of $f(x)$ in Y; hence, $f^{-1}(V)$ is a neighborhood of x in X by (a). So $f^{-1}(V)$ is open in X.

(b) $\Rightarrow$ (f). For any open neighborhood $V(f(x))$, the pre-image $f^{-1}(V)$ is open by (b) and so must contain some $A \in \mathcal{F}$ by the convergence of $\mathcal{F}$. Then V contains $f(A)$. This proves that $f(\mathcal{F})$ converges to $f(x)$.

(f) $\Rightarrow$ (a). Since the filterbase $\mathcal{U}(x)$ in X converges to x, then the filterbase $f(\mathcal{F})$ in Y converges to $f(x)$, by (f). Thus, f is continuous at x.

(b) $\Leftrightarrow$ (c). These are equivalent because the pre-image construction respects complements and open sets have closed complements, and vice versa.

(c) $\Rightarrow$ (d). Given $A \subseteq X$, we form the closed subset $\operatorname{CLS}_Y(f(A))$ of Y. Its pre-image $f^{-1}(\operatorname{CLS}_Y(f(A)))$ contains A and is closed in X by (c). By the characterization of closure in X, it follows that $\operatorname{CLS}_X(A) \subseteq f^{-1}(\operatorname{CLS}_Y(f(A)))$. This implies that $f(\operatorname{CLS}_X(A)) \subseteq \operatorname{CLS}_Y(f(A))$.

(d) $\Rightarrow$ (e). Given $B \subseteq Y$, form $A = f^{-1}(B)$. By (d)

$$f(\operatorname{CLS}_X(f^{-1}(B))) \subseteq \operatorname{CLS}_Y(f(f^{-1}(B))) \subseteq \operatorname{CLS}_Y(B).$$

Therefore, it follows that $\operatorname{CLS}_X(f^{-1}(B)) \subseteq f^{-1}(\operatorname{CLS}_Y(B))$.

$(e) \Rightarrow (c)$. Suppose that $F \subseteq Y$ is closed in Y. Then by (e),

$$\text{CLS}_X(f^{-1}(F)) \subseteq f^{-1}(\text{CLS}_Y(F)) = f^{-1}(F) \subseteq \text{CLS}_X(f^{-1}(F)).$$

Therefore, $f^{-1}(F) = \text{CLS}_X(f^{-1}(F))$, which makes $f^{-1}(F)$ closed in X. □

A function $f : X \to Y$ between topological spaces X and Y is **continuous** provided that the equivalent conditions in Theorem 4.1 hold. In short, f is continuous provided that $f^{-1}(V)$ is open in X whenever V is open in Y.

Example 2 By the characterizations of continuity in the metric setting (2.4.1), $f : X \to Y$ is a continuous function between the metric spaces (X, d) and (Y, m) if and only if it is a continuous function between the topological spaces $(X, \mathcal{T}_d)$ and $(Y, \mathcal{T}_m)$.

Example 3 Let $\mathcal{T}_1$ and $\mathcal{T}_2$ be a pair of topologies on X. Then the identity function $1_X : (X, \mathcal{T}_1) \to (X, \mathcal{T}_2)$ is continuous if and only if $\mathcal{T}_2 \subseteq \mathcal{T}_1$ in the power set $\mathcal{P}(X)$.

Example 4 Any function $g : X \to Y$ is continuous whenever either X has the discrete topology $\mathcal{P}(X)$ or Y has the indiscrete topology $\{\emptyset, Y\}$.

Example 5 A function $g : (X, \mathcal{T}(p)) \to (Y, \mathcal{T}(q))$ is continuous with respect to the particular point topologies if and only if $g(p) = q$ or g is constant on X. To argue this, recall that $V \in \mathcal{T}(q) \Leftrightarrow (V = \emptyset$ or $q \in V)$; similarly, $U \in \mathcal{T}(p) \Leftrightarrow (U = \emptyset$ or $p \in U)$.

A function g, such that either $g(p) = q$ or g is constant, is continuous: the preimage $g^{-1}(V)$ of any $V \in \mathcal{T}(q)$ is either $\emptyset$ or contains p.

Conversely, suppose that g is continuous but is not constant on X. Then, for any image point $g(x) \neq g(p)$ in Y, the open set $\{g(x), q\} \subseteq Y$ has a nonempty open preimage $g^{-1}(\{g(x), q\}) \subseteq X$, which therefore must contain p. Since $p \notin g^{-1}(\{g(x)\})$, it must be that $p \in g^{-1}(\{q\})$, so that $g(p) = q$.

Example 6 A function $g : (X, \mathcal{FC}) \to (Y, \mathcal{FC})$ is continuous with respect to the finite complement topologies on X and Y if and only if the pre-image of each co-finite set is co-finite or empty. The latter holds if and only if the pre-image of each finite set is finite or X, or equivalently, g is constant or the pre-image of each point is finite.

4.2 Theorem *The composite of continuous functions is continuous.*

Proof: Let $f : X \to Y$ and $g : Y \to Z$ be continuous. If $W \subseteq Z$ is open in Z, then $g^{-1}(W)$ is open in Y by the continuity of g. Then,

$$f^{-1}(g^{-1}(W)) = (g \circ f)^{-1}(W)$$

is open in X by the continuity of f. □

Often, two or more continuous functions defined on subspaces of a topological space can be assembled into a function on the entire space. Be aware that the continuity of the assembled function is not automatic. The following theorem gives two important situations in which continuity is assured.

4.3 Gluing Theorem *Let the topological space X be the union of a family, $\{A_\alpha : \alpha \in \mathcal{A}\}$, of open subspaces or finitely many closed subspaces. A family of continuous functions $\{h_\alpha : A_\alpha \to Y\}$ such that h_α and h_β agree on $A_\alpha \cap A_\beta$, for all $\alpha, \beta \in \mathcal{A}$, defines a continuous function $h : X \to Y$.*

Proof: A function h is well-defined by $h(x) = h_\alpha(x)$ for $x \in A_\alpha$ ($\alpha \in \mathcal{A}$), since $h_\alpha(x) = h_\beta(x)$ for all $x \in A_\alpha \cap A_\beta$ ($\alpha, \beta \in \mathcal{A}$). For any open (closed) subset $G \subseteq Y$, we have $h^{-1}(G) = \cup_\alpha h_\alpha^{-1}(G)$. Since the functions $\{h_\alpha\}$ are continuous, the pre-images $\{h_\alpha^{-1}(G)\}$ are relatively open (closed) subsets of the open (finitely many closed, respectively) subspaces $\{A_\alpha\}$. Hence, their union $h^{-1}(G) = \cup_\alpha h_\alpha^{-1}(G)$ is open (closed) in X. Thus, h is continuous. □

Intuition developed in the metric setting suggests that a continuous function $\mathbb{R} \to \mathbb{R}$ that is constant on the set $\mathbb{Q}$ of rationals is constant on all of $\mathbb{R}$. This is a consequence of the general fact that a continuous function into a Hausdorff space is completely determined by its values on any dense subset of its domain space. Here is a formulation of this result:

4.4 Coincidence Theorem *Continuous functions $f_1, f_2 : X \to Y$ into a Hausdorff space Y that coincide on a dense subset $D \subseteq X$ are equal on X.*

Proof: First, the set $A = \{x \in X : f_1(x) \neq f_2(x)\}$ is open in X. When $x \in A$, the distinct image points $f_1(x) \neq f_2(x)$ have disjoint open neighborhoods V_1 and V_2 in the Hausdorff space Y. Then $U = f_1^{-1}(V_1) \cap f_2^{-1}(V_2)$ is an open neighborhood of x that is contained in A as $f_1(U) \cap f_2(U) \subseteq V_1 \cap V_2 = \emptyset$.

So the coincidence set, $A^c = \{x \in X : f_1(x) = f_2(x)\}$, is a closed set in X. Because A^c contains the dense subset D, this coincidence set equals X. □

Example 7 Let X have the particular point topology $\mathcal{T}(p)$ (Example 1.9). Any continuous function $f : X \to Y$ into a Hausdorff space Y is constant. The reason is that $\{p\}$ is dense in $(X, \mathcal{T}(p))$, so that f must coincide with the continuous function that takes the constant value $f(p) \in Y$.

OPEN-CLOSED FUNCTIONS

Let $f : X \to Y$ be a function between topological spaces X and Y.

The function f is called **open** if the image of each open set in X is open in Y, and f is called **closed** if the image of each closed set in X is closed in Y.

Example 8 A function $f : (X, \mathcal{T}(p)) \to (Y, \mathcal{T}(q))$ is closed in the particular point topologies if and only if either f misses q, or f is onto and $f^{-1}(\{q\}) = \{p\}$.

A function $f : (X, \mathcal{T}(p)) \to (Y, \mathcal{T}(q))$ is open if and only if $f(p) = q$.

Example 9 A function $f : X \to Y$ is open in the finite complement topologies on X and Y (Example 1.5) if and only if the image of each co-finite set is co-finite.

A function $f : X \to Y$ is closed in the finite complement topologies on X and Y if and only if the image of f is finite or equals Y.

Example 10 The exponential function $e : \mathbb{R} \to \mathbb{S}^1$, $e(t) = \langle \cos 2\pi t, \sin 2\pi t \rangle$, is continuous, open, and closed. The restriction $e : [0, 1] \to \mathbb{S}^1$ is continuous and closed, but not open. For example, the image of the closed set $[\frac{1}{2}, 1] \subset [0, 1]$ is closed in $\mathbb{S}^1$, but the image of its complement $[0, \frac{1}{2})$ is not open in $\mathbb{S}^1$.

The concepts "open," "closed," and "continuous" are independent. The sampler below illustrates all eight combinations of these concepts. In each sample, the spaces are two- or three-point sets whose open subsets are enclosed and the function's assignments are depicted by arrows. Determine whether each of these functions is open, closed, or continuous.

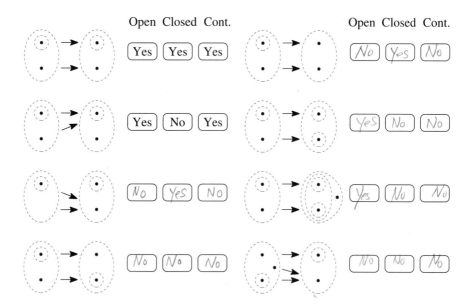

HOMEOMORPHISMS

Let X and Y be topological spaces. They are called *homeomorphic* spaces if there exists a pair of continuous inverse functions $f : X \to Y$ and $g : Y \to X$. Continuous inverse functions are called *homeomorphisms*.

4.5 Theorem *Let $f : X \to Y$ be a bijection between topological spaces. The following are equivalent conditions*:
(a) f is a homeomorphism.
(b) f is continuous and open: U is open in $X \Leftrightarrow f(U)$ is open in Y.
(c) f is continuous and closed: F is closed in $X \Leftrightarrow f(F)$ is closed in Y.
(d) For each subset $A \subseteq X$, we have $f(\text{CLS}_X(A)) = \text{CLS}_Y(f(A))$.
(e) For each subset $A \subseteq X$, we have $f(\text{INT}_X(A)) = \text{INT}_Y(f(A))$.

We leave the arguments for Theorem 4.5 as an exercise.

Homeomorphic spaces are indistinguishable to topologists; the set equivalence that underlies a homeomorphism gives bijective correspondences between the open sets and between the closed sets of the two spaces.

Example 11 Two discrete or two indiscrete spaces (Examples 1.2 and 1.3) are homeomorphic if and only if they are set equivalent.

Example 12 Stereographic projection $h : \mathbb{S}^n - \{p\} \to \mathbb{R}^n$ is a homeomorphism between the punctured n-sphere and real n-space. Find a coordinate expression for h.

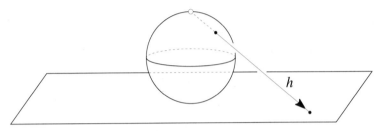

Example 12: Stereographic projection

TOPOLOGICAL INVARIANTS

The homeomorphism relation is denoted by $X \cong Y$. You may check that this relation between spaces is an equivalence relation:

- $\forall X: X \cong X$. *(reflexive)*
- $\forall X$ and $Y: (X \cong Y) \Rightarrow (Y \cong X)$. *(symmetric)*
- $\forall X, Y,$ and $Z: (X \cong Y) \wedge (Y \cong Z) \Rightarrow (X \cong Z)$. *(transitive)*

Therefore, the homeomorphism relation partitions any set of topological spaces into equivalences classes, called ***homeomorphism types***.

Example 13 There are just three distinct homeomorphism types of two-point spaces: Indiscrete, Sierpinski, and discrete two-point space (Examples 1.2-1.4).

A property $\mathcal{P}$ of topological spaces is a ***topological invariant*** if it respects homeomorphism types: if $X \cong Y$, then X has $\mathcal{P}$ if and only if Y has $\mathcal{P}$.

Any property that can ultimately be defined in terms of set relations and open sets is a topological invariant. *Topology* is a study of these invariants.

There are two ways to use topological invariants in furthering the classification of spaces into their homeomorphism types. To prove that $X \not\cong Y$, it suffices to find a topological invariant possessed by one of the two spaces X or Y but not the other. To prove that $X \cong Y$, it suffices to establish that one of the two spaces X or Y has the topological invariants that characterize the homeomorphism type of the other. The first use of topological invariants can be easy, but the second is usually quite difficult because of a lack of characterizing topological invariants for most homeomorphism types.

The following two examples utilize the topological invariants of metrizability and separability to distinguish between topological spaces.

Example 14 The real line $\mathbb{R}$ is not homeomorphic to $\mathbb{R}$ with the Fort topology (Example 2.12); the former is metrizable and the latter is not.

Example 15 The finite complement topology $\mathcal{FC}$ (Example 1.6) and the countable complement topology $\mathcal{CC}$ (Example 1.7) on an uncountable set X are not homeomorphic, since the former is separable but the latter is not.

Here is an example to illustrate the use of topological invariants to identify topological spaces.

Example 16 *A topological space that is compact, connected, metrizable, and has exactly two non-cut points is homeomorphic to the closed unit interval* $\mathbb{I} = [0, 1]$. These invariants are developed in subsequent chapters, and the stated characterization of $\mathbb{I}$ is a classical topological result, which is established in Section **7.5**.

Exercises

SECTION 1

1* Prove that, for any set X and point $p \in X$, the following are topologies on X:
   (a) Co-finite topology $\mathcal{FC} = \{\emptyset\} \cup \{U \subseteq X : U^c \text{ is finite}\}$ (Example 6).
   (b) Co-countable topology $\mathcal{CC} = \{\emptyset\} \cup \{U \subseteq X : U^c \text{ is countable}\}$ (Example 7).
   (c) Excluded point topology $\mathcal{D}(p) = \{X\} \cup \{U \subset X : p \notin U\}$ (Example 8).
   (d) Particular point topology $\mathcal{P}(p) = \{\emptyset\} \cup \{U \subseteq X : p \in U\}$ (Example 9).
   (e) Fort topology $\mathcal{F} = \{U \subseteq X : U^c \text{ is finite or } U^c \text{ includes } p\}$ (Example 2.12).

2* Prove: When X is any subset of a topological space $(Y, \mathcal{T})$, the collection
$$\mathcal{T}_X = \{V \subseteq X : \exists\, U \in \mathcal{T}, V = U \cap X\}$$
is a topology on X. This is the *subspace topology* on X of Example 10.

3* Prove: When $(X, \mathcal{T})$ and $(Y, \mathcal{R})$ are topological spaces, the collection,
$$\mathcal{T} \times \mathcal{R} = \{W = \cup_\alpha (U_\alpha \times V_\alpha) \subseteq X \times Y : U_\alpha \in \mathcal{T}, V_\alpha \in \mathcal{R}\},$$
of all arbitrary unions of products $U \times V \subseteq X \times Y$ of members $U \in \mathcal{T}$ and $V \in \mathcal{R}$ is a topology on $X \times Y$. This is the *product topology* on $X \times Y$ of Example 11.

4 (a) Show that the union of two topologies on a set X need not be a topology.
   (b) Prove that the intersection of any family of topologies on a set X is a topology.

5 Describe the convergent sequences in the following topological spaces and decide whether these sequences are adequate to determine the open and closed subsets:
   (a) Co-finite topology $\mathcal{FC} = \{\emptyset\} \cup \{U \subseteq X : U^c \text{ is finite}\}$ (Example 6).
   (b) Co-countable topology $\mathcal{CC} = \{\emptyset\} \cup \{U \subseteq X : U^c \text{ is countable}\}$ (Example 7).
   (c) Excluded point topology $\mathcal{D}(p) = \{X\} \cup \{U \subset X : p \notin U\}$ (Example 8).

SECTIONS 1-4   EXERCISES

(d) Particular point topology $\mathcal{P}(p) = \{\emptyset\} \cup \{U \subseteq X : p \in U\}$ (Example 9).

(e) Fort topology $\mathcal{F} = \{U \subseteq X : U^c \text{ is finite or } U^c \text{ includes } p\}$ (Example 2.12).

6 (a) Prove that in a metrizable space each neighborhood of a point contains a closed neighborhood of that point.

(b) A subset U of the closed upper half plane $\mathbb{R} \times [0, \infty) \subset \mathbb{R}^2$ is open in the **half-disc topology** if it is the union of open Euclidean discs $B \subset \mathbb{R} \times (0, \infty)$ and sets of the form $\{x\} \cup (B \cap (\mathbb{R} \times (0, \infty)))$, where B is an open Euclidean disc centered on some point $x \in \mathbb{R} \times \{0\}$. Prove that this defines a topology on $\mathbb{R} \times [0, \infty)$ that is not metrizable.

7 To construct the **line with two origins** $X = (\mathbb{R} - \{0\}) \cup \{0^+, 0^-\}$, delete the real number 0, and replace it with a pair of new points 0^+ and 0^-. To topologize X, retain each real open set $U \subseteq \mathbb{R} - \{0\}$, and replace each real open set U containing 0 by the pair of sets $U^+ = (U - 0) \cup \{0^+\}$ and $U^- = (U - 0) \cup \{0^-\}$. Prove that this defines a topology on X that is not metrizable.

8* A subset U of the closed upper half plane $\mathbb{R} \times [0, \infty) \subset \mathbb{R}^2$ is open in the **tangent-disc topology** if it is the union of open Euclidean discs $B \subset \mathbb{R} \times (0, \infty)$ and sets of the form $\{x\} \cup B$, where B is an open Euclidean disc in $\mathbb{R} \times (0, \infty)$ tangent to the axis $\mathbb{R} \times \{0\}$ at x. Prove that this defines a topology on $\mathbb{R} \times [0, \infty)$.

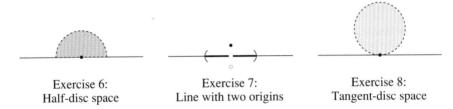

Exercise 6:   Exercise 7:   Exercise 8:
Half-disc space   Line with two origins   Tangent-disc space

Let X be a space and let $A \subseteq X$. The **derived set** of A, denoted by A', consists of the points $x \in X$ such that $N(x) \cap (A - \{x\}) \neq \emptyset$ for every neighborhood $N(x)$ of x.

9 Prove that $x \in A'$ if and only if there is a filterbase on $A - \{x\}$ that converges to x.

SECTION 2

1 Verify the following catalog of separation properties:

	T_5	T_4	T_3	T_2	T_1	T_0
Discrete topology $\mathcal{P}(X)$ (Example 1.2)	Yes	Yes	Yes	Yes	Yes	Yes
Indiscrete topology $\{\emptyset, X\}$ (Example 1.3)	Yes	Yes	Yes	No	No	No
Sierpinski topology $\mathcal{S}$ (Example 1.4)	Yes	Yes	No	No	No	Yes

2 Verify the following catalog of separation properties for non-discrete instances of the indicated topologies:

	T_5	T_4	T_3	T_2	T_1	T_0
Co-finite topology $\mathcal{FC}$ (Example 1.6)	No	No	No	No	Yes	Yes
Co-countable topology $\mathcal{CC}$ (Example 1.7)	No	No	No	No	Yes	Yes
Excluded point topology $\mathcal{D}(p)$ (Ex. 1.8)	Yes	Yes	No	No	No	Yes
Particular point topology $\mathcal{P}(p)$ (Ex. 1.9)	No	No	No	No	No	Yes

3 Prove that the topological product (Example 1.11) of two spaces is T_i if and only if both the factor spaces are T_i ($i = 0, 1, 2$).

4 Prove that a topological space X is Hausdorff (T_2) if and only if the diagonal subset $\Delta(x) = \{<x, x> \in X \times X : x \in X\}$ is a closed subset of the topological product $X \times X$.

5 (a) Prove that in a T_2 space each convergent sequence has a unique limit.
   (b) Prove that a space is T_1 if each convergent sequence has a unique limit.

6 Examine the separation properties of the half-disc space $\mathbb{R} \times [0, \infty)$ (Exercise 1.6) and decide whether it is metrizable or not.

7 Examine the separation properties of the Fort topology (Example 12).

8 Examine the separation properties of the line with two origins $(\mathbb{R} - \{0\}) \cup \{0^+, 0^-\}$ (Exercise 1.7) and decide whether it is metrizable or not.

9* Prove that the tangent-disc space $\mathbb{R} \times [0, \infty)$ (Exercise 1.8) is T_3, but neither T_4 nor T_5. Decide whether it is metrizable or not.

10* Prove that the interval space in Example 11 has no nonempty separated subsets, and hence it is T_5.

11 Prove that a topological space X is regular ($T_3 + T_0$) if and only if each neighborhood $N(x)$ of any point $x \in X$ contains a closed neighborhood $F(x)$ of x.

12 Prove that a topological space X is normal ($T_4 + T_1$) if and only if each neighborhood $N(A)$ of any closed subset $A \subseteq X$ contains a closed neighborhood $F(A)$ of A.

13 Prove that the closed unit interval $\mathbb{I}$ is a completely normal space.

14 Let X be a complete metric space. Use Baire's Category Theorem (2.5.6) to prove the following statements:
   (a) Any countable family of closed sets with union X has some member with nonempty interior.
   (b) The union of any countable family of nowhere dense sets has empty interior.

15 Let X be a space and let $A \subseteq X$. Prove that when X is a Hausdorff space the derived set A' (see Exercise 1.9) is always a closed set.

16 Investigate this claim: A space X is Hausdorff if and only if each convergent filterbase in X has a unique limit.

17 Prove that in any Hausdorff space X, the derived sets have these properties:
   (a) A' is closed in X.   (b) $(A')' \subseteq A'$.   (c) $(\overline{A})' = A'$.

SECTION 3

1 Prove that for any subset A of a topological space $(X, \mathcal{T})$,
   (a) $\text{BDY}(A) = \text{BDY}(A^c)$,
   (b) $\text{BDY}(A) = \text{CLS}(A) - \text{INT}(A)$, and
   (c) $\text{CLS}(A) = \text{INT}(A) \cup \text{BDY}(A)$.

2 Prove that for any subset A of a topological space $(X, \mathcal{T})$, the three sets $\text{INT}(A)$, $\text{BDY}(A)$, and $\text{INT}(A^c)$ are pairwise disjoint and have union X.

SECTIONS 1-4 EXERCISES 139

3. Determine whether the following conditions hold for all subsets $A \subseteq X$ of a topological space $(X, \mathcal{T})$:
   (a) BDY(BDY(A)) = BDY(A).
   (b) CLS(INT(A)) = CLS(A).
   (c) INT(CLS(A)) = INT(A).

4. Calculate CLS, INT, and BDY in the following topological spaces:
   (a) any set X with the excluded point topology $\mathcal{D}(p)$ (Example 1.8),
   (b) an uncountable set X with the co-countable topology $\mathcal{CC}$ (Example 1.7),
   (c) the topological product $\mathbb{R} \times \{0, 1\}$, where $\{0, 1\}$ has the discrete topology,
   (d) the interval space in Example 2.11,
   (e) the half-disc space (Exercise 1.6),
   (f) the line with two origins (Exercise 1.7), and
   (g) the tangent-disc space (Exercise 1.8).

5. Given a topology $\mathcal{T}$ on X, interior defines a function INT : $\mathcal{P}(X) \to \mathcal{P}(X)$. Establish the following conditions for all subsets $A, B \subseteq X$:
   (a) INT(A) $\subseteq A$.
   (b) INT(INT(A)) = INT(A).
   (c) INT($A \cap B$) = INT(A) $\cap$ INT(B).
   (d) INT(X) = X.

6. Consider any function $\iota : \mathcal{P}(X) \to \mathcal{P}(X)$ satisfying these properties:
   (a) $\iota(A) \subseteq A$.
   (b) $\iota(\iota(A)) = \iota(A)$.
   (c) $\iota(A \cap B) = \iota(A) \cap \iota(B)$.
   (d) $\iota(X) = X$.
   Prove that the collection $\mathcal{T}(\iota) = \{U \subseteq X : \iota(U) = U\}$ is a topology on X for which INT(A) = $\iota(A)$ for all $A \subseteq X$.

7. Given a topology $\mathcal{T}$ on X, closure defines a function CLS : $\mathcal{P}(X) \to \mathcal{P}(X)$. Establish the following conditions for all subsets $A, B \subseteq X$:
   (a) $A \subseteq$ CLS(A).
   (b) CLS(CLS(A)) = CLS(A).
   (c) CLS($A \cup B$) = CLS(A) $\cup$ CLS(B).
   (d) CLS($\emptyset$) = $\emptyset$.

8. Consider any function $\kappa : \mathcal{P}(X) \to \mathcal{P}(X)$ with the **Kuratowski closure properties**:
   (a) $A \subseteq \kappa(A)$.
   (b) $\kappa(\kappa(A)) = \kappa(A)$.
   (c) $\kappa(A \cup B) = \kappa(A) \cup \kappa(B)$.
   (d) $\kappa(\emptyset) = \emptyset$.
   Prove that the collection $\mathcal{T}(\kappa) = \{U \subseteq X : \kappa(U^c) = U^c\}$ is a topology on X for which CLS(A) = $\kappa(A)$ for all $A \subseteq X$.

9. For any subset A of a topological space X, let $\mathcal{C}(A) \subseteq \mathcal{P}(X)$ be the collection of subsets of X obtained from A by taking topological closures and set complements any number of times. So $\mathcal{C}(A) \subseteq \mathcal{P}(X)$ is the smallest collection of subsets of X such that (i) $A \in \mathcal{C}(A)$, (ii) $B \in \mathcal{C}(A) \Rightarrow$ CLS(B) $\in \mathcal{C}(A)$, and (iii) $B \in \mathcal{C}(A) \Rightarrow B^c \in \mathcal{C}(A)$.
   (a) Prove that $\mathcal{C}(A)$ has at most 14 distinct members.
   (b) Construct an example for which $\mathcal{C}(A)$ has 14 distinct members.

10. Let CLS$_X$, INT$_X$, and BDY$_X$ denote the closure, interior, and boundary operations in a subspace X (Example 1.10) of a space Y. Prove that for any subset $A \subseteq X$,
    (a) CLS$_X(A) = X \cap$ CLS$_Y(A)$,
    (b) INT$_X(A) \supseteq X \cap$ INT$_Y(A)$, and
    (c) BDY$_X(A) \subseteq X \cap$ BDY$_Y(A)$.
    Give examples that show that the inclusions in (b) and (c) can be proper.

11. Calculate the closure, interior, and boundary operations in the topological product $X \times Y$ (Example 1.11); for $A \subseteq X$ and $B \subseteq Y$, verify the following three properties:

(a) $\text{CLS}(A \times B) = \text{CLS}(A) \times \text{CLS}(B)$,
(b) $\text{INT}(A \times B) = \text{INT}(A) \times \text{INT}(B)$, and
(c) $\text{BDY}(A \times B) = (\text{BDY}(A) \times \text{CLS}(B)) \cup (\text{CLS}(A) \times \text{BDY}(B))$.

12 Let $\mathcal{D}(p)$ be the excluded point topology (Example 1.8) on a set X.
(a) Determine the proper dense subsets of the topological space $(X, \mathcal{D}(p))$.
(b) Prove that $(X, \mathcal{D}(p))$ is separable if and only if the set X is countable.

13 Let $\mathcal{CC}$ be the co-countable topology (Example 1.7) on an uncountable set X.
(a) Determine the proper dense subsets of $(X, \mathcal{CC})$.
(b) Decide whether the topological space $(X, \mathcal{CC})$ is separable.

14 Let $\mathcal{F}$ be Fort topology (Example 2.12) on an uncountable set X.
(a) Determine the proper dense subsets of $(X, \mathcal{F})$.
(b) Is the topological space $(X, \mathcal{F})$ separable?

15 Prove in detail that real n-space $\mathbb{R}^n$ is separable.

16 Prove that the topological product $X \times Y$ (Example 1.11) of two spaces X and Y is separable if and only if both factor spaces X and Y are separable.

17 Prove that an arbitrary subspace of a separable space need not be separable, but that an open subspace of a separable space is separable.

18 Prove that Hilbert space $\mathbb{H}$ (Exercise 2.3.16) is separable. Show that a countable dense set D is determined by the points $(x_k) \in \mathbb{H}$ whose non-zero coordinates $x_k \neq 0$ are rational and finite in number.

19 Consider two topologies $\mathcal{R} \subseteq \mathcal{S}$ on the set X. Compare the closures $\text{CLS}_\mathcal{R}(A)$ and $\text{CLS}_\mathcal{S}(A)$, the interiors $\text{INT}_\mathcal{R}(A)$ and $\text{INT}_\mathcal{S}(A)$, and the boundaries $\text{BDY}_\mathcal{R}(A)$ and $\text{BDY}_\mathcal{S}(A)$ of any subset $A \subseteq X$.

20 Let U be an open subset of a topological space X. Prove or disprove: for all $A \subseteq X$,
$$\text{CLS}(U \cup \text{CLS}(A)) = \text{CLS}(U \cap A).$$

21 Let X be a space and let $A \subseteq X$. Prove that $x \in \text{CLS}(A)$ if and only if there is a filterbase in A that converges in X to x.

A filterbase $\mathcal{A}$ in a space X *accumulates* at $x \in X$ if each neighborhood $U(x)$ intersects each $A \in \mathcal{A}$.

22 Prove that $\mathcal{A}$ accumulates at $x \in X$ if and only if $x \in \cap \{\text{CLS}(A) : A \in \mathcal{A}\}$.

23 Prove that the finite complement topology $\mathcal{FC}$ (Example 1.6) is separable.

SECTION 4

1 Prove that $f : (X, \mathcal{D}(p)) \to (Y, \mathcal{D}(q))$ is continuous with respect to the excluded point topologies (Example 1.8) if and only if either $f(p) = q$ or f is constant on X.

2 Find a pair of topological spaces X and Y such that every function $X \to Y$ is continuous, but no non-constant function $Y \to X$ is continuous.

3 Characterize the continuity of a function between two uncountable spaces with the co-countable topology (Example 1.7).

4 Decide whether either of the following two conditions is equivalent to the continuity of the function $f : X \to Y$ between topological spaces X and Y:
   (a) For every subset $B \subseteq Y$, we have $\text{INT}_X(f^{-1}(B)) \subseteq f^{-1}(\text{INT}_Y(B))$.
   (b) For every subset $A \subseteq X$, we have $f(\text{INT}_X(A)) \subseteq \text{INT}_Y(f(A))$.

5 Prove that the following properties are equivalent for any function $f : X \to Y$ between topological spaces X and Y:
   (a) f is an open map.
   (b) For every subset $A \subseteq X$, we have $f(\text{INT}_X(A)) \subseteq \text{INT}_Y(f(A))$.
   (c) For each neighborhood $N(x)$ in X, there exists a neighborhood $M(f(x))$ in Y such that $M \subseteq f(N)$.

6 Prove that the following properties are equivalent for any function $f : X \to Y$ between topological spaces X and Y:
   (a) f is a closed map.
   (b) For every subset $A \subseteq X$, we have $\text{CLS}_Y(f(A)) \subseteq f(\text{CLS}_X(A))$.

7 Let $f : X \to Y$ be a continuous function, where X is an infinite set with the finite complement topology (Example 1.6). Prove:
   (a) If Y is a Hausdorff space, then f is constant.
   (b) If Y is a T_1-space, then either f is constant or the pre-image of each point of Y is a finite subset of X.

8 Let $f : X \to Y$ and $g : W \to Z$ be continuous functions. Prove that the product function $f \times g : X \times W \to Y \times Z$, defined by $f \times g(<x, w>) = <f(x), g(w)>$, is continuous in the product topologies (Example 1.11).

9 Prove that if a function $f : X \to Y$ is continuous, then the graph

$$\Gamma(f) = \{<x, y> \in X \times Y : y = f(x)\}$$

is a closed subset of the topological product $X \times Y$. If $\Gamma(f)$ is closed in $X \times Y$, does it follow that f is continuous?

10 Determine the status of the eight functions in the sampler after Example 10.

11 Prove that the composite of open (closed) functions is open (closed, respectively).

12 The following sheet-like subspaces of real 3-space $\mathbb{R}^3$ are obtained from the first one by cutting the three vertical bars and re-connecting them according to the six possible permutations of the three ends.
    Classify these spaces up to homeomorphism using cut-and-glue techniques. State any topological claims that you invoke but that haven't been formally established at this point.

(a)   (b)   (c)   (d)   (e)   (f)

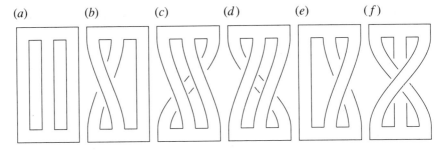

13 Let $(X, \mathcal{D}(p))$ and $(Y, \mathcal{D}(q))$ have excluded point topologies (Example 1.8).
(a) Prove that a function $f : (X, \mathcal{D}(p)) \to (Y, \mathcal{D}(q))$ is open if and only if either f misses q, or f is onto and $f^{-1}(\{q\}) = \{p\}$.
(b) Prove that $f : (X, \mathcal{D}(p)) \to (Y, \mathcal{D}(q))$ is closed if and only if $f(p) = q$.

14 Let X be finite and let Y be infinite. Determine the functions $f : X \to Y$ that are closed or open with respect to the finite complement topologies on X and Y.

15 Find all topologies on a three-point space $X = \{a, b, c\}$ and determine their homeomorphism types.

16 When is a subspace inclusion $i : X \to Y$ an open function; when is it closed?

17 The following sheet-like subspaces of real 3-space $\mathbb{R}^3$ are obtained from the first one by cutting one or more of the three horizontal bars, performing one or more half-twists, and re-connecting them without permuting the ends.

Classify these spaces up to homeomorphism using cut-and-glue techniques. State any topological claims that you invoke but that haven't been formally established at this point.

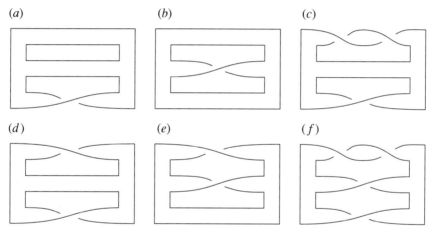

18 Determine whether the following 3-component subspaces of $\mathbb{R}^2$ are homeomorphic:

19 Let X be a Hausdorff space and let $f : X \to X$ be a continuous function. Prove that the fixed point set, $\{x \in X : f(x) = x\}$, is closed in X.

20 Let Ω be the first uncountable ordinal. Prove that any continuous real-valued function on the open ordinal subspace $[0, \Omega) \subset [0, \Omega]$ (Example 1.17) is eventually constant, i.e., constant on the interval (α, Ω) for some ordinal $\alpha < \Omega$.

5

CONSTRUCTION OF SPACES

This chapter presents several important topological constructions, including subspaces, identification spaces, product spaces, and adjunction spaces.

1 Subspaces and Identification Spaces

Given either an injective function and a topology on the codomain set or a surjective function and a topology on the domain set, we seek a natural topology on the other set for which the function is continuous.

SUBSPACE TOPOLOGY

Consider an inclusion function $i : X \subseteq Y$ and a topology $\mathcal{S}$ on the set Y. The discrete topology $\mathcal{P}(X)$ is the largest topology on X and it makes $i : (X, \mathcal{P}(X)) \to (Y, \mathcal{S})$ continuous. Is there a smallest topology $\mathcal{T}$ on X for which $i : (X, \mathcal{T}) \to (Y, \mathcal{S})$ is continuous?

1.1 Subspace Construction *Let $i : X \subseteq Y$ be an inclusion of a set X into the topological space $(Y, \mathcal{S})$. Then:*
(a) *The collection $i^*\mathcal{S} = \{i^{-1}(V) = V \cap X \subseteq X : V \in \mathcal{S}\}$ is a topology on X.*
(b) *$i : (X, \mathcal{T}) \to (Y, \mathcal{S})$ is continuous if and only if $\mathcal{T}$ contains $i^*\mathcal{S}$.*
Therefore, $i^*\mathcal{S}$ is the smallest topology on X for which i is continuous.

Proof: (a) A subset $U \subseteq X$ belongs to $i^*\mathcal{S}$ if and only if $U = i^{-1}(V) \subseteq X$ for some $V \in \mathcal{S}$. So the sets $\emptyset = i^{-1}(\emptyset)$ and $X = i^{-1}(Y)$ belong to $i^*\mathcal{S}$ because the sets $\emptyset$ and Y belong to $\mathcal{S}$. Also, an arbitrary union or finite intersection,

$$\cup_\alpha i^{-1}(V_\alpha) = i^{-1}(\cup_\alpha V_\alpha) \quad \text{or} \quad \cap_k i^{-1}(V_k) = i^{-1}(\cap_k V_k),$$

of members of $i^*\mathcal{S}$ is the pre-image of the member $\cup_\alpha V_\alpha$ or $\cap_k V_k$ of $\mathcal{S}$; hence, each one belongs to $i^*\mathcal{S}$.

(b) A topology $\mathcal{T}$ makes the inclusion $i : (X, \mathcal{T}) \to (Y, \mathcal{S})$ continuous if and only if $i^{-1}(V) \in \mathcal{T}$ for all $V \in \mathcal{S}$. Equivalently, $i^*\mathcal{S} \subseteq \mathcal{T}$. □

The topology $i^*\mathcal{S}$ is simply the **subspace topology** $\mathcal{S}_X$ (Example 4.1.10) that X acquires from $(Y, \mathcal{S})$: $i^*\mathcal{S} = \{V \cap X : V \in \mathcal{S}\} = \mathcal{S}_X$. The topological space $(X, i^*\mathcal{S})$ is called a **subspace** of $(Y, \mathcal{S})$.

Example 1 The inclusion $i : X = \{a, b\} \to \{a, b, c\} = Y$ is continuous with respect to the topologies $\mathcal{T} = \{\emptyset, \{b\}, X\}$ and $\mathcal{S} = \{\emptyset, Y\}$. But $\mathcal{T}$ is not the subspace topology, since it isn't the smallest topology, $i^*\mathcal{S} = \{\emptyset, X\} = \mathcal{S}_X$, for which i is continuous.

1.2 Transitivity Property *A subspace W of a subspace X of a space Y is a subspace of the space Y.*

Proof: Denote the inclusions by $j : W \subseteq X$ and $i: X \subseteq Y$. The main point is that $(i \circ j)^{-1}(V) = j^{-1}(i^{-1}(V))$ for all open subsets $V \subseteq Y$. □

1.3 Key Feature *Let X be a subspace of the topological space Y. For every space W, a function $k : W \to X$ is continuous if the composite function $i \circ k : W \to X \subseteq Y$ is continuous.*

Proof: If $i \circ k$ is continuous, then, for each open subset $V \subseteq Y$, the pre-image

$$(i \circ k)^{-1}(V) = k^{-1}(i^{-1}(V)) = k^{-1}(V \cap X)$$

is open in W. Thus, k is continuous, as every open subset of the subspace X has the form $V \cap X$ for some open subset $V \subseteq Y$. □

The subspace topology has this additional property:

1.4 Theorem *Let X be a subspace of Y. The restriction $g|_X : X \to Z$ of any continuous function $g : Y \to Z$ is continuous.*

Proof: The restriction $g|_X$ is the continuous composite $g \circ i : X \to Y \to Z$. □

By the Key Feature (1.3), any continuous function $h : W \to Y$ gives a continuous function $h : W \to h(W)$, provided that the image $h(W) \subseteq Y$ is given the subspace topology. The continuous function $h : W \to Y$ is called an **embedding** of W into Y provided that $h : W \to h(W)$ is a homeomorphism.

For a fixed space Y, the existence of an embedding $W \to Y$ is a topological invariant of W. Thus, topological spaces W_1 and W_2 are distinguished if there exists a space Y into which one, but not both, of W_1 and W_2 embed.

Example 2 The cylinder and the Möbius strip are not homeomorphic, since the cylinder, but not the Möbius strip, embeds into the 2-sphere $\mathbb{S}^2$.

To argue the latter claim, we view $\mathbb{S}^2$ as a subset that divides $\mathbb{R}^3$ into a bounded region and an unbounded region, which we fill with blue paint and red paint, respectively. If there were a copy of the Möbius strip embedded in $\mathbb{S}^2$, it would acquire a two-tone paint job, with the regions of the two colors meeting just along the edge of the Möbius strip. But this is an impossibility. If you start painting *one side* of the Möbius strip blue, you'll discover that the job is over, with all the exposed area painted blue, before you have a chance to use the red paint.

Example 2: Cylinder

Example 2: Möbius strip

Similarly, *for a fixed space W, the existence of an embedding $W \to Y$ is a topological invariant of Y*. So topological spaces Y_1 and Y_2 are distinguished if there exists a space W that embeds into one, but not both, of them.

Example 3 The wire handcuff space and the wire eight space (below) are not homeomorphic, because a cross × embeds into the wire eight but not into the wire handcuff. This non-embedding claim is intuitively obvious, but is fairly difficult to verify. Try it yourself after Chapter **6**.

Example 3: Wire handcuff

Example 3: Wire eight

IDENTIFICATION TOPOLOGY

Here is a problem dual to the subspace topology problem. Consider a surjection $p : X \to Y$ and a topology $\mathcal{T}$ on the domain set X. The indiscrete topology $\{\emptyset, Y\}$ is the smallest topology on Y and it makes the given function $p : (X, \mathcal{T}) \to (Y, \{\emptyset, Y\})$ continuous. Is there a largest topology $\mathcal{S}$ on Y for which $p : (X, \mathcal{T}) \to (Y, \mathcal{S})$ is continuous?

1.5 Identification Space Construction   Let $p : (X, \mathcal{T}) \to Y$ *be a surjection from the topological space* $(X, \mathcal{T})$ *to the set Y. Then,*
(a)   *The collection* $p_*\mathcal{T} = \{V \subseteq Y : p^{-1}(V) \in \mathcal{T}\}$ *is a topology on Y, and*
(b) $p : (X, \mathcal{T}) \to (Y, \mathcal{S})$ *is continuous if and only if $\mathcal{S}$ is contained in $p_*\mathcal{T}$. Therefore, $p_*\mathcal{T}$ is the largest topology on Y for which p is continuous.*

Proof: (a) By definition, a subset $V \subseteq Y$ belongs to $p_*\mathcal{T}$ if and only if its pre-image $p^{-1}(V) \subseteq X$ belongs to $\mathcal{T}$. Thus, the sets $\emptyset$ and Y belong to $p_*\mathcal{T}$ because their pre-images $p^{-1}(\emptyset) = \emptyset$ and $p^{-1}(Y) = X$ belong to $\mathcal{T}$. An arbitrary union $\cup_\alpha V_\alpha$ or a finite intersection $\cap_k V_k$ of members of $p_*\mathcal{T}$ is a member of $p_*\mathcal{T}$, because its pre-image is a union or finite intersection,

$$p^{-1}(\cup_\alpha V_\alpha) = \cup_\alpha p^{-1}(V_\alpha) \quad \text{or} \quad p^{-1}(\cap_k V_k) = \cap_k p^{-1}(V_k),$$

of members of the topology $\mathcal{T}$, accordingly.

(b) The topology $\mathcal{S}$ makes $p : (X, \mathcal{T}) \to (Y, \mathcal{S})$ continuous if and only if $p^{-1}(V) \in \mathcal{T}$, hence, $V \in p_*\mathcal{T}$ for all $V \in \mathcal{S}$. Equivalently, $\mathcal{S} \subseteq p_*\mathcal{T}$. □

The function $p : X \to Y$ *identifies* each point pre-image $p^{-1}(\{y\}) \subseteq X$ to the corresponding point $y \in Y$. For this reason, the topology $p_*\mathcal{T}$ on Y is called the **identification topology** determined by p and the topology $\mathcal{T}$ on X. The space $(Y, p_*\mathcal{T})$ is called an **identification space** of $(X, \mathcal{T})$, and the continuous surjection $p : (X, \mathcal{T}) \to (Y, p_*\mathcal{T})$ is called an **identification map**.

Thus: *p is an identification map provided that $V \subseteq Y$ is open (closed) if and only if its pre-image $p^{-1}(V) \subseteq X$ is open (closed, respectively)*.

Example 4 Let $X = \{1, 2, 3\}$, $\mathcal{T} = \{\emptyset, \{1\}, \{1, 2\}, \{1, 3\}, X\}$, and $Y = \{a, b\}$. Consider the function $p : X \to Y$ defined by $p(1) = a$, $p(2) = b$, and $p(3) = a$. The identification topology $p_*\mathcal{T}$ is $\{\emptyset, \{a\}, Y\}$.

Example 5 For each subset C of the interval $[0, 1]$, the characteristic function $\chi_C : [0, 1] \to \{0, 1\}$ is defined by $\chi_C(t) = 1$ if and only if $t \in C$.

The function χ_C is a surjection whenever $\emptyset \neq C \subset [0, 1]$. The identification topology on $\{0, 1\}$ determined by χ_C and the usual real subspace topology $\mathcal{T}$ on $[0, 1]$ is the indiscrete topology $\{\emptyset, \{0, 1\}\}$, if $C = \mathbb{Q} \cap [0, 1]$; it is the Sierpinski topology $\{\emptyset, \{1\}, \{0, 1\}\}$, if $C = [0, 1)$.

Although an identification map need not be an open function nor a closed function, each of these conditions is sufficient to imply that a continuous surjection is an identification map.

1.6 Theorem *A continuous surjection $p : X \to Y$ that is either open or closed is an identification map.*

Proof: When p is an open (closed) function, we consider a subset $V \subseteq Y$ whose pre-image $p^{-1}(V) \subseteq X$ is open (closed) in X. Then $V = p(p^{-1}(V)) \subseteq Y$ since p is a surjection, and $V = p(p^{-1}(V))$ is open (closed) in Y because p is an open (closed) function. Conversely, if $V \subseteq Y$ is open (closed) in Y, then the pre-image $p^{-1}(V) \subseteq X$ is open (closed) in X since p is continuous. This proves that the topology on Y is the identification topology. □

Example 6 The 1-sphere $\mathbb{S}^1$ is an identification space of the real line $\mathbb{R}$. The **exponential function** $e : \mathbb{R} \to \mathbb{S}^1$, $e(t) = <\cos 2\pi t, \sin 2\pi t>$, (opposite, left) wraps each unit interval $[k, k+1] \subset \mathbb{R}$ once around the unit circle $\mathbb{S}^1 \subset \mathbb{R}^2$. Since e is continuous and open with respect to the usual topologies on $\mathbb{R}$ and $\mathbb{S}^1$, then e is an identification map according to Theorem 1.6. Thus, $\mathbb{S}^1$ is the topological result of identifying every pair of real numbers in the real line $\mathbb{R}$ of integral distance apart.

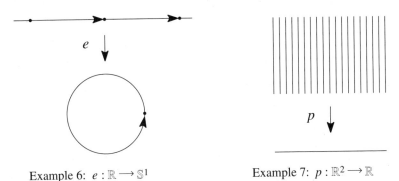

Example 6: $e: \mathbb{R} \to \mathbb{S}^1$

Example 7: $p: \mathbb{R}^2 \to \mathbb{R}$

Example 7 $\mathbb{R}$ *is a flattened identification space of* $\mathbb{R}^2$. The projection $p: \mathbb{R}^2 \to \mathbb{R}$, $p(<x, y>) = x$, (above right) carries vertical lines to their ordinate. Since p is a continuous and open function from the real plane $\mathbb{R}^2$ to the real line $\mathbb{R}$, it is an identification map. Thus the real line $\mathbb{R}$ is the topological result of identifying all points in the real plane $\mathbb{R}^2$ that have the same ordinate.

Example 8 *Creating the firmament from Cantor's dust.* The continuous surjection $\kappa: C \to \mathbb{I}$ (Example **2.4.8**) of Cantor's space C onto $\mathbb{I}$ is an identification map.

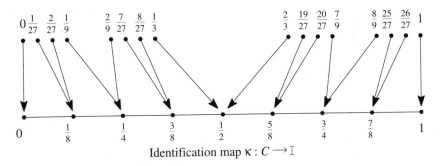

Identification map $\kappa: C \to \mathbb{I}$

To prove that κ is an identification map, it suffices by Theorem 1.6 to show that κ is a closed function. By definition, $\kappa((.t_1 t_2 t_3 \ldots t_k \ldots)_3) = (.b_1 b_2 b_3 \ldots b_k \ldots)_2$, where $b_k = \frac{1}{2} t_k$ for all $k \geq 1$. This shows that κ carries each intersection $T[t_1 t_2 t_3 \ldots t_k] \cap C$, where $T[t_1 t_2 t_3 \ldots t_k]$ is an *even* ternary interval, onto the corresponding binary interval $B[b_1 b_2 b_3 \ldots b_k] \subset \mathbb{I}$, with just the endpoints mapped to the endpoints. We denote the associated *open* ternary and binary intervals by $T(t_1 t_2 t_3 \ldots t_k)$ and $B(b_1 b_2 b_3 \ldots b_k)$, respectively. It follows that κ carries $C - T(t_1 t_2 t_3 \ldots t_k)$ onto $\mathbb{I} - B(b_1 b_2 b_3 \ldots b_k)$.

Consider any closed subset $F \subseteq C$ and any point $b \notin \kappa(F) \subseteq \mathbb{I}$.

If b is not a binary fraction, it is the image of a unique point t (not a ternary fraction) of the open set F^c. Choose k sufficiently large so that $T[t_1 t_2 t_3 \ldots t_k] \cap C \subseteq F^c$, where $T(t_1 t_2 t_3 \ldots t_k)$ is the kth stage open ternary interval containing t. The corresponding open binary interval $B(b_1 b_2 b_3 \ldots b_k) \subseteq \mathbb{I}$ contains b and lies in $\mathbb{I} - \kappa(F)$.

If b is a binary fraction, it is the image point of two ternary fractions $t < t'$ of the open set F^c. Choose k sufficiently large so that

$$C \cap (T[t_1t_2t_3\ldots t_k] \cup T[t_1't_2't_3'\ldots t_k']) \subseteq F^c,$$

where these are the kth stage ternary intervals that have t and t' as right and left-hand endpoints, respectively. The corresponding binary intervals have b as right- and left-hand endpoints, respectively, so that $B(b_1b_2b_3\ldots b_k] \cup B[b_1'b_2'b_3'\ldots b_k')$ is an open interval centered on b and contained in the complement of $\kappa(F)$.

This proves that κ is a closed function, hence, an identification map.

1.7 Transitivity Property *An identification space Z of an identification space Y of the space X is an identification space of X.*

Proof: Let $p : X \to Y$ and $q : Y \to Z$ be identification maps. Then a subset $V \subseteq Z$ is open in Z if and only if $q^{-1}(V) \subseteq Y$ is open in Y if and only if

$$p^{-1}(q^{-1}(V)) = (q \circ p)^{-1}(V)$$

is open in X. This proves that $q \circ p : X \to Z$ is an identification map. □

1.8 Key Feature *Let $p: X \to Y$ be an identification map. For every space Z, any function $k : Y \to Z$ is continuous if $k \circ p : X \to Z$ is continuous.*

Proof: Let $k \circ p: X \to Z$ be continuous. For each open subset $W \subseteq Z$,

$$(k \circ p)^{-1}(W) = p^{-1}(k^{-1}(W)) \subseteq X$$

is open in X; hence, $k^{-1}(W) \subseteq Y$ is open in the identification space Y. This proves that k is continuous. □

The Key Feature (1.8) of the identification topology has this application:

1.9 Transgression Theorem *Let $p : X \to Y$ be an identification map and let $g : X \to Z$ be any continuous function that respects the identifications of $p : X \to Y$ in this sense:*

for all $x, x' \in X$, $p(x) = p(x')$ implies $g(x) = g(x')$.

Then there is a unique continuous function $h : Y \to Z$ such that $h \circ p = g$.

Proof: By the hypothesis on g, the assignment $y \to g(p^{-1}(\{y\})$ is single valued. So it defines a function $h : Y \to Z$ such that $h \circ p = g$. Since g is continuous, so is h by the key feature of the identification topology. □

1.10 Corollary *Let $p : X \to Y$ and $q : X \to Z$ be identification maps that respect one another's identifications. Then Y and Z are homeomorphic.*

Proof: Twin applications of the Transgression Theorem yield continuous functions $h : Y \to Z$ and $k : Z \to Y$ such that $h \circ p = q$ and $k \circ q = p$. Then

$$k \circ h \circ p = k \circ q = p \quad \text{and} \quad h \circ k \circ q = h \circ p = q;$$

hence,
$$k \circ h = 1_Y \quad \text{and} \quad h \circ k = 1_Z,$$
by the surjectivity of p and q. Thus, $h : Y \cong Z : k$. □

QUOTIENT SPACES

By way of the identification topology, a topological space imposes a natural topology on each set image of itself. We can carry this story one step further by creating the image set, as well, from the topological space by means of the quotient set construction of Section **1**.5.

Any equivalence relation R on a set X determines a surjective function with domain X, as in Section **1**.5. Each $x \in X$ belongs to a unique *equivalence class* $[x]_R = \{z \in X : x \, R \, z\}$. The *quotient set of X modulo R* is the set, $X/R = \{[x]_R : x \in X\}$, of these classes. The *quotient function* $q_R : X \to X/R$, $q_R(x) = [x]_R$, sends each element $x \in X$ to its equivalence class $[x]_R \in X/R$.

Let $\mathcal{T}$ be a topology on X. The identification space $(X/R, q_{R*}\mathcal{T})$ is called the **quotient space of X modulo R**; the identification map
$$q_R : (X, \mathcal{T}) \to (X/R, q_{R*}\mathcal{T})$$
is called a **quotient map**.

The pre-image under q_R of a set $V \subseteq X/R$ of equivalence classes is
$$q_R^{-1}(V) = \cup \, \{[x]_R \subseteq X : [x]_R \in V\},$$
namely, the union in X of the equivalence classes in V. So $V \subseteq X/R$ is open in the quotient space X/R if and only if this union is open in the space X.

Example 9 *The cylinder $\mathbb{S}^1 \times \mathbb{I}$ is a quotient space of the unit square $\mathbb{I}^2$.* The edge point identifications $< 0, s > \sim < 1, s > (0 \leq s \leq 1)$ generate an equivalence relation $\sim$ on $\mathbb{I}^2$ whose quotient function $q : \mathbb{I}^2 \to \mathbb{I}^2/\sim$ glues the square into cylindrical form. Neighborhoods in $\mathbb{I}^2$ define neighborhoods in the quotient space $\mathbb{I}^2/\sim$ (below left) that make it *look like* the cylinder $\mathbb{S}^1 \times \mathbb{I}$.

Indeed, q makes the same identifications as the closed, continuous surjection
$$e \times 1_{\mathbb{I}} : \mathbb{I} \times \mathbb{I} \to \mathbb{S}^1 \times \mathbb{I}, \, g(<t, s>) = <e(t), s>,$$
which wraps $\mathbb{I} \times \mathbb{I}$ around the cylinder $\mathbb{S}^1 \times \mathbb{I}$, via the exponential function $e : \mathbb{I} \to \mathbb{S}^1$ (Example 4.4.10). So $\mathbb{I}^2/\sim$ is homeomorphic to the product $\mathbb{S}^1 \times \mathbb{I}$ by Corollary 1.10.

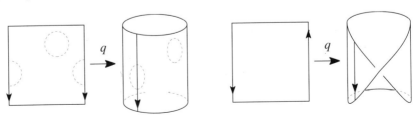

Example 9: Cylinder   Example 10: Möbius strip

Example 10 *The Möbius strip is a quotient space of the unit square $\mathbb{I}^2$. Let $\sim$ denote the equivalence relation on the closed unit square $\mathbb{I}^2 \subset \mathbb{R}^2$ generated by the edge point identifications $<0, s> \sim <1, 1-s>$, for all $0 \leq s \leq 1$. The quotient map $q : \mathbb{I}^2 \to \mathbb{I}^2/\sim$ (on p. 149) glues $\mathbb{I}^2$ into a topological Möbius strip.*

QUOTIENT MODULO A SUBSPACE

For any subset A of a topological space $(X, \mathcal{T})$, there is the equivalence relation $\sim_A$ defined by

$$x \sim_A z \text{ if and only if either } x = z \text{ or } \{x, z\} \subseteq A.$$

The quotient space of X modulo $\sim_A$ is denoted by X/A and is called the **quotient space of X modulo A**; it is a topological version of X in which the subset A is collapsed to a single point $[A]$. The quotient map is denoted by $q_A : X \to X/A$. The open subsets $V \subseteq X/A$ have the form $V = q_A(U) \subseteq X/A$, where U is open in X and either $U \subseteq X - A$ or $A \subseteq U$.

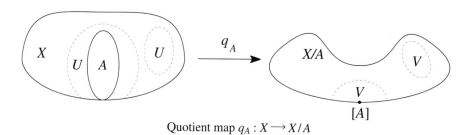

Quotient map $q_A : X \to X/A$

Example 11 $\mathbb{R}$ *is homeomorphic to its quotient space $\mathbb{R}/\mathbb{I}$ modulo $\mathbb{I}$.* Consider the closed continuous surjection $g : \mathbb{R} \to \mathbb{R}$ (below left) given by

$$g(x) = \begin{cases} x, & \text{if } x < 0 \\ 0, & \text{if } 0 \leq x \leq 1 \\ x - 1, & \text{if } 1 \leq x. \end{cases}$$

It makes the same identifications as the quotient map $q_\mathbb{I} : \mathbb{R} \to \mathbb{R}/\mathbb{I}$ that collapses $\mathbb{I}$ to a point. Therefore, $\mathbb{R}/\mathbb{I}$ is homeomorphic to $\mathbb{R}$, by Corollary 1.10.

Beware: $\mathbb{R}$ is not homeomorphic to its quotient space $\mathbb{R}/\mathbb{J}$ modulo the open interval $\mathbb{J} = (0, 1)$. In fact, $\mathbb{R}/\mathbb{J}$ is not even Hausdorff.

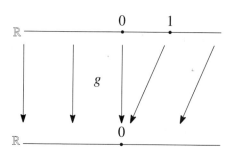
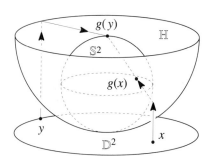

Example 11: $\mathbb{R}/\mathbb{I} \cong \mathbb{R}$ Example 12: $\mathbb{D}^2/\mathbb{S}^1 \cong \mathbb{S}^2$

Example 12 *The 2-sphere $\mathbb{S}^2$ is homeomorphic to the quotient space $\mathbb{D}^2/\mathbb{S}^1$ of the closed unit disc $\mathbb{D}^2$ modulo its boundary circle $\mathbb{S}^1$.* Here, let $\mathbb{S}^2 \subset \mathbb{R}^3$ denote the sphere of radius $\frac{1}{2}$, centered at $(0, 0, \frac{1}{2})$, and let $\mathbb{H}$ denote the lower hemisphere of the sphere of radius 1, centered at $(0, 0, 1)$. Then $\mathbb{S}^2$ and $\mathbb{H}$ are tangent to the closed unit disc $\mathbb{D}^2 \subset \mathbb{R}^2$ at the origin $\boldsymbol{O} \in \mathbb{R}^3$. A closed continuous surjection $g : \mathbb{D}^2 \to \mathbb{S}^2$ (opposite, right) sends each point $x \in \mathbb{D}^2$ to the intersection point of $\mathbb{S}^2$ with the line segment joining $(0, 0, 1)$ to the point on the hemisphere $\mathbb{H}$ directly above x. The function g makes the same identifications as the quotient map $q_{\mathbb{S}^1} : \mathbb{D}^2 \to \mathbb{D}^2/\mathbb{S}^1$ collapsing the boundary circle $\mathbb{S}^1 \subset \mathbb{D}^2$ of the closed unit disc. Therefore, $\mathbb{D}^2/\mathbb{S}^1$ is homeomorphic to $\mathbb{S}^2$ by Corollary 1.10.

2 Topological Bases

The local or global features of a topological space can be captured by certain sufficiently rich collections of the neighborhoods of points or members of the topology. These collections are called neighborhood bases, topological subbases, and topological bases. They are basic tools of the topological trade. The more economical the base that describes the topology, the easier it is to check various topological conditions of the space.

NEIGHBORHOOD BASE

Let x be a point of a topological space $(X, \mathcal{T})$. A **neighborhood base** at x is a set $\mathcal{U}_x$ of neighborhoods of x such that any other neighborhood of x contains at least one member of $\mathcal{U}_x$: $\forall\, V(x),\, \exists\, U \in \mathcal{U}_x,\, U \subseteq V$.

When a neighborhood base is available, local facts are easier to check:

- *If $\mathcal{U}_y$ is a neighborhood base at y in $(Y, \mathcal{S})$, then $f : (X, \mathcal{T}) \to (Y, \mathcal{S})$ is continuous on $f^{-1}(\{y\})$ if and only if $V \in \mathcal{U}_y$ implies $f^{-1}(V) \in \mathcal{T}$.*

- *If $\mathcal{U}_x$ is a neighborhood base at x in $(X, \mathcal{T})$, then $x \in \mathrm{CLS}(A)$ if and only if $U \cap A \neq \emptyset$ for all $U \in \mathcal{U}_x$.*

A topological space is **first countable** if there is a countable neighborhood base at each point.

Example 1 Any metric space (X, d) is first countable: the collections

$$\{B(x, r) : 0 < r \in \mathbb{Q}\} \quad \text{and} \quad \{B(x, 1/n) : n \in \mathbb{P}\}$$

are countable neighborhood bases at $x \in X$.

Example 2 An uncountable set X with the Fort topology is Hausdorff but not first countable (Example 4.2.12). Hence, X is not metrizable, in view of Example 1.

Example 3 For any point x in the particular-point topological space $(X, \mathcal{T}(p))$, the single set $\{x, p\}$ is a neighborhood base at x. Thus, $(X, \mathcal{T}(p))$ is first countable.

SUBBASE FOR A TOPOLOGY

It is possible to prescribe a topology on X by selecting any collection $\mathcal{S} \subseteq \mathcal{P}(X)$ of subsets of X and requiring that they be open sets:

2.1 Subbase Theorem *Given any family $\mathcal{S} \subseteq \mathcal{P}(X)$, there is a unique smallest topology $\mathcal{T}(\mathcal{S})$ containing $\mathcal{S}$; it consists precisely of $\emptyset$, X, all finite intersections of members of $\mathcal{S}$, and all unions of such finite intersections.*

Proof: The intersection of any collection of topologies on the same set X is a topology on X. So the intersection of all topologies on X containing $\mathcal{S}$,

$$\mathcal{T}(\mathcal{S}) = \cap \{\mathcal{T} : \mathcal{S} \subseteq \mathcal{T} \subseteq \mathcal{P}(X), \mathcal{T} \text{ is a topology}\},$$

is the smallest topology on X containing $\mathcal{S}$. The collection consisting of $\emptyset$, X, all finite intersections of members of $\mathcal{S}$, and all arbitrary unions of such finite intersections is a topology on X containing $\mathcal{S}$ and it is contained in any topology $\mathcal{T}$ containing $\mathcal{S}$. Thus, this collection equals $\mathcal{T}(\mathcal{S})$. □

Consider any topology $\mathcal{T}$ on X. A collection $\mathcal{S} \subseteq \mathcal{T}$ is called a ***subbase*** for $\mathcal{T}$ if $\mathcal{T} = \mathcal{T}(\mathcal{S})$, the topology consisting of $\emptyset$, X, all finite intersections of members of $\mathcal{S}$, and all arbitrary unions of such finite intersections.

Example 4 Given a family of topologies $\{\mathcal{T}_\alpha : \alpha \in \mathcal{A}\}$ on the set X, there is a unique smallest topology $\mathcal{T}$ on X containing each member of the family; namely, the topology with subbase $\cup_\alpha \mathcal{T}_\alpha$. This topology is called the ***supremum*** of the given family and is denoted by $sup\{\mathcal{T}_\alpha : \alpha \in \mathcal{A}\}$.

There is also a unique largest topology on X that is contained in each member of the family, namely, the intersection $\cap_\alpha \mathcal{T}_\alpha$. This topology is called the ***infimum*** of the given family and is denoted by $inf\{\mathcal{T}_\alpha : \alpha \in \mathcal{A}\}$.

Example 5 Let $(X, \leq)$ be a totally ordered set, as in Section 1.5. This means that the relation $\leq$ on X has these four properties:

- $\forall x \in X: x \leq x$. *(reflexive)*
- $\forall x, y \in X: (x \leq y) \wedge (y \leq x) \Rightarrow (x = y)$. *(antisymmetric)*
- $\forall x, y, z \in X: (x \leq y) \wedge (y \leq z) \Rightarrow (x \leq z)$. *(transitive)*
- $\forall x, y \in X: (x \leq y) \vee (y \leq x)$. *(chain condition)*

The strict relation $x < y$ means $x \neq y$ and $x \leq y$. The set of rays

$$(x, \rightarrow) = \{z \in X: x < z\} \quad \text{and} \quad (\leftarrow, y) = \{z \in X: z < y\}$$

for all $x, y \in X$ is a subbase $\mathcal{S}(\leq)$ for a topology $\mathcal{T}(\leq)$ on X, called the ***order topology***.

Thus, each ordinal acquires an order topology from its well-ordering $\leq$ (Theorem 1.5.11). The ordinal space $\alpha + 1 = [0, \alpha]$ has subbase $\{[0, \delta), (\gamma, \alpha] : 0 \leq \gamma, \delta \leq \alpha\}$. So $(\gamma, \delta] = (\gamma, \delta+1)$ is open, but $[\gamma, \delta)$ is open only if $\gamma = 0$ or γ is a non-limit ordinal $\gamma = \sigma + 1$, so $[\gamma, \delta) = (\sigma, \delta)$. The first uncountable ordinal Ω has no countable neighborhood base in $[0, \Omega]$: any countable subset $A = \{\gamma\} \subset [0, \Omega)$ has an upper bound $\beta \in [0, \Omega)$ (Theorem 1.5.12); so, Ω has a neighborhood $(\beta, \Omega] \subset (\gamma, \Omega]$ for all $\gamma \in A$.

BASE FOR A TOPOLOGY

A collection $\mathcal{B} \subseteq \mathcal{P}(X)$ is called a ***base*** for a topology $\mathcal{T}$ on X if $\mathcal{B} \subseteq \mathcal{T}$ and each member of $\mathcal{T}$ is a union of members of $\mathcal{B}$.

It is equivalent to require that $\mathcal{B} \subseteq \mathcal{T}$ *covers* X (i.e., $X = \cup\, \mathcal{B}$) and that, for each $x \in U \in \mathcal{T}$, there exists $B \in \mathcal{B}$ such that $x \in B \subseteq U$. The Subbase Theorem shows that every family $\mathcal{S} \subseteq \mathcal{P}(X)$ is the subbase for a topology on X. However, not every family $\mathcal{B} \subseteq \mathcal{P}(X)$ serves as the base for a topology:

2.2 Base Characterization *The following are equivalent conditions for any collection $\mathcal{B} \subseteq \mathcal{P}(X)$ that covers X:*
(a) *For each pair of subsets $U, V \in \mathcal{B}$, and each member $x \in U \cap V$, there exists $W \in \mathcal{B}$ such that $x \in W \subseteq U \cap V$.*
(b) *Finite intersections of members of $\mathcal{B}$ are unions of members of $\mathcal{B}$.*
(c) *$\mathcal{B}$ is a base for some topology on X.*

Proof: We have (a) $\Rightarrow$ (b) by induction, and (b) $\Rightarrow$ (a) by specialization. Statements (b) and (c) are equivalent by the definition of the topology $\mathcal{T}(\mathcal{B})$ with subbase $\mathcal{B}$ and the hypothesis that X is a union of members of $\mathcal{B}$.   □

Example 6 Because the collection $\mathcal{B}_d = \{B_d(x, r) : x \in X, r > 0\}$ of open balls in a metric space (X, d) has the neighborhood property (b), $\mathcal{B}_d$ serves as a subbase and as a base for the metric topology $\mathcal{T}_d$ on X.

Example 7 The collection of all rays $(-\infty, b)$ and $(a, +\infty)$ is a subbase, but not a base, for the usual topology of the real line $\mathbb{R}$. The collection of all finite open intervals (a, b) is a subbase and base for the same topology.

Example 8 The topology $\mathcal{T} \times \mathcal{R}$ on the product $X \times Y$ (Example **4**.1.11) has base $\{U \times V : U \in \mathcal{T}, V \in \mathcal{R}\}$ and subbase $\{U \times Y, X \times V : U \in \mathcal{T}, V \in \mathcal{R}\}$.

2.3 Theorem *Two bases define the same topology on X if and only if each member of either base is a union of members of the other base.*

Proof: The topologies $\mathcal{T}(\mathcal{B})$ and $\mathcal{T}(\mathcal{B}')$ with bases $\mathcal{B}$ and $\mathcal{B}'$ consist of all unions of members of $\mathcal{B}$ and $\mathcal{B}'$, respectively. So $\mathcal{T}(\mathcal{B}) = \mathcal{T}(\mathcal{B}')$ if and only if each member of $\mathcal{B}$ is a union of members of $\mathcal{B}'$, and vice versa.   □

Example 9 Two metrics d and m on the same space X are topologically equivalent (i.e., $\mathcal{T}_d = \mathcal{T}_m$) if each member of each base $\mathcal{B}_d$ and $\mathcal{B}_m$ of open balls is a union of members of the other base.

Example 10 The collection $\{(a, b] : a < b \text{ in } \mathbb{R}\}$ of all left-open right-closed intervals is a base for a new topology on $\mathbb{R}$, called the ***right interval topology***. Since each open interval is a union of left-open right-closed intervals, the right interval topology contains the usual topology on $\mathbb{R}$. Let $\mathbb{R}_{(\,]}$ denote $\mathbb{R}$ with this right interval topology.

Example 11 Let $\mathcal{OS}$ be the collection of all translates in $\mathbb{R}^2$ of the open squares:
$$(0, r) \times (0, r) \subset \mathbb{R} \times \mathbb{R} = \mathbb{R}^2 \qquad (r > 0).$$
Let $\mathcal{COS}$ be the collection of all translates in $\mathbb{R}^2$ of the closed-open squares:
$$[0, r) \times [0, r) \subset \mathbb{R} \times \mathbb{R} = \mathbb{R}^2 \qquad (r > 0).$$
$\mathcal{OS}$ and $\mathcal{COS}$ are bases for topologies on the plane $\mathbb{R}^2$, because the intersection of any two open or two closed-open squares is a union of the same kind of squares.

While each open square is the union of closed-open squares, no closed-open square is the union of open squares. Therefore, $\mathcal{OS}$ and $\mathcal{COS}$ are bases for distinct topologies $\mathcal{T}(\mathcal{OS}) \subset \mathcal{T}(\mathcal{COS})$ on $\mathbb{R}^2$.

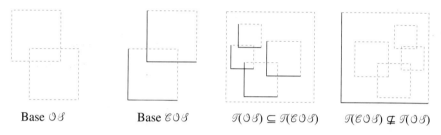

Base $\mathcal{OS}$ Base $\mathcal{COS}$ $\mathcal{T}(\mathcal{OS}) \subseteq \mathcal{T}(\mathcal{COS})$ $\mathcal{T}(\mathcal{COS}) \nsubseteq \mathcal{T}(\mathcal{OS})$

Example 11: Bases for distinct topologies

When bases $\mathcal{B}$ and $\mathcal{D}$ are available for the topological spaces $(X, \mathcal{T})$ and $(Y, \mathcal{R})$, respectively, certain topological facts are easier to check:

- $f: (X, \mathcal{T}) \to (Y, \mathcal{R})$ *is continuous if and only if* $f^{-1}(V) \in \mathcal{T}$ *for all* $V \in \mathcal{D}$.
- $x \in \mathrm{CLS}_X(A)$ *if and only if* $A \cap B \neq \emptyset$ *for all* $B \in \mathcal{B}$ *with* $x \in B$.
- $D \subseteq X$ *is dense in* $(X, \mathcal{T})$ *if and only if* $D \cap B \neq \emptyset$ *for all* $B \in \mathcal{B}$.

The more manageable the base, the more manageable these tasks become.

A topological space $(X, \mathcal{T})$ is called ***second countable*** if the topology $\mathcal{T}$ has a countable base $\mathcal{B}$.

Example 12 Real n-space $\mathbb{R}^n$ is second countable. In particular, the collection of all open Euclidean balls centered on points with rational coordinates and with rational radii is a countable base for the Euclidean metric topology.

Example 13 Let $P = \{2, 3, \ldots, p, \ldots\}$ be the set of primes. The subsets
$$V_n = \{p \in P : p \text{ does not divide } n\} \qquad (n = 0, 1, 2, 3, \ldots)$$
satisfy $V_0 = \emptyset$, $V_1 = P$, and $V_n \cap V_m = V_{nm}$. The latter condition is the contrapositive of this characterization of the primality of an integer p: p divides $mn \Leftrightarrow p$ divides m or p divides n. It follows that the collection $\{V_n : n = 0, 1, 2, 3, \ldots\}$ is a countable base for a topology on P; this topology is called the ***prime ideal topology***.

There are some connections between the second countability of a space X and the existence of a countable dense subset of X, i.e., its *separability*.

2.4 Theorem *Any second countable space is separable.*

Proof: Let $\mathcal{B}$ be any base for a space X, and let x_B denote a selected point from the member $\emptyset \neq B \in \mathcal{B}$. Then the subset $D_{\mathcal{B}} = \{x_B : \emptyset \neq B \in \mathcal{B}\}$ is dense in X. The reason is that $D_{\mathcal{B}}$ intersects each non-empty member of the base $\mathcal{B}$ by construction and each nonempty open subset $\emptyset \neq U \subseteq X$ is a union $\cup \{B \in \mathcal{B} : B \subseteq U\}$ of nonempty members $\emptyset \neq B \in \mathcal{B}$. Thus a space X with a countable base $\mathcal{B}$ contains a countable dense subset $D_{\mathcal{B}}$. In other words, a second countable space is separable. $\square$

2.5 Theorem *A separable metric space is second countable.*

Proof: Let C be a countable dense subset of the metric space (X, d). Then the family of open balls with rational radii and centered on points of C,

$$\mathcal{B} = \{B(c, r) : c \in C, 0 < r \in \mathbb{Q}\},$$

is a countable base for X. It suffices to show that, for each point x of an open set $U \subseteq X$, there is an open ball $B \in \mathcal{B}$ such that $x \in B \subseteq U$.

Since U is open, there exists some $0 < r \in \mathbb{Q}$ such that $B(x, 2r) \subseteq U$. If $x \in C$, we may take $B(x, r)$ as $B \in \mathcal{B}$. If $x \notin C$, we argue as follows. Since C is dense in X, the open ball $B(x, r)$ contains some $c \in C$. It follows from the triangle inequality (**D3**) for the metric d that $x \in B(c, r) \subseteq B(x, 2r) \subseteq U$. Then we may take the open ball $B(c, r)$ as $B \in \mathcal{B}$. $\square$

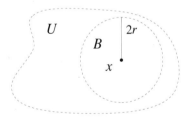   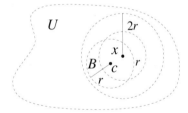

Theorem 2.5: $x \in C$ Theorem 2.5: $x \notin C$

2.6 Corollary *Second countability and separability are equivalent in the presence of metrizability.*

But, in general, *a separable space need not be second countable.*

Example 14 *The right interval space $\mathbb{R}_{(\,]}$ (Example 10) is not second countable.* When $\{(a_k, b_k] : k \in \mathbb{P}\}$ is a countable family of members of the base, there exists a real number $b \notin \{b_k : k \in \mathbb{P}\}$. Then no base interval with the form $(a, b]$ can be expressed as the union of members of the family.

But $\mathbb{R}_{(\,]}$ is separable and first countable: $\mathbb{Q}$, the set of rationals, is a countable dense subset and any x has the countable neighborhood base $\{(a, x] : a < x, a \in \mathbb{Q}\}$.

Example 15 Consider the finite complement topology $\mathcal{FC}$ on an infinite set X. This is separable, as any infinite subset is dense.

When X is second countable, with a countable base $\mathcal{B}$, then for any $x \in X$ we have

$$\{x\} \subseteq \cap \{B : x \in B \in \mathcal{B}\} \subseteq \cap \{X - \{y\} : y \neq x\} = \{x\}.$$

The containment relation holds, since each complement $X - \{y\}$, where $y \neq x$, is an open set that contains x and therefore contains some $B \in \mathcal{B}$ containing x. Thus $\{x\} = \cap \{B : x \in B \in \mathcal{B}\}$; so $X - \{x\}$ is a countable union $\cup \{B^c : x \in B \in \mathcal{B}\}$ of proper and closed, hence, finite sets. This proves that the set X is countable.

It follows that the finite complement topology $\mathcal{FC}$ on an uncountable set X is separable, but not second countable.

Example 16 *Any second countable space is first countable.* Therefore, the ordinal space $[0, \Omega]$, which is not first countable (Example 5), is not second countable.

3 Topological Unions and Products

This section is devoted to the construction of natural topologies on the disjoint union and the product of topological spaces.

DISJOINT UNION

Given an indexed family $\{X_\alpha : \alpha \in \mathcal{A}\}$ of sets, their **disjoint union** $\mathring{\cup}_\alpha X_\alpha$ is the set construction characterized by these properties:

(**DU1**) *There are functions* $i_\sigma : X_\sigma \to \mathring{\cup}_\alpha X_\alpha$ ($\sigma \in \mathcal{A}$), *called* **injections**.

(**DU2**) *For any set Z and any family $\{h_\alpha : X_\alpha \to Z\}$ of functions with common codomain set Z, there exists a unique function $h : \mathring{\cup}_\alpha X_\alpha \to Z$ such that $h \circ i_\alpha = h_\alpha$ for all $\alpha \in \mathcal{A}$.*

The inclusion functions $i_\sigma : X_\sigma \to \cup_\alpha X_\alpha$ ($\sigma \in \mathcal{A}$) into the usual set union $\cup_\alpha X_\alpha$ of subsets of a universe X satisfy condition (**DU2**) if and only if these subsets are pairwise disjoint (i.e., $X_\alpha \cap X_\sigma = \emptyset$ whenever $\alpha \neq \sigma$ in $\mathcal{A}$). Only then do arbitrary functions $\{h_\alpha : X_\alpha \to Z\}$ necessarily patch together to give a unique function $h : \cup_\alpha X_\alpha \to Z$ such that $h \circ i_\alpha = h_\alpha$ for all $\alpha \in \mathcal{A}$.

We can always replace the sets X_α ($\alpha \in \mathcal{A}$) by the pairwise disjoint sets:

$$X_\alpha \times \{\alpha\} \subseteq (\cup_\alpha X_\alpha) \times \mathcal{A} \qquad (\alpha \in \mathcal{A}).$$

Then the union $\cup_\alpha (X_\alpha \times \{\alpha\})$ can serve as the disjoint union $\mathring{\cup}_\alpha X_\alpha$.

TOPOLOGICAL UNION

Let $\{X_\alpha : \alpha \in \mathcal{A}\}$ be an indexed family of topological spaces. Putting a natural topology on their disjoint union $\mathring{\cup}_\alpha X_\alpha$ involves two opposing goals: the continuity of the injections in (**DU1**) is more likely for a small topology on

$\mathring{\cup}_\alpha X_\alpha$, while the continuity of the function h in (**DU2**) is more likely for a large topology on $\mathring{\cup}_\alpha X_\alpha$.

The ***disjoint union topology*** on $\mathring{\cup}_\alpha X_\alpha$ is the largest topology for which all the injections

$$i_\sigma : X_\sigma \to \mathring{\cup}_\alpha X_\alpha \qquad (\sigma \in \mathcal{A})$$

are continuous. This means that a subset $U \subseteq \mathring{\cup}_\alpha X_\alpha$ is *open* in $\mathring{\cup}_\alpha X_\alpha$ if and only if $i_\alpha^{-1}(U) = U \cap X_\alpha$ is open in X_α for all $\alpha \in \mathcal{A}$; and a subset $F \subseteq \mathring{\cup}_\alpha X_\alpha$ is *closed* in $\mathring{\cup}_\alpha X_\alpha$ if and only if $i_\alpha^{-1}(F) = F \cap X_\alpha$ is closed in X_α for all $\alpha \in \mathcal{A}$.

(In terms of Exercise 2.11, the disjoint union topology is *coinduced* by the collection of injections $\{i_\sigma : X_\sigma \to \mathring{\cup}_\alpha X_\alpha \, (\sigma \in \mathcal{A})\}$.)

When given the disjoint union topology, $\mathring{\cup}_\alpha X_\alpha$ is called the ***topological disjoint union*** with ***summands*** $\{X_\alpha : \alpha \in \mathcal{A}\}$.

The effect of this construction is to spread out the summands so that accumulation in the disjoint union occurs only in the individual summands.

Example 1 The subspace $\mathbb{R} - \mathbb{Z} \subset \mathbb{R}$ coincides with the topological disjoint union $\mathring{\cup}_n (n, n+1)$ of the interval subspaces $(n, n+1) \subset \mathbb{R}$. The reason is that a subset U is open in the subspace $\mathbb{R} - \mathbb{Z} \subset \mathbb{R}$ if and only if $U \cap (n, n+1)$ is open in the subspace $(n, n+1) \subset \mathbb{R}$ for all $n \in \mathbb{Z}$.

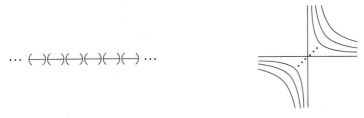

Example 1: $\mathbb{R} - \mathbb{Z}$ Example 2: $(\mathring{\cup}_k H_k) \mathring{\cup} A$

Example 2 The sets $H_k = \{<x, y> \in \mathbb{R}^2 : xy = 1/k\}$, indexed by $k \geq 1$, are disjoint hyperbolas that asymptotically approach the coordinate axes A in $\mathbb{R}^2$. The topological disjoint union, $(\mathring{\cup}_k H_k) \mathring{\cup} A$, of the subspaces $A, H_k \subset \mathbb{R}^2$ differs from the subspace $(\mathring{\cup}_k H_k) \mathring{\cup} A \subset \mathbb{R}^2$, since $(\mathring{\cup}_k H_k)$ is closed in the former space, but not in the latter.

3.1 Characteristic Feature *For any family $\{h_\alpha : X_\alpha \to Z\}$ of continuous functions with common codomain Z, there exists a unique continuous function $h : \mathring{\cup}_\alpha X_\alpha \to Z$ such that $h \circ i_\alpha = h_\alpha$ for all $\alpha \in \mathcal{A}$.*

Proof: The existence and uniqueness of the function h follows from the disjoint union set construction. For any open set $V \subseteq Z$, the pre-image

$$i_\alpha^{-1}(h^{-1}(V)) = (h \circ i_\alpha)^{-1}(V) = h_\alpha^{-1}(V)$$

is open X_α for each $\alpha \in \mathcal{A}$, by the continuity of the functions $\{h_\alpha \, (\alpha \in \mathcal{A})\}$. Thus, $h^{-1}(V)$ is open in the topological union $\mathring{\cup}_\alpha X_\alpha$. So h is continuous. □

SET PRODUCT

Given an indexed family $\{X_\alpha : \alpha \in \mathcal{A}\}$ of nonempty sets, their **product** $\prod_\alpha X_\alpha$ is the set construction characterized by these properties:

(**P1**)   *There are functions* $p_\sigma : \prod_\alpha X_\alpha \to X_\sigma$ ($\sigma \in \mathcal{A}$), *called* **projections**.

(**P2**)   *For any set W and any family* $\{g_\alpha : W \to X_\alpha\}$ *of functions with common domain W, there exists a unique function* $g : W \to \prod_\alpha X_\alpha$ *such that* $p_\alpha \circ g = g_\alpha$ *for all* $\alpha \in \mathcal{A}$.

For a finite indexed family $\{X_i : 1 \leq i \leq n\}$ of sets, the set $X_1 \times \ldots \times X_n$ of ordered n-tuples $(x_1, \ldots, x_n)$, with *coordinates* $x_i \in X_i$ ($1 \leq i \leq n$), serves as their product. The projections

$$p_i : X_1 \times \ldots \times X_n \to X_i \qquad (1 \leq i \leq n)$$

in (**P1**) are defined by $p_i(x_1, \ldots, x_n) = x_i$ for all $(x_1, \ldots, x_n)$, and the function

$$g : W \to X_1 \times \ldots \times X_n$$

in (**P2**) is defined by $g(w) = (g_1(w), \ldots, g_n(w))$ for all $w \in W$.

But what is the product of a family of sets indexed by an arbitrary, even uncountable, set $\mathcal{A}$? How can we define an element x with coordinates $x_\alpha \in X_\alpha$ for all $\alpha \in \mathcal{A}$ in that case?

Well, we can view an n-tuple $x = (x_1, \ldots, x_n) \in X_1 \times \ldots \times X_n$ as a function $x : \{1, \ldots, n\} \to \mathring{\cup}_i X_i$ such that $x(i) = x_i \in X_i$ for all $1 \leq i \leq n$. This suggests the following concept of an α-tuple in an arbitrary indexed family of sets.

For any indexed family $\{X_\alpha : \alpha \in \mathcal{A}\}$ of sets, an **α-tuple** $x : \mathcal{A} \to \mathring{\cup}_\alpha X_\alpha$ is a function with *coordinates* $x(\alpha) = x_\alpha \in X_\alpha$ for all $\alpha \in \mathcal{A}$. The **product set** $\prod_\alpha X_\alpha$ is defined to be the set of all α-tuples $x : \mathcal{A} \to \mathring{\cup}_\alpha X_\alpha$.

The Axiom of Choice (**1.5.4**) is equivalent to the existence of an α-tuple $x : \mathcal{A} \to \mathring{\cup}_\alpha X_\alpha$ for any indexed family $\{X_\alpha : \alpha \in \mathcal{A}\}$ of disjoint nonempty sets. Thus, by that axiom, the *product set* $\prod_\alpha X_\alpha$ *is nonempty*.

The product properties (**P1**) and (**P2**) are satisfied using the projections

$$p_\sigma : \prod_\alpha X_\alpha \to X_\sigma, \ p_\sigma(x) = x_\sigma, \qquad\qquad (\sigma \in \mathcal{A})$$

and the function

$$g : W \to \prod_\alpha X_\alpha, \ g(w)(\alpha) = g_\alpha(w).$$

TOPOLOGICAL PRODUCT

Now let $\{X_\alpha : \alpha \in \mathcal{A}\}$ be a family of topological spaces. A natural topology on the product $\prod_\alpha X_\alpha$ has two opposing goals: the continuity of the projections in (**P1**) is more likely for a large topology on $\prod_\alpha X_\alpha$, while that of the function g in (**P2**) is more likely for a small topology on $\prod_\alpha X_\alpha$.

We take as the **product topology** on $\prod_\alpha X_\alpha$ the smallest topology for which all the projections $p_\sigma : \prod_\alpha X_\alpha \to X_\sigma$ ($\sigma \in \mathcal{A}$) are continuous. Then $\prod_\alpha X_\alpha$ is called the ***topological product*** with *factor spaces* X_α ($\alpha \in \mathcal{A}$).

(In the terms of Exercise 2.10, the product topology is *induced* by the family of projections $\{p_\sigma : \prod_\alpha X_\alpha \to X_\sigma$ ($\sigma \in \mathcal{A}$)$\}$. By definition, the projection functions in (**P**1) are continuous. The proof that the function g in (**P**2) is continuous when the functions g_α are continuous is given in 3.2.)

One way to express the product topology is with its *subbase*:

$$\mathcal{S} = \{p_\sigma^{-1}(U_\sigma) \subseteq \prod_\alpha X_\alpha : \sigma \in \mathcal{A}, U_\sigma \text{ open in } X_\sigma\}.$$

Any $p_\sigma^{-1}(U_\sigma) \in \mathcal{S}$ consists of all points $x \in \prod_\alpha X_\alpha$ such that $p_\sigma(x) = x_\sigma \in U_\sigma$; so, $p_\sigma^{-1}(U_\sigma)$ is a product $\prod_\alpha V_\alpha$, where $V_\sigma = U_\sigma$ and $V_\alpha = X_\alpha$ for $\alpha \neq \sigma$. We call the subbase member $p_\sigma^{-1}(U_\sigma)$ a ***section*** of $\prod_\alpha X_\alpha$, since the coordinates of its points are restricted for only the one index $\sigma \in \mathcal{A}$.

Thus, a subset is open in $\prod_\alpha X_\alpha$ if and only if it is the union of finite intersections of sections $p_\sigma^{-1}(U_\sigma)$, where $\sigma \in \mathcal{A}$ and U_σ is open in X_σ.

We abbreviate each finite intersection of members of the subbase $\mathcal{S}$ by

$$\langle U_{\sigma(1)}, \ldots, U_{\sigma(k)} \rangle = p_{\sigma(1)}^{-1}(U_{\sigma(1)}) \cap \ldots \cap p_{\sigma(k)}^{-1}(U_{\sigma(k)}).$$

These finite intersections constitute a *base* for the product topology:

$$\mathcal{B} = \{\langle U_{\sigma(1)}, \ldots, U_{\sigma(k)} \rangle \subseteq \prod_\alpha X_\alpha : \sigma(i) \in \mathcal{A}, U_{\sigma(i)} \text{ open in } X_{\sigma(i)}, 1 \leq i \leq k\}.$$

The α–coordinate of any point $x \in \langle U_{\sigma(1)}, \ldots, U_{\sigma(k)} \rangle \in \mathcal{B}$ is restricted to the open set $U_\alpha \subseteq X_\alpha$ for the finitely many indices $\alpha \in \{\sigma(1), \ldots, \sigma(k)\}$ and is *unrestricted* for all other α. So for an infinite index set $\mathcal{A}$, a product $\prod_\alpha U_\alpha$ of *proper* open subsets $U_\alpha \subset X_\alpha$, for all $\alpha \in \mathcal{A}$, is *never* open in $\prod_\alpha X_\alpha$.

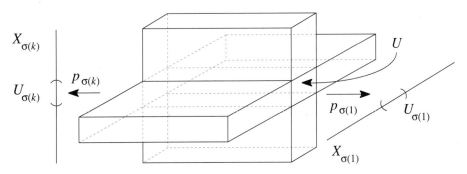

Member U of base $\mathcal{B}$ for product topology

FINITE TOPOLOGICAL PRODUCTS

Consider a finite indexed family $\{X_i : 1 \leq i \leq n\}$ of spaces. Each member of the subbase $\mathcal{S}$ for the product topology on $X_1 \times \ldots \times X_n$ is a *section*

$$p_i^{-1}(U_i) = X_1 \times \ldots \times U_i \times \ldots \times X_n \subseteq X_1 \times \ldots \times X_i \times \ldots \times X_n,$$

where U_i is open in X_i for some $1 \leq i \leq n$. Each member of the base $\mathcal{B}$, being a finite intersection of members of the subbase $\mathcal{S}$, is an *open product box*

$$U_1 \times \ldots \times U_n \subseteq X_1 \times \ldots \times X_n,$$

with open factors $U_i \subseteq X_i$ ($1 \leq i \leq n$), all of which may be proper subsets of the factor spaces (*unlike the infinite case*). In particular, the product topology on $X \times Y$ agrees with the one offered in Example **4.1.12**.

Example 3   The product $X_1 \times \ldots \times X_n$ of a finite family of discrete spaces is discrete because each point $x = (x_1, \ldots, x_n)$ equals an open product box $\{x_1\} \times \ldots \times \{x_n\}$. The product $\prod_\alpha X_\alpha$ of an infinite family of non-singleton spaces cannot be discrete; no point $x \in \prod_\alpha X_\alpha$ is an open set, because no member of the base $\mathcal{B}$ of the product topology restricts its points in more than finitely many coordinates.

Example 4   A product $X = X_1 \times \ldots \times X_n$ of the metric spaces (X_i, d_i) ($1 \leq i \leq n$) has the max-metric $m(x, y) = max\, \{d_i(x_i, y_i) : 1 \leq i \leq n\}$. The collection of open product boxes $U_1 \times \ldots \times U_n$ is a base for the metric topology induced by m, because an m-ball $B(x, r)$ in (X, m) is a product $B_1 \times \ldots \times B_n$ of the d_i-balls $B_i = B_{d_i}(x_i, r)$ in X_i ($1 \leq i \leq n$). Thus, the product metric topology and the product topology on $X_1 \times \ldots \times X_n$ coincide.

In particular, real n-space $\mathbb{R}^n$ is homeomorphic to the topological product $\mathbb{R} \times \ldots \times \mathbb{R}$ of n copies of the real line $\mathbb{R}$. Similarly, the n-cube $\mathbb{I}^n \subset \mathbb{R}^n$ is homeomorphic to the topological product $\mathbb{I} \times \ldots \times \mathbb{I}$ of n copies of the unit interval $\mathbb{I}$.

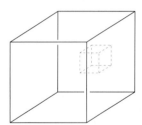
Example 4: $\mathbb{I}^n \cong \mathbb{I} \times \ldots \times \mathbb{I}$

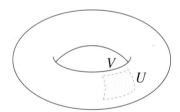
Example 5: $T^2 \cong \mathbb{S}^1 \times \mathbb{S}^1$

Example 5   The torus surface $T^2 \subset \mathbb{R}^3$ generated by rotating a circle about a disjoint coplanar line is homeomorphic to the topological product $\mathbb{S}^1 \times \mathbb{S}^1$. Curved rectangular patches on $T^2 \subset \mathbb{R}^3$ give a base for its subspace topology that corresponds to the base of open product boxes $U \times V \subseteq \mathbb{S}^1 \times \mathbb{S}^1$ for the product topology.

PROPERTIES OF THE TOPOLOGICAL PRODUCT

3.2   Characteristic Feature   *For any family $\{g_\alpha : W \to X_\alpha\}$ of continuous functions with common domain W, there exists a unique continuous function $g : W \to \prod_\alpha X_\alpha$ such that $p_\alpha \circ g = g_\alpha$ for all $\alpha \in \mathcal{A}$.*

Proof: The existence and uniqueness of the function g follow from the product set construction. For each subbase set $p_\sigma^{-1}(U_\sigma) \subseteq \prod_\alpha X_\alpha$, we have that $g^{-1}(p_\sigma^{-1}(U_\sigma)) = (p_\sigma \circ g)^{-1}(U_\sigma) = g_\sigma^{-1}(U_\sigma)$, which is open in W by the continuity of the given function g_σ. Therefore, g is continuous.   □

Here are some general consequences of the characteristic feature:

3.3 Theorem *Continuous functions $\{f_\alpha : W_\alpha \to X_\alpha \, (\alpha \in \mathcal{A})\}$ determine a continuous product function*

$$f = \prod_\alpha f_\alpha : \prod_\alpha W_\alpha \to \prod_\alpha X_\alpha, \quad (f(x))_\alpha = f_\alpha(x_\alpha).$$

Proof: By the Characteristic Feature (3.2) of the product topology, it is sufficient to observe the continuity of the composite,

$$p_\sigma \circ f = f_\sigma \circ p_\sigma : \prod_\alpha W_\alpha \to W_\sigma \to X_\sigma,$$

of $f : \prod_\alpha W_\alpha \to \prod_\alpha X_\alpha$ with each projection $p_\sigma : \prod_\alpha X_\alpha \to X_\sigma \, (\sigma \in \mathcal{A})$. □

Example 6 The exponential function $e : \mathbb{R} \to \mathbb{S}^1$ (Example 1.7) continuously wraps each closed unit interval $[k, k+1]$ in $\mathbb{R}$ once around the 1-sphere $\mathbb{S}^1$. Therefore, the product function $e \times e : \mathbb{R} \times \mathbb{R} \to \mathbb{S}^1 \times \mathbb{S}^1$ continuously wraps each closed unit square $[i, i+1] \times [k, k+1]$ in $\mathbb{R} \times \mathbb{R}$ once around the torus $\mathbb{S}^1 \times \mathbb{S}^1$.

Example 7 Peano's curve $P : \mathbb{I} \to \mathbb{I}^2 = \mathbb{I} \times \mathbb{I}$ (**3.5.1**) continuously stretches the closed unit interval $\mathbb{I}$ onto the entire closed unit square $\mathbb{I}^2$. The composite $(P \times P) \circ P : \mathbb{I} \to \mathbb{I} \times \mathbb{I} \to \mathbb{I}^2 \times \mathbb{I}^2$ continuously stretches $\mathbb{I}$ onto the entire closed 4-cube $\mathbb{I}^4 = \mathbb{I}^2 \times \mathbb{I}^2$.

3.4 Corollary *The topological product $\prod_\alpha W_\alpha$ of a family of subspaces $\{W_\alpha \subseteq X_\alpha : \alpha \in \mathcal{A}\}$ is a subspace of the topological product $\prod_\alpha X_\alpha$.*

Proof: By Theorem 3.3, the continuous inclusions $\{i_\alpha : W_\alpha \to X_\alpha\}$ induce a continuous inclusion $i = \prod_\alpha i_\alpha : \prod_\alpha W_\alpha \to \prod_\alpha X_\alpha$ of the topological products. Thus, the subspace topology on $\prod_\alpha W_\alpha \subseteq \prod_\alpha X_\alpha$ is contained in the product topology on $\prod_\alpha W_\alpha$.

Conversely, each member $p_\sigma^{-1}(W_\sigma \cap U_\sigma)$ $(U_\sigma \subseteq X_\sigma$ open) of the subbase for the product topology on $\prod_\alpha W_\alpha$ is a member $i^{-1}(p_\sigma^{-1}(U_\sigma))$ of the subspace topology on $\prod_\alpha W_\alpha \subseteq \prod_\alpha X_\alpha$. □

3.5 Corollary *The **slice** $\prod_\alpha W_\alpha \subseteq \prod_\alpha X_\alpha$ parallel to X_σ through a point $x = (x_\alpha)$ (i.e., $W_\sigma = X_\sigma$ and $W_\alpha = \{x_\alpha\}$ for all $\alpha \neq \sigma$) is a subspace of $\prod_\alpha X_\alpha$ homeomorphic to the factor space X_σ.*

3.6 Theorem *For any spaces X, Y, and Z, the natural set bijections*

$$X \times Y \to Y \times X \quad \text{and} \quad (X \times Y) \times Z \to X \times (Y \times Z),$$

given by

$$<x, y> \,\to\, <y, x> \quad \text{and} \quad <<x, y>, z> \,\to\, <x, <y, z>>,$$

are homeomorphisms of these product spaces.

Proof: These functions and their inverses are continuous by the characteristic feature of the product topology, because their composites with the projections onto the factor spaces are continuous.

Indeed, the interchange of coordinates, $<x, y> \leftrightarrow <y, x>$, is accomplished by the continuous functions

$$\tau: X \times Y \to Y \times X \quad \text{and} \quad \tau: Y \times X \to X \times Y$$

with $p_1 \circ \tau = p_2$ and $p_2 \circ \tau = p_1$.

Similarly, the association of coordinates, $<<x, y>, z> \to <x, <y, z>>$, is a continuous function since its ultimate projections to the factor spaces of $X \times (Y \times Z)$ are continuous compositions of projections of $(X \times Y) \times Z$. □

Example 8 *A space equalling its square.* Any topological product $Y = \prod_i X_i$ of an infinite sequence of copies of the same space $X_i = X$ has the bizarre property that it is homeomorphic to its topological square $Y \times Y$.

The ***shuffling function*** $s: Y \times Y \to Y$, defined by

$$s((x_1, x_2, \ldots, x_i, \ldots), (y_1, y_2, \ldots, y_i, \ldots)) = (x_1, y_1, x_2, y_2, \ldots, x_i, y_i, \ldots),$$

and the ***sorting function*** $u: Y \to Y \times Y$, defined by

$$u(z_1, z_2, z_3, z_4, \ldots, z_{2i}, z_{2i+1}, \ldots) = ((z_1, z_3, \ldots, z_{2i+1}, \ldots), (z_2, z_4, \ldots, z_{2i}, \ldots)),$$

are inverse functions. They are continuous by the characteristic feature of product topologies, because their composites with the projection functions to the ultimate factor spaces X_i are continuous.

Example 9 *Cantor's Dust as a topological product.* By Example 3, the topological product $\prod_k \{0, 2\}_k$ of a sequence of 2-point discrete spaces $\{0, 2\}_k$ ($k \geq 1$) is not discrete. In fact, $\prod_k \{0, 2\}_k$ is homeomorphic to Cantor's subspace $C \subset \mathbb{I}$. Recall that Cantor's space C consists precisely of points x that have a (necessarily unique) even ternary expansion $(.x_1 x_2 x_3 \ldots x_k \ldots)_3$, i.e., one whose ternary digits x_k are 0 or 2. Thus the ternary system provides the bijection $\kappa: C \to \prod_k \{0, 2\}_k$ defined by

$$\kappa((.x_1 x_2 x_3 \ldots x_k \ldots)_3) = (x_1, x_2, x_3, \ldots, x_k, \ldots).$$

For each $k \geq 1$, $C_k \subset \mathbb{I}$ denotes the set of points that have a ternary expansion whose first k ternary digits belong to $\{0, 2\}$. Equivalently, C_k is the union of the 2^k disjoint, even, k^{th} stage, closed ternary subintervals of $\mathbb{I}$. The kth ternary digit of members of $C \subset C_k$ is constantly 0 or 2 on each of the 2^k ternary subintervals in C_k. This proves that $\pi_k \circ \kappa: C \to \{0, 2\}_k$ is continuous for each $k \in \mathbb{P}$, so that κ is also.

To prove that κ is open, consider any open subset $U \subseteq C$ and any point $x \in U$. Then $B(x, r) \cap C \subseteq U$ for some $r > 0$. Choose k sufficiently large so that $1/3^k < r/3$. The even k^{th} stage ternary subinterval T in C_k containing x has width $1/3^k$ and so $T \cap C_k \subseteq B(x, r) \cap C \subseteq U$. Since $T \cap C_k$ consists of all points of C with the same first k ternary digits as x, then its image $\kappa(T \cap C_k)$ is the set

$$\{x_1\} \times \{x_2\} \times \ldots \times \{x_k\} \times \{0, 2\} \times \{0, 2\} \times \ldots,$$

which is an open subset of the topological product $\prod_k \{0, 2\}_k$. Thus, $\kappa(U)$ contains the open set $\kappa(T \cap C_k)$ about the point $\kappa(x)$. This proves that the continuous bijection κ is open; hence, κ is a homeomorphism.

This homeomorphism $\kappa: C \cong \prod_k \{0, 2\}_k$ implies that $C \cong C \times C$, by Example 8. This reproves the observation of Example 3.4.7.

3.7 Theorem *Each projection $p_\gamma : \prod_\alpha X_\alpha \to X_\gamma$ $(\gamma \in \mathcal{A})$ of a topological product $\prod_\alpha X_\alpha$ is a continuous open function.*

Proof: Each member $p_\sigma^{-1}(U_\sigma) \subseteq \prod_\alpha X_\alpha$ of the subbase $\mathcal{S}$ has images

$$p_\gamma(p_\sigma^{-1}(U_\sigma)) = \begin{cases} X_\gamma, & \text{if } \gamma \neq \sigma \\ U_\sigma, & \text{if } \gamma = \sigma. \end{cases}$$

Thus, each member $U = \cap_i\, p_{\sigma(i)}^{-1}(U_{\sigma(i)})$ of the base $\mathcal{B}$ has open images

$$p_\gamma(\cap_i\, p_{\sigma(i)}^{-1}(U_{\sigma(i)})) = \begin{cases} X_\gamma, & \text{if } \gamma \notin \{\sigma(1), \ldots, \sigma(k)\} \\ \cap_i \{U_{\sigma(i)} : \gamma = \sigma(i)\}, & \text{otherwise.} \end{cases} \qquad \square$$

Example 10 *Projections of a topological product need not be closed functions.* A closed subset $F \subseteq X \times Y$ may have a projection $p_Y(F) \subseteq Y$ that is not closed. In the following display, $X = \mathbb{R} = Y$ and F is one branch of a hyperbola; its projection $p_Y(F)$ is the non-closed ray $(0, \infty)$ in $\mathbb{R}$.

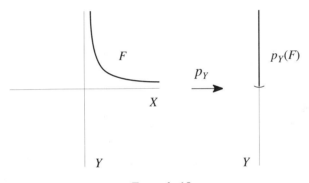

Example 10

4 Weak Topologies and Adjunction Spaces

We consider two more construction techniques. The first technique gives a set a *weak topology* relative a family of topologized subsets. The second technique assembles a topological space by *adjoining* one to another along a subspace.

WEAK TOPOLOGY

There are certain instances when it is possible to construct from a family of topologized subsets of a set a maximal topology on the set that has the family members as subspaces. This is usually referred to as the *weak topology* relative to the subspaces. The construction below is closely related to topological disjoint unions and identification spaces.

Let X be a set, and let $\{X_\alpha : \alpha \in \mathcal{A}\}$ be an indexed family of topological spaces, with each X_α a subset of X. To topologize X, we assume one of the two alternative pairs of hypotheses on the family: for all $\alpha \in \mathcal{A}$ and $\sigma \in \mathcal{A}$,

(**W**1) X_α and X_σ induce the same subspace topology on $X_\alpha \cap X_\sigma$, and
(**W**2) $X_\alpha \cap X_\sigma$ is closed (alternatively, open) in X_α and in X_σ.

Then the following collection is a topology on X, called the ***weak topology*** on X relative $\{X_\alpha : \alpha \in \mathcal{A}\}$:

$$\mathcal{T} = \{U \subseteq X : \forall\, \alpha \in \mathcal{A},\ U \cap X_\alpha \text{ is open in } X_\alpha\}.$$

Example 1 When $\{X_\alpha : \alpha \in \mathcal{A}\}$ is any family of pairwise disjoint spaces, the weak topology on their union X exists and is the disjoint union topology on $X = \mathring{\bigcup}_\alpha X_\alpha$.

4.1 Theorem *Let X have the weak topology relative a family of spaces $\{X_\alpha : \alpha \in \mathcal{A}\}$ that satisfy* (**W**1) *and a version of* (**W**2). *Then:*
(a) *A subset $V \subseteq X$ is closed in X if and only if $V \cap X_\alpha \subseteq X_\alpha$ is closed in X_α for all $\alpha \in \mathcal{A}$.*
(b) *Each space X_α is a closed (alternatively, open) subspace of X.*

Proof: (a) This restates the definition of the weak topology.
(b) Let $A \subseteq X_\alpha$ be closed (alternatively, open) in X_α. For each $\sigma \in \mathcal{A}$, $A \cap X_\sigma$ is closed (open) in the mutual subspace $X_\alpha \supseteq X_\alpha \cap X_\sigma \subseteq X_\sigma$ (**W**1), which is closed (open) in X_σ by (**W**2); hence, $A \cap X_\sigma$ is closed (open) in X_σ. So A is closed (open) in X by (a). This proves that X_α is a subspace of X and is also closed (open) in X. □

It is possible to form the weak topology as an alternative to a pre-existing topology on X. Let $\{X_\alpha : \alpha \in \mathcal{A}\}$ be any family of closed (alternatively, open) subspaces that cover a topological space X. Then the properties (**W**1) and (**W**2) are automatic. Let $w(X)$ denote the set X retopologized with the weak topology relative the family. In this situation, we have the following:

4.2 Theorem *The identity $1_X : w(X) \to X$ is a continuous function (alternatively, a homeomorphism, when the subspaces are open in X).*

Proof: The function is always continuous, because $U \subseteq X$ open in X implies $U \cap X_\alpha \subseteq X_\alpha$ open in the subspace X_α for all $\alpha \in \mathcal{A}$. When the subspaces are open in X, the converse holds by the topology axiom (**T**3) and the fact that $U = \bigcup_\alpha (U \cap X_\alpha)$ when the family covers X. □

Example 2 The weak topology relative a family $\{X_\alpha : \alpha \in \mathcal{A}\}$ of closed subspaces that cover X is usually strictly larger than the original topology. For example, the set $A = \{1/k : 1 \le k\}$ is closed in the weak topology on $X = [0, 1] = \mathbb{I}$ relative the family $\{\{0\}\} \cup \{[1/(k+1), 1/k] : 1 \le k\}$, but not in the usual real subspace topology on $\mathbb{I}$.

Example 3   The weak topology on $\mathbb{R}$ relative the family $\mathcal{F} = \{[n, n+1] : n \in \mathbb{Z}\}$ of closed intervals coincides with its original topology. Suppose that $F \subseteq \mathbb{R}$ is closed in $w(\mathbb{R})$ and that $x \notin F$. Then for each $n \in \mathbb{Z}$, there exists an open set V_n of $\mathbb{R}$ such that $V_n \cap [n, n+1] = [n, n+1] - F$. The point x has an open neighborhood $U(x)$ in $\mathbb{R}$ that intersects at most two members of $\mathcal{F}$. Then $W = \cap_n (U \cap V_n)$ is an intersection of at most two open neighborhoods of x in $\mathbb{R}$ that are disjoint from F; hence, W is an open neighborhood of x in $\mathbb{R}$ that is disjoint from F. This proves that F is closed in $\mathbb{R}$. (This family $\mathcal{F}$ exemplifies the *local finiteness property* involved in Exercise 3.)

Let X be any topological space, and let $\{X_\alpha : \alpha \in \mathcal{A}\}$ be a family of closed (alternatively, open) subspaces that cover X. Since the inclusions $i_\alpha : X_\alpha \to X$ ($\alpha \in \mathcal{A}$) are continuous, they determine a continuous surjection $q : \mathring{\cup}_\alpha X_\alpha \to X$ such that $q|_{X_\alpha} = i_\alpha$ for all $\alpha \in \mathcal{A}$, by the Characteristic Feature (3.1) of the topological disjoint union. In this situation, we have:

4.3   Theorem   *The space X has the weak topology if and only if the continuous surjection $q : \mathring{\cup}_\alpha X_\alpha \to X$, $q|_{X_\alpha} = i_\alpha$, is an identification map.*

Proof: The weak topology and the identification topology have the same characterization: $U \subseteq X$ is open in X if and only if $U \cap X_\alpha = q^{-1}(U) \cap X_\alpha$ is open in X_α for all $\alpha \in \mathcal{A}$, equivalently, $q^{-1}(U)$ is open in $\mathring{\cup}_\alpha X_\alpha$.   □

In the weak topology, open sets are tested against the subspaces. It follows that the continuity of a function defined on a space with the weak topology may be tested on the subspaces:

4.4   Theorem   *Let the space X have the weak topology relative the family of subspaces $\{X_\alpha : \alpha \in \mathcal{A}\}$. A function $g : X \to Y$ is continuous if and only if each restriction $g_\alpha = g|_{X_\alpha} : X_\alpha \to Y$ ($\alpha \in \mathcal{A}$) is continuous.*

Proof: By the subspace feature (1.4), the restrictions $g|_{X_\alpha}$ ($\alpha \in \mathcal{A}$) are continuous if g is continuous. The converse holds. Let $V \subseteq Y$ be open in Y. Then, for each $\alpha \in \mathcal{A}$, $g^{-1}(V) \cap X_\alpha = g_\alpha^{-1}(U)$ is open in X_α, by the continuity of g_α. Hence, $g^{-1}(V) \subseteq X$ is open in the weak topology on X.   □

ADJUNCTION SPACES

Special identification spaces called *adjunction spaces* arise when a space D is *attached* to a space A by means of a continuous function $\phi : S \to A$ defined on a closed subspace $S \subseteq D$, as follows.

The ***adjunction space of*** $D \supseteq S$ ***to*** A ***by*** $\phi : S \to A$, which is denoted by $X = A \cup_\phi D$, results from the topological disjoint union $A \mathring{\cup} D$ by identifying each point $z \in S \subseteq D$ with its image $\phi(z) \in A$, as indicated in the figure that follows.

We denote the identification map by $p : A \mathring{\cup} D \to A \cup_\phi D = X$.

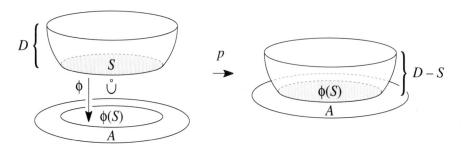

Adjunction space construction

4.5 Theorem *Let $p : A \mathbin{\mathring{\cup}} D \to A \cup_\phi D = X$ be the identification map that produces the adjunction space X. Then:*
(a) $p|_A : A \cong p(A)$ *is a homeomorphism onto a closed subspace* $p(A) \subseteq X$.
(b) $p|_{D-S} : D - S \cong p(D - S)$ *is a homeomorphism onto an open subspace* $p(D - S) \subseteq X$ *of the adjunction space X.*

Proof: (a) For each closed subset $F \subseteq A$, $p^{-1}(p(F)) \subseteq A \mathbin{\mathring{\cup}} D$ is the closed subset $F \cup \phi^{-1}(F) \subseteq A \mathbin{\mathring{\cup}} D$. Thus, $p(F) \subseteq p(A)$ is closed in the identification space $X = A \cup_\phi D$, and also in the subspace $p(A)$. This proves that $p(A)$ is a closed subspace of X and that the continuous bijection $p|_A : A \to p(A)$ is a closed function, hence, a homeomorphism.

(b) For each open subset $U \subseteq D - S$, the pre-image $p^{-1}(p(U)) \subseteq A \mathbin{\mathring{\cup}} D$ is the open subset $U \subseteq A \mathbin{\mathring{\cup}} D$. So $p(U) \subseteq p(D - S)$ is open in the identification space X, and also in the subspace $p(D - S)$. This proves that $p(U)$ is an open subspace of X and that the continuous bijection $p|_{D-S} : D - S \to p(D - S)$ is an open function, hence, a homeomorphism. □

Example 4 The adjunction space $X = \mathbb{D}_-^n \cup_\psi \mathbb{D}_+^n$ of one closed n-ball $\mathbb{D}_+^n \supset \mathbb{S}^{n-1}$ to a second one $\mathbb{D}_-^n$ by the inclusion $\psi: \mathbb{S}^{n-1} \subset \mathbb{D}_-^n$ is homeomorphic to the sphere $\mathbb{S}^n$. The n-balls $\mathbb{D}_-^n \subset X \supset \mathbb{D}_+^n$ correspond to the hemispheres $\mathbb{H}_-^n \subset \mathbb{S}^n \supset \mathbb{H}_+^n$.

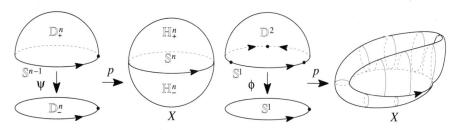

Example 4: Spherical adjunction space   Example 5: Dunce hat

Example 5 The ***dunce hat*** X results from a triangle by the identification of its three incompatibly oriented sides via linear homeomorphisms. Alternatively, X is the adjunction space of the disc $\mathbb{D}^2 \supset \mathbb{S}^1$ to $\mathbb{S}^1$ via an attaching map $\phi : \mathbb{S}^1 \to \mathbb{S}^1$ (above right) that wraps around $\mathbb{S}^1$ twice counter-clockwise, once clockwise.

Continuous functions *on* an adjunction space correspond to compatible pairs of continuous functions *on* the constituent spaces, as follows.

4.6 Adjunction Theorem   *Let $\phi : S \to A$, where $S \subseteq D$. Continuous functions $h : A \to Z$ and $k : D \to Z$ with $h \circ \phi = k|_S$ determine a unique continuous function*

$$l = <h, k> : A \cup_\phi D \to Z$$

such that $<h, k> \circ p|_A = h$ and $<h, k> \circ p|_D = k$.

Proof: As $h \circ \phi = k|_S$, the continuous function $A \cup_\phi D \to Z$ given by h and k respects the identifications of $p : A \mathring{\cup} D \to A \cup_\phi D$; hence, by the Transgression Theorem (1.9), it induces a continuous function $l : A \cup_\phi D \to Z$, as in the adjunction diagram, below left. The function l has the desired properties and it is unique by the surjectivity of p. Hereafter, l is denoted by $<h, k>$. □

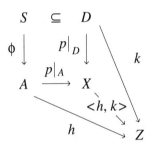
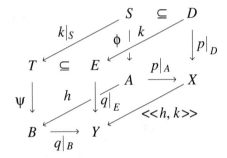

Adjunction Theorem 4.6   Corollary 4.7

4.7 Corollary   *Let $\phi : S \to A$, where $S \subseteq D$, and let $\psi : T \to B$, where $T \subseteq E$; consider the pair of adjunction spaces $X = A \cup_\phi D$ and $Y = B \cup_\psi E$. Continuous functions $h : A \to B$ and $k : (D, S) \to (E, T)$ such that $h \circ \phi = \psi \circ k|_S$ determine a unique continuous function*

$$<<h, k>> : A \cup_\phi D \to B \cup_\psi E$$

such that $<<h, k>> \circ p|_A = q|_B \circ h$ and $<<h, k>> \circ p|_D = q|_E \circ k$, where $p : A \mathring{\cup} D \to A \cup_\phi D$ and $q : B \mathring{\cup} E \to B \cup_\psi E$ are the identification maps.

Proof: Take $<<h, k>> = <q|_B \circ h, q|_E \circ k>$, as in the figure, above right.

Suppose that S is a closed subspace of both D and A. The adjunction space $X = A \cup_\phi D$ for $D \supseteq S$ and $\phi : S \subseteq A$ is a union of copies of A and D that share the closed subspace S and it has the weak topology relative them. In this case, we call X the ***amalgamated union*** of $D \supseteq S \subseteq A$ and we denote it by $A \cup_S D$. Any space X covered by a pair of closed subspaces A and D with intersection $S = A \cap D$ is the amalgamated union $A \cup_S D$ of $D \supseteq S \subseteq A$.

Example 6 The boundary of $\mathbb{D}^{n+1} \times \mathbb{D}^{n+1}$ in $\mathbb{R}^{n+1} \times \mathbb{R}^{n+1}$, which is homeomorphic to $\mathbb{S}^{2n+1}$, is the union of the closed subspaces $\mathbb{D}^{n+1} \times \mathbb{S}^n$ and $\mathbb{S}^n \times \mathbb{D}^{n+1}$ that have intersection $\mathbb{S}^n \times \mathbb{S}^n$. Thus, we may view the sphere $\mathbb{S}^{2n+1}$ as the amalgamated union $\mathbb{D}^{n+1} \times \mathbb{S}^n \cup_{\mathbb{S}^n \times \mathbb{S}^n} \mathbb{S}^n \times \mathbb{D}^{n+1}$. When $n = 0$, this describes $\mathbb{S}^1$ as the amalgamated union $\mathbb{D}^1 \times \mathbb{S}^0 \cup_{\mathbb{S}^0 \times \mathbb{S}^0} \mathbb{S}^0 \times \mathbb{D}^1$, namely, the four edges of the square $\mathbb{D}^1 \times \mathbb{D}^1$ glued together at the four corner points in $\mathbb{S}^0 \times \mathbb{S}^0$. When $n = 1$, this describes $\mathbb{S}^3$ as the amalgamated union $\mathbb{D}^2 \times \mathbb{S}^1 \cup_{\mathbb{S}^1 \times \mathbb{S}^1} \mathbb{S}^1 \times \mathbb{D}^2$, namely, the union of two solid torii that share the same torus boundary surface, with longitudes and meridians reversed.

In the case of amalgamated unions, the Adjunction Theorem (4.6) and its corollary (4.7) have these particularly straightforward expressions:

A continuous function $<h, k> : A \cup_S D \to Z$ arises from any pair of continuous functions $h : A \to Z$ and $k : D \to Z$ that coincide on the shared closed subspace S.

A continuous function $\ll h, k \gg : A \cup_S D \to B \cup_T E$ arises from any pair of continuous functions $h : (A, S) \to (B, T)$ and $k : (D, S) \to (E, T)$ that have the same restriction $S \to T$ on the shared subspaces S and T.

Example 7 Real multiplication defines a pair of continuous functions

$$m_+ : \mathbb{D}^1 \times \mathbb{S}^0 \to \mathbb{D}^1_+ \qquad m_- : \mathbb{S}^0 \times \mathbb{D}^1 \to \mathbb{D}^1_-$$

that have the same restriction $m : \mathbb{S}^0 \times \mathbb{S}^0 \to \mathbb{S}^0$. So they assemble into a continuous function $\mathbb{D}^1 \times \mathbb{S}^0 \cup_{\mathbb{S}^0 \times \mathbb{S}^0} \mathbb{S}^0 \times \mathbb{D}^1 \to \mathbb{D}^1_+ \cup_{\mathbb{S}^0} \mathbb{D}^1_-$, which, by Examples 4 and 6, we may view as $M : \mathbb{S}^1 \to \mathbb{S}^1$. You may check that M is a 2-fold wrap of $\mathbb{S}^1$ around $\mathbb{S}^1$.

Example 8 Complex multiplication defines a pair of continuous functions

$$m_+ : \mathbb{D}^2 \times \mathbb{S}^1 \to \mathbb{D}^2_+ \qquad m_- : \mathbb{S}^1 \times \mathbb{D}^2 \to \mathbb{D}^2_-$$

that have the same restriction $m : \mathbb{S}^1 \times \mathbb{S}^1 \to \mathbb{S}^1$. They assemble into a continuous function $\mathbb{D}^2 \times \mathbb{S}^1 \cup_{\mathbb{S}^1 \times \mathbb{S}^1} \mathbb{S}^1 \times \mathbb{D}^2 \to \mathbb{D}^2_+ \cup_{\mathbb{S}^1} \mathbb{D}^2_-$, abbreviated by $p : \mathbb{S}^3 \to \mathbb{S}^2$ and called ***Hopf's map***. The pre-images $p^{-1}(x)$, $x \in \mathbb{S}^2$, partition $\mathbb{S}^3$ into *circles*.

Exercises

SECTION 1

1. Determine the separation properties (T_i, $i = 0, 1, 2, 3, 4, 5$) of the quotient spaces $\mathbb{R}/(0, 1)$, $\mathbb{R}/[0, 1)$, $\mathbb{R}/(0, 1]$, and $\mathbb{R}/[0, 1]$ of $\mathbb{R}$ modulo the indicated intervals.

2. Prove that the quotient space $\mathbb{I}/\{0, 1\}$ is homeomorphic to the 1-sphere $\mathbb{S}^1$.

3. (a) Describe the quotient space $\mathbb{S}^2/\mathbb{S}^1$ of the 2-sphere modulo its equator.
   (b) Describe the quotient space $\mathbb{D}^n/\mathbb{S}^{n-1}$ of the closed n-ball modulo its boundary.

4. Let $f : X \to Y$ be a function from a set X to the topological space $(Y, \mathcal{S})$. Prove:
   (a) The collection $f^*\mathcal{S} = \{f^{-1}(V) \subseteq X : V \in \mathcal{S}\}$ is a topology on X.
   (b) The function $f : (X, \mathcal{T}) \to (Y, \mathcal{S})$ is continuous if and only if $\mathcal{T}$ contains $f^*\mathcal{S}$.
   So the ***induced topology***, $f^*\mathcal{S}$, is the smallest topology for which f is continuous.

5 (a) Prove that the topology $f^*\mathcal{S}$ on X induced by $f : X \to (Y, \mathcal{S})$, as in Exercise 4, has this key feature: for every space $(W, \mathcal{R})$, a function $k : (W, \mathcal{R}) \to (X, f^*\mathcal{S})$ is continuous if the composite function $f \circ k : (W, \mathcal{R}) \to (Y, \mathcal{S})$ is continuous.
   (b) Prove that $f^*\mathcal{S}$ is the unique topology on X for which f is continuous and the key feature in (a) is satisfied.

6 Prove this transitivity property of the induced topology defined in Exercise 4: for functions $g : W \to X$ and $f : X \to Y$ from the set W to the set X to the topological space $(Y, \mathcal{S})$, the induced topologies $(f \circ g)^*\mathcal{S}$ and $g^*(f^*\mathcal{S})$ on W coincide.

7 Let $f : X \to Y$ be a function from the topological space $(X, \mathcal{T})$ to the set Y. Prove:
   (a) The collection $f_*\mathcal{T} = \{V \subseteq Y : f^{-1}(V) \in \mathcal{T}\}$ is a topology on Y.
   (b) The function $f : (X, \mathcal{T}) \to (Y, \mathcal{S})$ is continuous if and only if $f_*\mathcal{T}$ contains $\mathcal{S}$.
   So the ***coinduced topology***, $f_*\mathcal{T}$, is the largest topology for which f is continuous.

8 (a) Prove that the topology $f_*\mathcal{T}$ on Y coinduced by $f : (X, \mathcal{T}) \to Y$, as in Exercise 7, has this key feature: for every space $(Z, \mathcal{R})$, a function $k : (Y, f_*\mathcal{T}) \to (Z, \mathcal{R})$ is continuous if the composite function $k \circ f : (X, \mathcal{T}) \to (Z, \mathcal{R})$ is continuous.
   (b) Prove that $f_*\mathcal{T}$ is the unique topology on Y for which f is continuous and the key feature in (a) is satisfied.

9 Prove this transitivity property of the coinduced topology defined in Exercise 7: for functions $f : X \to Y$ and $g : Y \to Z$ from the topological space $(X, \mathcal{T})$ to the set Y to the set Z, the coinduced topologies $(g \circ f)_*\mathcal{T}$ and $g_*(f_*\mathcal{T})$ on Z coincide.

10 Given a function $f : X \to (Y, \mathcal{S})$ from the set X to the topological space $(Y, \mathcal{S})$, relate the topologies $\mathcal{S}$ and $f_*(f^*\mathcal{S})$ on Y.

11 Given a function $f : (X, \mathcal{T}) \to Y$ from the topological space $(X, \mathcal{T})$ to the set Y, relate the topologies $\mathcal{T}$ and $f^*(f_*\mathcal{T})$ on X.

12 (a) Find an identification map that is not an open function.
    (b) Find an identification map that is not a closed function.

13 Find an equivalence relation $\sim$ on the real plane $\mathbb{R}^2$ such that the quotient space $\mathbb{R}^2/\sim$ is homeomorphic to the torus $\mathbb{S}^1 \times \mathbb{S}^1$ with the product metric topology.

14 Prove that there is a dense point in the quotient space $\mathbb{R}/\mathbb{Q}$ of $\mathbb{R}$ modulo $\mathbb{Q}$.

15 Prove that the quotient space of Cantor's space C modulo its subset of endpoints is homeomorphic to the quotient space of the closed unit interval $\mathbb{I}$ modulo its subset of binary fractions. What is this quotient space?

The ***projective plane*** $\mathbb{P}^2$ is the quotient space of the 2-sphere $\mathbb{S}^2$ modulo the antipodal equivalence relation: $<x_1, x_2, x_3> \sim <-x_1, -x_2, -x_3>$ for all $<x_1, x_2, x_3> \in \mathbb{S}^2$.

16 Prove that $\mathbb{P}^2$ is homeomorphic with the quotient space of the closed unit 2-ball $\mathbb{D}^2$ modulo the boundary antipodal equivalence relation:
$$<x_1, x_2> \sim <-x_1, -x_2> \text{ for all } <x_1, x_2> \in \mathbb{S}^1 \subset \mathbb{D}^2.$$

17 Decide whether any of the following spaces are homeomorphic:
    (a) the quotient space $\mathbb{R}/\mathbb{Z}$,
    (b) the union $K \subset \mathbb{R}^2$ of the circles C_k, centered on the points $<\pm\frac{1}{k}, 0>$ with radius $1/k$, for all integers $k \geq 1$, and
    (c) the quotient space $[-1, +1]/M$, where $M = \{\pm\frac{1}{k} : k \geq 1\}$.

170 CONSTRUCTION OF SPACES   CHAPTER 5

18 Give sketches that show that the quotient space $\mathbb{S}^2/\{<0, 0, -1>, <0, 0, +1>\}$ is homeomorphic to the pinched torus $(\mathbb{S}^1 \times \mathbb{S}^1)/\mathbb{S}^1 \times \{1\})$.

19 Let $q : X \to X/A$ be the quotient map that collapses a subspace $A \subseteq X$ to a point.
(a) Prove that when the subspace A is closed in X, q restricts to a homeomorphism $q : X - A \to q(X - A)$ of the subspaces $X - A \subseteq X$ and $q(X - A) \subseteq X/A$.
(b) Show that the restriction $q : X - A \to q(X - A)$ may not be a homeomorphism.
(c) Show that, for any space Y, the continuous functions $X \to Y$ that are constant on A are in bijective correspondence with the continuous functions $X/A \to Y$.

20 Prove that a quotient space X/R of a Hausdorff space X is Hausdorff if the relation $R \subseteq X \times X$ is closed in $X \times X$ and the quotient function $q_R : X \to X/R$ is open.

21 (a) Determine the quotient space of the torus surface $\mathbb{S}^1 \times \mathbb{S}^1$ modulo the longitudinal circle $\mathbb{S}^1 \times \{1\}$.
(b) Prove that $\mathbb{S}^2$ is homeomorphic to the quotient space of the torus surface $\mathbb{S}^1 \times \mathbb{S}^1$ modulo the 1-point union $\mathbb{S}^1 \vee \mathbb{S}^1 = (\mathbb{S}^1 \times \{1\}) \cup (\{1\} \times \mathbb{S}^1)$ of the longitudinal circle $\mathbb{S}^1 \times \{1\}$ and the meridional circle $\{1\} \times \mathbb{S}^1$.

22 Construct subspaces of $\mathbb{R}^3$ homeomorphic to the following quotient spaces:
(a) $\mathbb{R}^2/\mathbb{D}^2$, where $\mathbb{D}^2$ is the closed unit ball in $\mathbb{R}^2$,
(b) $\mathbb{R}^2/\mathbb{S}^1$, and
(c) $\mathbb{R}^2/(\mathbb{R}^2 - \mathbb{B}^2)$, where $\mathbb{B}^2$ is the open unit ball in $\mathbb{R}^2$.

23 Determine conditions on a topological space X and a subset $A \subseteq X$ that are necessary and sufficient for the quotient space X/A to be T_2, T_1, or T_0.

SECTION 2

1 Determine the topology on $\mathbb{R}^n$ that has the collection of all lines as its subbase.

2 Let $\mathcal{B}$ be a base for the topology $\mathcal{T}$ on Y. Prove that $\mathcal{B} \cap X = \{B \cap X : B \in \mathcal{B}\}$ is a base for the subspace topology $\mathcal{T}_X$ on $X \subseteq Y$. Deduce that a subspace of a second countable space is second countable. Are subspaces of a separable space separable?

3 (a) Let X be a topological space that contains a sequence of separable subspaces whose union is dense in X. Prove that X is separable.
(b) Prove that Hilbert space $\mathbb{H}$ (Exercise **2.3.16**) is a separable metric space.

4 Prove that if $\mathcal{U}_x$ is a neighborhood base at $x \in Y$ in the topological space $(Y, \mathcal{T})$, then $\mathcal{U}_x \cap X = \{U \cap X : U \in \mathcal{U}_x\}$ is a neighborhood base at x for the subspace $(X, \mathcal{T}_X)$. Deduce that a subspace of a first countable space is first countable.

5 Prove that the image of a first countable space is first countable if the function is continuous and open.

6 Let $\mathcal{OS}$ consist of all open squares $(a, a + r) \times (b, b + r) \subset \mathbb{R} \times \mathbb{R}$, and let $\mathcal{CS}$ consist of all closed squares $[a, a + r] \times [b, b + r] \subset \mathbb{R} \times \mathbb{R}$ ($a, b \in \mathbb{R}, r > 0$).
(a) Prove that the collection $\mathcal{CS}$ of closed squares is not a base for the topology $\mathcal{T}(\mathcal{CS})$ on $\mathbb{R} \times \mathbb{R}$ that has subbase $\mathcal{CS}$.
(b) Prove that the collection $\mathcal{OS}$ of open squares is a base for the topology $\mathcal{T}(\mathcal{OS})$ on $\mathbb{R} \times \mathbb{R}$ that has subbase $\mathcal{OS}$.
(c) Prove that $\mathcal{T}(\mathcal{CS})$ and $\mathcal{T}(\mathcal{OS})$ are distinct topologies on $\mathbb{R} \times \mathbb{R}$, one of which properly contains the other one.

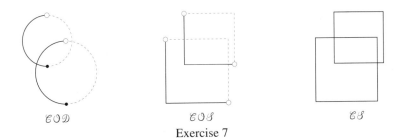

$\mathcal{COD}$ $\mathcal{COS}$ $\mathcal{CS}$

Exercise 7

7. Let $\mathcal{COD}$ and $\mathcal{COS}$ be the collections (above) of closed-open discs and squares.
   (a) Prove that the collection $\mathcal{COD}$ of closed-open discs is not a base for the topology $\mathcal{T}(\mathcal{COD})$ on $\mathbb{R} \times \mathbb{R}$ that has subbase $\mathcal{COD}$.
   (b) Prove that the collection $\mathcal{COS}$ of closed-open squares is a base for the topology $\mathcal{T}(\mathcal{COS})$ on $\mathbb{R} \times \mathbb{R}$ that has subbase $\mathcal{COS}$.
   (c) Prove that $\mathcal{T}(\mathcal{COD})$ and $\mathcal{T}(\mathcal{COS})$ are distinct topologies on $\mathbb{R} \times \mathbb{R}$, one of which properly contains the other one.

8. Determine the interior, closure, and boundary of the intervals $[a, b]$, $[a, b)$, and $(a, b]$ in the right-interval space $\mathbb{R}_{(\,]}$ (Example 10).

9. Prove that a second countable space is first countable, but not conversely.

10*. Consider a family $\{f_\alpha : X \to Y_\alpha\}$ of functions with common domain set X and various codomain spaces Y_α ($\alpha \in \mathcal{A}$).
    (a) Construct the smallest topology on X for which all the functions $\{f_\alpha\}$ are continuous. This is called the topology *induced* by the family $\{f_\alpha : X \to Y_\alpha\}$.
    (b) Prove that, for every space W and every function $h : W \to X$, h is continuous with respect to the induced topology on X if and only if $f_\alpha \circ h : W \to X \to Y_\alpha$ is continuous for all $\alpha \in \mathcal{A}$.
    (c) Prove that the topology on X induced by the family $\{f_\alpha : X \to Y_\alpha\}$ is the only topology on X for which the functions $\{f_\alpha\}$ are continuous and (b) holds.

11*. Consider a family $\{f_\alpha : X_\alpha \to Y\}$ of functions with common codomain set Y and various domain spaces X_α ($\alpha \in \mathcal{A}$).
    (a) Construct the largest topology on Y for which all the functions $\{f_\alpha\}$ are continuous. This is called the topology *coinduced* by the family $\{f_\alpha : X_\alpha \to Y\}$.
    (b) Prove that, for every space Z and every function $k : Y \to Z$, k is continuous with respect to the coinduced topology on Y if and only if $k \circ f_\alpha : X_\alpha \to Y \to Z$ is continuous for all $\alpha \in \mathcal{A}$.
    (c) Prove that the topology on X coinduced by the family $\{f_\alpha : X_\alpha \to Y\}$ is the only topology on X for which the functions $\{f_\alpha\}$ are continuous and (b) holds.

12. Prove that if a totally ordered set $(X, \leq)$ has no smallest element and no greatest element, then the family, $\mathcal{B}(\leq)$, consisting of all intervals $(x, y) = \{z \in X : x < z < y\}$, where $x, y \in X$, is a base for the order topology on X (Example 5).

13. Let $\mathcal{FC}$ be the finite complement topology on X and let $\mathcal{D}(p)$ be the excluded point topology with respect to $p \in X$. Prove that $sup\{\mathcal{FC}, \mathcal{D}(p)\}$ equals $\mathcal{F} = \mathcal{FC} \cup \mathcal{D}(p)$, the Fort topology on X. (See Example 4 for the definition of *sup*.)

14. Prove that the countable base $\{V_n : n = 0, 1, 2, 3, \dots \}$ equals the prime ideal topology on $P = \{2, 3, 5, \dots, p, \dots \}$ in Example 13. (Hint: $\cup_k V_{n_k} = V_{gcd(n_k)}$.)

15 Describe a countable neighborhood base at each point of the following spaces; determine whether or not these spaces are second countable:
  (a) the half-disc space (Exercise **4**.1.6),
  (b) the line with two origins (Exercise **4**.1.7), and
  (c) the tangent-disc space (Exercise **4**.1.8).

16 Prove that the usual topology on the real line $\mathbb{R}$ equals the order topology on $(\mathbb{R}, \leq)$.

17 Prove that a first countable space is Hausdorff if and only if each convergent sequence in the space has a unique limit.

18 (a) Let Ω be the first uncountable ordinal. Prove that the open ordinal interval $[0, \Omega)$, with its order topology, is first countable but not second countable.
  (b) Let α be any countable ordinal. Prove that both the ordinal intervals $[0, \alpha)$ and $[0, \alpha]$, with their order topologies, are first countable and second countable.

SECTION 3

1 Prove that, for any spaces X, Y, and Z, the natural set bijections
$$X \mathring{\cup} Y \to Y \mathring{\cup} X \quad \text{and} \quad (X \mathring{\cup} Y) \mathring{\cup} Z \to X \mathring{\cup} (Y \mathring{\cup} Z)$$
are homeomorphisms of these disjoint union spaces.

2 For functions $\{f_\alpha : X_\alpha \to W_\alpha \ (\alpha \in \mathcal{A})\}$, define their *union* $\mathring{\cup}_\alpha f_\alpha : \mathring{\cup}_\alpha X_\alpha \to \mathring{\cup}_\alpha W_\alpha$.
  (a) Prove that $\mathring{\cup}_\alpha f_\alpha$ is continuous if the functions $\{f_\alpha\}$ are continuous.
  (b) Prove that $\mathring{\cup}_\alpha f_\alpha$ is a homeomorphism if the functions $\{f_\alpha\}$ are homeomorphisms. Does the converse hold?

3 (a) Prove that the subspace topology on $\mathring{\cup}_k H_k \subset \mathbb{R}^2$ in Example 2 coincides with the disjoint union topology on $\mathring{\cup}_k H_k$ induced by the subspaces $H_k \subset \mathbb{R}^2$.
  (b) Find a subspace of $\mathbb{R}^2$ that is homeomorphic to the topological disjoint union $(\mathring{\cup}_k H_k) \mathring{\cup} A$ in Example 2.

4 Prove that the subspace $\Sigma_- \cup \mathbb{S}^1 \cup \Sigma_+ \subset \mathbb{R}^2$ in Example 2.2.5 is not homeomorphic to the topological disjoint union $\Sigma_- \mathring{\cup} \mathbb{S}^1 \mathring{\cup} \Sigma_+$.

5 Prove that the product properties (**P1**) and (**P2**) are satisfied by the product construction given in Section 3 for an arbitrary indexed family of sets.

6 Let A and B be subsets of topological spaces X and Y, respectively. Prove that the interior, closure, and boundary operations in the topological product $X \times Y$ and its factors satisfy the following properties:
  (a) $\text{INT}(A \times B) = \text{INT}(A) \times \text{INT}(B)$,
  (b) $\text{CLS}(A \times B) = \text{CLS}(A) \times \text{CLS}(B)$, and
  (c) $\text{BDY}(A \times B) = (\text{BDY}(A) \times \text{CLS}(B)) \cup (\text{CLS}(A) \times \text{BDY}(B))$.

7 Let $X = \Pi_\alpha X_\alpha$ be an arbitrary topological product; let $A_\alpha \subseteq X_\alpha$ for all $\alpha \in \mathcal{A}$. Prove:
  (a) $\text{INT}_X (\Pi_\alpha A_\alpha) \supseteq \Pi_\alpha \text{INT}_{X_\alpha}(A_\alpha)$, but equality fails in general.
  (b) $\text{CLS}_X (\Pi_\alpha A_\alpha) = \Pi_\alpha \text{CLS}_{X_\alpha}(A_\alpha)$.

8 Prove that a topological space X is Hausdorff if and only if the diagonal set,
$$\Delta(X) = \{<x, x> \in X \times X : x \in X\},$$
is a closed subset of the topological product $X \times X$.

9. Prove that a product function $\Pi_\alpha f_\alpha : \Pi_\alpha X_\alpha \to \Pi_\alpha W_\alpha$ is a homeomorphism of the product spaces if the given functions $\{f_\alpha : X_\alpha \to W_\alpha\}$ are homeomorphisms.

10. (a) Prove that a nonempty topological product of spaces that do not have the indiscrete topology is first (second) countable if and only if each factor space is first (second) countable and the index set is countable.
    (b) Prove that an arbitrary topological product is Hausdorff if and only if all the factor spaces are Hausdorff.

11. (a) Prove that a nonempty topological product of a countable family of separable spaces is separable.
    (b) Prove that if a family of spaces, each of which contains at least one pair of disjoint nonempty open sets, has a separable topological product, then the index set for the family injects into the set $\{0, 1\}^\mathbb{P}$ of binary sequences.
    (c) Construct a non-separable topological product of separable spaces.

12. Draw pictures to suggest a homeomorphism between these products of intervals:

    $$[0, 1] \times [0, 1) \quad \text{and} \quad (0, 1) \times [0, 1).$$

    Deduce the failure of cancellation in topology.

13. Show that the product topology on $\Pi_\alpha X_\alpha$ is the only topology with these properties:
    (a) The projection functions $\pi_\sigma : \Pi_\alpha X_\alpha \to X_\sigma$ are continuous for all $\sigma \in \mathcal{A}$.
    (b) Given any space W and any indexed family $\{g_\alpha : W \to X_\alpha\}$ of continuous functions with common domain space W, there exists a unique continuous function $g : W \to \Pi_\alpha X_\alpha$ such that $\pi_\alpha \circ g = g_\alpha$ for all $\alpha \in \mathcal{A}$.

14. Prove that Cantor's space C is homeomorphic to the topological product of any countable family of copies of itself.

15. Consider the annuli X and Y with *matched* and *mismatched fins*, below. Form their topological products $X \times \mathbb{I}$ and $Y \times \mathbb{I}$ with the closed unit interval $\mathbb{I} = [0, 1]$. Draw pictures to suggest a homeomorphism $X \times \mathbb{I} \cong Y \times \mathbb{I}$. Is $X \cong Y$?

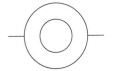

Exercise 15: X (matched fins)   Exercise 15: Y (mismatched fins)

16. Let W denote the torus surface with a single open disc removed (below left). Similarly, let Z denote the closed unit disc with two smaller open discs removed (below right). Form their topological products $Z \times \mathbb{I}$ and $W \times \mathbb{I}$ with the closed unit interval $\mathbb{I} = [0, 1]$. Draw pictures to suggest a homeomorphism $W \times \mathbb{I} \cong Z \times \mathbb{I}$.

Exercise 16: Punctured torus W Exercise 16: Doubly punctured disc Z

17 (a) Consider any indexed family $\{X_\alpha : \alpha \in \mathcal{A}\}$ of spaces. Prove that the set
$$\{\Pi_\alpha V_\alpha \subseteq \Pi_\alpha X_\alpha : V_\alpha \text{ open in } X_\alpha \text{ for all } \alpha \in \mathcal{A}\}$$
of all product boxes whose sides are open subsets of the factor spaces is a base for a topology on $\Pi_\alpha X_\alpha$. This is called the ***box topology***.
(b) Show that the topological product $\Pi_\alpha X_\alpha$ of non-indiscrete spaces has the box topology if and only if the index set $\mathcal{A}$ is finite.
(c) Are the projection maps of a product continuous relative to the box topology?

18 Prove that the n-cube $\mathbb{I}^n$ ($n \geq 1$) is an identification space of Cantor's space C.

19 Let $\{X_\alpha : \alpha \in \mathcal{A}\}$ be a family of subspaces of the space X. Form the topological product $X \times \mathcal{A}$, where $\mathcal{A}$ is given the discrete topology. Prove that the subspace $\cup_\alpha \{X_\alpha \times \{\alpha\} : \alpha \in \mathcal{A}\} \subseteq X \times \mathcal{A}$ is homeomorphic to the disjoint union $\mathring{\cup}_\alpha X_\alpha$.

20 Let $\Pi_\alpha X_\alpha$ be a topological product and let $z = (z_\alpha) \in \Pi X_\alpha$ be any point. Prove the density of the subset:
$$D = \{(x_\alpha) \in \Pi_\alpha X_\alpha : x_\alpha \neq z_\alpha \text{ for at most finitely many } \alpha \in \mathcal{A}\}.$$

21 Prove that for each pair of points x and y in Cantor's space C, there is a homeomorphism $h : C \to C$ such that $h(x) = y$ and $h(y) = x$.

22 Show that for any spaces X and Y, there exists a space Z such that $X \times Z \cong Y \times Z$.

SECTION 4

1 Let X have the weak topology relative the family $\{X_\alpha : \alpha \in \mathcal{A}\}$ of closed (alternatively, open) subspaces. Prove:
(a) The subspace topology on $X - \cup_\alpha X_\alpha$ is discrete.
(b) Any topology on X that has the family $\{X_\alpha : \alpha \in \mathcal{A}\}$ as subspaces is contained in the weak topology on X.

2 Prove that when the Hawaiian earrings subspace $HE \subset \mathbb{R}^2$ (Example 7.1.8) is retopologized with weak topology relative the family of tangent circles $\{C_k : 1 \leq k\}$, the expanding earrings subspace $EE \subset \mathbb{R}^2$ (Example 7.1.9) is produced.

A family $\{X_\alpha : \alpha \in \mathcal{A}\}$ of subsets of a space X is ***locally finite*** if each point $x \in X$ has a neighborhood $U(x)$ such that $U(x) \cap X_\alpha \neq \emptyset$ for at most finitely many $\alpha \in \mathcal{A}$.

3* Let $\{X_\alpha : \alpha \in \mathcal{A}\}$ be a locally finite family of closed subspaces that cover a topological space X. Prove that X has the weak topology relative $\{X_\alpha : \alpha \in \mathcal{A}\}$.

4 Determine the weak topology on $\mathbb{R}^2$ relative the family of all lines in $\mathbb{R}^2$.

5 Construct the 2-sphere $\mathbb{S}^2$ as an adjunction space $X = A \cup_\phi D$, where $(D, S) = (\mathbb{D}^2, \mathbb{S}^1)$ and A is a single point.

6 Construct the torus $\mathbb{S}^1 \times \mathbb{S}^1$ as an adjunction space $X = A \cup_\phi D$, where $(D, S) = (\mathbb{D}^2, \mathbb{S}^1)$ and A is the subspace $\mathbb{S}^1 \vee \mathbb{S}^1 = (\mathbb{S}^1 \times \{1\}) \cup (\{1\} \times \mathbb{S}^1)$ of $\mathbb{S}^1 \times \mathbb{S}^1$.

7 Use Corollary 4.7 and Exercises 5 and 6 above to construct a continuous surjection $\mathbb{S}^1 \times \mathbb{S}^1 \to \mathbb{S}^2$ that pinches the subspace $\mathbb{S}^1 \vee \mathbb{S}^1$ to a point.

8 Given Hopf's map $p : \mathbb{S}^3 \to \mathbb{S}^2$ (Example 8), form for each $x \in \mathbb{S}^2$ a neighborhood $U(x)$ and a homeomorphism $\lambda : \mathbb{S}^1 \times U \cong p^{-1}(U)$ with $p(\lambda(z, y)) = y$, $\forall\, y \in U$.

6

CONNECTEDNESS

In Chapter 3, a metric space with no nonempty, proper, simultaneously open and closed subset is called connected. Section 1 applies this concept to topological spaces. Section 2 investigates disconnected spaces and decomposes them into their maximal connected subspaces, called *components*. Section 3 develops the stronger concept of path-connectedness.

1 Connected Spaces

CONNECTEDNESS

Two subsets E and F of a topological space X are **separated in X** if

$$E \cap \mathrm{CLS}(F) = \emptyset = \mathrm{CLS}(E) \cap F.$$

A **separation** of a topological space X is a pair $\{E, F\}$ of nonempty separated subsets that have union X.

Notice that a disjoint pair $\{E \neq \emptyset \neq F\}$ with union X is a separation of X if and only if E and F are closed in X if and only if E and F are open in X.

The space X is called **disconnected** if there exists a separation $X = E \cup F$; otherwise, X is called **connected**.

Example 1 Any non-singleton discrete space X is disconnected, since every subset is open and closed in X. Any indiscrete space X is connected, since only $\emptyset$ and X are simultaneously open and closed in X.

Example 2 If $\mathcal{R} \subseteq \mathcal{T}$ are topologies on X, then $(X, \mathcal{R})$ is connected whenever $(X, \mathcal{T})$ is connected. Equivalently, a separation of $(X, \mathcal{R})$ is a separation of $(X, \mathcal{T})$.

Example 3 In a space X with the finite complement topology, the nonempty, proper, open-closed subsets are the finite-cofinite subsets. Thus, X is connected if and only if X is an infinite set or a singleton.

1.1 Lemma *A topological space X is disconnected if and only if X admits a continuous surjection to the two-point discrete space $\{0, 1\}$.*

Proof: Nonempty, disjoint sets $\{E, F\}$ with union $E \cup F = X$ determine a surjection $g: X \to \{0, 1\}$ by $g(E) = 0$ and $g(F) = 1$, and conversely. Also, g is continuous if and only if both E and F are open and closed in X. □

1.2 Theorem *The continuous image of a connected space is connected.*

Proof: Suppose that $f: X \to Y$ is a continuous surjection, with connected domain space X. If $g: Y \to \{0, 1\}$ is continuous, then $g \circ f: X \to \{0, 1\}$ is continuous, hence, constant by Lemma 1.1. Because f is surjective, g must also be constant. This proves that Y is connected by Lemma 1.1. □

1.3 Corollary *Connectedness is a topological invariant.*

> **Example 4** Any continuous real valued function on a denumerable set X with the finite complement topology is constant. X is connected by Example 3. But the connected subspaces of $\mathbb{R}$ are intervals and the countable nonempty ones are singletons.

CONNECTEDNESS OF SUBSPACES

In studying any space, whether connected or not, it is important to determine the connectedness status of its subspaces.

1.4 Lemma *A subspace $X \subseteq Y$ is disconnected if and only if X is the union of two nonempty subsets that are separated in Y.*

Proof: Two subsets E and F, with union X, are separated in X if and only if they are separated in Y, since $\mathrm{CLS}_X(A) = \mathrm{CLS}_Y(A) \cap X$ for $A \subseteq X$. □

> **Example 5** The hyperbola $H = \{<x, y> \in \mathbb{R}^2 : x^2 - y^2 = 1\}$ union its asymptotes $A = \{<x, y> \in \mathbb{R}^2 : y = \pm x\}$ is a disconnected subspace $X = H \cup A$ of $\mathbb{R}^2$. The complement in $\mathbb{R}^2$ of the hyperbola $\{<x, y> \in \mathbb{R}^2 : x^2 - y^2 = \frac{1}{2}\}$ has a separation $E \cup F$, with $H \subseteq E$ and $A \subseteq F$. So $H \cup A$ is a separation of X.

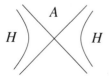
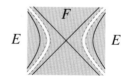

Example 5: Subspace $H \cup A \subseteq \mathbb{R}^2$ Example 5: Separation $E \cup F \supseteq H \cup A$

An extreme possibility is that the only connected subspaces of a space are singleton subspaces. The space is called ***totally disconnected*** in that case.

Example 6 Any discrete space is totally disconnected by Example 1, since every subspace of a discrete space is a discrete space.

Example 7 *A non-discrete totally disconnected space.* A product $\prod_\alpha X_\alpha$ of infinitely many non-singleton spaces $\{X_\alpha : \alpha \in \mathcal{A}\}$ is not discrete (Example **5.3.3**).

But $\prod_\alpha X_\alpha$ is totally disconnected when all the factor spaces X_α are discrete. For, if $x, x' \in \prod_\alpha X_\alpha$ are distinct points, then $x_\sigma \neq x'_\sigma$ in X_σ for some index $\sigma \in \mathcal{A}$. Since $\{x_\sigma\}$ and $X_\sigma - \{x_\sigma\}$ constitute a separation of the discrete space X_σ and the projection p_σ is continuous, then the pre-images $p_\sigma^{-1}(\{x_\sigma\})$ and $p_\sigma^{-1}(X_\sigma - \{x_\sigma\})$, which contain x and x', respectively, constitute a separation of $\prod_\alpha X_\alpha$.

Example 8 *Cantor's space C is totally disconnected.* Any connected subspace of C is a connected subspace of $\mathbb{R}$, hence, an interval, by the Interval Theorem (**3.1.3**). But any interval J contained within $C = \cap_k C_k$ has length zero, hence, is a singleton. The reason is that, for each $k \geq 1$, J is contained in C_k, the disjoint union of the 2^k, even, kth stage ternary intervals of width $1/3^k$. An alternate argument applies the conclusion of Example 7 to the product description $C \cong \prod_k \{0, 2\}_k$ in Example **5.3.10**.

1.5 Theorem *Let $X \subseteq Z \subseteq \mathrm{CLS}_Y(X)$ in the topological space Y. If X is a connected subspace of Y, then Z is a connected subspace of Y.*

Proof: We prove the contrapositive. Let Z be disconnected, so that there exists a continuous surjection $g : Z \to \{0, 1\}$ to two-point discrete space. The continuous restriction $g|_X : X \to \{0, 1\}$ is surjective, as the calculations

$$\{0, 1\} = g(Z) = g(Z \cap \mathrm{CLS}_Y(X)) = g(\mathrm{CLS}_Z(X)) \subseteq \mathrm{CLS}_{\{0, 1\}}(g(X)) = g(X)$$

hold for these reasons: (*i*) the surjectivity of g, (*ii*) the containment relation $Z \subseteq \mathrm{CLS}_Y(X)$, (*iii*) the subspace calculation $Z \cap \mathrm{CLS}_Y(X) = \mathrm{CLS}_Z(X)$, (*iv*) the continuity of g, and (*v*) the discreteness of $\{0, 1\}$.

Thus, the subspace X is disconnected, if the space Z is disconnected. □

Example 9 The polar coordinate spiral Σ_- (Example **2.2.5**), defined by $r = 1 - 1/\theta$ ($\pi \leq \theta$), is the continuous image of the connected ray $[\pi, \infty)$. Hence, Σ_- is a connected subspace of $\mathbb{R}^2$, by Theorem **1.2**. Its closure $\mathrm{CLS}(\Sigma_-) = \Sigma_- \cup \mathbb{S}^1$, called the ***closed polar spiral***, and any subspace $\Sigma_- \subseteq B \subseteq \Sigma_- \cup \mathbb{S}^1$ are connected, by Theorem **1.5**.

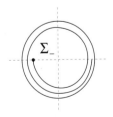
Example 9: Closed polar spiral

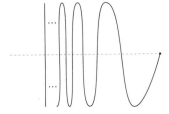
Example 10: Topologist's closed sine curve

Example 10 The topologist's sine curve (Exercise **3.4.14**), being the image of the continuous function $\langle x, \sin 1/x \rangle : (0, 1/\pi] \to \mathbb{R}^2$, is a connected subspace $S \subseteq \mathbb{R}^2$

according to Theorem 1.2. Hence, its closure $\text{CLS}(S) = S \cup (\{0\} \times [-1, +1])$ in $\mathbb{R}^2$, called the *topologist's closed sine curve*, is also connected by Theorem 1.5.

Example 11 Any subspace $Z \subseteq Y$ that contains a dense connected subspace X of Y is connected by Theorem 1.5, because $X \subseteq Z \subseteq Y = \text{CLS}_Y(X)$.

1.6 Theorem *The union $\cup_\alpha X_\alpha$ of any family $\{X_\alpha : \alpha \in \mathcal{A}\}$ of connected subspaces of any space X is connected, if there is some connected subspace $C \subseteq \cup_\alpha X_\alpha$ such that $X_\alpha \cap C \neq \emptyset$ for all $\alpha \in \mathcal{A}$.*

Proof: Let $f : \cup_\alpha X_\alpha \to \{0, 1\}$ be any continuous function to two-point discrete space. Its restrictions to the connected subspaces C and X_α ($\alpha \in \mathcal{A}$) are constant. Hence, all have the same value because $X_\alpha \cap C \neq \emptyset$ for all $\alpha \in \mathcal{A}$. This proves that f is constant. Thus, $\cup_\alpha X_\alpha$ is connected by Lemma 1.1. □

Theorem 1.6 is a generalization of Theorem 3.1.2 which implies the connectedness of convex and star-like metric subspaces of $\mathbb{R}^n$.

Example 12 The complement $\mathbb{R}^n - \{p\}$ of a point $p \in \mathbb{R}^n$ is not convex. But, when $n > 1$, the complement $\mathbb{R}^n - \{p\}$ is the union of line segments in $\mathbb{R}^n$ that intersect a line off p, and so it is a connected subspace of $\mathbb{R}^n$ by Theorem 1.6. Thus, $\mathbb{R} - \{0\}$ and $\mathbb{R}^n - \{p\}$ ($n > 1$) are not homeomorphic, and so neither are $\mathbb{R}$ and $\mathbb{R}^n$ ($n > 1$).

Example 13 The space $\mathbb{Q} \times \mathbb{Q}$ of pairs of rational numbers and the space $\mathbb{Q}^c \times \mathbb{Q}^c$ of pairs of irrational numbers are totally disconnected subspaces of the real plane. Yet their union, $Z = (\mathbb{Q} \times \mathbb{Q}) \cup (\mathbb{Q}^c \times \mathbb{Q}^c)$, is a connected subspace of $\mathbb{R}^2$ by Example 11 and Theorem 1.6. For Z contains all lines with non-zero rational slope through points of $\mathbb{Q} \times \mathbb{Q}$, and the union of all such lines is a dense connected subspace of $\mathbb{R} \times \mathbb{R}$.

CONNECTEDNESS OF PRODUCT SPACES

The next two results investigate the connectedness of topological products.

1.7 Lemma *If $X_1, \ldots, X_n$ are connected spaces, their topological product $X_1 \times \ldots \times X_n$ is connected.*

Proof: The product $X_1 \times X_2$ is the union of the connected slices $X_1 \times \{x_2\}$ ($x_2 \in X_2$), each of which has a nonempty intersection with each connected slice $\{x_1\} \times X_2$. Thus, $X_1 \times X_2$ is connected by Theorem 1.6. The lemma follows inductively, since $X_1 \times \ldots \times X_n \cong (X_1 \times \ldots \times X_{n-1}) \times X_n$. □

1.8 Theorem *A topological product $\prod_\alpha X_\alpha$ is connected if and only if all the factor spaces X_α ($\alpha \in \mathcal{A}$) are connected.*

Proof: Since each projection $p_\sigma : \prod_\alpha X_\alpha \to X_\sigma$ ($\sigma \in \mathcal{A}$) is a continuous surjection, each factor space X_σ is connected whenever $\prod_\alpha X_\alpha$ is connected.

Conversely, suppose that all the factor spaces X_α ($\alpha \in \mathcal{A}$) are connected. Select $x^* \in \prod_\alpha X_\alpha$. For each finite subset $\{\alpha_1, \ldots, \alpha_n\} \subseteq \mathcal{A}$, the subspace

$$\{x \in \prod_\alpha X_\alpha : x_\alpha = x^*_\alpha, \; \forall \, \alpha \notin \{\alpha_1, \ldots, \alpha_n\}\}$$

is homeomorphic to the finite product $X_{\alpha_1} \times \ldots \times X_{\alpha_n}$, hence, is connected by Lemma 1.7. The subspace D of all points $x \in \prod_\alpha X_\alpha$ that differ from x^* in at most finitely many coordinates $\alpha \in \mathcal{A}$ is the union of these connected subspaces of $\prod_\alpha X_\alpha$, all of which contain x^*. Therefore, D is connected by Theorem 1.6. Also, D is a dense subset of $\prod_\alpha X_\alpha$, since each member of the base for $\prod_\alpha X_\alpha$ restricts at most finitely many coordinates of its members and, hence, contains a member of D. Because $\prod_\alpha X_\alpha$ contains the dense connected subspace D, this product space is connected by Example 11. □

CUT POINTS

Let X be a connected space.

A subset $A \subseteq X$ is called a ***cut set*** if its complement $X - A$ is a disconnected subspace of X.

A point $x \in X$ is a ***cut point*** if its complement $X - \{x\}$ is a disconnected subspace of X. Otherwise, x is a ***non-cut point*** of X. A cut point x is a ***dispersion point*** if $X - \{x\}$ is a totally disconnected subspace of X.

Example 14 An infinite set X with the finite complement topology consists of non-cut points $x \in X$, because each subspace $X - \{x\}$ has the finite complement topology and so is connected by Example 3.

Example 15 The point $p \in X$ is a dispersion point of the space X with the particular-point topology $\mathcal{T}(p)$. Indeed, X is connected, while the subspace $X - \{p\} \subseteq X$ is discrete, hence, totally disconnected.

1.9 Theorem *Any homeomorphism carries cut points to cut points and non-cut points to non-cut points. Therefore,*
(a) the existence and number of cut points or non-cut points are topological invariants, and
(b) homeomorphic spaces have homeomorphic subspaces of cut points and homeomorphic subspaces of non-cut points.

Proof: Any homeomorphism $f : X \cong Y$ carries a separation $X - \{x\} = E \cup F$ to a separation $Y - \{f(x)\} = G \cup H$. Hence, f preserves both cut points and non-cut points. □

Example 16 The endpoints of the closed interval $\mathbb{I}$ are the only non-cut points of the interval by the Interval Theorem (**3.1.3**). Each point of the closed unit square $\mathbb{I}^2$ is a non-cut point of $\mathbb{I}^2$ by Theorem 1.6. Therefore, $\mathbb{I}$ and $\mathbb{I}^2$ are not homeomorphic, despite Cantor's set equivalence $\mathbb{I} \sim \mathbb{I}^2$ and Peano's continuous surjection $P : \mathbb{I} \to \mathbb{I}^2$.

Example 17 For each $n \geq 2$, let X_n be the union of the line segments in $\mathbb{R}^2$ joining the origin $\mathbf{O} = \,<0, 0>$ to the nth roots of unity $e^{2\pi i m/n} = \,<\cos 2\pi m/n, \sin 2\pi m/n>$ ($1 \leq m \leq n$) on the unit 1-sphere $\mathbb{S}^1$. Since X_n has exactly n non-cut points, these spaces represent different homeomorphism types: $X_n \cong X_m$ if and only if $n = m$.

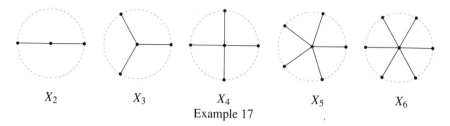

X_2 X_3 X_4 X_5 X_6

Example 17

Example 18 The wire handcuff (below left) has an open interval of cut points, and the wire eight (below right) has a single cut point. So they are not homeomorphic.

Example 18: Wire handcuff   Example 18: Wire eight

Example 19 Let $\mathbb{M} = \{0, 1, 1/2, \ldots, 1/k, \ldots\}$. The **harmonic broom** (below left) is the subspace $B \subset \mathbb{R}^2$ of all line segments joining $<0, 1> \in \mathbb{R}^2$ to the points in $\mathbb{M} \times \{0\}$. The **harmonic rake** (below right) is the subspace $R \subset \mathbb{R}^2$ of the upper semi-circles joining $<0, 0> \in \mathbb{R}^2$ to the points in $\mathbb{M} \times \{0\}$.

The non-cut points in B form the non-discrete space $(\{0\} \times [0, 1)) \cup (\mathbb{M} \times \{0\})$, while the non-cut points in R form the discrete space $(\mathbb{M} - \{0\}) \times \{0\}$. Therefore, B and R are not homeomorphic.

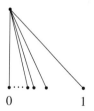
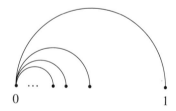

Example 19: Harmonic broom   Example 19: Harmonic rake

2 Components

Each topological space X can be viewed as the union of connected subspaces, namely, its singletons. A more interesting decomposition of X involves its maximal connected subspaces, i.e., connected subspaces that are contained in no larger connected subspace of X. They are the equivalence classes under the following relation on X.

CONNECTIVITY RELATION

The *connectivity relation* on the space X is defined as follows: two points $x, x' \in X$ are related, written $x \sim x'$, provided that there is a connected subspace $C \subseteq X$ such that $\{x, x'\} \subseteq C$.

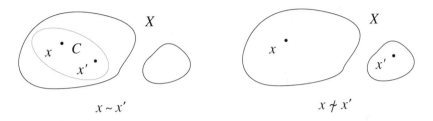

2.1 Lemma   *The connectivity relation is an equivalence relation on X.*

Proof: *Reflexivity:* $\forall\, x \in X,\, x \sim x$. The relation $x \sim x$ holds, as the subspace $C = \{x\}$ is connected.

Symmetry: $\forall\, x, x' \in X,\, x \sim x' \Leftrightarrow x' \sim x$. Both relations mean that there is a connected subspace $C \supseteq \{x, x'\}$.

Transitivity: $\forall\, x, x', x'' \in X,\, (x \sim x') \wedge (x' \sim x'') \Rightarrow (x \sim x'')$. The union of a connected subspace $C \supseteq \{x, x'\}$ and a connected subspace $C' \supseteq \{x', x''\}$ is a connected subspace $C \cup C' \supseteq \{x, x''\}$, by Theorem 1.6. □

COMPONENTS

The equivalence classes of X relative to the connectivity relation $\sim$ are called the *components* of X. For each point $x \in X$, let $C(x)$ denote the component, $\{x' \in X : x \sim x'\}$, containing x.

2.2 Theorem   *The components of a space X have these properties:*
(a)   *A connected subspace $C \subseteq X$ is a component if and only if C is **maximal** in this sense: if a connected subspace $A \subseteq X$ intersects C, then $A \subseteq C$.*
(b)   *The components of X are pairwise disjoint and have union X.*
(c)   *Each component $C(x)$ of X is a closed subset of X.*

Proof: (a) The component $C(x)$, being an equivalence class of the connectivity relation, is the union of all connected subspaces of X that contain x. It follows from Theorem 1.6 that $C(x)$ is itself a connected subspace of X containing x and that $C(x)$ contains any connected subspace of X containing one of its members. Conversely, a connected subspace C contains only related points, and one with the maximality property contains all points related to its members. So such a maximal connected subspace C is a component.

(b) Because the components are the equivalence classes of the connectivity relation on X, they form a partition of X by (**1.5.1**).

(c) By Theorem 1.5, the closure of a connected subspace is connected. So the maximality property (a) implies that a component is closed. □

Example 1 A space X is totally disconnected if and only if $C(x) = x$ for all $x \in X$.

Example 2 Let $Y_\pm \subseteq \mathbb{R}^2$ be the connected subspaces consisting of the straight line segments joining consecutive points of the sequences:

$$<\pm 1, 0>, <\pm \tfrac{1}{2}, 1>, <\pm \tfrac{1}{3}, 0>, \ldots, <\pm \tfrac{1}{2k}, 1>, <\pm \tfrac{1}{2k+1}, 0>, \ldots.$$

The subspace

$$Y = Y_- \cup \{O\} \cup Y_+$$

(below left) has a single component by Theorem 1.6, because the closures,

$$\mathrm{CLS}_Y(Y_-) = Y_- \cup \{O\} \quad \text{and} \quad \mathrm{CLS}_Y(Y_+) = Y_+ \cup \{O\},$$

are connected by Theorem 1.5, and they share the origin O.

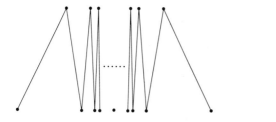
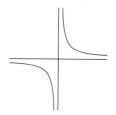

Example 2: $Y = Y_- \cup \{O\} \cup Y_+$ Example 3: $X = H \cup A$

Example 3 The subspace $X = H \cup A \subseteq \mathbb{R}^2$ (above right) consists of the Cartesian axes A and the hyperbola H with two branches. These three pieces of X are closed connected subspaces with union X. Hence, they are the maximal connected subspaces of X, i.e., its components.

Example 4 *Separation by components.* If a space X has just finitely many components C_i ($1 \leq i \leq n$), then the subspaces

$$E = \cup \{C_i : 1 \leq i \leq k\} \quad \text{and} \quad F = \cup \{C_i : k < i \leq n\}$$

are complementary closed subsets of X. Hence, $E \cup F$ is a separation of X. But, in general, not every partition of the components of a space gives a separation of that space. For example, $E = \{0\}$ and $F = \{1, 1/2, \ldots, 1/n, \ldots\}$ do not constitute a separation of their union $X = \{1, 1/2, \ldots, 1/n, \ldots, 0\}$. However, there is some separation of this space X by complementary unions of components.

Example 5 Since the connected subspaces of $\mathbb{R}$ are intervals, then the components of any subspace $X \subseteq \mathbb{R}$ are maximal intervals contained in X. When X is an open subset of $\mathbb{R}$, the components of X are maximal open intervals contained in X. So any open subset of $\mathbb{R}$ is the disjoint union of open intervals. There are only countably many components of an open subset of $\mathbb{R}$ because the components can be enumerated by a selection of a rational number from each of them.

COMPONENTS AND CONTINUOUS FUNCTIONS

Theorem 1.2 states that the continuous image of a connected space is connected. Thus, under a continuous function, the image of a connected subspace of the domain space is a connected subspace of the codomain space. This result has the following consequence.

2.3   Theorem   *Any continuous function $f : X \to Y$ carries each components of X into some component of Y: for all $x \in X$, $f(C(x)) \subseteq C(f(x))$.*

Proof: By Theorem 1.2, the continuous image of a connected subspace $C \supseteq \{x, x'\}$ of X is a connected subspace $f(C) \supseteq \{f(x), f(x')\}$ of Y. So $x \sim x'$ in X implies that $f(x) \sim f(x')$ in Y; equivalently, $f(C(x)) \subseteq C(f(x))$. □

2.4   Corollary   *In the topological product $\prod_\alpha X_\alpha$, the component $C(x)$ of the point x is the topological product $\prod_\alpha C(x_\alpha)$, where $C(x_\alpha)$ is the component of the coordinate $x_\alpha \in X_\alpha$, for all $\alpha \in \mathcal{A}$.*

Proof: The subspace $\prod_\alpha C(x_\alpha) \subseteq \prod_\alpha X_\alpha$ is the topological product of the connected subspaces $C(x_\alpha) \subseteq X_\alpha$, by Corollary 5.3.4. Thus, by Theorem 1.8, the topological product $\prod_\alpha C(x_\alpha)$ is a connected subspace of $\prod_\alpha X_\alpha$. Because $\prod_\alpha C(x_\alpha)$ contains x, then $\prod_\alpha C(x_\alpha) \subseteq C(x)$, by the maximality of $C(x)$.
   By Theorem 2.3, we have $p_\sigma(C(x)) \subseteq C(x_\sigma)$, hence, $C(x) \subseteq p_\sigma^{-1}(C(x_\sigma))$ for each projection $p_\sigma : \prod_\alpha X_\alpha \to X_\sigma$ ($\sigma \in \mathcal{A}$). So we also have

$$C(x) \subseteq \cap_\alpha \, p_\alpha^{-1}(C(x_\alpha)) = \prod_\alpha C(x_\alpha).$$

This proves that $C(x) = \prod_\alpha C(x_\alpha)$. □

2.5   Corollary   *A homeomorphism induces a bijective correspondence between the components of the domain and codomain spaces, and it restricts to homeomorphisms of corresponding components.*

Proof: Let $f : X \to Y$ and $g : Y \to X$ be continuous inverses. Theorem 2.3 provides continuous restrictions

$$f : C(x) \to C(f(x)) \quad \text{and} \quad g : C(y) \to C(g(y)),$$

which are inverses whenever $x = g(y)$ and $y = f(x)$. □

Example 6   *There is no topological Schröder-Bernstein Theorem* (1.4.4). *Even the existence of continuous bijections $X \to Y$ and $Y \to X$ does not imply that $X \cong Y$.*
   Consider the two subspaces of the real line $\mathbb{R}$:

$$X = [0, 1) \cup \{2\} \cup (3, 4) \cup \{5\} \cup (6, 7) \cup \{8\} \cup \ldots$$

and

$$Y = [0, 1] \cup \{2\} \cup (3, 4) \cup \{5\} \cup (6, 7) \cup \{8\} \cup \ldots .$$

There exists a continuous bijection $f : X \to Y$

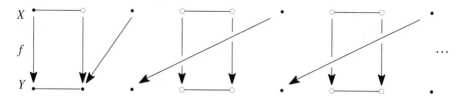

and there exists a continuous bijection $g : Y \to X$

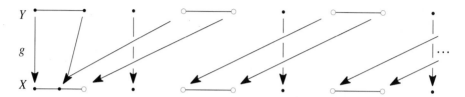

Yet $X \not\cong Y$. No component of X is homeomorphic to the component $[0, 1]$ of Y.

Example 7 The unions $X = \mathbb{S}^1 \cup ([0, 1] \times \{0\})$ and $Y = \mathbb{S}^1 \cup ([1, 2] \times \{0\})$ are homeomorphic subspaces of $\mathbb{R}^2$, and so are their complements $\mathbb{R}^2 - X$ and $\mathbb{R}^2 - Y$. But there is no homeomorphism $h : \mathbb{R}^2 \to \mathbb{R}^2$ that carries X to Y.
Otherwise, h would restrict to a homeomorphism $h : \mathbb{R}^2 - X \to \mathbb{R}^2 - Y$ pairing the components A_0 and A_∞ of $\mathbb{R}^2 - X$ with the components B_0 and B_∞ of $\mathbb{R}^2 - Y$. Since the bounded components A_0 and B_0 have sequentially compact closure, $\mathbb{D}^2$, in $\mathbb{R}^2$ and the unbounded components A_∞ and B_∞ do not, it would have to be that $h(A_0) = B_0$. Then this contradiction would follow:

$$Y = h(X) \subseteq h(\mathrm{CLS}(A_0)) \subseteq \mathrm{CLS}(h(A_0)) = \mathrm{CLS}(B_0) = \mathbb{D}^2.$$

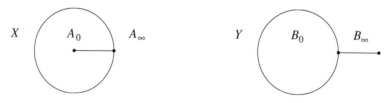

Example 7

Local Connectedness

Let X be a topological space and let $x \in X$. X is called ***locally connected at x*** if each neighborhood of x contains a connected neighborhood of x, and X is called ***locally connected*** if it is locally connected at each of its points $x \in X$.

Example 8 Any non-trivial discrete space is locally connected but not connected.

Example 9 The topologist's closed sine curve is connected (Example 1.10). Yet it is locally disconnected at each point $<0, y> (-1 \leq y \leq 1)$, because an open ball

$B = B_X(<0, y>, r)$ of radius $0 < r < 2$ contains no connected neighborhood of $<0, y>$.

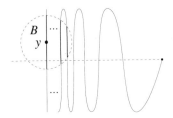 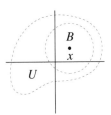

Example 9: Locally disconnected Example 10: Locally connected

Example 10 $\mathbb{R}^n$ is locally connected as each neighborhood $U(x)$ contains a convex, hence, connected open n-ball $B(x, r)$ for some $r > 0$.

We leave as an exercise the invariance of local connectedness.

2.6 Theorem *Local connectedness is a topological invariant, and any open subspace of a locally connected space is locally connected.* □

Example 11 The saw space $S \subset \mathbb{R}^2$ and the tapered saw space $T \subset \mathbb{R}^2$ (Examples 3.4.11-12) are connected spaces that are not homeomorphic. The saw space S fails to be locally connected at O, while the tapered saw space $T \cong [0, 1]$ is locally connected.

Most connectedness properties *do not* carry over to local connectedness:

Example 12 *The continuous image of a locally connected space need not be locally connected.* The identity function $1_X : (X, \mathcal{P}(X)) \to (X, \mathcal{T})$ from the discrete topology to an arbitrary topology $\mathcal{T}$ on X is continuous. The discrete space $(X, \mathcal{P}(X))$ is locally connected, but its continuous image $(X, \mathcal{T})$ need not be locally connected.

Example 13 *The closure of a locally connected subspace need not be locally connected.* The subspace $A = \{1, 1/2, \ldots, 1/k, \ldots\} \subset \mathbb{R}$ is discrete, hence, locally connected; but the closure of A in $\mathbb{R}$, $A \cup \{0\}$, is not locally connected at 0.

Example 14 *There exists a connected space that is locally connected at none of its points.* Let $0 < r < 1$ be an irrational number. The family of parallel lines

$$y = rx + kr + m \qquad (k, m \in \mathbb{Z})$$

with slope r constitute a dense subspace of the real plane $\mathbb{R} \times \mathbb{R}$, which is locally disconnected at each of its points. The function $e \times e : \mathbb{R} \times \mathbb{R} \to \mathbb{S}^1 \times \mathbb{S}^1$, wrapping the plane around the torus, carries the family of lines onto a subspace L of the torus $\mathbb{S}^1 \times \mathbb{S}^1$, which is locally disconnected at each of its points. But, because $e \times e$ is continuous and the entire subspace L is the image of each individual line in the family, this subspace $L \subset \mathbb{S}^1 \times \mathbb{S}^1$ is connected.

2.7 Theorem *A topological space X is locally connected if and only if each component of every open subspace $U \subseteq X$ is open in X.*

Proof: ($\Leftarrow$) The open components of open subsets provide the desired arbitrarily small connected neighborhoods of points.

($\Rightarrow$) Let C be a component of an open subspace $U \subseteq X$. For each $x \in C \subseteq X$, there exists a connected neighborhood $V(x) \subseteq U$ in X, by the local connectedness of X. By the maximality property, the component C of U contains the connected neighborhood $V(x)$. Thus, C is open. $\square$

Example 15 The harmonic broom space B (Example 1.19) is not locally connected at the points $<0, y>$ ($0 \leq y < 1$). By Theorem 2.7, the local disconnectedness of B is related to the existence of the non-open component $\{<0, y> : 0 \leq y < 1\}$ of the open subspace $B - \{<0, 1>\}$.

2.8 Corollary *In a locally connected space, components are open, as well as closed.*

Example 16 The ***harmonic target*** $HT \subset \mathbb{R}^2$ is the subspace consisting of the circles, centered on the origin O, with radii $0, 1, 1/2, \ldots, 1/k, \ldots$. The space HT is not locally connected, because O is a component that is not open in HT.

Example 17 *The topological product $\prod_\alpha X_\alpha$ of locally connected spaces X_α ($\alpha \in \mathcal{A}$) need not be locally connected.* This is illustrated by any infinite family $\{X_\alpha : \alpha \in \mathcal{A}\}$ of discrete, hence, locally connected spaces. The topological product $\prod_\alpha X_\alpha$ is totally disconnected (Example 1.7). So its components are singletons, which are not open by Example 5.3.3. Therefore, $\prod_\alpha X_\alpha$ is not locally connected according to Theorem 2.7.

3 Path-Connectedness

Since the closed unit interval $\mathbb{I} = [0, 1]$ is such a wholesome topological space, it is used as a standard by which other spaces are measured. Special topological features of a space X are detected when one tries to place a continuous image of $\mathbb{I}$ between given points of X, in the following sense.

PATHS AND PATH-CONNECTEDNESS

A continuous function $f : \mathbb{I} \to X$ is called a ***path in*** X. It is customary to view the parameter $t \in I$ as *time*, and to consider the path $f : \mathbb{I} \to X$ as traveling from ***initial point*** $f(0)$ to ***terminal point*** $f(1)$, with $f(t)$ as its position at time t. We say points $x_0, x_1 \in X$ are ***joined*** by the path f if $f(0) = x_0$ and $f(1) = x_1$, and we use the notation

$$f : (\mathbb{I}, 0, 1) \to (X, x_0, x_1)$$

to indicate the initial point x_0 and terminal point x_1 of the path f.

It can be challenging to determine whether a given pair of points of a space X are joined by a path in X. A space X is called ***path-connected*** if each pair of points of X are joined by some path in X.

Example 1 *Any convex subspace $X \subseteq \mathbb{R}^n$ is path-connected.* Given two points $a = (a_1, \ldots, a_n)$ and $b = (b_1, \ldots, b_n)$, the continuous function $f : \mathbb{I} \to \mathbb{R}^n$, defined by

$$f(t) = (1-t)(a_1, \ldots, a_n) + t(b_1, \ldots, b_n),$$

traces out the line segment between a and b. When a and b belong to the convex subset X, the function f is a path in X that joins a to b.

Example 2 *The subspace $Z = (\mathbb{Q} \times \mathbb{Q}) \cup (\mathbb{Q}^c \times \mathbb{Q}^c)$ of $\mathbb{R}^2$ is path-connected.* Z contains each non-vertical and non-horizontal line with rational slope through a point of $\mathbb{Q} \times \mathbb{Q}$. Thus, the origin O is joined in Z to $<x, y> \in \mathbb{Q} \times \mathbb{Q}$ by a path consisting of one or two line segments (below left). Also, the origin O is joined in Z to each point $<x, y> \in \mathbb{Q}^c \times \mathbb{Q}^c$ by a path consisting of infinitely many line segments in Z joining consecutive entries in a sequence

$$O = <0, 0>, <r_1, s_1>, <r_2, s_2>, \ldots, <r_n, s_n>, \ldots,$$

where $\{r_n\}$ and $\{s_n\}$ are strictly increasing sequences of rational numbers with limits x and y, respectively (below right).

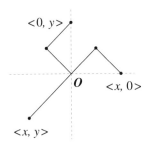 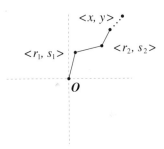

Example 2: $<x, y> \in \mathbb{Q} \times \mathbb{Q}$ Example 2: $<x, y> \in \mathbb{Q}^c \times \mathbb{Q}^c$

Example 3 *The Sierpinski topology $\{\emptyset, \{0\}, \{0, 1\}\}$ and the indiscrete topology $\{\emptyset, \{0, 1\}\}$ on $\{0, 1\}$ are path-connected.* In both cases, the characteristic function $\chi_{\{1\}} : \mathbb{I} \to \{0, 1\}$ ($\chi_{\{1\}}(t) = 1 \Leftrightarrow t = 1$) is continuous. The discrete topology on $\{0, 1\}$ is not path-connected; it admits no continuous surjection from the connected space $\mathbb{I}$.

Example 4 *The union $X \cup \{p\}$ of any space X and a point p is path-connected when it is topologized so that its nonempty open subsets are $U \cup \{p\}$, where U is open in X.* Any point $x \in X$ is joined to p by the path $f : \mathbb{I} \to X \cup \{p\}$ defined by

$$f(0) = x \quad \text{and} \quad f(t) = p \ (0 \ne t).$$

Any two points $x, y \in X$ are joined by the path $g : \mathbb{I} \to X \cup \{p\}$ defined by

$$g(0) = x, \quad g(t) = p \ (0 < t < 1), \quad \text{and} \quad g(1) = y.$$

3.1 Theorem   *A path-connected space is connected.*

Proof: If X is path-connected, it is the union of images of paths that join its points $x \in X$ to a selected point $x' \in X$. By Theorem 1.2, each path image is a connected subspace of X. Thus, X is connected by Theorem 1.6. $\square$

Example 5 *A connected space need not be path-connected.* The saw space S (Example **3.3.11**) is connected, but not path-connected. Were

$$f : I \to S \subset \mathbb{I}^2, f(t) = <f_1(t), f_2(t)>$$

a path in S joining $<0, 0>$ to $<1, 1>$, the second coordinate function would be a path $f_2 : \mathbb{I} \to \mathbb{I}$ joining 0 to 1. Then, by iterated applications of the Intermediate Value Theorem (**3.1.4**), there would be real numbers

$$0 \leq \ldots \leq t_n \leq \ldots \leq t_2 \leq t_1 = 1$$

such that $f_2(t_n) = 1/n$ for all $n \geq 1$. But there is a unique point $<x, 1/n> \in S$, for each $n \geq 1$, namely, $<0, 1/n>$ when n is even, and $<1, 1/n>$ when n is odd. Thus the first coordinate function would be a path $f_1 : (\mathbb{I}, 0, 1) \to (\mathbb{I}, 0, 1)$ with values $f_1(t_n) = 0$ for all even n, and $f_1(t_n) = 1$ for all odd n. This contradicts the continuity of f_1 at the limit $t^* = glb\{t_n\} \in \mathbb{I}$ of the strictly decreasing sequence $\{t_n\}$.

3.2 Theorem *The continuous image of a path-connected space is path-connected.*

Proof: If $f : \mathbb{I} \to X$ is a path that joins the points x_0 and x_1 in X and if $F : X \to Y$ is any continuous function, then $F \circ f : \mathbb{I} \to Y$ is a path in the image $F(X) \subseteq Y$ that joins the image points $F(x_0)$ and $F(x_1)$ in Y. Therefore, when X is a path-connected space, $F(X)$ is a path-connected subspace of Y. □

3.3 Corollary *Path-connectedness is a topological invariant.*

PATH-CONNECTIVITY RELATION

For each point $x \in X$, the **constant path** $x^* : \mathbb{I} \to X$ is defined by $x^*(\mathbb{I}) = x$.

For each path $f : \mathbb{I} \to X$, the **reverse path** $\bar{f} : \mathbb{I} \to X$ is defined by $\bar{f}(t) = f(1 - t)$ for all $t \in \mathbb{I}$.

For two paths $f : \mathbb{I} \to X$ and $g : \mathbb{I} \to X$ with $f(1) = g(0)$, the **product path** $k = f \cdot g : \mathbb{I} \to X$ is defined by $k(t) = f(2t)$ for $0 \leq t \leq \frac{1}{2}$ and $k(t) = g(2t - 1)$ for $\frac{1}{2} \leq t \leq 1$. The product path $k = f \cdot g$ is well-defined as $f(1) = k(\frac{1}{2}) = g(0)$; k is continuous on $\mathbb{I}$ by the Gluing Theorem (**4.4.3**), because it is continuous on the closed subsets $[0, \frac{1}{2}]$ and $[\frac{1}{2}, 1]$.

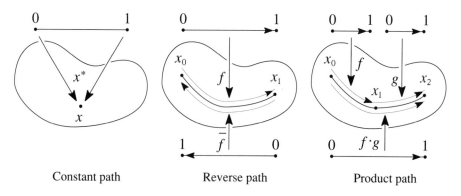

Constant path   Reverse path   Product path

If there is a path $f : (\mathbb{I}, 0, 1) \to (X, x_0, x_1)$, then we write $f : x_0 \sim x_1$, or simply $x_0 \sim x_1$. The relation $\sim$ is called the ***path-connectivity relation*** on X.

3.4 Lemma *Path-connectivity is an equivalence relation on X.*

Proof: *Reflexivity:* if $x \in X$, then $x^* : x \sim x$.
Symmetry: if $f : x_0 \sim x_1$, then $\bar{f} : x_1 \sim x_0$.
Transitivity: if $f : x_0 \sim x_1$ and $g : x_1 \sim x_2$, then $f \cdot g : x_0 \sim x_2$. □

The equivalence classes of a topological space X relative to the path-connectivity relation $\sim$ are called the ***path-components*** of X. For each point $x \in X$, let $P(x)$ denote the path-component, $\{x' \in X : x \sim x'\}$, containing x. The ***set of all path-components*** of the space X is denoted by $\pi_0(X)$.

3.5 Theorem *The path-components of a space X have these properties:*
*(a) A path-connected subspace $P \subseteq X$ is a path-component of X if and only if it is **maximal** in this sense: if a path-connected subspace $A \subseteq X$ intersects P, then $A \subseteq P$.*
(b) The path-components of X are pairwise disjoint and have union X.

Proof: (a) A subspace $P \subseteq X$ is a maximal path-connected subspace if and only if the path-connectivity relation holds among all of its points and it contains any point that is path-connected in X to one of its members. These conditions describe the equivalence classes under the path-connectivity relation, i.e., the path-components.
 (b) These are the properties of the equivalence classes of any equivalence relation, according to the Equivalence Class Properties (**1.5.2**). □

Example 6 *A path-component need not be a closed subset, and the closure of a path-connected subspace need not be path-connected.* The saw space S (Example 5) has two path-components, one of which consists of accumulation points of the other.

A space is ***totally path-disconnected*** if its path-components are singletons. Since path-connected spaces are connected, then each path-component is contained in a component. It follows that a totally disconnected space is totally path-disconnected. The converse is false:

Example 7 *There exists a connected, totally path-disconnected space.* Any infinite set X with the finite complement topology is connected (Example 1.3).
   When X is denumerable, it is totally path-disconnected. Any non-constant path $f : \mathbb{I} \to X$ would provide the denumerable collection $\{f^{-1}(x) : x \in f(\mathbb{I})\}$ of disjoint nonempty closed sets with union $\mathbb{I}$, which would contradict Sierpinski's Theorem (Exercise **7.5.8**).
   When X is set equivalent to $\mathbb{I}$, it is path-connected. For any pair of distinct points $a, b \in X$, there is a set bijection $f : (\mathbb{I}, 0, 1) \to (X, a, b)$. Since proper closed subsets

of X are finite, their pre-images under the bijection f are closed in $\mathbb{I}$. Thus, f is continuous; it is a path that joins a and b.

PATH-COMPONENTS AND CONTINUOUS FUNCTIONS

Continuous functions carry path-components into path-components.

3.6 Lemma *Any continuous function $F : X \to Y$ carries each path-component into some path-component: for all $x \in X$, $F(P(x)) \subseteq P(F(x))$.*

Proof: If the path $f : \mathbb{I} \to X$ joins x to x' in X, then the path $F \circ f : \mathbb{I} \to Y$ joins $F(x)$ to $F(x')$ in Y. Therefore, F carries the path-component $P(x)$ in X containing x into the path-component $P(F(x))$ in Y containing $F(x)$. □

3.7 Theorem *The path-component sets $\pi_0(X)$ have these properties:*
(a) Each continuous function $F : X \to Y$ induces a set function

$$F_\# : \pi_0(X) \to \pi_0(Y),$$

defined by $F_\#(P(x)) = P(F(x))$.
(b) *For continuous functions $F : X \to Y$ and $F : Y \to Z$, the composite function $G \circ F : X \to Z$ induces the composite set function:*

$$(G \circ F)_\# = G_\# \circ F_\# : \pi_0(X) \to \pi_0(Y) \to \pi_0(Z).$$

(c) *The identity function $1_X : X \to X$ induces the identity set function:*

$$(1_X)_\# = 1_{\pi_0(X)} : \pi_0(X) \to \pi_0(X).$$

Proof: (a) By Lemma 3.6, each function $F_\#$ is well-defined. (b) and (c) are immediate consequences of the definition of the induced set functions $F_\#$. □

3.8 Corollary *A homeomorphism induces a bijective correspondence between the path-components of the domain and codomain spaces, and it restricts to homeomorphisms of corresponding path-components.*

Proof: Let $F : X \to Y$ and $G : Y \to X$ be continuous inverses. Then the induced functions $F_\# : \pi_0(X) \to \pi_0(Y)$ and $G_\# : \pi_0(Y) \to \pi_0(X)$ are inverses by the parts (b) and (c) of Theorem 3.7:

$$G_\# \circ F_\# = (G \circ F)_\# = (1_X)_\# = 1_{\pi_0(X)} : \pi_0(X) \to \pi_0(X)$$

$$F_\# \circ G_\# = (F \circ G)_\# = (1_Y)_\# = 1_{\pi_0(Y)} : \pi_0(Y) \to \pi_0(Y).$$

Moreover, Lemma 3.6 provides the continuous restrictions

$$F : P(x) \to P(F(x)) \text{ and } G : P(y) \to P(G(y)),$$

which are inverses whenever $x = G(y)$ and $y = F(x)$. □

LOCAL PATH-CONNECTEDNESS

There is a local variation of path-connectedness that is strong enough to cause components and path-components to coincide.

3.9 Lemma *The following are equivalent properties for a space X:*
(a) Each path-component is open; hence, each is also closed.
(b) Each point has a path-connected neighborhood.

Proof: ($a \Rightarrow b$) If the path-components of X are open, they serve as path-connected neighborhoods of the points of X.

($b \Rightarrow a$) If each point of X has a path-connected neighborhood, then, by the maximality property (Theorem 3.5(a)), each path-component contains a path-connected neighborhood of each of its points, and so is open. □

3.10 Theorem *A space is path-connected if and only if it is connected and each point has a path-connected neighborhood.*

Proof: A path-connected space is connected and serves as a path-connected neighborhood of each of its points. Conversely, if X is connected and each point has a path-connected neighborhood, then, by Lemma 3.9, each path-component is open and closed in X. Thus, any component equals the connected space X. □

Thus, condition (b) of Lemma 3.9 is sufficient to cause connectedness and path-connectedness to coincide.

> **Example 8** An ***n*-dimensional manifold** is a second countable space in which each point has an open neighborhood homeomorphic to real n-space $\mathbb{R}^n$. By Theorem 3.10, a manifold is path-connected if and only if it is connected. For this reason, any open subspace of $\mathbb{R}^n$ or $\mathbb{S}^n$ is path-connected if and only if it is connected.

> **Example 9** Both the topologist's closed sine curve (Example 2.9) and the saw space (Example 5) show that path-connectedness and connectedness are not equivalent concepts for non-open subspaces of $\mathbb{R}^n$.

The following definition strengthens condition (b) of Lemma 3.9; it is enough to cause components and path-components to coincide in a space.

A space X is **locally path-connected** if each neighborhood of each point contains a path-connected neighborhood.

3.11 Theorem *In any locally path-connected space, components coincide with path-components and they are open as well as closed subsets.*

Proof: In any space X, $P(x) \subseteq C(x)$ for all $x \in X$, since the path-component

$P(x)$ is connected and the component $C(x)$ is a maximal connected subspace. Thus, each component is partitioned by the path-components that it contains. If X is locally path-connected, then its path-components are open by Lemma 3.9. It follows that each component $C \subseteq X$ is a path-component; otherwise, C would be separated by its distinct path-components. □

Exercises

SECTION 1

1. (a) Show that $\mathbb{Q} \times \mathbb{Q}$ and $\mathbb{Q}^c \times \mathbb{Q}^c$ are totally disconnected subspaces of $\mathbb{R}^2$, but that their union $(\mathbb{Q} \times \mathbb{Q}) \cup (\mathbb{Q}^c \times \mathbb{Q}^c)$ is a connected subspace of $\mathbb{R}^2$.
   (b) Prove that the complement $(\mathbb{Q} \times \mathbb{Q}^c) \cup (\mathbb{Q}^c \times \mathbb{Q})$ of $(\mathbb{Q} \times \mathbb{Q}) \cup (\mathbb{Q}^c \times \mathbb{Q}^c)$ in $\mathbb{R}^2$ is a totally disconnected subspace of $\mathbb{R}^2$.

2. Prove that an uncountable set with the countable complement topology is a connected space. Characterize its connected subspaces and its disconnected subspaces.

3. (a) Find a smallest connected subspace of $\mathbb{R}^2$ that contains the two polar coordinate spirals $\Sigma_\pm = \{(r, \theta) : r = 1 \pm 1/\theta \, (\pi \leq \theta)\}$ (Example **2.2.5**).
   (b) Find a smallest connected subspace of the plane that contains the doubled topologist's sine curve (Example **2.3.11**).

4. Determine whether or not the following are connected spaces:
   (a) the real line $\mathbb{R}_{()}$ with the right interval topology (Example **5.2.10**),
   (b) the half-disc space (Exercise **4.1.6**),
   (c) the line with two origins (Exercise **4.1.7**), and
   (d) the tangent disc space (Exercise **4.1.8**).

5. Prove that the 1-sphere $\mathbb{S}^1$ is not homeomorphic to any subspace of the real line $\mathbb{R}$.

6. Determine the homeomorphism types of connected spaces with three points.

7. Rework Exercise **3.4.12** by determining the subspaces of non-cut points of the doubled harmonic comb, the harmonic maze, and the harmonic filter.

8. Let x be a cut point of a connected Hausdorff space X, and let $X - \{x\} = E \cup F$ be a separation. Prove:
   (a) E and F are nonempty open subsets in X.
   (b) $\text{CLS}(E) = E \cup \{x\}$ and $\text{CLS}(F) = F \cup \{x\}$.
   (c) $\text{CLS}(E)$ and $\text{CLS}(F)$ are nontrivial connected subspaces of X.

9. (a) Let C be a connected subspace of a connected space X, and let $X - C = E \cup F$ be a separation. Prove that both $E \cup C$ and $F \cup C$ are connected.
   (b) Let A and B be closed subsets of a connected space $X = A \cup B$. Prove that if $A \cap B$ is connected, then A and B are connected.

10. Prove that a space is disconnected if and only if it is homeomorphic to the topological disjoint union of two nonempty subspaces of itself.

11. Prove that a space X is connected if and only if every family $\mathcal{U} = \{U\}$ of open sets

that cover X has this property: for each pair $U, V \in \mathcal{U}$, there exists a finite sequence $U = U_1, \ldots, U_n = V$ in $\mathcal{U}$ such that $U_k \cap U_{k+1} \neq \emptyset$ for all $1 \leq k < n$.

12 Prove that a connected Hausdorff space can have at most one dispersion point.

A space is ***reducible*** if it is homeomorphic to the topological product of two non-degenerate subspaces of itself; otherwise, it is ***prime***.

13 (a) Find all the reducible types of topologies on the four-point set $X = \{a, b, c, d\}$, and find some prime topology on X.
  (b) Prove that any interval in $\mathbb{R}$ is a prime space.
  (c) Prove that the 1-sphere $\mathbb{S}^1$ is a prime space.
  (d) Prove that the punctured plane $\mathbb{R}^2 - \{\boldsymbol{O}\}$ is a reducible space.

14 Consider a contraction $\Gamma : \mathfrak{H}(\mathbb{R}) \longrightarrow \mathfrak{H}(\mathbb{R})$, $\Gamma(A) = g_1(A) \cup g_2(A)$, associated with contractive linear maps $g_1(x) = a\,x + b$ and $g_2(x) = c\,x + d$ of $\mathbb{R}$ (see Theorem 3.3.2). Prove that the attractor P of Γ is either connected or totally disconnected.

15 Let $\sim$ be the connectivity relation on a topological space X. Prove that the quotient space $X/\sim$ is totally disconnected.

16 Let $f : X \longrightarrow Y$ be a closed continuous surjection. Prove that if Y is connected and if $f^{-1}(y)$ is a connected subspace of X for each $y \in Y$, then X is connected.

17 Let X be a connected space and let $f, g : X \longrightarrow [0, 1]$ be continuous functions. Prove that if f is surjective, there exists $x \in X$ such that $f(x) = g(x)$.

18 Classify up to homeomorphism the connected spaces that can be constructed from four intervals via identifications among their endpoints.

SECTION 2

1 Prove that any product of totally disconnected spaces is totally disconnected.

2 (a) Prove that the intersection of a decreasing sequence of connected subspaces of $\mathbb{R}$ is a connected subspace of $\mathbb{R}$.
  (b) Show that the intersection of a decreasing sequence of connected subspaces of $\mathbb{R}^2$ can be a disconnected subspace of $\mathbb{R}^2$.

3 Show that the complement of a dense open subset of $[0, 1]$ is totally disconnected.

4 Prove that, in any space, a nonempty connected subspace that is both open and closed is a component.

5 ***Cantor's Teepee*** $T \subseteq \mathbb{R}^2$ is the subspace of all points $<x, y> \in \mathbb{R}^2$ with rational height y on the line segments joining $<\frac{1}{2}, \frac{1}{2}>$ to the endpoints (i.e., ternary fractions) in Cantor's set $C \times \{0\}$, as well as, all points $<x, y>$ with irrational height y on the line segments joining $<\frac{1}{2}, \frac{1}{2}>$ to the non-endpoints in Cantor's set $C \times \{0\}$. Assume the fact that T is connected and prove that $<\frac{1}{2}, \frac{1}{2}>$ is a dispersion point of T.

6 Let S be the star-like set, $\{(r, \theta) \in \mathbb{R}^2 : 0 \leq r, \theta \in [0, 2\pi] \cap \mathbb{Q}\}$, of polar rays with rational radian angle.
  (a) Prove that S is a connected subspace of $\mathbb{R}^2$ that has the origin $\boldsymbol{O}$ as a cut point.
  (b) Determine the components of the open subspace $S - \boldsymbol{O}$.
  (c) Use the components of $S - \boldsymbol{O}$ to argue that S is not locally connected.

7 In the topologist's closed sine curve (Example 1.10), find an open subset that has a non-open component.

8 Determine the components of the topological product $(\mathbb{I} - M) \times (\mathbb{I} - M)$, where $\mathbb{I} = [0, 1]$ and $M = \{1/k : k \geq 1\}$.

9 Determine whether or not the harmonic rake (Example 1.19) is locally connected.

A cut point $x \in X$ of a connected space X has **order n** if the subspace $X - \{x\}$ has exactly n components.

10 (a) Prove that the orders of cut points are topological invariants.
    (b) Construct a space X with precisely two cut points, both of which have order 3.
    (c) Construct a space X with precisely the cut points $\{p_2, \ldots, p_n, \ldots\}$, where p_n has order n for each $n \geq 2$.

11 Determine whether the **doubled harmonic broom** (below left) and the **overlapping doubled harmonic broom** (below right) are homeomorphic subspaces of $\mathbb{R}^2$.

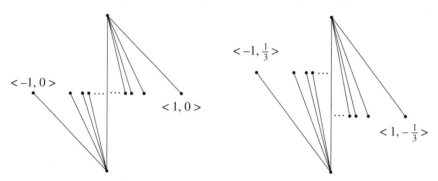

Exercise 11:
Doubled harmonic broom

Exercise 11:
Overlapping doubled harmonic broom

12 *Cantor's metropolis* M is the union of the copy $C \times \{0\} \subset \mathbb{R} \times \mathbb{R}$ of Cantor's space, together with the top edge and two vertical sides of the square on each open ternary interval deleted in the construction of C.
    (a) Prove that M is a connected subspace of $\mathbb{R}^2$.
    (b) Decide whether M is a locally connected subspace of $\mathbb{R}^2$.

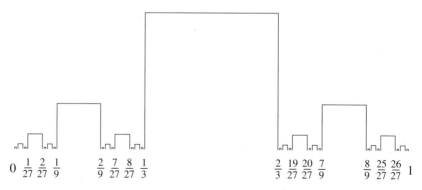

Exercise 12: Cantor's metropolis

SECTIONS 1-3
EXERCISES

13 *Cantor's uniform metropolis* UM is the union of the copy $C \times \{0\} \subset \mathbb{R} \times \mathbb{R}$ of Cantor's space, together with the top edge and two vertical sides of the one unit tall rectangle on each open ternary interval deleted in the construction of C.
  (a) Prove that UM is a connected subspace of $\mathbb{R}^2$.
  (b) Determine whether UM is a locally connected subspace of $\mathbb{R}^2$.

14 Let A be a connected subspace of a connected space X; let C be a component of the subspace $X - A$. Prove that the subspace $X - C$ is connected.

15 Prove that a topological product is locally connected if and only if all the factor spaces are locally connected and all but finitely many are connected.

16 Prove that a space is totally disconnected and locally connected if and only if it is discrete.

SECTION 3

1 Prove that the complement $\mathbb{R}^n - A$ of any countable subset $A \subset \mathbb{R}^n$ is a path-connected subspace of real n-space $\mathbb{R}^n$.

2 Prove that a topological product $\prod_\alpha X_\alpha$ is path-connected if and only if each factor space X_α is path-connected.

3 Let $A \subset \mathbb{I}^2$ be the union of the closed vertical line segments in $\mathbb{I}^2$ *above* these points:
$$\langle \tfrac{1}{2}, \tfrac{1}{2} \rangle, \langle \tfrac{1}{4}, \tfrac{1}{4} \rangle, \ldots, \langle \tfrac{1}{2k}, \tfrac{1}{2k} \rangle, \ldots,$$
as well as the closed vertical line segments in $\mathbb{I}^2$ *below* these points:
$$\langle \tfrac{1}{3}, \tfrac{2}{3} \rangle, \langle \tfrac{1}{5}, \tfrac{4}{5} \rangle, \ldots, \langle \tfrac{1}{2k+1}, \tfrac{2k}{2k+1} \rangle, \ldots.$$
Determine whether the subspaces $\mathbb{I}^2 - A$ and $\mathbb{J}^2 - A$ ($\mathbb{J} = (0, 1)$) are path-connected.

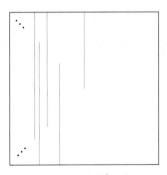

Exercise 3: $\mathbb{I}^2 - A$

4 Construct an original example of a connected space that is not path-connected.

5 Construct a path-connected space that is not locally path-connected.

6 Prove that the union of path-connected subspaces $\{P_\alpha : \alpha \in \mathscr{A}\}$ of a topological space X is path-connected if there is some path-connected subspace $P \subseteq \cup_\alpha P_\alpha$ such that $P \cap P_\alpha \neq \emptyset$ for each $\alpha \in \mathscr{A}$.

7 Show that the intersection of a sequence of nested path-connected subspaces of $\mathbb{R}^2$ need not be path connected.

8 Give the closed unit square $\mathbb{I}^2$ Peano's coordinate system described in Section **3.5**. Let $E \subset \mathbb{I}^2$ be the subspace of the closed unit square consisting of all points that have an ***even*** Peano expansion $.b_1b_2\ldots b_k\ldots$ (i.e., $b_k \in \{0, 2, 4, 6, 8\}$ for all $k \geq 1$). Prove that E is path-connected.
(Hint: Let E_k be the union of the 5^k even, k^{th} stage, Peano squares $S(b_1b_2\ldots b_k)$. Then $E = \cap_k E_k$, and the diagonals of the even Peano squares belongs to E.)

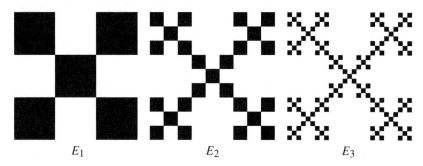

Exercise 8: Even Peano subspace $E \subset \mathbb{I}^2$

9 Prove that the complement, $(\mathbb{Q}^c \times \mathbb{Q}) \cup (\mathbb{Q} \times \mathbb{Q}^c)$, of the path-connected subspace $(\mathbb{Q} \times \mathbb{Q}) \cup (\mathbb{Q}^c \times \mathbb{Q}^c)$ of $\mathbb{R}^2$ is totally path-disconnected.
(Hint: If $f(t) = <f_1(t), f_2(t)>$ is a path in $(\mathbb{Q}^c \times \mathbb{Q}) \cup (\mathbb{Q} \times \mathbb{Q}^c)$, then real multiplication of its coordinate functions defines a path $g(t) = f_1(t)f_2(t)$ in $\{0\} \cup \mathbb{Q}^c$, which must be constant. Thus, $f(t)$ is a path in either $(\mathbb{Q}^c \times \{0\}) \cup (\{0\} \times \mathbb{Q}^c)$ or $(s\,\mathbb{Q} \times \mathbb{Q}) \cup (\mathbb{Q} \times s\,\mathbb{Q})$ for some $s \in \mathbb{Q}^c$.)

10 Prove that Cantor's Teepee $CT \subset \mathbb{R}^2$ (Exercise 2.5) is totally path-disconnected.

11 Determine whether or not Cantor's metropolis and uniform metropolis (Exercises 2.12 and 2.13) are path-connected.

12 Let $\{X_\alpha : \alpha \in \mathcal{A}\}$ be an indexed family of sets such that the intersection of any two of them is nonempty.
(a) Prove that if the sets $\{X_\alpha : \alpha \in \mathcal{A}\}$ are connected subspaces of a space X, then their union $\cup_\alpha X_\alpha$ is a connected subspace of X.
(b) If the sets $\{X_\alpha : \alpha \in \mathcal{A}\}$ are path-connected subspaces of a space X, must their union $\cup_\alpha X_\alpha$ be a path-connected subspace of X?

13 Show that Cantor's space C is not locally path-connected, yet its path-components and components coincide.

14 Let $A \subset \mathbb{I}$ be the union of the odd stage ternary intervals $T[0], T[0\,01], T[0\,21], T[2\,01], T[2\,21]$, etc. that are deleted from $\mathbb{I}$ in forming C and let $B \subset \mathbb{I}$ be the union of the even stage ternary intervals $T[01]$, etc. that are deleted from $\mathbb{I}$ in forming C. Form the subspace

$$X = (A \times \{0\}) \cup (C \times \mathbb{I}) \cup (B \times \{1\}) \subset \mathbb{I} \times \mathbb{I}.$$

Determine whether or not X is path-connected.

7

COMPACTNESS

In a metric space, sequential compactness and the Bolzano-Weierstrass property are equivalent to a property involving families of open subsets that cover the space. This chapter investigates this more general property, called compactness, in topological spaces.

1 Compact Spaces

COVERINGS AND COMPACTNESS

Let X be a set and let $\mathcal{U} = \{U\}$ be a family of subsets of X. The union $\cup\, \mathcal{U}$ is the subset $\cup\, \{U \subseteq X : U \in \mathcal{U}\}$ of X. The family $\mathcal{U}$ is called a ***covering*** of the subset $A \subseteq X$ if $A \subseteq \cup\, \mathcal{U}$, i.e., if each $a \in A$ belongs to at least one $U \in \mathcal{U}$.

To motivate the definition of compactness of a topological space, we consider coverings of the closed unit interval $\mathbb{I} = [0, 1] \subset \mathbb{R}$.

1.1 Theorem *For each covering $\mathcal{U}$ of $\mathbb{I}$ by open subsets of $\mathbb{I}$, there is a covering $\mathcal{W}$ of $\mathbb{I}$ consisting of finitely many members of $\mathcal{U}$.*

Proof: Let B be the set of $x \in \mathbb{I}$ such that the interval $[0, x]$ has a covering consisting of finitely many members of $\mathcal{U}$. B contains 0, since $[0, 0]$ belongs to some member of $\mathcal{U}$. Because $\mathcal{U}$ is a covering of $\mathbb{I}$ by open subsets, each point $x \in \mathbb{I}$ belongs to a relatively open interval $(x - r, x + r) \cap \mathbb{I}$ that is contained entirely in a member of $\mathcal{U}$. By the definition of B, any such interval in $\mathbb{I}$ is contained entirely in either B or in B^c. This proves that $B \neq \emptyset$ is an open and closed subset of $\mathbb{I}$; hence, $B = \mathbb{I}$ by connectedness. In particular, $1 \in B$. This means that $\mathbb{I} = [0, 1]$ has a covering by finitely many members of $\mathcal{U}$. □

Theorem 1.1 suggests the following definitions.

Let X be a topological space. A covering $\mathcal{U}$ of $A \subseteq X$ is an ***open covering*** if each member $U \in \mathcal{U}$ is an open subset of X. Another covering $\mathcal{W}$ of $A \subseteq X$ is a ***subcovering*** of $\mathcal{U}$ if each member of $\mathcal{W}$ is a member of $\mathcal{U}$; equivalently, $\mathcal{W} \subseteq \mathcal{U} \subseteq \mathcal{P}(X)$. $\mathcal{W}$ is ***finite*** if it consists of a finite number of subsets of X.

A topological space X is called ***compact*** if every open covering $\mathcal{U}$ of X has a finite subcovering $\mathcal{W}$. This is clearly a topological invariant.

Example 1 Any closed interval $[a, b]$ is homeomorphic to $\mathbb{I}$ and so is a compact space. Any non-closed interval, (a, b), $(a, b]$, or $[a, b)$, is a non-compact topological space. For example, there is no finite subcovering of this open covering of (a, b):
$$\mathcal{U} = \{(a + 1/k, b - 1/k) : k = 1, 2, \ldots\}.$$

Example 2 Any space with just finitely many open subsets is compact. In particular, any indiscrete space is compact. A discrete space X is compact if and only if X is finite, since the open covering of X by its singletons has no proper subcovering.

Example 3 Any space $X \neq \emptyset$ with the finite complement topology is compact. Given any open covering $\mathcal{U}$ of X, let $U \in \mathcal{U}$ be some nonempty member. Then U has a finite complement $F = X - U$. For each member $x \in F$, let U_x be a member of the covering $\mathcal{U}$ that contains x. Then $\{U\} \cup \{U_x : x \in F\}$ is a finite subcovering of $\mathcal{U}$.

Example 4 Each closed ordinal interval $[0, \beta] = \{\alpha : \alpha \text{ is an ordinal}, \alpha \leq \beta\}$ is compact in its order topology (Example **5.2.5**). Let $\mathcal{U} = \{U\}$ be any open covering of $[0, \beta]$. Then there exists for each ordinal $\alpha \leq \beta$ some $U_\alpha \in \mathcal{U}$ such that $\alpha \in U_\alpha$. By the definition of the order topology, there exists for each ordinal $0 \neq \alpha \leq \beta$ an ordinal $\phi(\alpha) < \alpha$ such that $(\phi(\alpha), \alpha] \subseteq U_\alpha$. So there is, by induction, a decreasing sequence of ordinals $\alpha_0 = \beta$, $\alpha_1 = \phi(\beta)$, $\ldots$, $\alpha_k = \phi(\alpha_{k-1})$, $\ldots$, which continues until some $\alpha_k = 0$. There must be some $\alpha_n = 0$; otherwise, $\{\alpha_k : 0 \leq k\}$ would be a nonempty set of ordinals that has no first element, contrary to their well-ordered feature (**1.5.11**). Then, the collection $\{U_0, U_{\alpha_0}, U_{\alpha_1}, \ldots, U_{\alpha_n}\}$ is a finite subcovering of the given covering $\mathcal{U}$.

There is a reformulation of compactness involving this notion:

A family $\mathcal{G}$ of subsets of X has the ***finite intersection property*** if each finite subfamily $\mathcal{F} \subseteq \mathcal{G}$ has nonempty intersection $\cap \mathcal{F} \neq \emptyset$.

1.2 FIP Theorem *The following are equivalent conditions for X.*
(a) *The space X is compact: for each family $\mathcal{U}$ of open sets in X with union $\cup \mathcal{U} = X$, there exists a finite subfamily $\mathcal{W} \subseteq \mathcal{U}$ with union $\cup \mathcal{W} = X$.*
(b) *For each family $\mathcal{G}$ of closed sets in X with empty intersection $\cap \mathcal{G} = \emptyset$, there exists a finite subfamily $\mathcal{F} \subseteq \mathcal{G}$ with empty intersection $\cap \mathcal{F} = \emptyset$.*
(c) *If a family $\mathcal{G}$ of closed sets in X has the finite intersection property, then $\mathcal{G}$ itself has nonempty intersection $\cap \mathcal{G} \neq \emptyset$.*
(d) *Each filterbase $\mathcal{A}$ in X accumulates in X, i.e., $\cap \{\text{CLS}(A) : A \in \mathcal{A}\} \neq \emptyset$.*

Proof: $(a) \Leftrightarrow (b)$. Take complements of the sets involved.
$(b) \Leftrightarrow (c)$. Take the contrapositive of the implications involved.
$(c) \Rightarrow (d)$. If $\mathcal{A} = \{A\}$ is a filterbase in X, then each finite intersection of

its members contains some member. Hence, $\mathcal{G} = \{\overline{A} : A \in \mathcal{A}\}$ has the finite intersection property. Therefore, (c) implies that $\cap \{\overline{A} : A \in \mathcal{A}\} = \cap \mathcal{G} \neq \emptyset$.

$(d) \Rightarrow (c)$. If the family $\mathcal{G} = \{G\}$ of closed sets has the finite intersection property, its members $G \in \mathcal{G}$ and their finite intersections form a filterbase $\mathcal{A}$ of closed sets in X. Then $\cap \mathcal{G} = \cap \mathcal{A} \neq \emptyset$, by construction and (d). $\square$

A useful corollary to the FIP Theorem is the following generalization of the Nested Intervals Theorem (**1.3.3**).

1.3 Cantor's Theorem   *In a compact space X, any sequence of nonempty nested closed subsets,*

$$F_1 \supseteq F_2 \supseteq \ldots \supseteq F_k \supseteq \ldots,$$

has a nonempty intersection $\cap_k F_k \neq \emptyset$.

Proof: The sequence of nested nonempty closed subsets has the finite intersection property, and so has nonempty intersection by the FIP Theorem. $\square$

SUBSPACES AND COMPACTNESS

By definition, the open subsets V of a subspace $A \subseteq X$ are the relatively open subsets of X, namely, the intersections $V = U \cap A$ with A of open subsets U of X. Therefore, the following lemma is immediate.

1.4 Lemma   *The following are equivalent for a subspace $A \subseteq X$:*
(a)   *The subspace topology on A is compact: each covering of A by relatively open subsets of X has a finite subcovering.*
(b)   *Each covering of A by open sets of X has a finite subcovering.*

Just as for sequential compactness of metric spaces, the closed subsets and the compact subspaces of a compact topological space are related:

1.5 Theorem
(a)   *A closed subset F of a compact space X is a compact subspace.*
(b)   *A compact subspace F of a Hausdorff space X is a closed subset of X.*
So a subspace of a compact Hausdorff space is compact if and only if it is closed.

Proof: (a) Let $\mathcal{U}$ be a covering of the closed set F by open subsets of the space X. Then $\mathcal{U} \cup \{X - F\}$ is an open covering of the compact space X, with some finite subcovering $\mathcal{W}$. Hence, $\mathcal{W} - \{X - F\}$ is a finite subcovering of the covering $\mathcal{U}$ of F. So F is a compact subspace of X, by Lemma 1.4.

(b) Fix $z \notin F$. For each $x \in F$, select disjoint open neighborhoods $V_x(z)$ and $U(x)$ of z and x in the Hausdorff space X. Then $\mathcal{U} = \{U(x) : x \in F\}$ is a covering of the subspace F by open subsets of X. By the compactness of F, there is a finite subcovering $\{U(x_i) : 1 \leq i \leq k\}$ of $\mathcal{U}$. Then the finite intersection $N(z) = \cap_i V_{x_i}(z)$ is an open neighborhood of z, disjoint from the union $\cup_i U(x_i)$, which contains F. This proves that every point $z \notin F$ has a neighborhood disjoint from F; hence, F is closed in X. □

Example 5 $\mathbb{Q} \cap [a, b]$ is not compact, since it is not closed in the interval $[a, b]$.

Example 6 *A compact subspace of a non-Hausdorff space need not be closed.* Any subspace A of a space X with the finite complement topology is compact, by Example 3. But when A is an infinite proper subset of X, it is not closed in X.

COMPACTNESS IN EUCLIDEAN SPACES

We now show that the compact topological subspaces of real n-space $\mathbb{R}^n$ are the sequentially compact metric subspaces of $\mathbb{E}^n$, namely, the subspaces that are closed in $\mathbb{R}^n$ and bounded in the Euclidean metric.

1.6 Lemma *A compact subspace of a metric space is bounded.*

Proof: Let X be any metric subspace of (Y, d). Since X is compact, its covering $\{B(x, 1) : x \in X\}$ by open unit balls in Y has a finite subcovering $\{B(x_i, 1) : 1 \leq i \leq m\}$. Set $M = diam\{x_k : 1 \leq k \leq m\}$. Any points $x, y \in X$ are within distance 1 of the set $\{x_k : 1 \leq k \leq m\}$, hence, within distance $2 + M$ of one another, according to the triangle inequality. So $diam\, X \leq 2 + M$. □

A reformulation of the proof of the Compact Cubes Theorem (**3.2.5**) establishes the compactness of Euclidean cubes, as follows.

1.7 Compact Cubes Theorem *Each cube is a compact subspace of $\mathbb{R}^n$.*

Proof: Let $\mathcal{U}$ be any open covering of a cube $B \subset \mathbb{R}^n$. Were there no finite subcovering of $\mathcal{U}$, there would exist a sequence of nested sub-cubes,

$$B = B_0 \supset B_1 \supset \ldots \supset B_{k-1} \supset B_k \supset \ldots,$$

such that the side length of each one is half that of its predecessor and each one fails to be covered by finitely many members of $\mathcal{U}$.

By the Nested Boxes Theorem (**3.2.4**), the intersection $\cap_k B_k$ would be a single point $x \in B$. Since $\mathcal{U}$ is an open covering of B, there would exist some member $U \in \mathcal{U}$ and a Euclidean ball $B(x, r) \subseteq U$, for some $r > 0$. Also, there would be an integer $k > 0$ such that $2^k > \sqrt{n}/r$; so r would exceed $(\frac{1}{2})^k \sqrt{n}$, the diameter of the cube B_k that contains x. Then B_k would be contained in $B(x, r)$ and thus would be covered by the single member $U \in \mathcal{U}$, contrary to the features of the nested boxes. So $\mathcal{U}$ must have a finite subcovering. □

1.8 Heine-Borel-Lebesgue Theorem

A topological subspace X of real n-space $\mathbb{R}^n$ is compact if and only if X is closed in $\mathbb{R}^n$ and bounded in the Euclidean metric.

Proof: Any compact subspace $X \subseteq \mathbb{R}^n$ is a closed subset of $\mathbb{R}^n$ by Theorem 1.5(b), and it is bounded in the Euclidean metric by Lemma 1.6.

Conversely, let X be closed in $\mathbb{R}^n$ and bounded in the Euclidean metric. As a closed subspace of some cube $[-k, +k]^n$, X is itself compact, by the Compact Cubes Theorem 1.7 and Theorem 1.5(b). □

In the next section, we shall show that, for all metric spaces, compactness coincides with sequential compactness, as defined in Section **3**.2.

Example 7 The harmonic broom and rake (Example **6**.1.19) are closed in $\mathbb{R}^2$ and bounded in the Euclidean metric; hence, they are compact subspaces of $\mathbb{R}^2$.

Example 8 The *Hawaiian earrings* subspace $HE \subset \mathbb{R}^2$ is the union of the mutually tangent circles C_k with center $<1/k, 0>$ and radius $1/k$, for each $k \geq 1$. Since HE is closed in $\mathbb{R}^2$ and bounded in the Euclidean metric, it is a compact subspace of $\mathbb{R}^2$.

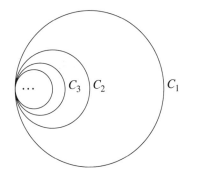
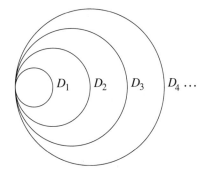

Example 8: Hawaiian earrings   Example 9: Expanding earrings

Example 9 The *expanding earrings* subspace $EE \subset \mathbb{R}^2$ is the union of the mutually tangent circles D_k, with center $<k, 0>$ and radius k, for all integers $k \geq 1$. The subspace EE is closed in $\mathbb{R}^2$, but not bounded in the Euclidean metric. So EE is a non-compact subspace of $\mathbb{R}^2$.

2 Topological Aspects of Compactness

COMPACTNESS AND CONTINUITY

Like sequential compactness for metric spaces (Section **3**.2), the compactness property is inherited by each continuous image of a compact topological space.

2.1 Compact Image Theorem

Let $f: X \to Y$ be a continuous function between topological spaces. If A is a compact subspace of X, then $f(A)$ is a compact subspace of Y.

Proof: Let $\mathcal{U}$ be a covering of $f(A) \subseteq Y$ by open subsets of Y. By the continuity of f, the family $f^{-1}(\mathcal{U}) = \{ f^{-1}(U) \subseteq X : U \in \mathcal{U} \}$ is an open covering of $A \subseteq X$. Since A is compact, it has a finite subcovering $\{ f^{-1}(U_i) : 1 \leq i \leq k \}$. So $A \subseteq \cup_i f^{-1}(U_i) = f^{-1}(\cup_i U_i)$ and $f(A) \subseteq \cup_i U_i$. Thus, the covering $\mathcal{U}$ of $f(A)$ has the finite subcovering $\{U_i : 1 \leq i \leq k\}$. By Lemma 1.4, this proves that $f(A)$ is a compact subspace of Y. $\square$

2.2 Corollary   *Compactness is a topological invariant.*

Example 1   Peano's curve provides a continuous surjection $\mathbb{I} \to \mathbb{I} \times \mathbb{I}$ of the closed unit interval onto the closed unit square. But there exists no continuous surjection $\mathbb{I} \to \mathbb{J} \times \mathbb{J}$, where $\mathbb{J} = (0, 1)$, nor continuous surjection $\mathbb{I} \to D$, where D is a dense proper subset of $\mathbb{I} \times \mathbb{I}$. The reason is that the open square $\mathbb{J} \times \mathbb{J}$ and a proper dense subset $D \subset \mathbb{I} \times \mathbb{I}$ are not compact.

Example 2   The ***doubled Hawaiian earrings*** subspace $DHE \subset \mathbb{R}^2$ consists of the mutually tangent circles C_k with center $< 1/k, 0 >$ and radius $1/|k|$, for $k \in \mathbb{Z} - \{0\}$. There is a continuous function $f: \mathbb{R} \to DHE$ that wraps the intervals $[-k, -k+1]$ and $[k-1, k]$ around the circles C_{-k} and C_k for each $k > 0$, identifying the set $\mathbb{Z} \subset \mathbb{R}$ of integers to the tangent point O in DHE. By the Transgression Theorem (**5.1.9**), f induces a continuous bijection $h: \mathbb{R}/\mathbb{Z} \to DHE$. Since DHE is compact while the quotient space $\mathbb{R}/\mathbb{Z}$ is not, h cannot be a homeomorphism.

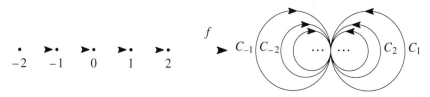

Example 2: Non-identification map $f: \mathbb{R} \to DHE$

2.3 Corollary   *Every continuous real-valued function $f: X \to \mathbb{R}$ on a nonempty compact space X attains a minimum value m and a maximum value M on X: there exist $x_m, x_M \in X$ such that, for all $x \in X$,*

$$m = f(x_m) \leq f(x) \leq f(x_M) = M.$$

Proof: The image $f(X) \subseteq \mathbb{R}$ is a compact subspace of $\mathbb{R}$; hence, it is closed in $\mathbb{R}$ and bounded in the Euclidean metric. Because $f(X)$ is bounded, the real numbers $m = \text{glb } f(X)$ and $M = \text{lub } f(X)$ exist, by the completeness of $\mathbb{R}$. Because $f(X)$ is closed, m and M belong to $f(X)$. Therefore, there exist $x_m, x_M \in X$ with values $m = f(x_m) \leq f(x) \leq f(x_M) = M$ for all $x \in X$. $\square$

2.4 Corollary *A continuous function $f : X \to Y$ with compact domain space X and Hausdorff codomain space Y is a closed function.*

Proof: If $F \subseteq X$ is a closed subset of the compact space X, then F is a compact subspace, by Theorem 1.5(*a*). In turn, the image $f(F) \subseteq Y$ is a compact subspace of Y, by the continuity of f and Theorem 2.1. Therefore, $f(F)$ is a closed subset of the Hausdorff space Y, by Theorem 1.5(*b*). □

THE HOMEOMORPHISM THEOREM

The exponential function $e : (0, 1] \to \mathbb{S}^1$, $e(t) = <\cos 2\pi t, \sin 2\pi t>$, and the function $h : \mathbb{R}/\mathbb{Z} \to DHE$ of Example 2 show that a continuous bijection need not be a homeomorphism. The following theorem shows that this phenomenon can't occur with compact Hausdorff spaces.

2.5 Homeomorphism Theorem *Any continuous bijection $f : X \to Y$ from a compact space X to a Hausdorff space Y is a homeomorphism.*

Proof: By Corollary 2.4, the continuous bijection f is a closed function. Hence, f is a homeomorphism by Theorem **4.4.5**. □

2.6 Corollary *Let $p : X \to Y$ be an identification map and let $g : X \to Z$ be any continuous function that makes the same identifications as p:*

$$\text{for all } x, x' \in X, \ g(x) = g(x') \text{ if and only if } p(x) = p(x').$$

If Y is compact and Z is Hausdorff, then the unique induced continuous function $h : Y \to Z$ such that $h \circ p = g$ is a homeomorphism.

Proof: The Transgression Theorem (**5.1.9**) provides the induced continuous bijection h such that $h \circ p = g$. Then Theorem 2.5 implies that h is a homeomorphism. □

Example 3 The continuous function $e \times e : \mathbb{I} \times \mathbb{I} \to \mathbb{S}^1 \times \mathbb{S}^1$ wraps the unit square $\mathbb{I} \times \mathbb{I}$ around the torus $\mathbb{S}^1 \times \mathbb{S}^1$, making the following identifications for all $x, y \in \mathbb{I}$:

$$<0, y> \sim <1, y> \quad \text{and} \quad <x, 0> \sim <x, 1>.$$

So the identification space $(\mathbb{I} \times \mathbb{I})/\sim$ of the square $\mathbb{I} \times \mathbb{I}$ modulo the relation $\sim$ is homeomorphic to the torus $\mathbb{S}^1 \times \mathbb{S}^1$, by Corollary 2.6.

Example 4 The Hawaiian earrings subspace $HE \subset \mathbb{R}^2$ (Example 1.8) is the union of the mutually tangent circles C_k with center $<1/k, 0>$ and radius $1/k$, for each $k \geq 1$. There is a continuous function $f : \mathbb{I} \to HE$ that wraps the interval $[1/(k+1), 1/k]$ around the circle C_k for each $k \geq 1$ and identifies $\mathbb{M} = \{0, 1, 1/2, \ldots, 1/k, \ldots\}$ to the tangent point O in HE. Corollary 2.6 provides a homeomorphism $h : \mathbb{I}/\mathbb{M} \to HE$ between the quotient space $\mathbb{I}/\mathbb{M}$ and the Hawaiian earrings space HE.

2.7 Corollary *If $\mathcal{T}$ is a compact Hausdorff topology on X, then*
(a) *there is no strictly smaller topology $\mathcal{R} \subset \mathcal{T}$ that is Hausdorff, and*
(b) *there is no strictly larger topology $\mathcal{T} \subset \mathcal{S}$ that is compact.*

Proof: (a) If $\mathcal{R} \subseteq \mathcal{T}$ is a Hausdorff topology, then the continuous bijection $1_X : (X, \mathcal{T}) \to (X, \mathcal{R})$ is a homeomorphism, by Theorem 2.5. So $\mathcal{R} = \mathcal{T}$.
  (b) If $\mathcal{S} \supseteq \mathcal{T}$ is a compact topology, then the continuous bijection $1_X : (X, \mathcal{S}) \to (X, \mathcal{T})$ is a homeomorphism, by Theorem 2.5. So $\mathcal{S} = \mathcal{T}$. □

Corollary 2.7 does not say that a set admits at most one compact Hausdorff topology, but only that distinct compact Hausdorff topologies on the same set are not comparable under the containment relation $\subseteq$.

COMPACTNESS OF PRODUCT SPACES

All examples of topological products of compact spaces that we have seen, such as the unit square $\mathbb{I} \times \mathbb{I}$ and the torus $\mathbb{S}^1 \times \mathbb{S}^1$, are themselves compact spaces. They illustrate the next theorem about topological products.

2.8 Lemma *A topological space X is compact if and only if each open covering by members of a base $\mathcal{B}$ for X has a finite subcovering.*

Proof: Let $\mathcal{U}$ be an open covering of X, and let $\mathcal{B}_\mathcal{U}$ denote the set of members B of the base $\mathcal{B}$, each of which is contained in a member of the open covering $\mathcal{U}$. Since $\mathcal{B}$ is a base, then $\mathcal{B}_\mathcal{U}$ is another open covering of X. By hypothesis, there is a finite subcovering $\{B_i : 1 \leq i \leq k\}$ of $\mathcal{B}_\mathcal{U}$. By the construction of $\mathcal{B}_\mathcal{U}$, each member B_i of the finite subcovering is contained in some member $U_i \in \mathcal{U}$. Then $\{U_i : 1 \leq i \leq k\}$ is a finite subcovering of $\mathcal{U}$. This establishes the non-trivial implication ($\Leftarrow$) in the lemma. □

2.9 Theorem *A topological product $X \times Y$ is compact if and only if the factor spaces X and Y are compact.*

Proof: Let $X \times Y$ be compact. Then the factor spaces X and Y are compact, being the continuous images of $X \times Y$ under the projections $p_X : X \times Y \to X$ and $p_Y : X \times Y \to Y$.
  For the converse, let X and Y be compact. The topological product $X \times Y$ of spaces X and Y has base $\mathcal{B}$ consisting of all products $U \times V$, where $U \subseteq X$ is open in X and $V \subseteq Y$ is open in Y. Let $\mathcal{W} = \{U_\alpha \times V_\alpha : \alpha \in \mathcal{A}\}$ be any open covering of $X \times Y$ by members of the base $\mathcal{B}$.
  For each $x \in X$, let $\mathcal{W}_x = \{U_\alpha \times V_\alpha : \alpha \in \mathcal{A}_x\}$ consist of all members $U_\alpha \times V_\alpha \in \mathcal{W}$ with $x \in U_\alpha$. Then $\mathcal{W}_x$ serves as an open covering of the slice $\{x\} \times Y$, which is a subspace of $X \times Y$ homeomorphic to the compact space Y. Therefore, $\{x\} \times Y$ has some finite subcovering $\{U_\alpha \times V_\alpha : \alpha \in \mathcal{F}_x \subseteq \mathcal{A}_x\}$.

The finite intersection $U_x = \cap \{U_\alpha : \alpha \in \mathcal{F}_x\}$ is an open subset of X containing x, and the finite union $\cup \{U_\alpha \times V_\alpha : \alpha \in \mathcal{F}_x\}$ contains $U_x \times Y$ (below).

The open covering $\{U_x : x \in X\}$ of the compact space X has a finite subcovering $\{U_x : x \in F \subseteq X\}$. Then $\{U_\alpha \times V_\alpha : \alpha \in \mathcal{F}_x, x \in F\}$ is a finite subcovering of $\mathcal{W}$, as it has union

$$\cup \{U_\alpha \times V_\alpha : \alpha \in \mathcal{F}_x, x \in F\} \supseteq \cup \{U_x \times Y : x \in F\} = X \times Y.$$

This proves that the topological product $X \times Y$ is compact, by Lemma 2.8. □

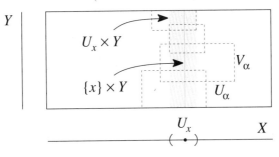

Theorem 2.9

2.10 Corollary *The topological product of a finite family of compact spaces is compact.*

> **Example 5** The topological product $\mathbb{S}^1 \times \ldots \times \mathbb{S}^1$ of n copies of the unit 1-sphere $\mathbb{S}^1$ is a compact space, called the ***n-dimensional torus***. It is not the n-sphere $\mathbb{S}^n$!

More generally, an arbitrary product $\prod_\alpha X_\alpha$ of topological spaces is compact if and only if each factor space X_α is compact. This theorem, established by A. Tychonoff in 1935, is equivalent to the axiom of choice (**1.5.4**) in set theory. We derive it using Zorn's lemma (**1.5.5**).

2.11 Tychonoff's Theorem *The topological product $X = \prod_\alpha X_\alpha$ of any family $\{X_\alpha : \alpha \in \mathcal{A}\}$ of compact spaces is compact.*

Proof: By the FIP Theorem (**1.2**), it suffices to prove that any family $\mathcal{G}$ of closed sets in $X = \prod_\alpha X_\alpha$ with the finite intersection property has nonempty intersection $\cap \mathcal{G} \neq \emptyset$. We consider the set $\mathcal{E}$ of all families $\mathcal{R} \subseteq \mathcal{P}(X)$ that contain $\mathcal{G}$ and that have the finite intersection property. We partially order $\mathcal{E}$ by inclusion, $\mathcal{R} \subseteq \mathcal{S} \subseteq \mathcal{P}(X)$, and we seek a maximal family in $\mathcal{E}$.

We can enlarge certain members of $\mathcal{E}$. If $\mathcal{R} \in \mathcal{E}$, then so is the set,

$$\mathcal{T} = \{\cap \mathcal{F} \subseteq X : \mathcal{F} \text{ finite and } \mathcal{F} \subseteq \mathcal{R}\},$$

of all finite intersections of members of $\mathcal{R}$. Notice that $\mathcal{T} \in \mathcal{E}$ and $\mathcal{T}$ contains all finite intersections of its own members. Also, if $A \subseteq X$ has $T \cap A \neq \emptyset$ for each $T \in \mathcal{T}$, then the set $\mathcal{S} = \mathcal{T} \cup \{A\}$ also belongs to $\mathcal{C}$.

Now each chain $\{\mathcal{R}_\gamma\}$ in $(\mathcal{C}, \subseteq)$ has an upper bound, namely, the union $\mathcal{R} = \cup_\gamma \mathcal{R}_\gamma$ in $\mathcal{P}(X)$ of the families $\mathcal{R}_\gamma$ in the chain. It is clear that $\mathcal{R} \in \mathcal{C}$ since any finite family $\mathcal{F} \subseteq \mathcal{R}$ is contained in some family $\mathcal{R}_\gamma$ of the chain.

Therefore, by Zorn's lemma (1.5.5), $\mathcal{C}$ has a maximal family, $\mathcal{S}$. Then $\mathcal{S}$ contains $\mathcal{G}$ and $\mathcal{S}$ cannot be enlarged in the manner above. Thus, $\mathcal{S}$ contains $\cap \mathcal{F}$, for each finite subfamily $\mathcal{F} \subseteq \mathcal{S}$, and $\mathcal{S}$ also contains any subset $A \subseteq X$ such that $S \cap A \neq \emptyset$ for each $S \in \mathcal{S}$.

For each $\alpha \in \mathcal{A}$, the family $\{\mathrm{CLS}_{X_\alpha}(p_\alpha(S)) : S \in \mathcal{S}\}$ in the compact space X_α has the finite intersection property, hence, a nonempty intersection. So there exists $x = (x_\alpha) \in X = \prod_\alpha X_\alpha$ with $x_\alpha \in \mathrm{CLS}_{X_\alpha}(p_\alpha(S))$ for all $S \in \mathcal{S}$ and all $\alpha \in \mathcal{A}$. For any subbase neighborhood $p_\alpha^{-1}(U_\alpha)$ of $x \in X$, the open set $U_\alpha \subseteq X_\alpha$ contains $x_\alpha \in \mathrm{CLS}_{X_\alpha}(p_\alpha(S))$, hence, intersects $p_\alpha(S)$, for all $S \in \mathcal{S}$. Therefore, $p_\alpha^{-1}(U_\alpha) \cap S \neq \emptyset$ for all $S \in \mathcal{S}$; hence, $p_\alpha^{-1}(U_\alpha) \in \mathcal{S}$ by the maximality of $\mathcal{S}$. Then, by a second application of maximality, $\mathcal{S}$ must also contain each base neighborhood $U(x) = \cap \{p_{\alpha(i)}^{-1}(U_{\alpha(i)}) : 1 \leq i \leq k\}$.

Because $\mathcal{S}$ contains $\mathcal{G}$ and has the finite intersection property, it follows that $U(x) \cap G \neq \emptyset$ for each $G \in \mathcal{G}$ and each base neighborhood $U(x)$ of x. Then $x \in \mathrm{CLS}_X G = G$ for each $G \in \mathcal{G}$. This proves that $\cap \mathcal{G} \supseteq \{x\} \neq \emptyset$. □

Example 6 Cantor's set C is a closed, hence, compact subspace of the closed unit interval $\mathbb{I} = [0, 1]$. For an alternate proof of the compactness of C, we can invoke Tychonoff's Theorem and the homeomorphism of C with the product $\prod_n \{0, 2\}_n$ of countably many copies of two-point discrete space $\{0, 2\}$ (Example 5.3.9).

Example 7 For any index set $\mathcal{A}$, the product $\prod_\alpha \mathbb{I}_\alpha$ of the family $\{\mathbb{I}_\alpha = \mathbb{I} : \alpha \in \mathcal{A}\}$ is a compact space, denoted by $\mathbb{I}^{\mathcal{A}}$ and called a ***topological cube***.

SEPARATION PROPERTIES

In a Hausdorff space, distinct points *can be separated* by disjoint neighborhoods. If the space is compact as well as Hausdorff, then any closed set and point outside it, and even disjoint closed sets, can be separated by disjoint neighborhoods. These are the separation properties T_3 and T_4, which when coupled with the Hausdorff property are called regularity and normality in Section 4.2. They are established for compact Hausdorff spaces, as follows.

2.12 Theorem *Any compact Hausdorff space is normal.*

Proof: Let F be a closed subset of a compact Hausdorff space X. Clearly, any points $x \in F$ and $q \in X - F$ have disjoint open neighborhoods $U_q(x)$ and $V_x(q)$ in X. Then the open covering $\{U_q(x) : x \in F\}$ of the closed, hence, compact subspace F has a finite subcovering $\{U_q(x_i) : 1 \leq i \leq k\}$. Then,

$$U_q(F) = \cup \{U_q(x_i) : 1 \leq i \leq k\} \quad \text{and} \quad V_F(q) = \cup \{V_{x_i}(q) : 1 \leq i \leq k\}$$

are disjoint open neighborhoods of F and $q \notin F$. This proves that the space X is regular.

Now let F and G be disjoint closed subsets of X. As above, disjoint open neighborhoods $V_F(q)$ and $U_q(F)$ of $q \in G$ and $F \subseteq X - G$, yield disjoint open neighborhoods of G and F. This proves that X is normal. □

3   Compactness in Metric Spaces

VARIATIONS OF COMPACTNESS

The following variations of compactness appear in the mathematical literature. They involve convergence of sequences and accumulation points of subsets, concepts defined for topological spaces just as for metric spaces.

- X is *compact* if each open covering of X has a finite subcovering.
- X is *countably compact* if each countable open covering of X has a finite subcovering.
- X is *sequentially compact* if each sequence in X has a convergent subsequence.
- X has the *Bolzano-Weierstrass property* if each infinite subset of X has an accumulation point in X.
- X is *Lindelöf* if each open covering of X has a countable subcovering.

There is this characterization of compactness of metrizable spaces:

3.1   Theorem   *Compactness, countable compactness, sequential compactness, and the Bolzano-Weierstrass property are equivalent for metrizable spaces.*

Our route to Theorem 3.1 passes through the following five results.

3.2   Theorem   *A Hausdorff space has the Bolzano-Weierstrass property if and only if it is countably compact.*

3.3   Theorem   *Any countably compact metric space is separable.*

3.4   Theorem   *Any separable metric space is second countable.*

3.5   Theorem   *Any second countable space is Lindelöf.*

3.6   Theorem   *Any countably compact Lindelöf space is compact.*

Theorems 3.2-3.6 imply Theorem 3.1, save the equivalence of sequential compactness and the Bolzano-Weierstrass property for metric spaces established in Theorem **3.2.3**. Here is a second way to sum Theorems 3.2-3.6:

- *Compactness, countable compactness, and the Bolzano-Weierstrass property are equivalent for second countable Hausdorff spaces.*

PROOFS OF THEOREMS 3.2-3.6

3.2 Theorem *A Hausdorff space has the Bolzano-Weierstrass property if and only if it is countably compact.*

Proof: First, suppose that the Hausdorff space X is not countably compact. Then some countable open covering $\mathcal{U} = \{U_k : k \geq 1\}$ of X contains no finite subcovering. So, for each $k \geq 1$, there exists a point $x_k \in X - (U_1 \cup \ldots \cup U_k)$. It follows that the set $A = \{x_k : k \geq 1\}$ is infinite, as no member of X belongs to the intersection $\cap_k (X - (U_1 \cup \ldots \cup U_k)) = X - \cup_k (U_1 \cup \ldots \cup U_k) = \emptyset$.

Now A has no accumulation point in $U_1 \cup \ldots \cup U_k$, since this open set contains at most the finitely many members x_i ($1 \leq i < k$) of A. It follows that A has no accumulation point in $X = \cup_k (U_1 \cup \ldots \cup U_k)$. Thus, X does not have the Bolzano-Weierstrass property.

Now suppose that X does not have the Bolzano-Weierstrass property. So there exists a denumerable set $A = \{a_k : k \geq 1\}$ that has no accumulation point in X. Then each $x \in X$ has an open neighborhood $U(x)$ that is disjoint from $A - \{x\}$. It follows that A is closed and that $\{U(a_k) : k \geq 1\} \cup \{X - A\}$ is a countable open covering of X that has no finite subcovering. This proves that X is not countably compact. □

3.3 Theorem *Any countably compact metric space is separable.*

Proof: Let X be a countably compact metric space. For each integer $n \geq 1$, there exists a finite covering $\mathcal{U}_n = \{B(x_{n,k}, 1/n) : 1 \leq k \leq m_n\}$ of X by open balls of radius $1/n$. Else, for some $n \geq 1$, we could inductively select points $x_{n,k} \in X$ for all $1 \leq k$ such that $x_{n,k} \notin \cup \{B(x_{n,i}, 1/n) : 1 \leq i \leq k-1\}$, hence, $d(x_{n,k}, x_{n,k'}) \geq 1/n$ for all $k \neq k'$. Then the set $A = \{x_{n,k} : 1 \leq k\}$ would be an infinite subset with no accumulation point in X, contrary to Theorem 3.2.

Then $D = \{x_{n,k} : 1 \leq n, 1 \leq k \leq m_n\}$ is a countable dense subset of X. For, given an open ball $B(x, r)$ about any point $x \in X$, there is an integer $n > 1/r$ and an open ball $B(x_{n,k}, 1/n) \in \mathcal{U}_n$ containing x. Because $d(x, x_{n,k}) < 1/n < r$, the ball $B(x, r)$ contains $x_{n,k} \in D$. This proves that X is separable. □

3.4 Theorem *Any separable metric space is second countable.*

Proof: This is Theorem **5**.2.5. □

3.5 Theorem *Any second countable space is Lindelöf.*

Proof: Let X be a second countable space, with countable base $\mathcal{B} = \{B\}$, and let $\mathcal{U}$ be an open covering of X. Each open set $U \in \mathcal{U}$ is a union of members $B \in \mathcal{B}$ of the base $\mathcal{B}$. Let $\mathcal{B}_\mathcal{U}$ be the set of members of the base $\mathcal{B}$ that are contained in some member of the open covering $\mathcal{U}$. For each $B \in \mathcal{B}_\mathcal{U}$, we select some $U_B \in \mathcal{U}$ with $B \subseteq U_B$. Then $\mathcal{W} = \{U_B : B \in \mathcal{B}_\mathcal{U}\}$ is a countable subcovering of the open covering $\mathcal{U}$ of X. If $x \in X$, then $x \in U$ for some $U \in \mathcal{U}$. Then $x \in B \subseteq U$ for some $B \in \mathcal{B}$, hence, $x \in B \subseteq U_B \in \mathcal{W}$. □

3.6 Theorem *Any countably compact Lindelöf space is compact.*

Proof: Let $\mathcal{U}$ be an open covering of X. Because X is Lindelöf, $\mathcal{U}$ has a countable subcovering $\mathcal{W}$. Because X is countably compact, the countable covering $\mathcal{W}$ has a finite subcovering. □

We've shown that second countability implies both separability (Theorem 5.2.2) and the Lindelöf property (Theorem 3.5). But neither converse holds.

Example 1 *The right interval space $\mathbb{R}_{(\,]}$ is separable and Lindelöf, but not second countable.* In view of Example 5.2.14, it remains only to show that $\mathbb{R}_{(\,]}$ is Lindelöf. Let $\mathcal{U} = \{U\}$ be an open covering of $\mathbb{R}_{(\,]}$. The union, V, of the real interiors $\mathrm{INT}_\mathbb{R}(U)$ of the members $U \in \mathcal{U}$ is covered by countably many of them, because the real subspace $V \subset \mathbb{R}$ is Lindelöf by Theorem 3.5. Another countable collection of $U \in \mathcal{U}$ covers $\mathbb{R} - V$, because for any point $b \in \mathbb{R} - V$ there must be a nonempty open interval (a, b) contained in V. The intervals that arise this way are necessarily *disjoint*, so there are countably many of them (indexed by a rational number from each of them).

3.7 Theorem *In metric spaces, the properties of second countability, separability, and Lindelöf are equivalent.*

Proof: A separable metric space is second countable, by Theorem 3.4. A second countable space is Lindelöf, by Theorem 3.5. Finally, we show that a Lindelöf metric space is separable. Given $k \geq 1$, let $\{B(x_{i,k}, 1/k) : i \geq 1\}$ be a countable subcover of the cover $\{B(x, 1/k) : x \in X\}$ of the metric space (X, d). Then $D = \{x_{i,k} : k, i \geq 1\}$ is a countable dense subset. □

LEBESGUE COVERING LEMMA

This lemma is concerned with this question: given an open covering $\mathcal{U}$ of a metric space (X, d), what diameter set can be pushed everywhere in X and be entirely contained at each instance within some member of $\mathcal{U}$? Recall that the *diameter* of a subset A of a metric space (X, d) is defined as

$$diam(A) = lub\, \{d(x, y) : x, y \in A\}.$$

Example 2 In the open covering $\mathcal{U} = \{(n-1, n+1) : n \in \mathbb{Z}\}$ of $\mathbb{R}$, consecutive intervals overlap in an open interval $(n-1, n)$ of length 1. So any subset $A \subset \mathbb{R}$ with $diam(A) < 1$ is contained within some member of $\mathcal{U}$.

Example 3 In the open covering $\mathcal{W} = \{(n - 1/n, n + 1 + 1/(n+1)) : n \geq 1\}$ of $(0, \infty)$ consecutive intervals overlap in an interval $(n - 1/n, n + 1/n)$ of width $2/n$. When any set $A \subset (0, \infty)$ of positive diameter is positioned about an integer $n \geq 1$ so large that $1/n < diam(A)$, then it is contained within no member of the open covering $\mathcal{W}$.

Let $\mathcal{U}$ be an open covering of a metric space (X, d). A real number $\lambda > 0$ is called a ***Lebesgue number*** of the covering $\mathcal{U}$ of X if, for each subset $A \subseteq X$ with $diam(A) < \lambda$, there exists some member $U \in \mathcal{U}$ such that $A \subseteq U$.

In Examples 2 and 3, $\mathcal{U}$ has a Lebesgue number $\lambda = 1$, while $\mathcal{W}$ has none.

3.8 Lebesgue Covering Lemma *Each open covering $\mathcal{U}$ of a compact metric space (X, d) has a Lebesgue number $\lambda > 0$.*

Proof: For each $x \in X$, there exists some neighborhood $U(x) \in \mathcal{U}$ and some open ball $B(x, r(x)) \subseteq U(x)$. Let $\{B(x_k, \frac{1}{2}r(x_k)) : 1 \leq k \leq m\}$ be a finite subcovering of the open covering $\{B(x, \frac{1}{2}r(x)) : x \in X\}$ of the compact space X.

To show that $\lambda = min\{\frac{1}{2}r(x_k) : 1 \leq k \leq m\} > 0$ is a Lebesgue number for $\mathcal{U}$, we let $\emptyset \neq A \subseteq X$ have $\delta = diam(A) < \lambda$. Since $A \cap B(x_k, \frac{1}{2}r(x_k)) \neq \emptyset$ for some $1 \leq k \leq m$, then $A \subseteq B(x_k, r(x_k)) \subseteq U(x_k)$, by the triangle inequality. $\square$

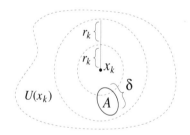
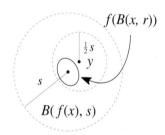

Lebesgue Covering Lemma   Uniform continuity application

The Lebesgue Covering Lemma has numerous applications in subsequent chapters. Our first application concerns a strong version of continuity.

A function $f : (X, d) \to (Y, m)$ between metric spaces is ***uniformly continuous*** if for each $s > 0$, there exists $r > 0$ such that, for all $x \in X$,

$$f(B_d(x, r)) \subseteq B_m(f(x), s).$$

3.9 Theorem *Any continuous function $f : X \to Y$ from a compact metric space (X, d) to a metric space (Y, m) is uniformly continuous.*

Proof: We must construct a radius $r > 0$ that works for the given $s > 0$ at all points $x \in X$. First, we select any Lebesgue number λ for the open covering $\mathcal{U} = \{ f^{-1}(B_m(y, \frac{1}{2}s)) : y \in Y \}$ of X, and then we take $r < \frac{1}{2}\lambda$. For each $x \in X$, $\operatorname{diam}(B_d(x, r)) < \lambda$; hence, $B_d(x, r) \subseteq f^{-1}(B_m(y, \frac{1}{2}s))$, for some $y \in Y$. So,

$$f(B_d(x, r)) \subseteq B_m(y, \tfrac{1}{2}s) \subseteq B_m(f(x), s),$$

by the triangle inequality (**D**3) for the metric m. □

Uniform continuity can fail when the domain space is non-compact.

Example 4 The continuous function $f : \mathbb{R} - \{0\} \to \mathbb{R}$, defined by $f(x) = 1/x$, is not uniformly continuous. There exists no $r > 0$ such that $f(B(x, r)) \subseteq B(f(x), 1)$ for all $x \neq 0$. For any radius $r > 0$, the ball $B(r, r) = (0, 2r)$ centered on $x = r$ has image $f(B(r, r)) = (1/(2r), \infty)$, which isn't contained in the ball $B(f(r), 1) = (1/r - 1, 1/r + 1)$.

4   Locally Compact Spaces

Some spaces, like $\mathbb{R}^n$, that fail to be compact are locally compact!

LOCAL COMPACTNESS

A topological space X is **locally compact** if, for each $x \in X$ and each neighborhood $N(x)$, there is a compact neighborhood $K(x)$ with $K \subseteq N$.

Example 1 The closed discs $D(x, r) = \{ y \in \mathbb{R}^n : d(x, y) \leq r \}$ provide arbitrarily small compact neighborhoods in real n-space $\mathbb{R}^n$. So $\mathbb{R}^n$ is locally compact.

Example 2 A discrete space is locally compact. It is compact if and only if finite.

4.1   Lemma   *These are equivalent conditions for a Hausdorff space X:*
(a)   *X is a locally compact space.*
(b)   *For each $x \in X$ and each neighborhood $N(x)$, there is an open neighborhood $V(x)$ such that $\operatorname{CLS}(V)$ is compact and $\operatorname{CLS}(V) \subseteq N$.*
(c)   *Each $x \in X$ has a compact neighborhood $C(x)$ in X.*

Proof: $(b) \Rightarrow (a)$. Take $K = \operatorname{CLS}(V)$.
  $(a) \Rightarrow (c)$. Take $N(x) = X$ and $C(x) = K(x) \subseteq N$.
  $(c) \Rightarrow (b)$. Given $x \in X$, there exists a compact, hence, closed neighborhood $C(x)$. Given any neighborhood $U(x)$, the regularity property of the compact Hausdorff subspace C provides a relatively open neighborhood $W(x)$ in C such that $W \subseteq \operatorname{CLS}_C(W) \subseteq U \cap C$. It follows that $W = R \cap C$ for some open neighborhood $R(x)$ in X. Then $V = R \cap \operatorname{INT}_X(C) \subseteq W$ is a neighborhood of x in X, and $\operatorname{CLS}_X(V)$ is a compact neighborhood $K(x) \subseteq U$, as

$$\operatorname{CLS}_X(V) \subseteq \operatorname{CLS}_X(W) \cap C = \operatorname{CLS}_C(W) \subseteq U \cap C.$$ □

4.2 Corollary *Any compact Hausdorff space is locally compact.*

Example 3 An infinite set X with the finite complement topology is compact, but not Hausdorff. Nevertheless, X is locally compact because every subspace has the finite complement topology and therefore is compact.

4.3 Theorem (a) *Any locally compact subspace A of a Hausdorff space X has the form $A = V \cap F$, where V is open and F is closed in X.*
(b) *If U is open and F is closed in a locally compact Hausdorff space X, then the subspace $A = U \cap F$ is locally compact.*

Proof: (a) Let $a \in A$. The closure in A, $\overline{V(a)} \cap A$, of some relative neighborhood $V(a) \cap A$ of a is compact, hence, closed in X. Therefore, $V(a) \cap A$ is closed in $V(a)$. Then A is closed in the union $V = \cup \{V(a) : a \in A\}$, which is an open neighborhood of A. So $A = V \cap F$ for some closed set F in X.
(b) Each $a \in A = U \cap F$ has an open neighborhood $V(a)$ in X with compact closure $\overline{V} \subseteq U$ in X. In A, the relative neighborhood $V \cap A$ of a has closure $\overline{V} \cap A = \overline{V} \cap (U \cap F) = \overline{V} \cap F$, which is closed in $\overline{V}$, hence, compact. □

Example 4 The intersection, $\mathbb{B}^2 \cap (\mathbb{R} \times [0, \infty))$, of the open unit ball $\mathbb{B}^2$ and the closed upper-half plane $\mathbb{R} \times [0, \infty)$ is a locally compact, non-compact, subspace of the (locally compact Hausdorff) real plane $\mathbb{R}^2$, according to Theorems 1.5(b) and 4.3(b).

Example 5 *A Hausdorff continuous image of a locally compact Hausdorff space may not be locally compact.* The identity function $1_\mathbb{Q} : (\mathbb{Q}, \mathcal{D}) \to (\mathbb{Q}, \mathcal{T})$ from the discrete topology to the usual topology on the space of rationals gives an example.

Locally compact spaces are particularly well-behaved with respect to product space and identification space constructions, according to the following theorem of J. H. C. Whitehead (1949):

4.4 Whitehead's Theorem *Let Z be a locally compact Hausdorff space. If $p : X \to Y$ is an identification map, then so is the product function*

$$p \times 1_Z : X \times Z \to Y \times Z.$$

Proof: This means that $W \subseteq Y \times Z$ is open in $Y \times Z$ whenever the pre-image $(p \times 1_Z)^{-1}(W) \subseteq X \times Z$ is open in $X \times Z$. To show that W contains a neighborhood of each point $<y_0, z_0> \in W$, we select a point $x_0 \in p^{-1}(y_0)$ and a basic open set $<x_0, z_0> \in U \times V \subseteq (p \times 1_Z)^{-1}(W)$. Since Z is a locally compact Hausdorff space, there exists an open neighborhood $L(z_0)$ with compact closure $\overline{L} = \text{CLS}_Z(L) \subseteq V$. Then $U \times \overline{L} \subseteq (p \times 1_Z)^{-1}(W)$. Let

$$M = \{x \in X : \{x\} \times \overline{L} \subseteq (p \times 1_Z)^{-1}(W)\},$$

so that $M \supseteq U$ is the maximal subset of X such that $M \times \overline{L} \subseteq (p \times 1_Z)^{-1}(W)$.

M is an open neighborhood of $x_0 \in X$. First, $\{x\} \times \bar{L} \subseteq (p \times 1_Z)^{-1}(W)$ for $x \in M$, and any covering of the compact space $\{x\} \times \bar{L}$ by basic open sets

$$<x, z> \in U_z \times V_z \subseteq (p \times 1_Z)^{-1}(W)$$

indexed by $z \in \bar{L}$ has a finite subcovering $\{U_{z_i} \times V_{z_i} : 1 \leq i \leq k\}$. Then $U_x = \cap \{U_{z_i} : 1 \leq i \leq k\}$ is an open neighborhood of x contained in M, since

$$U_x \times \bar{L} \subseteq U_x \times V \subseteq (p \times 1_Z)^{-1}(W),$$

where $V = \cup \{V_{z_i} : 1 \leq i \leq k\}$.

$p(M)$ *is an open neighborhood of* $y_0 = p(x_0) \in Y$. Since $p(M) \times \bar{L} \subseteq W$,

$$p^{-1}(p(M)) \times \bar{L} \subseteq (p \times 1_Z)^{-1}(W).$$

Hence, $p^{-1}(p(M)) \subseteq M$ by the maximality of M. As $M \subseteq p^{-1}(p(M))$, then $M = p^{-1}(p(M))$. This implies that $p(M)$ is open in Y, because p is an identification map and M is open in X.

So W contains the basic open neighborhood $p(M) \times L$ of $<y_0, z_0> \in W$. Thus, W is open in $Y \times Z$. This proves that $p \times 1_Z$ is an identification map. □

4.5 Corollary *If* $p : X \to Y$ *and* $q : W \to Z$ *are identification maps and if the domain of one and the codomain of the other are locally compact Hausdorff spaces, then* $p \times q : X \times W \to Y \times Z$ *is an identification map.*

Proof: If, say, Y and W are locally compact Hausdorff spaces, then

$$p \times 1_W : X \times W \to Y \times W \quad \text{and} \quad 1_Y \times q : Y \times W \to Y \times Z$$

are identification maps, by Whitehead's Theorem (4.4). So their composite, $p \times q$, is an identification map by the Transitivity Property 5.1.7. □

Example 6 The exponential function $e : \mathbb{R} \to \mathbb{S}^1$ (Example 5.1.6) is an identification map that wraps each unit interval $[k, k+1]$ in $\mathbb{R}$ once around the 1-sphere $\mathbb{S}^1$. Therefore, $e \times e : \mathbb{R} \times \mathbb{R} \to \mathbb{S}^1 \times \mathbb{S}^1$ is an identification map that wraps each unit square $[i, i+1] \times [k, k+1]$ in the real plane $\mathbb{R} \times \mathbb{R}$ once around the torus $\mathbb{S}^1 \times \mathbb{S}^1$.

BAIRE SPACES

Locally compact Hausdorff spaces have the following useful property proved for complete metric spaces in Baire's Category Theorem (2.5.6).

A space is a **Baire space** if the intersection of any countable family of open dense subsets is dense in the space.

4.6 Theorem *Any locally compact Hausdorff space is a Baire space.*

Proof: We mimic the proof of (2.5.6). Consider a countable family $\{A_m\}$ of open dense sets in a locally compact Hausdorff space X. We must show that each open set U intersects $\cap_m A_m$. Because A_1 is an open dense set, $U \cap A_1$ is nonempty and open, hence by Lemma 4.1, contains the compact closure $\bar{B}_1$ of some nonempty open set B_1. Similarly, $B_1 \cap A_2$ contains the compact closure $\bar{B}_2$ of some nonempty open set B_2. By induction, there exists an open set B_m with compact closure $\bar{B}_m \subseteq B_{m-1} \cap A_m$ for all $m \geq 1$. By Cantor's Theorem (1.3), the sequence $\{\bar{B}_m\}$ of nested closed subsets of the compact space $\bar{B}_1$ has $\cap_m \bar{B}_m \neq \emptyset$. But, by this construction, $U \cap (\cap_m A_m) \supseteq \cap_m \bar{B}_m$. Thus, U intersects $\cap_m A_m$. This proves that $\cap_m A_m$ is dense in X. ∎

In analysis, the intersection of a countable family of open sets is called a G_δ *set* and the union of a countable family of closed sets is called an F_σ *set*. For example, in $\mathbb{R}^n$ each open set and each closed set is both a G_δ and an F_σ set (Exercise 25). Theorem 4.6 precludes some sets from being G_δ sets.

Example 7 A countable dense set D of accumulation points of a Baire space X can't be a G_δ set in X. Otherwise, we would have $D = \cap_k A_k$ for some countable family $\{A_k\}$ of open sets, and $\{A_k\} \cup \{X - \{d\} : d \in D\}$ would be a countable family of open dense sets with intersection $D \cap (X - D) = \emptyset$, contrary to Theorem 4.6.

Example 8 A function $f : X \to Y$ between metric spaces is continuous at precisely those points $x \in X$ whose neighborhoods have images in Y of arbitrarily small diameter. Thus, the set of continuity of f is the G_δ set $\cap_k U_k$ in X, where U_k consists of all $x \in X$ that have a neighborhood $N(x)$ for which $diam\, f(N) < 1/k$. By Example 7, this set of continuity cannot be a dense countable set of accumulation points of a complete metric space X. For example, $\mathbb{Q}$ cannot be the set of continuity of any $f : \mathbb{R} \to Y$.

As in Section 4.3, any countable union of nowhere dense sets in a space X is called a *first category set in X* and any set that is not a first category set is called a *second category set in X*.

4.7 Theorem *Let X be a Baire space. Then:*
(a) Any countable family of closed sets that cover X has some member with nonempty interior.
(b) In X, any first category set has empty interior.
(c) In X, the complement of a first category set is a second category set.

Proof: (a) Let $\{F_k\}$ be a countable cover of X by closed sets. As $\cup_k F_k = X$, then $\cap_k (F_k)^c = \emptyset$. Since X is a Baire space, at least one of the open sets $(F_k)^c$ is not dense. Then $(\text{INT } F_k)^c = \text{CLS } (F_k)^c \neq X$; hence, $\text{INT } F_k \neq \emptyset$.

(b) Let $\{A_k\}$ be a countable family of nowhere dense sets in X, so that the open set $(\bar{A}_k)^c$ is dense in X for each $k \geq 1$. Then, by the Baire property, $\cap_k (\bar{A}_k)^c$ is dense in X. For any open set $U \subseteq \cup_k A_k \subseteq \cup_k \bar{A}_k$, we deduce that $\cap_k (\bar{A}_k)^c \subseteq (\cup_k \bar{A}_k)^c \subseteq U^c$. This makes U^c a dense closed set, i.e., equal to X. In other words, $U = \emptyset$. This proves that $\cup_k A_k$ is nowhere dense.

(c) The proof of Theorem 4.3.5 invokes just the Baire property of complete metric spaces (Theorem 2.5.6) and so applies in any Baire space. ∎

Example 9 A countable Hausdorff space X with no open points, such as $\mathbb{Q}$, cannot be a Baire space; the family $\{\{x\} : x \in X\}$ would contradict Theorem 4.7(a).

Example 10 Any first category set $A = \cup_k A_k$ in X enlarges to the first category set $B = \cup_k \mathrm{CLS}(A_k)$ in X, since nowhere dense sets $\{A_k : 1 \leq k\}$ in X have nowhere dense closures in X. Strangely, such a first category set in X may be dense in X. Any countable dense subset of a non-trivial Hausdorff space, such as $\mathbb{Q} \subseteq \mathbb{R}$, is an example.

Example 11 A first category set in a Baire space has no interior, yet the complementary second category set may also have no interior; e.g., $\mathbb{Q}$ and $\mathbb{R} - \mathbb{Q}$.

K-SPACES

In any Hausdorff space X, the compact subspaces are closed and they cover X. Therefore, as in Theorems 5.4.1-2, there is a weak topology on X relative the family $\{C_\alpha \subseteq X : \alpha \in \mathcal{A}\}$ of all compact subspaces of X. It is called the ***compactly generated*** topology and it converts X into a space $k(X)$ in which a subset is closed (open) if and only if it intersects each compact subspace C_α of X in a closed (open) subset of C_α. Also, $k(X)$ has all the compact spaces of X as subspaces and the identity function $1_X : k(X) \to X$ is continuous.

A Hausdorff space X is called a ***k-space*** if X has its compactly generated topology; equivalently, $1_X : k(X) \to X$ is a homeomorphism.

4.8 Theorem *Any locally compact Hausdorff space is a k-space.*

Proof: Let U be open in $k(X)$, where X is locally compact and Hausdorff. Each $x \in U$ has an open neighborhood V with compact closure $\overline{V} = \mathrm{CLS}(V)$ in X. Then $U \cap \overline{V}$ is open in the subspace $\overline{V}$ that X and $k(X)$ share; hence, $U \cap \overline{V} \cap V = U \cap V$ is open in the open subspace V of X and so in X as well. The neighborhoods $U \cap V(x)$ in X show that U is open in X. So $X = k(X)$. □

4.9 Theorem *Let X be a Hausdorff space. Then X is a k-space if and only if X is an identification space of a locally compact space.*

Proof: Let X be a k-space, with compact subspaces $\{C_\alpha : \alpha \in \mathcal{A}\}$. Then the surjection $q : \mathring{\cup}_\alpha C_\alpha \to X$, $q|_{C_\alpha} = i_\alpha$, is an identification map by Theorem 5.4.3. This gives half the argument, since $\mathring{\cup}_\alpha C_\alpha$ is locally compact.

Now let $p : W \to X$ be an identification map; let W be a locally compact space, hence, a k-space by Theorem 4.8; and let U be open in $k(X)$. For each compact subspace $K \subseteq W$, the image $p(K) \subseteq X$ is compact. Thus $U \cap p(K)$ is open in $p(K)$, so that $U \cap p(K) = V \cap p(K)$ for some open subset $V \subseteq X$. Then $p^{-1}(U) \cap K = p^{-1}(V) \cap K$ is open in K. Thus, $p^{-1}(U)$ is open in the k-space W, so that U is open in the identification space X. This proves that $X = k(X)$. □

4.10 Corollary *Any identification space of a k-space is a k-space.*

Proof: By the Transitivity Property **5.1.7**, the composition of identification maps is an identification map. So Theorem 4.9 yields the corollary. □

4.11 K-Factory *For each Hausdorff space X, the space k(X) is a k-space; and each continuous function $g : X \to Y$ between Hausdorff spaces is a continuous function $g : k(X) \to k(Y)$ between their k-spaces.*

Proof: The space $k(X)$ retains the compact subspaces of X as subspaces, by Theorem **5.4.2**. Conversely, each compact subspace of $k(X)$ is a compact subspace of X, by Theorem 2.1 and the continuity of $1_X : k(X) \to X$. Thus, $k(X)$ and X have the same compact subspaces and $k(X)$ is a k-space.

A continuous function $g : X \to Y$ is continuous on each compact subspace $k(X) \supseteq C \subseteq X$ and takes values in the compact subspace $Y \supseteq g(C) \subseteq k(Y)$. Thus, $g : k(X) \to k(Y)$ has a continuous restriction $g|_C : C \to k(Y)$, for each compact subspace $C \subseteq X$. Hence, $g : k(X) \to k(Y)$ is itself continuous by Theorem **5.4.4**. □

COMPACTIFICATIONS

Often, a non-compact topological space can be embedded in a compact one.

If X embeds as a dense subspace of a compact Hausdorff topological space X^*, then X^* is called a (Hausdorff) ***compactification*** of X.

When the complement $X^* - X$ consists of n points, the space X^* is called a (Hausdorff) ***n-point compactification*** of X.

Example 12 The closed unit interval $\mathbb{I} = [0, 1]$ and the 1-sphere $\mathbb{S}^1$ are compactifications of the open unit interval $\mathbb{J} = (0, 1)$. $\mathbb{J}$ densely embeds in $\mathbb{I}$ via the inclusion $i : \mathbb{J} \to \mathbb{I}$, and it also densely embeds in $\mathbb{S}^1$ via the exponential function $e : \mathbb{J} \to \mathbb{S}^1$. $\mathbb{S}^1$ is a one-point compactification of $\mathbb{J}$, while $\mathbb{I}$ is a 2-point compactification of $\mathbb{J}$.

Example 13 *The closed ordinal interval $[0, \beta]$ is a one-point compactification of the open ordinal interval $[0, \beta)$, whenever β is a limit ordinal.* $[0, \beta)$ is a subspace of $[0, \beta]$, in the order topology (Example **5.2.5**). $[0, \beta]$ is compact, by Example 1.4. Finally, $[0, \beta)$ is dense in $[0, \beta]$: any basic neighborhood $(\alpha, \beta]$ of $\beta \in [0, \beta]$ is determined by an ordinal $\alpha < \beta$ and so contains the non-limit ordinal $\alpha + 1 < \beta$ in $[0, \beta)$.

Example 14 Both the closed square $\mathbb{I} \times \mathbb{I}$ and the torus $\mathbb{S}^1 \times \mathbb{S}^1$ are compactifications of the open square $\mathbb{J} \times \mathbb{J}$, via the product embeddings $i \times i : \mathbb{J} \times \mathbb{J} \to \mathbb{I} \times \mathbb{I}$ and $e \times e : \mathbb{J} \times \mathbb{J} \to \mathbb{S}^1 \times \mathbb{S}^1$, respectively. What is a one-point compactification of $\mathbb{J} \times \mathbb{J}$?

Suppose that X admits a (Hausdorff) one-point compactification X_∞. Let $\{\infty\}$ denote the one-point complement $X_\infty - X$. Because X_∞ is a Hausdorff space, the point ∞ is closed; hence, the complement $X = X_\infty - \{\infty\}$ is open in X_∞. Because X_∞ is a compact Hausdorff space, its open subspace X is locally compact and Hausdorff by Theorem 4.3. Because X is a proper dense subspace of the Hausdorff space X_∞, it is neither closed in X_∞ nor compact.

SECTION 4 · LOCALLY COMPACT SPACES · 217

Thus, a space X that admits a (Hausdorff) one-point compactification X_∞ must be Hausdorff, non-compact, and locally compact. Also, each open neighborhood U in X_∞ of the point ∞ has a closed, hence, compact complement $X_\infty - U = K \subseteq X$. These observations provide guidelines for the construction of Hausdorff one-point compactifications.

4.12 Alexandroff's Theorem *Any non-compact, locally compact, Hausdorff space has a Hausdorff one-point compactification.*

Proof: Let $(X, \mathcal{T})$ be a non-compact, locally compact, Hausdorff space. Let $X_\infty = X \cup \{\infty\}$, where $\infty \notin X$, and let $\mathcal{T}_\infty$ denote the set of all $W \subseteq X_\infty$ such that either $W = U \in \mathcal{T}$ or $W = X_\infty - K$, where K is a compact subspace of X.

Then: (*a*) $\mathcal{T}_\infty$ is a topology on X_∞, (*b*) $(X, \mathcal{T})$ is a subspace of $(X_\infty, \mathcal{T}_\infty)$, (*c*) $(X_\infty, \mathcal{T}_\infty)$ is compact, (*d*) $(X_\infty, \mathcal{T}_\infty)$ is Hausdorff, and (*e*) X is dense in X_∞.

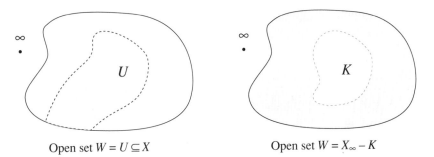

Open set $W = U \subseteq X$ · · · Open set $W = X_\infty - K$

(*a*) This part is left to Exercise 9.

(*b*) For $W \in \mathcal{T}_\infty$, the intersection $W \cap X$ equals either $U \cap X = U \in \mathcal{T}$ or $(X_\infty - K) \cap X = X - K \in \mathcal{T}$, where K is a compact, hence, closed subspace of the Hausdorff space X. Thus, the subspace topology on X is $\mathcal{T}$.

(*c*) Let $\mathcal{W}$ be an open covering of X_∞. Then some member $W \in \mathcal{W}$ contains ∞ and so equals $X_\infty - K$, for some compact subspace K of X. The covering $\mathcal{W}$ contains a finite covering $\mathcal{U}$ of this compact subspace K, hence, also the finite covering $\mathcal{U} \cup \{W\}$ of $K \cup (X_\infty - K) = X_\infty$.

(*d*) Because X is Hausdorff and open in X_∞, it suffices to construct disjoint neighborhoods of $x \in X$ and ∞ in X_∞. Since X is locally compact, it contains a compact neighborhood $K(x)$. Then K and $X_\infty - K$ are disjoint neighborhoods of x and ∞ in X_∞.

(*e*) Let $\emptyset \neq W \in \mathcal{T}_\infty$. If $W \in \mathcal{T}$, then $W \cap X = W \neq \emptyset$. If $W = X_\infty - K$, where K is a compact, hence, proper subspace of the non-compact space X, then $W \cap X = X - K \neq \emptyset$. This proves that X meets each nonempty open subset of X_∞; it is dense in X_∞. □

Example 15 The n-pronged space X_n (Example 6.1.17) is a one-point compactification of the topological disjoint union of n half-open intervals.

Example 16 The unit sphere $\mathbb{S}^n$ in $\mathbb{R}^{n+1}$ is a one-point compactification of real space $\mathbb{R}^n$. The inverse of stereographic projection (Example 4.4.12) embeds $\mathbb{R}^n$ as the dense complement of the north pole in $\mathbb{S}^n$.

Example 17 Consider the pinched torus $(\mathbb{S}^1 \times \mathbb{S}^1)/(\mathbb{S}^1 \times \{1\})$ and the pin-cushion space $\mathbb{S}^2/\mathbb{S}^0$, in which the poles in $\mathbb{S}^0 = \{-1, +1\}$ are identified. Both of these spaces are one-point compactifications of the punctured plane $\mathbb{R}^2 - \{\boldsymbol{O}\}$. The following theorem implies that $\mathbb{S}^2/\mathbb{S}^0$ and $(\mathbb{S}^1 \times \mathbb{S}^1)/(\mathbb{S}^1 \times \{1\})$ are homeomorphic.

4.13 Theorem *The (Hausdorff) one-point compactification of a non-compact, locally compact, Hausdorff space is unique.*

Proof: We show that $\mathcal{T}_\infty$ equals any compact Hausdorff topology $\mathcal{R}$ on $X_\infty = X \cup \{\infty\}$ that has $(X, \mathcal{T})$ as a subspace. Let $R \in \mathcal{R}$.

If $R \subseteq X$, then $R = R \cap X$ belongs to the subspace topology $\mathcal{T}$ on X, hence, also to $\mathcal{T}_\infty$.

If $R \not\subseteq X$, then $K = X_\infty - R \subseteq X$ is a closed, hence, compact subspace of $(X_\infty, \mathcal{R})$ and $(X, \mathcal{T})$. Therefore, $R = X_\infty - K$ belongs to $\mathcal{T}_\infty$ in this case also.

This proves that $\mathcal{R} \subseteq \mathcal{T}_\infty$. Hence, $\mathcal{R} = \mathcal{T}_\infty$ by Corollary 2.7. □

5 A Characterization of Arcs

The closed interval $\mathbb{I} = [0, 1] \subset \mathbb{R}$ is such an important topological space that it is worth considering what topological properties completely determine $\mathbb{I}$.

Hereafter, any topological space homeomorphic to $\mathbb{I}$ is called an ***arc***.

TOPOLOGICAL PROPERTIES OF ARCS

Here is a summary of some previously observed properties of arcs:

5.1 Theorem *Each arc is a compact, connected, metrizable space with exactly two non-cut points.*

The following three spaces take measure of the four properties of arcs cited in Theorem 5.1; each space has all but one of the properties.

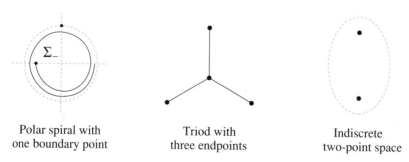

Polar spiral with one boundary point   Triod with three endpoints   Indiscrete two-point space

The polar spiral fails to be compact, the triod fails to have exactly two non-cut points, and the indiscrete two-point space fails to be metrizable.

We call any compact, connected, metrizable space a ***compactum***.

Despite the above examples, Theorem 5.1 has a valid converse that is this section's goal: *any compactum with exactly two non-cut points is an arc*.

CUT POINT SEPARATION

Here is a quick review of some pertinent terminology from Chapter **6**.

Let X be a topological space. Two subsets E and F are ***separated in X*** if

$$E \cap \mathrm{CLS}_X(F) = \emptyset = \mathrm{CLS}_X(E) \cap F.$$

A subspace $A \subseteq X$ is ***disconnected*** if A has a ***separation*** $A = E \cup F$, i.e., a decomposition into two nonempty separated subsets $E, F \subset X$. In a connected space X, a point $x \in X$ is a ***cut point*** if the point complement $X - \{x\}$ is a disconnected space, i.e., if there is a separation $X - \{x\} = E \cup F$.

5.2   Proposition   *Let x be a cut point of a connected Hausdorff space X, and let $X - \{x\} = E \cup F$ be a separation. Then,*
(a)   *E and F are nonempty open subsets in X,*
(b)   *$\overline{E} = \mathrm{CLS}_X(E) = E \cup \{x\}, \overline{F} = \mathrm{CLS}_X(F) = F \cup \{x\}$, and*
(c)   *$\overline{E}$ and $\overline{F}$ are non-degenerate, connected subspaces of X.*

Proof: (a) Since $E \cap \overline{F} = \emptyset = \overline{E} \cap F$ and since $\{x\}$ is closed in X, then E and F are the complements of the closed sets $\overline{F} \cup \{x\}$ and $\overline{E} \cup \{x\}$, respectively.

(b) Since E is open in X, its complement $X - E = F \cup \{x\}$ is closed in X. Hence, $F \subseteq \overline{F} \subseteq F \cup \{x\}$. Because X is connected, the nonempty open set F cannot also be closed. So $\overline{F} = F \cup \{x\}$. Similarly, $\overline{E} = E \cup \{x\}$.

(c) Were $\overline{E} = E \cup \{x\}$ to have a separation $U \cup V$, with, say, $U \subseteq E$ and $x \in V$, then X would have the separation $X = U \cup (F \cup V)$, as

$$\overline{U} \cap (F \cup V) = (\overline{U} \cap F) \cup (\overline{U} \cap V) \subseteq (\overline{E} \cap F) \cup (\overline{U} \cap V) = \emptyset$$

and

$$U \cap \overline{(F \cup V)} = U \cap (\overline{F} \cup \overline{V}) = (U \cap \overline{F}) \cup (U \cap \overline{V}) \subseteq (E \cap \overline{F}) \cup (U \cap \overline{V}) = \emptyset.$$

Since X is connected, then $\overline{E}$, and similarly $\overline{F}$, must be connected. $\square$

5.3   Proposition   *Let x and z be distinct cut points of a connected Hausdorff space X, and let $X - \{x\} = E_x \cup F_x$ and $X - \{z\} = E_z \cup F_z$ be separations. Then there are four mutually exclusive possibilities:*
(a)   *$x \in E_z$ and $z \in F_x$; hence, $\overline{E}_x \subseteq E_z$ and $\overline{F}_z \subseteq F_x$.*
(b)   *$x \in F_z$ and $z \in E_x$; hence, $\overline{F}_x \subseteq F_z$ and $\overline{E}_z \subseteq E_x$.*
(c)   *$x \in F_z$ and $z \in F_x$; hence, $\overline{E}_x \subseteq F_z$ and $\overline{E}_z \subseteq F_x$.*
(d)   *$x \in E_z$ and $z \in E_x$; hence, $\overline{F}_x \subseteq E_z$ and $\overline{F}_z \subseteq E_x$.*

Proof: There are two possibilities for the location of either cut point x or z in the given separation of the complement of the other cut point. The cases listed above record the four possibilities for the location of the pair $\{x, z\}$. The containment relations stated in each case are verified this way.

In (a), $x \in E_z$ implies that $\overline{F}_z = F_z \cup \{z\} = X - E_z$ contains z but not x. Thus $\overline{F}_z = \overline{F}_z \cap (X - \{x\})$ is the union of the two separated sets $(\overline{F}_z \cap E_x)$ and $(\overline{F}_z \cap F_x)$. Also $z \in F_x$ implies that the second of the two separated sets, $\overline{F}_z \cap F_x$, contains z. Because $\overline{F}_z$ is connected by Proposition 5.2(c), it must be that the first separated set, $\overline{F}_z \cap E_x$, is empty. Then we have $E_x \subseteq X - \overline{F}_z = E_z$; hence, $\overline{E}_x = E_x \cup \{x\} \subseteq E_z$, in Case (a). It follows by set complementation that the containment relation $\overline{F}_z \subseteq F_x$ also holds in this case. □

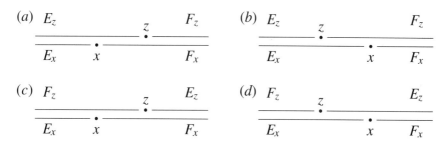

Four cases in Proposition 5.3

Existence of Non-Cut Points

The next result shows how the topological properties of compactness, connectedness, and metrizability conspire to guarantee non-cut points in a non-degenerate space, i.e., a space with more than one point.

5.4 Existence Theorem *Any compact, connected, non-degenerate, metrizable space X has at least two non-cut points.*

Proof: By Section 3, a compact metrizable space is separable and Hausdorff. Therefore, the theorem follows from the next lemma. □

5.5 Lemma *Any connected, non-degenerate, separable, Hausdorff space that has at most one non-cut point is non-compact.*

Proof: Let D be a countable dense subset of X, and let N be the (empty or singleton) set of non-cut points of X. Because X is non-degenerate and Hausdorff, the dense set D contains at least two points, so that $D - N$ is non-empty and consists of cut points. Let x_1 be the first member of (some fixed enumeration of) $D - N$. Let $X - \{x_1\} = E_1 \cup F_1$ be a separation, labeled so that $N \subseteq F_1$. So E_1 is disjoint from N. But, it contains members of the dense set D, because it is open by Proposition 5.2. Let x_2 be the first member of

$E_1 \cap (D - N)$, and let $X - \{x_2\} = E_2 \cup F_2$ be a separation, labeled so that $x_2 \in E_1$ and $x_1 \in F_2$. Then $\bar{E}_2 \subseteq E_1$ and $\bar{F}_1 \subseteq F_2$, by Proposition 5.3; E_2 is open, by Proposition 5.2; and $N \cup \{x_1\} \subseteq F_2$ so that $E_2 \cap (D - (N \cup \{x_1\}))$ is nonempty, by construction. An iteration of this argument provides a sequence of separations $X - \{x_k\} = E_k \cup F_k$, for $1 \le k$, with these properties:

$$\ldots \subseteq \bar{E}_{k+1} \subseteq E_k \subseteq \bar{E}_k \subseteq \ldots \subseteq E_2 \subseteq \bar{E}_2 \subseteq E_1,$$

$$N \subseteq F_1 \subseteq \bar{F}_1 \subseteq F_2 \subseteq \ldots \subseteq F_k \subseteq \bar{F}_k \subseteq F_{k+1} \subseteq \ldots,$$

$$\bar{E}_k = E_k \cup \{x_k\}, F_k = F_k \cup \{x_k\}, \text{ and } \{x_1, x_2, \ldots, x_k\} \subseteq F_{k+1}.$$

Because the nested sets $\bar{F}_k$ contain x_1 and are connected by Proposition 5.2, their union $C = \cup_k \bar{F}_k$ is connected, by Theorem **6**.1.6. Also, C contains $N \cup D$ by construction. Then it follows from the containment relation $N \subseteq C$ that any $x \in X - C$ is a cut point of X, and it follows from the containment relations $C \subseteq X - \{x\} \subseteq X = \bar{C}$ (in view of the density of $D \subseteq C$) that $X - \{x\}$ is connected. It must be that $X - C = \emptyset$, so that $X = C$.

The nested family of nonempty closed sets $\{\bar{E}_k : 1 \le k\}$ has intersection:

$$\cap_k \bar{E}_k = \cap_k E_k = \cap_k (X - \bar{F}_k) = X - \cup_k \bar{F}_k = X - C = \emptyset.$$

Therefore, X is non-compact, by Cantor's Theorem (1.3). □

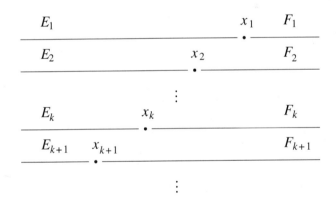

Construction in Lemma 5.5

5.6 Corollary *Let X be a compact, connected, non-degenerate, metrizable space. Let $x \in X$ be any cut point of X, and let $X - \{x\} = E_x \cup F_x$ be any separation. Then,*
(a) *each of the sets E_x and F_x contains at least one non-cut point of X; and*
(b) *each non-cut point of $\bar{E}_x$ or $\bar{F}_x$ distinct from x is a non-cut point of X.*

Proof: By the Existence Theorem 5.4, the compact, connected, non-degenerate, metrizable subspace $\bar{E}_x$ has at least two non-cut points. So $\bar{E}_x$ contains some non-cut point $z \in \bar{E}_x = E_x \cup \{x\}$. Then z is a non-cut point of X.

Otherwise, there would be a separation $X - \{z\} = E_z \cup F_z$. By Proposition 5.3, we may label so that $x \in F_z$, $z \in E_x$, $\overline{F}_x \subseteq F_z$, and $\overline{E}_z \subseteq E_x$. Then the separated sets $(\overline{E}_x \cap E_z)$ and $(\overline{E}_x \cap F_z)$, which contain E_z and x, respectively, would give a separation

$$\overline{E}_x - \{z\} = (\overline{E}_x \cap E_z) \cup (\overline{E}_x \cap F_z),$$

contradicting the fact that z is a non-cut point of $\overline{E}_x$. This proves the claims for E_x and $\overline{E}_x$, and similar arguments apply to F_x and $\overline{F}_x$. □

5.7 Corollary *Let X be a compactum with exactly two non-cut points p and q. Let $x \in X$ be any cut point of X, and let $X - \{x\} = E_x \cup F_x$ be any separation. Then we may label so that:*
(a) $\overline{E}_x = E_x \cup \{x\}$ *is a compactum with exactly two non-cut points p and x.*
(b) $\overline{F}_x = F_x \cup \{x\}$ *is a compactum with exactly two non-cut points x and q.*

Proof: By the Existence Theorem 5.4, each one of $\overline{E}_x$ and $\overline{F}_x$ is a compactum with at least two non-cut points. By Corollary 5.6, $\overline{E}_x$ has at most two non-cut points, x and one of the non-cut points, p, of X. Likewise, $\overline{F}_x$ has at most two non-cut points, x and the other non-cut point, q, of X. □

SEPARATION ORDER RELATION

For the remainder of Section 5, we work with a compactum X that has exactly two non-cut points p and q. Then, for each $x \in X - \{p, q\}$, there exists a separation $X - \{x\} = E_x \cup F_x$. We label so that $p \in E_x$ and $q \in F_x$, as we may by Corollary 5.7. This convention eliminates two of the four notational possibilities in Proposition 5.3.

For distinct points $x, z \in X - \{p, q\}$, there are just two mutually exclusive possibilities:
(a) $x \in E_z, z \in F_x, \overline{E}_x \subseteq E_z, \overline{F}_z \subseteq F_x$:

$$\begin{array}{ccccc}
& E_z & & F_z & \\
p & \underline{} & z & \underline{} & q \\
\bullet & & \bullet & & \bullet \\
\bullet & \underline{} & \bullet & \underline{} & \bullet \\
p & E_x & x & F_x & q
\end{array}$$

(b) $x \in F_z, z \in E_x, \overline{F}_x \subseteq F_z, \overline{E}_z \subseteq E_x$:

$$\begin{array}{ccccc}
& E_z & & F_z & \\
p & \underline{} & z & \underline{} & q \\
\bullet & & \bullet & & \bullet \\
\bullet & \underline{} & & \underline{} & \bullet \\
p & E_x & x & F_x & q
\end{array}$$

We say $x < z$ or $z < x$ in the ***separation order*** on $X - \{p, q\}$ if (a) or (b) holds, respectively. Notice that the case that holds can be expressed independently in terms of the relationship of x to the separation $E_z \cup F_z$, or the relationship of z to the separation $E_x \cup F_x$. Thus, the separation order depends upon just the cut points x and z, and not on the specific separations.

5.8 Lemma   *The separation order $<$ on $X - \{p, q\}$ is a simple order:*
(a) *Trichotomy: for $x, z \in X - \{p, q\}$, exactly one of the relations holds:*

$$x < z, \qquad x = z, \quad \text{or} \quad z < x.$$

(b) *Transitivity: for $x, y, z \in X - \{p, q\}$, $x < y$ and $y < z$ imply $x < z$.*

Proof: (a) By Proposition 5.3 and the labeling convention, either $x < z$ or $z < x$, but not both.
 (b) If $x < y$ and $y < z$, then $\bar{E}_x \subseteq E_y \subseteq \bar{E}_y \subseteq E_z$, hence, $x < z$. □

The separation order $<$ on $X - \{p, q\}$ extends to a simple order on X by letting $p < q$ and $p < x < q$ for all $x \neq p, q$. Let $x \leq z$ mean $x < z$ or $x = z$. Then $\leq$ is a total order, as in Example 5.2.5.

For any pair of points $x, z \in X$, there are the intervals:

$$(x, z), \qquad (x, z], \qquad [x, z), \quad \text{and} \quad [x, z].$$

As usual, they are the sets of all points $y \in X$ satisfying these conditions:

$$x < y < z, \qquad x < y \leq z, \qquad x \leq y < z, \quad \text{and} \quad x \leq y \leq z.$$

These are called ***separation subintervals*** of $X = [p, q]$.

SEPARATION ORDER TOPOLOGY

The ***separation order topology*** on X has a subbase consisting of the intervals $[p, x)$ and $(z, q]$, where $x, z \in X$. Thus, each interval $(x, z) = [p, z) \cap (x, q]$ is open and each interval $[x, z] = (X - [p, x)) \cap (X - (z, q])$ is closed in the separation order topology on X.

5.9 Theorem   *The separation order topology on the compactum X with exactly two non-cut points coincides with the original topology on X.*

Proof: The separation order topology is Hausdorff. For $x < z$ in X, either the interval (x, z) is empty or contains some point $x < y < z$. Therefore, either $[p, z)$ and $(x, q]$ or $[p, y)$ and $(y, q]$ are disjoint neighborhoods of x and z, respectively. (The next proposition shows that only the latter case is possible.) The subbase sets for the separation order topology have the forms:

$$[p, x) = \begin{cases} \emptyset, & \text{if } p = x \\ E_x, & \text{if } p < x < q \\ X - \{q\}, & \text{if } x = q. \end{cases} \qquad (z, q] = \begin{cases} X - \{p\}, & \text{if } p = z \\ F_z, & \text{if } p < z < q \\ \emptyset, & \text{if } z = q. \end{cases}$$

These sets are open in the compactum X, by Proposition 5.2. It follows that the identity function on X is a continuous bijection from the original compact topology to the Hausdorff separation order topology. Consequently, this identity function is a homeomorphism, by Theorem 2.5. □

Thus, each open interval (x, z) is open and each closed interval $[x, z]$ is closed in the original topology on the compactum X.

5.10 Proposition   *If $x < z$ in the separation order on X, then $(x, z) \neq \emptyset$.*

Proof: Using the above expressions for $[p, x)$ and $(z, q]$, it is straightforward to express the separation interval (x, z) using the separations for x and z:

$$(x, z) = \begin{cases} E_z - \{p\}, & p = x < z < q \\ E_z \cap F_x, & p < x < z < q \\ X - \{p, q\}, & p = x < z = q \\ F_x - \{q\}, & p < x < z = q. \end{cases}$$

The compacta X, $\overline{E}_z = E_z \cup \{z\}$, and $\overline{F}_x = F_x \cup \{x\}$ cannot equal their sets of non-cut points, $\{p, q\}$, $\{p, z\}$ and $\{x, q\}$. So the interval (x, z) is nonempty in the first and last two cases, above. It is also nonempty in the second case; otherwise, the relations

$$E_z \cap F_x = (X - \overline{F}_z) \cap (X - \overline{E}_x) = X - (\overline{F}_z \cup \overline{E}_x)$$

would provide the separation $X = \overline{F}_z \cup \overline{E}_x$ of the connected space X. □

BINARY COORDINATE SYSTEM IN $\mathbb{I}$

We are now prepared to prove the main theorem of this section; we show that the compactum X with exactly two non-cut points has a coordinate system that matches the binary coordinate system on $\mathbb{I}$.

The binary system involves an iteration of the process of subdividing intervals into two equal subintervals, beginning with $\mathbb{I}$.

Each of the rational subdivision points that arise in $\mathbb{I}$ via this inductive process has the form $A/2^k$ and is called a **binary fraction**.

The subintervals that arise are called **binary intervals** and they are labeled as follows. The interval $\mathbb{I}$ is subdivided into two equal first stage binary subintervals $B[0] = [0, 1/2]$ and $B[1] = [1/2, 1]$. Then inductively, each $(k-1)$st stage binary interval $B[b_1 b_2 \ldots b_{k-1}]$ is subdivided into two equal binary subintervals $B[b_1 b_2 \ldots b_{k-1} 0]$ and $B[b_1 b_2 \ldots b_{k-1} 1]$, in order.

The real number $0 \leq r \leq 1$ has the **binary expansion** $(.b_1 b_2 \ldots b_k \ldots)_2$, with digits $b_k \in \{0, 1\}$, if and only if $r = \cap_k B[b_1 b_2 \ldots b_k]$.

Each binary fraction has two binary expansions, one ending in *zeros*, the other ending in *ones;* every other real number has a unique binary expansion.

SEPARATION COORDINATE SYSTEM

We now create a similar coordinate system on the compactum X with exactly two non-cut points p and q. We begin with any countable dense subset $D \supseteq \{p, q\}$ of the compact, hence, separable metric space X. Then we iterate the process of subdividing separation intervals at members of D, beginning with $X = [p, q]$. The open and closed separation subintervals that arise are called **D-intervals**. They are labeled as follows.

Because D is dense in the compactum X and the interval (p, q) is open and nonempty, there exists a first member x_1 of (some fixed enumeration of) $D \cap (p, q)$. The point x_1 subdivides the compactum $X = [p, q]$ into two 1st stage D-intervals $D[0] = [p, x_1]$ and $D[1] = [x_1, q]$. They are analogues of the two binary intervals $B[0] = [0, \frac{1}{2}]$ and $B[1] = [\frac{1}{2}, 1]$ of $\mathbb{I} = [0, 1]$.

Because D is dense in the compactum X and the open D-intervals $D(0) = (p, x_1)$ and $D(1) = (x_1, q)$ are nonempty, there exists a first member $x_{01} \in D \cap D(0)$ and a first member $x_{11} \in D \cap D(1)$. They subdivide the intervals $D[0] = [p, x_1]$ and $D[1] = [x_1, q]$ into four 2nd stage closed D-intervals:

$$D[00] = [p, x_{01}], \quad D[01] = [x_{01}, x_1], \quad D[10] = [x_1, x_{11}], \quad D[11] = [x_{11}, q].$$

Inductively, each $(k-1)$st stage D-interval $D[.b_1 b_2 \ldots b_{k-1}]$ contains a first member $x_{b_1 b_2 \ldots b_{k-1} 1} \in D \cap D(b_1 b_2 \ldots b_{k-1})$, which subdivides $D[b_1 b_2 \ldots b_{k-1}]$ into two kth stage D-intervals $D[b_1 b_2 \ldots b_{k-1} 0]$ and $D[b_1 b_2 \ldots b_{k-1} 1]$, in order.

Then each sequence $b_1, b_2, \ldots, b_k, \ldots$ of digits $b_k \in \{0, 1\}$ determines a nested sequence of closed D-intervals

$$X \supset D[b_1] \supset D[b_1 b_2] \supset \ldots \supset D[b_1 b_2 \ldots b_k] \supset \ldots .$$

Because X is compact and these nested, closed, D-intervals are nonempty, Cantor's Theorem (1.3) implies that $\cap_k D[b_1 b_2 \ldots b_k]$ is nonempty. Furthermore, this intersection is a single point of X. Otherwise, $\cap_k D[b_1 b_2 \ldots b_k]$ would contain distinct points $x < z$, the nonempty open separation interval (x, z) between them, and, therefore, some member y of the dense subset D. But were y the kth member in the enumeration of D, it would be an endpoint of a kth stage D-interval. It couldn't belong to the open kth stage D-interval $D(b_1 b_2 \ldots b_k)$ that contains (x, z).

A point $x \in X$ has **D-expansion** $(b_1 b_2 \ldots b_k \ldots)_D$ if $x = \cap_k D[b_1 b_2 \ldots b_k]$.

Each member $D - \{p, q\}$ has two D-expansions, one ending in *zeros*, the other endings in *ones*. Every other member of X belongs to a unique kth stage D-interval for each $k \geq 1$; hence, it has a unique D-expansion.

5.11 Arc Characterization *A topological space is an arc if and only if it is a compactum with exactly two non-cut points.*

Proof: By Theorem 5.9, a compactum X with exactly two non-cut points p and q has the separation order topology. Let $h : \mathbb{I} \to X$ be the function that sends the point $r \in \mathbb{I}$ with binary expansion $(.b_1b_2\ldots b_k\ldots)_2$ to the point $h(r) \in X$ with D-expansion $(b_1b_2\ldots b_k\ldots)_D$.

h is well-defined. Any $r \in \mathbb{I}$ with distinct binary expansions is a binary fraction shared by consecutive binary intervals; one expansion terminates in *ones* and another terminates in *zeros*. The corresponding D-expansions that terminate in *ones* and *zeros* describe the same point $h(r) \in X$, a subdivision point in $D - \{p, q\}$ shared by consecutive D-intervals.

h is surjective. Any point $x \in X$ belongs to at least one nested sequence of closed D-intervals:

$$X \supset D[b_1] \supset D[b_1b_2] \supset \ldots \supset D[b_1b_2\ldots b_k] \supset \ldots .$$

Then x is the image of the point $r = (.b_1b_2\ldots b_k\ldots)_2 \in \mathbb{I}$.

h is one-one. Two D-expansions that represent the same member of X terminate in *zeros* and *ones* and describe a subdivision point $x \in D - \{p, q\}$. The corresponding binary expansions represent the same binary fraction.

h is a homeomorphism. By definition, h sends each binary interval $B[b_1b_2\ldots b_k]$ bijectively onto the corresponding D-interval $D[b_1b_2\ldots b_k]$. If $A \subseteq \mathbb{I}$ is arbitrarily near r, then A enters any open binary interval that contains r and also any pair of consecutive binary intervals that share r as an endpoint. Therefore, the image $h(A)$ enters every open D-interval that contains $h(r)$ and also every pair of consecutive D-intervals that share $h(r)$ as an endpoint. Since D is dense in X and consists of the endpoints of the D-intervals, it follows that $h(A)$ enters every open separation interval that contains $h(r)$. Thus, the image set $h(A)$ is arbitrarily near the image point $h(r)$. This establishes the continuity of h. Because $\mathbb{I}$ is compact and X is Hausdorff, h is a homeomorphism by Theorem 2.5. □

Similar techniques give the following characterization of the 1-sphere $\mathbb{S}^1$ (Exercises 5 and 6).

5.12 Circle Characterization *A topological space is a simple closed curve (i.e., homeomorphic to the 1-sphere $\mathbb{S}^1$) if and only if it is a nondegenerate compactum in which each pair of distinct points is a cut set.*

Exercises

SECTION 1

1. Give a reformulation of the proof of Theorem 1.1 that appeals directly to the order completeness of $(\mathbb{R}, \leq)$ to prove that the set A equals $\mathbb{I}$.

2. Decide whether the following are compact topological spaces:
   (a) an infinite set with the particular point topology (Example 4.1.9),
   (b) $\mathbb{N}$ with the Fort topology with respect to $p = 0 \in \mathbb{N}$ (Example 4.2.12),

(c) an uncountable set with the countable complement topology (Example **4**.1.7),
(d) the half-disc space (Exercise **4**.1.6),
(e) the line with two origins (Exercise **4**.1.7), and
(f) the tangent disc space (Exercise **4**.1.8).

3 Investigate the following two claims:
(a) Every compact subspace of a topological space has compact closure.
(b) No compact subspace of a topological space has compact interior.

4 (a) Prove that the quotient space $\mathbb{R}/\mathbb{Z}$, obtained from the real line $\mathbb{R}$ by identifying the set $\mathbb{Z}$ of integers to a point, is non-compact.
(b) Prove that the quotient space $\mathbb{R}^2/L^2$, obtained from the real plane $\mathbb{R}^2$ by identifying the lattice subset $L = (\mathbb{Z} \times \mathbb{R}) \cup (\mathbb{R} \times \mathbb{Z})$ to a point, is non-compact.

A subspace $A \subseteq X$ of a topological space X is ***co-compact*** if the complementary subspace $X - A$ is compact.

5 Prove that the quotient space $\mathbb{R}^2/A$ of the real plane $\mathbb{R}^2$, obtained by identifying a co-compact subspace A to a point, is compact.

6 Decide whether a connected, non-compact, Hausdorff space has any nonempty compact open subspaces.

7 Describe all the dense compact subspaces of a Hausdorff space.

8 Prove Cantor's Theorem (1.3).

9 (a) Consider the obvious bijection $HE \longleftrightarrow EE$ that matches the tangent circle C_k in the Hawaiian earrings HE (Example 8) with the tangent circle D_k in the expanding earrings space EE (Example 9) for each $k \geq 1$. Is it continuous in either direction?
(b) Is either DE or EE is homeomorphic to the doubled Hawaiian earrings space DHE (Example 2.2).

10 (a) Prove that the quotient space $\mathbb{R}/(\mathbb{R} - [0, 1])$ is compact but not Hausdorff.
(b) Prove that the quotient space $\mathbb{R}/(\mathbb{R} - (0, 1))$ is compact and Hausdorff.

11 Let $X = \mathbb{N} - \{1\}$ and let $U_n = \{x \in X : x \text{ divides } n\}$ for each $n \in X$. Let X have the topology with subbase $\{U_n : n \in X\}$. Prove:
(a) X is T_0 and T_4, but not T_1.
(b) X is not compact, but every point of X has a compact neighborhood.

A metric space (X, d) is ***well-chained*** if, for each pair $x, y \in X$ and each $r > 0$, there exist finitely many points $x = x_0, x_1, \ldots x_n = y$ such that $d(x_{k-1}, x_k) < r$ for $1 \leq k \leq n$.

12 Prove:
(a) A connected metric space is well-chained, but not conversely.
(b) A compact, well-chained, metric space is connected.

13 Let $\{A_\alpha : \alpha \in \mathcal{A}\}$ be a family of compact subspaces of a Hausdorff space X such that each finite intersection $\cap \{A_\alpha : \alpha \in \mathcal{F} \subseteq \mathcal{A}\}$ is a connected subspace of X. Prove that $\cap \{A_\alpha : \alpha \in \mathcal{F} \subseteq \mathcal{A}\}$ is a connected subspace of X.

14 Let Ω be the first uncountable ordinal and let $A \subseteq [0, \Omega)$ be a closed uncountable subset whose complement in $[0, \Omega)$ is also uncountable. Prove:
(a) The quotient map $[0, \Omega) \to [0, \Omega)/A$ is a continuous closed surjection.
(b) The quotient space $[0, \Omega)/A$ is compact but not first countable.

SECTION 2

1   (a)   Prove that any topology contained in a compact topology is itself compact.
  (b)   Prove that any topology containing a Hausdorff topology is itself Hausdorff.
  (c)   Prove that distinct compact Hausdorff topologies on a set X are not related by the containment relation in $\mathcal{P}(X)$.
  (d)   Prove that any set $X \sim \mathbb{R}$ admits distinct compact Hausdorff topologies.

2   Prove that a continuous function $f : \mathbb{I} \to \mathbb{I}$ is a homeomorphism if and only if either
  (a) $f(0) = 0$, $f(1) = 1$, and f is strictly increasing ($s < t \Rightarrow f(s) < f(t)$), or
  (b) $f(0) = 1$, $f(1) = 0$, and f is strictly decreasing ($s < t \Rightarrow f(s) > f(t)$).

3   Prove that a subspace $X \subset \mathbb{R}$ is compact if and only if every continuous function $f : X \to \mathbb{R}$ is bounded and attains a maximum value on X.

4   Prove that the quotient space $\mathbb{D}^n / \mathbb{S}^{n-1}$ is homeomorphic to $\mathbb{S}^n$.

5   Let X be compact and let Y be Hausdorff. Prove:
  (a)   The projection $p_Y : X \times Y \to Y$ is a closed continuous function.
  (b)   A function $f : X \to Y$ is continuous if and only if its graph,
$$\Gamma(f) = \{<x, y> \in X \times Y : y = f(x), x \in X\},$$
is closed in $X \times Y$.

6   Let X be a compact Hausdorff space and let $f : X \to X$ be a continuous function. Prove that there exists a nonempty closed subset $B \subseteq X$ such that $f(B) = B$. (Hint: Try X as a first approximation to B.)

7   Complete the proof of the normality property for a compact Hausdorff space X (Theorem 2.12). Show that if $F, G \subseteq X$ are disjoint closed sets, then there exist disjoint open sets U and V containing F and G, respectively.

8   Let F and G be compact subspaces of a metric space (X, d). Prove:
  (a) $\text{diam } F = \text{lub}\{d(x, y) : x, y \in F\}$ equals $d(x, y)$ for some $x, y \in F$.
  (b) $d(F, G) = \text{lub}\{d(x, y) : x \in F, y \in G\}$ equals $d(x, y)$ for some $x \in F$ and $y \in G$.

9   Let X be a compact Hausdorff space and let R be a closed equivalence relation on X (i.e., $R \subseteq X \times X$ is closed in the topological product). Prove that the quotient space X/R is Hausdorff.

10   Prove that any regular space with a countable base is normal.

11   Let $f : X \to Y$ be a continuous closed surjection. Prove that if Y is compact and if each pre-image $f^{-1}(y)$, $y \in Y$, is a compact subspace of X, then X is compact.

12   Let A and B be compact subspaces of a space X, and let N be an open neighborhood of $A \times B$ in $X \times X$. Construct an open box neighborhood $A \times B \subseteq U \times V \subseteq N$.

SECTION 3

1   (a)   Verify the Limit-Accumulation Properties (2.3.2) in any first countable space.
  (b)   Prove that a first countable topological space has the Bolzano-Weierstrass Property if and only if it is sequentially compact.

2   Prove this partial converse of Theorem 3.5: any metrizable Lindelöf space is second countable.

3. Prove: Any continuous image of a countably compact space is countably compact.

4. (a) Prove that any continuous surjection from a compact space onto a Hausdorff space has compact point pre-images.
(b) Prove that any Hausdorff space that is the continuous image of a compact metric space is second countable.

5. Let ∞ denote a point that doesn't belong to real n-space $\mathbb{R}^n$. Call a subset V of the union $\mathbb{R}^n_\infty = \mathbb{R}^n \cup \{\infty\}$ *open* if either (i) $V = U \subseteq \mathbb{R}^n$, where U is open in $\mathbb{R}^n$, or (ii) $V = \mathbb{R}^n_\infty - K$, where $K \subseteq \mathbb{R}^n$ is a compact subspace. Prove that these sets constitute a compact Hausdorff topology on $\mathbb{R}^n_\infty$ that has real n-space $\mathbb{R}^n$ as a subspace.

6. Prove:
(a) A compact metric space is complete.
(b) Not every complete metric space is compact.

7. (a) Prove that a compact metric space has a finite diameter.
(b) Prove that every metric space has finite diameter with respect to some topologically equivalent metric.

8. Find a Lebesgue number for each of the following coverings:
(a) the covering $\mathcal{U}_r = \{(n-r, n+r) \subseteq \mathbb{R}: n \in \mathbb{Z}\}$, where $r > 0$, of $\mathbb{R}$,
(b) the covering $\mathcal{U}_r = \{B(x, r) \subseteq \mathbb{R}^2: x \in \mathbb{Z} \times \mathbb{Z}\}$, where $r > 1$, of $\mathbb{R}^2$,
(c) the covering $\{X - x : x \in X\}$ of an arbitrary compact metric space (X, d), and
(d) the covering $\mathcal{U}_r = \{B(x, r) \subseteq X: x \in X\}$, where $r > 0$, of an arbitrary compact metric space (X, d).

9. Let (X, d) be a compact metric space and let $\mathcal{F} = \{F\}$ be a family of closed sets with empty intersection $\cap \mathcal{F} = \emptyset$. Prove that there is a constant $\lambda > 0$ such that every subset $A \subseteq X$ with $diam(A) < \lambda$ is disjoint from at least one $F \in \mathcal{F}$.

10. Prove that the open ordinal interval $[0, \Omega)$ is countably compact but not compact.

11. Prove that the open ordinal interval $[0, \Omega)$ is not Lindelöf, but that the closed ordinal interval $[0, \Omega]$ is Lindelöf.

SECTION 4

1. Determine whether the right interval space $\mathbb{R}_{(\,]}$ (Example 5.2.10) is locally compact.

2. Consider the topologist's sine curve $X = \{< x, \sin 1/x > \in \mathbb{R}^2 : 0 < x \leq 1/\pi\}$. Determine whether $X \cup \{(0, 0)\}$ and $\text{CLS}(X)$ are locally compact subspaces of $\mathbb{R}^2$.

3. Show that the collection $\{\emptyset\} \cup \{(0, 1)\} \cup \{(1/n, 1) : n = 2, 3, \ldots\}$ of intervals is a topology on the open interval $\mathbb{J} = (0, 1)$. Prove that with this topology $\mathbb{J}$ is locally compact, but that every neighborhood has a non-compact closure.

4. Prove that the image of a locally compact space under a continuous open function is locally compact.

5. Prove that the quotient space $\mathbb{R}/\mathbb{Z}$ is the Hausdorff image of a first countable, locally compact, Hausdorff space under a continuous closed function, yet it is neither first countable nor locally compact.

6   Prove that an open subspace of a Baire space is a Baire space.

7   Prove that in a Baire space the intersection of a countable family of open dense sets is a second category set, i.e., is not the countable union of nowhere dense sets.

8   Prove that $\mathbb{R}_{(\,]}$, the real line $\mathbb{R}$ with the right interval topology (Example 5.2.10), is a Baire space.

9* Let $(X, \mathcal{T})$ be a non-compact, locally compact, Hausdorff space. Let $\mathcal{T}_\infty$ consist of all subsets $W \subseteq X_\infty = X \cup \{\infty\}$ ($\infty \notin X$) such that either $W = U \in \mathcal{T}$ or $W = X_\infty - K$, where K is a compact subspace of X. Prove that $\mathcal{T}_\infty$ is a topology on X_∞.

10 Prove or disprove: A compact Hausdorff space X is the one-point compactification of the complement $X - \{x\}$ of each of its points $x \in X$.

11 Describe the Hausdorff one-point compactifications of these non-compact, locally compact, Hausdorff spaces:
    (a) the open unit ball $\mathbb{B}^n$ in Euclidean n-space,
    (b) the complement $\mathbb{R} - \{0, 1\}$ of the two points 0, 1 of the real line $\mathbb{R}$,
    (c) the topologist's sine curve $\{<x, \sin 1/x> \in \mathbb{R}^2 : 0 < x \le 1/\pi\}$, and
    (d) the doubled topologist's sine curve $\{<x, \sin 1/x> \in \mathbb{R}^2 : 0 < |x| \le 1/\pi\}$.

12 Determine whether the one-point compactification of the subspace $\mathbb{R} - \mathbb{Z}$ of the real line $\mathbb{R}$ is homeomorphic to the doubled Hawaiian earrings subspace $DHE \subseteq \mathbb{R}^2$ (Example 2.2).

13 Define embeddings into the real line $\mathbb{R}$ of the one-point compactifications of the discrete spaces $\mathbb{N}$ and $\mathbb{Z}$ of natural numbers and integers.

14 Prove: If one does not require the Hausdorff property of a compactification, any space has a one-point compactification.

15 Let X_∞ and Y_∞ denote one-point compactifications of two non-compact, locally compact, Hausdorff spaces. Prove that $X \cong Y$ implies $X_\infty \cong Y_\infty$, but not conversely.

16 Show: A compact Hausdorff space has only itself as a compactification.

17 Prove that any totally disconnected, locally compact, Hausdorff space can be embedded in a product of discrete two-point spaces.

18 Let X be a locally compact, locally connected, Hausdorff space and let $a, b \in X$. Prove that there is a compact connected subspace $A \subseteq X$ such that $\{a, b\} \subseteq A$.

19 Prove that Hilbert space (Exercise 2.2.16) is not locally compact but it is separable.

20 Determine whether the topological product of any denumerable family of locally compact spaces is locally compact.

21 Let Z be a k-space and let X be a Hausdorff space. Establish a bijective correspondence between the continuous functions $X \to Z$ on X and the continuous functions $k(X) \to Z$ on the k-space associated with X, as in the K-Factory (4.11).

22 Prove: In the ordinal space $[0, \Omega]$, the point $\{\Omega\}$ is a closed set that is not a G_δ set.

23 Construct a Baire space X and a second category set in X whose complement is not a first category set in X.

24 Prove that a compact Hausdorff space is a second category set in itself.

25* (a) Prove that in $\mathbb{R}^n$ each open set and each closed set is both a G_δ and an F_σ set.
(b) Does the statement in (a) hold in every metric space, in every complete metric space, in every locally compact metric space?

26 (a) Prove: The set $\mathbb{Q}$ of rational numbers cannot be the set of zeros, $g^{-1}(0)$, of a continuous function $g : \mathbb{R} \to \mathbb{R}$.
(b) Construct a continuous function $g : \mathbb{R} \to \mathbb{R}$ whose set of zeros, $g^{-1}(0)$, is the set $\mathbb{R} - \mathbb{Q}$ of irrational numbers.

SECTION 5

1 Construct your own examples of spaces that have exactly three of the four topological properties (compactness, connectedness, metrizability, existence of exactly two non-cut points) that characterize arcs.

2 Let $(X, \leq)$ be a set with a simple order such that there is no maximum element and no minimum element and each interval (x, y), for $x < y$, is nonempty. Establish the following relations in the order topology:
(a) $(x, y) = \text{INT}[x, y]$.
(b) $[x, y] = \text{CLS}(x, y)$.

3 Let X be a compactum with exactly two non-cut points p and q.
(a) Prove that the function $h : \mathbb{I} \to X$ defined to carry $(.b_1 b_2 \ldots b_k \ldots)_2 \in \mathbb{I}$ to $(b_1 b_2 \ldots b_k \ldots)_D \in X$ is order-preserving; show that if $r < s$ in usual order on $\mathbb{I} \subseteq \mathbb{R}$, then $h(r) < h(s)$ in the separation order on X.
(b) Prove that any order-preserving function $h : \mathbb{I} \to X$ that $h(0) = p$ and $h(1) = q$ is a homeomorphism.

4 Let D be a dense subset of a compactum X with exactly two non-cut points p and q. Let $\prod_k \{0,1\}_k$ be the topological product of a sequence of 2-point discrete spaces. Consider the functions $q_X : \prod_k \{0,1\}_k \to X$ and $q_\mathbb{I} : \prod_k \{0,1\}_k \to \mathbb{I}$ defined by
$$q_X(b_1, b_2, \ldots, b_k, \ldots) = (b_1 b_2 \ldots b_k \ldots)_D = \cap_k D[b_1 b_2 \ldots b_k] \in X$$
and
$$q_\mathbb{I}(b_1, b_2, \ldots, b_k, \ldots) = (.b_1 b_2 \ldots b_k \ldots)_2 = \cap_k B[b_1 b_2 \ldots b_k] \in \mathbb{I}.$$
(a) Prove: q_X and $q_\mathbb{I}$ are identification maps that make the same identifications.
(b) Deduce that the assignment $(.b_1 b_2 \ldots b_k \ldots)_2 \to (b_1 b_2 \ldots b_k \ldots)_D$ defines a homeomorphism $\mathbb{I} \to X$.

5* Let X be a compact, connected, non-degenerate metrizable space in which each pair of distinct points is a cut set. Prove:
(a) X has no cut points.
(b) If $x \neq z$ and $X - \{x, z\} = E \cup F$ is a separation, then $E, F \subseteq X$ are open, $\bar{E} = E \cup \{x, z\}$ and $\bar{F} = F \cup \{x, z\}$, and the latter are connected subspaces.
(c) If $x \neq z$ and $X - \{x, z\} = E \cup F$ is a separation, then $\bar{E} = E \cup \{x, z\}$ and $\bar{F} = F \cup \{x, z\}$ are arcs with non-cut points x and z.

6* Use Exercise 5.5 to prove the Circle Characterization (5.12): a topological space is a simple closed curve if and only if it is a compact, connected, non-degenerate, metrizable space in which each pair of distinct points is a cut set.

7   Let A be a connected cut set of a connected space X, and let $X - A = E \cup F$ be a separation. Prove that both $E \cup A$ and $F \cup A$ are connected subspaces of X.

A ***continuum*** is a compact connected Hausdorff space.

8* Use Cantor's Theorem (1.3) to prove Sierpinski's Theorem:
No continuum X is the union of a countable collection of disjoint closed subsets, at least two of which are nonempty.
(Hint: Consider a sequence $\{A_k : k \geq 1\}$ of disjoint closed subsets of X, at least two of which are nonempty, and construct a sequence $\{C_k : k \geq 1\}$ of nested nonempty continua such that $C_k \cap A_k = \emptyset$ for all $k \geq 1$. To begin, take $C_1 = X$ if $A_1 = \emptyset$; otherwise, take C_1 to be a component of a closed neighborhood N of some $A_m \neq \emptyset$ ($m \neq 1$) with $N \cap A_1 = \emptyset$ and $C_1 \cap A_m \neq \emptyset$. Then show that $\{C_1 \cap A_k : k \geq 1\}$ is a sequence of disjoint closed subsets of C_1, at least two of which are nonempty.)

8

SEPARATION & COVERING PROPERTIES

In this final chapter of Part One, we consider the separation properties of regularity, normality, and complete regularity, and we apply them to the metrization of topological spaces. We also consider the covering property of paracompactness and study its applications.

1   Normal Spaces

Hausdorff, regular, and normal spaces are defined in Section **4**.2. They have successively stronger separation properties. Regularity and normality are such important properties that we investigate them more thoroughly in this section. Their definitions are rigged to ensure the Hausdorff property.

REGULAR SPACES

A *regular* ($T_3 + T_0$) space is a Hausdorff space in which each point and closed set not containing the point have disjoint neighborhoods.

1.1   Lemma   *For a T_0-space X, the following conditions are equivalent:*
(*a*)   The space X is T_3, thus, regular: for each point $x \in X$ and closed set A not containing x, there exist disjoint open neighborhoods V(x) and W(A).
(*b*)   For each point $x \in X$ and closed set A not containing x, there exists an open neighborhood V(x) whose closure $\overline{V}$ is disjoint from A, i.e., $V \subseteq \overline{V} \subseteq A^c$.
(*c*)   For each point $x \in X$ and open neighborhood U(x), there exists an open neighborhood V(x) such that $V \subseteq \overline{V} \subseteq U$.

Proof: (*a*) $\Rightarrow$ (*b*). Given disjoint open neighborhoods V(x) and W(A), the closed complement W^c contains V, hence, also $\overline{V}$. Thus, $\overline{V} \cap A = \varnothing$.

233

$(b) \Rightarrow (c)$. Given an open neighborhood $U(x)$, we apply (b) to the closed set $A = U^c$ to obtain an open neighborhood $V(x)$ such that $V \subseteq \overline{V} \subseteq A^c = U$.

$(c) \Rightarrow (a)$. Given a point $x \in X$ and a closed set A not containing x, we apply (c) to the open neighborhood $U(x) = A^c$ to obtain an open neighborhood $V(x)$ such that $V \subseteq \overline{V} \subseteq U = A^c$. Then $W = (\overline{V})^c$ is an open neighborhood of A disjoint from V. □

1.2 Theorem *Regularity has these invariance features:*
(a) *Each subspace of a regular space is regular.*
(b) *A product is regular if and only if all the factor spaces are regular.*
(c) *A quotient of a regular space modulo a closed subspace is Hausdorff.*

Proof: (a) This is part of Theorem **4.2.2**.

(b) By (a) and Corollary **5.3.5**, each factor space X_α ($\alpha \in \mathcal{A}$) is regular whenever the topological product $\Pi_\alpha X_\alpha$ is regular.

For the converse, suppose that the factor spaces X_α ($\alpha \in \mathcal{A}$) are regular. Let $U \subseteq X = \Pi_\alpha X_\alpha$ be any neighborhood of a point $x \in \Pi_\alpha X_\alpha$. There exists some basic neighborhood $\Pi_\alpha U_\alpha \subseteq U$, where U_α is a neighborhood of the coordinate x_α for each $\alpha \in \mathcal{A}$ and $U_\alpha = X_\alpha$ for all but finitely many $\alpha \in \mathcal{A}$. By Lemma 1.1, the regular space X_α contains an open neighborhood V_α of x_α such that $V_\alpha \subseteq \overline{V_\alpha} \subseteq U_\alpha$, for each $\alpha \in \mathcal{A}$. When $U_\alpha = X_\alpha$, we take $V_\alpha = X_\alpha$, so that $\Pi_\alpha V_\alpha$ is a basic neighborhood of x. Then we have:

$$\overline{\Pi_\alpha V_\alpha} = \Pi_\alpha \overline{V_\alpha} \subseteq \Pi_\alpha U_\alpha \subseteq U.$$

Therefore, $\Pi_\alpha X_\alpha$ is a regular space by Lemma 1.1.

(c) Let A be a closed subset of a regular space X. Then $X - A$ is an open subset in both X and X/A and its two subspace topologies agree. Thus, points in $X - A \subseteq X/A$ are different from $[A]$ and have disjoint neighborhoods as X is Hausdorff. Finally, for $x \in X - A$, there exist disjoint open neighborhoods $V(x)$ and $W(A)$. Their images, $q(V)$ and $q(W)$, are disjoint open neighborhoods of x and $[A]$ in X/A, because $V = q^{-1}(q(V))$ and $W = q^{-1}(q(W))$ are disjoint open sets in X. □

NORMAL SPACES

A *normal* ($T_4 + T_1$) space is a Hausdorff space in which disjoint closed subsets have disjoint open neighborhoods.

1.3 Lemma *For a T_1 space X, the following conditions are equivalent:*
(a) *The space X is T_4, hence, normal: disjoint closed subsets A and B have disjoint open neighborhoods $M(A)$ and $N(B)$.*
(b) *For each open neighborhood $U(A)$ of a closed set $A \subseteq X$, there is an open neighborhood $V(A)$ with $A \subseteq V \subseteq \overline{V} \subseteq U$.*
(c) *Disjoint closed sets have disjoint closed neighborhoods.*

Proof: $(a) \Rightarrow (b)$. Consider a closed set $A \subseteq X$ and an open neighborhood $U(A)$. By the normality property, the closed sets A and $X - U$ have disjoint open neighborhoods V and W. Then $V \subseteq \overline{V} \subseteq X - W \subseteq U$.

$(b) \Rightarrow (c)$. Given disjoint closed sets A and B, apply (b) twice to obtain open sets V and W such that $A \subseteq V \subseteq \overline{V} \subseteq W \subseteq \overline{W} \subseteq X - B$. So $\overline{V}$ and $X - W$ are disjoint closed neighborhoods of A and B, respectively.

$(c) \Rightarrow (a)$. This is trivial, by the definition of neighborhoods. □

Example 1 *If the partially ordered set $(X, \leq)$ is well-ordered, then X is normal in the order topology* (Example **5.2.5**). Each neighborhood of any $x \neq \min X$ contains an interval of the form $(z, x]$, which is the open interval $(z, \rightarrow)$ when x has no successor, and otherwise is the open interval (z, s_x) where s_x is the immediate successor of x. Each neighborhood of $x = \min X$ contains the open interval $[x, x] = [x, s_x)$!

Let A and B be disjoint closed sets in X. Then each $a \in A$ has an open neighborhood $(z_a, a]$ ($= [a, a]$ if $a = \min X$) disjoint from B and each $b \in B$ has an open neighborhood $(w_b, b]$ ($= [b, b]$ if $b = \min X$) disjoint from A. Therefore,

$$U = \cup \{(z_a, a] : a \in A\} \quad \text{and} \quad V = \cup \{(w_b, b] : b \in B\}$$

are open neighborhoods of A and B, respectively. They are disjoint; otherwise, there would exist $a \in A$, $b \in B$, and $x \in (z_a, a] \cap (w_b, b]$. And then the two options $a < b$ and $b < a$ would yield the contradictions $a \in (w_b, b]$ and $b \in (z_a, a]$, respectively.

1.4 Theorem *Normality has these invariance features:*
(a) *Each closed subspace of a normal space is normal.*
(b) *If a topological product is normal, so are all the factor spaces.*

Proof: (a) Closed sets A and B in a closed subspace $X \subseteq Y$ are closed in Y. When Y is normal, there are disjoint open neighborhoods $U(A)$ and $V(B)$ in Y. Then $U \cap X$ and $V \cap X$ are disjoint open neighborhoods in X for A and B.

(b) Because each factor space is homeomorphic to a closed slice of the topological product, (b) follows from (a). □

Example 2 *The topological product of normal spaces need not be normal.* The argument employed in Example 1 shows that the real line $\mathbb{R}_{(]}$ with the right interval topology (Example **5.2.10**) is normal. But $\mathbb{R}_{(]} \times \mathbb{R}_{(]}$ is not normal:

$\mathbb{R}_{(]} \times \mathbb{R}_{(]}$ has as subbase the sets $S(<x, y>, r) = (x - r, x] \times (y - r, y]$. Their shape shows that the cross diagonal line Δ of points $<x, y>$ with $y = -x$ is a closed discrete subspace in $\mathbb{R}_{(]} \times \mathbb{R}_{(]}$. Therefore, the disjoint subsets $A \subseteq \Delta$ and $B \subseteq \Delta$ consisting of points $<x, -x>$ with $x \in \mathbb{Q}$ and $x \in \mathbb{R} - \mathbb{Q}$, respectively, are both closed in $\mathbb{R}_{(]} \times \mathbb{R}_{(]}$.

Let U be a neighborhood of B in $\mathbb{R}_{(]} \times \mathbb{R}_{(]}$. Now Δ is the countable union of the points of A and the sets G_k of points $<x, -x> \in B$ with $S(<x, -x>, 1/k) \subseteq U$, for $k \geq 1$. As a real subspace $\Delta \subseteq \mathbb{R}^2$, Δ is a Baire space. By Theorem **7.4.7**(b), one of the sets G_k must fail to be nowhere dense in the subspace $\Delta \subseteq \mathbb{R}$ because each point of A is nowhere dense. So some G_k has closure with nonempty real interior so that $\overline{G_k}$ contains some interval J of the line Δ. Since $S(<x, -x>, 1/k)$ is contained in U for all points $<x, -x> \in G_k$, no point of $A \cap J$ can have a base neighborhood disjoint from U. So the closed sets A and B do not have disjoint open neighborhoods in $\mathbb{R}_{(]} \times \mathbb{R}_{(]}$.

Example 3 While topological products of normal spaces need not be normal, topological products of compact Hausdorff spaces are compact by Tychonoff's Theorem (**7.2.11**), hence, normal by Theorem **7.2.12**.

Example 4 *An arbitrary subspace of a normal space need not be normal.* Let ω be the first infinite ordinal, and let Ω be the first uncountable ordinal. The product $T = [0, \Omega] \times [0, \omega]$, called the ***Tychonoff plank***, is normal by Example 3. But the point complement $T_\infty = T - \{<\Omega, \omega>\}$ is not normal: $A = \{<\Omega, n> : 0 \leq n < \omega\}$ and $B = \{<\alpha, \omega> : 0 \leq \alpha < \Omega\}$ are closed but cannot be separated in T_∞. In fact, any neighborhood U of A contains a segment $(\alpha_n, \Omega] \times \{n\}$ with endpoint $<\Omega, n> \in A$, for each $n < \omega$. The countable set $\{\alpha_n \in \Omega : n < \omega\}$ has an upper bound α in $[0, \Omega)$ by Theorem **1.5.12**. Then $(\alpha, \Omega] \times [0, \omega) \subseteq U$. Therefore, any neighborhood of $<\alpha+1, \omega> \in B$ intersects U.

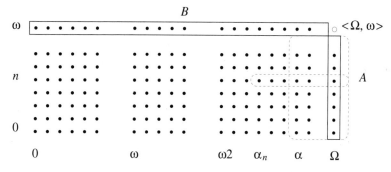

Example 4: Deleted Tychonoff Plank $T_\infty = T - \{<\Omega, \omega>\}$

To place the normality property in perspective, we record these previously established relationships:

- *A normal space is regular* (Theorem **4.2.1**).

- *A metrizable space is normal* (Theorem **2.3.5**).

- *A compact Hausdorff space is normal* (Theorem **7.2.12**).

The first statement has the following partial converse, proved by Tychonoff in 1925. Recall from Section **7.3** that a Lindelöf space is a topological space in which each open covering has a countable subcovering. Also, recall that each second countable space is Lindelöf by Theorem **7.3.5**.

1.5 Tychonoff's Lemma *Each regular Lindelöf space is normal. In particular, each regular second countable space is normal.*

Proof: Let A and B be disjoint closed subsets of a regular Lindelöf space X. Because X is regular, Lemma 1.1(*b*) shows that each of the closed sets A and B is covered by open sets that have closure disjoint from the other closed set. These closed subspaces A and B of the Lindelöf space X are Lindelöf; so A

has a countable covering $\{U_k : 1 \leq k\}$ by open sets with closure $\overline{U}_k \subseteq B^c$ and B has a countable covering $\{V_k : 1 \leq k\}$ by open sets with closure $\overline{V}_k \subseteq A^c$.

By replacing each U_k and V_k with the union of its predecessors in the indexing, we obtain coverings of A and B by nested open sets:

$$\{U_1 \subseteq U_2 \subseteq \ldots \subseteq U_k \subseteq \ldots\} \quad \text{and} \quad \{V_1 \subseteq V_2 \subseteq \ldots \subseteq V_k \subseteq \ldots\}.$$

These still have the closure properties $\overline{U}_k \subseteq B^c$ and $\overline{V}_k \subseteq A^c$, because the closure of a finite union of subsets is the union of their closures.

Then $U = \cup \{U_k - \overline{V}_k : 1 \leq k\}$ and $V = \cup \{V_k - \overline{U}_k : 1 \leq k\}$ are open neighborhoods of A and B, respectively. Moreover, $U \cap V = \emptyset$, because we have

$$U_k \cap (X - \overline{U}_n) \supseteq (U_k - \overline{V}_k) \cap (V_n - \overline{U}_n) \subseteq (X - \overline{V}_k) \cap V_n,$$

where $U_k \cap (X - \overline{U}_n) = \emptyset$, when $k \leq n$, and $(X - \overline{V}_k) \cap V_n = \emptyset$, when $n \leq k$. □

NORMALITY AND URYSOHN FUNCTIONS

On some spaces, the only continuous real-valued functions possible are constant (Example **6**.1.4). Normal spaces are much better behaved, according to the following two characterizations of normality, due to P. Urysohn (1898-1924) and H. Tietze. Urysohn's characterization, established in 1924, shows that any two disjoint closed subsets of a normal space can be made to receive distinct constant values under some continuous function on the space. Tietze's characterization shows that any continuous real-valued function on a closed subspace of a normal space extends to a continuous real-valued function on the entire space.

For use in both characterizations of normality, we point out that a continuous function $u : X \to \mathbb{I}$ on any space X, together with the binary fractions $0 < d = k/2^n \leq 1$, give rise to open subsets $V_d = u^{-1}([0, d))$ of X such that $\overline{V}_d \subseteq V_{d'}$ whenever $d < d'$. More importantly, we have the converse:

1.6 Lemma *Let X be any topological space and let $V_d \subseteq X = V_1$ be open sets that are indexed by the binary fractions $0 < d = k/2^n \leq 1$ such that $\overline{V}_d \subseteq V_{d'}$ whenever $0 < d < d' \leq 1$.*

Then the function $u : X \to \mathbb{I}$, $u(x) = \mathrm{glb}\,\{d : x \in V_d\}$, is continuous.

Proof: By the nested property, $d < d' \Rightarrow \overline{V}_d \subseteq V_{d'}$, we have:

(i) $x \notin V_d \Rightarrow (x \notin \overline{V}_{d'}$ for all $d' < d) \Rightarrow d \leq u(x)$.

(ii) $x \in \overline{V}_d \Rightarrow (x \in V_{d'}$ for all $d < d') \Rightarrow u(x) \leq d$.

To establish the continuity of u, let $x \in X$ and $u(x) = t$. Consider any interval neighborhood $(t - r, t + r)$ of $u(x) = t$ with radius $r > 0$.

When $0 < t$ or $t < 1$, we have binary fractions d' or d'' such that:

$$0 < t - r < d' < t \quad \text{or} \quad t < d'' < t + r < 1.$$

When $0 < t < 1$, both binary fractions $d' < t < d''$ exist. Then $W = V_{d''} - \overline{V}_{d'}$ is an open neighborhood of x, as $u(x) = t < d''$ implies $x \in V_{d''}$ by (i) and $d' < t = u(x)$ implies $x \notin \overline{V}_{d'}$ by (ii). And $u(W) \subseteq (t - r, t + r)$, as $z \in V_{d''}$ implies $u(z) \leq d'' < t + r$ by (ii) and $z \notin \overline{V}_{d'}$ implies $t - r < d' \leq u(z)$ by (i).

When $t = 0$ or 1, the binary fraction d'' or d' exists. Then either $W = V_{d''}$ or $W = X - \overline{V}_{d'}$ is an open neighborhood of x with $u(W) \subseteq (t - r, t + r)$.

This proves that u is continuous at each $x \in X$. $\square$

1.7 Urysohn's Theorem
Let X be a Hausdorff space. Then the following properties are equivalent:
(a)   X is normal.
*(b)   For each pair of disjoint closed subsets A and B in X, there exists a continuous function $u : X \to \mathbb{I}$, called a **Urysohn function** for A and B, such that $u(A) = \{0\}$ and $u(B) = \{1\}$.*

Proof: $(b) \Rightarrow (a)$. A Urysohn function, $u : X \to \mathbb{I}$, for A and B provides disjoint open neighborhoods $U = u^{-1}([0, \frac{1}{2}))$ and $V = u^{-1}((\frac{1}{2}, 1])$ of A and B.

$(a) \Rightarrow (b)$. For any pair $C \subseteq U$ in X, where C is closed and U is open, Lemma 1.3 provides an open set V such that $C \subseteq V \subseteq \overline{V} \subseteq U$. This produces two pairs $C \subseteq V$ and $\overline{V} \subseteq U$ on which to apply Lemma 1.3 again.

We iterate this procedure, starting with the pair $A \subseteq X - B$, for disjoint closed subsets A and B of X. We obtain subsets that we index, as follows:

$$A \subseteq V_{1/2} \subseteq \overline{V}_{1/2} \subseteq X - B \subseteq X = V_1$$
$$A \subseteq V_{1/4} \subseteq \overline{V}_{1/4} \subseteq V_{1/2} \subseteq \overline{V}_{1/2} \subseteq V_{3/4} \subseteq \overline{V}_{3/4} \subseteq X - B \subseteq X = V_1$$
$$\vdots$$

In this way we introduce, for each binary fraction $0 < d = k/2^n \leq 1$, an open subset V_d such that (i) $A \subseteq V_d \subseteq \overline{V}_d \subseteq X - B$ and (ii) $d < d' \Rightarrow \overline{V}_d \subseteq V_{d'}$.

Then the function $u : X \to \mathbb{I}$, $u(x) = glb\{d : x \in V_d\}$, is continuous by Lemma 1.6. Also, $u(A) = 0$ because $a \in V_d$ for all $a \in A$ and all binary fraction $0 < d < 1$, and $u(B) = 1$ because $b \in B \cap V_d$ only for $d = 1$. $\square$

The Urysohn function satisfies $u(A) = 0$; but, its *set of zeros*, $u^{-1}(0)$, may be larger than A. Recall from Section **7.4** that any countable intersection of open sets in a topological space X, such as the set of zeros of a continuous function $u : X \to \mathbb{R}$, is called a G_δ set. In a normal space X, any G_δ set arises as the set of zeros of a continuous real-valued function:

1.8   Corollary
For disjoint closed subsets A and B in a normal space X, there is a Urysohn function $u : X \to \mathbb{I}$ such that $A = u^{-1}\{0\}$ or $B = u^{-1}\{1\}$ if and only if A or B, respectively, is a G_δ set.

Proof: If $u : X \to \mathbb{I}$ is a Urysohn function with $A = u^{-1}\{0\}$ or $B = u^{-1}\{1\}$, then $A = \cap_n u^{-1}([0, 1/n))$ or $B = \cap_n u^{-1}((1-1/n, 1])$ is a G_δ set.

Conversely, when A is a G_δ set $\cap_n A_n$, we repeat the preceding construction of a Urysohn function for A and B, after replacing each $V_{1/2^n}$ by its intersection with A_n. The resulting Urysohn function $a : X \to \mathbb{I}$ (for A and B) has $A = a^{-1}(0)$, as $A \subseteq \cap_n V_{1/2^n} \subseteq \cap A_n = A$. Similarly, when B is a G_δ set, there exists a Urysohn function $b : X \to \mathbb{I}$ (for B and A) such that $B = b^{-1}(0)$. When both A and B are G_δ sets, Urysohn functions a and b are available. Then $u = a/(a + b)$ is the desired Urysohn function. □

Example 5 Given any pair of disjoint closed subsets A and B in real n-space $\mathbb{R}^n$, there is a continuous function $v : \mathbb{R}^n \to \mathbb{R}$ that takes prescribed values c and d exactly on A and B: $v^{-1}\{c\} = A$ and $v^{-1}\{d\} = B$. Take $v(x) = (d - c)\, u(x) + c$, where u is provided by Urysohn's Theorem (1.7).

NORMALITY AND EXTENSION

Let X be a topological space and let $A \subseteq X$ be a closed subspace.

When continuous functions $g : A \to Y$ and $G : X \to Y$ with the same codomain space Y satisfy the relationship $g = G \circ i : A \subseteq X \to Y$, then g is called the ***restriction*** of G to A and G is called an ***extension*** of g over X.

A general extension problem in topology asks what continuous functions on a subspace $A \subseteq X$ admit an extension to a continuous function on X with the same codomain space Y.

Example 6 *A continuous extension need not exist.* In particular, the identity $1_A : A \to A$ on a closed subspace $A \subseteq X$ has a continuous extension $X \to A$ if and only if A is a *retract* of X. For example, the closed subspace $\{0, 1\} \subseteq \mathbb{I}$ is not a retract of $\mathbb{I}$, as the disconnected space $\{0, 1\}$ cannot be the image of the connected interval $\mathbb{I}$.

Example 7 *The extension problem becomes trivial when the codomain space may be enlarged.* For any closed subspace $A \subseteq X$ and any continuous function $g : A \to Y$, there is the adjunction space $Y \cup_g X$ and the identification map $p : Y \mathring{\cup} X \to Y \cup_g X$ identifying each $a \in A \subseteq X$ with $g(a) \in Y$. Then the composite $j \circ g : A \to Y \subseteq Y \cup_g X$ admits the extension $G = p|_X : X \subseteq Y \mathring{\cup} X \to Y \cup_g X$.

Urysohn's Theorem (1.7) expresses the existence of a continuous extension over the normal space X of the function $A \cup B \to \mathbb{R}$ with values 0 and 1 on any disjoint closed sets A and B, respectively. So Urysohn's Theorem is a special case of the following general extension theorem of H. Tietze.

1.9 Tietze's Extension Theorem
Let X be a Hausdorff space. Then the following properties are equivalent:
(a) X *is normal.*
(b) *For every closed subset $A \subseteq X$, each continuous function $g : A \to \mathbb{I}$ has a continuous extension $G : X \to \mathbb{I}$.*

Proof: $(b) \Rightarrow (a)$. Urysohn functions are special cases of the functions produced by (b). Thus, X is normal by Urysohn's Theorem (1.7).

$(a) \Rightarrow (b)$. Given a continuous function $g : A \to \mathbb{I}$, we refine the proof of Urysohn's Theorem, as follows. We set $V_1 = X$ and then seek disjoint open subsets V_d and U_d of X (for all binary fractions $0 < d = k/2^n < 1$) such that

$$A \cap V_d = g^{-1}([0, d)) \quad \text{and} \quad A \cap U_d = g^{-1}((d, 1]),$$

and also $\overline{V}_d \subseteq V_{d'}$ and $\overline{U}_{d'} \subseteq U_d$, whenever $d < d'$.

We construct V_d and U_d by induction on the lexicographic order $\ll$ of the binary fractions $d = k/2^n$ ($1 \leq n$ and odd $1 \leq k \leq 2^n$):

$$d = \tfrac{1}{2} \ll (\tfrac{1}{2})^2 \ll 3(\tfrac{1}{2})^2 \ll (\tfrac{1}{2})^3 \ll 3(\tfrac{1}{2})^3 \ll 5(\tfrac{1}{2})^3 \ll 7(\tfrac{1}{2})^3 \ll \dots .$$

We use the normality of X and the fact that $g^{-1}([0, d])$ and $g^{-1}([d, 1])$ are closed subsets of the closed subspace A, hence, are closed in X for all d.

Suppose that $d = \tfrac{1}{2}$ or that V_e and U_e have been constructed for all binary fractions $e \ll d = k/2^n$ in the order $\ll$. Of the binary fractions e that preceed d, let e' be the largest one $e' < d$ and let e'' be the smallest one $d < e''$. For $m \geq 1$, the closed set $\overline{V}_{e'} \cup g^{-1}([0, \ d - (\tfrac{1}{2})^{n+m}])$ has an open neighborhood $R_m \subseteq \overline{R}_m \subseteq X - (g^{-1}([d, 1]) \cup \overline{U}_{e''})$; the closed set $g^{-1}([d + (\tfrac{1}{2})^{n+m}, 1]) \cup \overline{U}_{e''}$ has an open neighborhood $S_m \subseteq \overline{S}_m \subseteq X - (\overline{V}_{e'} \cup g^{-1}([0, d]))$. Then, as in the proof of Tychonoff's Lemma (1.5), we make each of family $\{R_m : 1 \leq m\}$ and $\{S_m : 1 \leq m\}$ nested and then form the disjoint open sets

$$V_d = \cup \{R_m - \overline{S}_m : 1 \leq m\} \quad \text{and} \quad U_d = \cup \{S_m - \overline{R}_m : 1 \leq m\}.$$

Then V_d is an open neighborhood of $\overline{V}_{e'}$ and $A \cap V_d = g^{-1}([0, d))$; similarly, U_d is an open neighborhood of $\overline{U}_{e''}$ and $A \cap U_d = g^{-1}((d, 1])$. Because disjoint open subsets are necessarily separated in X, we also have

$$\overline{V}_d \subseteq X - U_d \subseteq X - \overline{U}_{e''} \subseteq V_{e''} \quad \text{and} \quad \overline{U}_d \subseteq X - V_d \subseteq X - \overline{V}_{e'} \subseteq U_{e'}.$$

By induction, V_d and U_d have the desired properties.

By Lemma 1.6, the function $G : X \to \mathbb{I}$, $G(x) = glb \{d : x \in V_d\}$, is continuous. Finally, $g(a) = G(a)$ for all $a \in A$, because, by construction, we have

$$g(a) < d \Rightarrow (a \in g^{-1}([0, d)) \subseteq V_d) \Rightarrow G(a) \leq d$$

and

$$G(a) < d \Rightarrow (a \in A \cap V_d = g^{-1}([0, d))) \Rightarrow g(a) < d.$$

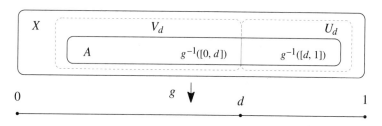

$\square$

Tietze's Extension Theorem generalizes to allow other intervals in $\mathbb{R}$ as the codomain space (Exercise 17):

1.10 Corollary *For every closed subset $A \subseteq X$ of a normal space X, each continuous function $g : A \to \mathbb{R}$ has a continuous extension $G : X \to \mathbb{R}$. If $g(A) \subseteq (a, b)$, there is an extension G with $G(X) \subseteq (a, b)$.*

Certain spaces can be substituted for $\mathbb{I}$ and $\mathbb{R}$ in Tietze's Theorem (1.10).

A space Y is called an ***absolute retract*** (for normal spaces) if, for every normal space X and closed subspace $A \subseteq X$, each continuous function $g : A \to Y$ has a continuous extension $G : X \to Y$.

The terminology is appropriate: an *absolute retract* Y is a retract of any normal space X in which it appears as a closed subspace, since the identity $1_Y : Y \to Y$ extends to a continuous function $r : X \to Y$, i.e., to a retraction.

1.11 Theorem
(a) Any retract of an absolute retract is an absolute retract.
(b) A topological product is an absolute retract if and only if each factor space is an absolute retract.

Proof: (a) Let Y be an absolute retract and let $r : Y \to B$ be a retraction, i.e., $1_B = r \circ j : B \subseteq Y \to B$. Let $g : A \to B$ be a continuous function defined on a closed subspace $A \subseteq X$ of a normal space X. Because Y is an absolute retract, $j \circ g : A \to B \subseteq Y$ has a continuous extension $G : X \to Y$. Then $r \circ G : X \to B$ is an extension of g because $r \circ G \circ i = r \circ j \circ g = g$.

(b) Because the factor spaces Y_α ($\alpha \in \mathcal{A}$) are retracts of $\prod_\alpha Y_\alpha$ (Corollary 5.3.5), they are absolute retracts whenever $Y = \prod_\alpha Y_\alpha$ is one, by (a).

Now let each Y_α ($\alpha \in \mathcal{A}$) be an absolute retract, and let $g : A \to Y$ be a continuous function on a closed subspace A of a normal space X. Because Y_α is an absolute retract, $p_\alpha \circ g : A \to Y \to Y_\alpha$ has a continuous extension $G_\alpha : X \to Y_\alpha$. Then $G : X \to Y$, $p_\alpha \circ G = G_\alpha$, is the desired extension of g. □

Example 8 $\mathbb{R}^n$ and $\mathbb{I}^n$ are absolute retracts by Theorem 1.11(b). Thus, they are retracts of any normal spaces in which they appear as closed subspaces. In particular, $\mathbb{R}^m$ retracts onto any embedding of $\mathbb{I}^n$ or $\mathbb{R}^n$ as a closed subspace in $\mathbb{R}^m$.

2 Completely Regular Spaces

We seek to characterize subspaces of normal spaces. They need not be normal, but they have the following interesting feature in common.

A Hausdorff space X is ***completely regular*** if, for each point $q \in X$ and closed set A not containing q, there exists a continuous function $u : X \to \mathbb{I}$ such that $u(A) = \{0\}$ and $u(q) = 1$.

It is equivalent to require that, for each neighborhood $U(q)$, there exists a continuous function $u : X \to \mathbb{I}$ such that $u(U^c) = \{0\}$ and $u(q) = 1$.

RELATIONS AND INVARIANCE

Complete regularity has these relations to other separation properties.

2.1   Proposition
(a)   Every subspace of a normal space is completely regular.
(b)   Every completely regular space is regular.

Proof: (a) Let $X \subseteq Y$ be a subspace of the space Y. If $A \subseteq X$ is closed and doesn't contain $q \in X$, then $A = X \cap F$ where $F \subseteq Y$ is closed in Y and doesn't contain q. When Y is normal, Urysohn's Theorem (1.7) provides a Urysohn function $u|_X : X \subseteq Y \to \mathbb{I}$ with $u(A) = \{0\}$ and $u(q) = 1$. This proves that X is completely regular.

(b) Let $A \subseteq X$ be a closed subset not containing the point q. Any continuous function $u : X \to \mathbb{I}$ with $u(A) = \{0\}$ and $u(q) = 1$ determines disjoint open neighborhoods $u^{-1}([0, \frac{1}{2}))$ and $u^{-1}((\frac{1}{2}, 1])$ of A and q, respectively.   □

Complete regularity is better behaved than normality relative to subspace and product space constructions.

2.2   Theorem   *Complete regularity has these invariance properties:*
(a)   Every subspace of a completely regular space is completely regular.
(b)   A topological product is completely regular if and only if each factor space is completely regular.

Proof: (a) The proof of 2.1(a) suffices here.

(b) Consider any given point $q = (q_\alpha) \in \Pi_\alpha X_\alpha$ of a topological product of completely regular spaces and consider any given base neighborhood of q:

$$U = \langle U_{\sigma(1)}, \ldots, U_{\sigma(k)} \rangle = \cap \{ p_{\sigma(i)}^{-1}(U_{\sigma(i)}) : 1 \le i \le k \}.$$

There exist continuous functions $u_i : X_{\sigma(i)} \to \mathbb{I}$ of the factor spaces such that

$$u_i(X_{\sigma(i)} - U_{\sigma(i)}) = 0 \quad \text{and} \quad u_i(q_{\sigma(i)}) = 1,$$

for all $1 \le i \le k$, by complete regularity. The function $u : \Pi_\alpha X_\alpha \to \mathbb{I}$,

$$u(x) = \min \{ u_{\sigma(i)} \circ p_{\sigma(i)}(x) : 1 \le i \le k \},$$

is continuous and satisfies $u_i(X - U) = 0$ and $u(q) = 1$. Therefore, the product is completely regular. The converse holds by (a).   □

Example 1   By Theorem 2.2(a), any non-normal subspace of a normal space, such as the space in Example 1.4, shows that a complely regular space need not be normal.

Any subspace of a normal space is competely regular, by 2.1(a). To prove that, conversely, any completely regular space is the subspace of a normal space, we consider special cubes, i.e., topological products of $\mathbb{I}$.

For any space X, we denote the set of all continuous functions $g : X \to \mathbb{I}$ by $(X, \mathbb{I})$. For $g \in (X, \mathbb{I})$, let $\mathbb{I}_g$ denote a copy of the closed unit interval $\mathbb{I}$.

The topological product $\Pi_g \mathbb{I}_g$ of the closed intervals $\mathbb{I}_g$, $g \in (X, \mathbb{I})$, is called the ***cube*** bounding X and is denoted by $\mathbb{I}^{(X, \mathbb{I})}$.

2.3 Embedding Theorem *When X is completely regular, the function*

$$\lambda : X \to \mathbb{I}^{(X, \mathbb{I})}, \; \lambda(x)_g = g(x),$$

embeds X into its cube $\mathbb{I}^{(X, \mathbb{I})}$.

Proof: λ *is continuous.* The projections $h = p_h \circ \lambda : X \to \Pi_g \mathbb{I}_g \to \mathbb{I}_h$, $h \in (X, \mathbb{I})$, are continuous; therefore, so is λ by the Characteristic Feature (**5.3.2**) of the product topology.

λ *is injective.* For distinct points $x \neq z$ in the Hausdorff space X, there is a neighborhood U of x not containing z. By the complete regularity of X, there is a function $u : X \to \mathbb{I}$ such that $u(x) = 1$ and $u(z) \in u(U^c) = \{0\}$. Then the image points $\lambda(x)$ and $\lambda(z)$ differ in their u^{th} coordinates: $u(x) = 1 \neq 0 = u(z)$.

When X is completely regular, the open sets $V_g = g^{-1}((0, 1])$, $g \in (X, \mathbb{I})$, constitute a base for its topology. Given any point $q \in X$ and neighborhood $U(q)$, there is, by complete regularity, a continuous function $u : X \to \mathbb{I}$ such that $u(q) = 1$ and $u(U^c) = \{0\}$. Thus, $q \in V_u = u^{-1}((0, 1]) \subseteq U$.

$\lambda : X \to \lambda(X)$ *is open.* The image under λ of any basic set V_h in X is

$$\lambda(V_h) = \lambda(X) \cap \{t \in \Pi_g \mathbb{I}_g : t_h > 0\}.$$

This image is the portion of the open slice $p_h^{-1}((0, 1])$ of the topological product $\Pi_g \mathbb{I}_g$ that is contained in the image subspace $\lambda(X) \subseteq \mathbb{I}^{(X, \mathbb{I})}$.

So $\lambda : X \to \lambda(X)$ is an embedding of X onto its image in its cube $\mathbb{I}^{(X, \mathbb{I})}$. □

The construction of cubes has the following naturality property:

2.4 Corollary *Any continuous function $w : X \to Y$ of completely regular spaces extends to a continuous function $W : \mathbb{I}^{(X, \mathbb{I})} \to \mathbb{I}^{(Y, \mathbb{I})}$ of their cubes:*

$$\begin{array}{ccc} X & \xrightarrow{w} & Y \\ \lambda_X \downarrow & & \downarrow \lambda_Y \\ \overline{\lambda(X)} & \xrightarrow{W} & \overline{\lambda(Y)} \\ \cap & & \cap \\ \mathbb{I}^{(X, \mathbb{I})} & \xrightarrow{W} & \mathbb{I}^{(Y, \mathbb{I})} \end{array}$$

Moreover, W has a continuous restriction $W : \overline{\lambda(X)} \to \overline{\lambda(Y)}$.

Proof: The sets $(X, \mathbb{I})$ and $(Y, \mathbb{I})$ of continuous functions index the factors in the cubes $\mathbb{I}^{(X, \mathbb{I})}$ and $\mathbb{I}^{(Y, \mathbb{I})}$. Because w is continuous, each index $g \in (Y, \mathbb{I})$ determines an index $g \circ w \in (X, \mathbb{I})$. We define $W : \mathbb{I}^{(X, \mathbb{I})} \to \mathbb{I}^{(Y, \mathbb{I})}$ to be the continuous function between the cubes that has its projection

$$p_g \circ W : \mathbb{I}^{(X, \mathbb{I})} \to \mathbb{I}^{(Y, \mathbb{I})} \to \mathbb{I}_g$$

given by

$$p_{g \circ w} : \mathbb{I}^{(X, \mathbb{I})} \to \mathbb{I}_{g \circ w} \equiv \mathbb{I}_g,$$

for all $g \in (Y, \mathbb{I})$. The function W is continuous by the Characteristic Feature (**5.3.2**) for topological products.

For each $x \in X$, the two images $\lambda_Y(w(x))$ and $W(\lambda_X(x))$ have the same coordinates, namely, $g(w(x))$ for each index $g \in (Y, \mathbb{I})$. This proves the commutative feature, $\lambda_Y \circ w = W \circ \lambda_X$, of the display in Corollary **2.4**.

It follows that $W : \mathbb{I}^{(X, \mathbb{I})} \to \mathbb{I}^{(Y, \mathbb{I})}$ has a restriction $W : \lambda_X(X) \to \lambda_Y(Y)$ that we may identify with the given function $w : X \to Y$. Because W is continuous, it also restricts to a continuous function of the closures:

$$W : \overline{\lambda(X)} \to \overline{\lambda(Y)}. \qquad \square$$

The Embedding Theorem (**2.3**) provides the following succinct characterization of subspaces of a normal space.

2.5 Theorem *The following are equivalent properties for a space X:*
(a) *X is completely regular.*
(b) *X is a subspace of a cube.*
(c) *X is a subspace of a compact Hausdorff space.*
(d) *X is a subspace of a normal space.*

Proof: The implications $(a) \Rightarrow (b)$, $(b) \Rightarrow (c)$, $(c) \Rightarrow (d)$, and $(d) \Rightarrow (a)$ follow from the Embedding Theorem (**2.3**), Tychonoff's Theorem (**7.2.11**), the Normality Property (**7.2.12**), and Urysohn's Theorem (**1.7**). $\qquad \square$

STONE-ČECH COMPACTIFICATION

As in Section **7.4**, an embedding of a space X as a dense subspace of a compact Hausdorff space X^* is called a *compactification* of X. We usually identify X with its embedded image in any of its compactifications X^* to produce a subspace inclusion $X \subseteq X^*$ of X as a dense subspace of X^*.

The one-point compactification of Section **7.4** applies only to locally compact Hausdorff spaces. We now consider a compactification technique that applies to all completely regular spaces.

Let X be a completely regular space. By the Embedding Theorem (**2.3**), there is the embedding

$$\lambda : X \to \lambda(X) \subseteq \mathbb{I}^{(X, \mathbb{I})}, \ \lambda(x)_g = g(x).$$

By Theorem 7.1.5, the closure $\overline{\lambda(X)} \subseteq \mathbb{I}^{(X, \mathbb{D})}$, which we denote by $\beta(X)$, is a compact subspace of the compact cube. Thus the embedding

$$\lambda : X \to \lambda(X) \subseteq \overline{\lambda(X)} = \beta(X)$$

shows that $\beta(X)$ is a compactification of X, its **Stone-Čech compactification**.

This compactification $\beta(X)$, developed independently in 1937 by M. H. Stone and E. Čech, is the largest compactification of X:

2.6 Theorem *Let $\lambda : X \to \overline{\lambda(X)} = \beta(X)$ be the Stone-Čech compactification of a completely regular space X. Then:*
(a) For every compact Hausdorff space Y, each continuous function $w : X \to Y$ extends uniquely to a continuous function $W : \beta(X) \to Y$, i.e., there is a continuous function W such that $w = W \circ \lambda : X \to \beta(X) \to Y$.
(b) Every compactification $\hat{X}$ of X is an identification space of $\beta(X)$ preserving X.

Proof: (a) We apply the Stone-Čech compactification to both X and Y. By Corollary 2.4, the function $w : X \to Y$ has this extension:

$$\begin{array}{ccc} X & \to & Y \\ \lambda_X \downarrow & & \downarrow \lambda_Y \\ \beta(X) = \overline{\lambda(X)} & \xrightarrow{W} & \overline{\lambda(Y)} = \beta(Y) \end{array}$$

Since Y is a compact Hausdorff space, the embedding $\lambda_Y : Y \to \beta(Y)$ is a homeomorphism by which we may identify Y and $\beta(Y)$. Then W becomes the desired continuous extension $W : \beta(X) \to Y$ of $w : X \to Y$. The extension is unique by the Coincidence Theorem (4.4.4).

(b) By (a), the embedding $w : X \subseteq \hat{X}$ into any compactification has a continuous extension $W : \beta(X) \to \hat{X}$. Since $\beta(X)$ is compact, the image subspace $W(\beta(X)) \subseteq \hat{X}$ is compact, hence, closed. Because $W(\beta(X))$ contains the dense subset $X \subseteq \hat{X}$, it equals $\hat{X}$. Since $\beta(X)$ is compact and $\hat{X}$ is Hausdorff, the continuous surjection $W : \beta(X) \to \hat{X}$ is a closed function by Corollary 7.2.4. Hence, W is also an identification map by Theorem 5.1.6. □

Example 2 *The Stone-Čech compactification $\beta((0, 1])$ appends an uncountable set to compactify $(0, 1]$. The continuous function $w : (0, 1] \to [-1, +1]$, $w(x) = \sin 1/x$, extends to a continuous function $W : \beta((0, 1]) \to [-1, +1]$, by Theorem 2.6. For each $y \in [-1, +1]$, the pre-image $w^{-1}(y) \subseteq (0, 1]$ is infinite but does not accumulate in $(0, 1]$. But each pre-image $w^{-1}(y)$ must have an accumulation point in the compactification $\beta((0, 1])$, by Theorem 7.3.2. These accumulation points for different pre-images $w^{-1}(y)$ and $w^{-1}(y')$ are distinct points of $\beta((0, 1])$, for they have the distinct images $y \neq y'$ in $[-1, +1]$ under W by continuity.*

Property (a) in Theorem 2.6 characterizes the Stone-Čech compactification and also provides the following corollary (Exercise 7):

2.7 Corollary *If $X \cong Y$, then $\beta(X) \cong \beta(Y)$.*

Two compactifications $X \subseteq \hat{X}_1$ and $X \subseteq \hat{X}_2$ of X are called **equivalent** if there is a homeomorphism $\hat{X}_1 \to \hat{X}_2$ that is the identity on X.

2.8 Theorem
(a) Compactifications $X \subseteq \hat{X}_1$ and $X \subseteq \hat{X}_2$ are equivalent if there are continuous functions $f : \hat{X}_1 \to \hat{X}_2$ and $g : \hat{X}_2 \to \hat{X}_1$ extending the identity 1_X on X.
(b) A compactification $X \subseteq \hat{X}$ is equivalent to the Stone-Čech compactification $\beta(X)$ if there is a continuous function $\hat{X} \to \beta(X)$ extending 1_X.

Proof: (a) Because $g \circ f : \hat{X}_1 \to \hat{X}_1$ agrees with the identity $1_{\hat{X}_1} : \hat{X}_1 \to \hat{X}_1$ on the dense subspace $X \subseteq \hat{X}_1$, then $g \circ f = 1_{\hat{X}_1}$, by the Coincidence Theorem (4.4.4). Similarly, $f \circ g : \hat{X}_2 \to \hat{X}_2$ equals the identity $1_{\hat{X}_2} : \hat{X}_2 \to \hat{X}_2$.
(b) Theorem 2.6(b) provides a continuous function $\beta(X) \to \hat{X}$ extending 1_X. Therefore, $\hat{X}$ and $\beta(X)$ are equivalent by (a). □

Example 3 Let $X \subseteq \hat{X}$ be a compactification of X such that each continuous function $g : X \to \mathbb{I}$ has a continuous extension $G : \hat{X} \to \mathbb{I}$. Then the evaluation embedding $\lambda : X \to \prod_g \mathbb{I}_g = \mathbb{I}^{(X, \mathbb{I})}$, $\lambda(x) = (g(x))$, has the continuous extension $\Lambda : \hat{X} \to \mathbb{I}^{(X, \mathbb{I})}$, $\Lambda(x) = (G(x))$. Since X is dense in $\hat{X}$ and Λ is continuous, then $\Lambda(\hat{X}) = \beta(X) \subseteq \mathbb{I}^{(X, \mathbb{I})}$, so that Λ defines a continuous function $\Lambda : \hat{X} \to \beta(X)$. Therefore, $\hat{X}$ is equivalent to the Stone-Čech compactification $\beta(X)$ by Theorem 2.8(b).

Example 4 *The Stone-Čech compactification $\beta([0, \Omega))$ of the open ordinal interval $[0, \Omega)$ is the closed ordinal interval $[0, \Omega]$.* $[0, \Omega]$ is compact by Example 7.1.4. Also every continuous function $g : [0, \Omega) \to \mathbb{I}$ has a continuous extension $G : [0, \Omega] \to \mathbb{I}$; hence, $[0, \Omega] \cong \beta([0, \Omega))$, by Example 3. In fact, every continuous function g is eventually constant on $[0, \Omega)$, i.e., has a constant value on (α, Ω) for some ordinal α, and so extends continuously by assigning that constant value to Ω. To show why g is eventually constant, we construct an increasing sequence (α_n) in $[0, \Omega)$ such that the intervals (α_n, Ω) have *diam* $g((\alpha_n, \Omega)) < 1/n$ for all $n \geq 1$. Such a sequence exists; otherwise, for some $n \geq 1$ there would be an increasing sequence (γ_i) in $[0, \Omega)$ such that $|g(\gamma_i) - g(\gamma_{i-1})| > 1/n$ for all $i \geq 1$. Because (γ_i) would converge to its least upper bound in $[0, \Omega)$ (Example 4.1.17) and the sequence $(g(\gamma_i))$ would not converge in $\mathbb{I}$, the continuity of g would be contradicted. Since (γ_i) can't exist, (α_n) must exist. It has a least upper bound α in $[0, \Omega)$; by construction, *diam* $g((\alpha, \Omega)) = 0$, as desired.

2.9 Theorem *Let X be a Hausdorff space. Then:*
(a) *X is completely regular if and only if it admits a compactification X^*.*
(b) *X is locally compact if and only if it admits a compactification and is an open set in each of its compactifications.*
(c) *X is compact if and only if X is its only compactification.*

Proof: (a) Because a compact Hausdorff space X^* is normal (**5.2.12**), any space X that admits a compactification X^* must be completely regular, according to Theorem 2.5. Conversely, a completely regular space has its Stone-Čech compactification.

(b) When X is an open subset of a compactification X^*, then X is locally compact by Theorem **7.4.3**(b). Conversely, if X is locally compact and X^* is a compactification, then by Theorem **7.4.3**(a) we have $X = V \cap F$, where V is open and F is closed in X^*. Because the closed set F contains the dense subset X of X^*, then $F = X^*$ and, therefore, $X = V \cap F = V$ is open in X^*.

(c) When X is a compact Hausdorff space, its embedding $X \subseteq X^*$ into any compactification X^* is a homeomorphism: Since X is compact and X^* is Hausdorff, then X is a closed and dense in X^*. Hence, $X = \overline{X} = X^*$. □

Statement 2.9(c) does not characterize the compact Hausdorff spaces as those Hausdorff spaces that admit a unique compactification. Example 4 and Theorem 2.6(b) show that the non-compact ordinal interval $[0, \Omega)$ has a unique compactification, namely, $\beta([0, \Omega)) = [0, \Omega]$!

3 Metrization Theorems

A topological space $(X, \mathcal{T})$ is said to be ***metrizable*** if there exists a metric d on X that induces the given topology $\mathcal{T}$ on X.

We have seen various purely topological consequences of metrizability, namely, Hausdorff, regularity, and normality. We now seek topological properties that imply metrizability.

We give two metrization theorems in this section. The first, a theorem dating to 1925 and due to Urysohn, gives a necessary and sufficient condition that a space be metrizable and separable. The second, established in 1950, independently, by J. Nagata and Y. Smirnov, gives necessary and sufficient conditions that a space be metrizable.

In the proofs of both theorems, a space acquires a metric via an embedding of it into a (pseudo-) metrizable topological product. (A ***pseudometrizable space*** has its topology induced by a pseudo-metric, as in Exercise **2.3.21**.) We begin with an investigation of such topological products.

PRODUCTS OF METRIC SPACES

We now characterize those topological products of non-trivial spaces that are metrizable. The same arguments are valid for pseudo-metrizable spaces.

3.1 Theorem *Let $\prod_\alpha X_\alpha$ be a topological product of spaces, none of which is a single point. If $\prod_\alpha X_\alpha$ is metrizable, then each factor space X_α is metrizable and the index set $\mathcal{A} = \{\alpha\}$ is countable.*

Proof: Because $\prod_\alpha X_\alpha$ is metrizable and because each factor X_α is homeomorphic to a slice of $\prod_\alpha X_\alpha$ (Corollary **5.3.5**), each factor space is metrizable. Moreover, $\prod_\alpha X_\alpha$ is first countable, by Example **5.2.1**. To show that $\mathcal{A}$ is countable, consider $x \in \prod_\alpha X_\alpha$ and a countable neighborhood base $\mathcal{U}_x$ at x. For each index $\alpha \in \mathcal{A}$, we select a proper open neighborhood $V_\alpha \subset X_\alpha$ of $x_\alpha \in X_\alpha$ and then select some $U \in \mathcal{U}_x$ such that $U \subseteq p_\alpha^{-1}(V_\alpha)$. The function $\rho : \mathcal{A} \to \mathcal{U}_x$ defined by such choices, $\rho(\alpha) = U \subseteq p_\alpha^{-1}(V_\alpha)$, is finite-to-one, because each neighborhood $U \in \mathcal{U}_x$ in the product topology restricts at most finitely many coordinates of its members. Therefore, $\mathcal{A}$ is countable. □

3.2 Product Metrization Theorem

Let $\{(X_k, d_k) : 1 \leq k\}$ be a denumerable family of metric spaces with diameters $diam(X_k)$. On the product set $X \times X$, where $X = \prod_k X_k$, we define

$$m((x_k), (y_k)) = lub\{d_k(x_k, y_k) : 1 \leq k\}.$$

(a) m is a metric on $X = \prod_k X_k$ if $\{diam(X_k) : 1 \leq k\}$ is bounded in $\mathbb{E}^1$.
(b) m metrizes the product topology on $\prod_k X_k$ if and only if $diam(X_k) \to 0$.

Proof: (a) The bounded diameters show that m is a function $m : X \times X \to \mathbb{R}$. The metric axioms (**D1**)-(**D3**) for m are consequences of these same axioms for the metrics $\{d_k : 1 \leq k\}$ and the basic properties of the *lub* in $(\mathbb{R} \leq)$.

(b) Suppose that m metrizes the product topology. Were there a radius $r > 0$ and a subsequence (k_n) such that $diam(X_{k_n}) \geq r$ for all $n \geq 1$, then any metric ball $B_m(x, \frac{1}{2}r)$ in $\prod_k X_k$ would restrict the coordinates y_k of its members y for the infinitely many indices k that appear in the subsequence (k_n). This ball wouldn't be a neighborhood of x in the product topology on $\prod_k X_k$. Thus, no such radius and subsequence exist. Equivalently, $diam(X_k) \to 0$.

Suppose next that $diam(X_k) \to 0$. We show that each m-ball $B_m(x, r)$ about any point $x \in \prod_k X_k$ contains some base neighborhood

$$U = \langle B_1(x_1, r_1), \ldots, B_n(x_n, r_n) \rangle,$$

and conversely. This will prove that m metrizes the product topology.

Given $r > 0$, choose $n \geq 1$ such that $diam(X_k) \leq \frac{1}{2}r$ for all $k \geq n$. Then

$$U = \langle B_1(x_1, \tfrac{1}{2}r), \ldots, B_n(x_k, \tfrac{1}{2}r) \rangle \subseteq B_m(x, r),$$

since $y \in U$ has $d_k(x_k, y_k) \leq \frac{1}{2}r$ for all $1 \leq k$, so that $m(x, y) < r$. Conversely, given U, we choose $r = min\{r_k : 1 \leq k \leq n\}$. Then,

$$B_m(x, r) \subseteq \langle B_1(x_1, r_1), \ldots, B_n(x_n, r_n) \rangle = U,$$

because $y \in B_m(x, r)$ has $d_k(x_k, y_k) \leq r \leq r_k$, for all $1 \leq k \leq n$. □

For each metric space (X, d) and bound $M > 0$, there is the topologically equivalent metric $b(x, y) = M \, min\{d(x, y), 1\}$ for which b-$diam \, X \leq M$. Therefore, the Product Metrization Theorem has the following consequence:

3.3 Corollary *The topological product $\prod_k X_k$ of a countable family of metrizable spaces $\{(X_k, d_k) : 1 \leq k\}$ is metrizable.*

Proof: We give each factor space X_k ($1 \leq k$) its topologically equivalent metric $b_k = (\frac{1}{2})^k \min\{d_k, 1\}$. Because $b_k\text{-}diam(X_k) \leq (\frac{1}{2})^k \to 0$, the topological product $\prod_k X_k$ has the metric $m = lub\{b_k : 1 \leq k\}$ by Theorem 3.2. □

EMBEDDING THEOREM

We shall consider two situations involving a space X and a family $\mathcal{F}$ of continuous functions $g : X \to Y_g$. We ask whether the space X can be reconstructed from the knowledge of its images in the spaces Y_g under the continuous functions $g \in \mathcal{F}$. This is comparable to trying to identify an object in a product set from the knowledge of the coordinates of its points. But here we have to construct a product set large enough to contain X.

As in Section 2, a family $\mathcal{F} = \{g : X \to Y_g\}$ determines an ***evaluation function***

$$\lambda : X \to \prod_g Y_g, \; (\lambda(x))_g = g(x).$$

We say that the family $\mathcal{F}$ ***distinguishes points and closed sets*** if, for each point $x \in X$ and closed set $A \subseteq X$ not containing x, there exists $g \in \mathcal{F}$ such that $g(x) \notin \text{CLS } g(A) \subseteq Y_g$.

We say that the family $\mathcal{F}$ ***distinguishes points*** if, for each pair of distinct points $x \neq y$ in X, there exists some $g \in \mathcal{F}$ such that $g(x) \neq g(y)$. In a space with closed points (i.e., T_1-space), the first property implies the second.

When the spaces $\{Y_g : g \in \mathcal{F}\}$ are metrizable and the family $\mathcal{F}$ is countable, the topological product $\prod_g Y_g$ is metrizable, by Corollary 3.3. Then the metrizability of X follows whenever the evaluation function λ is an embedding. Hence, our interest in the following:

3.4 Embedding Theorem *Let $\mathcal{F} = \{g : X \to Y_g\}$ be a family of continuous functions on a space X and let $\lambda : X \to \prod_g Y_g$ be the evaluation function.*
(a) λ is a continuous function into the topological product;
(b) λ is an open function of X onto $\lambda(X) \subseteq \prod_g Y_g$ whenever $\mathcal{F}$ distinguishes points and closed sets in X; and
(c) λ is injective if and only if $\mathcal{F}$ distinguishes points in X.
Therefore, λ is an embedding when the conditions in (b) and (c) hold.

Proof: (a) The projections $h = p_h \circ \lambda : X \to \prod_g Y_g \to Y_h$ ($h \in \mathcal{F}$) are the given continuous functions; therefore, λ is continuous by the Characteristic Feature (5.3.2) of the product topology.

(b) To show that $\lambda : X \to \lambda(X)$ is open, we prove that, for each $x \in X$ and each open neighborhood $U(x)$ in X, the image $\lambda(U)$ contains $\lambda(X) \cap V$ for some neighborhood V of $\lambda(x) \in \prod_g Y_g$. Because $\mathcal{F}$ distinguishes points and

closed sets, there exists $h \in \mathcal{F}$ such that $h(x) \notin \mathrm{CLS}(h(U^c)) \subseteq Y_h$. Then
$$V = p_h^{-1}(Y_h - \mathrm{CLS}(h(U^c))) \subseteq \prod_g Y_g$$
is a neighborhood of $\lambda(x)$, since $p_h(\lambda(x)) = h(x) \in Y_h - \mathrm{CLS}(h(U^c))$. Finally, we have $\lambda(X) \cap V \subseteq \lambda(U)$, since
$$\lambda(z) \in V \;\Rightarrow\; p_h(\lambda(z)) = h(z) \notin \mathrm{CLS}(h(U^c)) \;\Rightarrow\; z \in U.$$

(c) This is clear from the definitions. □

URYSOHN METRIZATION THEOREM

Given a space X and a continuous function $g : X \to \mathbb{I}$, we can use the Euclidean distance between image points $g(x), g(z) \in \mathbb{I}$ as a coarse metric on X: $d_g(x, z) = |g(x) - g(z)|$. This function d_g is technically a *pseudo-metric* on X; it satisfies all the properties (**D**1)-(**D**3) of a metric, except possibly the requirement $d_g(x, z) = 0 \Rightarrow x = z$. The reason is that distinct points $x, z \in X$ may *project the same shadow* in $\mathbb{I}$ under g. Urysohn's idea is to catalog sufficiently many functions $g : X \to \mathbb{I}$ that allow the reconstruction of X from the shadows that it projects, but few enough to allow the pseudo-metrics d_g to coalesce into a metric on X.

Let $\mathbb{I}^\omega$ denote the cube $\prod_k \mathbb{I}_k$ obtained as the topological product of copies $\mathbb{I}_k = \mathbb{I}$ ($1 \leq k$) of the closed unit interval. By Corollary 3.3, the product topology on the cube $\mathbb{I}^\omega$ is metrizable.

3.5 Urysohn's Metrization Theorem
The following properties of a topological space X are equivalent:
(a) X is a separable metric space;
(b) X is a second countable regular space; and
(c) X is homeomorphic to a subspace of the cube $\mathbb{I}^\omega$.

Proof: (a) $\Rightarrow$ (b). Any metrizable space is regular, even normal (Theorem **2**.3.5); a separable metric space is second countable (Theorem **5**.2.5).

(b) $\Rightarrow$ (c). Each second countable regular space is normal (Tychonoff's Lemma 1.6), so we may apply Urysohn's Theorem (1.7) to the given space X. Let $\mathcal{B}$ be a countable base for the topology on X, and let
$$\mathcal{A} = \{(U, V) \in \mathcal{B} \times \mathcal{B} : \overline{U} \subseteq V\}.$$

For each $(U, V) \in \mathcal{A}$, let $u = u_{(U, V)} : X \to \mathbb{I}$ be a Urysohn function such that $u(\overline{U}) = \{0\}$ and $u(V^c) = \{1\}$. To show that the countable family $\mathcal{F} = \{u_{(U, V)}\}$ indexed by $\mathcal{A}$ distinguishes points and closed sets, we consider any point $x \in X$ and closed set $A \subseteq X$ not containing x. Some base member $V \in \mathcal{B}$ satisfies $x \in V \subseteq A^c$. Then, by regularity, there exists a second base member $U \in \mathcal{B}$ such that $x \in U \subseteq \overline{U} \subseteq V \subseteq A^c$. The Urysohn function $u = u_{(U, V)}$ for the pair $(U, V) \in \mathcal{A}$ satisfies $u(x) \in u(\overline{U}) = \{0\}$ and $u(A) \subseteq u(V^c) = \{1\}$.

Hence, $u(x) \notin \mathrm{CLS}(u(A))$. This proves that $\mathcal{F}$ distinguishes x and A.

Thus, by the Embedding Theorem 3.4, $\lambda : X \to \prod_u \mathbb{I}_u$ gives a homeomorphism between X and a subspace of the countable topological product $\prod_u \mathbb{I}_u$, which we view as a subspace of the denumerable cube $\mathbb{I}^\omega$.

$(c) \Rightarrow (a)$. The denumerable cube $\mathbb{I}^\omega$ is metrizable by Corollary 3.3. Because $\mathbb{I}$ has a countable base $\mathcal{D}$, the denumerable cube $\mathbb{I}^\omega$ is also second countable, with countable base

$$\mathcal{B} = \{ \langle U_1, \ldots, U_n \rangle \subseteq \prod_k \mathbb{I}_k : 1 \leq n,\ U_i \in \mathcal{D},\ \forall\ 1 \leq i \leq n \}.$$

Since these topological properties of $\mathbb{I}^\omega$ are hereditary, the subspace X is also metrizable and second countable, hence, separable by Theorem **5.2.5**. □

NAGATA-SMIRNOV METRIZATION THEOREM

The Nagata-Smirnov metrization of spaces that are not necessarily separable builds on Urysohn's techniques, but involves some new ideas:

A family $\mathcal{G} = \{G\}$ of subsets of a topological space X is ***locally finite*** if each $x \in X$ has a neighborhood that intersects only finitely many $G \in \mathcal{G}$.

An accumulation point of the union $\cup\, \mathcal{G}$ of a locally finite family $\mathcal{G}$ must be an accumulation point of some member $G \in \mathcal{G}$. Thus, the closure of the union $\cup\, \{G : G \in \mathcal{G}\}$ equals the union, $\cup\, \{\overline{G} : G \in \mathcal{G}\}$, of the closures.

A family $\mathcal{G}$ is σ-***locally finite*** if it is the union $\mathcal{G} = \cup\, \{\mathcal{G}_k : 1 \leq k\}$ of a denumerable collection of locally finite families $\mathcal{G}_k$.

A covering $\mathcal{V}$ of a set X is a ***refinement*** of a covering $\mathcal{U}$ of X if each member of $\mathcal{V}$ is a subset of a member of $\mathcal{U}$.

Some of these concepts are involved in the topological property of paracompactness, a generalization of compactness developed by J. Dieudonné in 1944. We shall consider paracompactness in Section 4. In 1949, A. H. Stone showed that each metrizable space is paracompact using the following technique of decomposing an open covering into a denumerable collection of locally finite families:

3.6 Stone's Theorem   *Each open covering $\mathcal{U}$ of a metric space (X, d) has a σ-locally finite open refinement $\mathcal{V} = \cup\, \{\mathcal{V}_k : 1 \leq k\}$.*

Proof: Let $\mathcal{U}$ be a given open covering of the metric space (X, d). For each positive integer $1 \leq k$ and each member $U \in \mathcal{U}$, we form the subset

$$U_k = \{x \in X : d(x, X - U) \geq (\tfrac{1}{2})^k\} \subseteq U.$$

By the triangle inequality, $d(U_k, X - U_{k+1}) \geq (\tfrac{1}{2})^k - (\tfrac{1}{2})^{k+1} = (\tfrac{1}{2})^{k+1}$.

We invoke the Well-Ordering Principle (**1.5.6**) to choose a well-ordering $\ll$ of the set $\mathcal{U}$. For each $1 \leq k$ and each member $U \in \mathcal{U}$, we form the subset

$$U'_k = U_k - \cup\, \{V_{k+1} : V \in \mathcal{U},\ V \ll U\} \subseteq U_k.$$

In the well-ordering $\ll$ of $\mathcal{U}$, any pair $U \neq V$ in $\mathcal{U}$ satisfies either $V \ll U$ or $U \ll V$, but not both. So either $U'_k \subseteq X - V_{k+1}$ or $V'_k \subseteq X - U_{k+1}$; hence, either

$$d(U'_k, V'_k) \geq d(X - V_{k+1}, V_k) \geq (\tfrac{1}{2})^{k+1}$$

or

$$d(U'_k, V'_k) \geq d(U_k, X - U_{k+1}) \geq (\tfrac{1}{2})^{k+1}.$$

Then $U''_k = \{x \in X : d(x, U'_k) \leq (\tfrac{1}{2})^{k+3}\}$ is contained in U. And, for $U \neq V$,

$$d(U''_k, V''_k) \geq (\tfrac{1}{2})^{k+1} - 2(\tfrac{1}{2})^{k+3} = (\tfrac{1}{2})^{k+2}.$$

Thus each family $\mathcal{V}_k = \{U''_k : U \in \mathcal{U}\}$ ($k \geq 1$) is locally finite. Indeed, for each $x \in X$, the open ball $B(x, (\tfrac{1}{2})^{k+3})$ intersects *at most one* member of $\mathcal{V}_k$.

The family $\mathcal{V} = \cup \{\mathcal{V}_k : 1 \leq k\}$ is a covering of X. Each $x \in X$ belongs to some $U \in \mathcal{U}$, hence, to U_k for some $k \geq 1$. Then $x \in U'_k \subseteq U''_k$, when U is the first member of $(\mathcal{U}, \ll)$ that contains x. Since U''_k is open and $U''_k \subseteq U$, this proves that $\mathcal{V}$ is a σ-locally finite open refinement of $\mathcal{U}$. $\square$

For the Nagata-Smirnov Metrization Theorem, we require the following generalization of Tychonoff's Lemma (1.5), which incidentally was itself invoked in Urysohn's Metrization Theorem (3.5).

3.7 Lemma *Each regular space whose topology has a σ-locally finite base $\mathcal{B} = \cup \{\mathcal{B}_k : 1 \leq k\}$ is normal.*

Proof: Let A and B be disjoint closed subsets of the space X. Then A has a covering $\mathcal{U} \subseteq \mathcal{B}$ by open sets that have closure disjoint from B, and B has a covering $\mathcal{V} \subseteq \mathcal{B}$ by open sets that have closure disjoint from A.

We express $\mathcal{U}$ and $\mathcal{V}$ as $\mathcal{U} = \cup \{\mathcal{U}_k : 1 \leq k\}$ and $\mathcal{V} = \cup \{\mathcal{V}_k : 1 \leq k\}$, where $\mathcal{U}_k \subseteq \mathcal{B}_k$ and $\mathcal{V}_k \subseteq \mathcal{B}_k$ for all $k \geq 1$. For each $k \geq 1$, let

$$U_k = \cup \{U : U \in \mathcal{U}_k\} \quad \text{and} \quad V_k = \cup \{V : V \in \mathcal{V}_k\}.$$

By local finiteness of $\mathcal{B}_k$,

$$\overline{U}_k = \cup \{\overline{U} : U \in \mathcal{U}_k\} \quad \text{and} \quad \overline{V}_k = \cup \{\overline{V} : V \in \mathcal{V}_k\}.$$

Thus, A has the countable covering $\{U_k : 1 \leq k\}$ by open sets with closure $\overline{U}_k \subseteq B^c$ and B has the countable covering $\{V_k : 1 \leq k\}$ by open sets with closure $\overline{V}_k \subseteq A^c$. This is precisely the middle point in the proof of Tychonoff's Lemma 1.5, which proceeds to convert the members of these countable coverings into nested sets and then construct the disjoint neighborhoods

$$U = \cup \{U_k - \overline{V}_k : 1 \leq k\} \quad \text{and} \quad V = \cup \{V_k - \overline{U}_k : 1 \leq k\}. \quad \square$$

We now have the tools necessary for the metrization theorem established independently by J. Nagata (1950) and Yu. M. Smirnov (1951). Stone's Theorem (3.6) gives half of the following:

3.8 Nagata-Smirnov Metrization Theorem

The following properties of a topological space X are equivalent:
(a)   X is metrizable.
(b)   X is regular and has a σ-locally finite base $\mathcal{B} = \cup \{\mathcal{B}_k : 1 \leq k\}$.

Proof: (a) $\Rightarrow$ (b). For each $n \geq 1$, the covering $\mathcal{U}(n)$ of X by open balls of radius $1/n$ has a σ-locally finite open refinement $\mathcal{B}(n) = \cup \{\mathcal{B}_k(n) : 1 \leq k\}$, by Stone's Theorem (3.6). Then $\mathcal{B} = \cup \{\mathcal{B}_k(n) : 1 \leq n, 1 \leq k\}$ is a σ-locally finite base for the metric topology on X.

(b) $\Rightarrow$ (a). Let $\mathcal{B} = \cup \{\mathcal{B}_k : 1 \leq k\}$ be a σ-locally finite base for X. For each pair $<m, n> \in \mathbb{P} \times \mathbb{P}$ and each $U \in \mathcal{B}_m$, let $G = \cup \{V \in \mathcal{B}_n : \overline{V} \subseteq U\}$. Because $\mathcal{B}_n$ is locally finite, $\overline{G} = \cup \{\overline{V} \in \mathcal{B}_n : V \subseteq U\}$; hence, $\overline{G} \subseteq U$.

By Lemma 3.7, X is normal. We choose a Urysohn function $u_U : X \to \mathbb{I}$ for which $u_U(\overline{G}) = 1$ and $u_U(U^c) = 0$. Then we consider the function

$$d_{m,n} : X \times X \to \mathbb{R}, \; d_{m,n}(x, y) = \Sigma \{ |u_U(x) - u_U(y)| : U \in \mathcal{B}_m\}.$$

Because $\mathcal{B}_m$ is a locally finite family, each $<x, y> \in X \times X$ has a neighborhood on which the summation in $d_{m,n}(x, y)$ is finite, hence, continuous. Thus, $d_{m,n}$ is continuous for each pair $<m, n> \in \mathbb{P} \times \mathbb{P}$.

Now each $d_{m,n}$ is a metric on X in every respect except that the implication $d_{m,n}(x, y) = 0 \Rightarrow x = y$ may fail. It is a pseudo-metric and induces a pseudo-metric topology with base consisting of the pseudo-metric open balls $B_{m,n}(x, r) = \{y \in X : d_{m,n}(x, y) < r\}$ (see Exercise 2.3.21). Let $Y_{m,n}$ denote X with this pseudo-metric topology and let $g_{m,n} : X \to Y_{m,n}$ denote the identity set function. The function $g_{m,n}$ is continuous, because

$$g_{m,n}^{-1}(B_{m,n}(x, r)) = d_{m,n}(x, -)^{-1}([0, r))$$

is open in X by the continuity of the restriction $d_{m,n}(x, -) : \{x\} \times X \to \mathbb{R}$.

To show that the denumerable family $\mathcal{F} = \{g_{m,n} : X \to Y_{m,n}\}$ separates points and closed sets, we consider any point $x \in X$ and closed set $A \subseteq X$ not containing x. There exists a member $U \in \mathcal{B}_m \subseteq \mathcal{B}$ of the base such that $x \in U \subseteq X - A$; and by regularity, there exists a second member $V \in \mathcal{B}_n \subseteq \mathcal{B}$ of the base such that $x \in V \subseteq \overline{V} \subseteq U \subseteq X - A$. Then the pseudo-metric $d_{m,n}$ satisfies $d_{m,n}(x, A) \geq 1$, since $u_U(\overline{V}) = 1$ and $u_U(A) \subseteq u_U(U^c) = 0$. Thus, $g_{m,n}(x) = x$ does not belong to the $d_{m,n}$-closure of $g_{m,n}(A) = A$.

Therefore, by the Embedding Theorem (3.4), the evaluation function $\lambda : X \to \prod_{m,n} Y_{m,n}$ is an embedding of X onto a subspace of the countable product of pseudo-metrizable spaces. By the proofs of Theorem 3.2 and Corollary 3.3, the topological product $\prod_{m,n} Y_{m,n}$ is pseudo-metrized by $m((x_{m,n}), (y_{m,n})) = lub\{(\frac{1}{2})^{m+n} d_{m,n}(x, y) : 1 \leq k\}$. In fact, m is a metric on the image $\lambda(X)$, which consists of constant sequences $\lambda(x) = (x_{m,n} = x)$. Indeed, $m((x_{m,n}), (y_{m,n})) = 0$ implies $x = y$, for there exists a pair $<m, n> \in \mathbb{P} \times \mathbb{P}$ with $d_{m,n}(x, y) \geq 1$ whenever $x \neq y$ (using $A = \{y\}$).   □

4 Paracompact Spaces

In 1944, J. Dieudonné introduced a generalization of compactness called *paracompactness*. This is an extremely useful covering property; it postulates special refinements for all open coverings of a space.

REFINEMENTS AND PARACOMPACTNESS

The definition of paracompactness involves the following terminology, most of which has been introduced previously:

A family $\mathcal{U} = \{U_\alpha : \alpha \in \mathcal{A}\}$ of subsets of a topological space X is **locally finite** if each point has a neighborhood that intersects U_α for only finitely many $\alpha \in \mathcal{A}$. The family $\mathcal{U}$ is called a **covering** if $X = \cup\, \mathcal{U}$, and it is called an **open** (**closed**) **covering** if each $U_\alpha \in \mathcal{U}$ is open (closed) in X.

A covering $\mathcal{V} = \{V_\beta : \beta \in \mathcal{B}\}$ is called a **refinement** of another covering $\mathcal{U} = \{U_\alpha : \alpha \in \mathcal{A}\}$ if there is a function $\rho : \mathcal{B} \to \mathcal{A}$ such that $B_\beta \subseteq A_{\rho(\beta)}$ for all $\beta \in \mathcal{B}$. $\mathcal{V}$ is a **precise refinement** of $\mathcal{U}$ if $\mathcal{A} = \mathcal{B}$ and $B_\alpha \subseteq A_\alpha$ for all $\alpha \in \mathcal{A}$.

For most purposes, a given refinement can be converted into a precise refinement, using the following construction:

4.1 Lemma *When the covering $\mathcal{V}$ is a refinement of the covering $\mathcal{U}$, the family $\mathcal{W} = \{W_\alpha : \alpha \in \mathcal{A}\}$ of subsets, given by*

$$W_\alpha = \cup\, \{B_\beta \in \mathcal{V} : \rho(\beta) = \alpha\},$$

is a covering that is a precise refinement of $\mathcal{U}$. When $\mathcal{V}$ is a locally finite open covering, $\mathcal{W}$ is a locally finite open covering. □

A Hausdorff space X is **paracompact** if each open covering of X has an open locally finite refinement. By Lemma 4.1, the required refinement may be assumed to be a precise refinement.

Example 1 *Any compact Hausdorff space is paracompact*, because a finite subcovering is a locally finite refinement.

The converse of the implication in Example 1 fails; paracompactness is an honest generalization of compactness. We shall see in Example 4 that real n-space $\mathbb{R}^n$ gives a non-compact example of paracompactness. There is however this partial converse:

4.2 Theorem *A countably compact, paracompact space is compact.*

Proof: Let $\mathcal{U} = \{U_\alpha : \alpha \in \mathcal{A}\}$ be any open covering of a paracompact space X. Then $\mathcal{U}$ has a locally finite, precise refinement $\mathcal{V} = \{V_\alpha : \alpha \in \mathcal{A}\}$.

We order the subcoverings of $\mathcal{V}$ by inclusion. The intersection of a chain of subcoverings of $\mathcal{V}$ is a subcovering of $\mathcal{V}$; because each $x \in X$ belongs to just finitely many $V_\alpha \in \mathcal{V}$, one such V_α containing x must belong to all members of the chain of subcoverings, hence, to their intersection.

Therefore, by Zorn's Lemma (**1.5.5**), there is a minimal subcovering $\mathcal{W} = \{V_\beta : \beta \in \mathcal{B} \subseteq \mathcal{A}\}$ of $\mathcal{V}$. Because $\mathcal{W}$ has no subcovering, each $V_\beta \in \mathcal{W}$ contains some $x_\beta \in X$ that is contained in no other member of $\mathcal{W}$.

Then $\mathcal{B}$ is finite; otherwise, $\{x_\beta : \beta \in \mathcal{B}\}$ would be an infinite set that has no accumulation point, contrary to the Bolzano-Weierstrass property (Theorem **7.3.2**) for the countably compact Hausdorff space X.

It follows that $\{U_\beta : \beta \in \mathcal{B} \subseteq \mathcal{A}\}$ is a finite subcovering of the original covering $\mathcal{U}$. $\square$

Example 2   Each closed ordinal space $[0, \beta]$ is compact (Example **7.1.4**), hence, paracompact by Example 1. The non-closed ordinal interval $[0, \Omega)$ is not paracompact, since it is countably compact but not compact.

4.3   Theorem   *Every paracompact space is normal.*

Proof: As in the proof of the normality property (**7.2.12**) for compact Hausdorff spaces, the argument proceeds in two stages.

First, we prove that the paracompact space X is regular. Let F be a closed subset and consider any point $q \in X - F$. Because X is Hausdorff, each $x \in F$ has an open neighborhood U_x such that $q \notin \overline{U}_x$. By paracompactness and Lemma 4.1, the open covering

$$\{U_x : x \in F\} \cup \{X - F\}$$

of X has an open locally finite refinement $\{V_x : x \in F\} \cup \{W\}$. Because this refinement is a locally finite open covering and because $W \subseteq X - F$, then $V = \cup \{V_x : x \in F\}$ is open, contains F, and has closure $\overline{V} = \cup \{\overline{V}_x : x \in F\}$. Since $q \notin \overline{U}_x \supseteq \overline{V}_x$ for all $x \in F$, then $q \notin \overline{V}$. Thus, V and $X - \overline{V}$ are disjoint open neighborhoods of F and q, respectively. This proves that X is regular.

Now let F and G be disjoint closed subsets. By the regularity of X, each $x \in F$ has an open neighborhood U_x such that $G \cap \overline{U}_x = \emptyset$. By the previous reasoning, with q replaced by G, there exist disjoint open neighborhoods of F and G. This proves that X is normal. $\square$

Example 3   *A normal space need not be paracompact.* As Example 1.1 and Example 2 show, the ordinal interval $[0, \Omega)$ is normal but not paracompact.

VARIATIONS ON A THEME

As in Section 3, we say that a family is **σ-*locally finite*** if it is the union of a denumerable collection of locally finite families.

The following connections between open or closed, locally finite or σ-locally finite, refinements were established by E. Michael in 1953.

4.4 Michael's Theorem
For a regular space X, the following properties are equivalent:
(a)   *X is paracompact.*
(b)   *Each open covering of X has an open σ-locally finite refinement.*
(c)   *Each open covering of X has a locally finite refinement.*
(d)   *Each open covering of X has a closed locally finite refinement.*

Proof: $(a) \Rightarrow (b)$. This is trivial.

$(b) \Rightarrow (c)$. Let $\mathcal{U} = \{U\}$ be an open covering of X; let $\mathcal{V} = \cup \{\mathcal{V}_k : 1 \leq k\}$ be a σ-locally finite refinement of $\mathcal{U}$. We form the union $V_k = \cup \mathcal{V}_k$ for each of the locally finite families $\mathcal{V}_k$. Then $\{V_k : 1 \leq k\}$ is an open covering of X; hence, the sets $W_k = V_k - \cup \{V_i : i < k\}$ $(k \geq 1)$ also cover X. The covering $\{W_k : 1 \leq k\}$ is locally finite, because no member V_i of the open covering $\{V_k : 1 \leq k\}$ intersects any of the W_k for $k > i$. Then $\{W_k \cap V : V \in \mathcal{V}_k, 1 \leq k\}$ is a locally finite refinement of $\mathcal{U}$. Indeed, each $x \in X$ has a neighborhood that intersects W_k for just finitely many k and for each k the point x has a neighborhood that intersects just finitely many members of $\mathcal{V}_k$.

$(c) \Rightarrow (d)$. Let $\mathcal{U} = \{U\}$ be any open covering of X. For each $x \in X$, we select a neighborhood $U_x \in \mathcal{U}$. Because X is regular, there exists for each $x \in X$ an open neighborhood $V_x \subseteq \overline{V}_x \subseteq U_x$. The family $\mathcal{V} = \{V_x : x \in X\}$ is an open covering; $\mathcal{V}$ has a locally finite refinement $\mathcal{G} = \{G\}$, by (c). Then the closed family $\overline{\mathcal{G}} = \{\overline{G}\}$ is also locally finite. Moreover, $\overline{\mathcal{G}}$ refines $\mathcal{U}$, because, for each G, we have that $\overline{G} \subseteq \overline{V}_x \subseteq U_x$ for some $x \in X$.

$(d) \Rightarrow (a)$. Let $\mathcal{U} = \{U\}$ be any open covering of X. By (d), $\mathcal{U}$ has a closed locally finite refinement $\mathcal{F} = \{F\}$. For each $x \in X$, we select a neighborhood V_x that intersects at most finitely many $F \in \mathcal{F}$. By (d), the open covering $\mathcal{V} = \{V_x : x \in X\}$ has a closed locally finite refinement $\mathcal{G} = \{G\}$. For each $F \in \mathcal{F}$, we form the set $R(F) = X - \cup \{G \in \mathcal{G} : F \cap G = \emptyset\}$. $R(F)$ contains F, and it is open because $\mathcal{G}$ is a closed locally finite covering.

Then $\{R(F) : F \in \mathcal{F}\}$ is locally finite. Any $x \in X$ has a neighborhood W that intersects just finitely many $G \in \mathcal{G}$ because $\mathcal{G}$ is locally finite. Then W intersects just finitely many $R(F)$, $F \in \mathcal{F}$. For $G \cap R(F) \neq \emptyset \Leftrightarrow F \cap G \neq \emptyset$ and $\mathcal{G}$ refines $\mathcal{V}$, whose members intersect at most finitely many $F \in \mathcal{F}$.

Now we select, for each $F \in \mathcal{F}$, one member $U(F) \in \mathcal{U}$ containing F. Then $\{R(F) \cap U(F) : F \in \mathcal{F}\}$ is an open locally finite refinement of $\mathcal{U}$. □

4.5 Corollary *Paracompactness appears in the following ways:*
(a)   *Any metrizable space is paracompact.*
(b)   *Any regular Lindelöf space is paracompact.*
(c)   *Any regular second countable space is paracompact.*

Proof: (a) By Stone's Theorem (3.6), each open covering of a metric space has a σ-locally finite open refinement.

(b) In a Lindelöf space, each open covering $\mathcal{U}$ has a countable subcovering $\mathcal{V} = \{V_k : 1 \leq k\}$, which we can express as a σ-locally finite refinement $\mathcal{V} = \cup \{\mathcal{V}_k : 1 \leq k\}$ of $\mathcal{U}$ using the singleton families $\mathcal{V}_k = \{V_k\}$ ($1 \leq k$).

(c) A second countable space is Lindelöf (Theorem 7.3.5); hence, part (b) applies to establish paracompactness. □

Example 4 *Real n-space $\mathbb{R}^n$ is non-compact but paracompact.*

Example 5 *A topological product of paracompact spaces need not be paracompact.* By Example 7.3.1 and Example 1.2, the right interval space $\mathbb{R}_{(]}$ is Lindelöf (but not second countable) and regular. Hence, $\mathbb{R}_{(]}$ is paracompact by Corollary 4.5(b). According to Theorem 4.3, the topological product $\mathbb{R}_{(]} \times \mathbb{R}_{(]}$ cannot be paracompact because it is not normal (Example 1.2).

Example 6 *A subspace of a paracompact space need not be paracompact.* The Tychonoff plank $T = [0, \Omega] \times [0, \omega]$ is compact, hence, paracompact. The deleted Tychonoff plank $T_\infty = T - \{(\Omega, \omega)\}$ is not normal (Example 1.4), hence, not paracompact according to Theorem 4.3.

4.6 Theorem *Paracompactness has these invariance features:*
(a) *A closed subspace of a paracompact space is paracompact.*
(b) *The factors spaces of a paracompact product are paracompact.*

Proof: (a) Let $A \subseteq X$ be a closed subspace of a paracompact space X. Because $X - A$ is open in X, any covering of A by relatively open subsets has the form $\mathcal{V} = \{V = U \cap A\}$ for some open covering $\mathcal{U} = \{U\}$ of X. An open locally finite refinement of $\mathcal{U}$ determines one for $\mathcal{V}$.

(b) The factors spaces are homeomorphic to closed slices in the topological product. Therefore, (a) applies to establish their paracompactness. □

PARTITIONS OF UNITY

Given a space X, we consider **unity** $1 : X \to \mathbb{R}$, by which we mean the continuous function that takes the constant value 1.

Whenever X is paracompact, it is possible to partition unity into a family of continuous functions supported on an arbitrarily fine family of subspaces of X. This terminology and result are developed next.

The **support** of a continuous function $g: X \to \mathbb{R}$ is the closed set

$$\text{supp}(g) = \text{CLS}(\{x \in X : g(x) \neq 0\}).$$

A family $\{g_\alpha : \alpha \in \mathcal{A}\}$ of continuous functions $g_\alpha : X \to \mathbb{R}$ is called a **partition of unity** on X provided that the following two properties hold:

- The supports, $\text{supp}(g_\alpha)$, form a locally finite closed covering of X.
- For each $x \in X$, the values $g_\alpha(x) \in \mathbb{R}$ ($\alpha \in \mathcal{A}$) have sum $\sum_\alpha g_\alpha(x) = 1$.

Example 7 The following diagram gives the graphs of a partition of unity $\{g_-, g_+\}$ on $\mathbb{R}$ with supports $\{supp(g_-) = (-\infty, 1], supp(g_+) = [-1, \infty)\}$:

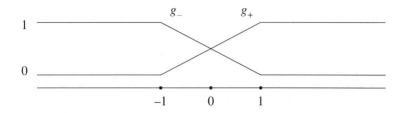

Example 7: Partition of unity on $\mathbb{R}$

We shall show that in a paracompact space there are partitions of unity supported on arbitrarily fine closed coverings. For this purpose, we establish the ability to *shrink* locally finite coverings in a normal space.

4.7 Shrinkability Lemma *In a normal space X, any locally finite covering $\mathcal{U} = \{U_\alpha : \alpha \in \mathcal{A}\}$ has a precise open refinement $\mathcal{V} = \{V_\alpha : \alpha \in \mathcal{A}\}$ such that $\overline{V}_\alpha \subseteq U_\alpha$ for all $\alpha \in \mathcal{A}$.*

Proof: Given $\mathcal{U}$, we will make a transfinite construction of the desired refinement $\mathcal{V}$. We invoke the Well-Ordering Principle (**1.5.6**), in order to choose well-orderings of the indexing set $\mathcal{A}$ and the topology $\mathcal{T}$ on X. We wish to construct a function $\lambda : \mathcal{A} \longrightarrow \mathcal{T}$, denoted by $\lambda(\alpha) = V_\alpha \in \mathcal{T}$, such that:

(i_α) $\overline{V}_\alpha \subseteq U_\alpha$, and
(ii_α) $\{V_\sigma : \sigma \leq \alpha\} \cup \{U_\tau : \tau > \alpha\}$ is a covering of X.

Suppose that $\lambda(\beta)$ has been defined satisfying (i_β) and (ii_β) for all $\beta < \alpha$. Then the set

$$G = X - [\cup \{V_\sigma : \sigma < \alpha\}) \cup (\cup \{U_\tau : \tau > \alpha\})]$$

is closed and contained in U_α, by (ii_β) for all $\beta < \alpha$. By normality, there exists a member $V \in \mathcal{T}$ such that $G \subseteq V \subseteq \overline{V} \subseteq U_\alpha$; let $\lambda(\alpha) = V_\alpha \in \mathcal{T}$ be the first such member in the well-ordering of $\mathcal{T}$. Then (i_α) and (ii_α) are satisfied.

The resulting family $\{V_\alpha : \alpha \in \mathcal{A}\}$ is a covering of X; because $\mathcal{U}$ is a locally finite covering, any $x \in X$ belongs to just finitely many members of $\mathcal{U}$, hence, to some complement $X - \cup \{U_\tau : \tau > \alpha\} = \cup \{V_\sigma : \sigma \leq \alpha\}$. □

4.8 Theorem *Let X be a paracompact space. For each open covering $\mathcal{U} = \{U_\alpha : \alpha \in \mathcal{A}\}$, there is a partition of unity $\{g_\alpha : \alpha \in \mathcal{A}\}$ whose supports constitute a precise refinement of $\mathcal{U}$: $supp(g_\alpha) \subseteq U_\alpha$ for all $\alpha \in \mathcal{A}$.*

Proof: Because the paracompact space X is regular, a precise locally finite refinement of the open covering $\mathcal{U} = \{U_\alpha : \alpha \in \mathcal{A}\}$ has successive refinements $\mathcal{V} = \{V_\alpha : \alpha \in \mathcal{A}\}$ and $\mathcal{R} = \{R_\alpha : \alpha \in \mathcal{A}\}$ such that, for all $\alpha \in \mathcal{A}$,

$$R_\alpha \subseteq \overline{R}_\alpha \subseteq V_\alpha \subseteq \overline{V}_\alpha \subseteq U_\alpha.$$

Urysohn's Theorem (1.7) provides $u_\alpha : X \to \mathbb{I}$ such that $u_\alpha(\overline{R}_\alpha) = 1$ and $u_\alpha(X - V_\alpha) = 0$. Since $\mathcal{V}$ is an open locally finite covering, each $x \in X$ has a neighborhood on which the summation $\sum_\alpha g_\alpha$ is finite, hence, continuous. Since continuity is a local property, the function $u = \sum_\alpha g_\alpha : X \to \mathbb{R}$, defined by $u(x) = \sum_\alpha g_\alpha(x)$, is continuous. Since each $x \in X$ belongs to at least one member of $\mathcal{R}$, the value $u(x) \in \mathbb{R}$ is always positive.

Then $\{g_\alpha = u_\alpha/u : \alpha \in \mathcal{A}\}$ is a suitable partition of unity on X. □

A partition of unity $\{g_\alpha : \alpha \in \mathcal{A}\}$ on X gives a way to blend any family $\{f_\alpha : \alpha \in \mathcal{A}\}$ of continuous functions $f_\alpha : X \to \mathbb{R}$ into a single continuous function, as follows:

4.9 Theorem *Let $\{g_\alpha : \alpha \in \mathcal{A}\}$ be a partition of unity on X and let $\{f_\alpha : X \to \mathbb{R} \ (\alpha \in \mathcal{A})\}$ be a family of continuous functions. Then $f : X \to \mathbb{R}$,*

$$f(x) = \sum_\alpha g_\alpha(x) f_\alpha(x),$$

is a continuous function.

Proof: The supports of the continuous functions $\{g_\alpha : \alpha \in \mathcal{A}\}$ form a locally finite covering of X. Therefore, each point of X has some neighborhood on which f is a finite sum of continuous functions. Hence, f is continuous. □

Exercises

SECTION 1

1. Prove that a quotient space $X|R$ of a space X is T_1 if and only if each equivalence class is closed in X.

2. Prove that continuous functions $f : X \to Y$ and $g : Y \to X$ that fix all points of a dense subspace D shared by X and Y are inverse homeomorphisms.

3. Prove that each F_σ set in a normal space is a normal subspace.

4. Prove that in a regular space distinct points have disjoint closed neighborhoods. Find a space that is not regular that has this property.

    A space X is called *completely Hausdorff* if distinct points of X have disjoint *closed* neighborhoods.

5. (a) Must a regular Hausdorff space be completely Hausdorff?
   (b) Do subspaces and product spaces inherit the complete Hausdorff property?

6. Prove that a regular Lindelöf space is normal.

7   Let R be an equivalence relation on a regular space X. Prove:
    (a) If the quotient map $q_R : X \rightarrow X/R$ is closed, then $R \subseteq X \times X$ is closed.
    (b) If $q_R : X \rightarrow X/R$ is an open and closed function, then X/R is Hausdorff.

8   Prove: In a regular space, any closed set equals the intersection of all its open neighborhoods.

9   Prove that the image of a normal space under a continuous closed surjection is normal. Does there exist a non-normal continuous image of a normal space?

10 Determine whether or not there exists a denumerable, connected, normal space.

11 Consider an increasing sequence of topological spaces $X_1 \subseteq X_2 \subseteq \ldots$. Give the union $X = \cup \{X_k : 1 \leq k\}$ the weak topology with respect to the family $\{X_k : 1 \leq k\}$. Prove: If each X_k is normal and closed in X_{k+1}, then X is normal.

12 Prove that a space is completely normal if and only if every subspace is normal.

13 Prove: If a partially ordered set $(X, \leq)$ is well-ordered, then the order topology on X is completely normal. So all ordinal intervals are completely normal.

14 Prove: The quotient space of a normal space modulo a closed subspace is normal.

15 Let R be an equivalence relation on a normal space X. Prove: If the quotient map $q_R : X \rightarrow X/R$ is an open and closed function, then X/R is normal.

16 Let X be a locally compact Hausdorff space, let $C \subseteq X$ be a compact subspace, and let U be an open neighborhood of C in X. Prove: There exists a continuous function $f : X \rightarrow \mathbb{I}$ such that $f(C) = 1$ and $f(X - C) = 0$.

17* Use Urysohn's Theorem (1.7) and Tietze's Extension Theorem (1.9) to establish Corollary 1.10. Consider, in order, the cases in which the codomain space is the interval $(-k, +k)$, the real line $\mathbb{R}$, and the general interval (a, b).

SECTION 2

1   Let U be an open neighborhood of a point p in a completely regular space X. Prove that the singleton $\{p\}$ is a G_δ set in X if and only if there is a continuous function $u : X \rightarrow \mathbb{I}$ such that $u^{-1}(1) = \{p\}$ and $u^{-1}(0) = U^c$.

2   Prove that a completely regular space is connected if and only if its Stone-Čech compactification is connected.

3   Use the box topology (Exercise 5.3.17) on the product $\mathbb{R}^\omega$ of a denumerable family of copies of the real line $\mathbb{R}$. Determine whether this product is Hausdorff, regular, and completely regular.

4   Let $\beta(\mathbb{N})$ be the Stone-Čech compactification of the discrete space $\mathbb{N}$. Prove:
    (a) $\beta(\mathbb{N})$ is separable.
    (b) $\beta(\mathbb{N})$ is totally disconnected.
    (c) $\beta(\mathbb{N})$ is uncountable.

5   Let the space X have the property that every continuous function $f : X \rightarrow [0, 1]$ is constant off some compact subspace $K_f \subseteq X$. Prove that the one-point compactification and the Stone-Čech compactification of X are equivalent.

SECTIONS 1-4

6 Let $X \subseteq \hat{X}$ be a compactification of X with this property: for each continuous function $w : X \rightarrow Y$ *to a compact Hausdorff space* Y, there exists a unique continuous function $W : \hat{X} \rightarrow Y$ such that $w = W|_X : X \subseteq \hat{X} \rightarrow Y$. Prove that $\hat{X}$ is equivalent to the Stone-Čech compactification $\beta(X)$.

7* Prove that homeomorphic spaces have homeomorphic Stone-Čech compactifications.

8 (a) Prove that a locally compact, second countable, Hausdorff space is metrizable.
   (b) Let X be a second countable, regular, Hausdorff space and let D be a countable subset. Show that X is metrizable in a new topology with subbase consisting of the original open sets and the points of D.

9 Prove: The one-point compactification X_∞ of a locally compact Hausdorff space X is metrizable if and only if X is second countable.

10 Let X be a Hausdorff space; consider the relation $xRy \Leftrightarrow g(x) = g(y)$ for all continuous functions $g : X \rightarrow \mathbb{I}$. Prove:
   (a) R is an equivalence relation on X.
   (b) The quotient space X/R is completely regular.

11 Prove that a non-trivial connected and completely regular space is not denumerable.

SECTION 3

1 Consider a locally finite family $\mathcal{G}$ of subsets of a topological space X. Prove that the family $\overline{\mathcal{G}} = \{\overline{G} : G \in \mathcal{G}\}$ of closures in X is also locally finite.

2 Prove that a T_0-space that is the union of a locally-finite family of closed metrizable subspaces is itself metrizable.

3 Prove that a metrizable space X has a totally bounded metric if and only if X is second countable.

4 Is the one-point compactification of an uncountable discrete space metrizable?

5 If a space is the union of two metrizable subspaces, must the space be metrizable?

6 Let X be the topological product $\mathbb{R}^\mathbb{R}$ of copies of the real line $\mathbb{R}$ indexed by $\mathbb{R}$. Assume the fact that X is normal. Determine whether or not X is compact, locally compact, connected, metrizable, separable, Lindelöf, or completely regular.

SECTION 4

For a neighborhood V of the diagonal in $X \times X$, let $V(x) = \{y \in X : <x, y> \in V\}$. A covering $\mathcal{U}$ of a topological space is called an **even** covering if there is a neighborhood V of the diagonal in $X \times X$ such that $\mathcal{V} = \{V(x) : x \in X\}$ refines $\mathcal{U}$.

1 Prove that a regular space is paracompact if and only if each open covering is even.

2 Prove that if each open set in a paracompact space X is a paracompact subspace, then every subspace of X is paracompact.

3 Prove that each F_σ set in a paracompact space is a paracompact subspace.

4. Prove that a Hausdorff space X is normal if there exists for each locally finite open covering $\mathcal{U} = \{U_\alpha : \alpha \in \mathcal{A}\}$ some precise open refinement $\mathcal{V} = \{V_\alpha : \alpha \in \mathcal{A}\}$ such that $\overline{V}_\alpha \subseteq U_\alpha$ for all $\alpha \in \mathcal{A}$.

5. Prove that the quotient of a paracompact space modulo a closed subset is paracompact.

6. Prove that a Hausdorff space X is paracompact whenever there exists for each open covering $\mathcal{U} = \{U_\alpha : \alpha \in \mathcal{A}\}$ some partition of unity $\{g_\alpha : \alpha \in \mathcal{A}\}$ whose supports constitute a precise refinement of $\mathcal{U}$: $supp(g_\alpha) \subseteq U_\alpha$ for each $\alpha \in \mathcal{A}$.

7. Prove: In paracompact spaces, countable compactness is equivalent to compactness.

8. Let the space X have the weak topology relative the closed covering $\{A_\alpha : \alpha \in \mathcal{A}\}$. Prove that X is normal if each subspace A_α ($\alpha \in \mathcal{A}$) is normal.

9. Let X be a regular space, and let $\{A_\alpha : \alpha \in \mathcal{A}\}$ be a locally finite family of closed paracompact subspaces. Prove that $\cup_\alpha A_\alpha$ is a paracompact subspace of X.

Part Two: Homotopy

9

FUNDAMENTAL GROUP

In Section **6**.2, paths are used to partition each topological space X into path-components. The set $\pi_0(X)$ of path-components records the topological features of X related to the existence of paths joining prescribed points of X. In this chapter, *homotopies* between paths are used to partition the set $P(X)$ of all paths in the space X into path-homotopy classes. The resulting set, $\pi(X)$, of path-homotopy classes measures more subtle topological features of X related to the variety of paths joining prescribed points of X.

1   The Fundamental Groupoid

PATH-HOMOTOPY

Interesting features of a space are revealed when one tries to continuously deform one path into another one with the same endpoints, as follows.

Let X be any space and let $f, g : \mathbb{I} \to X$ be paths in X that have the same endpoints $f(0) = x_0 = g(0)$ and $f(1) = x_1 = g(1)$. A *path-homotopy* from the path f to the path g is a continuous function $H : \mathbb{I} \times \mathbb{I} \to X$ such that, for all path and homotopy parameter values $0 \leq t \leq 1$ and $0 \leq r \leq 1$, we have:

$$H(t, 0) = f(t) \qquad H(t, 1) = g(t) \qquad H(0, r) = x_0 \qquad H(1, r) = x_1.$$

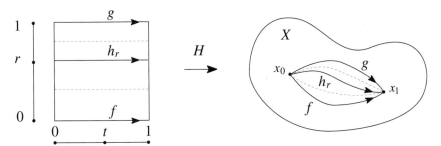

The parametrized family of paths $h_r : \mathbb{I} \to X$ ($0 \leq r \leq 1$), defined by $h_r(t) = H(t, r)$ for $0 \leq t \leq 1$, is called a ***continuous family of paths*** in X from f to g. Two paths f and g that are related by a path-homotopy H are called ***path-homotopic***. The path-homotopy relationship between f and g is denoted by $f \simeq g$, or by $H : f \simeq g$ when the path-homotopy H is to be emphasized.

Example 1 Consider the annulus $A \subseteq \mathbb{R}^2$ consisting of all points $x \in \mathbb{R}^2$ of distance $1 \leq \|x\| \leq 2$ from the origin $O \in \mathbb{R}^2$. There is a variety of paths in A joining the antipodal points $1 = \langle 1, 0 \rangle$ and $-1 = \langle -1, 0 \rangle$. Below left, are three paths f, g, and h in A; f is a clockwise path, g is a counter-clockwise path, and h is a deformed version of g. In the complex notation $re^{i\theta} = \langle r \cos\theta, r \sin\theta \rangle$, their definitions are:

$$f(t) = e^{-\pi i t}, \qquad g(t) = e^{\pi i t}, \qquad \text{and} \qquad h(t) = (1 + \sin \pi t)\, e^{\pi i t}.$$

The paths g and h enclose only points of A, and there exists a continuous family of paths in A from g to h. Specifically, for each $0 \leq r \leq 1$, there is the path $h_r : \mathbb{I} \to A$ given by $h_r(t) = (1 + r \sin \pi t)\, e^{\pi i t}$. Notice that $h_0 = g$, $h_1 = h$, and the paths h_r vary continuously in r. On the other hand, the two paths f and g enclose the hole of the annulus A and there appears to be no continuous family of paths in A between f and g.

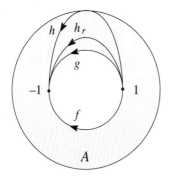
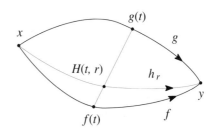

Example 1: Homotopy $g \simeq h$ in annulus A

Example 2: Straight-line homotopy in $\mathbb{R}^n$

Example 2 In $\mathbb{R}^n$, the vector space addition $+ : \mathbb{R}^n \times \mathbb{R}^n \to \mathbb{R}^n$ and scalar multiplication $\cdot : \mathbb{R}^1 \times \mathbb{R}^n \to \mathbb{R}^n$ (Section **2**.2) are continuous. In terms of these operations, the line segment between points $a, b \in \mathbb{R}^n$ is written $r \cdot a + (1 - r) \cdot b$, where $0 \leq r \leq 1$. Thus, these operations can be used to construct a path-homotopy $H : \mathbb{I} \times \mathbb{I} \to \mathbb{R}^n$ between any two paths $f, g : (\mathbb{I}, 0, 1) \to (\mathbb{R}^n, x, y)$ with common endpoints, as follows: $H(t, r) = (1 - r) \cdot f(t) + r \cdot g(t)$. The same technique obviously applies in any convex subspace $X \subseteq \mathbb{R}^n$.

A space X is called ***simply connected*** provided that it is path-connected and any two paths with the same endpoints are path-homotopic in X.

Example 2 shows that any convex subspace $X \subseteq \mathbb{R}^n$, such as the open unit ball $\mathbb{B}^n \subset \mathbb{R}^n$ or the closed unit ball $\mathbb{D}^n \subset \mathbb{R}^n$, is simply connected. Example 1 suggests that the situation is richer for the 1-sphere $\mathbb{S}^1 \subset \mathbb{R}^2$: the clockwise path f and the counter-clockwise path g joining the antipodal points

$1 = \langle 1, 0 \rangle$ and $-1 = \langle -1, 0 \rangle$ do not appear to be path-homotopic in $\mathbb{S}^1$. Section 2 verifies this and shows that the path-homotopy relation detects the difference between $\mathbb{S}^1$ and $\mathbb{D}^2$.

Here is a less expected example:

Example 3 *Two-point Sierpinski space S is simply connected.* The Sierpinski topology $\{\emptyset, \{a\}, \{a, b\}\}$ on $S = \{a, b\}$ is path-connected (Example 6.3.3). Moreover, two paths $f, g : \mathbb{I} \to S$ with the same endpoints are path-homotopic in S. A path-homotopy $H : \mathbb{I} \times \mathbb{I} \to S$ from f to g can be defined to take the constant value a on the interior of the unit square $\mathbb{I} \times \mathbb{I}$. This function H is continuous, because the pre-image $H^{-1}(a)$ of the open subset $\emptyset \neq \{a\} \subset S$ is the union of the open square $(0, 1) \times (0, 1)$, the relatively open subsets $f^{-1}(a) \subseteq \mathbb{I} \times \{0\}$ and $g^{-1}(a) \subseteq \mathbb{I} \times \{1\}$, as well as the sides $\{0\} \times \mathbb{I}$ or $\{1\} \times \mathbb{I}$ whenever $g(0) = a = f(0)$ or $g(1) = a = f(1)$.

PATH-HOMOTOPY CLASSES

For any space X, let $P(X)$ denote the set of all paths in X. We now study the path homotopy relation $\simeq$ on $P(X)$.

1.1 Lemma *The path-homotopy relation $\simeq$ is an equivalence relation.*

Proof: *Reflexivity*: If $f \in P(X)$, we have $S : f \simeq f$ via the *stationary* path-homotopy S given by $S(t, r) = f(t)$ for all $0 \leq r \leq 1$ and $0 \leq t \leq 1$.

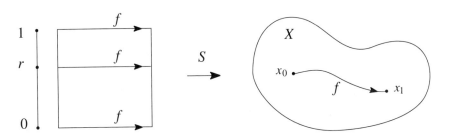

Stationary path-homotopy

Symmetry: If $H : f \simeq g$, then $K : g \simeq f$, where the path-homotopy K is the *reverse* of H. That is, $K(t, r) = H(t, 1 - r)$ for all $0 \leq r \leq 1$ and $0 \leq t \leq 1$.

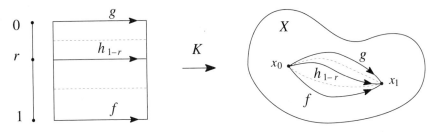

Reverse path-homotopy K

Transitivity: If $H : f \simeq g$ and $L : g \simeq h$, then $M : f \simeq h$, where the path-homotopy M is obtained by *stacking L atop H*:

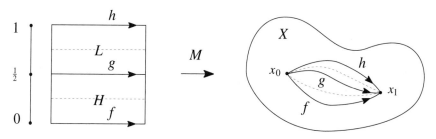

Stacked path-homotopy M

Since $H(t, 1) = g(t) = L(t, 0)$ for all $0 \leq t \leq 1$, M is well-defined by

$$M(t, r) = \begin{cases} L(t, 2r - 1), & \text{if } \tfrac{1}{2} \leq r \leq 1 \\ H(t, 2r), & \text{if } 0 \leq r \leq \tfrac{1}{2}. \end{cases}$$

The function M is continuous by the Gluing Theorem (**4.4.3**). □

The equivalence class in $P(X)$ of any path $f : \mathbb{I} \to X$ with respect to the path-homotopy relation $\simeq$ is called a **path-homotopy class**; it is denoted by

$$[f] = \{g \in P(X) : g \simeq f\}.$$

The quotient set, $P(X)/\simeq$, of all path-homotopy classes of paths in X is denoted by $\pi(X)$ and is called the **fundamental set** of X.

Example 4   Let $\tau : \pi(X) \to X \times X$ denote the function that associates to each path-homotopy class $[f] \in \pi(X)$ the ordered pair of its endpoints $< f(0), f(1) > \in X \times X$. Clearly, τ is onto $\Leftrightarrow X$ is path-connected and τ is bijective $\Leftrightarrow X$ is simply connected.

INDUCED FUNCTIONS

Any continuous function $F : X \to Y$ converts a path $f : \mathbb{I} \to X$ into a path $F \circ f : \mathbb{I} \to Y$. So F induces a set function

$$F_\# : P(X) \to P(Y), F_\#(f) = F \circ f.$$

The function $F_\#$ respects the path-homotopy relation on $P(X)$ and $P(Y)$:

1.2 Naturality Property   *Let $F : X \to Y$ be a continuous function. If the paths $f, g : \mathbb{I} \to X$ are path-homotopic in the domain space X, then the paths $F \circ f, F \circ g : \mathbb{I} \to Y$ are path-homotopic in the codomain space Y.*

Proof: If $f \simeq g$ in X via the path-homotopy $H : \mathbb{I} \times \mathbb{I} \to X$, then $F \circ f \simeq F \circ g$ in Y via the path-homotopy $F \circ H : \mathbb{I} \times \mathbb{I} \to X \to Y$. □

1.3 Theorem
The fundamental sets satisfy these functorial properties:

(a) Each continuous function $F : X \to Y$ induces a set function:
$$F_\# : \pi(X) \to \pi(Y), \quad F_\#([f]) = [F \circ f].$$

(b) For any two continuous functions $F : X \to Y$ and $G : Y \to Z$, the composite function $G \circ F : X \to Z$ induces the composite set function:
$$(G \circ F)_\# = G_\# \circ F_\# : \pi(X) \to \pi(Y) \to \pi(Z).$$

(c) The identity function $1_X : X \to X$ induces the identity set function:
$$(1_X)_\# = 1_{\pi(X)} : \pi(X) \to \pi(X).$$

Proof: (a) The set function $F_\#$ is well-defined by the Naturality Property (1.2). The proofs of (b) and (c) are left to Exercise 1. □

MULTIPLICATION IN P(X) AND Π(X)

The set $P(X)$ of all paths in a space X admits a sometimes-defined multiplication: for two paths
$$f : (\mathbb{I}, 0, 1) \to (X, x_0, x_1) \quad \text{and} \quad g : (\mathbb{I}, 0, 1) \to (X, x_1, x_2),$$

the **product path** $f \cdot g : (\mathbb{I}, 0, 1) \to (X, x_0, x_2)$ is defined in Section **6**.3 by
$$f \cdot g(t) = \begin{cases} g(2t - 1), & \text{if } \tfrac{1}{2} \le t \le 1 \\ f(2t), & \text{if } 0 \le t \le \tfrac{1}{2}. \end{cases}$$

The algebraic structure of $(P(X), \cdot)$ has many failings; see Exercise 2. The following results show that life gets better in the fundamental set $\pi(X)$.

1.4 Lemma
If $f \simeq f'$ and $g \simeq g'$ and if the product $f \cdot g$ is defined, then the product $f' \cdot g'$ is defined and $f \cdot g \simeq f' \cdot g'$.

Proof: As $f(1) = g(0)$, any two path-homotopies $H : f \simeq f'$ and $K : g \simeq g'$ glue side-by-side to form a path-homotopy $L : f \cdot g \simeq f' \cdot g'$:

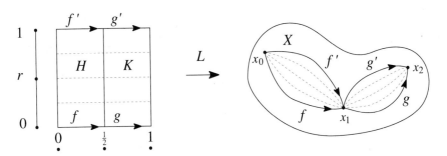

Side-by-side path-homotopy L

Since the relations $H(1, r) = f(1) = g(0) = K(0, r)$ hold for all $0 \leq r \leq 1$, the function L is well-defined by

$$L(t, r) = \begin{cases} K(2t - 1, r), & \text{if } \frac{1}{2} \leq t \leq 1 \\ H(2t, r), & \text{if } 0 \leq t \leq \frac{1}{2}. \end{cases}$$

Also, L is continuous by the Gluing Theorem (**4.4.3**). □

Thus, the multiplication for the set $P(X)$ of paths in a space X determines a multiplication for the fundamental set $\pi(X)$ of path-homotopy classes of paths in X: $[f] \cdot [g] = [f \cdot g]$, whenever the product path $f \cdot g$ is defined. This induced algebraic structure of $\pi(X)$ is more pleasant than that on $P(X)$ because the path-homotopy relation $\simeq$ eliminates the distinction between two paths that differ merely in their parametrization of the unit interval.

1.5 Lemma *Two paths $f, g : (\mathbb{I}, 0, 1) \to (X, x_0, x_1)$ that differ by just a continuous change of parameter $p : (\mathbb{I}, 0, 1) \to (\mathbb{I}, 0, 1)$ are path-homotopic.*

Proof: Let $g = f \circ p : (\mathbb{I}, 0, 1) \to (X, x_0, x_1)$. Because $\mathbb{I}$ is simply connected space $\mathbb{I}$, there is a path-homotopy $1_{\mathbb{I}} \simeq p : (\mathbb{I}, 0, 1) \to (\mathbb{I}, 0, 1)$. Hence, in the space X, there is a path-homotopy $f = f \circ 1_{\mathbb{I}} \simeq f \circ p = g : (\mathbb{I}, 0, 1) \to (X, x_0, x_1)$, by the Naturality Property (**1.2**). □

1.6 Theorem *The fundamental set $\pi(X)$ of path-homotopy classes of paths in a space X has the algebraic structure of a groupoid, as follows.*
(a) *The sometimes-defined multiplication in $\pi(X)$ is associative:*

$$[f] \cdot ([g] \cdot [h]) = ([f] \cdot [g]) \cdot [h], \text{ whenever either triple product is defined.}$$

(b) *Every element $[f] \in \pi(X)$ has a left unit and a right unit:*

$$[f] \cdot [x_1^*] = [f] = [x_0^*] \cdot [f], \text{ if } f(0) = x_0 \text{ and } f(1) = x_1.$$

(c) *Every element $[f] \in \pi(X)$ has a two-sided inverse:*

$$[f] \cdot [\bar{f}] = [x_0^*)] \text{ and } [\bar{f}] \cdot [f] = [x_1^*], \text{ if } f(0) = x_0 \text{ and } f(1) = x_1.$$

Proof: (a) For three elements $[f], [g], [h] \in \pi(X)$ that can be multiplied in the given order, the associativity relation $[f] \cdot ([g] \cdot [h]) = ([f] \cdot [g]) \cdot [h]$ follows from Lemma 1.5, since $f \cdot (g \cdot h) = ((f \cdot g) \cdot h) \circ p$ for the piecewise-linear change of parameter $p : (\mathbb{I}, 0, 1/2, 3/4, 1) \to (\mathbb{I}, 0, 1/4, 1/2, 1)$.

(b) The constant path at any point $x \in X$ is denoted by $x^* : \mathbb{I} \to X$. For any path $f : (\mathbb{I}, 0, 1) \to (X, x_0, x_1)$, the relations $[x_0^*] \cdot [f] = [f] = [f] \cdot [x_1^*]$ follow from Lemma 1.5, since $x_0^* \cdot f = f \circ q$ and $f \cdot x_1^* = f \circ r$, where q and r are the piecewise-linear changes of parameter

$$q : (\mathbb{I}, 0, 1/2, 1) \to (\mathbb{I}, 0, 0, 1) \quad \text{and} \quad r : (\mathbb{I}, 0, 1/2, 1) \to (\mathbb{I}, 0, 1, 1).$$

Thus, the elements $[x_1^*]$ and $[x_0^*]$ serve as left and right units, respectively, for the path-homotopy class $[f]$, assuming $f(0) = x_0$ and $f(1) = x_1$.

(c) For any path $f : (\mathbb{I}, 0, 1) \to (X, x_0, x_1)$, there is the reverse path $\bar{f} : (\mathbb{I}, 0, 1) \to (X, x_1, x_0)$, defined by $\bar{f}(t) = f(1-t)$. The product path $f \cdot \bar{f}$ equals the path $f \circ u : (\mathbb{I}, 0, 1) \to (X, x_0, x_0)$, where u is the piecewise-linear change of parameter $u : (\mathbb{I}, 0, \tfrac{1}{2}, 1) \to (\mathbb{I}, 0, 1, 0)$. By the Naturality Property (1.2), any path-homotopy $u \simeq 0^* : (\mathbb{I}, 0, 1) \to (\mathbb{I}, 0, 0)$ in the simply connected space $\mathbb{I}$ provides a path-homotopy in the space X:

$$f \cdot \bar{f} = f \circ u \simeq f \circ 0^* = x_0^* : (\mathbb{I}, 0, 1) \to (X, x_0, x_0).$$

Similarly, $\bar{f} \cdot f = f \circ v : (\mathbb{I}, 0, 1) \to (X, x_0, x_0)$, where v is the piecewise-linear change of parameter $v : (\mathbb{I}, 0, \tfrac{1}{2}, 1) \to (\mathbb{I}, 1, 0, 1)$. Therefore, there is a path-homotopy $\bar{f} \cdot f = f \circ v \simeq f \circ 1^* = x_1^* : (\mathbb{I}, 0, 1) \to (X, x_1, x_1)$.

These path-homotopies imply that $[f] \cdot [\bar{f}] = [x_0^*]$ and $[\bar{f}] \cdot [f] = [x_1^*]$. In short, the path-homotopy classes $[f]$ and $[\bar{f}]$ are inverses in $\pi(X)$. □

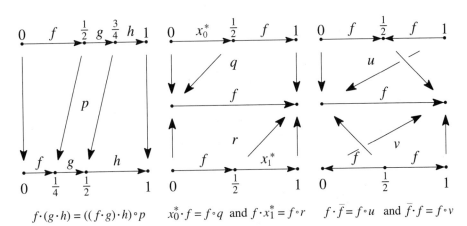

$f \cdot (g \cdot h) = ((f \cdot g) \cdot h) \circ p$ $\quad x_0^* \cdot f = f \circ q$ and $f \cdot x_1^* = f \circ r$ $\quad f \cdot \bar{f} = f \circ u$ and $\bar{f} \cdot f = f \circ v$

1.7 Theorem *For each continuous function $F : X \to Y$, the induced set function $F_\# : \pi(X) \to \pi(Y)$, $F_\#([f]) = [F \circ f]$, is a homomorphism of the fundamental groupoids: for all products $[f] \cdot [g] \in \pi(X)$, we have*

$$F_\#([f] \cdot [g]) = F_\#([f]) \cdot F_\#([g]) \in \pi(Y).$$

Proof: Whenever the product path $f \cdot g : \mathbb{I} \to X$ is defined, so is the product path $(F \circ f) \cdot (F \circ g) : \mathbb{I} \to Y$. Furthermore, $(F \circ f) \cdot (F \circ g) = F \circ (f \cdot g)$ in $P(Y)$, since these paths agree pointwise. Then the definition of the induced function $F_\#$ and the definition of the multiplications in $\pi(X)$ and $\pi(Y)$ yield the following calculations:

$$\begin{aligned} F_\#([f] \cdot [g]) &= F_\#([f \cdot g]) = [F \circ (f \cdot g)] = [(F \circ f) \cdot (F \circ g)] \\ &= [F \circ f] \cdot [F \circ g] = F_\#([f]) \cdot F_\#([g]). \end{aligned}$$
□

We summarize Theorems 1.3, 1.6, and 1.7 by saying that π is a *groupoid-valued functor*: π associates to each space X a groupoid $\pi(X)$, and to each continuous function $F: X \to Y$ a groupoid homomorphism $F_\# : \pi(X) \to \pi(Y)$, satisfying the *functorial properties* $(G \circ F)_\# = G_\# \circ F_\#$ and $(1_X)_\# = 1_{\pi(X)}$. The value of these functorial properties is indicated in Exercise 1(c).

2 Path-Homotopy in Spheres

PATH-HOMOTOPY IN THE 1-SPHERE

It is easy to define paths in the 1-sphere $\mathbb{S}^1$ that join specified endpoints and that wrap around $\mathbb{S}^1$, in either direction any number of times. Clearly, the path-homotopy relationships in $\mathbb{S}^1$ depend not only upon each path's endpoints, but also its *essential length*. The latter is a real number that measures its net direction and length of travel; it is defined as follows.

The removal of any pair $\{x, -x\}$ of antipodal points separates the circle $\mathbb{S}^1$ into two connected open subspaces, called *open hemispheres* of $\mathbb{S}^1$. For two points $a, b \in \mathbb{S}^1$ that lie in an open hemisphere of $\mathbb{S}^1$, the *directed distance* from a to b, denoted by $D(a \to b)$, is the length $-\pi < \varphi < \pi$ of the circular arc from a to b in the open hemisphere, with sign + or − accordingly as the arc from a to b is traversed counter-clockwise or clockwise. Using the complex numbers $e^{i\varphi} = \langle \cos 2\pi\varphi, \sin 2\pi\varphi \rangle \in \mathbb{S}^1$, the directed distance $D(e^{i\alpha} \to e^{i\beta})$ is the unique real number $-\pi < \varphi < \pi$ such that $e^{i\alpha} e^{i\varphi} = e^{i\beta}$.

For three points $a, b, c \in \mathbb{S}^1$ that lie in an open hemisphere of $\mathbb{S}^1$, there is a *triangle equality*: $D(a \to b) + D(b \to c) = D(a \to c)$ (Exercise 1).

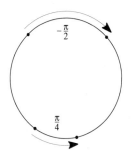
Directed distance

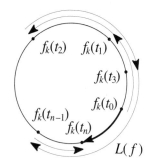
Essential length

Given a path $f : \mathbb{I} \to \mathbb{S}^1$, we say that a partition of $\mathbb{I}$,

$$\mathcal{P} : 0 = t_0 < t_1 < \ldots < t_{n-1} < t_n = 1,$$

is *f-admissible* if each subinterval $[t_i, t_{i+1}] \subseteq \mathbb{I}$ is carried by f into some open hemisphere of $\mathbb{S}^1$. An arbitrary partition $\mathcal{P}$ is f-admissible if its *mesh*, $\max \{|t_{i+1} - t_i| : 0 \leq i < n\}$, is less than a Lebesgue number of a covering of

$\mathbb{I}$ by the pre-images under f of open hemispheres of $\mathbb{S}^1$. So an admissible partition $\mathcal{P}$ exists for each path f, by the Lebesgue Covering Lemma (7.3.8).

The *essential length* $L(f)$ of a path f is defined to be the sum,

$$\sum_i D(f(t_i) \to f(t_{i+1})),$$

of the directed distances between the images under f of consecutive points in an f-admissible partition $\mathcal{P}$ of $\mathbb{I}$.

$L(f)$ is independent of the choice of the f-admissible partition $\mathcal{P}$. By the triangle equality, $L(f)$ is unaltered by the insertion of any single new partition point in an admissible partition; and any two admissible partitions can be made to coincide by the insertion of each one's points into the other.

So $L(f)$ measures the net direction and length of the path f, independently of the partition $\mathcal{P}$. In fact, $L(f)$ is an invariant of the path class $[f]$.

2.1 Path-Homotopy Invariance   *If $f \simeq g : \mathbb{I} \to \mathbb{S}^1$, then $L(f) = L(g)$.*

Proof: Let $H : \mathbb{I} \times \mathbb{I} \to \mathbb{S}^1$ be a path-homotopy from f to g. By a Lebesgue number argument, there exist partitions

$$\mathcal{P} : 0 = t_0 < t_1 < \ldots < t_{n-1} < t_n = 1 \quad \text{and} \quad \mathcal{R} : 0 = r_0 < r_1 < \ldots < r_{m-1} < r_m = 1$$

such that H carries each sub-rectangle $R_{i,k} = [t_i, t_{i+1}] \times [r_k, r_{k+1}]$ of $\mathbb{I} \times \mathbb{I}$ into an open hemisphere of $\mathbb{S}^1$. Therefore, the partition $\mathcal{P}$ is admissible for each path $h_k = H|_{\mathbb{I} \times \{r_k\}} : \mathbb{I} \to \mathbb{S}^1$ $(0 \le k \le m)$ and may be used to calculate $L(h_k)$:

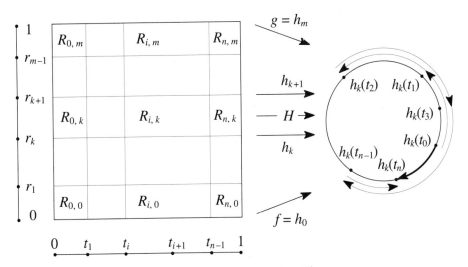

Path homotopy $H : \mathbb{I} \times \mathbb{I} \to \mathbb{S}^1$

Because H carries the sub-rectangles $R_{i,k} = [t_i, t_{i+1}] \times [r_k, r_{k+1}]$ $(0 \le i < n)$ of $\mathbb{I} \times \mathbb{I}$ into an open hemisphere of $\mathbb{S}^1$, n applications of the *triangle equality* show that the sum, $L(h_k)$, of the directed distances between the consecutive image points $h_k(t_i)$ $(0 \leq i \leq n)$ equals the sum, $L(h_{k+1})$, for $h_{k+1}(t_i)$ $(0 \leq i \leq n)$. (Recall that the ends $\{0\} \times [r_k, r_{k+1}]$ and $\{1\} \times [r_k, r_{k+1}]$ are collapsed by H.) Thus, by induction on $0 \leq k \leq m$, $L(f) = L(h_0)$ equals $L(g) = L(h_m)$. □

Example 1 The clockwise and counter-clockwise paths $f(t) = e^{-\pi i t}$ and $g(t) = e^{\pi i t}$ in $\mathbb{S}^1$ are not path-homotopic. The partition $0 < \frac{1}{2} < 1$ is admissible for both paths f and g; their essential lengths, $L(f)$ and $L(g)$, are given by these calculations:

$$L(f) = D(e^0 \to e^{-\frac{1}{2}\pi i}) + D(e^{-\frac{1}{2}\pi i} \to e^{-\pi i}) = (-\tfrac{1}{2})\pi + (-\tfrac{1}{2})\pi = -\pi, \text{ and}$$

$$L(g) = D(e^0 \to e^{\frac{1}{2}\pi i}) + D(e^{\frac{1}{2}\pi i} \to e^{\pi i}) = \tfrac{1}{2}\pi + \tfrac{1}{2}\pi = +\pi.$$

Since their essential lengths differ, f and g are not path-homotopic in $\mathbb{S}^1$.

There is a valid converse to the previous result: paths in $\mathbb{S}^1$ that share the same endpoints and have the same essential length are path-homotopic. We prove this converse, using the following ideas.

The **exponential function** $e : \mathbb{R}^1 \to \mathbb{S}^1$, $e(t) = e^{2\pi i t}$, wraps each interval $[n, n+1]$ in $\mathbb{R}^1$ with consecutive integer endpoints once around the 1-sphere $\mathbb{S}^1$, counter-clockwise. The real line $\mathbb{R}^1$ can be arranged as a helix aligned about the z-axis in xyz-space so that the exponential function becomes the projection, parallel to the z-axis, of the helix onto the unit circle $\mathbb{S}^1$ in the xy-plane. The pre-image $e^{-1}(x) \subset \mathbb{R}^1$, called the *fiber* over $x = e(\hat{x}) \in \mathbb{S}^1$, is the translated copy $\hat{x} + \mathbb{Z}$ of the set $\mathbb{Z}$ of integers. If U is any open hemisphere of $\mathbb{S}^1$, then its pre-image $e^{-1}(U)$ has denumerably many path components $\hat{U}$, each of which is carried by e homeomorphically onto U.

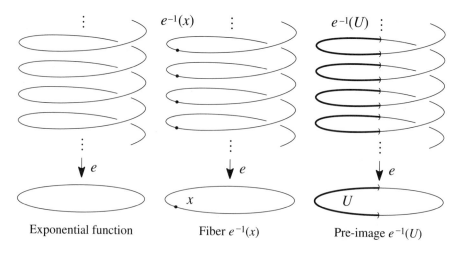

Exponential function   Fiber $e^{-1}(x)$ Pre-image $e^{-1}(U)$

Let $x \in \mathbb{S}^1$ and $\hat{x} \in e^{-1}(x) \subset \mathbb{R}^1$. If $f : \mathbb{I} \to \mathbb{S}^1$ is a path with initial point $f(0) = x$, there exists a path $\hat{f} : \mathbb{I} \to \mathbb{R}^1$ with initial point $\hat{f}(0) = \hat{x}$ such that $f = e \circ \hat{f} : \mathbb{I} \to \mathbb{R}^1 \to \mathbb{S}^1$. We say that $\hat{f}$ *covers* f or is a *lifting* of f at $\hat{x}$.

To visualize the lifting $\hat{f}$, imagine a *knight* traveling on the unit circle in the xy-plane, carrying an upraised *lance* that snags a small *ring* threaded on the helix coiled above him. As the *knight* travels the continuous path $f(t)$ on the unit circle, the *ring* travels a continuous path $\hat{f}(t)$ on the helix that covers $f(t)$ for all $0 \le t \le 1$.

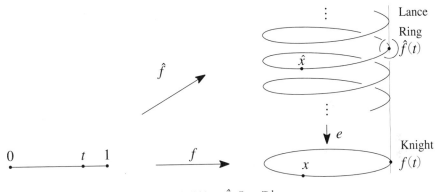

Path lifting $\hat{f} : \mathbb{I} \to \mathbb{R}^1$

More formally, the path lifting $\hat{f}$ can be constructed inductively on consecutive sub-intervals of an f-admissible partition,

$$\mathcal{P} : 0 = t_0 < t_1 < \ldots < t_{n-1} < t_n = 1,$$

of the unit interval $\mathbb{I}$, as follows. Suppose that $\hat{f} : [0, t_i] \to \mathbb{R}^1$ is a path lifting at $\hat{x}$ of the path $f : [0, t_i] \to \mathbb{S}^1$. Let U be a hemisphere of $\mathbb{S}^1$ containing the image $f([t_i, t_{i+1}])$; let $\hat{U}$ be the path-component of the pre-image $e^{-1}(U)$ containing the point $\hat{f}(t_i)$; and let $\varphi : U \to \hat{U}$ be the continuous inverse of the exponential function $e : \hat{U} \to U$. Then $\varphi \circ f : [t_i, t_{i+1}] \to U \to \hat{U}$ is the lifting at $\hat{f}(t_i)$ of the path $f : [t_i, t_{i+1}] \to U$. The liftings $\hat{f}$ on $[0, t_i]$ and $\varphi \circ f$ on $[t_i, t_{i+1}]$ glue into a lifting $\hat{f} : [0, t_{i+1}] \to \mathbb{R}^1$ at $\hat{x}$ of the path $f : [0, t_{i+1}] \to \mathbb{S}^1$. By induction, we get a lifting $\hat{f} : [0, 1] \to \mathbb{R}^1$ at $\hat{x}$ of the path $f : [0, 1] \to \mathbb{S}^1$.

2.2 Lifting Lemma *Each path $f : (\mathbb{I}, 0, 1) \to (\mathbb{S}^1, x, y)$ has a path lifting $\hat{f} : (\mathbb{I}, 0, 1) \to (\mathbb{R}^1, \hat{x}, \hat{y})$. Moreover, $L(f) = 2\pi\hat{y} - 2\pi\hat{x}$.*

Proof: The path lifting $\hat{f} : (\mathbb{I}, 0, 1) \to (\mathbb{R}^1, \hat{x}, \hat{y})$ is path-homotopic in the simply connected space $\mathbb{R}^1$ to the linear path $\hat{h} : (\mathbb{I}, 0, 1) \to (\mathbb{R}^1, \hat{x}, \hat{y})$ given by $\hat{h}(t) = (1-t)\hat{x} + t\hat{y}$. Then $f = e \circ \hat{f}$ and $h = e \circ \hat{h}$ are path-homotopic in $\mathbb{S}^1$, by the Naturality Property (1.2). Hence, $L(f) = L(h) = 2\pi\hat{y} - 2\pi\hat{x}$, by the Path-Homotopy Invariance (2.1) of L and a direct calculation of $L(h)$. □

2.3 Path-Homotopy Classification *Let the paths $f, g : \mathbb{I} \to \mathbb{S}^1$ have the same initial point x. Then, $L(f) = L(g)$ if and only if $f \simeq g$.*

Proof: When $L(f) = L(g)$, the path liftings $\hat{f}, \hat{g} : \mathbb{I} \to \mathbb{R}^1$ at the same initial point $\hat{x}$ have the same terminal point $\hat{y}$, by the Lifting Lemma (2.2). Then the liftings $\hat{f}, \hat{g} : (\mathbb{I}, 0, 1) \rightarrow (\mathbb{R}^1, \hat{x}, \hat{y})$ are path-homotopic in the simply connected space $\mathbb{R}^1$. It follows that $f = e \circ \hat{f}, g = e \circ \hat{g} : (\mathbb{I}, 0, 1) \rightarrow (\mathbb{S}^1, x, y)$ have the same terminal point $y = e(\hat{y})$ and are path-homotopic in $\mathbb{S}^1$, by the Naturality Property (1.2). The converse implication, $f \simeq g \Rightarrow L(f) = L(g)$, is the Path-Homotopy Invariance (2.1). □

Thus, a path-homotopy class $[f] \in \pi(\mathbb{S}^1)$ is completely determined by its initial point $x = f(0) \in \mathbb{S}^1$ and its essential length $L(f) \in \mathbb{R}$. This fact can be expressed as a groupoid isomorphism $\pi(\mathbb{S}^1) \approx \mathbb{S}^1 \times \mathbb{R}^1$ (Exercise 3).

BROUWER FIXED POINT THEOREM

The homotopy classification of paths in the 1-sphere is a powerful result with the following important topological consequences.

A subspace $A \subseteq X$ is a **retract** of X if there exists a continuous function $r : X \rightarrow A$ that is pointwise fixed on A, i.e., whose composite with the inclusion $i_A : A \subseteq X$ is the identity function $1_A = r \circ i_A : A \subseteq X \rightarrow A$. Such a continuous function $r : X \rightarrow A$ is called a **retraction** of X onto A.

Example 2 The subspace $\mathbb{Q}$ of rationals is not a retract of $\mathbb{R}^1$, since a disconnected space cannot be the continuous image of a connected one. The closed unit ball $\mathbb{D}^n$ is a retract of real n-space $\mathbb{R}^n$, but the open unit ball $\mathbb{B}^n$ is not (see Exercise 5(c)).

2.4 Theorem *The sphere $\mathbb{S}^1$ is not a retract of the closed disc $\mathbb{D}^2$.*

Proof: The clockwise and counter-clockwise paths $f, g : \mathbb{I} \rightarrow \mathbb{S}^1$, joining the antipodal points $<1, 0>$ and $<-1, 0>$, with essential lengths $L(f) = -\pi$ and $L(g) = +\pi$, respectively, are not path-homotopic in $\mathbb{S}^1$. But their composites $i \circ f, i \circ g : \mathbb{I} \rightarrow \mathbb{S}^1 \subset \mathbb{D}^2$ are path-homotopic, since $\mathbb{D}^2$ is a convex, hence, simply connected subspace of $\mathbb{R}^2$. There can be no retraction $r : \mathbb{D}^2 \rightarrow \mathbb{S}^1$; otherwise, any path-homotopy $i \circ f \simeq i \circ g$ in $\mathbb{D}^2$ would yield a path-homotopy $f = r \circ i \circ f \simeq r \circ i \circ g = g$ in $\mathbb{S}^1$, by the Naturality Property (1.2).

Another way to express the contradiction provided by a retraction $r : \mathbb{D}^2 \rightarrow \mathbb{S}^1$ is to consider the induced groupoid homomorphisms:

$$1_{\pi(\mathbb{S}^1)} = (1_{\mathbb{S}^1})_\# = (r \circ i)_\# = r_\# \circ i_\# : \pi(\mathbb{S}^1) \rightarrow \pi(\mathbb{D}^2) \rightarrow \pi(\mathbb{S}^1).$$

Since $[f] \neq [g]$ in $\pi(\mathbb{S}^1)$ and $i_\#[f] = i_\#[g]$ in $\pi(\mathbb{D}^2)$, the homomorphism $r_\# : \pi(\mathbb{D}^2) \rightarrow \pi(\mathbb{S}^1)$ with $1_{\pi(\mathbb{S}^1)} = r_\# \circ i_\#$ cannot exist. □

The negative result in the previous theorem might be considered of little use, but it has a very important positive reformulation. To state it we use the fixed point terminology from Section 2.5. Recall that we say that a point $x \in X$ is a **fixed point** of a function $F : X \rightarrow X$ if $F(x) = x$. Thus, a function $F : X \rightarrow X$ is called **fixed point free** (*f.p.f.*) if $F(x) \neq x$ for all $x \in X$.

Example 3 It is often easy to find fixed point free continuous functions $F : X \to X$ of a non-compact space X; e.g., translations of real n-space are fixed point free.

Some compact spaces also admit fixed point free continuous functions; e.g., the antipodal function $\alpha : \mathbb{S}^n \to \mathbb{S}^n$, $\alpha(x) = -x$, on a sphere $\mathbb{S}^n$ ($n \geq 0$) is fixed point free.

2.5 Fixed Point Theorem *Each continuous function $F : \mathbb{D}^2 \to \mathbb{D}^2$ of the closed unit disc $\mathbb{D}^2$ has at least one fixed point.*

Proof: Were $F : \mathbb{D}^2 \to \mathbb{D}^2$ a fixed point free map, then, for each $x \in \mathbb{D}^2$, the continuation of the straight-line segment in $\mathbb{R}^2$ from $F(x)$ to x would meet the boundary 1-sphere $\mathbb{S}^1$ in a well-defined point $r(x)$. The result would be a retraction $r : \mathbb{D}^2 \to \mathbb{S}^1$; the continuity of F and the geometry of the situation would imply the continuity of r and the relation $r(x) = x$ for each $x \in \mathbb{S}^1$. Since a retraction $r : \mathbb{D}^2 \to \mathbb{S}^1$ cannot exist according to Theorem 2.4, then neither can a fixed point free map $F : \mathbb{D}^2 \to \mathbb{D}^2$. □

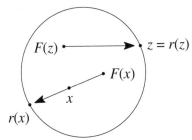
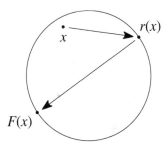

Retraction r from a f.p.f. map F F.p.f. map F from a retraction r

Thus, when a copy of a paper disc is picked up, crumpled, then repositioned over the original, some point of the copy covers the identical point of the original. When a wind storm stirs the particles of an oil slick on a puddle, it passes on, having returned some particle to its original position. These observations are guaranteed by the Fixed Point Theorem, provided the paper disc and puddle are homeomorphic to closed discs and the crumpling and wind motion are continuous. These claims fail if either the disc and puddle are annular or they are subject to tearing and wind shearing.

In 1911, L. E. J. Brouwer (1881-1967) established the fixed point theorem for the closed disc $\mathbb{D}^{n+1}$ in each dimension. The proof of Theorem 2.5 works in all dimensions once it is shown that the boundary sphere $\mathbb{S}^n$ is not a retract of $\mathbb{D}^{n+1}$. This requires a study of higher dimensional connectedness beyond the reach of the fundamental group. What is needed are the higher homotopy groups π_n ($n \geq 2$); they are introduced in Section **15**.2. The non-triviality of the group $\pi_n(\mathbb{S}^n)$ detects the non-retractibility of $\mathbb{D}^{n+1}$ onto $\mathbb{S}^n$.

The fixed point theorems have applications to numerous mathematical disciplines. They provide, for example, the existence of solutions to systems of equations and the existence of solutions of differential equations.

SIMPLE-CONNECTEDNESS OF $\mathbb{S}^n$

We now consider paths in the unit sphere $\mathbb{S}^n$ of dimension $n \geq 2$. The sphere $\mathbb{S}^2$ appears to be simply connected; it is path-connected and any belt slips right off it! Indeed, any two paths in $\mathbb{S}^2$ that share their endpoints are homotopic in the simply connected complement $\mathbb{S}^2 - \{z\} \cong \mathbb{R}^2$ of any point z that they miss. The problem with this argument in general is that no such point z exists when either path is a sphere-covering curve provided by Peano.

We need a preliminary argument to create points free of the path images.

2.6 Free Point Lemma *Let $n \geq 2$ and let $z \in \mathbb{S}^n - \{x, y\}$. Any path $f : (\mathbb{I}, 0, 1) \to (\mathbb{S}^n, x, y)$ is path-homotopic to some path $g : \mathbb{I} \to \mathbb{S}^n$ that misses the point z.*

Proof: (a) The complements of any two points $w, z \in \mathbb{S}^n - \{x, y\}$ define an open covering of $\mathbb{S}^n$. So the Lebesgue Covering Lemma provides a partition,

$$\mathcal{P} : 0 = t_0 < t_1 < \ldots < t_{n-1} < t_n = 1,$$

of $\mathbb{I}$ such that the restrictions $f_i = f : [t_i, t_{i+1}] \to \mathbb{S}^n$ ($0 \leq i < n$) take values alternately in the two complements $\mathbb{S}^n - \{w\}$ and $\mathbb{S}^n - \{z\}$. Then all of the partition points t_i ($0 \leq i \leq n$) are sent into the complement $\mathbb{S}^n - \{w, z\}$.

Now $\mathbb{S}^n - \{w, z\} \cong \mathbb{R}^n - O$ is path-connected as $n \geq 2$; so there is a path $g_i : [t_i, t_{i+1}] \to \mathbb{S}^n - \{w, z\} \subset \mathbb{S}^n$ joining $f(t_i)$ to $f(t_{i+1})$. Since $\mathbb{S}^n - \{w\} \cong \mathbb{R}^n$ and $\mathbb{S}^n - \{z\} \cong \mathbb{R}^n$ are simply connected by Example 1.2, there exist path-homotopies $H_i : f_i \simeq g_i : [t_i, t_{i+1}] \to \mathbb{S}^n$ ($0 \leq i < n$). These path-homotopies

$$H_0 : [t_0, t_1] \times \mathbb{I} \to \mathbb{S}^n, \ H_1 : [t_1, t_2] \times \mathbb{I} \to \mathbb{S}^n, \ \ldots, \ H_{n-1} : [t_{n-1}, t_n] \times \mathbb{I} \to \mathbb{S}^n$$

glue side-by-side to form a path-homotopy $H : \mathbb{I} \times \mathbb{I} \to \mathbb{S}^n$, $H(t, r) = H_i(t, r)$ ($t_i \leq t \leq t_{i+1}$), from f to a path g that is contained entirely in $\mathbb{S}^n - \{w, z\}$. □

2.7 Theorem *Each sphere $\mathbb{S}^n$ of dimension $n \geq 2$ is simply connected.*

Proof: By the Free Point Lemma, the simple connectedness of the complement $\mathbb{S}^n - \{z\}$, and the transitivity of the path-homotopy relation, we conclude that two paths with the same endpoints are path-homotopic in $\mathbb{S}^n$. □

BORSUK-ULAM THEOREM

Let $\alpha_n : \mathbb{S}^n \to \mathbb{S}^n$ denote the antipodal function $\alpha_n(x) = -x$ on the n-sphere $\mathbb{S}^n$ ($n \geq 0$). A function $G : \mathbb{S}^n \to \mathbb{S}^m$ is called ***antipodal-preserving*** if it carries antipodal points to antipodal points: $\alpha_m(G(x)) = G(\alpha_n(x))$ for all $x \in \mathbb{S}^n$.

Example 4 The inclusion $\mathbb{S}^{n-1} \subset \mathbb{S}^n$ of the equator $\mathbb{S}^{n-1}$ into an n-sphere is antipodal preserving. The continuous function $d_k : \mathbb{S}^1 \to \mathbb{S}^1$, $d_k(e^{i\theta}) = e^{ik\theta}$, is antipodal-preserving if and only if the integer k is odd.

2.8 Lemma *When a continuous function $F : \mathbb{S}^1 \to \mathbb{S}^1$ is antipodal-preserving, the induced homomorphism $F_\# : \pi(\mathbb{S}^1) \to \pi(\mathbb{S}^1)$ carries the class $[e] \in \pi(\mathbb{S}^1)$ of the exponential path $e : \mathbb{I} \to \mathbb{S}^1$ to a non-unit.*

Proof: The antipodal function $\alpha : \mathbb{S}^1 \to \mathbb{S}^1$ rotates the 1-sphere through 180°; so it interchanges the counter-clockwise paths $g, h : \mathbb{I} \to \mathbb{S}^1$, $g(t) = e^{\pi i t}$ and $h(t) = e^{\pi i (t+1)}$, which have antipodal endpoints 1 and -1. Since F is antipodal preserving, α interchanges the paths $F \circ g, F \circ h : \mathbb{I} \to \mathbb{S}^1$, which have antipodal endpoints $F(1)$ and $F(-1)$. Thus, $F \circ g$ and $F \circ h$ have the same non-trivial essential length L. Since the exponential path $e : \mathbb{I} \to \mathbb{S}^1$, $e(t) = e^{2\pi i t}$, is the path product $g \cdot h$, then $F \circ e$ is the path product $(F \circ g) \cdot (F \circ h)$, which has non-trivial essential length $2L$. So, $F_\#([e]) = [F \circ e]$ is not a unit. □

2.9 Theorem *There is no continuous antipodal-preserving function*

$$G : \mathbb{S}^2 \to \mathbb{S}^1.$$

Proof: Any continuous antipodal-preserving function $G : \mathbb{S}^2 \to \mathbb{S}^1$ would restrict to the equator $\mathbb{S}^1 \subset \mathbb{S}^2$ to give a continuous antipodal preserving function $F = G \circ i : \mathbb{S}^1 \subset \mathbb{S}^2 \to \mathbb{S}^1$. Since $\mathbb{S}^2$ is simply connected, the resulting factorization $F_\# = G_\# \circ i_\# : \pi(\mathbb{S}^1) \to \pi(\mathbb{S}^2) \to \pi(\mathbb{S}^1)$ would imply that $F_\#([e])$ is the unit $[F(1)^*]$, contrary to Lemma 2.8. □

This negative result has a positive reformulation involving this terminology: any continuous function $J : \mathbb{S}^2 \to \mathbb{R}^2$ is called a *flattening* of $\mathbb{S}^2$.

2.10 Borsuk-Ulam Theorem *Any flattening $J : \mathbb{S}^2 \to \mathbb{R}^2$ of the sphere $\mathbb{S}^2$ identifies some pair of antipodal points.*

Proof: Were $J : \mathbb{S}^2 \to \mathbb{R}^2$ a flattening that identifies no pair of antipodal points of $\mathbb{S}^2$, then $J(x) - J(\alpha(x))$ would be in $\mathbb{R}^2 - \{O\}$ for all $x \in \mathbb{S}^2$. Hence,

$$G(x) = \frac{J(x) - J(\alpha(x))}{\| J(x) - J(\alpha(x)) \|}$$

would define a continuous antipodal-preserving function $G : \mathbb{S}^2 \to \mathbb{S}^1$.

Since a continuous antipodal-preserving function $G : \mathbb{S}^2 \to \mathbb{S}^1$ cannot exist according to Theorem 2.9, then neither can a flattening $J : \mathbb{S}^2 \to \mathbb{R}^2$ that identifies no pair of antipodal points of $\mathbb{S}^2$. □

The following application of the Borsuk-Ulam Theorem takes the Pancake Theorem (**3.1.6**) one dimension higher (Exercise 9).

2.11 Ham-Sandwich Theorem *Any three open subsets of real space $\mathbb{R}^3$ are simultaneously bisected by some plane.*

3 The Fundamental Group

It is cumbersome to deal with the sometimes defined product in the fundamental groupoid $\pi(X)$ of a space X. This section introduces a way to obtain subgroupoids of $\pi(X)$ in which multiplication of two elements is always defined and which, therefore, are groups.

FUNDAMENTAL GROUP AT A BASEPOINT

Consider any space X and any selected point $x \in X$, called a **basepoint**. A **loop in X based at x** is a path $f : \mathbb{I} \to X$ that has initial and terminal point x. The loops in X based at x constitute a subset $\Omega(X, x) \subset P(X)$ and their path-homotopy classes constitute a subset $\pi_1(X, x) \subset \pi(X)$.

The path product of two *loops* $f, g \in \Omega(X, x)$ is always defined and the result is another *loop* $f \cdot g \in \Omega(X, x)$. Thus, the sometimes defined multiplication on $\pi(X)$ restricts to a binary operation $\pi_1(X, x) \times \pi_1(X, x) \to \pi_1(X, x)$.

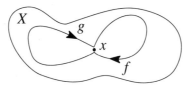

Product of loops at x

Therefore, Theorem 1.6 has the following specialization:

3.1 Theorem *The set $\pi_1(X, x)$ of path-homotopy classes of loops in X based at x has the algebraic structure of a group:*
(a) *The multiplication of path-homotopy classes of loops based at x is always defined and is associative:*

$$[f] \cdot ([g] \cdot [h]) = ([f] \cdot [g]) \cdot [h], \text{ for all } [f], [g], [h] \in \pi_1(X, x).$$

(b) *The path-homotopy class $[x^*]$ of the constant loop x^* is an identity:*

$$[f] \cdot [x^*] = [f] = [x^*] \cdot [f], \text{ for all } [f] \in \pi_1(X, x).$$

(c) *The path-homotopy class $[\bar{f}]$ of the reverse loop $\bar{f}$ is a two-sided inverse of the path-homotopy class $[f]$ of the loop f based at x:*

$$[f] \cdot [\bar{f}] = [x^*] = [\bar{f}] \cdot [f], \text{ for all } [f] \in \pi_1(X, x).$$

In view of these facts, $\pi_1(X, x)$ is called the **fundamental group** of X at x.

For a continuous function $F : X \to Y$ and loop $f : \mathbb{I} \to X$ in X based at x, the composite $F \circ f : \mathbb{I} \to Y$ is a loop in Y based at $F(x)$.

Thus, Theorems 1.3 and 1.7 have the following specialization:

3.2 Theorem *The fundamental groups have these functorial properties:*
(a)   For each continuous function $F : X \to Y$, the induced function

$$F_\# : \pi_1(X, x) \to \pi_1(Y, F(x)), \quad F_\#([f]) = [F \circ f],$$

is a group homomorphism.
(b)   For continuous functions $F : X \to Y$ and $G : Y \to Z$, the composite function $G \circ F : X \to Z$ induces the composite homomorphism:

$$(G \circ F)_\# = G_\# \circ F_\# : \pi_1(X, x) \to \pi_1(Y, F(x)) \to \pi_1(Z, G(F(x))).$$

(c)   The identity function $1_X : X \to X$ induces the identity homomorphism:

$$(1_X)_\# = 1_{\pi_1(X, x)} : \pi_1(X, x) \to \pi_1(X, x).$$

In short, π_1 is a *group-valued functor*. Thus, it is a *topological invariant*:

3.3 Corollary *Any homeomorphism $F : X \to Y$ induces an isomorphism,*

$$F_\# : \pi_1(X, x) \to \pi_1(Y, F(x)),$$

between the fundamental groups at corresponding basepoints.

Proof: Inverse continuous functions $F : (X, x) \leftrightarrow (Y, y) : G$ induce inverse group homomorphisms $F_\# : \pi_1(X, x) \leftrightarrow \pi_1(Y, y) : G_\#$; for, by Theorem 3.2,

$$1_{\pi_1(X, x)} = (1_X)_\# = (G \circ F)_\# = G_\# \circ F_\# : \pi_1(X, x) \to \pi_1(Y, y) \to \pi_1(X, x)$$

and

$$1_{\pi_1(Y, y)} = (1_Y)_\# = (F \circ G)_\# = F_\# \circ G_\# : \pi_1(Y, y) \to \pi_1(X, x) \to \pi_1(Y, y). \quad \square$$

THE ROLE OF THE BASEPOINT

When f is a loop in X based at x_0 and h is a path in X from x_0 to x_1, the path product $\bar{h} \cdot (f \cdot h)$ is a loop in X based at x_1.

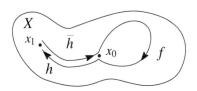

Change of basepoint

Moreover, the path-homotopy class $[\bar{h} \cdot (f \cdot h)]$ depends only upon the path-homotopy classes $[f]$ and $[h]$. So the path-homotopy class $[h] \in \pi(X)$ of $h : (\mathbb{I}, 0, 1) \to (X, x_0, x_1)$ determines a ***change of basepoint*** function:

$$\omega_{[h]} : \pi_1(X, x_0) \to \pi_1(X, x_1), \quad \omega_{[h]}([f]) = [\bar{h}] \cdot [f] \cdot [h].$$

The indicated multiplication occurs in the *fundamental groupoid* $\pi(X)$.

3.4 Theorem   *The change of basepoint functions have these properties:*
(a)   *Inverse classes* $[h]$ *and* $[\bar{h}]$ *give inverse homomorphisms*

$$\omega_{[h]} : \pi_1(X, x_0) \to \pi_1(X, x_1) \quad \text{and} \quad \omega_{[\bar{h}]} : \pi_1(X, x_1) \to \pi_1(X, x_0).$$

(b)   *The two change of basepoint isomorphisms*

$$\omega_{[h]} : \pi_1(X, x_0) \to \pi_1(X, x_1) \quad \text{and} \quad \omega_{[k]} : \pi_1(X, x_0) \to \pi_1(X, x_1)$$

differ in $\pi_1(X, x_1)$ *by conjugation by the group element* $[\bar{h} \cdot k] \in \pi_1(X, x_1)$.

Proof: (a) The homomorphism property for the function $\omega_{[h]}$ is the relation $\omega_{[h]}([f] \cdot [g]) = \omega_{[h]}([f]) \cdot \omega_{[h]}([g])$ for $[f], [g] \in \pi_1(X, x_0)$. This relation,

$$[\bar{h}] \cdot ([f] \cdot [g]) \cdot [h] = ([\bar{h}] \cdot [f] \cdot [h]) \cdot ([\bar{h}] \cdot [g] \cdot [h]),$$

holds since $[h] \cdot [\bar{h}] = [x_0^*]$ is the identity in the group $\pi_1(X, x_0)$.

The homomorphisms $\omega_{[h]}$ and $\omega_{[\bar{h}]}$ are inverses, i.e., the relations

$$[h] \cdot ([\bar{h}] \cdot [f] \cdot [h]) \cdot [\bar{h}] = [f] \quad \text{and} \quad [\bar{h}] \cdot ([h] \cdot [d] \cdot [\bar{h}]) \cdot [h] = [d]$$

hold for $[f] \in \pi_1(X, x_0)$ and $[d] \in \pi_1(X, x_1)$. The reason is that $[h] \cdot [\bar{h}]$ is the identity $[x_0^*] \in \pi_1(X, x_0)$ and $[\bar{h}] \cdot [h]$ is the identity $[x_1^*] \in \pi_1(X, x_1)$.

(b) This statement summarizes the relation:

$$\omega_{[h]}([f]) = [\bar{h} \cdot k] \cdot ([\bar{k}] \cdot [f] \cdot [k]) \cdot [\bar{h} \cdot k]^{-1} = [\bar{h} \cdot k] \cdot (\omega_{[k]}([f])) \cdot [\bar{h} \cdot k]^{-1}.$$

The latter triple product is called the ***conjugate*** of $\omega_{[k]}([f]) \in \pi_1(X, x_1)$ by the group element $[\bar{h} \cdot k] \in \pi_1(X, x_1)$. □

3.5 Corollary   *The fundamental groups $\pi_1(X, x_0)$ and $\pi_1(X, x_1)$ are isomorphic if the basepoints x_0 and x_1 lie in the same path-component of X.*

Thus, in a path-connected space X, all the fundamental groups $\pi_1(X, x)$, $x \in X$, are isomorphic via the change of basepoint isomorphisms $\omega_{[g]}$ associated with the path-classes $[g] \in \pi(X)$. There is no danger in defining the fundamental group $\pi_1(X)$ of a path-connected space X to be a copy of these isomorphic groups, provided that we remember that the change of basepoint isomorphisms between them are not unique, and that different choices for these isomorphisms differ by conjugation.

TRIVIALITY OF THE FUNDAMENTAL GROUP

3.6 Theorem   *For any space X and any basepoint $x \in X$, the following conditions are equivalent:*
(a) $\pi_1(X, x)$ *is a trivial group, i.e., contains just the identity element* $[x^*]$.
(b)   *Each loop f in X based at x is **null-homotopic**, i.e., $f \simeq x^*$.*
(c)   *Any two paths in X with common endpoints that lie in the path component $P(x)$ of the basepoint x are path-homotopic.*

Proof: (a) ⇔ (b). This is clear from the definition of $\pi_1(X, x)$ as the group of path-homotopy classes of loops in X based at x.

(c) ⇒ (b). Since paths in X with common endpoints in $P(x)$ include loops in X based at x, statement (c) includes statement (b).

(a) ⇒ (c). Let the two paths $f, g : (\mathbb{I}, 0, 1) \to (X, x_0, x_1)$ have endpoints $x_0, x_1 \in P(x)$. As $\pi_1(X, x)$ is trivial, so is the group $\pi_1(X, x_0)$, by Corollary 3.5. So the class $[f \cdot \bar{g}]$ of the loop $f \cdot \bar{g}$ based at x_0 is the identity element of $\pi_1(X, x_0)$. Then we have $[g] = [f \cdot \bar{g}] \cdot [g] = [f] \cdot [\bar{g} \cdot g] = [f]$ in the fundamental groupoid $\pi(X)$, i.e., $f \simeq g : (\mathbb{I}, 0, 1) \to (X, x_0, x_1)$. □

3.7 Corollary *A space is simply connected if and only if it is path-connected and has a trivial fundamental group.*

WINDING NUMBER AND DEGREE

Let $1 = <1, 0>$ be the basepoint of the 1-sphere $\mathbb{S}^1 \subset \mathbb{R}^2$. By Section 2, a loop $f : \mathbb{I} \to \mathbb{S}^1$ based at 1 has a lifting $\hat{f} : \mathbb{I} \to \mathbb{R}$ through the exponential function $e : \mathbb{R}^1 \to \mathbb{S}^1$, $e(t) = e^{2\pi i t}$. The difference, $\hat{f}(1) - \hat{f}(0)$, of the initial and terminal points $\hat{f}(0), \hat{f}(1) \in e^{-1}(1) = \mathbb{Z}$ is an integer, denoted by $w(f)$ and called the **winding number** of f. The essential length $L(f)$ of f is given by $L(f) = 2\pi(\hat{f}(1) - \hat{f}(0)) = 2\pi w(f)$, according to the Lifting Lemma (2.2).

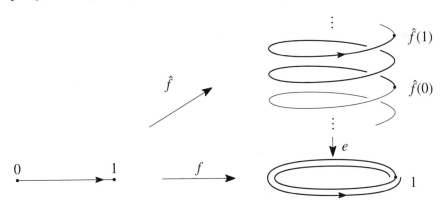

Winding number $w(f) = \hat{f}(1) - \hat{f}(0) \in \mathbb{Z}$

Since homotopic paths have the same essential length, then homotopic loops have the same winding number. In other words, the winding number is a well-defined function $w : \pi_1(\mathbb{S}^1, 1) \to \mathbb{Z}$. Since the essential length of a product of two paths is given by the relation $L(f \cdot g) = L(f) + L(g)$ (Exercise 2.3), $w : \pi_1(\mathbb{S}^1, 1) \to \mathbb{Z}$ is a homomorphism from the fundamental group of the 1-sphere $\mathbb{S}^1$ to the additive group $\mathbb{Z}$ of integers. The homomorphism w is injective, because any two loops that have the same essential length are path-homotopic. The homomorphism w is surjective; for each integer $n \in \mathbb{Z}$, there exists the loop $d_n : \mathbb{I} \to \mathbb{S}^1$, $d_n(t) = e^{2\pi n i t}$, whose essential length is

$L(d_n) = 2\pi n$ and whose winding number is $w(d_n) = n$.

Thus, $w : \pi_1(\mathbb{S}^1, 1) \to \mathbb{Z}$ is a group isomorphism under which the path-homotopy class $[e]$ of the exponential loop $e : \mathbb{I} \to \mathbb{S}^1$, $e(t) = e^{2\pi i t}$, corresponds to $1 \in \mathbb{Z}$. The following statement summarizes these observations.

3.8 Theorem $\pi_1(\mathbb{S}^1, 1)$ *is an infinite cyclic group generated by* $[e]$.

The *degree* of a continuous function $F : (\mathbb{S}^1, 1) \to (\mathbb{S}^1, 1)$ is the integer $Deg(F)$ that records the value of the induced homomorphism

$$F_\# : \pi_1(\mathbb{S}^1, 1) \to \pi_1(\mathbb{S}^1, 1)$$

on the generator $[e]$, i.e., $F_\#([e]) = [e]^{Deg(F)}$. If we use the isomorphism, $w : \pi_1(\mathbb{S}^1) \to \mathbb{Z}$ to convert to the additive group $\mathbb{Z}$, we see that a continuous function $F : (\mathbb{S}^1, 1) \to (\mathbb{S}^1, 1)$ with degree n is one that induces the homomorphism $\mathbb{Z} \to \mathbb{Z}$ that multiplies by n. Equivalently, $Deg(F)$ is the winding number of the path $F \circ e : \mathbb{I} \to \mathbb{S}^1 \to \mathbb{S}^1$.

Example 1 The function $F_n : (\mathbb{S}^1, 1) \to (\mathbb{S}^1, 1)$, $F_n(e^{i\theta}) = e^{in\theta}$, has $Deg(F_n) = n$, since the path $F_n \circ e = d_n$ has winding number $w(d_n) = n$.

THE FUNDAMENTAL GROUP OF A TOPOLOGICAL PRODUCT

The topological product $X \times Y$ of two spaces X and Y is characterized by the property that, for each space W, there is a bijective correspondence between continuous functions $h : W \to X \times Y$ into the product and pairs of continuous functions $f : W \to X$ and $g : W \to Y$ into the factors. Specifically, h corresponds to the pair $f = p_X \circ h$ and $g = p_Y \circ h$.

So there is a bijective correspondence between paths $h : \mathbb{I} \to X \times Y$ in the product and pairs of paths in the factors:

$$f = p_X \circ h : \mathbb{I} \to X \quad \text{and} \quad g = p_Y \circ h : \mathbb{I} \to Y.$$

Similarly, there is a bijective correspondence between path-homotopies $H : \mathbb{I} \times \mathbb{I} \to X \times Y$ in the product and pairs of path-homotopies in the factors:

$$F = p_X \circ H : \mathbb{I} \times \mathbb{I} \to X \quad \text{and} \quad G = p_Y \circ H : \mathbb{I} \times \mathbb{I} \to Y.$$

When H is a path-homotopy $h_0 \simeq h_1$ in the product $X \times Y$, then $F = p_X \circ H$ and $G = p_Y \circ H$ are path-homotopies in the factors X and Y, respectively:

$$f_0 = p_X \circ h_0 \simeq p_X \circ h_1 = f_1 \quad \text{and} \quad g_0 = p_Y \circ h_0 \simeq p_Y \circ h_1 = g_1.$$

This shows that there is a bijection

$$\eta : P(X \times Y) \to P(X) \times P(Y), \eta(h) = \ <f = p_X \circ h, g = p_Y \circ h>,$$

that respects the path-homotopy relations in $P(X \times Y)$, $P(X)$, and $P(Y)$.

Thus, there results a bijection:

$$\eta : \pi(X \times Y) \to \pi(X) \times \pi(Y), \eta([h]) = <[p_X \circ h], [p_Y \circ h]>.$$

This bijection is a groupoid isomorphism, provided that $\pi(X) \times \pi(Y)$ is given the coordinate-wise groupoid multiplication:

$$<[f_1], [g_1]> \cdot <[f_2], [g_2]> = <[f_1 \cdot f_2], [g_1 \cdot g_2]>.$$

When basepoints are considered, the following theorem results.

3.9 Theorem *The fundamental group $\pi_1(X \times Y, <x, y>)$ of a product space $X \times Y$ is isomorphic to the direct product $\pi_1(X, x) \times \pi_1(Y, y)$ of the fundamental groups of the factor spaces X and Y.*

Proof: A path $h : \mathbb{I} \to X \times Y$ is a loop based at $<x, y>$ if and only if the paths $f = p_X \circ h : \mathbb{I} \to X$ and $g = p_Y \circ h : \mathbb{I} \to Y$ are loops based at x and y, respectively. Thus, the groupoid isomorphism described above restricts to a group isomorphism $\eta : \pi_1(X \times Y, <x, y>) \to \pi_1(X, x) \times \pi_1(Y, y)$. □

Example 2 The 2-dimensional torus T^2 is the topological product $\mathbb{S}^1 \times \mathbb{S}^1$ of two copies of the 1-sphere $\mathbb{S}^1$. By Theorem 3.9, the fundamental group $\pi_1(T^2)$ at the basepoint $<1, 1>$ is isomorphic to the direct product $\pi_1(\mathbb{S}^1) \times \pi_1(\mathbb{S}^1)$ of two copies of the fundamental group $\pi_1(\mathbb{S}^1) \approx \mathbb{Z}$ at the basepoint $1 \in \mathbb{S}^1$.

By the topological invariance of π_1 (Corollary 3.3), this calculation proves that the simply connected 2-sphere $\mathbb{S}^2$ and the torus T^2 are not homeomorphic. Despite being intuitively obvious, this is a non-trivial result. These spaces are indistinguishable by most topological invariants, such as connectedness, compactness, metrizability, etc.

The path-homotopy classes, $[l]$ and $[m]$, of the **longitudinal** and **meridional loops**

$$l : \mathbb{I} \to \mathbb{S}^1 \times \mathbb{S}^1 \quad \text{and} \quad m : \mathbb{I} \to \mathbb{S}^1 \times \mathbb{S}^1,$$

given by $l(t) = <e^{2\pi i t}, 1>$ and $m(t) = <1, e^{2\pi i t}>$, generate $\pi_1(T^2)$ and they satisfy the commutativity relation $[l] \cdot [m] = [m] \cdot [l]$, because these facts hold for the corresponding elements $<1, 0>$ and $<0, 1>$ in the isomorphic group $\mathbb{Z} \times \mathbb{Z}$.

4 Poincaré Index Formula

The winding number appears in the work of Henri Poincaré [1854-1912] on the qualitative theory of differential equations. The connections Poincaré developed during 1881-1886 in his "Mémoire" involve the indices of critical points of vector fields. This section gives a brief development of these ideas.

VECTOR FIELDS

Let X be an open path-connected subset of the real plane $\mathbb{R}^2$. A **continuous vector field** V on X is simply a continuous function $V : X \to \mathbb{R}^2$.

A path $\lambda : [a, b] \to X$ is an **integral curve** for the vector field V if its velocity vector, $\lambda'(t) = (x'(t), y'(t))$, equals the vector $V(\lambda(t))$ at each point

$\lambda(t) \in X$ ($a \le t \le b$) of the curve. If we view the vector $V(p)$ anchored with its tail at $p \in X$, the integral curve through p is tangent to $V(p)$ and its speed equals the length of $V(p)$. If we write $V(x, y) = (F(x, y), G(x, y)) \in \mathbb{R}^2$, then an integral curve is a solution of the system of differential equations determined by the coordinate functions $F(x, y)$ and $G(x, y)$:

$$\frac{dx}{dt} = F(x, y) \qquad \frac{dy}{dt} = G(x, y).$$

When F and G are analytic functions, V is called **analytic**. The existence and uniqueness theorems for analytic systems imply that there is exactly one integral curve for V through each point $p \in X$ at which $V(p)$ is not zero. The collection of all integral curves constitutes the **phase portrait** of the system.

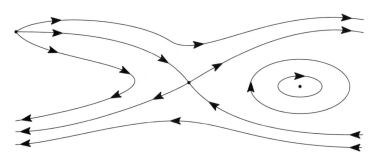

Phase portrait

INDEX OF CRITICAL POINTS

The phase portrait of a system of differential equations is determined by the exceptional points that do not lie on an integral curve. These are the **singular points** $p \in X$ of the vector field V, that is, the points at which $V(p) = O$. Non-singular points are called **regular points**. The behavior of a continuous vector field V is measured near each singular point by means of a numerical index, defined as follows.

Consider any path $g : \mathbb{I} \to X$, not necessarily an integral curve, of regular points. Then $V_g = V \circ g : \mathbb{I} \to \mathbb{R}^2 - \{O\}$ reads the direction and magnitude of the vector field along g; the normalization $v_g = V_g / \| V_g \| : \mathbb{I} \to \mathbb{S}^1$ reads just its direction or *angular variation*. If $g : \mathbb{I} \to X$ is a loop, so is $v_g : \mathbb{I} \to \mathbb{S}^1$.

The path-homotopy class $[v_g] \in \pi_1(\mathbb{S}^1, v_g(0))$ has winding number $w(v_g) \in \mathbb{Z}$, called the **index of V along g** and denoted by $Index(V, g)$.

Think of $Index(V, g)$ as the number of twists in one's neck that result when one walks the loop g, keeping one's chest parallel to the x-axis and head continuously pointed in the vector field's direction at each point of the walk. Counter-clockwise twists are positive; clockwise twists are negative.

By the path-homotopy invariance and the additive property of the winding number, the index has the following basic features.

SECTION 4   POINCARÉ INDEX FORMULA   287

4.1 Index Features (a) *Loops $f, g : \mathbb{I} \to X - V^{-1}(O)$ for which there is a path-homotopy $f \simeq g$ or $\bar{h} \cdot f \cdot h \simeq g$ in $X - V^{-1}(O)$ have equal index:*

$$Index(V, f) = Index(V, g).$$

(b) *A product $k \cdot l : \mathbb{I} \to X - V^{-1}(O)$ of loops has index:*

$$Index(V, k \cdot l) = Index(V, k) + Index(V, l).$$

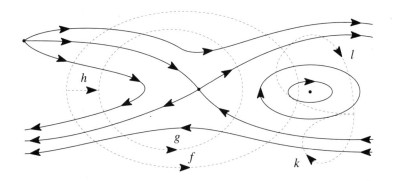

A singular point $p \in X$ of the vector field V is **isolated** if it has a closed disc neighborhood $D(p, r)$ containing no other singular point of V. In that case, let $g : \mathbb{I} \to X$ denote a loop that traverses the boundary circle $S(p, r)$, counter-clockwise. The **index of V at p**, denoted by $Index(V, p)$, is defined to be the index of V along the loop g that circumscribes p. By 4.1(a), this index at p does not depend upon the radius of the disc neighborhood.

Example 1   When $\alpha\delta - \beta\gamma \neq 0$, the the linear system

$$\frac{dx}{dt} = \alpha x + \beta y \qquad \frac{dy}{dt} = \gamma x + \delta y \qquad (\alpha, \beta, \gamma, \delta \in \mathbb{R})$$

has a single singular point, $O = \langle 0, 0 \rangle$. And an affine transformation of coordinates puts the system into one of three elementary forms, depending upon whether the matrix of coefficients has 2, 1, or 0 distinct real eigenvalues. The phase portraits for these three cases follow. In each case, the singular point O has Index +1, except for the saddle point, which has Index −1.

I: $\dfrac{dx}{dt} = \alpha x$ $\dfrac{dy}{dt} = \beta y$ Solution: $x = A e^{\alpha t} \quad y = B e^{\beta t}$

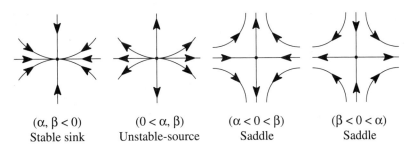

$(\alpha, \beta < 0)$ $(0 < \alpha, \beta)$ $(\alpha < 0 < \beta)$ $(\beta < 0 < \alpha)$
Stable sink   Unstable-source   Saddle   Saddle

II: $\dfrac{dx}{dt} = \alpha x \qquad \dfrac{dy}{dt} = x + \alpha y$ Solution: $x = A e^{\alpha t} \qquad y = (Bt + C) e^{\beta t}$

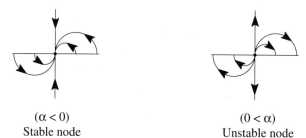

($\alpha < 0$)
Stable node

($0 < \alpha$)
Unstable node

III: $\dfrac{dx}{dt} = \alpha x - \beta y \qquad \dfrac{dy}{dt} = \beta x + \alpha y$ Solution: $x = A e^{\alpha t} \cos \beta t \qquad y = B e^{\alpha t} \sin \beta t$

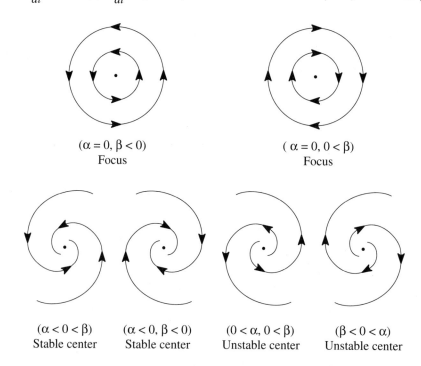

($\alpha = 0, \beta < 0$)
Focus

($\alpha = 0, 0 < \beta$)
Focus

($\alpha < 0 < \beta$)
Stable center

($\alpha < 0, \beta < 0$)
Stable center

($0 < \alpha, 0 < \beta$)
Unstable center

($\beta < 0 < \alpha$)
Unstable center

Example 2 Other singular points arise for non-linear systems:

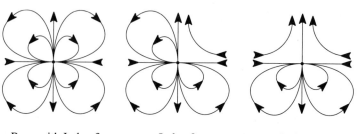

Rose with Index 3   Index 2   Index 1

INDEX FORMULA

One of Poincaré's discoveries relates the index of a continuous vector field V along the boundary path of a disc in X with the indices of the singular points contained in the interior of the disc.

4.2   Poincaré's Index Formula   *Let V be a continuous vector field on an embedded disc $D \subset \mathbb{R}^2$ whose singularities are isolated and lie in the interior of D. Then V has finitely many singularities p_k $(1 \leq k \leq n)$ and the counter-clockwise oriented boundary path g of D has index:*

$$Index(V, g) = Index(V, p_1) + \ldots + Index(V, p_n).$$

Proof: Because D is a compact subspace of $\mathbb{R}^2$, any infinite subset of D has an accumulation point in D. Since the set $V^{-1}(O)$ of singular points is closed and its members are isolated, $V^{-1}(O)$ must be a finite set $\{p_k : 1 \leq k \leq n\}$.

For suitably small radius $r > 0$, the closed discs $D(p_k, r)$ $(1 \leq k \leq n)$ are disjoint and contained in the interior of D. Let the loops g_k $(1 \leq k \leq n)$ be counter-clockwise oriented boundary paths of these discs $D(P_k, r)$. We may choose paths h_k $(1 \leq k \leq n)$ so that there is a path-homotopy in $X - V^{-1}(O)$:

$$g \simeq (\bar{h}_1 \cdot g_1 \cdot h_1) \cdot \ldots \cdot (\bar{h}_n \cdot g_n \cdot h_n).$$

To visualize this path-homotopy in the figure below, imagine the embedded disc D sliced open along the paths h_k $(1 \leq k \leq n)$ and the open discs $B(p_k, r)$ $(1 \leq k \leq n)$ removed. Then deform the path g across the remainder.

Then, by Lemma 4.1, we have these calculations:

$$\begin{aligned} Index(V, g) &= Index(V, \bar{h}_1 \cdot g_1 \cdot h_1) + \ldots + Index(V, \bar{h}_n \cdot g_n \cdot h_n) \\ &= Index(V, g_1) + \ldots + Index(V, g_n) \\ &= Index(V, p_1) + \ldots + Index(V, p_n). \end{aligned}$$ $\square$

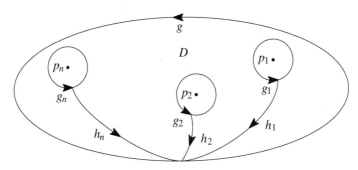

Index Formula 4.2

Poincaré investigated the behavior at infinity of a system defined throughout the plane. He completed the plane to a *projective plane* by the addition of a *line at infinity*. And he extended the system on the plane to a system on the projective plane, usually with new singular points on the line at infinity. He then developed the index formula into an equation between the sum of the indices of the singular points on the projective plane and the *Euler characteristic* of the projective plane. Section **13**.3 defines the Euler characteristic of surfaces and proves Poincaré's Index Formula for continuous vector fields and foliations on all surfaces.

PERIODIC SOLUTIONS

An integral curve $\lambda : [0, b] \to X$ for the continuous vector field V is called *closed* if $\lambda(0) = \lambda(b)$. A closed integral curve represents a periodic solution to the system of differential equations defined by V. The value b is called the *period* of the closed curve λ if $\lambda(0) \neq \lambda(t)$ for all $0 < t < b$.

A closed integral curve $\lambda : [0, b] \to X$ with period $b > 0$ is a simple (i.e., non-self-intersecting) closed path consisting of regular points; hence, $Index(V, g)$ is defined. When an embedded disc D that has λ as its boundary path is contained in X, the Index Formula applies: $Index(V, \lambda)$ equals the sum of the indices of the singular points contained in D. Therefore, the following calculation of Poincaré explains why every hurricane has an eye. (Incidentally, D exists by the Jordan Curve Theorem, which we don't prove.)

4.3 Integral Curve Index *Any counter-clockwise oriented, closed, integral curve of an analytic vector field has index 1.*

Proof: Because the vector field V is analytic, the closed integral curve λ is *rectifiable*, that is, all its segments have finite arc length. Without affecting $Index(V, \lambda)$, we normalize V to consist of unit length vectors on λ and we scale the plane so that λ has unit arc length. Then λ is parametrized by arc length $0 \leq t \leq 1$. Also, λ remains in the upper half plane above its horizontal tangent line at its southern-most point P, which we take as $\lambda(0) = \lambda(1)$.

Using a device of Heinz Hopf, we consider a new *st*-coordinate plane and form a unit vector field U on the right triangle $\triangledown = \{<s, t> : 0 \leq s \leq t \leq 1\}$. For a point $<s, t> \in \triangledown$ such that $\lambda(s) \neq \lambda(t)$, let $U(s, t)$ be the unit vector in the direction of the chord in $\mathbb{R}^2$ from $\lambda(s)$ to $\lambda(t)$:

$$U(s, t) = \frac{\lambda(t) - \lambda(s)}{\| \lambda(t) - \lambda(s) \|}.$$

Upon approach of $<t, t>$ and $<0, 1>$ in $\triangledown$, the values of U have the limits $\lambda'(t) = V(\lambda(t))$ and $-\lambda'(1) = -V(\lambda(1))$. Therefore, U extends continuously over $\triangledown$ by $U(t, t) = V(\lambda(t))$, $0 \leq t \leq 1$, and $U(0, 1) = -V(\lambda(1))$.

Because the vector field U on the triangle $\triangledown$ has no singular points, its index along the boundary of $\triangledown$ is zero by Theorem 4.2. Therefore, the winding number that arises on a counter-clockwise traverse of its hypothenuse, namely, *Index*(V, λ), equals the winding number that arises on a clockwise traverse of its other two edges (below right). The latter equals 1: the vectors $U(0, t)$, $0 \leq t \leq 1$, (below left) record the direction of the moving point $\lambda(t)$ as viewed from $P = \lambda(0)$; when anchored at P, these vectors sweep out the counter-clockwise straight angle above the tangent line at P. The vectors $U(s, 1)$, $0 \leq s \leq 1$, (below left) record the direction of $P = \lambda(1)$ as viewed from the moving point $\lambda(s)$; when anchored at P, these vectors sweep out the counter-clockwise straight angle below the tangent line at P. □

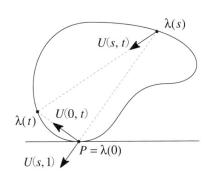
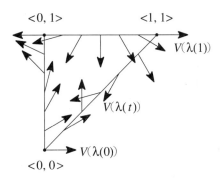

Theorem 4.3: Closed integral curve λ   Theorem 4.3: Vector field U on $\triangledown$

4.4 Corollary (*a*) *A closed integral curve surrounds at least one singular point with positive index.*
(*b*) *A closed integral curve that surrounds only elementary singular points surrounds one fewer saddle point than nodes, foci, and centers.*

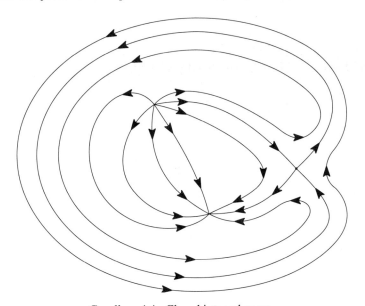

Corollary 4.4: Closed integral curve

In 1901, Ivar Bendixson studied some finer points concerning local phase-portraits around a singular point. Under reasonable hypotheses, the region around a singular point decomposes into parabolic, elliptic, and hyperbolic sectors that have the following characteristic features:

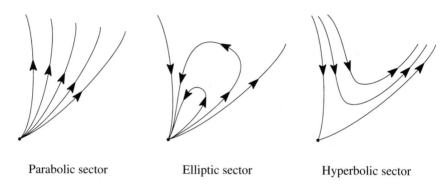

Parabolic sector   Elliptic sector   Hyperbolic sector

One of Bendixson's results is the following (Exercise 7):

4.5 Local Index Formula   *For any singular point p of a vector field V,*

$$\text{Index}(V, p) = \tfrac{1}{2}(e - h + 2),$$

where e and h are respectively the numbers of elliptic and hyperbolic sectors surrounding p.

Exercises

SECTION 1

1* Prove the functorial properties in Theorem 1.3:
   (a) For any two continuous functions $F : X \to Y$ and $G : Y \to Z$, the composite function $G \circ F : X \to Z$ induces the composite groupoid homomorphism:
   $$(G \circ F)_\# = G_\# \circ F_\# : \pi(X) \to \pi(Y) \to \pi(Z).$$
   (b) Any identity function $1_X : X \to X$ induces the identity homomorphism:
   $$(1_X)_\# = 1_{\pi(X)} : \pi(X) \to \pi(X).$$
   (c) A homeomorphism $X \cong Y$ induces a groupoid isomorphism $\pi(X) \approx \pi(Y)$.

2* (a) Prove that three paths $f, g, h \in P(X)$ associate, i.e., $f \cdot (g \cdot h) = (f \cdot g) \cdot h$, if and only if there are infinite factorizations:
   $$f = f \cdot g = (f \cdot g) \cdot g = ((f \cdot g) \cdot g) \cdot g = \dots$$
   and
   $$h = g \cdot h = g \cdot (g \cdot h) = g \cdot (g \cdot (g \cdot h)) = \dots .$$
   (b) Prove that a path $e \in P(X)$ serves as a left or right unit for a path $g \in P(X)$ if

and only if g has an infinite factorization

$$g = e \cdot g = e \cdot (e \cdot g) = e \cdot (e \cdot (e \cdot g)) = \ldots$$

or

$$g = g \cdot e = (g \cdot e) \cdot e = ((g \cdot e) \cdot e) \cdot e = \ldots .$$

3   Let X be a simply connected space. Introduce a sometimes defined multiplication on the product set $X \times X$ and define a groupoid isomorphism $\pi(X) \approx X \times X$.

4   Let the projection functions $p_X : X \times Y \to X$ and $p_Y : X \times Y \to Y$ for $X \times Y$ induce groupoid homomorphisms $p_{X\#} : \pi(X \times Y) \to \pi(X)$ and $p_{Y\#} : \pi(X \times Y) \to \pi(Y)$.
   (a)   Verify that a bijection $\eta : \pi(X \times Y) \to \pi(X) \times \pi(Y)$ is defined by

$$\eta([h]) = <p_{X\#}([h]), p_{Y\#}([h])>.$$

   (b)   Introduce a sometimes defined multiplication on the product set $\pi(X) \times \pi(Y)$ such that the set bijection η in (a) is a groupoid isomorphism.

5   Prove that the particular or excluded point topology on a space is simply connected.

SECTION 2

1*   Prove the triangle equality $D(a \to b) + D(b \to c) = D(a \to c)$ relating the directed distances between any three points $a, b, c \in \mathbb{S}^1$ that lie in an open hemisphere of $\mathbb{S}^1$.

2   For each integer $n \in \mathbb{Z}$, there is the path $d_n : \mathbb{I} \to \mathbb{S}^1$, $d_n(t) = e^{2\pi i n t}$. Calculate $L(d_n)$ directly via a d_n-admissible partition and indirectly using a path lifting $\hat{d}_n : \mathbb{I} \to \mathbb{R}^1$.

3*   (a)   Prove that the essential length of a product of two paths in $\mathbb{S}^1$ is the sum of their essential lengths: $L(f \cdot g) = L(f) + L(g)$.
   (b)   Establish a groupoid isomorphism $\pi(\mathbb{S}^1) \approx \mathbb{S}^1 \times \mathbb{R}^1$.

4   The straight-line path in $\mathbb{R}^{n+1}$ joining non-antipodal points of $\mathbb{S}^n$ misses the origin O and so can be radially projected to a ***curvilinear path*** in $\mathbb{S}^n$. Prove that any path in $\mathbb{S}^n$ ($n \geq 1$) is path-homotopic to a path product of finitely many curvilinear paths. Is such a path product non-surjective, hence, inessential?

5*   Prove:
   (a)   A retract A of a connected space X is a connected subspace.
   (b)   A retract A of a compact space X is a compact subspace.
   (c)   A retract A of a Hausdorff space X is closed in X.

6   Prove that any retract of a simply connected space is simply connected.

7   Use Exercise 6 to deduce that the following spaces are not simply connected:
   (a) $\mathbb{R}^2 - \{O\}$.   (b) $\mathbb{S}^1 \times \mathbb{S}^1$.

8   Prove that on a continuous globe there is some pair of antipodal points with the same temperature and elevation.

9*   Use the Borsuk-Ulam Theorem for flattenings of the 2-sphere to prove the Ham Sandwich Theorem (2.11):
   (a)   Rescale to make the subsets A, B, and C lie in the open unit ball $\mathbb{B}^3$.
   (b)   For each of the subsets $D = A, B, C$ and for each $x \in \mathbb{S}^2$, let $t_D(x)$ be the scalar $-1 < t \leq +1$ such that the plane Π_D passing through tx and perpendicular to the ray from O through x bisects the set D.

(c) Prove that the function $g : \mathbb{S}^2 \to \mathbb{R}^2$, $g(x) = <t_A(x) - t_B(x), t_B(x) - t_C(x)>$, is continuous and satisfies $g(-x) = -g(x)$ for all $x \in \mathbb{S}^2$.
(d) Apply the Borsuk-Ulam Theorem (2.10) to $g : \mathbb{S}^2 \to \mathbb{R}^2$.

SECTION 3

1. (a) Determine all possible group homomorphisms $h : \mathbb{Z} \to \mathbb{Z}$ from the additive group of integers to itself.
   (b) Prove that, for each group homomorphism $h : \mathbb{Z} \to \mathbb{Z}$, there is a continuous function $F : (\mathbb{S}^1, 1) \to (\mathbb{S}^1, 1)$ such that $h = F_\# : \pi_1(\mathbb{S}^1, 1) \to \pi_1(\mathbb{S}^1, 1)$.

2. Prove that the degree of maps $F : (\mathbb{S}^1, 1) \to (\mathbb{S}^1, 1)$ is multiplicative:
$$Deg(G \circ F) = Deg(G) \cdot Deg(F).$$

3. Prove that the inverse of the group isomorphism
$$\eta : \pi_1(X \times Y, <x, y>) \to \pi_1(X, x) \times \pi_1(Y, y), \eta([h]) = <p_{X\#}[h], p_{Y\#}[h]>,$$
can be described in these two ways:
   (a) $<[f], [g]> \to [h]$, where $h(t) = <f(t), g(t)>$ for all $t \in \mathbb{I}$.
   (b) $<[f], [g]> \to i_{X\#}([f]) \cdot i_{Y\#}([g])$, where the two inclusion maps $i_X : X \to X \times Y$ and $i_Y : Y \to X \times Y$ are defined by $i_X(x') = <x', y>$ and $i_Y(y') = <x, y'>$.

4. Use the identification map $\mathbb{I} \times \mathbb{I} \to T^2$ that produces the torus from the unit square to define a path-homotopy $l \cdot m \simeq m \cdot l$. This gives a geometric argument that the generators $[l], [m] \in \pi_1(T^2)$ satisfy the commutativity relation $[l] \cdot [m] = [m] \cdot [l]$.

5. Calculate the fundamental group of the n-dimensional torus $\mathbb{S}^1 \times \ldots \times \mathbb{S}^1$ (n copies).

6. Prove that the inclusion $j : \mathbb{S}^1 \to \mathbb{R}^2 - \{O\}$ and any retraction $r : \mathbb{R}^2 - \{O\} \to \mathbb{S}^1$ induce inverse fundamental group homomorphisms.

7. Compare and contrast the fundamental groups of the following spaces:
   (a) the 2-dimensional torus T^2,
   (b) the doubly-punctured plane $\mathbb{R}^2 - \{<1, 0>, <-1, 0>\}$, and
   (c) the pinched torus (Exercise 5.1.24).

8. Let $a, b, c, d \in \mathbb{Z}$. Calculate the group homomorphism $\mathbb{Z} \times \mathbb{Z} \to \mathbb{Z} \times \mathbb{Z}$ induced on fundamental groups by the continuous function $F : \mathbb{S}^1 \times \mathbb{S}^1 \to \mathbb{S}^1 \times \mathbb{S}^1$, given by
$$F(<e^{i\theta}, e^{i\varphi}>) = <e^{i(a\theta + b\varphi)}, e^{i(c\theta + d\varphi)}>.$$

9. Prove:
   (a) Each factor of a topological product is a retract of that product.
   (b) Any non-surjective continuous function $F : X \to \mathbb{S}^1$ induces a trivial homomorphism of the fundamental groups.
   (c) The 1-sphere $\mathbb{S}^1$ is not homeomorphic to a product of non-degenerate spaces.

10. In complex analysis, the winding number of a closed path λ in the punctured complex plane $\mathbb{C} - O$ is defined by the integral
$$w(\lambda) = \frac{1}{2\pi i} \int_\lambda \frac{1}{z} dz.$$

Relate this to the winding number definition given in Section 3.

SECTION 4

1. Give detailed arguments for the Index Features (4.1).

2. Sketch each of the following vector fields $V(x, y)$ on the plane $\mathbb{R}^2$, and calculate the index of V along the loop $e(t) = <\cos 2\pi t, \sin 2\pi t> \ (0 \leq t \leq 1)$:
   (a) $V(x, y) = (x, y)$,
   (b) $V(x, y) = (-x, -y)$,
   (c) $V(x, y) = (-x, y)$,
   (d) $V(x, y) = (-x + y, x - y)$, and
   (e) $V(x, y) = (x + y, x - y)$.

3. Sketch the vector field $V(x, y) = (x - xy, -y + xy)$ on the plane $\mathbb{R}^2$, and calculate the index of each of its two critical points, $<0, 0>$ and $<1, 1>$.

4. Determine the index for each of the singular points indicated below:

   (a)

   (b)

   (c)

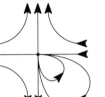
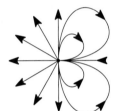

5. Consider a continuous mountain range on a disc whose topographic features consist of a finite number of *pits*, *peaks*, and saddle-like *passes*.
   (a) Prove that a closed contour curve (i.e., a curve of constant elevation) on this mountainous disc must enclose at least one pit or peak.
   (b) Determine the *topographic count*:

   $$\#(pits) - \#(passes) + \#(peaks).$$

6. Consider a disc covered by an archipelago consisting of mountainous, lake-strewn islands. Verify J. Clerk Maxwell's observation that the *topological count*:

   $$\#(islands) - \#(lakes)$$

   equals the *topographic count*:

   $$\#(pits) - \#(passes) + \#(peaks).$$

7.* Prove Bendixson's Index Formula: For a singular point p of a vector field,

   $$\text{Index}(V, p) = \tfrac{1}{2}(e - h + 2)$$

   where e and h are respectively the numbers of elliptic and hyperbolic sectors surrounding p.

8. Consider any non-vanishing vector field V on the boundary $\mathbb{S}^1$ of the closed unit disc $\mathbb{D}^2$ such that the index of V along the boundary curve $e : I \to \mathbb{S}^1$ is zero. Construct an extension of V to a continuous non-vanishing vector field on $\mathbb{D}^2$.

10

HOMOTOPY

In Chapter **9**, path-homotopies are introduced in the definition of the fundamental group. To facilitate calculations of the fundamental group, this chapter presents general homotopical concepts. We leave the familiar topological discipline of spaces and continuous functions for a new homotopical world of spaces and homotopy classes of continuous functions. Technically speaking, these two mathematical settings are called *categories* and a proper transition from one category to another is a *functor*. For convenience, we begin this chapter with a discussion of categories and functors.

1 Categories and Functors

Most mathematical disciplines involve *objects* that are sets equipped with extra structure and *morphisms* that are set functions that respect the extra structure. This framework is the archetype for a *category*.

CATEGORIES

Consider a class $\mathcal{C}$ of ***objects***, denoted by A, B, C, etc. Suppose that, for each ordered pair A and B of objects in $\mathcal{C}$, there is a set $(A, B)_\mathcal{C}$, and suppose that, for each ordered triple A, B, and C of objects in $\mathcal{C}$, there is a function

$$\circ : (A, B)_\mathcal{C} \times (B, C)_\mathcal{C} \to (A, C)_\mathcal{C}, \text{ denoted by } \circ(f, g) = g \circ f.$$

A member $f \in (A, B)_\mathcal{C}$ is called a ***morphism*** from A to B and is denoted by $f: A \to B$ even though it may not be a function, and $\circ$ is called ***composition***. Such a system $\mathcal{C}$ is called a ***category*** provided that these two axioms hold:

- For all morphisms $f \in (A, B)_\mathcal{C}$, $g \in (B, C)_\mathcal{C}$, and $h \in (C, D)_\mathcal{C}$,

$$h \circ (g \circ f) = (h \circ g) \circ f. \qquad (\textbf{\textit{associativity}})$$

- *For each object $B \in \mathcal{C}$, there is a morphism $1_B \in (B, B)_\mathcal{C}$ such that*
$$g \circ 1_B = g \quad \text{and} \quad 1_B \circ f = f \qquad \text{(\textbf{identity})}$$
for all morphisms $g \in (B, C)_\mathcal{C}$ and $f \in (A, B)_\mathcal{C}$.

Example 1 The class of sets and their set functions, with the usual composition, form a category $\mathcal{S}$. The class of topological spaces and their continuous functions, with the usual composition, form a category $\mathcal{T}$. The categories $\mathcal{G}$ and $\mathcal{A}b$ of groups and abelian groups and their group homomorphisms are described in Chapter **11**.

Example 2 Any category $\mathcal{C}$ gives rise to a (synthetic) *dual category* $\mathcal{C}^*$. The objects $A^* \in \mathcal{C}^*$ and the morphisms $f^* \in (B^*, A^*)_{\mathcal{C}^*}$ of $\mathcal{C}^*$ are in bijective correspondence with the objects $A \in \mathcal{C}$ and morphisms $f \in (A, B)_\mathcal{C}$ of $\mathcal{C}$. The composition function in $\mathcal{C}^*$, $\circ : (C^*, B^*)_{\mathcal{C}^*} \times (B^*, A^*)_{\mathcal{C}^*} \to (C^*, A^*)_{\mathcal{C}^*}$, is defined by $f^* \circ g^* = (g \circ f)^*$

Morphisms $f : A \to B$ and $g : B \to A$ are called *inverses* in $\mathcal{C}$ if $g \circ f = 1_A$ and $f \circ g = 1_B$. A morphism $f : A \to B$ is called an *equivalence* in $\mathcal{C}$ (or an *invertible morphism*) if it has an inverse $g : B \to A$ in $\mathcal{C}$.

The axioms for a category $\mathcal{C}$ imply the following (Exercise 1):

- *The **identity morphism** $1_B \in (B, B)_\mathcal{C}$ is unique for each object $B \in \mathcal{C}$.*
- *When an inverse morphism $g : B \to A$ exists for $f : A \to B$, it is unique and therefore may be denoted by $f^{-1} : B \to A$.*
- *The inversion rule $(h \circ f)^{-1} = f^{-1} \circ h^{-1}$ holds for invertible morphisms.*

Objects $A, B \in \mathcal{C}$ are *equivalent* in $\mathcal{C}$, written $A \equiv B$, if there are inverse morphisms $f : A \leftrightarrow B : g$ in $\mathcal{C}$. The composite of equivalences in $\mathcal{C}$ is an equivalence in $\mathcal{C}$, so $\equiv$ is an equivalence relation on any set of objects of $\mathcal{C}$.

Example 3 Equivalence in the three categories $\mathcal{S}$, $\mathcal{T}$, and $\mathcal{G}$ is called set equivalence, homeomorphism, and isomorphism, respectively.

INITIAL AND TERMINAL OBJECTS

An object $I \in \mathcal{C}$ is *initial* in $\mathcal{C}$ if, for each object $B \in \mathcal{C}$, there exists a single morphism $I \to B$ in $\mathcal{C}$, i.e., $(I, B)_\mathcal{C}$ is a singleton.

An object $T \in \mathcal{C}$ is *terminal* in $\mathcal{C}$ if, for each object $B \in \mathcal{C}$, there exists a single morphism $B \to T$ in $\mathcal{C}$, i.e., $(B, T)_\mathcal{C}$ is a singleton.

Example 4 In the category $\mathcal{S}$ of sets, the empty set is an initial object but not a terminal object, and the singleton sets are terminal objects but not initial objects.

In the category of (abelian) groups $\mathcal{G}$ ($\mathcal{A}b$), the trivial group $\{1\}$ (or $\{0\}$) is both an initial object and a terminal object.

Example 5 Initial and terminal are dual concepts: I is initial in $\mathcal{C}$ if and only if I^* is terminal in $\mathcal{C}^*$; T is terminal in $\mathcal{C}$ if and only if T^* is initial in $\mathcal{C}^*$.

The axioms for a category $\mathcal{C}$ imply the following:

1.1 Theorem *When an initial object I or a terminal object T in a category $\mathcal{C}$ exists, it is unique up to equivalence in $\mathcal{C}$.*

Proof: For example, when I and J are initial objects in $\mathcal{C}$, then there are unique morphisms $f : I \to J$ and $g : J \to I$. The composites $g \circ f : I \to I$ and $f \circ g : J \to J$ must be the unique elements $1_I \in (I, I)_\mathcal{C}$ and $1_J \in (J, J)_\mathcal{C}$. □

SUBCATEGORIES

A category $\mathcal{B}$ is called a ***subcategory*** of a category $\mathcal{C}$ if each object in $\mathcal{B}$ is an object in $\mathcal{C}$, each morphism $A \to B$ in $\mathcal{B}$ is a morphism in $\mathcal{C}$, and the composition of morphisms in $\mathcal{B}$ agrees with their composition in $\mathcal{C}$. We denote the subcategory relationship by $\mathcal{B} \subseteq \mathcal{C}$.

The subcategory $\mathcal{B}$ is a ***proper*** subcategory of $\mathcal{C}$, written $\mathcal{B} \subset \mathcal{C}$, if there is some object of $\mathcal{C}$ that does not belong to $\mathcal{B}$. The subcategory $\mathcal{B}$ is a ***full*** subcategory of $\mathcal{C}$ if $(A, B)_\mathcal{B} = (A, B)_\mathcal{C}$ for all objects $A, B \in \mathcal{B}$.

Example 6 The category $\mathcal{A}b$ of abelian groups is a full and proper subcategory of the category $\mathcal{G}$ of groups. The Hausdorff spaces, the compact Hausdorff spaces, and the k-spaces form full and proper subcategories $\mathcal{T}_h$, $\mathcal{T}_c$, and $\mathcal{T}_k$ of the category $\mathcal{T}$ of topological spaces. Moreover, $\mathcal{T}_c \subset \mathcal{T}_k \subset \mathcal{T}_h \subset \mathcal{T}$.

PRODUCTS IN A CATEGORY

Let $\{A_\alpha \in \mathcal{C} : \alpha \in \mathcal{A}\}$ be an indexed family of objects in a category $\mathcal{C}$. Their ***product*** $(P, \{p_\alpha\})$ in $\mathcal{C}$ consists of an object $P \in \mathcal{C}$ and an indexed family of morphisms $\{p_\alpha : P \to A_\alpha \ (\alpha \in \mathcal{A})\}$, called ***projections***, satisfying the universal property that has these equivalent formulations:

- $\forall X \in \mathcal{C}, \forall \{g_\alpha : X \to A_\alpha \ (\alpha \in \mathcal{A})\}, \exists !$ *(unique)* $g : X \to P$ such that $g_\alpha = p_\alpha \circ g : X \to P \to A_\alpha, \forall \alpha \in \mathcal{A}$.

- $\forall X \in \mathcal{C}$, *the function* $(X, P)_\mathcal{C} \to \prod_\alpha (X, A_\alpha)_\mathcal{C}, g \to (p_\alpha \circ g)$, *is bijective.*

- $(P, \{p_\alpha\})$ is a terminal object in the category $\mathcal{C} // \{A_\alpha\}$ ***above*** $\{A_\alpha\}$. An object in $\mathcal{C} // \{A_\alpha\}$ is a pair $(X, \{g_\alpha\})$, consisting of an object $X \in \mathcal{C}$ and an indexed family of morphisms $\{g_\alpha : X \to A_\alpha \ (\alpha \in \mathcal{A})\}$ with domain X, and a morphism $f : (X, \{g_\alpha\}) \to (Y, \{h_\alpha\})$ in $\mathcal{C} // \{A_\alpha\}$ is a morphism $f : X \to Y$ in $\mathcal{C}$ such that $h_\alpha \circ f = g_\alpha$ for all $\alpha \in \mathcal{A}$.

When a product $(P, \{p_\alpha\})$ exists in $\mathcal{C}$ for the family $\{A_\alpha \in \mathcal{C} : \alpha \in \mathcal{A}\}$, it is unique up to equivalence in the category $\mathcal{C} // \{A_\alpha\}$ above $\{A_\alpha\}$, according to Theorem 1.1. This means that P is unique up to an equivalence in $\mathcal{C}$ that respects the family of projections in the product. Whenever the product $(P, \{p_\alpha\})$ exists, we shall denote the product object P by $\prod_\alpha A_\alpha$.

Example 7 The set product $(A = \prod_\alpha A_\alpha, \{p_\alpha : A \to A_\alpha\})$ of an indexed family of sets $\{A_\alpha \in \mathcal{S} : \alpha \in \mathcal{A}\}$ is defined in Section **5**.3. It qualifies as their product in the category $\mathcal{S}$ of sets, according to the product properties (**P1**) and (**P2**) in Section **5**.3.

Example 8 The topological product $(X = \prod_\alpha X_\alpha, \{p_\alpha : X \to X_\alpha\})$ of an indexed family of spaces $\{X_\alpha \in \mathcal{T} : \alpha \in \mathcal{A}\}$ is defined in Section **5**.4. It qualifies as their product in the topological category $\mathcal{T}$, according to the Characteristic Feature (**5**.3.2).

Example 9 *The product in a category may or may not qualify as the product in a subcategory.* By Tychonoff's Theorem (**7**.2.11), the product of compact spaces in $\mathcal{T}$ qualifies as the product in the subcategory $\mathcal{T}_c$ of compact spaces. By Example **8**.1.2, the product of normal spaces in $\mathcal{T}$ may not qualify as their product in the subcategory $\mathcal{T}_n$ of normal spaces.

Example 10 In the categories $\mathcal{G}$ and $\mathcal{A}b$ of groups and abelian groups, the ***direct product*** $(\mathbb{G} = \prod_\alpha \mathbb{G}_\alpha, \{p_\alpha : \mathbb{G} \to \mathbb{G}_\alpha\})$ serves as product (Exercise 6). The direct product is the set product with the group structure of coordinatewise multiplication.

SUMS IN A CATEGORY

Let $\{A_\alpha \in \mathcal{C} : \alpha \in \mathcal{A}\}$ be an indexed family of objects in a category $\mathcal{C}$. Their ***sum*** $(S, \{i_\alpha\})$ in $\mathcal{C}$ consists of an object $S \in \mathcal{C}$ and an indexed family of morphisms $\{i_\alpha : A_\alpha \to S \, (\alpha \in \mathcal{A})\}$, called ***injections***, satisfying the universal property that has these equivalent formulations:

- $\forall X \in \mathcal{C}, \forall \{g_\alpha : A_\alpha \to X \, (\alpha \in \mathcal{A})\}, \exists \, ! \, (unique) \, g : S \to X$ such that $g_\alpha = g \circ i_\alpha : A_\alpha \to S \to X, \forall \, \alpha \in \mathcal{A}$.

- $\forall X \in \mathcal{C}$, the function $(S, X)_\mathcal{C} \to \prod_\alpha (A_\alpha, X)_\mathcal{C}, g \to (g \circ i_\alpha)$, is bijective.

- $(S, \{i_\alpha\})$ is an initial object in the category $\mathcal{C} \setminus\!\setminus \{A_\alpha\}$ ***below*** $\{A_\alpha\}$. An object in $\mathcal{C} \setminus\!\setminus \{A_\alpha\}$ is a pair $(X, \{g_\alpha\})$, consisting of an object $X \in \mathcal{C}$ and an indexed family of morphisms $\{g_\alpha : A_\alpha \to X \, (\alpha \in \mathcal{A})\}$ with codomain X, and a morphism $f : (X, \{g_\alpha\}) \to (Y, \{h_\alpha\})$ in $\mathcal{C} \setminus\!\setminus \{A_\alpha\}$ is a morphism $f : X \to Y$ in $\mathcal{C}$ such that $f \circ g_\alpha = h_\alpha$ for all $\alpha \in \mathcal{A}$.

Thus, if there exists a sum $(S, \{i_\alpha\})$ in $\mathcal{C}$ for the family $\{A_\alpha \in \mathcal{C} : \alpha \in \mathcal{A}\}$, it is unique up to equivalence in the category $\mathcal{C} \setminus\!\setminus \{A_\alpha\}$ below $\{A_\alpha\}$, by Theorem 1.1. This means that S is unique up to an equivalence in $\mathcal{C}$ that respects the family of injections in the sum. When the sum $(S, \{i_\alpha\})$ exists in $\mathcal{C}$, we shall denote the sum object S by $\Sigma_\alpha A_\alpha$.

Example 11 In the category $\mathcal{S}$ of sets, any family $\{A_\alpha \in \mathcal{S} : \alpha \in \mathcal{A}\}$ of sets has the sum $(S = \mathring{\cup}_\alpha A_\alpha, \{i_\alpha : A_\alpha \to S\})$, according to the properties (**DU1**) and (**DU2**) of the disjoint union construction (Section **5**.3).

Example 12 In the category $\mathcal{T}$ of topological spaces, any family $\{X_\alpha \in \mathcal{T} : \alpha \in \mathcal{A}\}$ of spaces has the sum $(X = \mathring{\cup}_\alpha X_\alpha, \{i_\alpha : X_\alpha \to X\})$, according to the Characteristic Feature (**5**.3.1) of the topological disjoint union.

Example 13 *Sums in a subcategory may differ from sums in the larger category.* In the category $\mathcal{G}$ of groups, any family $\{\mathbb{G}_\alpha \in \mathcal{G} : \alpha \in \mathcal{A}\}$ has the sum $(\mathbb{G} = *_\alpha \mathbb{G}_\alpha, \{i_\alpha : \mathbb{G}_\alpha \to \mathbb{G}\})$, where $*_\alpha \mathbb{G}_\alpha$ is the ***free product*** of the groups (see Section **11.3**).

In the category $\mathcal{A}b$ of (additive) abelian groups, any family $\{A_\alpha \in \mathcal{G} : \alpha \in \mathcal{A}\}$ has the sum $(\mathbb{A} = \oplus_\alpha A_\alpha, \{i_\alpha : A_\alpha \to \mathbb{A}\})$, called the ***direct sum*** (Exercises **11.**3.5-6).

FUNCTORS BETWEEN CATEGORIES

Let $\mathcal{C}$ and $\mathcal{D}$ be any two categories. A ***covariant functor*** $\Phi : \mathcal{C} \to \mathcal{D}$ assigns to each object $A \in \mathcal{C}$ an object $\Phi(A) \in \mathcal{D}$ and to each morphism $f : A \to B$ in $\mathcal{C}$ a morphism $\Phi(f) : \Phi(A) \to \Phi(B)$ in $\mathcal{D}$ such that:

$$\Phi(1_A) = 1_{\Phi(A)} : \Phi(A) \to \Phi(A)$$

and

$$\Phi(g \circ f) = \Phi(g) \circ \Phi(f) : \Phi(A) \to \Phi(B) \to \Phi(C).$$

Contravariant functors are comparable but reverse morphism directions.

For each category $\mathcal{C}$, there is an identity functor $1_\mathcal{C} : \mathcal{C} \to \mathcal{C}$. Covariant functors $\Psi : \mathcal{B} \to \mathcal{C}$ and $\Phi : \mathcal{C} \to \mathcal{D}$ yield a covariant functor $\Phi \circ \Psi : \mathcal{B} \to \mathcal{D}$.

Example 14 If $\mathcal{B} \subseteq \mathcal{C}$, there is a covariant *inclusion functor* $I : \mathcal{B} \to \mathcal{C}$.

Example 15 Each Hausdorff space Y can be retopologized as a k-space $k(Y)$ using its compactly generated topology. According to the K-Factory (**7.**4.11), this defines a covariant functor $K : \mathcal{T}_h \to \mathcal{T}_k$ from the category of Hausdorff spaces to the category of k-spaces; K is a left-inverse of the inclusion functor $I : \mathcal{T}_k \to \mathcal{T}_h$. Because we have $(I(X), Y)_{\mathcal{T}_h} = (X, k(Y))_{\mathcal{T}_k}$ for any k-space X and Hausdorff space Y, the functor K carries a product $(\prod_\alpha Y_\alpha, \{p_\alpha\})$ in $\mathcal{T}_h$ into a product $(k(\prod_\alpha Y_\alpha), \{p_\alpha\})$ in $\mathcal{T}_k$ (Exercise 7).

1.2 Theorem *Let $\Phi : \mathcal{C} \to \mathcal{D}$ be a covariant functor. If $f : A \to B$ is an equivalence in $\mathcal{C}$, then $\Phi(f) : \Phi(A) \to \Phi(B)$ is an equivalence in $\mathcal{D}$.*

Proof: If the morphisms $f : A \leftrightarrow B : g$ are inverses in $\mathcal{C}$, then the morphisms $\Phi(f) : \Phi(A) \leftrightarrow \Phi(B) : \Phi(g)$ are inverses in $\mathcal{D}$. The functorial properties give

$$\Phi(g) \circ \Phi(f) = \Phi(g \circ f) = \Phi(1_A) = 1_{\Phi(A)}$$

and

$$\Phi(f) \circ \Phi(g) = \Phi(f \circ g) = \Phi(1_B) = 1_{\Phi(B)}. \qquad \square$$

By Theorem 1.2, objects in $\mathcal{C}$ that are carried to inequivalent objects in some category $\mathcal{D}$ by some functor Φ are necessarily inequivalent in $\mathcal{C}$. This is the main thrust of *algebraic topology*, whose beginning we have charted in Chapter **9**:

Example 16 By Section **9.**1, the fundamental set π is a functor from the category $\mathcal{T}$ of topological spaces to the category of groupoids. Thus, spaces with non-isomorphic fundamental groupoids are not homeomorphic.

2 Basic Homotopy Concepts

We now introduce the basic homotopy relation between continuous functions, and we consider specialized construction techniques for homotopies.

THE HOMOTOPY RELATION

Let X and Y be spaces. For $0 \leq t \leq 1$, define $i_t : X \rightarrow X \times \mathbb{I}$ by $i_t(x) = <x, t>$.

Continuous functions $f, g : X \rightarrow Y$ are **homotopic** if there exists a continuous function $H : X \times \mathbb{I} \rightarrow Y$ such that $H \circ i_0 = f$ and $H \circ i_1 = g$. Then H is called a **homotopy from f to g**; it is defined **on** X and takes place **in** Y.

The homotopy relation is denoted by $f \simeq g$ or by $H : f \simeq g$ to specify H.

Example 1 For a singleton space $X = \{*\}$, a continuous function $f : X \rightarrow Y$ is essentially just a point of Y and a homotopy $H : X \times \mathbb{I} \rightarrow Y$ is simply a path between two points of Y. So a space Y is path-connected if and only if any two continuous functions $f, g : X \rightarrow Y$ of the singleton space $X = \{*\}$ are homotopic in Y.

Example 2 Each $z \in \mathbb{S}^1$ determines a **meridional** embedding $m : \mathbb{S}^1 \rightarrow \mathbb{S}^1 \times \mathbb{S}^1$, $m(x) = <z, x>$, and a **longitudinal** embedding $l : \mathbb{S}^1 \rightarrow \mathbb{S}^1 \times \mathbb{S}^1$, $l(x) = <x, z>$. Any two meridional or two longitudinal embeddings are homotopic, via a homotopy carrying $\mathbb{S}^1 \times \mathbb{I}$ onto an annular region that they bound in $\mathbb{S}^1 \times \mathbb{S}^1$, as below:

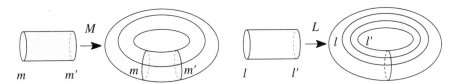

Example 2: $M : m \simeq m'$ Example 2: $L : l \simeq l'$

A continuous function is **null-homotopic**, or **inessential**, if it is homotopic to a constant function; otherwise, it is **essential**. A homotopy $H : f \simeq c$ from f to a constant function c is called a **null-homotopy** of f.

Example 3 The identity function $1_{\mathbb{S}^1} : \mathbb{S}^1 \rightarrow \mathbb{S}^1$ is essential, since homotopic maps $f \simeq g : \mathbb{S}^1 \rightarrow \mathbb{S}^1$ have the same degree (Exercise 3) and $Deg(1_{\mathbb{S}^1}) = 1 \neq 0 = Deg(c)$.

The inclusion $i : \mathbb{S}^1 \rightarrow \mathbb{D}^2$ is inessential; there is a null-homotopy $H : \mathbb{S}^1 \times \mathbb{I} \rightarrow \mathbb{D}^2$ of i to the constant function c_O given by $H(x, t) = (1-t) \cdot x$, for all $x \in \mathbb{S}^1$ and $t \in \mathbb{I}$.

HOMOTOPIES IN REAL SUBSPACES

In real space $\mathbb{R}^n$, any two points x and z are joined by a line segment:

$$[x, z] = \{t \cdot x + (1-t) \cdot z \in \mathbb{R}^n : 0 \leq t \leq 1\}.$$

So it is easy to construct a homotopy *in* $\mathbb{R}^n$ between continuous functions defined *on* any space X. More generally, we have the following result:

2.1 Theorem *Two continuous functions $f, g : X \to Y \subseteq \mathbb{R}^n$ are homotopic in Y when Y contains the line segment $[f(x), g(x)]$ for each $x \in X$.*

Proof: Since the vector space operations of addition and scalar multiplication in $\mathbb{R}^n$ are continuous, the homotopy $H : X \times \mathbb{I} \to \mathbb{R}^n$, given by

$$H(x, t) = t \cdot g(x) + (1-t) \cdot f(x), \qquad (\forall\, x \in X, t \in \mathbb{I})$$

is continuous. In addition, H takes values within Y, by the hypothesis on this subspace; hence, H is a homotopy $H : X \times \mathbb{I} \to Y$ in Y from f to g. □

The homotopy H in Theorem 2.1 is called a ***straight-line homotopy*** in Y.

Example 4 The function $i \circ \rho : \mathbb{R}^n - O \to \mathbb{S}^{n-1} \subset \mathbb{R}^n - O$, where ρ is the radial retraction defined by $\rho(x) = x/\|x\|$ and i is the inclusion function, is homotopic to the identity function $1_{\mathbb{R}^n - O} : \mathbb{R}^n - O \to \mathbb{R}^n - O$ via the straight line homotopy in $\mathbb{R}^n - O$.

A subset $C \subseteq \mathbb{R}^n$ is ***convex*** if it contains each line segment joining any two of its points. Clearly, Theorem 2.1 applies to every pair of continuous functions $f, g : X \to C$ into a convex subset $C \subseteq \mathbb{R}^n$. This gives:

2.2 Corollary *Any two continuous functions into a convex subspace C of $\mathbb{R}^n$ are homotopic in C.*

Example 5 The n-disc $\mathbb{D}^n$ and the n-cube $\mathbb{I}^n$ are convex subspaces of $\mathbb{R}^n$. Thus, any two continuous functions into either of these spaces are homotopic. In particular, the inclusion $i : \mathbb{S}^{n-1} \subset \mathbb{D}^n$ is inessential.

When $Y \subseteq \mathbb{R}^n$ does not satisfy the hypothesis on line segments in Theorem 2.1, the straight-line homotopy $H : X \times \mathbb{I} \to \mathbb{R}^n$ associated with the continuous functions $f, g : X \to Y$ is not confined to Y and so doesn't qualify as a homotopy from f to g in Y. Sometimes a modification of the straight-line homotopy does qualify. Here is one such situation.

2.3 Corollary *Two continuous functions $f, g : X \to \mathbb{S}^{n-1}$, such that $f(x)$ and $g(x)$ are never antipodal for $x \in X$, are homotopic in $\mathbb{S}^{n-1}$. In particular, any non-surjective continuous function $f : X \to \mathbb{S}^{n-1}$ is inessential.*

Proof: Because $f(x)$ and $g(x)$ are never antipodal points, the straight-line homotopy $H : X \times \mathbb{I} \to \mathbb{R}^n$ from f to g never passes through the origin O. Then H and the radial retraction $\rho : \mathbb{R}^n - O \to \mathbb{S}^{n-1}$, $\rho(x) = x/\|x\|$, define a homotopy $K = \rho \circ H : X \times \mathbb{I} \to \mathbb{S}^{n-1}$ from f to g in $\mathbb{S}^{n-1}$. □

Warning: Don't attempt to construct a straight-line homotopy in a space that isn't a subspace of $\mathbb{R}^n$. Other situations require other techniques.

HOMOTOPIES ON IDENTIFICATION SPACES

A homotopy *on* a space may induce a homotopy *on* an identification space:

2.4 Homotopy Transgression Theorem *Let $K : W \times \mathbb{I} \to Y$ be a homotopy that respects the identification map $p : W \to X$ in this sense:*

for all $w, w' \in W$ and $t \in \mathbb{I}$, $p(w) = p(w')$ implies $K(w, t) = K(w', t)$.

Then K induces a unique homotopy $H : X \times \mathbb{I} \to Y$ such that $H \circ (p \times 1_\mathbb{I}) = K$.

Proof: By Whitehead's Theorem (7.4.4), $p \times 1_\mathbb{I} : W \times \mathbb{I} \to X \times \mathbb{I}$ is an identification map. Since K respects the identifications of $p \times 1_\mathbb{I}$, this theorem is a special case of the standard Transgression Theorem (5.1.9). □

Theorem 2.4 applies to a homotopy $K : W \times \mathbb{I} \to Y$ that respects the identification map $p : W \to X$ and a continuous function $q : Y \to Z$, in this sense:

for all $w, w' \in W$ and $t \in \mathbb{I}$, $p(w) = p(w')$ implies $q(K(w, t)) = q(K(w', t))$.

There results a homotopy $H : X \times \mathbb{I} \to Z$ such that $H \circ (p \times 1_\mathbb{I}) = q \circ K$.

Example 6 Consider the diagonal path $\Delta : \mathbb{I} \to \mathbb{I} \times \mathbb{I}$, $\Delta(x) = <x, x>$, and the piecewise-linear path $f : \mathbb{I} \to \mathbb{I} \times \mathbb{I}$ from $<0, 0>$, thru $<0, 1>$, to $<1, 1>$. The straight-line homotopy $K : \mathbb{I} \times \mathbb{I} \to \mathbb{I} \times \mathbb{I}$ from Δ to f respects the identification map $e : \mathbb{I} \to \mathbb{S}^1$ and the function $e \times e : \mathbb{I} \times \mathbb{I} \to \mathbb{S}^1 \times \mathbb{S}^1$. K induces a homotopy from the diagonal embedding $\Delta : \mathbb{S}^1 \to \mathbb{S}^1 \times \mathbb{S}^1$, $\Delta(x) = <x, x>$, to a continuous function $\mu : \mathbb{S}^1 \to \mathbb{S}^1 \times \mathbb{S}^1$ that traces the one-point union $\mathbb{S}^1 \vee \mathbb{S}^1 = (\mathbb{S}^1 \times \{1\}) \cup (\{1\} \times \mathbb{S}^1)$.

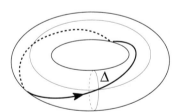
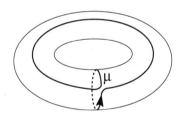

Example 6: $H : \Delta \simeq \mu$

HOMOTOPIES ON ADJUNCTION SPACES

Elaborate but extremely useful applications of the Homotopy Transgression Theorem involve *adjunction spaces*, the special identification spaces introduced in Section 5.4. Recall, the adjunction space $X = A \cup_\phi D$ arises when the space D is *attached* to the space A by means of a continuous function $\phi : S \to A$ defined on a closed subspace $S \subseteq D$. In detail, X results from the disjoint union $A \overset{\circ}{\cup} D$ when each point $z \in S \subseteq D$ is identified with its image $\phi(z) \in A$. We denote the identification map by $p : A \overset{\circ}{\cup} D \to A \cup_\phi D = X$.

We may view $(A \cup_\phi D) \times \mathbb{I}$ as the adjunction space $(A \times \mathbb{I}) \cup_{\phi \times 1_\mathbb{I}} (D \times \mathbb{I})$

by Corollary **5.1.10** to the Transgression Theorem. For the obvious homeomorphism $(A \,\mathring{\cup}\, D) \times \mathbb{I} \cong (A \times \mathbb{I}) \,\mathring{\cup}\, (D \times \mathbb{I})$ respects the identification maps

$$p \times 1_{\mathbb{I}} : (A \,\mathring{\cup}\, D) \times \mathbb{I} \to (A \cup_\phi D) \times \mathbb{I}$$

and

$$(A \times \mathbb{I}) \,\mathring{\cup}\, (D \times \mathbb{I}) \to (A \times \mathbb{I}) \cup_{\phi \times 1_{\mathbb{I}}} (D \times \mathbb{I}).$$

So the Adjunction Theorem (**5.4.6**) can be used to construct a homotopy on an adjunction space out of a pair of homotopies on the constituent spaces.

2.5   Homotopy Adjunction Theorem   *Let $\phi : S \to A$, where $S \subseteq D$ is a closed subspace. A pair of homotopies $H : A \times \mathbb{I} \to Z$ and $K : D \times \mathbb{I} \to Z$ such that $H \circ (\phi \times 1_{\mathbb{I}}) = K|_{S \times \mathbb{I}}$ induce a homotopy:*

$$L = <H, K> : (A \cup_\phi D) \times \mathbb{I} \to Z.$$

If $H : h_0 \simeq h_1$ and $K : k_0 \simeq k_1$, then $<H, K> : <h_0, k_0> \simeq <h_1, k_1>$.

Proof: As $(A \cup_\phi D) \times \mathbb{I} \cong (A \times \mathbb{I}) \cup_{\phi \times 1_{\mathbb{I}}} (D \times \mathbb{I})$, we may produce L from an application of the Adjunction Theorem **5.4.6**, as in this adjunction diagram:

$$\begin{array}{ccc} S \times \mathbb{I} & \subseteq & D \times \mathbb{I} \\ & \searrow^{K} & \\ \phi \times 1_{\mathbb{I}} \downarrow \quad & Z \quad & \downarrow (p|_D) \times 1_{\mathbb{I}} \\ & \nearrow^{H} \quad \searrow^{L} & \\ A \times \mathbb{I} & \xrightarrow[(p|_A) \times 1_{\mathbb{I}}]{} & (A \cup_\phi D) \times \mathbb{I} \end{array}$$

$\square$

2.6   Corollary   *Let $\phi : S \to A$ and $\psi : T \to B$, where $S \subseteq D$ and $T \subseteq E$. Two homotopies $H : A \times \mathbb{I} \to B$ and $K : (D \times \mathbb{I}, S \times \mathbb{I}) \to (E, T)$ such that $H \circ (\phi \times 1_{\mathbb{I}}) = \psi \circ K|_{S \times \mathbb{I}}$ induce a homotopy:*

$$<<H, K>> : (A \cup_\phi D) \times \mathbb{I} \to B \cup_\psi E.$$

If $H : h_0 \simeq h_1$ and $K : k_0 \simeq k_1$, then $<<H, K>> : <<h_0, k_0>> \simeq <<h_1, k_1>>$.

Proof: We take $<<H, K>> = <q|_B \circ H, q|_E \circ K>$. $\square$

RELATIVE HOMOTOPIES

Let $f, g : X \to Y$ be continuous functions that agree on a subspace $A \subseteq X$. They are **homotopic relative to A** if there exists a homotopy $H : X \times \mathbb{I} \to Y$ from f to g such that $f(a) = H(a, t) = g(a)$ for all $a \in A$ and $t \in \mathbb{I}$. Then H is called a **homotopy relative to A** from f to g.

When $f, g : X \to Y$ are homotopic relative to A, we write $f \simeq g \; rel \; A$, or even $H : f \simeq g \; rel \; A$ when a specific homotopy H is to be emphasized.

Example 7 A path-homotopy $H: f \simeq g$ (Section **9**.1) between paths $f, g : \mathbb{I} \to Y$ fixes their common endpoints and so is a relative homotopy $H: \mathbb{I} \times \mathbb{I} \to Y \, rel \, \{0, 1\}$.

Example 8 The straight-line homotopy in $Y \subseteq \mathbb{R}^n$ between continuous functions $f, g: X \to Y$ is a homotopy relative to any subspace $A \subseteq X$ on which f and g agree.

Example 9 Consider the harmonic broom $B \subseteq \mathbb{R}^2$ (Example **6**.1.19). Every function $f: B \to B$ is homotopic to the constant function $c: B \to B$ at the apex point $<0, 1>$ via a straight-line homotopy. By the transitivity of the homotopy relation $\simeq$, the horizontal projection $p: B \to B$, $p(<x, y>) = <0, y>$, onto the limiting straw $A = \{0\} \times \mathbb{I}$ is homotopic to the identity function $1_B: B \to B$. Although p and 1_B coincide on A, there is no homotopy $H: 1_B \simeq p \, rel \, A$. (Exercise 8).

3 Homotopy Category

We now construct the homotopy category $\mathfrak{H}$ of spaces and homotopy classes of continuous functions. In this new category, the problem of classifying spaces up to homeomorphism type is replaced by the simpler problem of classifying spaces up to the coarser relation of homotopy type.

HOMOTOPY CLASSES

For each pair of topological spaces X and Y, let (X, Y) denote the set of all continuous functions $X \to Y$.

3.1 Theorem *Homotopy is an equivalence relation $\simeq$ on (X, Y).*

Proof: *Reflexive:* For all $f: X \to Y$, we have $f \simeq f$ via the *stationary* homotopy $S: X \times \mathbb{I} \to Y$, defined by $S(x, t) = f(x)$ for all $x \in X$ and $t \in \mathbb{I}$.

Symmetric: If $H: f \simeq g$, then $G: g \simeq f$ via the *reverse* homotopy $G: X \times \mathbb{I} \to Y$, defined by $G(x, t) = H(x, 1 - t)$.

Transitive: If $H: f \simeq g$ and $K: g \simeq h$, then $L: f \simeq h$ via the homotopy $L: X \times \mathbb{I} \to Y$, obtained by stacking K atop H. Since $H(x, 1) = g(x) = L(x, 0)$ for all $x \in X$, L is well-defined by

$$L(x, t) = \begin{cases} K(x, 2t - 1), & \text{if } \tfrac{1}{2} \leq t \leq 1 \\ H(x, 2t), & \text{if } 0 \leq t \leq \tfrac{1}{2}. \end{cases}$$

In addition, L is continuous by the Gluing Theorem (**4**.4.3). □

Thus, the homotopy relation $\simeq$ partitions (X, Y) into **homotopy classes:**

$$[f] = \{g \in (X, Y) : f \simeq g\}.$$

Let $[X, Y]$ denote the quotient set, $(X, Y)/\simeq$, of these homotopy classes.

Example 1 The path-homotopy investigations of Section 9.2 imply that $[\mathbb{S}^1, \mathbb{S}^1]$ is $\{[d_n] : n \in \mathbb{Z}\}$, while $[\mathbb{S}^1, \mathbb{S}^n]$ is the single class of the constant function whenever $n \geq 2$. In other words, the 1-sphere $\mathbb{S}^1$ has infinitely many homotopically distinct styles of wearing a circular belt, while any higher dimensional sphere $\mathbb{S}^n$ has but one.

3.2 Theorem *Composites of homotopic functions are homotopic: If $f \simeq f' : X \to Y$ and $g \simeq g' : Y \to Z$, then $g \circ f \simeq g' \circ f' : X \to Z$.*

Proof: If $H : f \simeq f'$ and $K : g \simeq g'$, then
$$g \circ H : g \circ f \simeq g \circ f' \text{ and } K \circ (f' \times 1_\mathbb{I}) : g \circ f' \simeq g' \circ f'.$$
Hence, $g \circ f \simeq g' \circ f'$, by the transitivity property in Theorem 3.1. □

Thus, for any three spaces X, Y, and Z, a ***composition function***
$$\circ : [X, Y] \times [Y, Z] \to [X, Z]$$
is well-defined by $[g] \circ [f] = [g \circ f]$ for all $[f] \in [X, Y]$ and $[g] \in [Y, Z]$.

The composition is associative: $[h] \circ ([g] \circ [f]) = ([h] \circ [g]) \circ [f])$. Also, the homotopy classes $[1_X]$ and $[1_Y]$ act as identities under composition with all homotopy classes $[f] \in [X, Y]$: $[1_Y] \circ [f] = [f] = [f] \circ [1_X]$.

So the composition operation makes the class of spaces and homotopy classes of continuous functions into a category $\mathfrak{H}$, called the ***homotopy category***. In $\mathfrak{H}$, the morphism set $(X, Y)_\mathfrak{H}$ for any ordered pair of spaces X and Y is the homotopy set $[X, Y]$.

The equivalence relation in $\mathfrak{H}$ between spaces is called *homotopy equivalence*. This new relation is defined below and investigated more thoroughly in Section 5.

HOMOTOPY EQUIVALENCE

Two spaces X and Y are ***homotopy equivalent***, denoted by $X \simeq Y$, if they are equivalent as objects in the homotopy category $\mathfrak{H}$.

In detail, $X \simeq Y$ provided that there are homotopy classes $[f] \in [X, Y]$ and $[g] \in [Y, X]$ such that $[g] \circ [f] = [1_X]$ in $[X, X]$ and $[f] \circ [g] = [1_Y]$ in $[Y, Y]$. Equivalently, there are continuous functions $f : X \leftrightarrow Y : g$ such that
$$g \circ f \simeq 1_X \quad \text{and} \quad f \circ g \simeq 1_Y.$$
Such continuous functions f and g are called ***homotopy equivalences***; each is called a ***homotopy inverse*** of the other.

Because homotopy equivalence is the relation of equivalence between objects in the homotopy category $\mathfrak{H}$, it is an equivalence relation on any set of spaces. Therefore, the homotopy equivalence relation partitions any set of topological spaces into classes called ***homotopy types***.

Homeomorphic spaces are homotopy equivalent, because each homeomorphism is a homotopy equivalence. But there are homotopy equivalent spaces that are not homeomorphic. The partition into homotopy types is simply coarser than the partition into homeomorphism types.

Example 2 The annulus $A = \{x \in \mathbb{R}^2 : 1 \leq \|x\| \leq 2\}$ and the 1-sphere $\mathbb{S}^1$ are homotopy equivalent via the inclusion $i : \mathbb{S}^1 \subset A$ and the radial retraction $\rho : A \to \mathbb{S}^1$, $\rho(x) = x/\|x\|$. For $\rho \circ i = 1_{\mathbb{S}^1}$ and $i \circ \rho \simeq 1_A$ via the straight-line homotopy in A:
$$H : A \times \mathbb{I} \to A, H(x, t) = t \cdot x + (1-t) \cdot \rho(x).$$

Example 3 Punctured n-space $\mathbb{R}^n - \{O\}$ is homotopy equivalent to the $(n-1)$-sphere $\mathbb{S}^{n-1}$. The inclusion $i : \mathbb{S}^{n-1} \subset \mathbb{R}^n - \{O\}$ and the radial retraction
$$\rho : \mathbb{R}^n - \{O\} \to \mathbb{S}^{n-1}, \rho(x) = x/\|x\|,$$
are homotopy inverses.

CONTRACTIBLE SPACES

A space Y is **contractible** if the identity function $1_Y : Y \to Y$ is inessential, i.e., if there is a homotopy $H : Y \times \mathbb{I} \to Y$ between the identity function 1_Y and some function $c_y : Y \to Y$ with a constant value $c_y(Y) = y$.

Any homotopy $H : 1_Y \simeq c_y$ is called a **contraction** of the space Y to $y \in Y$.

Contractible spaces are the simplest possible spaces in the homotopy category $\mathfrak{H}$, according to the following characterization.

3.3 Theorem *The following are equivalent conditions on any space Y:*
(a) *Y is contractible.*
(b) *Y is homotopy equivalent to a singleton space.*
(c) *Any two continuous functions of any space X into Y are homotopic.*

Proof: $(a) \Rightarrow (b)$. Let $H : 1_Y \simeq c_y$ be a contraction. Then $i : \{y\} \hookrightarrow Y : k$ are homotopy inverses, as $k \circ i = 1_{\{y\}}$ and $H : 1_Y \simeq c_y = i \circ k$.

$(b) \Rightarrow (c)$. Let $h : * \hookrightarrow Y : k$ be homotopy inverses. Then, $1_Y \simeq h \circ k = c_y$, where $y = h(*)$. So, for any $f, g : X \to Y$, there are the homotopy relations:
$$f = 1_Y \circ f \simeq c_y \circ f = c_y \circ g \simeq 1_Y \circ g = g.$$

$(c) \Rightarrow (a)$. In particular, $1_Y \simeq c_y : Y \to Y$ for any $y \in Y$. □

3.4 Corollary *A contractible space is path-connected and there exists a contraction to any of its points.*

Example 4 The closed and open n-balls $\mathbb{D}^n$ and $\mathbb{B}^n$ are contractible to the origin O via the contractions $H : X \times \mathbb{I} \to X$ ($X = \mathbb{D}^n, \mathbb{B}^n$) defined by $H(x, t) = (1-t) \cdot x$. By the transitivity of the homotopy equivalence relation, these contractible spaces $\mathbb{D}^n$ and $\mathbb{B}^n$ are homotopy equivalent to each other: $\mathbb{D}^n \simeq O \simeq \mathbb{B}^n$. So the homotopy equivalence relation does not respect the basic topological property of compactness.

Example 5 S^0 is not contractible as it is not path connected; S^1 is not contractible as $1, c : S^1 \to S^1$ have different degree and so can't be homotopic (see Exercise 3). It seems obvious that each n-sphere S^n ($n \geq 2$) is non-contractible, but it is surprisingly difficult to establish this fact. The non-contractibility of higher dimensional spheres is established in the text: J. Dugundji, *Topology*, Boston: Allyn and Bacon, Inc., 1966.

It can be challenging to determine the contractibility status of even very simple spaces. The harmonic comb (Exercise **3**.4.11) and the doubled harmonic comb (Exercise **3**.4.12) are two interesting examples.

Example 6 *The harmonic comb HC is a contractible space.* The homotopy

$$H(<x, y>, t) = \begin{cases} <x, 2t + (1-2t)y>, & 0 \leq t \leq \frac{1}{2} \\ <(2-2t)x, 1>, & \frac{1}{2} \leq t \leq 1 \end{cases}$$

defines a contraction $H : HC \times \mathbb{I} \to HC$ that pushes the teeth in HC vertically up to the spine, and then contracts the spine horizontally to the point $<0, 1>$.

Example 7 *The doubled harmonic comb DHC is not a contractible space.* Were $H : DHC \times \mathbb{I} \to DHC$ a contraction to the origin $\mathbf{O}$, then

(1) there would exist a time $0 < t < 1$ when $H(\mathbf{O}, t) = <0, 1>$;
(2) for each time $0 < t < 1$ when $H(\mathbf{O}, t) = <0, 1>$, there would exist an earlier time $0 < s < t < 1$ when $H(\mathbf{O}, s) = <0, -1>$; and
(3) for each time $0 < s < 1$ when $H(\mathbf{O}, s) = <0, -1>$, there would exist an earlier time $0 < t < s < 1$ when $H(\mathbf{O}, t) = <0, 1>$.

Assuming (1-3), there would be a monotonic sequence,

$$0 < \ldots s_k < t_k < \ldots < s_2 < t_2 < s_1 < t_1 < 1,$$

such that $H(\mathbf{O}, s_k) = <0, -1>$ and $H(\mathbf{O}, t_k) = <0, 1>$ for all $k \geq 1$. Then at the common limit, $\lim s_k = r = \lim t_k$, we would have $<0, -1> = H(\mathbf{O}, r) = <0, 1>$, by the continuity of H.

By this contradiction, it suffices to establish the three claims (1-3).

(1) The contraction H of DHC to the origin $\mathbf{O} = <0, 0>$ would restrict to a path $H(<1/n, 0>, -) : \mathbb{I} \to DHC$ joining points $<1/n, 0>$ and $\mathbf{O}$ of DHC that are separated in $DHC - \{<1/n, 1>\}$. By the connectedness of $\mathbb{I}$, this path would have the value $<1/n, 1>$ at some time a_n. Then, by the sequential compactness of $\mathbb{I}$, the resulting sequence (a_n) would have a convergent subsequence (a_{n_i}) with limit t, at which time we would have

$$H(\mathbf{O}, t) = \lim H(<1/n_i, 0>, a_{n_i}) = \lim <1/n_i, 1> = <0, 1>.$$

(2) Suppose it were the case that $H(\mathbf{O}, t) = <0, 1>$. Since $\lim <-1/n, 0> = \mathbf{O}$, we would have $\lim H(<-1/n, 0>, t) = H(\mathbf{O}, t) = <0, 1>$. So for all sufficiently large n, the path $H(<-1/n, 0>, -) : [0, t] \to DHC$ would join $<-1/n, 0>$ and a point near $<0, 1>$, which are points that are separated in $DHC - \{<-1/n, -1>\}$. So this path would take the value $<-1/n, -1>$ at some time $b_n < t$. The sequence (b_n) in $\mathbb{I}$ would have a convergent subsequence (b_{n_i}) with limit $s < t$, at which time we would have

$$H(\mathbf{O}, s) = \lim H(<-1/n_i, 0>, b_{n_i}) = \lim <-1/n_i, -1> = <0, -1>.$$

(3) This follows from (2), by symmetry.

SECTION 3   HOMOTOPY CATEGORY   309

CONE CONSTRUCTION

Let X be any topological space. The ***cone CX over X*** is the quotient space, $(X \times \mathbb{I})/(X \times \{1\})$, obtained from the topological product $X \times \mathbb{I}$ by collapsing the end $X \times \{1\}$ to a single point $*$, called the ***apex*** of CX.

Let $q : X \times \mathbb{I} \to CX$ denote the quotient map. The quotient topology is characterized by this property: $F \subseteq CX$ is closed if and only if $q^{-1}(F)$ is closed in $X \times \mathbb{I}$.

Example 8   *The cone $C\mathbb{S}^{n-1}$ over the sphere $\mathbb{S}^{n-1}$ is homeomorphic to the closed n-ball $\mathbb{D}^n$.* The continuous surjection $f : \mathbb{S}^{n-1} \times \mathbb{I} \to \mathbb{D}^n$, $f(x, t) = (1-t) \cdot x$, makes the same identifications as the quotient map $q : \mathbb{S}^{n-1} \times \mathbb{I} \to C\mathbb{S}^{n-1}$. By the Transgression Theorem (5.1.9), f induces a continuous bijection $g : C\mathbb{S}^{n-1} \to \mathbb{D}^n$. Since the cone $C\mathbb{S}^{n-1}$ is compact and the disc $\mathbb{D}^n$ is Hausdorff, then g is a homeomorphism by the Homeomorphism Theorem (5.2.5).

When $X \subset \mathbb{R}^n$ is a compact subspace, the cone CX over X is homeomorphic to the *geometric cone over X* consisting of all line segments in $\mathbb{R}^{n+1}$ joining points $x \in X$ to $<0, 1>$ (Exercise 2).

3.5 Theorem   *Any space embeds as a closed subspace of its cone.*

Proof: The image $q(X \times \{0\})$ under the quotient map $q : X \times \mathbb{I} \to CX$ is a closed subspace of CX since the pre-image $q^{-1}(q(X \times \{0\})) = X \times \{0\}$ is a closed subspace of $X \times \mathbb{I}$. The restriction

$$q|_{X \times \{0\}} : X \times \{0\} \to q(X \times \{0\})$$

is a homeomorphism because it is a closed continuous bijection. Hereafter, we let $X \subset CX$ denote the embedding $q|_{X \times \{0\}} : X \times \{0\} \to CX$.   □

3.6 Theorem   *Every cone is contractible.*

Proof: By the Homotopy Transgression Theorem (2.4), the homotopy

$$H : X \times \mathbb{I} \times \mathbb{I} \to X \times \mathbb{I},$$

given by

$$H(x, s, t) = (x, (1-t) \cdot s + t \cdot 1),$$

induces a homotopy $K : CX \times \mathbb{I} \to CX$, as in this commutative diagram:

$$\begin{array}{ccc} X \times \mathbb{I} \times \mathbb{I} & \xrightarrow{H} & X \times \mathbb{I} \\ q \times 1_{\mathbb{I}} \downarrow & & \downarrow q \\ CX \times \mathbb{I} & \xrightarrow{K} & CX \end{array}$$

The definition
$$K(q(x, s), t) = q(H(x, s, t)) = q(x, (1 - t) \cdot s + t \cdot 1)$$
shows that K is a homotopy from the identity function 1_{CX} to the constant function at the apex, $* = q(X \times \{1\})$, of the cone CX. Thus, $CX \simeq *$. □

The cone construction shows that every space appears as a closed subspace of some contractible space. This makes most topological properties useless for homotopical purposes, as we now show.

A topological property $\mathcal{P}$ is a **homotopy property** provided that it respects homotopy types: whenever $X \simeq Y$, then X has $\mathcal{P}$ if and only if Y has $\mathcal{P}$.

For example, path connectedness is a homotopy property.

3.7 Theorem *A topological property possessed by each singleton, but not all spaces, and inherited by closed subspaces isn't a homotopy property.*

Proof: Such a topological property $\mathcal{P}$ cannot be a homotopy property. Otherwise, the homotopy equivalence $* \simeq CX$ (Theorem 3.6) for any space X and the property that $*$ has $\mathcal{P}$ would imply that CX has $\mathcal{P}$. Then the hereditary property would imply that the closed subspace $X \subset CX$ (Theorem 3.5) has $\mathcal{P}$, contrary to the fact that not all spaces have $\mathcal{P}$. □

HOMOTOPY INVARIANCE OF Π_1

Theorem 3.7 shows that most topological invariants, such as compactness, Hausdorff, metrizability, first countability, and second countability, are useless for investigations in the homotopy category.

However, since the fundamental group is defined in terms of special homotopies, the path-homotopies, π_1 provides a homotopy invariant. To prove this, we begin with the following lemma.

3.8 Lemma *Homotopic maps $F, G : X \to Y$ induce homomorphisms*
$$F_\# : \pi_1(X, x) \to \pi_1(Y, F(x)) \quad \text{and} \quad G_\# : \pi_1(X, x) \to \pi_1(Y, G(x))$$
that correspond under the change of basepoint isomorphism
$$\omega_{[h]} : \pi_1(Y, F(x)) \to \pi_1(Y, G(x)), \quad \omega_{[h]}([f]) = [\bar{h}] \cdot [f] \cdot [h],$$
where the path $h = H|_{\{x\} \times \mathbb{I}} : (\mathbb{I}, 0, 1) \to (X, F(x), G(x))$ is obtained from any homotopy $H : F \simeq G$.

Proof: For any loop $f : \mathbb{I} \to X$ based at x, the composite function
$$K = H \circ (f \times 1_\mathbb{I}) : \mathbb{I} \times \mathbb{I} \to X \times \mathbb{I} \to Y$$
restricts to the path $h = H|_{\{x\} \times \mathbb{I}}$ on the sides $\{0\} \times \mathbb{I}$ and $\{1\} \times \mathbb{I}$, to the path

$F \circ f$ on the bottom $\mathbb{I} \times \{0\}$, and to the path $G \circ f$ on the top $\mathbb{I} \times \{1\}$. K and a quotient map $q : \mathbb{I} \times \mathbb{I} \to \mathbb{I} \times \mathbb{I}$ that collapses $\{0\} \times \mathbb{I}$ and $\{1\} \times \mathbb{I}$ can be used to define a path-homotopy $q \circ K : \bar{h} \cdot (F \circ f) \cdot h \simeq G \circ f$ in Y. In other words,

$$G_\# = \omega_{[h]} \circ F_\# : \pi_1(X, x) \to \pi_1(Y, F(x)) \to \pi_1(Y, G(x)). \quad \square$$

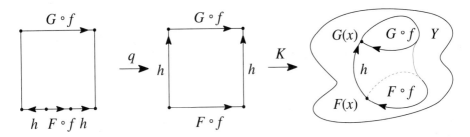

Lemma 3.8

3.9 Theorem *Any homotopy equivalence $F : X \to Y$ induces a fundamental group isomorphism for each $x \in X$:*

$$F_\# : \pi_1(X, x) \to \pi_1(Y, F(x)).$$

Proof: Let $F : X \leftrightarrow Y : G$ be homotopy inverses. Because $1_X \simeq G \circ F$ and $1_Y \simeq F \circ G$, Lemma 3.8 provides paths $h : \mathbb{I} \to X$ and $k : \mathbb{I} \to Y$ such that

$$(G \circ F)_\# = \omega_{[h]} : \pi_1(X, x) \to \pi_1(X, G(F(x)))$$

and

$$(F \circ G)_\# = \omega_{[k]} : \pi_1(Y, F(x)) \to \pi_1(Y, F(G(F(x)))).$$

These combine to give a commutative diagram:

$$\begin{array}{ccc} \pi_1(X, x) & \xrightarrow{\omega_{[h]}} & \pi_1(X, G(F(x))) \\ F_\# \downarrow & {}^{G_\#} \nearrow & \downarrow F_\# \\ \pi_1(Y, F(x)) & \xrightarrow[\omega_{[k]}]{} & \pi_1(Y, F(G(F(x)))) \end{array}$$

Since $\omega_{[h]}$ and $\omega_{[k]}$ are isomorphisms, so are all $F_\#$ and $G_\#$ in the diagram. $\square$

3.10 Corollary *Any contractible space has a trivial fundamental group; hence, any contractible space is simply connected.*

Example 9 The cone over the Hawaiian earrings is contractible (Theorem 3.6), hence, has trivial fundamental group. Two copies of this cone joined at their apex points yield a contractible space; but two copies joined at the tangency points in their bases form a space with non-trivial fundamental group, hence, a non-contractible one.

4 Fibrations and Cofibrations

The twin problems of lifting homotopies through a continuous surjection and extending homotopies defined on a subspace lead to special functions called *fibrations* and *cofibrations*. In this section, we introduce these functions.

HOMOTOPY LIFTING PROPERTY

Let $p : Y \to B$ be a continuous surjection. A continuous function $k : X \to Y$ is called a *lifting* of a continuous function $h : X \to B$ if $p \circ k = h$. A homotopy $K : X \times \mathbb{I} \to Y$ is a *lifting* of a homotopy $H : X \times \mathbb{I} \to B$ if $p \circ K = H$.

The existence of a lifting of h is an invariant of the homotopy class of h when p satisfies the following *homotopy lifting property*.

The function $p : Y \to B$ has the **homotopy lifting property** for a space X if, for each continuous function $k : X \to Y$, each homotopy $H : X \times \mathbb{I} \to B$ of $p \circ k$ ($H \circ i_0 = p \circ k$) has a lifting to a homotopy $K : X \times \mathbb{I} \to Y$ of k ($K \circ i_0 = k$), and K is constant on $\{x\} \times \mathbb{I}$ whenever H is constant on $\{x\} \times \mathbb{I}$.

The lifting property is expressed by this commutative diagram:

$$\begin{array}{ccc} X & \xrightarrow{k} & Y \\ i_0 \downarrow & \nearrow K & \downarrow p \\ X \times \mathbb{I} & \xrightarrow{H} & B \end{array} \quad (\forall k \text{ and } H, \exists K)$$

A continuous function $p : Y \to B$ is a (**regular Hurewicz**) *fibration* if it has the homotopy lifting property for every space X.

Example 1 Any topological projection $p_B : B \times F \to B$ is a fibration. Given any continuous function $k : X \to B \times F$ and homotopy $H : X \times \mathbb{I} \to B$ of $H \circ i_0 = p_B \circ k$, there is the lifting $K : X \times \mathbb{I} \to B \times F$, defined by $K(x, t) = (H(x, t), p_F(k(x)))$.

Example 2 Let $Y = (\mathbb{I} \times \{0\}) \cup (\{0\} \times \mathbb{I}) \subseteq \mathbb{I} \times \mathbb{I}$. The restriction $p : Y \to \mathbb{I}$ of the projection fibration $p : \mathbb{I} \times \mathbb{I} \to \mathbb{I}$, $p(t, r) = t$, from Example 1 is not a fibration. For any non-empty space X, the homotopy $H : X \times \mathbb{I} \to \mathbb{I}$, $H(x, t) = t$, has no lifting to a homotopy $K : X \times \mathbb{I} \to Y$ of the constant function $k : X \to Y$ at $k(X) = <0, 1>$.

There are continuous surjections that locally have the structure of a projection $p_B : B \times F \to B$. A *fiber bundle* (Y, p, B, F) involves a continuous surjection $p : Y \to B$ and a space F such that there exists an open covering $\mathcal{U}$ of B and, for each $U \in \mathcal{U}$, a homeomorphism $\varphi_U : U \times F \to p^{-1}(U)$ such that $p \circ \varphi_U : U \times F \to p^{-1}(U) \to U$ is projection onto the first factor. The space F is called the *fiber*, each homeomorphism φ_U is called a *coordinate chart*, and p is called an *F-bundle*, with *total space* Y and *base space* B.

Example 3 There are two $\mathbb{I}$-bundles with base space $\mathbb{S}^1$: the *trivial* product bundle $p_{\mathbb{S}^1} : \mathbb{S}^1 \times \mathbb{I} \to \mathbb{S}^1$ and the *twisted* Möbius bundle $p : M \to \mathbb{S}^1$.

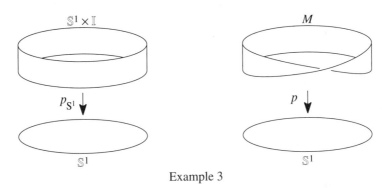

Example 3

Example 4 Hopf's map $p : \mathbb{S}^3 \to \mathbb{S}^2$, assembled in Example 5.4.8 from complex multiplications $m_+ : \mathbb{S}^1 \times \mathbb{D}^2 \to \mathbb{D}^2_+$ and $m_- : \mathbb{D}^2 \times \mathbb{S}^1 \to \mathbb{D}^2_-$, is an $\mathbb{S}^1$-bundle.

In 1955, W. Hurewicz proved that a local fibration $p : Y \to B$ (i.e., a continuous surjection for which there is open covering $\mathcal{U} = \{U\}$ of B such that $p|_U : p^{-1}(U) \to U$ is a fibration for each $U \in \mathcal{U}$) is a fibration whenever the base space B is paracompact. The argument utilizes partitions of unity on the paracompact base space. Because a fiber bundle projection is a local fibration by Example 1, Hurewicz's result implies that any fiber bundle projection onto a paracompact base space is a fibration. We assume this result.

Here are two ways to generalize Hopf's $\mathbb{S}^1$-bundle $p : \mathbb{S}^3 \to \mathbb{S}^2$.

Example 5 The pair of quaternionic multiplications $m_- : \mathbb{S}^3 \times \mathbb{D}^4 \to \mathbb{D}^4_-$ and $m_+ : \mathbb{D}^4 \times \mathbb{S}^3 \to \mathbb{D}^4_+$ assemble into a bundle projection $p : \mathbb{S}^8 \to \mathbb{S}^4$ with fiber $\mathbb{S}^3$, by use of Corollary 5.4.7 of the Adjunction Theorem. Similarly, Cayley multiplication determines a bundle projection $p : \mathbb{S}^{15} \to \mathbb{S}^8$ with fiber $\mathbb{S}^7$ (Exercise 2).

Example 6 *Hopf's $\mathbb{S}^1$-bundle $p : \mathbb{S}^{2n+1} \to \mathbb{P}_n(\mathbb{C})$.* Let $\mathbb{C}^{n+1}$ be $(n+1)$-dimensional complex space. View $\mathbb{S}^{2n+1}$ as the sphere of points $z \in \mathbb{C}^{n+1}$ with norm $\| z \| = 1$. Scalar multiplication $\lambda (z_0, \ldots, z_n) = (\lambda z_0, \ldots, \lambda z_n)$ by $\lambda \in \mathbb{S}^1 \subseteq \mathbb{C}$ introduces an equivalence relation $z \sim \lambda z$ on $\mathbb{S}^{2n+1}$, with equivalence classes $[z] = [z_0, \ldots, z_n]$. The resulting quotient map, $p : \mathbb{S}^{2n+1} \to \mathbb{P}_n(\mathbb{C})$, is a fiber bundle projection with fiber:

$$p^{-1}([1, 0, \ldots, 0]) = \{(\lambda 1, \lambda 0, \ldots, \lambda 0) \in \mathbb{S}^{2n+1}\} = \mathbb{S}^1.$$

For this, we use the open covering $U_i = \{[z] \in \mathbb{P}_n(\mathbb{C}) : z_i \neq 0\}$ ($0 \leq i \leq n$) of $\mathbb{P}_n(\mathbb{C})$ and the open hemispheres $V_i = \{w \in \mathbb{S}^{2n+1} : 0 < w_i \text{ (real)}\}$ ($0 \leq i \leq n$) of $\mathbb{S}^{2n+1}$. Whenever $0 \neq z_i \in \mathbb{C}$, there is a unique $\lambda \in \mathbb{S}^1$ with $0 < \lambda z_i = w_i$ (real). So $p(V_i) = U_i$ and $\varphi_{U_i} : U_i \times \mathbb{S}^1 \to p^{-1}(U_i)$, where $([w], \lambda) \to \lambda^{-1} w$ for $w \in V_i$, form a coordinate chart.

Example 7 A fiber bundle (Y, p, B, F) whose fiber F is discrete is called a *covering space*. The exponential function $e : \mathbb{R}^1 \to \mathbb{S}^1$, $e(t) = e^{2\pi i t}$, is a prime example. Covering spaces and their homotopy lifting property are developed in Chapter **13**.

The fibration property is propagated in a *fibered product construction* that is the categorical dual to the adjunction space construction (Section **5**.4).

The *fibered product* of continuous functions $p : Y \to B$ and $g : D \to B$ is the subspace $P_g = \{<d, y> \in D \times Y : g(d) = p(y)\}$ of the product $D \times Y$.

4.1 Induced Fibration Theorem   Let $p : Y \to B$ be a fibration and let $g : D \to B$ any continuous function. Then the projection $p_g : P_g \to D$, $p_g(<d, y>) = d$, is a fibration, called the fibration **induced from** p **by** g.

Proof: Given any space X, any continuous function $k : X \to P_g$, and any homotopy $H : X \times \mathbb{I} \to D$ of $p_g \circ k$ ($H \circ i_0 = p_g \circ k$), we form the continuous function $p_Y \circ k : X \to Y$ and the homotopy $g \circ H : X \times \mathbb{I} \to B$ of $p \circ p_Y \circ k$ ($g \circ H \circ i_0 = p \circ p_Y \circ k$). Since p is a fibration, the homotopy $g \circ H$ has a lifting through p to a homotopy $J : X \times \mathbb{I} \to Y$ of $p_Y \circ k$ ($J \circ i_0 = p_Y \circ k$). Then H has a lifting through p_g to a homotopy $K : X \times \mathbb{I} \to P_g$, $K(x, t) = <H(x, t), J(x, t)>$, of k (below). Whenever H is constant on $\{x\} \times \mathbb{I}$, so are J and H. □

$$\begin{array}{ccccc}
X & \xrightarrow{k} & P_g & \xrightarrow{p_Y} & Y \\
i_0 \downarrow & \nearrow K & p_g \downarrow & & \downarrow p \\
X \times \mathbb{I} & \xrightarrow{H} & D & \xrightarrow{g} & B
\end{array}$$

Induced Fibration Theorem (4.1)

In Section 5, we shall examine the homotopy features of fibrations and show that they provide interesting examples of essential functions.

HOMOTOPY EXTENSION PROPERTY

Let $i : A \subseteq X$ be the inclusion of a closed subspace. A continuous function $k : X \to Y$ is called an *extension* of a continuous function $h : A \to Y$ provided that $k \circ i = h$. A homotopy $K : X \times \mathbb{I} \to Y$ is called an *extension* of a homotopy $H : A \times \mathbb{I} \to Y$ provided that $K \circ (i \times 1_{\mathbb{I}}) = H$.

The existence of an extension of h is an invariant of the homotopy class of h when i has a *homotopy extension property*. To express this property, we view $(A \times \mathbb{I}) \cup (X \times \{0\})$ as the union of two closed subspaces of $X \times \mathbb{I}$.

A continuous function $H : (A \times \mathbb{I}) \cup (X \times \{0\}) \to Y$ is called a *partial homotopy* of $h : X \to Y$ if $H \circ i_0$ and h agree on A. By the Gluing Theorem (**4**.4.3), a homotopy $H : A \times \mathbb{I} \to Y$ such that $H|_{A \times \{0\}} = h|_{A \times \{0\}}$ determines a partial homotopy $H : (A \times \mathbb{I}) \cup (X \times \{0\}) \to Y$ of $h : X \times \{0\} \to Y$.

The inclusion $i : A \subseteq X$ has the **homotopy extension property** (**HEP**) for the space Y if every partial homotopy $H : (A \times \mathbb{I}) \cup (X \times \{0\}) \to Y$ of every continuous function $h : X \times \{0\} \to Y$ extends to a homotopy $K : X \times \mathbb{I} \to Y$.

A subspace inclusion $i : A \subseteq X$ is called a **cofibration** provided that it has the homotopy extension property for all spaces Y.

Example 8 The inclusion $i : \mathbb{S}^{n-1} \subset \mathbb{D}^n$ is a cofibration. First, there is a retraction
$$R : \mathbb{D}^n \times \mathbb{I} \to (\mathbb{S}^{n-1} \times \mathbb{I}) \cup (\mathbb{D}^n \times \{0\}) \subset \mathbb{D}^n \times \mathbb{I}.$$
(As indicated below left, R can be viewed in $\mathbb{R}^{n+1} = \mathbb{R}^n \times \mathbb{R}$ as radial retraction away from the point $((0, \ldots, 0), 2)$ centered above the top end $\mathbb{D}^n \times \{1\}$ of the solid cylinder $\mathbb{D}^n \times \mathbb{I}$. Below right, R is projection of $\mathbb{D}^n \times \mathbb{I} \cong \{x \in \mathbb{R}^n \times [0, \infty) : 1 \leq \|x\| \leq 2\}$ onto $\{x \in \mathbb{R}^n \times [0, \infty) : \|x\| = 1 \text{ or } x_{n+1} = 0\}$.) Using the retraction R, a partial homotopy $H : (\mathbb{S}^{n-1} \times \mathbb{I}) \cup (\mathbb{D}^n \times \{0\}) \to Y$ of any map $h : \mathbb{D}^n \times \{0\} \to Y$ extends to the homotopy $K = H \circ R : \mathbb{D}^n \times \mathbb{I} \to (\mathbb{S}^{n-1} \times \mathbb{I}) \cup (\mathbb{D}^n \times \{0\}) \to Y$.

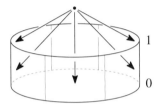 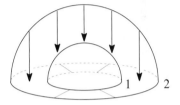

Example 8: Two views of $R : \mathbb{D}^n \times \mathbb{I} \to (\mathbb{S}^{n-1} \times \mathbb{I}) \cup (\mathbb{D}^n \times \{0\})$

The technique in Example 8 leads to this characterization (Exercise 4):

4.2 Theorem *The following are equivalent for a closed subspace $A \subseteq X$.*
(a) *The inclusion $i : A \subseteq X$ is a cofibration.*
(b) *$i : A \subseteq X$ has the homotopy extension property for $(A \times \mathbb{I}) \cup (X \times \{0\})$.*
(c) *The subspace $(A \times \mathbb{I}) \cup (X \times \{0\})$ is a retract of $X \times \mathbb{I}$.*

Example 9 $\{0\} \subseteq \mathbb{I}$ is a cofibration since $(\{0\} \times \mathbb{I}) \cup (\mathbb{I} \times \{0\})$ is a retract of $\mathbb{I} \times \mathbb{I}$.

Example 10 By Theorem 4.2, the inclusion $i : \mathbb{S}^{n-1} \subset \mathbb{S}^n$ ($n \geq 1$) is a cofibration. To prove this, we view $\mathbb{S}^n$ as the amalgamated union $\mathbb{D}^n_+ \cup_{\mathbb{S}^{n-1}} \mathbb{D}^n_-$ of two hemispherical n-balls $\mathbb{D}^n_+$ and $\mathbb{D}^n_-$ that share the equator $\mathbb{S}^{n-1}$ as their boundary. Then we view $\mathbb{S}^n \times \mathbb{I}$ as the amalgamated union $\mathbb{D}^n_+ \times \mathbb{I} \cup_{\mathbb{S}^{n-1} \times \mathbb{I}} \mathbb{D}^n_- \times \mathbb{I}$ and we glue two copies of the retraction in Example 8 into a retraction:
$$R : \mathbb{S}^n \times \mathbb{I} \to (\mathbb{S}^{n-1} \times \mathbb{I}) \cup (\mathbb{S}^n \times \{0\}) \subset \mathbb{S}^n \times \mathbb{I}.$$

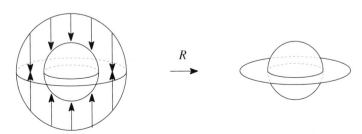

Example 10

Example 11 The inclusion $\{0\} \subset \{0, 1, 1/2, 1/3, \ldots\} = M$ is not a cofibration. There is no retraction of $M \times \mathbb{I}$ onto $(\{0\} \times \mathbb{I}) \cup (M \times \{0\})$.

Cofibrations can arise from the adjunction space construction in Section 5.4. For any continuous function $\phi : S \to A$ defined on a closed subspace $S \subseteq D$, the adjunction space $X = A \cup_\phi D$ results from the topological disjoint union $A \mathring{\cup} D$ when each point $z \in S \subseteq D$ is identified with its image $\phi(z) \in A$. By Theorem 5.4.4, the space A embeds as a closed subspace of X.

4.3 Induced Cofibration Theorem *When $j : S \subseteq D$ is a cofibration and $\phi : S \to A$ is any attaching map, the inclusion $i : A \subseteq X = A \cup_\phi D$ is a cofibration, called the **cofibration induced from j by ϕ**.*

Proof: By Theorem 4.2, the hypothesis and conclusion are equivalent to the existence of a retraction $r : D \times \mathbb{I} \to (S \times \mathbb{I}) \cup (D \times \{0\}) \subseteq D \times \mathbb{I}$

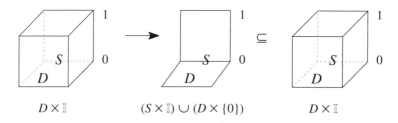

and a retraction $R : X \times \mathbb{I} \to (A \times \mathbb{I}) \cup (X \times \{0\}) \subseteq X \times \mathbb{I}$.

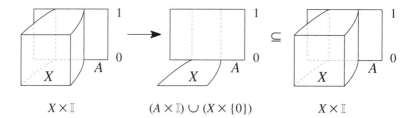

To construct R from r, we view $X \times \mathbb{I} = (A \cup_\phi D) \times \mathbb{I}$ as the adjunction space $B \cup_\psi E = (A \times \mathbb{I}) \cup_{\phi \times 1_\mathbb{I}} (D \times \mathbb{I})$ in Corollary 2.6. The two homotopies

$$1_{A \times \mathbb{I}} : A \times \mathbb{I} \to A \times \mathbb{I} \quad \text{and} \quad r : (D \times \mathbb{I}, S \times \mathbb{I}) \to (D \times \mathbb{I}, S \times \mathbb{I})$$

satisfy $1_{A \times \mathbb{I}} \circ (\phi \times 1_\mathbb{I}) = (\phi \times 1_\mathbb{I}) \circ r|_{S \times \mathbb{I}}$, and so induce the desired retraction:

$$R = \ll 1_{A \times \mathbb{I}}, r \gg \, : (A \cup_\phi D) \times \mathbb{I} \to (A \cup_\phi D) \times \mathbb{I}. \qquad \square$$

Example 12 In $\mathbb{D}^3$, the union $\mathbb{S}^2 \cup \mathbb{D}^2$ of the boundary 2-sphere and the equatorial 2-ball is the adjunction space of $\mathbb{S}^2 \supset \mathbb{S}^1$ to $\mathbb{D}^2$ by the inclusion $\phi : \mathbb{S}^1 \subset \mathbb{D}^2$. As the inclusion $i : \mathbb{S}^1 \subset \mathbb{S}^2$ is a cofibration by Example 10, the inclusion $j : \mathbb{D}^2 \subset \mathbb{S}^2 \cup \mathbb{D}^2$ is a cofibration by the Induced Cofibration Theorem 4.3.

5 Homotopy Equivalences

This section presents techniques for recognizing and constructing homotopy equivalences.

DEFORMATION RETRACTIONS

Inclusions and retractions are often homotopy equivalences. Here is the specialized vocabulary devised for such situations.

Let $i : A \subseteq X$ be an inclusion of a subspace A into a space X.

A *retraction* of X onto A is a continuous function $r : X \to A$ such that $1_A = r \circ i : A \subseteq X \to A$. When a retraction r exists, A is called a *retract* of X.

A *deformation* of X into A is a homotopy $K : X \times \mathbb{I} \to X$ from $1_X : X \to X$ to $i \circ d : X \to A \subseteq X$, where $d : X \to A$ is any continuous function. When a deformation K exists, X is said to be *deformable* into A.

Example 1 Each slice $\{x\} \times Y$ is a retract of the topological product $X \times Y$. The product $X \times Y$ is deformable into the slice $\{x\} \times Y$ if and only if X is contractible.

A *deformation retraction* of X into A is a homotopy $K : X \times \mathbb{I} \to X$ from $1_X : X \to X$ to $i \circ r : X \to A \subseteq X$, where r is a retraction of X onto A. When a deformation retraction K exists, A is called a *deformation retract* of X.

5.1 Theorem *A subspace $A \subseteq X$ is a deformation retract of X if and only if A is a retract of X and X is deformable into A. The inclusion and retraction for a deformation retract are inverse homotopy equivalences.*

Proof: Let $1_X \simeq i \circ d : X \to X$ be a deformation of X into A, and let $r : X \to A$ be a retraction. Then, $r = r \circ 1_X \simeq r \circ i \circ d = 1_A \circ d = d$; hence, $1_X \simeq i \circ d \simeq i \circ r$ is a deformation retraction of X into A. So, i and r are homotopy inverses. □

Example 2 The $(n-1)$-sphere $\mathbb{S}^{n-1}$ is a deformation retract of the hollowed n-space $\mathbb{R}^n - \mathbb{B}^n$. The straight-line homotopy in $\mathbb{R}^n - \mathbb{B}^n$ between the identity and the radial retraction $\rho : \mathbb{R}^n - \mathbb{B}^n \to \mathbb{S}^{n-1}$, $\rho(x) = x/\|x\|$, is a deformation retraction.

A *strong deformation retraction* of X into A is a deformation retraction relative to A, that is, a homotopy $K : X \times \mathbb{I} \to X$ rel A from $1_X : X \to X$ to $i \circ r : X \to A \subseteq X$, where $r : X \to A$ is a retraction of X onto A. Then A is called a *strong deformation retract* of X.

Example 3 The one-point union $\mathbb{S}^1 \vee \mathbb{S}^1 = (\mathbb{S}^1 \times \{1\}) \cup (\{1\} \times \mathbb{S}^1)$ is a strong deformation retract of the punctured torus $\mathbb{S}^1 \times \mathbb{S}^1 - \{<-1,-1>\}$. By the Homotopy Transgression Theorem (2.4), a strong deformation retraction is induced by one of the punctured square $\mathbb{I} \times \mathbb{I} - \{<\frac{1}{2}, \frac{1}{2}>\}$ onto its boundary. For the latter strong deformation retraction exists and respects the identifications of $e \times e : \mathbb{I} \times \mathbb{I} \to \mathbb{S}^1 \times \mathbb{S}^1$.

Example 4 The limiting tooth $\{0\} \times \mathbb{I}$ in the harmonic comb HC is a deformation retract of HC, but it is not a strong deformation retract of HC (compare Example 3.7).

Example 5 In $\mathbb{R}^3$, the union $B = (\mathbb{R} \times \mathbb{R} \times \{0\}) \cup (\{0\} \times \mathbb{R} \times \mathbb{R})$ is a *book with four pages* that share the spine $\{0\} \times \mathbb{R} \times \{0\}$. The complement, $B - \{O\}$, of the spine point $O \in B$ has the strong deformation retract $(\mathbb{S}^1 \times \{0\}) \cup (\{0\} \times \mathbb{S}^1)$, which is an amalgamated union $\mathbb{S}^1 \cup_{\mathbb{S}^0} \mathbb{S}^1$ of two 1-spheres that share their equator $\mathbb{S}^0$. B isn't homeomorphic to the *two page book* $\mathbb{R}^2$, else we would have $B - \{O\} \cong \mathbb{R}^2 - \{O\}$ and $\mathbb{S}^1 \cup_{\mathbb{S}^0} \mathbb{S}^1 \simeq \mathbb{S}^1$, by Theorem 5.1. As $\pi_1(\mathbb{S}^1) = \mathbb{Z} \neq \pi_1(\mathbb{S}^1 \cup_{\mathbb{S}^0} \mathbb{S}^1)$, the homotopy equivalence is impossible by the homotopy invariance of π_1 (Theorem 10.3.9).

A strong deformation retract can arise as an adjunction space whose constituents include a strong deformation retract, as follows. Given closed subspaces $S \subseteq B \subseteq D$ and a continuous function $\phi : S \to A$, the adjunction space $A \cup_\phi B$ is a closed subspace of the adjunction space $A \cup_\phi D$:

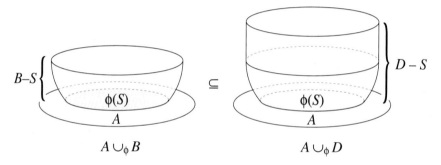

5.2 Deformation Theorem *If $B \subseteq D$ is a strong deformation retract of D, then $A \cup_\phi B$ is a strong deformation retract of $A \cup_\phi D$.*

Proof: Let $K : D \times \mathbb{I} \to D$ rel B be a strong deformation retraction, with $K : 1_D \simeq i \circ r : D \to B \subseteq D$, where $r : D \to B$ is a retraction. Then K and the stationary homotopy $H : A \times \mathbb{I} \to A$ at 1_A satisfy $H \circ (\phi \times 1_\mathbb{I}) = \phi \circ K|_{S \times \mathbb{I}}$. By Corollary 2.6 to the Homotopy Adjunction Theorem (2.5), they induce a homotopy:
$$\ll H, K \gg \, : (A \cup_\phi D) \times \mathbb{I} \to A \cup_\phi D \text{ rel } A \cup_\phi B.$$
This is a homotopy
$$\ll H, K \gg \, : \, \ll 1_A, 1_D \gg \, \simeq \, \ll 1_A, i \circ r \gg \, = \, \ll 1_A, i \gg \circ \ll 1_A, r \gg.$$

Now $\ll 1_A, 1_D \gg$ is the identity on $A \cup_\phi D$; $\ll 1_A, i \gg$ is the inclusion $A \cup_\phi B \subseteq A \cup_\phi D$, and $\ll 1_A, r \gg$ is a retraction $A \cup_\phi D \to A \cup_\phi B$. So the displayed homotopy is a deformation retraction of $A \cup_\phi D$ into $A \cup_\phi B$. □

Example 6 For a continuous function $f : X \to Y$, the *mapping cylinder* M_f is the adjunction space of $X \times \mathbb{I} \supset X \times \{0\}$ to Y by $f : X \times \{0\} \to Y$. Since $X \times \{0\} \subset X \times \mathbb{I}$ is a strong deformation retract of $X \times \mathbb{I}$, then Y is a strong deformation retract of M_f.

FIBER HOMOTOPY EQUIVALENCES

Let the continuous surjection $p : Y \to B$ be a fibration. Each point $b \in B$ is the image of its fiber $p^{-1}(b) \subseteq Y$. The homotopy lifting property of the fibration makes it possible to relate the homotopy types of the fibers and Y.

5.3 Fibration Theorem *Let $p : Y \to B$ be a fibration.*
(a) If p is inessential, Y is deformable into any fiber $F = p^{-1}(b) \subseteq Y$.
(b) If $\{b\} \subseteq B$ is a strong deformation retract of B, then $F = p^{-1}(b) \subseteq Y$ is a strong deformation retract of Y.

Proof: (a) Let $H : Y \times \mathbb{I} \to B$ be a null-homotopy $H : p \simeq c_b$. Since H is a homotopy of $p \circ 1_Y$, it has a lifting to a homotopy $K : Y \times \mathbb{I} \to Y$ of 1_Y. Since $p \circ K \circ i_1 = H \circ i_1 = c_b$, then $K \circ i_1 : Y \to p^{-1}(b) \subseteq Y$. So K is a deformation of Y into the fiber $F = p^{-1}(b)$.
(b) Let $H : B \times \mathbb{I} \to B$ rel $\{b\}$ be a strong deformation of B into $\{b\}$. Since $H \circ (p \times 1_\mathbb{I}) : Y \times \mathbb{I} \to B$ is a homotopy relative to $F = p^{-1}(b)$ of $p \circ 1_Y$, it has a lifting to a homotopy $K : Y \times \mathbb{I} \to Y$ rel F of 1_Y. This K is a strong deformation retraction of Y into the fiber $F = p^{-1}(b)$. □

> **Example 7** *The Hopf fibration $p : \mathbb{S}^3 \to \mathbb{S}^2$ (Example 4.4) is essential.* Otherwise, $\mathbb{S}^3$ would be deformable into the fiber $\mathbb{S}^1$, by Theorem 5.3(a). Then, $1_{\mathbb{S}^3}$ would be homotopic to a non-surjective function, hence, would be inessential by Corollary 2.3. This is contrary to the non-contractibility of $\mathbb{S}^3$, a fact that we recorded but didn't prove in Example 3.5.

5.4 Corollary *Let $p : Y \to B$ be a fibration. Fibers over points in the same path component of B are homotopy equivalent.*

Proof: Let $g : \mathbb{I} \to B$ be a path joining b_0 to b_1. The fibration $p_g : P_g \to \mathbb{I}$ induced from p by g (Theorem 4.1) has fibers $p_g^{-1}(i) = p^{-1}(b_i) = F_i$ ($i = 0, 1$). Because each endpoint $\{i\} \subset \mathbb{I}$ is a strong deformation retract of $\mathbb{I}$, each fiber F_i is a strong deformation retract of the induced fibration P_g, by Theorem 5.3(b). Thus $F_0 \simeq P_g \simeq F_1$. □

QUOTIENT HOMOTOPY EQUIVALENCES

To what extent is a contractible subspace $A \subseteq X$ negligible in analyzing the homotopy type of X? The following example and theorem show that the answer depends critically upon the way the subspace A is situated in X.

> **Example 8** By Examples 3.6 and 3.7, the harmonic comb HC is a contractible subspace of the non-contractible double harmonic comb DHC. But DHC is not homotopy equivalent to its quotient space $DHC/HC \cong HC$, since the latter is contractible. According to the following theorem, the failure of $DHC \to DHC/HC$ to be a homotopy equivalence implies that $HC \subseteq DHC$ is not a cofibration.

5.5 Cofibration Theorem *Let the inclusion $i : A \subseteq X$ be a cofibration. If $A \subseteq X$ is a contractible subspace of X, then the quotient map $q : X \to X/A$ collapsing A to a point is a homotopy equivalence.*

Proof: If $H : A \times \mathbb{I} \to A$ is a contraction, then $i \circ H : A \times \mathbb{I} \to A \subseteq X$ defines a partial homotopy of the identity function $1_X : X \times \{0\} \to X$. Since the inclusion $i : A \subseteq X$ is a cofibration, this partial homotopy extends to a full homotopy $K : X \times \mathbb{I} \to X$ of $1_X : X \to X$.

Since $K|_{X \times \{1\}} = i \circ H|_{X \times \{1\}}$ is constant on $A \times \{1\}$, then, by the Transgression Theorem **5.1.9**, $K|_{X \times \{1\}}$ factors as $k \circ q : X \to X/A \to X$, for some continuous function $k : X/A \to X$. Therefore, we have

$$k \circ q = K|_{X \times \{1\}} \simeq K|_{X \times \{0\}} = 1_X.$$

Since $K : X \times \mathbb{I} \to X$ extends $i \circ H : A \times \mathbb{I} \to A \subseteq X$, then $q \circ K : X \times \mathbb{I} \to X/A$ is constant on $A \times \mathbb{I}$. By the Homotopy Transgression Theorem (2.5), the homotopy $q \circ K$ factors as

$$q \circ K = G \circ (q \times 1_{\mathbb{I}}) : X \times \mathbb{I} \to (X/A) \times \mathbb{I} \to X/A,$$

for some homotopy $G : (X/A) \times \mathbb{I} \to X/A$. Then $G : 1_{X/A} \simeq q \circ k$, since

$$G|_{(X/A) \times \{0\}} \circ (q \times \{0\}) = q \circ K|_{X \times \{0\}} = q \circ 1_X = 1_{X/A} \circ q$$

and

$$G|_{(X/A) \times \{1\}} \circ (q \times \{1\}) = q \circ K|_{X \times \{1\}} = q \circ k \circ q.$$

This proves that $q : X \to X/A$ and $k : X/A \to X$ are homotopy inverses. □

Example 9 By the Cofibration Theorem and Example 4.10, the subspace $\mathbb{S}^2 \cup \mathbb{D}^2$ of $\mathbb{R}^3$ is homotopy equivalent to its quotient space $(\mathbb{S}^2 \cup \mathbb{D}^2)/\mathbb{D}^2 \cong \mathbb{S}^2/\mathbb{S}^1 \cong \mathbb{S}^2 \vee \mathbb{S}^2$.

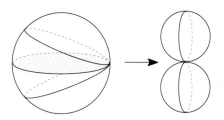

Example 9: $\mathbb{S}^2 \cup \mathbb{D}^2 \simeq \mathbb{S}^2 \vee \mathbb{S}^2$

When A is a closed subspace of X, the cone CA over A is a closed subspace of the cone CX over X. Then the subspace $CA \cup X \subseteq CX$ can be viewed as the adjunction space of $X \supseteq A$ to CA by the inclusion $\phi : A \subset CA$.

The quotient space X/A is a topological method of trivializing the subspace A; the adjunction space $CA \cup X$ is a homotopical method of trivializing A. The two methods are homotopically equivalent when $A \subseteq X$ is a cofibration, according to the Cofibration Theorem (5.5):

5.6 Corollary *When the inclusion $i : A \subseteq X$ is a cofibration, there is a quotient homotopy equivalence*

$$q : CA \cup X \longrightarrow (CA \cup X)/CA \cong X/A.$$

Proof: Now $CA \subset CA \cup X$ is the cofibration coinduced from the cofibration $i : A \subseteq X$ by $\phi : A \subset CA$, as in Theorem 4.3. Since CA is contractible, then $q : CA \cup X \longrightarrow (CA \cup X)/CA$ is a homotopy equivalence by (5.5). □

Example 10 Because the inclusion $\mathbb{S}^0 \subset \mathbb{S}^1$ is a cofibration, the theta curve $\theta = C\mathbb{S}^0 \cup \mathbb{S}^1$ is homotopically equivalent to the figure-of-eight curve $\infty = \mathbb{S}^1/\mathbb{S}^0$.

ATTACHED CELLS

Let $\phi : \mathbb{S}^{n-1} \longrightarrow A$ be a continuous function. To attach the *n*-ball $\mathbb{D}^n \supset \mathbb{S}^{n-1}$ to A by ϕ, we form the adjunction space $X = A \cup_\phi \mathbb{D}^n$, as in Section 5.4. The space X is the topological disjoint union $A \mathbin{\mathring{\cup}} \mathbb{D}^n$, modulo the identification of each point $x \in \mathbb{S}^{n-1} \subset \mathbb{D}^n$ with its image $\phi(x) \in A$. By Theorem 5.4.5, the identification map $p : A \mathbin{\mathring{\cup}} \mathbb{D}^n \longrightarrow X$ embeds A as a closed subspace of X and it embeds the open *n*-ball $\mathbb{B}^n = \mathbb{D}^n - \mathbb{S}^{n-1}$ as the open subspace $X - A$.

The open *n*-ball $X - A \cong \mathbb{B}^n$ is denoted by c^n and is called an ***n*-cell** c^n; $\phi : \mathbb{S}^{n-1} \longrightarrow A$ is called the ***attaching map*** for c^n; and X is denoted by $A \cup_\phi c^n$.

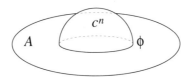

Adjunction space $X = A \cup_\phi c^n$

It is sometimes possible to identify a given space X as an adjunction space $A \cup_\phi c^n$ if one can find a closed subspace $A \subset X$ whose complementary subspace $X - A$ is homeomorphic to the open *n*-ball $\mathbb{B}^n = \mathbb{D}^n - \mathbb{S}^{n-1}$.

Example 11 In the pinched torus $(\mathbb{S}^1 \times \mathbb{S}^1)/(\mathbb{S}^1 \times \{1\})$, the complement of $\{1\} \times \mathbb{S}^1$ is homeomorphic to an open 2-ball $\mathbb{B}^2$ that is adjoined to $\{1\} \times \mathbb{S}^1$ by the inessential map $\phi : \mathbb{S}^1 \longrightarrow \{1\} \times \mathbb{S}^1$ depicted below, right. So, $(\mathbb{S}^1 \times \mathbb{S}^1)/(\mathbb{S}^1 \times \{1\}) \cong \mathbb{S}^1 \cup_\phi c^2$.

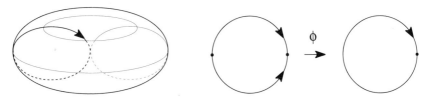

Example 11: Pinched torus   Example 11: Attaching map

The attached cell viewpoint aids analysis of homotopy types, since homotopic attaching maps yield homotopy equivalent adjunction spaces.

To prove this, we use the solid cylinder $\mathbb{D}^n \times \mathbb{I}$ and its boundary cylinder $\mathbb{S}^{n-1} \times \mathbb{I}$. Any homotopy $H : \mathbb{S}^{n-1} \times \mathbb{I} \to A$ determines an adjunction space $A \cup_H (\mathbb{D}^n \times \mathbb{I})$. When $H : \phi \simeq \psi$, the adjunction spaces $A \cup_\phi c^n$ and $A \cup_\psi c^n$ arise as these closed subspaces of $A \cup_H (\mathbb{D}^n \times \mathbb{I})$:

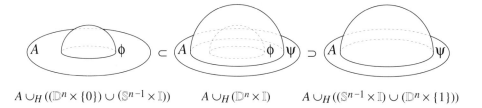

$A \cup_H ((\mathbb{D}^n \times \{0\}) \cup (\mathbb{S}^{n-1} \times \mathbb{I}))$ $A \cup_H (\mathbb{D}^n \times \mathbb{I})$ $A \cup_H ((\mathbb{S}^{n-1} \times \mathbb{I}) \cup (\mathbb{D}^n \times \{1\}))$

5.7 Cell Sliding Theorem *Homotopic attaching maps $\phi, \psi : \mathbb{S}^{n-1} \to A$ produce homotopy equivalent adjunction spaces $A \cup_\phi c^n$ and $A \cup_\psi c^n$.*

Proof: The closed subspaces

$$(\mathbb{D}^n \times \{0\}) \cup (\mathbb{S}^{n-1} \times \mathbb{I}) \subset \mathbb{D}^n \times \mathbb{I} \supset (\mathbb{S}^{n-1} \times \mathbb{I}) \cup (\mathbb{D}^n \times \{1\})$$

are strong deformation retracts of the solid cylinder $\mathbb{D}^n \times \mathbb{I}$. All three contain the closed boundary cylinder $\mathbb{S}^{n-1} \times \mathbb{I}$, by which they can be attached to A via a homotopy $H : \phi \simeq \psi$. By the Deformation Theorem (5.2) and the preliminary observations, those subspaces give strong deformation retracts:

$$A \cup_\phi c^n \subset A \cup_H (\mathbb{D}^n \times \mathbb{I}) \supset A \cup_\psi c^n.$$

Thus, the adjunction spaces $A \cup_\phi c^n$ and $A \cup_\psi c^n$ are homotopy equivalent. □

Example 12 The attaching map $\phi : \mathbb{S}^1 \to \mathbb{S}^1$ for the 2-cell in the pinched torus in Example 11 is inessential. Thus, by the Cell Sliding Theorem (5.7), the pinched torus $\mathbb{S}^1 \cup_\phi c^2$ is homotopy equivalent to the adjunction space $\mathbb{S}^1 \cup_* c^2$ for the constant map $* : \mathbb{S}^1 \to \mathbb{S}^1$. The latter is homeomorphic to the one-point union $\mathbb{S}^1 \vee \mathbb{S}^2$.

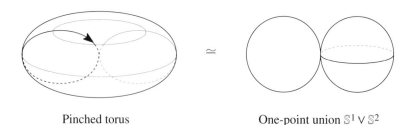

Pinched torus   One-point union $\mathbb{S}^1 \vee \mathbb{S}^2$

Example 13 Any two continuous functions $\phi, \psi : (\mathbb{S}^1, 1) \to (\mathbb{S}^1, 1)$ with the same degree are homotopic (Exercise 2.3); hence, their adjunction spaces $A \cup_\phi c^n$ and $A \cup_\psi c^n$ are homotopy equivalent, by the Cell Sliding Theorem (5.7).

Exercises

SECTION 1

1* Prove that the following hold in any category $\mathcal{C}$:
   (a) The identity morphism $1_A : A \to A$ is unique for each object $A \in \mathcal{C}$.
   (b) If an inverse for a morphism $f : A \to B$ exists, it is unique.
   (c) The inversion rule $(g \circ f)^{-1} = f^{-1} \circ g^{-1}$ holds when f and g are invertible in $\mathcal{C}$.

2 Prove: Equivalence of objects is an equivalence relation on any set of objects in any category $\mathcal{C}$.

3 Let S be any set and view its power set $\mathcal{P}(S)$ as a category whose morphisms are subset inclusions $A \subseteq B$ in S between its members $A, B \in \mathcal{P}(S)$. Thus, each morphism set $(A, B)_{\mathcal{P}(S)}$ is either empty or a singleton. Prove that every indexed family $\{A_\alpha \in \mathcal{P}(S) : \alpha \in \mathcal{A}\}$ has a sum and a product in $\mathcal{P}(S)$.

4 Prove that the direct sum $(\mathbb{A} = \oplus_\alpha \mathbb{A}_\alpha, \{i_\alpha : \mathbb{A}_\alpha \to \mathbb{A}\})$ qualifies as the sum of the family $\{\mathbb{A}_\alpha \in \mathcal{A}b : \alpha \in \mathcal{A}\}$ in the category $\mathcal{A}b$ of (additive) abelian groups.

5 Prove that the direct sum $(\mathbb{A} = \oplus_\alpha \mathbb{A}_\alpha, \{i_\alpha : \mathbb{A}_\alpha \to \mathbb{A}\})$ of any family of non-trivial abelian groups $\{\mathbb{A}_\alpha \in \mathcal{A}b : \alpha \in \mathcal{A}\}$ is not their sum in the category $\mathcal{G}$ of all groups.

6* Prove that the direct product $(\mathbb{G} = \prod_\alpha \mathbb{G}_\alpha, \{p_\alpha : \mathbb{G} \to \mathbb{G}_\alpha\})$ qualifies as the product of the family $\{\mathbb{G}_\alpha \in \mathcal{G} : \alpha \in \mathcal{A}\}$ in the category $\mathcal{G}$ of groups.

7* Let $I : \mathcal{T}_k \to \mathcal{T}_h$ and $K : \mathcal{T}_h \to \mathcal{T}_k$ be the functors in Example 15. Prove:
   (a) $(I(X), Y)_{\mathcal{T}_h} = (X, k(Y))_{\mathcal{T}_k}$ for each ordered pair of objects $X \in \mathcal{T}_k$ and $Y \in \mathcal{T}_h$.
   (b) $K : \mathcal{T}_h \to \mathcal{T}_k$ converts the product $(\prod_\alpha Y_\alpha, \{p_\alpha\})$ of the family $\{Y_\alpha : \alpha \in \mathcal{A}\}$ in $\mathcal{T}_h$ into a product $(k(\prod_\alpha Y_\alpha), \{p_\alpha\})$ for the family $\{k(Y_\alpha) : \alpha \in \mathcal{A}\}$ in $\mathcal{T}_k$.

View a directed set $(\mathcal{D}, \leq)$ as a category $\mathcal{D}$ with each element $a \in \mathcal{D}$ as an object and each relation $a \leq b$ as a unique morphism. Then a covariant functor $T : \mathcal{D} \to \mathcal{C}$ is called a ***direct system*** in $\mathcal{C}$. Define:

$$(T(\mathcal{D}), X)_\mathcal{C} = \{(g_a) \in \prod_a (T(a), X)_\mathcal{C} : g_b \circ T(a \leq b) = g_a, \forall\, a \leq b \text{ in } \mathcal{D}\}.$$

A pair $(L, (u_a))$, where $L \in \mathcal{C}$ and $(u_a) \in (T(\mathcal{D}), X)_\mathcal{C}$, is a ***direct limit*** of the directed system T if the function $(L, X)_\mathcal{C} \to (T(\mathcal{D}), X)_\mathcal{C},\, g \to (g \circ u_a)$, is a bijection $\forall\, X \in \mathcal{C}$.

8 (a) Prove that every direct system in the category $\mathcal{S}$ of sets has a direct limit.
   (b) Prove that every direct system in the category $\mathcal{T}$ of spaces has a direct limit.

View a directed set $(\mathcal{D}, \leq)$ as a category $\mathcal{D}$ with each element $a \in \mathcal{D}$ as an object and each relation $a \leq b$ a unique morphism. Then a contravariant functor $T : \mathcal{D} \to \mathcal{C}$ is called a ***inverse system*** in $\mathcal{C}$. Define:

$$(X, T(\mathcal{D}))_\mathcal{C} = \{(g_a) \in \prod_a (X, T(a))_\mathcal{C} : T(a \leq b) \circ g_b = g_a, \forall\, a \leq b \text{ in } \mathcal{D}\}.$$

A pair $(L, (v_a))$, where $L \in \mathcal{C}$ and $(v_a) \in (X, T(\mathcal{D}))_\mathcal{C}$, is an ***inverse limit*** of the inverse system T if the function $(X, L)_\mathcal{C} \to (X, T(\mathcal{D}))_\mathcal{C},\, g \to (u_a \circ g)$, is a bijection $\forall\, X \in \mathcal{C}$.

9 (a) Investigate inverse limits of inverse systems in the category $\mathcal{S}$ of sets.
   (b) Investigate inverse limits of inverse systems in the category $\mathcal{T}$ of spaces.
   (c) Investigate inverse limits of inverse systems in the category $\mathcal{T}_c$ of compact Hausdorff spaces.

SECTION 2

1. Prove that any two continuous functions into a starlike subspace of $\mathbb{R}^n$ (see p. 89) are homotopic in that subspace.

2. Prove that two continuous functions $f, g : X \to Y$ into a convex subspace $Y \subseteq \mathbb{R}^n$ that agree on a subspace $A \subseteq X$ are homotopic relative to A.

3*. (a) Prove that $f \simeq g : (\mathbb{S}^1, 1) \to (\mathbb{S}^1, 1)$ if and only if $Deg\ f = Deg\ g$.
   (b) Prove that no meridional embedding $m : \mathbb{S}^1 \to \mathbb{S}^1 \times \mathbb{S}^1$ and longitudinal embedding $l : \mathbb{S}^1 \to \mathbb{S}^1 \times \mathbb{S}^1$ are homotopic.

4. Prove or disprove: Inessential continuous functions between any two spaces are homotopic.

5*. Let Y be any space and let $f : \mathbb{S}^n \to Y$ be any continuous function. Show that f is inessential if and only if f admits an extension $F : \mathbb{D}^{n+1} \to Y$.

6. Prove: $f \simeq g : X \to \prod_\alpha Y_\alpha$ if and only if $p_\alpha \circ f \simeq p_\alpha \circ g : X \to Y_\alpha$ for all $\alpha \in \mathscr{A}$, where $p_\sigma : \prod_\alpha Y_\alpha \to Y_\sigma$ ($\sigma \in \mathscr{A}$) is the projection function.

7. Prove that if $f_\alpha \simeq g_\alpha : X_\alpha \to Y_\alpha$ for all $\alpha \in \mathscr{A}$, then
   (a) $\mathring{\cup}_\alpha f_\alpha \simeq \mathring{\cup}_\alpha g_\alpha : \mathring{\cup}_\alpha X_\alpha \to \mathring{\cup}_\alpha Y_\alpha$, and
   (b) $\prod_\alpha f_\alpha \simeq \prod_\alpha g_\alpha : \prod_\alpha X_\alpha \to \prod_\alpha Y_\alpha$.

A *topological pair* (X, A) consists of a space X and a subspace $A \subseteq X$. A *continuous function* $f : (X, A) \to (Y, B)$ of topological pairs is a continuous function $f : X \to Y$ such that $f(A) \subseteq B$.

Continuous functions $f, g : (X, A) \to (Y, B)$ are **homotopic** provided that there is a continuous function $H : (X \times \mathbb{I}, A \times \mathbb{I}) \to (Y, B)$ such that $H(x, 0) = f(x)$ and $H(x, 1) = g(x)$ for all $x \in X$. The homotopy relation, denoted by $f \simeq g$, partitions the set $(X, A; Y, B)$ of continuous functions $(X, A) \to (Y, B)$ into **homotopy classes** $[f]$. Let $[X, A; Y, B]$ denotes the *set of homotopy classes*.

8*. Let HB be the harmonic broom (Example **6**.1.19). Both the identity function $1_{HB} : HB \to HB$ and the collapsing function $k : HB \to HB$, $k(<x, y>) = <0, y>$, are constant on the limiting straw $S = \{0\} \times \mathbb{I}$. Prove:
   (a) $k, 1_{HB} : (HB, S) \to (HB, S)$ are homotopic as continuous functions of pairs.
   (b) $k, 1_{HB} : HB \to HB$ are not homotopic relative to S.

9*. Fundamental group revisited:
   (a) Construct a bijection $\eta : \pi_1(X, x) = [\mathbb{I}, \{0, 1\}; X, x] \to [\mathbb{S}^1, 1; X, x]$.
   (b) Construct a based continuous function $\mu : (\mathbb{S}^1, 1) \to (\mathbb{S}^1 \vee \mathbb{S}^1, 1)$ such that

$$\eta([g] \cdot [h]) = [<G, H> \circ \mu]$$

for all $[g], [h] \in \pi_1(X, x)$, where $\eta([g]) = [G]$, $\eta([h]) = [H] \in [\mathbb{S}^1, 1; X, x]$.

SECTION 3

1. Prove that the following are contractible spaces:
   (a) the harmonic broom (Example **6**.1.19),
   (b) the harmonic rake (Example **6**.1.19), and
   (c) the space $E \subset \mathbb{I}^2$ of points with even Peano expansion (Exercise **6**.3.8).

2* Prove that when $X \subseteq \mathbb{R}^n$ is a compact subspace, the cone CX is homeomorphic to the *geometric cone* of all line segments in $\mathbb{R}^{n+1}$ joining points $x \in X$ to $<O, 1>$.

3 Consider the *geometric cone* over the interval $X = [0, 1)$, as defined in Exercise 2. Prove that the cone CX is not homeomorphic to the geometric cone under the correspondence of $q(x, t) \in CX$ with $<(1-t)x, t> \in \mathbb{R}^2$.

If continuous functions $f : X \leftrightarrow Y : g$ satisfy $g \circ f \simeq 1_X$, then f is called a *right homotopy inverse* for g and g is called a *left homotopy inverse* for f.

4 Prove that if $f : X \to Y$ has a left homotopy inverse $g : Y \to X$ and a right homotopy inverse $h : Y \to X$, then f has a (two-sided) homotopy inverse.

5 Prove that the following are homotopy equivalences:
  (a) any continuous function homotopic to a homotopy equivalence, and
  (b) any continuous function between contractible spaces.

6 Show that a homotopy equivalence between two spaces provides a bijective correspondence between their path-components and it restricts to homotopy equivalences of corresponding path components.

7 Prove that if $f_\alpha : X_\alpha \to Y_\alpha$ ($\alpha \in \mathcal{A}$) are homotopy equivalences, then so are
$$\mathring{\cup}_\alpha f_\alpha : \mathring{\cup}_\alpha X_\alpha \to \mathring{\cup}_\alpha Y_\alpha \quad \text{and} \quad \Pi_\alpha f_\alpha : \Pi_\alpha X_\alpha \to \Pi_\alpha Y_\alpha.$$

8 Prove that spaces that have homeomorphic *thickenings* are homotopy equivalent, that is, $X \times \mathbb{I} \cong Y \times \mathbb{I} \Rightarrow X \simeq Y$.

9 (a) Are the Sierpinski and indiscrete two-point spaces homotopy equivalent?
  (b) Are the indiscrete and discrete two-point spaces homotopy equivalent?

10 Classify the homeomorphism types of three-point spaces into homotopy types.

11 Construct a homotopy equivalence between these wire-like subspaces of $\mathbb{R}^2$:

 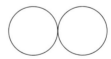

12 Prove that a topological product is contractible if and only all its factor spaces are.

13 (a) Is the doubled harmonic broom (Exercise **6.2.11**) contractible?
   (b) Is the doubled harmonic rake (Example **6.1.19**) contractible?

14 Prove that sums and products exist in the homotopy category $\mathfrak{H}$.

SECTION 4

1 (a) Prove that the composition of two fibrations is a fibration.
  (b) Prove that the composition of two cofibrations is a cofibration.

2* Prove:
  (a) Hopf's map $p : \mathbb{S}^3 \to \mathbb{S}^2$ (Example **5.4.8**) is an $\mathbb{S}^1$-bundle.
  (b) The quaternionic multiplications $m_+ : \mathbb{S}^3 \times \mathbb{D}^4 \to \mathbb{D}^4_+$ and $m_- : \mathbb{D}^4 \times \mathbb{S}^3 \to \mathbb{D}^4_-$ assemble into an $\mathbb{S}^3$-bundle $p : \mathbb{S}^8 \to \mathbb{S}^4$.
  (c) Cayley multiplication determines an $\mathbb{S}^7$-bundle $p : \mathbb{S}^{15} \to \mathbb{S}^8$.

3. Prove that, when $n = 2$, Hopf's $\mathbb{S}^1$-bundle $p : \mathbb{S}^{2n+1} \to \mathbb{P}_n(\mathbb{C})$ in Example 6 can be identified with Hopf's map $p : \mathbb{S}^3 \to \mathbb{S}^2$ in Example **5.4.8**.

4*. Establish the characterization of cofibrations stated in Theorem 4.2.

5. Prove that the inclusion $X \subset CX$ of any space X into its cone is a cofibration.

6. Prove that the inclusion of the origin O into the harmonic comb space HC (see Exercise **3.3.11**) is not a cofibration.

7. Prove that the inclusion of the interval $\{0\} \times [-1, +1]$ into the topologist's closed sine curve (Example **6.1.10**) is not a cofibration.

8. (a) Prove that the standard inclusion $\mathbb{R} \subseteq \mathbb{R}^2$ is a cofibration.
   (b) Is every embedding of $\mathbb{R}$ into $\mathbb{R}^2$ a cofibration?

SECTION 5

1. Prove that the following three spaces are homotopy equivalent:
   (a) the union of the three circles in $\mathbb{R}^2$ of radius 1, centered on $0, 2, 4 \in \mathbb{R}$,
   (b) the triply punctured plane $\mathbb{R}^2 - \{<0, 0>, <2, 0>, <4, 0>\}$, and
   (c) the one-point union $\mathbb{S}^1 \vee \mathbb{S}^1 \vee \mathbb{S}^1$ of three copies of the 1-sphere $\mathbb{S}^1$.

2. (a) Prove that any retraction is an identification map.
   (b) Prove that any retract of a contractible space is contractible.

3. Prove:
   (a) A is a retract of X if and only if, for every space Y, each continuous function $f : A \to Y$ admits an extension $F : X \to Y$.
   (b) Let X be deformable into A. Then, for every space Y, extensions $F, G : X \to Y$ of homotopic continuous functions $f, g : A \to Y$ are homotopic.

4. Let $f : X \to Y$ be a continuous function and let M_f be its mapping cylinder. Prove:
   (a) The inclusion $i : X \equiv X \times \{1\} \subset M_f$ is homotopic to $j \circ f : X \to Y \subset M_f$.
   (b) $f : X \to Y$ factors into the inclusion $i : X \equiv X \times \{1\} \subset M_f$ followed by a strong deformation retracton $r : M_f \to Y$.
   (c) $X \equiv X \times \{1\}$ is a retract of M_f if and only if f has a left homotopy inverse.
   (d) M_f is deformable into X if and only if f has a right homotopy inverse.
   (e) X is a deformation retract of M_f if and only if f is a homotopy equivalence.

5. Prove that any retract $A \subseteq X$ of a convex subspace $X \subseteq \mathbb{R}^n$ is a strong deformation retract of X.

6. Prove that the torus $\mathbb{S}^1 \times \mathbb{S}^1$ is a deformation retract of $(\mathbb{S}^1 \times \mathbb{D}^2) - (\mathbb{S}^1 \times \{O\})$.

7. Prove: The origin O is a deformation retract of the harmonic broom space B (Example **6.1.19**), but it is not a strong deformation retract of B.

8. Analyze the homotopy type of the complement of the coordinate axes in $\mathbb{R}^3$.

The **mapping torus** T_f of a continuous function $f : X \to X$ is the quotient space obtained from $X \times \mathbb{I}$ by identifying $<x, 1>$ with $<f(x), 0>$ for all $x \in X$.

9. Prove these features of the mapping torus construction:
   (a) If $f \simeq g : X \to X$, then $T_f \simeq T_g$.
   (b) If $r : X \to A \subseteq X$ is a retraction, then $A \times \mathbb{S}^1$ is a deformation retract of T_r.

10 (a) Prove that the quotient map $q : W \to \mathbb{S}^1 \vee \mathbb{S}^1$ collapsing the connecting interval in the wire handcuff W is a homotopy equivalence:

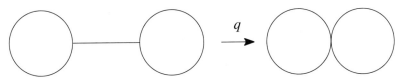

   (b) Construct an explicit homotopy inverse $g : \mathbb{S}^1 \vee \mathbb{S}^1 \to W$ for $q : W \to \mathbb{S}^1 \vee \mathbb{S}^1$.

11 Prove: When the attaching map $\phi : \mathbb{S}^{n-1} \to A$ is constant, the adjunction space $A \cup_\phi \mathbb{D}^n$ is homeomorphic to a one-point union $A \vee \mathbb{S}^n$.

12 (a) Prove that the quotient map $q : \mathbb{R}^n \to \mathbb{R}^n / \mathbb{D}^n$ is a homotopy equivalence.
   (b) Is the quotient map $q : \mathbb{R}^n \to \mathbb{R}^n / \mathbb{B}^n$ a homotopy equivalence?

13 Prove that the dunce hat (Example 5.4.5) is contractible.

14 Prove that the torus-with-membrane subspace $(\mathbb{S}^1 \times \mathbb{S}^1) \cup (\{1\} \times \mathbb{D}^2) \subset \mathbb{S}^1 \times \mathbb{D}^2$ is homotopy equivalent to the one-point union $\mathbb{S}^1 \vee \mathbb{S}^2$.

Let $f : X \to Y$ be a continuous function. The ***mapping cone*** C_f is the adjunction space $Y \cup_f CX$ of $CX \supset X$ to Y via $f : X \to Y$. Let $H : X \times \mathbb{I} \to Y$ be a homotopy. The ***mapping wedge*** W_H is the adjunction space $Y \cup_H (CX \times \mathbb{I})$ of $CX \times \mathbb{I} \supset X \times \mathbb{I}$ to Y via $H : X \times \mathbb{I} \to Y$.

15 Prove these features of the mapping cone construction:
   (a) The identification map $Y \mathring{\cup} CX \to C_f$ embeds the space Y as a closed subspace of C_f, and it embeds the *open cone* $CX - X$ as the open subspace $C_f - Y$.
   (b) If $H : f \simeq g$, then C_f and C_g are strong deformation retracts of W_H, so $C_f \simeq C_g$.

16 Let A be a subspace comprised of n distinct points of the 2-sphere $X = \mathbb{S}^2$. Prove that the quotient space X/A obtained from X by collapsing A to a single point is homotopy equivalent to a one-point union of $\mathbb{S}^2$ and $n-1$ copies of the 1-sphere $\mathbb{S}^1$.

17 Prove:
   (a) Any five of the six faces of a 3-cube $\mathbb{I}^3$ form a strong deformation retract of $\mathbb{I}^3$.
   (b) Prove that the ***house with two rooms*** (below, right) is a strong deformation retract of $\mathbb{I}^3$. The house is constructed with floors, ceilings, panels, and four complete walls so that the lower room is accessible only from the roof and the upper room is accessible only from the basement.

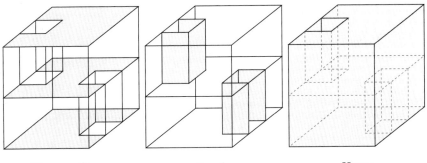

Floors-ceilings　　　　　　Panels　　　　　　House

11

GROUP THEORY

The topological and homotopical invariance of the fundamental group of a space make it an important tool in topology. Calculations of fundamental groups are made in the next chapter using the Seifert-Van Kampen Theorem. In practice, this theorem expresses the fundamental group in combinatorial terms, giving generators and relators for it. This chapter develops the combinatorial approach to groups and shows how to distinguish some groups that are presented in these terms.

1 Overview

This overview is adequate preparation for Chapter **12**. You may choose to skip the remainder of this chapter until you feel the need for its results.

GROUP AXIOMS

Let G be any set. A binary operation on G is a function $\cdot : G \times G \to G$. It assigns to each ordered pair $<g, h>$ of elements of G an element $g \cdot h$ of the same set G. The element $g \cdot h$ is called the ***product*** of the elements $g, h \in G$.

A ***group*** $(G, \cdot)$ is a set G equipped with a binary operation $\cdot : G \times G \to G$ that satisfies these three axioms:

- $\forall\, g, h, k \in G,\ g \cdot (h \cdot k) = (g \cdot h) \cdot k.$ (***associativity***)
- $\exists\, 1 \in G$ such that $g \cdot 1 = g = 1 \cdot g,\ \forall\, g \in G.$ (***existence of identity***)
- $\forall\, g \in G, \exists\, h \in G$ such that $g \cdot h = 1 = h \cdot g.$ (***existence of inverses***)

The simplest example of an infinite group is the group of integers,

$$\mathbb{Z} = \{\,\ldots, -n, \ldots, -1, 0, 1, \ldots, n, \ldots\,\},$$

under addition $+$. A multiplicative version of $\mathbb{Z}$ is the *infinite cyclic group*

SECTION 1 OVERVIEW 329

$\langle x \rangle$ generated by a *letter* x; $\langle x \rangle$ consists of the integral powers of x:

$$\ldots, x^{-n}, \ldots, x^{-1}, x^0 = 1, x^1 = x, \ldots, x^n, \ldots.$$

Multiplication in $\langle x \rangle$ is given by the usual exponential law: $x^k x^m = x^{k+m}$. With this law, the group axioms in $\mathbb{Z}$ provide the group axioms in $\langle x \rangle$:

$$x^k(x^m x^n) = (x^k x^m)x^n \qquad x^k 1 = x^k = 1 \, x^k \qquad x^k x^{-k} = 1 = x^{-k} x^k$$

$$\text{Associativity} \qquad \text{Existence of identity} \qquad \text{Existence of inverse}$$

The basic example of a finite group is the group of integers modulo n,

$$\mathbb{Z}_n = \{0, 1, \ldots, n-1\},$$

with the operation of addition modulo n. A multiplicative version of $\mathbb{Z}_n$ is the *cyclic group of order* n, $\langle x : x^n \rangle$, generated by a *letter* x whose powers are subject to just the group axioms and the relation $x^n = 1$. This group $\langle x : x^n \rangle$ consists of these powers of x:

$$x^0 = 1, x^1 = x, \ldots, x^{n-1}.$$

Multiplication in $\langle x : x^n \rangle$ is given by addition modulo n of the integral exponents. The relation $x^n = 1$ in the finite cyclic group $\langle x : x^n \rangle$ fails to hold in the infinite cyclic group $\langle x \rangle$. So it is not a consequence of the group axioms.

FREE GROUPS

There is a group $\langle x, z \rangle$ generated by two *letters* x and z whose products are subject to just the group axioms and the exponential law for powers. This group $\langle x, z \rangle$ consists of all *reduced words* w in x and z, i.e., expressions

$$w = x^{k_1} z^{m_1} x^{k_2} z^{m_2} \ldots x^{k_n} z^{m_n}$$

with non-zero integral exponents, save possibly $k_1 = 0$ or $m_n = 0$. The empty word $\varnothing$ is allowed and it is denoted by 1. The product wv of two reduced words w and v in $\langle x, z \rangle$ is the reduced word obtained by simplifying their concatenation, using just the group axioms and the usual exponential law for powers. For example, $(zx^2 zx^{-1})(xzxz) = zx^2 z^2 xz$.

Because the products in the groups $\langle x \rangle$ and $\langle x, z \rangle$ are evaluated using just the group axioms and the exponential law, $\langle x \rangle$ and $\langle x, z \rangle$ are called *free groups* with *free bases* $\{x\}$ and $\{x, z\}$, respectively.

For any set $X = \{x_\alpha : \alpha \in \mathcal{A}\}$, there is a *free group* $\langle X \rangle$ with *free basis* X; $\langle X \rangle$ consists of *reduced words* in the *alphabet* X. They are expressions

$$x_{\alpha_1}^{k_1} x_{\alpha_2}^{k_2} \ldots x_{\alpha_n}^{k_n}$$

of length $n \geq 0$, in non-zero powers of the x_α, with indices $\alpha_i \neq \alpha_{i+1}$ in $\mathcal{A}$. Two reduced words in $\langle X \rangle$ are equal if and only if they have the same length, the same indices, and the same exponents. The multiplication in $\langle X \rangle$ is concatenation, followed by reduction using just the group axioms and the exponential law. The empty word $\varnothing$ is denoted by 1 and serves as the identity in the group $\langle X \rangle$; inverse elements in the group $\langle X \rangle$ have this form:

$$x_{\alpha_1}^{k_1} x_{\alpha_2}^{k_2} \ldots x_{\alpha_n}^{k_n} \quad \text{and} \quad x_{\alpha_n}^{-k_n} \ldots x_{\alpha_2}^{-k_2} x_{\alpha_1}^{-k_1}.$$

GROUPS WITH RELATIONS

As the finite cyclic group $\langle x : x^n \rangle$ illustrates, a given group $\mathbb{G}$ need not be free. It is always possible to select a *set of generators* $\{g_\alpha \in \mathbb{G} : \alpha \in \mathcal{A}\}$ for $\mathbb{G}$, i.e., elements whose products $g_{\alpha_1}^{k_1} g_{\alpha_2}^{k_2} \ldots g_{\alpha_n}^{k_n}$ constitute $\mathbb{G}$. But typically, some products of the generators are equal in $\mathbb{G}$ to the identity element 1, even though they do not reduce to 1 using just the group axioms and the usual exponential law for powers. Any such equality in $\mathbb{G}$,

$$g_{\alpha_1}^{k_1} g_{\alpha_2}^{k_2} \ldots g_{\alpha_n}^{k_n} = 1,$$

is called a *relation* among the generators $\{g_\alpha : \alpha \in \mathcal{A}\}$ of $\mathbb{G}$.

There is no restriction on the relations that are possible in groups.

Given any set of generators, $X = \{x_\alpha : \alpha \in \mathcal{A}\}$, and any set of reduced words, $R = \{r_\omega(x) = x_{\alpha_1}^{k_1} x_{\alpha_2}^{k_2} \ldots x_{\alpha_n}^{k_n} : \omega \in \mathcal{W}\}$, in the free group $\langle X \rangle$, we can form the smallest equivalence relation $\sim$ on $\langle X \rangle$ satisfying these properties:

$$r_\omega(x) \sim 1 \qquad (\omega \in \mathcal{W})$$

and

$$w \sim w' \text{ and } v \sim v' \text{ imply } wv \sim w'v' \qquad (w, w', v, v' \in \langle X \rangle).$$

The set $\mathbb{G}$ of equivalence classes, $\underline{w}$, of the reduced words $w \in \langle X \rangle$ under the equivalence relation $\sim$ inherits a group structure from $\langle X \rangle$ in the obvious manner; namely, the product is given by $\underline{u}\,\underline{v} = \underline{u\,v}$. As $r_\omega(x) \sim 1$, we have $\underline{r}_\omega(x) = \underline{1}$ in $\mathbb{G}$ for all $\omega \in \mathcal{W}$.

This group $\mathbb{G}$ is denoted by $\langle X : R \rangle$, where $R = \{r_\omega(x) \in \langle X \rangle : \omega \in \mathcal{W}\}$ is called a set of *relators* for $\mathbb{G}$. $\mathbb{G}$ has the generators $\{g_\alpha = \underline{x}_\alpha : \alpha \in \mathcal{A}\}$ whose products are subject only to the group axioms and to the prescribed relations:

$$r_\omega(g) = g_{\alpha_1}^{k_1} g_{\alpha_2}^{k_2} \ldots g_{\alpha_n}^{k_n} = \underline{x}_{\alpha_1}^{k_1} \underline{x}_{\alpha_2}^{k_2} \ldots \underline{x}_{\alpha_n}^{k_n} = \underline{r}_\omega(x) = \underline{1} \qquad (\omega \in \mathcal{W}).$$

An example is the group $\langle x, z : xzx^{-1}z^{-1} \rangle$ with the relation $\underline{x}\,\underline{z}\,\underline{x}^{-1}\underline{z}^{-1} = \underline{1}$. Here a commutativity law holds: $\underline{z}\,\underline{x} = \underline{1}\,\underline{z}\,\underline{x} = (\underline{x}\,\underline{z}\,\underline{x}^{-1}\,\underline{z}^{-1})\underline{z}\,\underline{x} = \underline{x}\,\underline{z}$.

In general, it is extremely difficult to fully analyze the structure of a group presented by generators and relators. But the Seifert-Van Kampen Theorem (**12.1.3**) provides just such a description for the fundamental group of various topological spaces. For this reason, we are forced to consider such combinatorial presentations of groups.

For example, the wire eight space, or one-point union $\mathbb{S}_a^1 \vee \mathbb{S}_b^1$ of two copies of the 1-sphere $\mathbb{S}^1$, has free fundamental group $\pi_1(\mathbb{S}_a^1 \vee \mathbb{S}_b^1) = \langle a, b \rangle$, generated by the path-homotopy classes $i = [e_i]$ of the two exponential loops $e_i : \mathbb{I} \to \mathbb{S}_a^1 \vee \mathbb{S}_b^1$ ($i = a, b$). The torus $\mathbb{S}_a^1 \times \mathbb{S}_b^1$ has fundamental group

$$\pi_1(\mathbb{S}_a^1 \times \mathbb{S}_b^1) = \langle a, b : aba^{-1}b^{-1} \rangle,$$

generated by the path-homotopy classes of the two exponential loops on the subspace $\mathbb{S}_a^1 \vee \mathbb{S}_b^1 \subset \mathbb{S}_a^1 \times \mathbb{S}_b^1$. The commutativity relation $aba^{-1}b^{-1} = 1$ derives from the identification map $e \times e : \mathbb{I} \times \mathbb{I} \to \mathbb{S}_a^1 \times \mathbb{S}_b^1$.

The remainder of this chapter is devoted to a formal treatment of the presentation of groups by generators and relators, and to the admittedly meager techniques available for distinguishing between presented groups.

2 Groups and Homomorphisms

This section introduces basic concepts in group theory.

GROUPS

As defined in Section 1, a ***group*** $(\mathbb{G}, \cdot)$ is a set $\mathbb{G}$ equipped with a binary operation $\cdot : \mathbb{G} \times \mathbb{G} \to \mathbb{G}$ that satisfies these three axioms: associativity, existence of identity, and existence of inverses.

A group $\mathbb{G}$ is called ***abelian*** (in honor of the Norwegian mathematician Niels Abel (1802-1829)) if it satisfies this extra axiom:

- $\forall\, g, k \in \mathbb{G},\ g \cdot k = k \cdot g.$ \hfill (***commutativity***)

Often, the group operation in an abelian group is called addition and is denoted by +. Then the identity element is called ***zero*** and is denoted by 0, and the inverse of an element g is called its ***negative*** and is denoted by $-g$.

Example 1 Integer addition gives an infinite abelian group $(\mathbb{Z}, +)$, in which 0 is the identity and the inverse of an integer n is its *negative* $-n$.

The number of elements of a finite group $\mathbb{G}$ is denoted by $|\mathbb{G}|$ and is called its ***order***.

Example 2 For $n > 0$, the congruence relation $k \equiv l \pmod{n}$ on $\mathbb{Z}$ means that $k - l$ is a multiple of n; it partitions $\mathbb{Z}$ into n congruence classes,

$$[0]_n, [1]_n, [2]_n, \ldots, [n-1]_n,$$

called ***integers modulo n***. These classes form a group $\mathbb{Z}_n$ under addition modulo n:

$$[k]_n +_n [l]_n = [k+l]_n.$$

The congruence class $[0]_n$ is the identity, and the inverse of a congruence class $[k]_n$ is the class $[n-k]_n$. So $(\mathbb{Z}_n, +_n)$ is an abelian group of order n.

Example 3 For each modulus $n > 0$, there is the multiplicative group,

$$\mathbb{Z}_n^* = \{[k]_n \in \mathbb{Z}_n : \exists\, [m]_n \in \mathbb{Z}_n, [k]_n \cdot [m]_n = [1]_n\},$$

of ***invertible*** integers modulo n. $\mathbb{Z}_n^*$ has order $\varphi(n)$, given by Euler's phi-function.

LIFE IN A GROUP

Because the product operation in a group $\mathbb{G}$ is associative, all products formed from any list of members $g_1, g_2, \ldots, g_n$ by multiplying them in order are equal and may be denoted by $g_1 \cdot g_2 \cdot \ldots \cdot g_n$. The following theorem gives other basic features of life in a group.

2.1 Theorem *Let $(\mathbb{G}, \cdot)$ be a group. Then,*
(a) *the identity element $1 \in \mathbb{G}$ is unique;*
(b) *the inverse $g^{-1} \in \mathbb{G}$ of any member $g \in \mathbb{G}$ is unique; and*
(c) *the inversion rule $(g \cdot h)^{-1} = h^{-1} \cdot g^{-1}$ holds for each pair $g, h \in \mathbb{G}$.*

Proof: (a) Let 1 and e be identity elements in $\mathbb{G}$. Then $e = e \cdot 1 = 1$.
(b) Suppose that both h and k are inverses in $\mathbb{G}$ for the member g. Then, by the associativity and identity axioms, we have

$$h = 1 \cdot h = (k \cdot g) \cdot h = k \cdot (g \cdot h) = k \cdot 1 = k.$$

So the inverse is unique and may safely be denoted by g^{-1}.
(c) The calculations

$$g \cdot h \cdot h^{-1} \cdot g^{-1} = g \cdot 1 \cdot g^{-1} = g \cdot g^{-1} = 1$$

and

$$h^{-1} \cdot g^{-1} \cdot g \cdot h = h^{-1} \cdot 1 \cdot h^{-1} = h \cdot h^{-1} = 1$$

prove that $g \cdot h$ and $h^{-1} \cdot g^{-1}$ are inverse members of $\mathbb{G}$. □

SUBGROUPS

A subset $\mathbb{K} \subseteq \mathbb{G}$ is called a ***subgroup*** of a group $(\mathbb{G}, \cdot)$ if it contains the product $k_1 \cdot k_2$ in $\mathbb{G}$ of each pair of its members $k_1, k_2 \in \mathbb{K}$, as well as the inverse k^{-1} in $\mathbb{G}$ of each member $k \in \mathbb{K}$. Then $(\mathbb{K}, \cdot)$ is a group. The ***subgroup relationship*** is denoted by $\mathbb{K} \leq \mathbb{G}$.

The intersection of any collection of subgroups of a group $\mathbb{G}$ is a subgroup of $\mathbb{G}$. Thus, the intersection of all subgroups of $\mathbb{G}$ that contain a given subset $S \subseteq \mathbb{G}$ is a subgroup of $\mathbb{G}$, denoted by $\langle S \rangle$ and called the ***subgroup generated by*** S. This group $\langle S \rangle$ is the set of all products in $\mathbb{G}$ formed from the members $g \in S$ and their inverses $g^{-1} \in \mathbb{G}$. If $\langle S \rangle = \mathbb{G}$, then S is called a ***set of generators*** of the group $\mathbb{G}$.

Any analysis of the structure of a group begins with a determination of a convenient set of generators, but such a set is usually not unique.

Example 4 $(\mathbb{Z}, +)$ is generated by $S = \{1\}$ and it is also generated by any pair $\{k, m\}$ of relatively prime integers.

$(\mathbb{Z}_n, +_n)$ is generated by $\{[1]_n\}$ and it is also generated by $[k]_n$ whenever k and n are relatively prime.

A determination of a set of generators for $(\mathbb{Z}_n^*, \cdot_n)$ is an interesting problem in number theory. For example, $\mathbb{Z}_5^* = \{[1]_5, [2]_5, [3]_5, [4]_5\}$ is generated by the single invertible class $[2]_5$, since this class has these powers:

$$[2]_5{}^1 = [2]_5, \ [2]_5{}^2 = [4]_5, \ [2]_5{}^3 = [3]_5, \ [2]_5{}^4 = [1]_5.$$

But $\mathbb{Z}_{12}^* = \{[1]_{12}, [5]_{12}, [7]_{12}, [11]_{12}\}$ is generated by no single member.

The subgroup $\langle g \rangle$ generated by an element $g \in \mathbb{G}$ consists of its **powers**:

$$\ldots \ g^{-2} = g^{-1} \cdot g^{-1}, \ g^{-1}, \ g^0 = 1, \ g, \ g^2 = g \cdot g, \ \ldots .$$

If these powers are distinct, then $g \in \mathbb{G}$ is said to have **infinite order**. Otherwise, there is some duplication $g^l = g^k$ for some $l < k$; hence, $g^{k-l} = 1$ for the positive exponent $k - l$. Then the least positive exponent $n > 0$ such that $g^n = 1$ is denoted by $|g|$ and is called the **order** of g. When g has finite order $n = |g|$, the subgroup $\langle g \rangle$ generated by g is $\{1, g, g^2, \ldots, g^{n-1}\}$, in view of the relation $g^n = 1$ and the minimality of n.

When $\mathbb{G} = \langle g \rangle$ for some member $g \in \mathbb{G}$, the group $\mathbb{G}$ is called a **cyclic group**; its order $|\mathbb{G}|$ is the order $|g|$ of g.

Example 5 There is the cyclic group $(\mathbb{Z}, +)$ of integers, generated by the element 1 of infinite order; there is the cyclic group $(\mathbb{Z}_n, +_n)$ of integers modulo n, generated by the class $1 = [1]_n$ of finite order n.

HOMOMORPHISMS

Let $(\mathbb{G}, \cdot)$ and $(\mathbb{H}, \square)$ be groups. A **homomorphism** $\varphi : (\mathbb{G}, \cdot) \to (\mathbb{H}, \square)$ is a function $\varphi : \mathbb{G} \to \mathbb{H}$ such that $\varphi(g_1 \cdot g_2) = \varphi(g_1) \square \varphi(g_2)$ for all $g_1, g_2 \in \mathbb{G}$.

Example 6 For any group element $h \in \mathbb{H}$, there is a group homomorphism $\varphi_h : (\mathbb{Z}, +) \to (\mathbb{H}, \square)$ defined by $\varphi_h(k) = h^k$ for all $k \in \mathbb{Z}$. The homomorphism property $\varphi_h(k + l) = \varphi_h(k) \square \varphi_h(l)$ is the exponential rule $h^{k+l} = h^k \square h^l$ for powers in $\mathbb{H}$.

Example 7 Multiplication by $k \in \mathbb{Z}$ defines a function $\mu_k : \mathbb{Z}_m \to \mathbb{Z}_n$, $\mu_k([a]_m) = [k \, a]_n$, if and only if $k \, m \equiv 0 \pmod{n}$. When μ_k exists, it is a homomorphism from $(\mathbb{Z}_m, +_m)$ to $(\mathbb{Z}_n, +_n)$. We have the homomorphism property,

$$[k \, (a_1 + a_2)]_n = [k \, a_1]_n + [k \, a_2]_n,$$

by the distributive law for integers.

2.2 Theorem *If $\varphi : (\mathbb{G}, \cdot) \to (\mathbb{H}, \square)$ is a group homomorphism, then*
(a) *$\varphi(1) = 1$, and*
(b) *$\varphi(g)^{-1} = \varphi(g^{-1})$ for all $g \in \mathbb{G}$.*

Proof: (a) Since $1 = 1 \cdot 1$ in $\mathbb{G}$ and φ is a homomorphism, then we have $\varphi(1) = \varphi(1) \square \varphi(1)$ in $\mathbb{H}$. So $1 = \varphi(1)$ by right multiplication in $\mathbb{H}$ by $\varphi(1)^{-1}$.
(b) Since $1 = \varphi(1) = \varphi(g \cdot g^{-1}) = \varphi(g) \square \varphi(g^{-1})$, then $\varphi(g)^{-1} = \varphi(g^{-1})$ by left multiplication in $\mathbb{H}$ by $\varphi(g)^{-1}$. □

The ***composition*** of two group homomorphisms $\varphi : (\mathbb{G}, \cdot) \to (\mathbb{H}, \square)$ and $\gamma : (\mathbb{H}, \square) \to (\mathbb{K}, \times)$ is the group homomorphism $\gamma \circ \varphi : (\mathbb{G}, \cdot) \to (\mathbb{K}, \times)$ defined by $\gamma \circ \varphi(g) = \gamma(\varphi(g))$ for $g \in \mathbb{G}$.

For any group $(\mathbb{H}, \square)$, the identity function $1_\mathbb{H} : (\mathbb{H}, \square) \to (\mathbb{H}, \square)$ is a homomorphism satisfying $\gamma \circ 1_\mathbb{H} = \gamma$ and $1_\mathbb{H} \circ \varphi = \varphi$ for all homomorphisms φ and γ. Also, $\delta \circ (\gamma \circ \varphi) = (\delta \circ \gamma) \circ \varphi$ whenever these composites exist.

Thus, the class of groups and their homomorphisms, with the usual composition operation, constitute a category $\mathcal{G}$, the ***category of groups***.

Inverse homomorphisms $\varphi : (\mathbb{G}, \cdot) \to (\mathbb{H}, \square)$ and $\gamma : (\mathbb{H}, \square) \to (\mathbb{G}, \cdot)$ in $\mathcal{G}$ (i.e., $\gamma \circ \varphi = 1_\mathbb{G}$ and $\varphi \circ \gamma = 1_\mathbb{H}$) are called group ***isomorphisms***.

Equivalent groups $(\mathbb{G}, \cdot)$ and $(\mathbb{H}, \square)$ in $\mathcal{G}$ are called ***isomorphic***; the isomorphism relation is denoted by $\mathbb{G} \approx \mathbb{H}$.

Example 8 Any cyclic group $\mathbb{G} = \langle g \rangle$ of infinite order is isomorphic to the additive group $(\mathbb{Z}, +)$ of integers, via the correspondence $g^k \to k$. Any cyclic group $\mathbb{G} = \langle g \rangle$ of finite order n is isomorphic to the additive group $(\mathbb{Z}_n, +_n)$ of integers modulo n, via the correspondence $g^k \to [k]_n$.

Any surjective homomorphism is called an ***epimorphism***; any injective homomorphism is called a ***monomorphism***.

2.3 Theorem *If $\varphi : (\mathbb{G}, \cdot) \to (\mathbb{H}, \square)$ is a group homomorphism, then*
(a) *the subsets*

$$Ker\ \varphi = \{g \in \mathbb{G} : \varphi(g) = 1 \in \mathbb{H}\} \quad and \quad Im\ \varphi = \{\varphi(g) \in \mathbb{H} : g \in \mathbb{G}\}$$

*are subgroups of $\mathbb{G}$ and $\mathbb{H}$, called the **kernel** and **image of** φ, respectively,*
(b) *φ is an epimorphism if and only if $Im\ \varphi = \mathbb{H}$,*
(c) *φ is a monomorphism if and only if $Ker\ \varphi = \{1\}$, and*
(d) *φ is an isomorphism if and only if $Ker\ \varphi = \{1\}$ and $Im\ \varphi = \mathbb{H}$.*

Proof: (a) If $g_1, g_2 \in Ker\ \varphi$, then $\varphi(g_1 \cdot g_2) = \varphi(g_1) \square \varphi(g_2) = 1 \square 1 = 1$ in $\mathbb{H}$; hence, $g_1 \cdot g_2 \in Ker\ \varphi$. If $g \in Ker\ \varphi$, then $\varphi(g^{-1}) = \varphi(g)^{-1} = 1^{-1} = 1$ in $\mathbb{H}$; hence, $g^{-1} \in Ker\ \varphi$. Thus, $Ker\ \varphi$ is a subgroup of $\mathbb{G}$.

If $h_1 = \varphi(g_1)$, $h_2 = \varphi(g_2) \in Im\ \varphi$, then $h_1 \square h_2 = \varphi(g_1) \square \varphi(g_2) = \varphi(g_1 \cdot g_2)$; hence, $h_1 \cdot h_2 \in Im\ \varphi$. If $h = \varphi(g) \in Im\ \varphi$, then $h^{-1} = \varphi(g)^{-1} = \varphi(g^{-1})$; hence, $h^{-1} \in Im\ \varphi$. Thus, $Im\ \varphi$ is a subgroup of $\mathbb{H}$.

(b) This follows immediately from the definition of $Im\ \varphi$.

(c) Because $\varphi(1) = 1$ in $\mathbb{H}$, then $Ker\ \varphi = \{1\}$ whenever φ is injective. Conversely, if $Ker\ \varphi = \{1\}$, then φ is injective. For $\varphi(g_1) = \varphi(g_2)$ means $\varphi(g_1 \cdot g_2^{-1}) = 1$ in $\mathbb{H}$, and this implies that $g_1 \cdot g_2^{-1} = 1$ so $g_1 = g_2$ in $\mathbb{G}$.

(d) A group homomorphism $\varphi : (\mathbb{G}, \cdot) \to (\mathbb{H}, \square)$ satisfies $Ker\ \varphi = \{1\}$ and $Im\ \varphi = \mathbb{H}$ if and only if φ is bijective (by (b) and (c)), and so is invertible in $\mathcal{G}$ (by the Inversion Theorem **1.4.3**). Because an inverse function to a group homomorphism is necessarily a group homomorphism, (d) holds. □

3 Products and Sums of Groups

This section is devoted to the construction of sums and products in the category $\mathcal{G}$ of groups. We describe the concepts by way of their universal (i.e., characterizing) property discussed in Section **10**.1, and then we make an explicit construction possessing the universal property.

PRODUCT OF GROUPS

Let $\{\mathbb{G}_\alpha : \alpha \in \mathcal{A}\}$ be an indexed family of groups and let $\{p_\alpha : \mathbb{P} \to \mathbb{G}_\alpha\}$ be an indexed family of group homomorphisms with common domain group $\mathbb{P}$.

The group $\mathbb{P}$ is a ***product of the groups*** $\{\mathbb{G}_\alpha : \alpha \in \mathcal{A}\}$, ***with projections*** $\{p_\alpha : \mathbb{P} \to \mathbb{G}_\alpha\}$, if the pair $(\mathbb{P}, \{p_\alpha\})$ satisfies the universal property that has these equivalent formulations:

- *For every group $\mathbb{H}$ and indexed family of group homomorphisms $\{\varphi_\alpha : \mathbb{H} \to \mathbb{G}_\alpha\}$ with domain $\mathbb{H}$, there is a unique homomorphism $\varphi : \mathbb{H} \to \mathbb{P}$ such that $\varphi_\alpha = p_\alpha \circ \varphi : \mathbb{H} \to \mathbb{P} \to \mathbb{G}_\alpha$ for each $\alpha \in \mathcal{A}$.*

- *For every group $\mathbb{H}$, the function $(\mathbb{H}, \mathbb{P})_\mathcal{G} \to \prod_\alpha (\mathbb{H}, \mathbb{G}_\alpha)_\mathcal{G}$, defined by $\varphi \to (p_\alpha \circ \varphi)$, is bijective.*

As in any category, Theorem **10**.1.1 implies that products (that exist) in $\mathcal{G}$ are unique.

DIRECT PRODUCT

The following construction of the *direct product* of groups puts a group structure on their product in $\mathcal{S}$, as a candidate for their product in $\mathcal{G}$.

For any indexed family $\{\mathbb{G}_\alpha : \alpha \in \mathcal{A}\}$ of groups, their set product $\prod_\alpha \mathbb{G}_\alpha$ is a group under coordinatewise multiplication: if $x = (x_\alpha)$ and $z = (z_\alpha)$ in $\prod_\alpha \mathbb{G}_\alpha$, then $x \cdot z = (x_\alpha \cdot z_\alpha)$ in $\prod_\alpha \mathbb{G}_\alpha$. $\prod_\alpha \mathbb{G}_\alpha$ has identity element (1_α), where 1_α is the identity element of $\mathbb{G}_\alpha$ for each $\alpha \in \mathcal{A}$, and any element $x = (x_\alpha)$ in $\prod_\alpha \mathbb{G}_\alpha$ has inverse $x^{-1} = (x_\alpha^{-1})$, where x_α^{-1} is the inverse of x_α in $\mathbb{G}_\alpha$ for each $\alpha \in \mathcal{A}$. $\prod_\alpha \mathbb{G}_\alpha$ is called the ***direct product*** of the family $\{\mathbb{G}_\alpha : \alpha \in \mathcal{A}\}$.

The direct product of the finite indexed family $\{\mathbb{G}_i : 1 \leq i \leq k\}$ is the set $\mathbb{G}_1 \times \ldots \times \mathbb{G}_k$ of k-tuples $x = (x_1, \ldots, x_k)$, with coordinatewise multiplication:

$$(x_1, \ldots, x_k) \cdot (z_1, \ldots, z_k) = (x_1 \cdot z_1, \ldots, x_k \cdot z_k).$$

The identity element of $\mathbb{G}_1 \times \ldots \times \mathbb{G}_k$ is the k-tuple $1 = (1_1, \ldots, 1_k)$; the inverse of $x = (x_1, \ldots, x_k)$ in $\mathbb{G}_1 \times \ldots \times \mathbb{G}_k$ is $x^{-1} = (x_1^{-1}, \ldots, x_k^{-1})$.

Example 1 The product $\mathbb{Z}_2 \times \mathbb{Z}_2$ of two copies of $(\mathbb{Z}_2, +_2)$ consists of the four elements $<0, 0>, <1, 0>, <0, 1>$, and $<1, 1>$, with coordinatewise addition modulo 2. $\mathbb{Z}_2 \times \mathbb{Z}_2$ is isomorphic to $\mathbb{Z}_{12}^*$ (Example 2.3) under these correspondences:

$<0, 0> \leftrightarrow [1]_{12}$ $<1, 0> \leftrightarrow [5]_{12}$ $<0, 1> \leftrightarrow [7]_{12}$ $<1, 1> \leftrightarrow [11]_{12}$.

We state without proof the following:

3.1 Fundamental Theorem *Any finitely generated abelian group* $\mathbb{G}$ *is isomorphic to a finite direct product*

$$\mathbb{Z} \times \ldots \times \mathbb{Z} \times \mathbb{Z}_{m_1} \times \ldots \times \mathbb{Z}_{m_k}$$

of cyclic groups, where the modulii m_i $(1 \leq i \leq k)$ *are greater than 1 and each one divides the next. Such a factorization of* $\mathbb{G}$ *is unique; the number of infinite cyclic factors is called the free-abelian* **rank** *of* $\mathbb{G}$ *and the modulii,*

$$1 < m_1 \mid m_2 \mid \ldots \mid m_k,$$

are called the **torsion coefficients** *of* $\mathbb{G}$.

Example 2 The group $\mathbb{G} = \mathbb{Z}_2 \times \mathbb{Z}_3 \approx \mathbb{Z}_6$ has torsion coefficient 6. The group $\mathbb{G} = \mathbb{Z}_2 \times \mathbb{Z}_2$ has torsion coefficients $\{2, 2\}$, and so is not isomorphic to $\mathbb{Z}_4$.

EXISTENCE OF PRODUCTS

We now show that the direct product of groups serves as their product in $\mathcal{G}$.

3.2 Lemma *Consider the direct product* $\mathbb{P} = \prod_\alpha \mathbb{G}_\alpha$. *The projection functions* $p_\alpha : \mathbb{P} \to \mathbb{G}_\alpha$, $p_\alpha(x) = x_\alpha$, *are homomorphisms.*

Proof: The α-coordinate of the product $x \cdot z$ in $\mathbb{P} = \prod_\alpha \mathbb{G}_\alpha$ is the product $x_\alpha \cdot z_\alpha$ in $\mathbb{G}_\alpha$ of the α-coordinates of x and z. Thus, for each $\alpha \in \mathcal{A}$, we have

$$p_\alpha(x \cdot z) = x_\alpha \cdot z_\alpha = p_\alpha(x) \cdot p_\alpha(z). \qquad \square$$

3.3 Existence of Products *The direct product* $\mathbb{P} = \prod_\alpha \mathbb{G}_\alpha$ *is a product of the groups* $\{\mathbb{G}_\alpha : \alpha \in \mathcal{A}\}$, *with projections* $\{p_\alpha : \mathbb{P} \to \mathbb{G}_\alpha\}$.

Proof: For any group $\mathbb{H}$ and indexed family of group homomorphisms $\{\varphi_\alpha : \mathbb{H} \to \mathbb{G}_\alpha\}$, the unique set function $\varphi : \mathbb{H} \to \mathbb{P}$ such that $p_\alpha \circ \varphi = \varphi_\alpha$ for each $\alpha \in \mathcal{A}$ (i.e., $\varphi(h) = (\varphi_\alpha(h))$) is a group homomorphism. For we have

$$\varphi(h \cdot k) = (\varphi_\alpha(h \cdot k)) = (\varphi_\alpha(h) \cdot \varphi_\alpha(k)) = (\varphi_\alpha(h)) \cdot (\varphi_\alpha(k)) = \varphi(h) \cdot \varphi(k),$$

for all $h, k \in \mathbb{H}$, by the homomorphism property for each φ_α ($\alpha \in \mathcal{A}$). $\qquad \square$

SUM OF GROUPS

Let $\{\mathbb{G}_\alpha : \alpha \in \mathcal{A}\}$ be an indexed family of groups and let $\{\psi_\alpha : \mathbb{G}_\alpha \to \mathbb{S}\}$ be an indexed family of group homomorphisms with codomain group $\mathbb{S}$.

The group $\mathbb{S}$ is a **sum of the groups** $\{\mathbb{G}_\alpha : \alpha \in \mathcal{A}\}$, **with injections** $\{\psi_\alpha : \mathbb{G}_\alpha \to \mathbb{S}\}$, if the pair $(\mathbb{S}, \{\psi_\alpha\})$ satisfies the universal property that has these equivalent formulations:

- *For every group $\mathbb{H}$ and indexed family of group homomorphisms $\{\lambda_\alpha : \mathbb{G}_\alpha \to \mathbb{H}\}$ with codomain $\mathbb{H}$, there is a unique group homomorphism $\lambda : \mathbb{S} \to \mathbb{H}$ such that $\lambda \circ \psi_\alpha = \lambda_\alpha$ for each $\alpha \in \mathcal{A}$.*
- *For every group $\mathbb{H}$, the function $(\mathbb{S}, \mathbb{H})_{\mathcal{G}} \to \prod_\alpha (\mathbb{G}_\alpha, \mathbb{H})_{\mathcal{G}}$, defined by $\lambda \to (\lambda \circ \psi_\alpha)$, is bijective.*

Theorem 10.1.1 implies that sums (that exist) in $\mathcal{G}$ are unique.

FREE PRODUCT

The universal property for the sum of groups is the group-theoretic analogue of the relationship between a disjoint set union and the individual sets. Although the disjoint union of groups isn't a group, there is a free product construction that builds a sum from the disjoint union.

Consider any indexed family $\{\mathbb{G}_\alpha : \alpha \in \mathcal{A}\}$ of groups. Their disjoint union $\mathring{\cup}_\alpha \mathbb{G}_\alpha$ generates a group, as follows. A *word* in the **alphabet** $\mathring{\cup}_\alpha \mathbb{G}_\alpha$ is a finite sequence $w = (g_1, \ldots, g_n)$, of arbitrary length $n \geq 0$, of elements from the disjoint union $\mathring{\cup}_\alpha \mathbb{G}_\alpha$. The sequence $\emptyset$ of length 0 is called the **empty word**; a word g of length 1 is called a **letter** in the alphabet $\mathring{\cup}_\alpha \mathbb{G}_\alpha$. The set of all words in the alphabet $\mathring{\cup}_\alpha \mathbb{G}_\alpha$ is denoted by $\mathbb{W}(\mathring{\cup}_\alpha \mathbb{G}_\alpha)$.

Juxtaposition or **concatenation** of words,

$$\mathbb{W}(\mathring{\cup}_\alpha \mathbb{G}_\alpha) \times \mathbb{W}(\mathring{\cup}_\alpha \mathbb{G}_\alpha) \to \mathbb{W}(\mathring{\cup}_\alpha \mathbb{G}_\alpha),$$

is defined by

$$(g_1, \ldots, g_n)(h_1, \ldots, h_m) = (g_1, \ldots, g_n, h_1, \ldots, h_m);$$

it is an associative binary operation, with identity $\emptyset$, on the set $\mathbb{W}(\mathring{\cup}_\alpha \mathbb{G}_\alpha)$. But $\mathbb{W}(\mathring{\cup}_\alpha \mathbb{G}_\alpha)$ is not a group since inverses fail to exist.

Two words $w, w' \in \mathbb{W}(\mathring{\cup}_\alpha \mathbb{G}_\alpha)$ are **equivalent**, written $w \sim w'$, if they are related by a sequence of these **elementary replacements**:

(R1) $(g_1, \ldots, g_{i-1}, g_i, g_{i+1}, \ldots, g_n) \leftrightarrow (g_1, \ldots, g_{i-1}, g_{i+1}, \ldots, g_n)$, where $g_i = 1 \in \mathbb{G}_{\alpha_i}$.

(R2) $(g_1, \ldots, g_{i-1}, g_i, g_{i+1}, g_{i+2}, \ldots, g_n) \leftrightarrow (g_1, \ldots, g_{i-1}, g, g_{i+2}, \ldots, g_n)$, where g_i and g_{i+1} belong to $\mathbb{G}_\alpha$ ($\alpha_i = \alpha = \alpha_{i+1}$) and $g = g_i \cdot g_{i+1} \in \mathbb{G}_\alpha$.

Let $*_\alpha \mathbb{G}_\alpha$ be the set $[w]$ of equivalence classes of words $w \in \mathbb{W}(\mathring{\cup}_\alpha \mathbb{G}_\alpha)$ under the relation $\sim$. Concatenation is compatible with the relation $\sim$. So an associative binary operation, with identity $1 = [\emptyset]$, results on $*_\alpha \mathbb{G}_\alpha$. Since

$$(g_1, \ldots, g_n, g_n^{-1}, \ldots, g_1^{-1}) \sim \emptyset \sim (g_n^{-1}, \ldots, g_1^{-1}, g_1, \ldots, g_n),$$

the classes $[g_1, \ldots, g_n]$ and $[g_n^{-1}, \ldots, g_1^{-1}]$ are inverses in $*_\alpha \mathbb{G}_\alpha$. So $*_\alpha \mathbb{G}_\alpha$ is a group, called the ***free product*** of the groups $\{\mathbb{G}_\alpha : \alpha \in \mathcal{A}\}$. For a finite indexed family of groups $\{\mathbb{G}_i : 1 \leq i \leq k\}$, the free product $*_i \mathbb{G}_i$ is usually denoted by $\mathbb{G}_1 * \ldots * \mathbb{G}_k$, with their order of appearance unimportant.

3.4 Normal Form Lemma *Every element of $*_\alpha \mathbb{G}_\alpha$ is represented uniquely by a word $w = (g_1, \ldots, g_n)$ in $\overset{\circ}{\bigcup}_\alpha \mathbb{G}_\alpha$ that is **reduced** in this sense:*
(a) *no entry g_i is the identity element $1 \in \mathbb{G}_{\alpha_i}$, and*
(b) *consecutive entries g_i and g_{i+1} belong to distinct groups $\mathbb{G}_{\alpha_i} \neq \mathbb{G}_{\alpha_{i+1}}$.*

Proof (van der Waerden 1945): Let $\mathbb{R}(\overset{\circ}{\bigcup}_\alpha \mathbb{G}_\alpha)$ denote the set of reduced words. It suffices to show that $\mathbb{R}(\overset{\circ}{\bigcup}_\alpha \mathbb{G}_\alpha) \subset \mathbb{W}(\overset{\circ}{\bigcup}_\alpha \mathbb{G}_\alpha) \to *_\alpha \mathbb{G}_\alpha$ is bijective. Each word w is equivalent to a reduced word, so that this assignment is surjective. Right multiplication in $\mathbb{W}(\overset{\circ}{\bigcup}_\alpha \mathbb{G}_\alpha)$ by $g_\alpha \neq 1_\alpha \in \mathbb{G}_\alpha$ determines a permutation ϕ_{g_α} of the set $\mathbb{R}(\overset{\circ}{\bigcup}_\alpha \mathbb{G}_\alpha)$ of reduced words:

$\phi_{g_\alpha}(g_1, \ldots, g_n) = (g_1, \ldots, g_n, g_\alpha)$, if $\alpha_n \neq \alpha$;

$\phi_{g_\alpha}(g_1, \ldots, g_n) = (g_1, \ldots, g_{n-1})$, if $\alpha_n = \alpha$ and $g_n \cdot g_\alpha = 1_\alpha \in \mathbb{G}_\alpha$; and

$\phi_{g_\alpha}(g_1, \ldots, g_n) = (g_1, \ldots, g_{n-1}, g_n \cdot g_\alpha)$, if $\alpha_n = \alpha$ and $g_n \cdot g_\alpha \neq 1_\alpha \in \mathbb{G}_\alpha$.

For an identity element $1_\alpha \in \mathbb{G}_\alpha$, let ϕ_{1_α} be the identity permutation of $\mathbb{R}(\overset{\circ}{\bigcup}_\alpha \mathbb{G}_\alpha)$. The multiplicative extension $\phi : \mathbb{W}(\overset{\circ}{\bigcup}_\alpha \mathbb{G}_\alpha) \to \text{Perm}(\mathbb{W}(\overset{\circ}{\bigcup}_\alpha \mathbb{G}_\alpha))$ of the assignment ϕ has these features:

- $w \sim w' \Rightarrow \phi_w = \phi_{w'}$;
- $\phi_w(\emptyset) = w_0 \Rightarrow w_0$ *is reduced and* $w \sim w_0$; *and*
- w *reduced* $\Rightarrow \phi_w(\emptyset) = w$.

Thus, equivalent reduced words are equal in $\mathbb{W}(\overset{\circ}{\bigcup}_\alpha \mathbb{G}_\alpha)$ and the restriction of the function $\mathbb{W}(\overset{\circ}{\bigcup}_\alpha \mathbb{G}_\alpha) \to *_\alpha \mathbb{G}_\alpha$ to $\mathbb{R}(\overset{\circ}{\bigcup}_\alpha \mathbb{G}_\alpha) \subset \mathbb{W}(\overset{\circ}{\bigcup}_\alpha \mathbb{G}_\alpha)$ is injective. □

By this lemma, the function $\psi_\sigma : \mathbb{G}_\sigma \to *_\alpha \mathbb{G}_\alpha$, $\psi_\sigma(g) = [g]$, is an injection for each $\sigma \in \mathcal{A}$. Once each letter $g \in \overset{\circ}{\bigcup}_\alpha \mathbb{G}_\alpha$ is identified with its equivalence class $[g] \in *_\alpha \mathbb{G}_\alpha$, then each group element $[w] \in *_\alpha \mathbb{G}_\alpha$ can be represented uniquely as a reduced product $g_1 g_2 \ldots g_n$ of letters $g_i \in \mathbb{G}_{\alpha_i}$ in $*_\alpha \mathbb{G}_\alpha$. Multiplication of two reduced products is achieved by concatenation, followed by reduction via the elementary replacements (**R1**) and (**R2**).

Example 3 The free product of the finite cyclic groups $\mathbb{Z}_3 = \{x^0, x^1, x^2\}$ and $\mathbb{Z}_4 = \{z^0, z^1, z^2, z^3\}$ is the infinite group $\mathbb{Z}_3 * \mathbb{Z}_4$ of reduced words

$$x^{k_1} z^{m_1} x^{k_2} z^{m_2} \ldots x^{k_n} z^{m_n},$$

where $n \geq 0$, $1 \leq k_i \leq 2$ and $1 \leq m_i \leq 3$, except possibly $k_1 = 0$ or $m_n = 0$.

EXISTENCE OF SUMS

We now show that the free product of groups serves as their sum in $\mathcal{G}$.

3.5 Lemma *For the free product $\mathbb{S} = *_\alpha \mathbb{G}_\alpha$, each inclusion function $\psi_\alpha : \mathbb{G}_\alpha \to \mathbb{S}$, $\psi_\alpha(g) = [g]$, is a homomorphism.*

Proof: For $g_1, g_2 \in \mathbb{G}_\alpha$, the word $(g_1 \cdot g_2)$ of length 1 is equivalent to the word (g_1, g_2) of length 2, which equals the product $(g_1)(g_2)$ in $\mathbb{W}(\mathring{\cup}_\alpha \mathbb{G}_\alpha)$. These facts imply the homomorphism property for $\psi_\alpha : \mathbb{G}_\alpha \to \mathbb{S}$, as follows:

$$\psi_\alpha(g_1 \cdot g_2) = [g_1 \cdot g_2] = [g_1][g_2] = \psi_\alpha(g_1)\psi_\alpha(g_2). \qquad \square$$

3.6 Existence Theorem *The free product $\mathbb{S} = *_\alpha \mathbb{G}_\alpha$ is a sum of the groups $\{\mathbb{G}_\alpha : \alpha \in \mathcal{A}\}$, with injections $\{\psi_\alpha : \mathbb{G}_\alpha \to \mathbb{S}\}$.*

Proof: This requires that, for any group $\mathbb{H}$ and indexed family of group homomorphisms $\{\lambda_\alpha : \mathbb{G}_\alpha \to \mathbb{H}\}$, there is a unique group homomorphism $\lambda : \mathbb{S} \to \mathbb{H}$ such that $\lambda \circ \psi_\alpha = \lambda_\alpha$ for each $\alpha \in \mathcal{A}$.

The function $\lambda : \mathbb{S} \to \mathbb{H}$,

$$\lambda([g_1, \ldots, g_n]) = \lambda_{\alpha_1}(g_1) \cdot \ldots \cdot \lambda_{\alpha_n}(g_n),$$

is well-defined, because the displayed element of $\mathbb{H}$ is unaffected by the elementary replacements (**R**1) and (**R**2) of the word $(g_1, \ldots, g_n)$. And λ is a homomorphism by its definition. Since $\lambda([g]) = \lambda_\alpha(g)$ when $g \in \mathbb{G}_\alpha$, the homomorphism λ satisfies the condition $\lambda \circ \psi_\alpha = \lambda_\alpha$, for each $\alpha \in \mathcal{A}$. Since $\mathbb{S} = *_\alpha \mathbb{G}_\alpha$ is generated by these elements $\psi_\alpha(g) = [g]$ ($g \in \mathbb{G}_\alpha, \alpha \in \mathcal{A}$), then λ is the unique homomorphism satisfying the conditions. $\qquad \square$

FREE GROUPS

Let X be any subset of a group $\mathbb{F}$. Then $\mathbb{F}$ is called a *free group* with *free basis* X provided that $\mathbb{F}$ has this universal property:

- *For any group $\mathbb{G}$, any set function $\Lambda : X \to \mathbb{G}$ extends uniquely to a group homomorphism $\lambda : \mathbb{F} \to \mathbb{G}$.*

By Examples 2.6 and 2.8, an infinite cyclic group $\mathbb{G} = \langle g \rangle$ generated by $g \in \mathbb{G}$ is a free group with singleton free basis $\{g\}$. We abbreviate the relationship between a free group $\mathbb{F}$ and a free basis X by writing $\mathbb{F}(X)$.

3.7 Uniqueness of Free Groups *Let $\mathbb{F}$ and $\mathbb{D}$ be free groups with free basis X. Then there exist inverse group isomorphisms $\mathbb{F} \leftrightarrow \mathbb{D}$ that fix X.*

Proof: The inclusions $X \subset \mathbb{D}$ and $X \subset \mathbb{F}$ extend uniquely to homomorphisms $\mathbb{F} \to \mathbb{D}$ and $\mathbb{D} \to \mathbb{F}$, by the universal property for $\mathbb{F}$ and $\mathbb{D}$. Since the composite $\mathbb{F} \to \mathbb{D} \to \mathbb{F}$ extends the inclusion $X \subset \mathbb{F}$, it equals $1_\mathbb{F}$, by the universal property for $\mathbb{F}$. Similarly, the composite $\mathbb{D} \to \mathbb{F} \to \mathbb{D}$ extends the inclusion $X \subset \mathbb{D}$; hence, it equals $1_\mathbb{D}$, by the universal property for $\mathbb{D}$. $\qquad \square$

3.8 Existence of Free Groups *For any set X, there is a free group $\langle X \rangle$ with free basis X.*

Proof: Suppose that $X = \{x_\alpha : \alpha \in \mathcal{A}\}$. Let $\langle X \rangle$ be the free product $*_\alpha \langle x_\alpha \rangle$ of the infinite cyclic groups $\langle x_\alpha \rangle$ for $\alpha \in \mathcal{A}$, and let $\{\psi_\alpha : \langle x_\alpha \rangle \to \langle X \rangle\}$ be the injections. $\langle X \rangle$ consists of reduced words $x_{\alpha_1}^{k_1} x_{\alpha_2}^{k_2} \ldots x_{\alpha_n}^{k_n}$ in powers of the generators x_α. Given any group $\mathbb{G}$ and function $\Lambda : X \to \mathbb{G}$, the family of homomorphisms $\{\lambda_\alpha : \langle x_\alpha \rangle \to \mathbb{G}\}$, uniquely defined by $\lambda_\alpha(x_\alpha) = \Lambda(x_\alpha)$, determines a homomorphism $\lambda : \langle X \rangle \to \mathbb{G}$ such that $\lambda \circ \psi_\alpha = \lambda_\alpha$ for each $\alpha \in \mathcal{A}$, by the universal property for the sum $\langle X \rangle = *_\alpha \langle x_\alpha \rangle$. Thus, λ extends the function $\Lambda : X \to \mathbb{G}$; λ is unique, since $\langle X \rangle$ is generated by X. $\square$

Example 4 By definition, there is a unique homomorphism on a free group extending any group-valued assignments for the members of its free basis. Thus the assignments $a \to c$ and $b \to cd$ define a homomorphism $\lambda : \langle a, b \rangle \to \langle c, d \rangle$; similarly, the assignments $c \to a$ and $d \to a^{-1}b$ define a homomorphism $\psi : \langle c, d \rangle \to \langle a, b \rangle$.

Since $\psi \circ \lambda(a) = \psi(c) = a$ and $\psi \circ \lambda(b) = \psi(cd) = a(a^{-1}b) = b$, then $\psi \circ \lambda = 1_{\langle a, b \rangle}$. Since $\lambda \circ \psi(c) = \lambda(a) = c$ and $\varphi \circ \psi(d) = \varphi(a^{-1}b) = c^{-1}(cd) = d$, then $\varphi \circ \psi = 1_{\langle c, d \rangle}$. In other words, ψ and λ are inverse isomorphisms.

4 Quotient Groups

Certain equivalence relations on a group yield a quotient set that inherits the group structure. This section studies such *quotient groups*.

COSETS

Let $\mathbb{K} \leq \mathbb{G}$, and let $x \in \mathbb{G}$. The function $x \cdot - : \mathbb{G} \to \mathbb{G}$ obtained by left multiplication by x carries the subgroup $\mathbb{K}$ into the subset

$$x \cdot \mathbb{K} = \{x \cdot k \in \mathbb{G} : k \in \mathbb{K}\}.$$

This subset $x \cdot \mathbb{K}$ is called the **left coset** of $\mathbb{K}$ in $\mathbb{G}$ determined by $x \in \mathbb{G}$. (Right multiplication by x determines a **right coset** $\mathbb{K} \cdot x$ of $\mathbb{K}$ in $\mathbb{G}$.)

Since $\mathbb{K}$ is a subgroup, it contains all products of its members and their inverses. Thus, $k \cdot \mathbb{K} = \mathbb{K}$ if and only if $k \in \mathbb{K}$. In fact, $x \cdot \mathbb{K} \cap z \cdot \mathbb{K} \neq \emptyset \Leftrightarrow x \cdot \mathbb{K} = z \cdot \mathbb{K} \Leftrightarrow x^{-1} \cdot z \in \mathbb{K}$, according to the following lemma.

4.1 Coset Partition Lemma Let $\mathbb{K} \leq \mathbb{G}$.
(a) The relation $\equiv_\mathbb{K}$ on $\mathbb{G}$, defined by $x \equiv_\mathbb{K} z$ if and only if $x^{-1} \cdot z \in \mathbb{K}$, is an equivalence relation whose equivalence classes are the left cosets of $\mathbb{K}$.
(b) The distinct left cosets of $\mathbb{K}$ in $\mathbb{G}$ constitute a partition of $\mathbb{G}$.

Proof: (a) The reflexive property, $x \equiv_\mathbb{K} x$, holds as $x^{-1} \cdot x = 1 \in \mathbb{K}$ for all $x \in \mathbb{G}$. The symmetry property, $x \equiv_\mathbb{K} z \Leftrightarrow z \equiv_\mathbb{K} x$, holds since $x^{-1} \cdot z \in \mathbb{K} \Leftrightarrow z^{-1} \cdot x = (x^{-1} \cdot z)^{-1} \in \mathbb{K}$. The transitive property, $x \equiv_\mathbb{K} y$ and $y \equiv_\mathbb{K} z \Rightarrow x \equiv_\mathbb{K} z$, holds since $x^{-1} \cdot y \in \mathbb{K}$ and $y^{-1} \cdot z \in \mathbb{K} \Rightarrow x^{-1} \cdot z = (x^{-1} \cdot y) \cdot (y^{-1} \cdot z) \in \mathbb{K}$. Thus, $\equiv_\mathbb{K}$ is an equivalence relation on $\mathbb{G}$. As $x \equiv_\mathbb{K} z \Leftrightarrow x^{-1} \cdot z \in \mathbb{K} \Leftrightarrow z \in x \cdot \mathbb{K}$, the equivalence class $[x]$ in $\mathbb{G}$ equals the coset $x \cdot \mathbb{K}$.

(*b*) The equivalence classes of an equivalence relation always partition the set, by (1.5.1). So, $x \cdot \mathbb{K} \cap z \cdot \mathbb{K} \neq \emptyset \Leftrightarrow x \cdot \mathbb{K} = z \cdot \mathbb{K} \Leftrightarrow x^{-1} \cdot z \in \mathbb{K}$. □

Example 1 The subgroup $\langle x^n \rangle$ of powers of x^n in the infinite cyclic group $\langle x \rangle$ has n distinct cosets: $\langle x^n \rangle, x \cdot \langle x^n \rangle, \ldots, x^{n-1} \cdot \langle x^n \rangle$.

Let $\mathbb{G}/\mathbb{K}$ denote the set of all distinct left cosets $x \cdot \mathbb{K}$ of $\mathbb{K}$ in $\mathbb{G}$. The number of distinct left cosets of $\mathbb{K}$ in $\mathbb{G}$ is called the ***index of*** $\mathbb{K}$ ***in*** $\mathbb{G}$ and is denoted by $[\mathbb{G}, \mathbb{K}]$. The Coset Partition Lemma (4.1) implies the following.

4.2 Lagrange's Theorem *For* $\mathbb{K} \leq \mathbb{G}$ *(finite)*, $|\mathbb{G}| = |\mathbb{K}| [\mathbb{G}, \mathbb{K}]$.

QUOTIENT GROUPS

A subgroup $\mathbb{N}$ of $\mathbb{G}$ is ***normal*** if $x \cdot \mathbb{N} = \mathbb{N} \cdot x$ for every $x \in \mathbb{G}$; equivalently, $x \cdot \mathbb{N} \cdot x^{-1} \subseteq \mathbb{N}$ for every $x \in \mathbb{G}$. Thus, a normal subgroup $\mathbb{N}$ contains all the ***conjugates*** $x \cdot n \cdot x^{-1} \in \mathbb{G}$ of its members $n \in \mathbb{N}$ by the members $x \in \mathbb{G}$.

Given a subset $S \subseteq \mathbb{G}$, the set $\mathbb{N}(S)$ of all products of all conjugates $x \cdot s^{\pm 1} \cdot x^{-1} \in \mathbb{G}$ of the members of S and their inverses is the smallest normal subgroup of $\mathbb{G}$ containing the subset S (Exercise 3). This subgroup $\mathbb{N}(S)$ is called the ***normal closure*** in $\mathbb{G}$ of the subset $S \subseteq \mathbb{G}$.

When $\mathbb{N}$ is a normal subgroup, the set $\mathbb{G}/\mathbb{N}$ of distinct left cosets $x \cdot \mathbb{N}$ of $\mathbb{N}$ in $\mathbb{G}$ acquires a well-defined multiplication:

$$(x \cdot \mathbb{N}) \cdot (z \cdot \mathbb{N}) = x \cdot (\mathbb{N} \cdot z) \cdot \mathbb{N} = x \cdot (z \cdot \mathbb{N}) \cdot \mathbb{N} = (x \cdot z) \cdot \mathbb{N}.$$

4.3 Theorem *Let* $\mathbb{N}$ *be a normal subgroup of the group* $\mathbb{G}$. *Then:*
(*a*) $\mathbb{G}/\mathbb{N}$ *is a group with the multiplication* $(x \cdot \mathbb{N}) \cdot (z \cdot \mathbb{N}) = (x \cdot z) \cdot \mathbb{N}$, *and*
(*b*) $q : \mathbb{G} \to \mathbb{G}/\mathbb{N}$, $q(x) = x \cdot \mathbb{N}$, *is a homomorphism with* $\operatorname{Ker} q = \mathbb{N}$.

Proof: (*a*) The associativity of multiplication in $\mathbb{G}$ implies the associativity of multiplication in $\mathbb{G}/\mathbb{N}$. The coset $\mathbb{N} = 1 \cdot \mathbb{N}$ is an identity in $\mathbb{G}/\mathbb{N}$ and the coset $x \cdot \mathbb{N}$ has inverse $x^{-1} \cdot \mathbb{N}$.

(*b*) Since $(x \cdot z) \cdot \mathbb{N} = (x \cdot \mathbb{N}) \cdot (z \cdot \mathbb{N})$, the quotient function q satisfies the homomorphism property: $q(x \cdot z) = q(x) \cdot q(z)$ for all $x, z \in \mathbb{G}$. □

4.4 Theorem *For any group homomorphism* $\varphi : (\mathbb{G}, \cdot) \to (\mathbb{H}, \square)$,
(*a*) $\mathbb{K} = \operatorname{Ker} \varphi$ *is a normal subgroup of* $\mathbb{G}$, *and*
(*b*) φ *has a factorization* $\varphi = \overline{\varphi} \circ q : \mathbb{G} \to \mathbb{G}/\mathbb{K} \to \mathbb{H}$, *where the homomorphism* $\overline{\varphi} : \mathbb{G}/\mathbb{K} \to \mathbb{H}$ *is given by* $\overline{\varphi}(x \cdot \mathbb{K}) = \varphi(x)$.

Proof: (*a*) For each $x \in \mathbb{G}$, we have the calculation:

$$\varphi(x^{-1} \cdot \mathbb{K} \cdot x) = \varphi(x)^{-1} \square \varphi(\mathbb{K}) \square \varphi(x) = \varphi(x)^{-1} \square \varphi(x) = 1.$$

This implies the containment relation $x^{-1} \cdot \mathbb{K} \cdot x \subseteq \mathbb{K}$ for all $x \in \mathbb{G}$, which is equivalent to the normality property.

(b) The function $\overline{\varphi} : \mathbb{G}/\mathbb{K} \to \mathbb{H}$, $\overline{\varphi}(x \cdot \mathbb{K}) = \varphi(x)$, is well defined, because

$$x \cdot \mathbb{K} = z \cdot \mathbb{K} \Rightarrow (z \in x \cdot k \text{ for some } k \in \mathbb{K} = Ker\ \varphi) \Rightarrow \varphi(x) = \varphi(z).$$

Also, $\overline{\varphi}$ is a homomorphism, since the relations $(x \cdot \mathbb{K}) \cdot (y \cdot \mathbb{K}) = (x \cdot y) \cdot \mathbb{K}$ and

$$\overline{\varphi}((x \cdot y) \cdot \mathbb{K}) = \varphi(x \cdot y) = \varphi(x) \square \varphi(y) = \overline{\varphi}(x \cdot \mathbb{K}) \square \overline{\varphi}(y \cdot \mathbb{K})$$

hold by the definition of $\overline{\varphi}$ and the homomorphism property for φ. Finally, $\varphi = \overline{\varphi} \circ q$ by the definitions $\overline{\varphi}(x \cdot \mathbb{K}) = \varphi(x)$ and $q(x) = x \cdot \mathbb{K}$. □

4.5 Corollary *Any group homomorphism $\varphi : (\mathbb{G}, \cdot) \to (\mathbb{H}, \square)$ induces a group isomorphism $\overline{\varphi} : \mathbb{G}/Ker\ \varphi \to Im\ \varphi$.*

Proof: The function $\overline{\varphi} : \mathbb{G}/Ker\ \varphi \to Im\ \varphi$ is an epimorphism, because $Im\ \overline{\varphi} = Im\ \varphi$. Also, $\overline{\varphi}$ is a monomorphism, because $\overline{\varphi}(x \cdot \mathbb{K}) = \varphi(x) = 1$ in $\mathbb{H}$ implies that $x \in Ker\ \varphi = \mathbb{K}$, so that $x \cdot \mathbb{K} = \mathbb{K} = 1$ in $\mathbb{G}/\mathbb{K}$. □

Example 2 The epimorphism $\varphi : \langle x, z \rangle \to \mathbb{Z} \times \mathbb{Z}$, given by

$$\varphi(x) = <1, 0> \quad \text{and} \quad \varphi(z) = <0, 1>,$$

induces an isomorphism $\overline{\varphi} : \langle x, z \rangle / Ker\ \varphi \to \mathbb{Z} \times \mathbb{Z}$. $Ker\ \varphi$ consists of all reduced words $x^{k_1} z^{m_1} \ldots x^{k_n} z^{m_n}$ in the free group $\langle x, z \rangle$ that have exponent sums $\sum_i k_i = 0$ and $\sum_i m_i = 0$. It follows that $Ker\ \varphi$ equals the normal closure in $\langle x, z \rangle$ of the **commutator element** $xzx^{-1}z^{-1}$ (Exercise 4).

The previous theorem has the following extension (Exercise 5):

4.6 Homomorphism Theorem *Let $\mathbb{N}$ be a normal subgroup of $\mathbb{G}$. A group homomorphism $\varphi : (\mathbb{G}, \cdot) \to (\mathbb{H}, \square)$ induces a unique homomorphism $\overline{\varphi} : \mathbb{G}/\mathbb{N} \to \mathbb{H}$ with $\varphi = \overline{\varphi} \circ q_\mathbb{N} : \mathbb{G} \to \mathbb{G}/\mathbb{N} \to \mathbb{H}$ if and only if $Ker\ \varphi \supseteq \mathbb{N}$.*

PRESCRIBED GENERATORS AND RELATORS

It is possible to construct a group with a specified set of generators and a specified set of laws or relations among the generators. We simply combine the free group construction with the quotient group construction.

Example 3 To construct a group $\mathbb{G}$ generated by two elements $\underline{x}$ and $\underline{z}$ for which the relations $\underline{x}^2 = \underline{z}^3 = \underline{xzxz} = \underline{1}$ hold, we begin with the free group $\mathbb{F} = \langle x, z \rangle$. By the Normal Form Lemma 3.4, none of the elements x^2, z^3, and $xzxz$ equals the identity $1 \in \mathbb{F}$. Then we form the normal closure $\mathbb{N}$ in $\mathbb{F}$ of the subset $\{x^2, z^3, xzxz\}$; $\mathbb{N}$ consists of all products of all conjugates of the three elements x^2, z^3, and $xzxz$ and their inverses. In the quotient group $\mathbb{G} = \mathbb{F}/\mathbb{N}$, the cosets $x \cdot \mathbb{N}$, $y \cdot \mathbb{N}$, and $\mathbb{N}$, which may be abbreviated by $\underline{x}$, $\underline{z}$, and $\underline{1}$, satisfy the desired laws $\underline{x}^2 = \underline{z}^3 = \underline{xzxz} = \underline{1}$. For example, $\underline{x}^2 = x^2 \cdot \mathbb{N} = \mathbb{N} = \underline{1}$, because $x^2 \in \mathbb{N}$.

Here is the most general version of this construction process.

Let $\mathbb{F} = \mathbb{F}(X)$ be a free group with free basis $X = \{x_\alpha : \alpha \in \mathcal{A}\}$. Any nonempty reduced word $r = x_{\alpha_1}^{k_1} x_{\alpha_2}^{k_2} \ldots x_{\alpha_n}^{k_n}$ in powers of the generators is an element of $\mathbb{F}$ different from the identity element $1 \in \mathbb{F}$, according to the Normal Form Lemma 3.4. In other words, $\mathbb{F}$ is free of all unnecessary relationships between the members of its free basis.

Given any indexed set $R = \{r_\omega \in \mathbb{F} : \omega \in \mathcal{W}\}$ of nonempty reduced words, let $\mathbb{N}$ be its normal closure $\mathbb{N}(R)$ in $\mathbb{F}$. To introduce the relations $r_\omega = 1$ ($\omega \in \mathcal{W}$), we form the quotient group $\mathbb{F}/\mathbb{N}$ of $\mathbb{F} = \mathbb{F}(X)$ modulo the normal closure $\mathbb{N} = \mathbb{N}(R)$. For each $\omega \in \mathcal{W}$, the coset $\underline{r}_\omega = r_\omega \cdot \mathbb{N}$ equals the identity coset $\underline{1} = 1 \cdot \mathbb{N}$ in $\mathbb{F}/\mathbb{N}$ because $r_\omega \in \mathbb{N}$.

The quotient group $\mathbb{F}/\mathbb{N}$, denoted by $\langle X : R \rangle$, is called the **group with generators** $X = \{x_\alpha : \alpha \in \mathcal{A}\}$ and **relators** $R = \{r_\omega : \omega \in \mathcal{W}\}$, or the **group presented by** $\{X : R\}$.

Example 4 The group $\langle x : \rangle$ is the infinite cyclic group $\langle x \rangle$ generated by x.
The group $\langle z : z^n \rangle$ is a finite cyclic group of order n generated by z.
The group $\langle a, b : a^2, b^3 \rangle$ is isomorphic to the free product $\langle a : a^2 \rangle * \langle b : b^3 \rangle$ of finite cyclic groups of order 2 and 3.
By Example 2, the group $\langle x, z : xzx^{-1}z^{-1} \rangle$ is isomorphic to the direct product $\langle x \rangle \times \langle z \rangle$ of infinite cyclic groups $\langle x \rangle$ and $\langle z \rangle$.

Here is a way to summarize the construction of the group $\langle X : R \rangle$ with generators $X = \{x_\alpha : \alpha \in \mathcal{A}\}$ and relators $R = \{r_\omega \in \langle X \rangle : \omega \in \mathcal{W}\}$. $\langle X : R \rangle$ is the group of equivalence classes of words in the alphabet $\{x_\alpha^{\pm 1} : \alpha \in \mathcal{A}\}$, under the smallest equivalence relation $\sim$ such that: (i) $x_\alpha x_\alpha^{-1} \sim \emptyset \sim x_\alpha^{-1} x_\alpha$, (ii) $r_\omega \sim \emptyset$, and (iii) $(w \sim w'$ and $v \sim v') \Rightarrow wv \sim w'v'$.

Regardless of the definition used for the group $\langle X : R \rangle$, it is difficult in general to determine whether a word represents the identity or whether two words represent the same group element. It is even more difficult to determine whether two presented groups are isomorphic.

Example 5 The two groups $\langle a, b : aba^{-1}b \rangle$ and $\langle x, y : xxyy \rangle$ are isomorphic. The homomorphisms $\varphi : \langle a, b \rangle \longleftrightarrow \langle x, y \rangle : \psi$, given by

$$\varphi(a) = x, \quad \varphi(b) = xy, \quad \psi(x) = a, \quad \text{and} \quad \psi(y) = a^{-1}b,$$

are inverses that exchange $aba^{-1}b$ and $xxyy$ (Exercise 8). By the Homomorphism Theorem (4.6), they induce inverse homomorphisms:

$$\overline{\varphi} : \langle a, b : aba^{-1}b \rangle \longleftrightarrow \langle x, y : xxyy \rangle : \overline{\psi}.$$

The same technique shows that, for any word w in an alphabet J disjoint from $\{a, b, x, y\}$, there is an isomorphism $\langle a, b, J : aba^{-1}bw \rangle \approx \langle x, y, J : xxyyw \rangle$.

Example 6 The two groups $\langle a, b, c : aba^{-1}b^{-1}cc \rangle$ and $\langle x, y, z : xyx^{-1}yzz \rangle$ are isomorphic. There is a homomorphism $\varphi : \mathbb{F}(a, b, c) \to \mathbb{F}(x, y, z)$ given by

$$\varphi(a) = xyx^{-1}, \quad \varphi(b) = yzx^{-1}, \quad \text{and} \quad \varphi(c) = yz.$$

And there is a homomorphism $\psi : \mathbb{F}(x, y, z) \to \mathbb{F}(a, b, c)$ given by:
$$\psi(x) = b^{-1}c, \quad \psi(y) = c^{-1}bab^{-1}c, \quad \text{and} \quad \psi(z) = c^{-1}ba^{-1}b^{-1}cc.$$
They are inverses that exchange $aba^{-1}b^{-1}cc$ and $xyx^{-1}yzz$ (Exercise 9). So they induce inverse isomorphisms $\overline{\varphi} : \langle a, b, c : aba^{-1}b^{-1}cc \rangle \leftrightarrow \langle x, y, z : xyx^{-1}yzz \rangle : \overline{\psi}$.

Example 7   By Examples 5 and 6,
$$\langle a, b, c : aba^{-1}b^{-1}cc \rangle \approx \langle x, y, z : xyx^{-1}yzz \rangle \approx \langle x, y, z : xxyyzz \rangle.$$

The source of these algebraic isomorphisms is some topological subdivide-join operations on surfaces that are described in Chapter **13**.

PRESENTATIONS OF GROUPS

Any group $\mathbb{G}$ can be presented by a set of generators and relators, as follows.

First, select an indexed set of generators $\{g_\alpha : \alpha \in \mathcal{A}\}$ of $\mathbb{G}$ and form the free group $\mathbb{F} = \mathbb{F}(X)$ with a similarly indexed free basis $X = \{x_\alpha : \alpha \in \mathcal{A}\}$. Then form the homomorphism $\varphi : \mathbb{F} \to \mathbb{G}$, $\varphi(x_\alpha) = g_\alpha$ for all $\alpha \in \mathcal{A}$. Let $\mathbb{N} = Ker\ \varphi$. The induced homomorphism $\overline{\varphi} : \mathbb{F}/\mathbb{N} \to \mathbb{G}$, $\overline{\varphi}(x \cdot \mathbb{N}) = \varphi(x)$, is an isomorphism by Corollary 4.5. Some set $R = \{r_\omega : \omega \in \mathcal{W}\}$ of members of $\mathbb{N}$ has normal closure $\mathbb{N}(R) = \mathbb{N}$ in $\mathbb{F}$. Thus, $\mathbb{F}/\mathbb{N} = \langle X : R \rangle$. The isomorphism $\overline{\varphi} : \mathbb{F}/\mathbb{N} \approx \mathbb{G}$ is called a ***presentation*** of $\mathbb{G}$.

Here is a way to interpret this description of $\mathbb{G}$. If all relationships in $\mathbb{G}$ between the products of the generators $\{g_\alpha\}$ are lost, the result is the free group $\mathbb{F}$ with the free basis $\{x_\alpha\}$. A sufficiently rich collection of relations $g_{\alpha_1}^{k_1} g_{\alpha_2}^{k_2} \ldots g_{\alpha_n}^{k_n} = 1$ in $\mathbb{G}$ among the generators $\{g_\alpha\}$ are encoded in relators $\{r_\omega \in \mathbb{F} : \omega \in \mathcal{W}\}$ chosen to have normal closure $\mathbb{N}$. Then $\mathbb{G}$ is recovered from $\mathbb{F}$ by forming the quotient group $\mathbb{F}/\mathbb{N}$.

Example 8   The abelian group $\mathbb{Z}_m \times \mathbb{Z}_n$ is generated by $<[1]_m, [0]_n>$ and $<[0]_m, [1]_n>$. The epimorphism $\varphi : \langle x, z \rangle \to \mathbb{Z}_m \times \mathbb{Z}_n$, defined by
$$\varphi(x) = <[1]_m, [0]_n> \quad \text{and} \quad \varphi(z) = <[0]_m, [1]_n>,$$
has kernel $Ker\ \varphi = \mathbb{N}(\{x^m, z^n, xzx^{-1}z^{-1}\})$. Thus, φ gives the presentation:
$$\overline{\varphi} : \langle x, z : x^m, z^n, xzx^{-1}z^{-1} \rangle \approx \mathbb{Z}_m \times \mathbb{Z}_n.$$

ABELIANIZATION

Let $\mathbb{G}$ be any group. A homomorphism $a : \mathbb{G} \to \mathbb{A}$ to an abelian group $\mathbb{A}$ is an ***abelianization*** of $\mathbb{G}$ if the following universal property holds:

- *For each abelian group $\mathbb{B}$ and homomorphism $\varphi : \mathbb{G} \to \mathbb{B}$, there is a unique homomorphism $\overline{\varphi} : \mathbb{A} \to \mathbb{B}$ such that $\varphi = \overline{\varphi} \circ a : \mathbb{G} \to \mathbb{A} \to \mathbb{B}$.*

4.7   Uniqueness of Abelianizations   *If $a : \mathbb{G} \to \mathbb{A}$ and $c : \mathbb{G} \to \mathbb{C}$ are abelianizations of $\mathbb{G}$, there exist inverse group isomorphisms $\overline{c} : \mathbb{A} \to \mathbb{C}$ and $\overline{a} : \mathbb{C} \to \mathbb{A}$ such that $c = \overline{c} \circ a$ and $a = \overline{a} \circ c$.*

Proof: The homomorphism $\bar{c} : A \to C$ such that $c = \bar{c} \circ a : G \to A \to C$ exists by the universal property for $a : G \to A$; the homomorphism $\bar{a} : C \to A$ such that $a = \bar{a} \circ c : G \to C \to A$ exists by the universal property for $c : G \to C$. The composite $\bar{a} \circ \bar{c} : A \to C \to A$ equals $1_A : A \to A$ by the universal property for $a : G \to A$, as $\bar{a} \circ \bar{c} \circ a = \bar{a} \circ c = a = 1_A \circ a$. Similarly, the composite $\bar{c} \circ \bar{a} : C \to A \to C$ equals $1_C : C \to C$ by the universal property for $c : G \to C$, as $\bar{c} \circ \bar{a} \circ c = \bar{c} \circ a = c = 1_C \circ c$. □

For any group G, the normal closure in G of the commutator elements

$$[x, z] = xzx^{-1}z^{-1} \qquad (x, z \in G)$$

is called the *commutator subgroup* of G and is denoted by $[G, G]$.

Let $Ab(G)$ denote the quotient group $G/[G, G]$ and let

$$a_G : G \to Ab(G) = G/[G, G]$$

denote the quotient homomorphism given by $a_G(g) = g \cdot [G, G]$.

4.8 Existence of Abelianizations *For any group G, the quotient homomorphism $a_G : G \to Ab(G)$ is an abelianization.*

Proof: $Ab(G)$ is abelian: $xzx^{-1}z^{-1} \in [G, G] \Rightarrow xz \cdot [G, G] = zx \cdot [G, G]$ in $Ab(G)$ for all $x, z \in G$. For any abelian group A and group homomorphism $\varphi : G \to A$, we have that $\varphi([x, z]) = [\varphi(x), \varphi(z)] = 1$ for all $x, z \in G$. Hence, $[G, G] \le Ker\, \varphi$. By the Homomorphism Theorem (4.6), φ induces a unique homomorphism $\bar{\varphi} : Ab(G) \to A$ such that $\varphi = \bar{\varphi} \circ a_G : G \to Ab(G) \to A$. Thus, a_G is an abelianization of G. □

Example 9 For any two groups G and H, we have

$$[G \times H, G \times H] = [G, G] \times [H, H]$$

in $G \times H$, because $[<g, h>, <g', h'>] = <[g, g'], [h, h']>$ in $G \times H$ for all elements $<g, h>, <g', h'> \in G \times H$. Therefore, the homomorphism

$$a = a_G \times a_H : G \times H \to Ab(G) \times Ab(H)$$

induces an isomorphism $\bar{a} : Ab(G \times H) \approx Ab(G) \times Ab(H)$. Hence, $a = a_G \times a_H$ is an abelianization of $G \times H$.

Example 10 An abelianization $a : G * H \to Ab(G) \times Ab(H)$ for any free product $G * H$ is defined by $a(g) = <a_G(g), 1>$ and $a(h) = <1, a_H(h)>$ for all $g \in G$ and $h \in H$. For, given any homomorphism $\varphi : G * H \to A$ to an abelian group A, the restrictions $\varphi_G : G \to A$ and $\varphi_H : H \to A$ induce homomorphisms $\bar{\varphi}_G : Ab(G) \to A$ and $\bar{\varphi}_H : Ab(H) \to A$. The latter, in turn, define a unique homomorphism $\bar{\varphi} : Ab(G) \times Ab(H) \to A$ such that $\varphi = \bar{\varphi} \circ a$, namely,

$$\bar{\varphi}(<a_G(g), a_H(h)>) = \bar{\varphi}_G(a_G(g)) + \bar{\varphi}_H(a_H(h)) = \varphi(g) + \varphi(h)$$

for all $g \in G$ and $h \in H$.

Examples 9 and 10 illustrate these general results (Exercise 11):

$$Ab(\mathbb{G}_1 \times \ldots \times \mathbb{G}_n) \approx Ab(\mathbb{G}_1) \times \ldots \times Ab(\mathbb{G}_n)$$

and

$$Ab(*_\alpha \mathbb{G}_\alpha) \approx \oplus_\alpha Ab(\mathbb{G}_\alpha).$$

In particular, the abelianization of the free group $\mathbb{F}(X) = *_\alpha \langle x_\alpha \rangle$ with free basis $X = \{x_\alpha\}$ is isomorphic to the free-abelian group $\mathbb{A}(X) = \oplus_\alpha \langle x_\alpha \rangle$ with free-abelian basis $\{x_\alpha\}$. (See the Exercises of Section 3 for the direct sum $\oplus$ and the free-abelian terminology.)

4.9 Theorem *Any quotient group $\mathbb{G}/\mathbb{N}$ has $Ab(\mathbb{G}/\mathbb{N}) \approx Ab(\mathbb{G})/a_\mathbb{G}(\mathbb{N})$.*

Proof: Let $a : \mathbb{G}/\mathbb{N} \to Ab(\mathbb{G})/a_\mathbb{G}(\mathbb{N})$ be the homomorphism induced by the abelianization $a_\mathbb{G} : \mathbb{G} \to Ab(\mathbb{G})$. For an abelian group $\mathbb{A}$ and homomorphism $\varphi : \mathbb{G}/\mathbb{N} \to \mathbb{A}$, the homomorphism $\psi = \varphi \circ q_\mathbb{N} : \mathbb{G} \to \mathbb{G}/\mathbb{N} \to \mathbb{A}$ induces a unique homomorphism $\overline{\psi} : Ab(\mathbb{G}) \to \mathbb{A}$ such that $\overline{\psi} \circ a_\mathbb{G} = \psi = \varphi \circ q_\mathbb{N}$. Therefore, $Ker\ \overline{\psi}$ contains $a_\mathbb{G}(\mathbb{N})$, and $\overline{\psi}$ induces a unique homomorphism $\overline{\varphi} : Ab(\mathbb{G})/a_\mathbb{G}(\mathbb{N}) \to \mathbb{A}$ such that $\varphi = \overline{\varphi} \circ a$. This proves that the homomorphism $a : \mathbb{G}/\mathbb{N} \to Ab(\mathbb{G})/a_\mathbb{G}(\mathbb{N})$ is an abelianization. □

Example 11 The group $\langle x, z : xzx^{-1}z \rangle = \mathbb{F}(x,z)/\mathbb{N}(xzx^{-1}z)$ has abelianization $\mathbb{Z} \oplus \mathbb{Z}_2 = \mathbb{A}(x,z)/\langle z^2 \rangle$.

In Chapter **13**, the following 1-relator groups appear as the fundamental groups of surfaces:

$$\mathbb{H}_g = \langle a_1, b_1, \ldots, a_g, b_g : a_1 b_1 a_1^{-1} b_1^{-1} \ldots a_g b_g a_g^{-1} b_g^{-1} \rangle \quad (g \geq 1)$$

$$\mathbb{C}_g = \langle a_1, \ldots, a_g : a_1 a_1 \ldots a_g a_g \rangle \quad (g \geq 1).$$

4.10 Theorem $\mathbb{H}_g$ *and* $\mathbb{C}_g$ *have non-isomorphic abelianizations:*

$$Ab(\mathbb{H}_g) \approx (\mathbb{Z} \oplus \mathbb{Z}) \oplus \ldots \oplus (\mathbb{Z} \oplus \mathbb{Z}) \quad and \quad Ab(\mathbb{C}_g) \approx \mathbb{Z}_2 \oplus \mathbb{Z} \oplus \ldots \oplus \mathbb{Z}.$$

The first abelianization is free-abelian with rank $2g$ and the second has rank $g - 1$ and torsion coefficient 2.

Proof: The abelianization $a : \mathbb{F}(\{a_i, b_i\}) \to \mathbb{A}(\{a_i, b_i\})$ trivializes the relator $\prod a_i b_i a_i^{-1} b_i^{-1}$ of $\mathbb{H}_g$. Thus, by Theorem 4.9, $Ab(\mathbb{H}_g)$ is isomorphic to the free-abelian group $\mathbb{A}(\{a_i, b_i\}) = (\mathbb{Z} \oplus \mathbb{Z}) \oplus \ldots \oplus (\mathbb{Z} \oplus \mathbb{Z})$ with rank $2g$.

The abelianization $a : \mathbb{F}(\{a_i\}) \to \mathbb{A}(\{a_i\}) = \mathbb{Z} \oplus \mathbb{Z} \oplus \ldots \oplus \mathbb{Z}$ sends a_i to the generator $x_i = (0, \ldots, 0, 1, 0, \ldots, 0)$ in $\mathbb{A}(\{a_i\})$ and, so, $a_1 a_1 \ldots a_g a_g$ to $(2, \ldots, 2)$. Therefore, $Ab(\mathbb{C}_g) \approx (\mathbb{Z} \oplus \mathbb{Z} \oplus \ldots \oplus \mathbb{Z})/\langle (2, \ldots, 2) \rangle$.

Consider the homomorphism

$$\varphi : \mathbb{Z} \oplus \mathbb{Z} \oplus \ldots \oplus \mathbb{Z} \to \mathbb{Z} \oplus \mathbb{Z} \oplus \ldots \oplus \mathbb{Z}$$

given by
$$\varphi(x_1) = x_1 - x_2 - x_3 - \ldots - x_g$$
and
$$\varphi(x_i) = x_i, \qquad (i > 1)$$

and the homomorphism
$$\psi : \mathbb{Z} \oplus \mathbb{Z} \oplus \ldots \oplus \mathbb{Z} \to \mathbb{Z} \oplus \mathbb{Z} \oplus \ldots \oplus \mathbb{Z}$$

given by
$$\psi(x_1) = x_1 + x_2 + x_3 + \ldots + x_g$$
and
$$\psi(x_i) = x_i \qquad (i > 1).$$

They are inverse homomorphisms interchanging $(2, \ldots, 2)$ and $(2, 0, \ldots, 0)$. Thus, they induce an isomorphism
$$(\mathbb{Z} \oplus \mathbb{Z} \oplus \ldots \oplus \mathbb{Z})/\langle (2, \ldots, 2) \rangle \approx (\mathbb{Z} \oplus \mathbb{Z} \oplus \ldots \oplus \mathbb{Z})/\langle (2, 0, \ldots, 0) \rangle,$$
which is the desired isomorphism $Ab(\mathbb{C}_g) \approx \mathbb{Z}_2 \oplus \mathbb{Z} \oplus \ldots \oplus \mathbb{Z}$.

By the Fundamental Theorem (3.1) classifying finitely generated abelian groups, none of the abelianizations $Ab(\mathbb{H}_g)$ and $Ab(\mathbb{C}_g)$ ($g \geq 1$) are isomorphic. □

Exercises

SECTION 2

1. Prove that in any group $(\mathbb{G}, \cdot)$ the equations
$$g \cdot x = h \qquad \text{and} \qquad y \cdot g = h$$
have unique solutions $x, y \in \mathbb{G}$ for every pair $g, h \in \mathbb{G}$.

2. Prove that in any group $(\mathbb{G}, \cdot)$ there are these cancellation laws:
$$g \cdot x = h \cdot x \Rightarrow g = h \qquad \text{and} \qquad x \cdot g = x \cdot h \Rightarrow g = h.$$

3. Find a set of generators for each of these groups:
   (a) the multiplicative group $\mathbb{Z}_{15}^*$ of invertible congruence classes modulo 15,
   (b) the additive group $(\mathbb{Q}, +)$ of rational numbers, and
   (c) the multiplicative group $\mathbb{Z}_{18}^*$ of invertible congruence classes modulo 18.

4. Find a single element that generates the group $\mathbb{Z}_3 \times \mathbb{Z}_5$ of ordered pairs $< [k]_3, [l]_5 >$ with coordinatewise addition:
$$< [k]_3, [l]_5 > + < [m]_3, [n]_5 > = < [k+m]_3, [l+n]_5 >.$$

5. Find a set of generators for the group $\mathbb{Z}_3 \times \mathbb{Z}_6$ of ordered pairs $< [k]_3, [l]_6 >$ with coordinatewise addition:
$$< [k]_3, [l]_6 > + < [m]_3, [n]_6 > = < [k+m]_3, [l+n]_6 >.$$

6. Determine the following subgroups of the infinite cyclic group $(\mathbb{Z}, +)$.
   (a) $\langle 3 \rangle$ (b) $\langle 6, 8 \rangle$ (c) $\langle 3, 7 \rangle$ (d) $\langle 4, 9 \rangle$

7. Determine the following subgroups of the cyclic group $(\mathbb{Z}_9, +_9)$ of order 9.
   (a) $\langle [3]_9 \rangle$ (b) $\langle [2]_9 \rangle$ (c) $\langle [3]_9, [2]_9 \rangle$ (d) $\langle [5]_9 \rangle$

8. Let $\mathbb{G} = \langle g \rangle$ and $\mathbb{H} = \langle h \rangle$ be finite cyclic groups of order m and n, respectively. Prove that there is a homomorphism $\varphi : \mathbb{G} \to \mathbb{H}$ satisfying $\varphi(g) = h^k$ if and only if n divides mk.

9. Construct a group isomorphism $\varphi : (\mathbb{Z}_7^*, \cdot_7) \to (\mathbb{Z}_6, +_6)$.

SECTION 3

1. Construct an isomorphism $\mathbb{Z}_p \times \mathbb{Z}_q \approx \mathbb{Z}_{pq}$ when p and q are relatively prime modulii.

2. Prove that the groups $\mathbb{Z}_3 \times \mathbb{Z}_3$ and $\mathbb{Z}_9$ are not isomorphic.

3. Prove that any homomorphism $\varphi : \mathbb{G} * \mathbb{H} \to \mathbb{J} \times \mathbb{K}$ is completely determined by the four homomorphisms obtained by composing it with the two projections

$$\pi_\mathbb{J} : \mathbb{J} \times \mathbb{K} \to \mathbb{J} \quad \text{and} \quad \pi_\mathbb{K} : \mathbb{J} \times \mathbb{K} \to \mathbb{K}$$

   and the two injections

$$\psi_\mathbb{G} : \mathbb{G} \to \mathbb{G} * \mathbb{H} \quad \text{and} \quad \psi_\mathbb{H} : \mathbb{H} \to \mathbb{G} * \mathbb{H}.$$

4. Prove that a free product of free groups is a free group.

   An abelian group $\mathbb{S}$ is an ***abelian sum of the abelian groups*** $\{\mathbb{G}_\alpha: \alpha \in \mathcal{A}\}$, ***with injections*** $\{\psi_\alpha : \mathbb{G}_\alpha \to \mathbb{S}\}$, if the following universal property holds: *for any abelian group $\mathbb{H}$ and indexed family of homomorphisms $\{\varphi_\alpha : \mathbb{G}_\alpha \to \mathbb{H}\}$, there is a unique homomorphism $\varphi : \mathbb{S} \to \mathbb{H}$ such that $\varphi \circ \psi_\alpha = \varphi_\alpha$ for all $\alpha \in \mathcal{A}$.*

5*. Establish a uniqueness property for abelian sums of abelian groups.

   The ***direct sum*** $\oplus_\alpha \mathbb{G}_\alpha$ of the groups $\{\mathbb{G}_\alpha : \alpha \in \mathcal{A}\}$ is the subgroup of the product $\prod_\alpha \mathbb{G}_\alpha$ consisting of all α-tuples $x = (x_\alpha)$ with $x_\alpha = 1_\alpha \in \mathbb{G}_\alpha$ for all but finitely many $\alpha \in \mathcal{A}$.

6*. Establish the existence of abelian sums of abelian groups: prove that the direct sum $\mathbb{S} = \oplus_\alpha \mathbb{G}_\alpha$ of abelian groups $\{\mathbb{G}_\alpha : \alpha \in \mathcal{A}\}$ serves as their abelian sum.

   Let X be any subset of an abelian group $\mathbb{A}$. Then $\mathbb{A}$ is called a ***free-abelian group*** with ***free-abelian basis*** X if $\mathbb{A}$ has this universal property: *for any abelian group $\mathbb{G}$, any function $\varphi : X \to \mathbb{G}$ extends uniquely to a group homomorphism $\varphi : \mathbb{A} \to \mathbb{G}$.*
   When this holds, we abbreviate the relationship by writing $\mathbb{A}(X)$.

7*. Establish the following:
   (a) (Uniqueness of Free-Abelian Groups) If $\mathbb{A}$ and $\mathbb{B}$ be free-abelian groups with free-abelian basis X, then there exist inverse group isomorphisms $\mathbb{A} \leftrightarrow \mathbb{B}$ fixing X.
   (b) (Existence of Free-Abelian Groups) For any set X, there is a free-abelian group $\mathbb{A}$ with free-abelian basis X.

8. Assume the Fundamental Theorem (3.1) of finitely generated abelian groups. Prove that the rank of a free basis of a finitely generated free group is an invariant of the group.

SECTION 4

1. Prove that the following are equivalent for a subgroup $\mathbb{K}$ of a group $\mathbb{G}$:
   (a) For all $x, x', z, z' \in \mathbb{G}$, $x \equiv_\mathbb{K} x'$ and $z \equiv_\mathbb{K} z'$ imply $x \cdot z \equiv_\mathbb{K} x' \cdot z'$.
   (b) For every $x \in \mathbb{G}$, we have $x \cdot \mathbb{K} \cdot x^{-1} \subseteq \mathbb{K}$.
   (c) For every $x \in \mathbb{G}$, we have $x \cdot \mathbb{K} = \mathbb{K} \cdot x$.

2. Determine the subgroup $\langle n-1 \rangle$ of $\mathbb{Z}_{2n}$ generated by $n-1$, and form the cosets of $\langle n-1 \rangle$ in $\mathbb{Z}_{2n}$. (There are two cases, depending upon whether n is even or odd.)

3*. Prove that the normal closure in a group $\mathbb{G}$ of a subset $S \subseteq \mathbb{G}$ has these descriptions:
   (a) the set $\mathbb{N}(S)$ of all products in $\mathbb{G}$ of all conjugates $x \cdot s^{\pm 1} \cdot x^{-1} \in \mathbb{G}$ of the members of S and their inverses,
   (b) the smallest normal subgroup $\mathbb{N}(S)$ of $\mathbb{G}$ containing the subset S, and
   (c) the intersection $\mathbb{N}(S)$ of all normal subgroups of $\mathbb{G}$ containing the subset S.

4*. Prove that the normal closure of the commutator $xzx^{-1}z^{-1}$ in the free group $\langle x, z \rangle$ consists of all reduced words $x^{k_1} z^{m_1} x^{k_2} z^{m_2} \ldots x^{k_n} z^{m_n}$ in the free group $\langle x, z \rangle$ that have exponent sums $\sum_i k_i = 0$ and $\sum_i m_i = 0$.

5*. Prove that a group homomorphism $\varphi : (\mathbb{G}, \cdot) \to (\mathbb{H}, \square)$ factors through the quotient homomorphism $q : \mathbb{G} \to \mathbb{G}/\mathbb{N}$ for a normal subgroup $\mathbb{N}$ if and only if $\operatorname{Ker} \varphi \supseteq \mathbb{N}$.

6. Construct presentations for each of the following groups.
   (a) $\mathbb{Z}$
   (b) $\mathbb{Z} \times \mathbb{Z} \times \mathbb{Z}$
   (c) $\mathbb{Z}_m \times \mathbb{Z}_n \times \mathbb{Z}_r$
   (d) $\mathbb{Z}_m * \mathbb{Z}_n$
   (e) $\mathbb{Z}_2 * \mathbb{Z}_2 * \mathbb{Z}_2$
   (f) $\mathbb{F}(\{x_1, \ldots, x_n\})$

7. Construct presentations for each of the following groups.
   (a) $\langle x_\alpha : r_\beta \rangle \times \langle y_\gamma : s_\delta \rangle$
   (b) $\langle x_\alpha : r_\beta \rangle * \langle y_\gamma : s_\delta \rangle$

8*. Prove that the homomorphisms $\varphi : \langle a, b \rangle \leftrightarrow \langle x, y \rangle : \psi$, given by
$$\varphi(a) = x, \quad \varphi(b) = xy, \quad \psi(x) = a, \text{ and } \psi(y) = a^{-1}b,$$
are inverses that exchange $aba^{-1}b$ and $xxyy$.

9*. Prove that the homomorphisms $\varphi : \mathbb{F}(a, b, c) \leftrightarrow \mathbb{F}(x, y, z) : \psi$, given by
$$\varphi(a) = xyx^{-1} \qquad \varphi(b) = yzx^{-1} \qquad \varphi(c) = yz,$$
and
$$\psi(x) = b^{-1}c \qquad \psi(y) = c^{-1}bab^{-1}c \qquad \psi(z) = c^{-1}ba^{-1}b^{-1}cc,$$
are inverses that exchange $aba^{-1}b^{-1}cc$ and $xyx^{-1}yzz$.

10. Let $\mathbb{C}_g = \langle a_1, \ldots, a_g : a_1 a_1 \ldots a_g a_g \rangle$.
   (a) Prove that when $g = 2n + 1$,
$$\mathbb{C}_g \approx \langle x_1, y_1, \ldots, x_n, y_n, z : [x_1, y_1] \ldots [x_n, y_n] zz \rangle.$$
   (b) Prove that when $g = 2n$,
$$\mathbb{C}_g \approx \langle x_1, y_1, \ldots, x_n, y_n, : [x_1, y_1] \ldots [x_{n-1}, y_{n-1}] x_n y_n x_n^{-1} y_n \rangle.$$
   (c) Use (a) and (b) to determine the abelianization of $\mathbb{C}_g$.

11. Prove:
   (a) $\operatorname{Ab}(\mathbb{G}_1 \times \ldots \times \mathbb{G}_n) \approx \operatorname{Ab}(\mathbb{G}_1) \oplus \ldots \oplus \operatorname{Ab}(\mathbb{G}_n)$.
   (b) $\operatorname{Ab}(*_\alpha \mathbb{G}_\alpha) \approx \oplus_\alpha \operatorname{Ab}(\mathbb{G}_\alpha)$. (See Section 3, Exercises, for the direct sum $\oplus$.)

12

CALCULATIONS OF Π_1

Because of the topological invariance of the fundamental group, spaces with non-isomorphic fundamental groups are not homeomorphic. To distinguish topological spaces in this way, we need to be able to calculate their fundamental groups. This chapter shows how to present the fundamental group of a wide variety of spaces by generators and relators, and also applies techniques of Chapter **11** to distinguish among them.

1 The Seifert-Van Kampen Theorem

The Seifert-Van Kampen Theorem determines the fundamental group of a topological space X that is given as the union of open subspaces U_α ($\alpha \in \mathcal{A}$). The algebraic relationship between the fundamental group of X and those of the open subspaces mimics their topological relationship.

One can reconstruct the topological space X from the subspaces $\{U_\alpha\}$ if one knows the intersections $U_\sigma \cap U_\tau$ and the inclusions $i_{\sigma\tau} : U_\sigma \cap U_\tau \subseteq U_\sigma$ and $i_{\tau\sigma} : U_\tau \cap U_\sigma \subseteq U_\tau$, for each pair $\sigma, \tau \in \mathcal{A}$. X is an *amalgamated union* of this directed system of subspace inclusions, and it is characterized by these properties (Exercise 1):

- There exist continuous functions $i_\alpha : U_\alpha \to X$ ($\alpha \in \mathcal{A}$) such that
  $$i_\sigma \circ i_{\sigma\tau} : U_\sigma \cap U_\tau \subseteq U_\sigma \to X \text{ and } i_\tau \circ i_{\tau\sigma} : U_\tau \cap U_\sigma \subseteq U_\tau \to X$$
  coincide for all $\sigma, \tau \in \mathcal{A}$.

- If $f_\alpha : U_\alpha \to Y$ ($\alpha \in \mathcal{A}$) are any continuous functions such that
  $$f_\sigma \circ i_{\sigma\tau} : U_\sigma \cap U_\tau \subseteq U_\sigma \to Y \text{ and } f_\tau \circ i_{\tau\sigma} : U_\tau \cap U_\sigma \subseteq U_\tau \to Y$$
  coincide for all $\sigma, \tau \in \mathcal{A}$, then there exists a unique continuous function $f : X \to Y$ such that $f_\alpha = f \circ i_\alpha : U_\alpha \to X \to Y$ for all $\alpha \in \mathcal{A}$.

SECTION 1 — THE SEIFERT–VAN KAMPEN THEOREM

AMALGAMATED SUMS OF GROUPS

There is a similar amalgamation relationship for groups. Consider an indexed family of groups $\{\mathbb{G}_\alpha, \alpha \in \mathcal{A}\}$ such that, for each pair $\sigma \neq \tau \in \mathcal{A}$, there is a group $\mathbb{G}_{\{\sigma, \tau\}}$ and a pair of homomorphisms

$$\mathbb{G}_\sigma \xleftarrow{\varphi_{\sigma\tau}} \mathbb{G}_{\{\sigma, \tau\}} \xrightarrow{\varphi_{\tau\sigma}} \mathbb{G}_\tau \qquad (\sigma \neq \tau \in \mathcal{A}).$$

We call the family $\{\varphi_{\sigma\tau}\}$ a ***bonding system*** of homomorphisms.

Suppose that $\mathbb{G}$ is a group and there are homomorphisms $\rho_\alpha : \mathbb{G}_\alpha \to \mathbb{G}$ ($\alpha \in \mathcal{A}$) such that the two composite homomorphisms $\rho_\sigma \circ \varphi_{\sigma\tau}$ and $\rho_\tau \circ \varphi_{\tau\sigma}$ coincide for all $\sigma \neq \tau \in \mathcal{A}$, as in this commutative diagram:

$$\begin{array}{ccc} & \mathbb{G}_{\{\sigma,\tau\}} & \\ \varphi_{\sigma\tau}\swarrow & & \searrow\varphi_{\tau\sigma} \\ \mathbb{G}_\sigma & & \mathbb{G}_\tau \\ \rho_\sigma\searrow & & \swarrow\rho_\tau \\ & \mathbb{G} & \end{array} \qquad (\sigma \neq \tau \in \mathcal{A})$$

Then the family $\{\rho_\alpha : \mathbb{G}_\alpha \to \mathbb{G}\}$ is called an ***amalgamating system*** for the bonding system $\{\varphi_{\sigma\tau}\}$.

We call $(\mathbb{G}, \{\psi_\alpha\})$ an ***amalgamated sum of the bonding system*** $\{\varphi_{\sigma\tau}\}$ provided that $\{\psi_\alpha : \mathbb{G}_\alpha \to \mathbb{G}\}$ is an amalgamating system satisfying the universal property that has these equivalent formulations:

- If $\{\rho_\alpha : \mathbb{G}_\alpha \to \mathbb{H}\}$ *is any amalgamating system for the given bonding system* $\{\varphi_{\sigma\tau}\}$, *there exists a unique homomorphism* $\rho : \mathbb{G} \to \mathbb{H}$ *such that* $\rho_\alpha = \rho \circ \psi_\alpha : \mathbb{G}_\alpha \to \mathbb{G} \to \mathbb{H}$ *for all* $\alpha \in \mathcal{A}$.

- $\forall \, \mathbb{H} \in \mathcal{G}$, *the function to the set of amalgamating systems*

  $$(\mathbb{G}, \mathbb{H})_\mathcal{G} \to \{(\rho_\alpha) \in \textstyle\prod_\alpha (\mathbb{G}_\alpha, \mathbb{H})_\mathcal{G} : \rho_\sigma \circ \varphi_{\sigma\tau} = \rho_\tau \circ \varphi_{\tau\sigma}\ (\forall\ \sigma \neq \tau)\},$$

  defined by $\varphi \to (\rho_\alpha \circ \varphi)$, *is a bijection*.

- $(\mathbb{G}, \{\psi_\alpha\})$ *is an initial object in the category* $\mathcal{G}\backslash\backslash\{\varphi_{\sigma\tau}\}$ ***below*** $\{\varphi_{\sigma\tau}\}$. *An object in* $\mathcal{G}\backslash\backslash\{\varphi_{\sigma\tau}\}$ *is a pair* $(\mathbb{H}, \{\rho_\alpha\})$, *consisting of a group* $\mathbb{H}$ *and an amalgamating system* $\{\rho_\alpha : \mathbb{G}_\alpha \to \mathbb{H}\}$ *for the bonding system* $\{\varphi_{\sigma\tau}\}$, *and a morphism* $\kappa : (\mathbb{H}, \{\rho_\alpha\}) \to (\mathbb{K}, \{\lambda_\alpha\})$ *in* $\mathcal{G}\backslash\backslash\{\varphi_{\sigma\tau}\}$ *is a homomorphism* $\kappa : \mathbb{H} \to \mathbb{K}$ *such that* $\kappa \circ \rho_\alpha = \lambda_\alpha$ *for all* $\alpha \in \mathcal{A}$.

By Theorem 10.1.1, any amalgamated sum (that exists) in $\mathcal{G}$ is unique:

1.1 Theorem *An amalgamated sum of a bonding system is unique up to a group isomorphism that respects the amalgamating homomorphisms.*

The universal property satisfied by the amalgamated sum is expressed in this ***amalgamated sum diagram***:

$$\begin{array}{c} \mathbb{G}_{\{\sigma,\tau\}} \\ {}^{\varphi_{\sigma\tau}}\swarrow \quad \searrow{}^{\varphi_{\tau\sigma}} \\ \mathbb{G}_\sigma \xrightarrow{\psi_\sigma} \mathbb{G} \xleftarrow{\psi_\tau} \mathbb{G}_\tau \\ {}_{\rho_\sigma}\searrow \; \exists!\downarrow\rho \; \swarrow{}_{\rho_\tau} \\ \mathbb{H} \end{array} \qquad (\forall\, \sigma\neq\tau \in \mathcal{A})$$

Example 1 By the Homomorphism Theorem (**11.4.6**), a quotient group $\mathbb{G}/\mathbb{N}$ is an amalgamated sum of the bonding system $\{\{1\} \leftarrow \mathbb{N} \leq \mathbb{G}\}$.

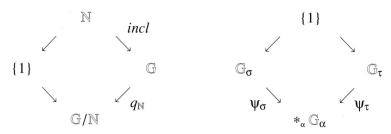

Example 1: Quotient group   Example 2: Free product

Example 2 By Theorem **11.3.6**, the free product $(\mathbb{S} = *_\alpha \mathbb{G}_\alpha, \{\psi_\alpha : \mathbb{G}_\alpha \to \mathbb{S}\})$ is an amalgamated sum of the trivial bonding system $\{\mathbb{G}_\sigma \leftarrow \{1\} \to \mathbb{G}_\tau\,(\sigma\neq\tau\in\mathcal{A})\}$.

1.2 Theorem *Each bonding system in $\mathcal{G}$ has an amalgamated sum.*

Proof: An amalgamated sum $(\mathbb{G}, \{\psi_\alpha\})$ for a bonding system

$$\mathbb{G}_\sigma \xleftarrow{\varphi_{\sigma\tau}} \mathbb{G}_{\{\sigma,\tau\}} \xrightarrow{\varphi_{\tau\sigma}} \mathbb{G}_\tau \qquad (\sigma\neq\tau\in\mathcal{A})$$

can be constructed as the quotient group $\mathbb{G}$ of the sum $*_\alpha \mathbb{G}_\alpha$, modulo the normal closure $\mathbb{N}$ of the set of difference elements

$$\varphi_{\sigma\tau}(g)\,\varphi_{\tau\sigma}(g)^{-1} \in *_\alpha \mathbb{G}_\alpha,$$

where $g \in \mathbb{G}_{\{\sigma,\tau\}}$ and $\sigma\neq\tau\in\mathcal{A}$. The universal property for $\mathbb{G}$ follows from the universal property for the sum $*_\alpha \mathbb{G}_\alpha$ (Section **11.3**) and an application of the Homomorphism Theorem (**11.4.6**) to the quotient group $\mathbb{G}$. □

Example 3 The bonding system $\langle a: \rangle \leftarrow \langle c: \rangle \to \langle b: \rangle$, defined by $a^2 \leftarrow c \to b^3$, has amalgamated sum $\langle a, b : a^2 = b^3 \rangle$.

SEIFERT-VAN KAMPEN THEOREM

Let $\mathcal{U} = \{U_\alpha : \alpha \in \mathcal{A}\}$ be any covering of a topological space X by *path-connected open subspaces* that contain the basepoint x. Also, let $\mathcal{U}$ contain the intersection $U_{\{\sigma,\tau\}} = U_\sigma \cap U_\tau$ of any pair of its members. The inclusions

$$U_\sigma \overset{i_{\sigma\tau}}{\supseteq} U_\sigma \cap U_\tau \overset{i_{\tau\sigma}}{\subseteq} U_\tau \qquad (\sigma \neq \tau \in \mathcal{A})$$

induce a bonding system of homomorphisms:

$$\pi_1(U_\sigma) \overset{\varphi_{\sigma\tau}}{\longleftarrow} \pi_1(U_\sigma \cap U_\tau) \overset{\varphi_{\tau\sigma}}{\longrightarrow} \pi_1(U_\tau) \qquad (\sigma \neq \tau \in \mathcal{A}).$$

And the inclusions $i_\alpha : U_\alpha \subseteq X$ ($\alpha \in \mathcal{A}$) induce an amalgamating system of homomorphisms $\{\psi_\alpha : \pi_1(U_\alpha) \to \pi_1(X)\ (\alpha \in \mathcal{A})\}$, as in this diagram:

$$\begin{array}{ccc} & \pi_1(U_\sigma \cap U_\tau) & \\ \varphi_{\sigma\tau} \swarrow & & \searrow \varphi_{\tau\sigma} \\ \pi_1(U_\sigma) & & \pi_1(U_\tau) \qquad (\forall\, \sigma \neq \tau \in \mathcal{A}) \\ \psi_\sigma \searrow & & \swarrow \psi_\tau \\ & \pi_1(X) & \end{array}$$

1.3 Seifert-Van Kampen Theorem *Under the above hypotheses on the open covering* $\mathcal{U} = \{U_\alpha : \alpha \in \mathcal{A}\}$ *of the space X, the bonding system*

$$\pi_1(U_\sigma) \overset{\varphi_{\sigma\tau}}{\longleftarrow} \pi_1(U_\sigma \cap U_\tau) \overset{\varphi_{\tau\sigma}}{\longrightarrow} \pi_1(U_\tau) \qquad (\sigma \neq \tau \in \mathcal{A})$$

has amalgamated sum $(\pi_1(X), \{\psi_\alpha : \pi_1(U_\alpha) \to \pi_1(X)\})$.

Proof: The following six steps show that for each group $\mathbb{H}$ and each amalgamating system $\{\rho_\alpha : \pi_1(U_\sigma) \to \mathbb{H}\ (\alpha \in \mathcal{A})\}$ for the bonding system, there exists a unique homomorphism $\rho : \pi_1(X) \to \mathbb{H}$ such that, for all $\alpha \in \mathcal{A}$,

$$\rho_\sigma = \rho \circ \psi_\sigma : \pi_1(U_\sigma) \to \pi_1(X) \to \mathbb{H}.$$

In the following discussion, a loop $f : \mathbb{I} \to X$ based at x is called a $\mathcal{U}$-**loop** if the image $f(\mathbb{I})$ is contained entirely in some member of the open covering $\mathcal{U} = \{U_\alpha : \alpha \in \mathcal{A}\}$. A factorization

$$f = i_\alpha \circ f_\alpha : \mathbb{I} \to U_\alpha \subseteq X$$

of a $\mathcal{U}$-loop f through a specific member U_α of $\mathcal{U}$ is called an α-**loop**.

Step 1 *For any $\mathcal{U}$-loop f, the element $\rho(f) = \rho_\alpha([f_\alpha]) \in \mathbb{H}$ is independent of the factorization $f = i_\alpha \circ f_\alpha : \mathbb{I} \to U_\alpha \subseteq X$ of f as an α-loop.*

Proof: If $f = i_\sigma \circ f_\sigma : \mathbb{I} \to U_\sigma \subseteq X$ and $f = i_\tau \circ f_\tau : \mathbb{I} \to U_\tau \subseteq X$ are factorizations expressing f as a σ-loop and a τ-loop, then there are the factorizations

$$f_\sigma = i_{\sigma\tau} \circ f_{\sigma\tau} : \mathbb{I} \to U_\sigma \cap U_\tau \subseteq U_\sigma \quad \text{and} \quad f_\tau = i_{\tau\sigma} \circ f_{\sigma\tau} : \mathbb{I} \to U_\sigma \cap U_\tau \subseteq U_\tau.$$

So, $\rho_\sigma([f_\sigma]) = \rho_\sigma(\varphi_{\sigma\tau}([f_{\sigma\tau}])) = \rho_\tau(\varphi_{\tau\sigma}([f_{\sigma\tau}])) = \rho_\tau([f_\tau])$ by the compatibility property, $\rho_\sigma \circ \varphi_{\sigma\tau} = \rho_\tau \circ \varphi_{\tau\sigma}$, of the amalgamating homomorphisms $\{\rho_\alpha\}$. □

Step 2 *If an α-loop f is path-homotopic in U_α to a product $f_1 \cdot f_2$ of two α-loops f_1 and f_2, then the elements $\rho(f)$, $\rho(f_1)$, and $\rho(f_2) \in \mathbb{H}$ defined in Step 1 are related this way:*

$$\rho(f) = \rho(f_1) \cdot \rho(f_2) \in \mathbb{H}.$$

Proof: By hypothesis, $[f_\alpha] = [f_{1\alpha}] \cdot [f_{2\alpha}]$ in $\pi_1(U_\alpha)$. Since $\rho_\alpha : \pi_1(U_\alpha) \to \mathbb{H}$ is a homomorphism, we have the evaluation:

$$\rho_\alpha([f_\alpha]) = \rho_\alpha([f_{1\alpha}] \cdot [f_{2\alpha}]) = \rho_\alpha([f_{1\alpha}]) \cdot \rho_\alpha([f_{2\alpha}]).$$

This is the relation $\rho(f) = \rho(f_1) \cdot \rho(f_2)$ in $\mathbb{H}$, by Step 1. □

Consider any path $f : \mathbb{I} \to X$. A partition of the interval $\mathbb{I}$,

$$\mathcal{P} : 0 = t_0 < t_1 < \ldots < t_{n-1} < t_n = 1,$$

is called *f-admissible* if f carries each subinterval $[t_{i-1}, t_i] \subset \mathbb{I}$ into some member U_{α_i} of the open covering $\mathcal{U}$. Such a partition exists by the Lebesgue Covering Lemma. Let $f_i : (\mathbb{I}, 0, 1) \to (X, f(t_{i-1}), f(t_i))$ denote a reparametrization of $f : [t_{i-1}, t_i] \to X$, for each $0 < i \leq n$. Then f is path-homotopic in X to an iterated product $f_1 \cdot f_2 \cdot \ldots \cdot f_n$ of these n paths. The image $f(t_i)$ of the partition point $t_i = [t_{i-1}, t_i] \cap [t_i, t_{i+1}]$ belongs to $U_{\alpha_i} \cap U_{\alpha_{i+1}}$, which is a path-connected subspace containing x. Let $h_i : (\mathbb{I}, 0, 1) \to (X, x, f(t_i))$ be a path in $U_{\alpha_i} \cap U_{\alpha_{i+1}}$ joining x to $f(t_i)$, for each $0 < i < n$. The process of dragging each point $f(t_i)$ ($0 < i < n$) backwards along the path h_i gives a path-homotopy,

$$f_1 \cdot f_2 \cdot \ldots \cdot f_i \cdot \ldots \cdot f_n \simeq (f_1 \cdot \overline{h}_1) \cdot (h_1 \cdot f_2 \cdot \overline{h}_2) \cdot \ldots \cdot (h_{i-1} \cdot f_i \cdot \overline{h}_i) \cdot \ldots \cdot (h_{n-1} \cdot f_n),$$

that deforms f into a product of $\mathcal{U}$-loops:

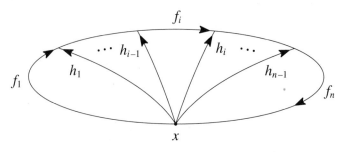

Factorization into $\mathcal{U}$-loops

Step 3 *For each loop $f : \mathbb{I} \to X$ and f-admissible partition $\mathcal{P}$, the element*

$$\rho(f, \mathcal{P}) = \rho(f_1 \cdot \bar{h}_1) \cdot \rho(h_1 \cdot f_2 \cdot \bar{h}_2) \cdot \ldots \cdot \rho(h_{n-1} \cdot f_n) \in \mathbb{H}$$

is independent of the paths h_i in $U_{\alpha_i} \cap U_{\alpha_{i+1}}$ joining x to $f(t_i)$ $(0 < i < n)$.

Proof: Let k_i $(0 < i < n)$ be other paths in $U_{\alpha_i} \cap U_{\alpha_{i+1}}$ joining x to $f(t_i)$. The product paths $h_i \cdot \bar{k}_i$ and $k_i \cdot \bar{h}_i$ are reverse loops in $U_{\alpha_i} \cap U_{\alpha_{i+1}}$ at x. Therefore, $\rho(h_i \cdot \bar{k}_i) \cdot \rho(k_i \cdot \bar{h}_i) = 1 \in \mathbb{H}$, for each $0 < i < n$, by Step 2. Insertion of these trivial products between the consecutive factors of $\rho(f, \mathcal{P})$ yields:

$$\rho(f_1 \cdot \bar{h}_1) \cdot \rho(h_1 \cdot f_2 \cdot \bar{h}_2) \cdot \ldots \cdot \rho(h_{n-1} \cdot f_n)$$
$$= \rho(f_1 \cdot \bar{h}_1) \cdot \rho(h_1 \cdot \bar{k}_1) \cdot \rho(k_1 \cdot \bar{h}_1) \cdot \rho(h_1 \cdot f_2 \cdot \bar{h}_2) \cdot \ldots$$
$$= \rho(f_1 \cdot \bar{h}_1 \cdot h_1 \cdot \bar{k}_1) \cdot \rho(k_1 \cdot \bar{h}_1 \cdot h_1 \cdot f_2 \cdot \bar{h}_2 \cdot h_2 \cdot \bar{k}_2) \cdot \ldots$$
$$= \rho(f_1 \cdot \bar{k}_1) \cdot \rho(k_1 \cdot f_2 \cdot \bar{k}_2) \cdot \ldots \cdot \rho(k_{n-1} \cdot f_n). \qquad \square$$

Step 4 *For each loop $f : \mathbb{I} \to X$, the element $\rho(f) = \rho(f, \mathcal{P}) \in \mathbb{H}$ defined in Step 3 is independent of the f-admissible partition $\mathcal{P}$.*

Proof: Let an f-admissible partition $\mathcal{P}$ be refined by the addition of a new partition point $t_{i-1} < t < t_i$. Then $\rho(f, \mathcal{P}) \in \mathbb{H}$ is unaltered, except its factor $\rho(h_{i-1} \cdot f_i \cdot \bar{h}_i)$ is replaced by the product $\rho(h_{i-1} \cdot f_i' \cdot \bar{h}) \cdot \rho(h \cdot f_i'' \cdot \bar{h}_i)$, where h is a path in U_{α_i} joining x to $f(t)$, and f_i' and f_i'' are reparametrizations of the restrictions of f to $[t_{i-1}, t]$ and $[t, t_i]$, respectively. The α_i-loop $h_{i-1} \cdot f_i \cdot \bar{h}_i$ is path-homotopic in U_{α_i} to the product $(h_{i-1} \cdot f_i' \cdot \bar{h}) \cdot (h \cdot f_i'' \cdot \bar{h}_i)$ of two α_i-loops. So the factor $\rho(h_{i-1} \cdot f_i \cdot \bar{h}_i)$ equals its replacement by Step 2. Therefore, $\rho(f, \mathcal{P}) \in \mathbb{H}$ is unaltered by the addition of any partition point to $\mathcal{P}$. Since any two f-admissible partitions have a common refinement obtained by the addition of the points of each one to the other one, the proof is complete. $\square$

Step 5 *If the loops $f, g : \mathbb{I} \to X$ are path-homotopic in X, then the group elements $\rho(f)$ and $\rho(g)$ defined in Step 4 are equal: $\rho(f) = \rho(g)$ in $\mathbb{H}$.*

Proof: We suppose first that there is an f-, g-admissible partition,

$$\mathcal{P} : 0 = t_0 < t_1 < \ldots < t_{n-1} < t_n = 1,$$

and a path-homotopy $H : \mathbb{I} \times \mathbb{I} \to X$ from f to g that carries the sub-rectangle $[t_{i-1}, t_i] \times \mathbb{I} \subset \mathbb{I} \times \mathbb{I}$ into some $U_{\alpha_i} \in \mathcal{U}$, for each $0 < i \leq n$. For $0 < i \leq n$, let

$$f_i : (\mathbb{I}, 0, 1) \to (X, f(t_{i-1}), f(t_i)) \quad \text{and} \quad g_i : (\mathbb{I}, 0, 1) \to (X, g(t_{i-1}), g(t_i))$$

denote reparametrizations of the restrictions $f, g : [t_{i-1}, t_i] \to X$. Also, for each $0 < i < n$, let the path $d_i : (\mathbb{I}, 0, 1) \to (U_{\alpha_i}, f(t_i), g(t_i))$ denote the restriction $H|_{\{t_i\} \times \mathbb{I}}$ of the path-homotopy H and let $h_i : (\mathbb{I}, 0, 1) \to (X, x, f(t_i))$ be a path in the intersection $U_{\alpha_i} \cap U_{\alpha_{i+1}}$. These paths appear in the next display.

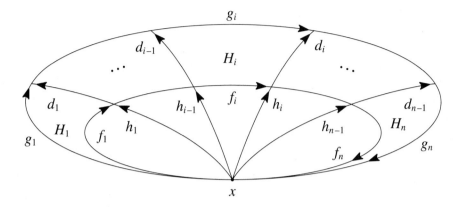

Then the restrictions $H_i : [t_{i-1}, t_i] \times \mathbb{I} \to U_{\alpha_i}$ $(0 < i \le n)$ of H can be incorporated into these path homotopies in U_{α_i} of α_i-loops:

$$f_1 \cdot \bar{h}_1 \simeq g_1 \cdot \bar{d}_1 \cdot \bar{h}_1 \text{ in } U_{\alpha_1}$$
$$\vdots$$
$$h_{i-1} \cdot f_i \cdot \bar{h}_i \simeq h_{i-1} \cdot \bar{d}_{i-1} \cdot g_i \cdot \bar{d}_i \cdot \bar{h}_i \text{ in } U_{\alpha_i}$$
$$\vdots$$
$$h_{n-1} \cdot f_n \simeq h_{n-1} \cdot \bar{d}_{n-1} \cdot g_n \text{ in } U_{\alpha_n}.$$

Therefore, by Steps 1 and 3,

$$\rho(f, \mathcal{P}) = \rho(f_1 \cdot \bar{h}_1) \cdot \ldots \cdot \rho(h_{i-1} \cdot f_i \cdot \bar{h}_i) \cdot \ldots \cdot \rho(h_{n-1} \cdot f_n)$$

equals

$$\rho(g, \mathcal{P}) = \rho(g_1 \cdot \bar{d}_1 \cdot \bar{h}_1) \cdot \ldots \cdot \rho(h_{i-1} \cdot \bar{d}_{i-1} \cdot g_i \cdot \bar{d}_i \cdot \bar{h}_i) \cdot \ldots \cdot \rho(h_{n-1} \cdot \bar{d}_{n-1} \cdot g_n).$$

By Step 4, $\rho(f) = \rho(f, \mathcal{P}) = \rho(g, \mathcal{P}) = \rho(g)$.

We now consider any two path-homotopic loops $f, g : \mathbb{I} \to X$ and any path-homotopy $H : \mathbb{I} \times \mathbb{I} \to X$ from f to g. By the Lebesgue Covering Lemma, there exist partitions,

$$\mathcal{P} : 0 = t_0 < t_1 < \ldots < t_{n-1} < t_n = 1 \text{ and } \mathcal{R} : 0 = r_0 < r_1 < \ldots < r_{m-1} < r_m = 1,$$

of the path parameter t and the homotopy parameter r such that the path-homotopy H carries each sub-rectangle $[t_{i-1}, t_i] \times [r_{k-1}, r_k] \subset \mathbb{I} \times \mathbb{I}$ into some $U_{\alpha_{i,k}} \in \mathcal{U}$. For each $0 \le k \le m$, the restriction $H|_{\mathbb{I} \times \{r_k\}}$ of the homotopy H is a loop $l_k : \mathbb{I} \to X$ at x. Furthermore, the restriction $H : \mathbb{I} \times [r_{k-1}, r_k] \to X$ is a homotopy from l_{k-1} to l_k of the special sort considered above. This implies that $\rho(l_{k-1}) = \rho(l_k)$. Then, by induction on k, the loops $f = l_0$ and $l_m = g$ have the same ρ assignment. $\square$

Step 6 *For loops $f, g : \mathbb{I} \to X$, the elements $\rho(f), \rho(g), \rho(f \cdot g) \in \mathbb{H}$ defined in Step 4 satisfy $\rho(f \cdot g) = \rho(f) \cdot \rho(g)$ in $\mathbb{H}$.*

Proof: Let

$$\mathcal{P} : 0 = t_0 < t_1 < \ldots < t_{n-1} < t_n = 1 \quad \text{and} \quad \mathcal{R} : 0 = s_0 < s_1 < \ldots < s_{m-1} < s_m = 1$$

be an f-admissible and a g-admissible partition, respectively. Then,

$$\mathcal{S} : 0 = \tfrac{1}{2}t_0 < \tfrac{1}{2}t_1 < \ldots < \tfrac{1}{2}t_{n-1} < \tfrac{1}{2} < \tfrac{1}{2}s_1 + \tfrac{1}{2} < \ldots < \tfrac{1}{2}s_{m-1} + \tfrac{1}{2} < \tfrac{1}{2}s_m + \tfrac{1}{2} = 1$$

is an $f \cdot g$-admissible partition for which $\rho(f \cdot g, \mathcal{S}) = \rho(f, \mathcal{P}) \cdot \rho(g, \mathcal{R})$. □

Conclusion *There is a unique homomorphism* $\rho : \pi_1(X) \to \mathbb{H}$ *such that* $\rho_\alpha = \rho \circ \psi_\alpha : \pi_1(U_\alpha) \to \pi_1(X) \to \mathbb{H}$ *for all* $\alpha \in \mathcal{A}$.

Proof: The assignment $\rho([f]) = \rho(f) \in \mathbb{H}$ is well-defined, by Steps 1-5; it is a homomorphism, by Step 6; and it satisfies $\rho_\alpha = \rho \circ \psi_\alpha$ for all $\alpha \in \mathcal{A}$, by Steps 1 and 4. The argument before Step 3 shows that any loop in X at x is path-homotopic to a product of α-loops. So $\pi_1(X)$ is generated by the images of the inclusion-induced homomorphisms $\psi_\alpha : \pi_1(U_\alpha) \to \pi_1(X)$. Thus, ρ is unique and the Seifert-Van Kampen Theorem is proved. □

> **Example 4** The hypothesis that the open sets have path-connected intersections is crucial to the Seifert-Van Kampen Theorem: although the complements U_- and U_+ of the north and south poles of $\mathbb{S}^1$ are path-connected open spaces that cover $\mathbb{S}^1$, the group $\pi_1(\mathbb{S}^1) = \mathbb{Z}$ cannot be an amalgamated sum of their trivial fundamental groups.

2 Polygonal Complexes

ONE-POINT UNIONS AND FREE PRODUCTS

2.1 Theorem *If each pair of distinct members of the open covering $\mathcal{U}$ in the Seifert-Van Kampen Theorem has simply connected intersection in X, then the fundamental group $\pi_1(X)$ is the sum of the groups $\{\pi_1(U_\alpha) : \alpha \in \mathcal{A}\}$.*

Proof: The pair $(\pi_1(X), \{\psi_\alpha : \pi_1(U_\alpha) \to \pi_1(X)\})$ is a sum in $\mathcal{G}$ because, by hypothesis, $\pi_1(U_\sigma \cap U_\tau) = \{1\}$ for each pair $\sigma \neq \tau \in \mathcal{A}$ and so in this case the bonding system consists of trivial homomorphisms:

$$\pi_1(U_\sigma) \xleftarrow{\varphi_{\sigma\tau}} \{1\} \xrightarrow{\varphi_{\tau\sigma}} \pi_1(U_\tau). \qquad \square$$

By Example 2, this implies that $\pi_1(X) \approx *_\alpha \pi_1(U_\alpha)$.

Let $\{(X_\alpha, b_\alpha) : \alpha \in \mathcal{A}\}$ be any collection of Hausdorff spaces with preferred basepoints. Their ***one-point union*** $\vee_\alpha X_\alpha$ is the space obtained from their disjoint union $\mathring{\cup}_\alpha X_\alpha$ by identifying all their basepoints $b_\alpha \in X_\alpha$ ($\alpha \in \mathcal{A}$). The identification point serves as the basepoint b for $\vee_\alpha X_\alpha$. The identification map, $\mathring{\cup}_\alpha X_\alpha \to \vee_\alpha X_\alpha$, carries each space X_σ homeomorphically onto a closed subspace of $\vee_\alpha X_\alpha$. And a subset is closed in $\vee_\alpha X_\alpha$ if and only if it meets each X_α in a closed subset.

The basepoint $b \in \vee_\alpha X_\alpha$ is called **non-degenerate** if it has a simply connected open neighborhood U in $\vee_\alpha X_\alpha$ such that each inclusion $X_\alpha \subseteq X_\alpha \cup U$ ($\alpha \in \mathcal{A}$) induces an isomorphism $\pi_1(X_\alpha) \approx \pi_1(X_\alpha \cup U)$.

2.2 Corollary *If $\vee_\alpha X_\alpha$ has a non-degenerate basepoint, $\pi_1(\vee_\alpha X_\alpha)$ is a sum of the groups $\{\pi_1(X_\alpha) : \alpha \in \mathcal{A}\}$.*

Proof: Since distinct members of the open covering $\mathcal{U} = \{X_\alpha \cup U : \alpha \in \mathcal{A}\}$ of $\vee_\alpha X_\alpha$ have the simply connected intersection U, Theorem 2.1 applies. Therefore, $\pi_1(\vee_\alpha X_\alpha)$ is a sum of the groups $\{\pi_1(X_\alpha) \approx \pi_1(X_\alpha \cup U) : \alpha \in \mathcal{A}\}$, with injections $\{\psi_\alpha : \pi_1(X_\alpha) \approx \pi_1(X_\alpha \cup U) \to \pi_1(\vee_\alpha X_\alpha)\}$. □

Any one-point union $\vee_\alpha \mathbb{S}^1_\alpha$ of copies of the 1-sphere $\mathbb{S}^1$ has a non-degenerate basepoint 1. The complement of the set of points in $\vee_\alpha \mathbb{S}^1_\alpha$ antipodal to the basepoint 1 is a simply connected open neighborhood U, as required. By Theorem **9**.3.8 and Corollary 2.2, $\pi_1(\vee_\alpha \mathbb{S}^1_\alpha)$ is a free group $\mathbb{F}(\{x_\alpha\})$ with free basis $\{x_\alpha = [e_\alpha] : \alpha \in \mathcal{A}\}$ consisting of the path-homotopy classes of the exponential loops $\{e_\alpha : \mathbb{I} \to \mathbb{S}^1_\alpha\}$. Each reduced word

$$w = x_{\alpha_1}^{k_1} x_{\alpha_2}^{k_2} \ldots x_{\alpha_n}^{k_n} \in \mathbb{F}(\{x_\alpha\})$$

corresponds to the path-homotopy class of the loop that *spells w*:

$$f_w = e_{\alpha_1}^{k_1} \cdot e_{\alpha_2}^{k_2} \cdot \ldots \cdot e_{\alpha_n}^{k_n} : \mathbb{I} \to \vee_\alpha \mathbb{S}^1_\alpha.$$

Example 1 $\pi_1(\mathbb{S}^1_a \vee \mathbb{S}^1_b)$ is a free group $\mathbb{F}(a, b)$ with free basis $\{a = [e_a], b = [e_b]\}$. The word $w = aba^{-1}b^{-1} \in \mathbb{F}(a, b)$ corresponds to the loop $f_w = e_a \cdot e_b \cdot e_a^{-1} \cdot e_b^{-1}$ traversing $\mathbb{S}^1_a$ then $\mathbb{S}^1_b$, counter-clockwise, and $\mathbb{S}^1_a$ then $\mathbb{S}^1_b$, clockwise.

The fundamental group of a one-point union is not always isomorphic to the free product of the fundamental groups of the original spaces. The union point may fail to be a non-degenerate basepoint. The following example shows how bad the situation can be.

Example 2 *The one-point union of simply connected spaces need not be simply connected.* The geometric cone C over the Hawaiian earrings $HE \subset \mathbb{R}^2$ (Example 7.1.8) is the subspace of $\mathbb{R}^3$ consisting of all line segments joining $<0, 0, 1>$ to points of HE. C is path-connected and each loop at $<0, 0, 1>$ is path-homotopic to the constant loop at $<0, 0, 1>$ via the straight-line homotopy in C. Thus, the geometric cone C is simply connected.

However, the one-point union $C \vee C$ of two copies of C, with the origin O as basepoint, is not simply connected. Any loop at O that circumscribes infinitely many of the base circles in a manner that alternates between the two copies of C is not path-homotopic to the constant loop at O.

ATTACHING 2-CELLS

Let A be a path-connected topological space, with selected basepoint x. Loops $\lambda : \mathbb{I} \to A$ based at x and continuous functions $\phi : (\mathbb{S}^1, 1) \to (A, x)$ are in bijective correspondence via the relation $\lambda = \phi \circ e : \mathbb{I} \to \mathbb{S}^1 \to A$, according to the Transgression Theorem (**5.1.9**).

It is possible to enlarge the space A so that any specific loop λ becomes path-homotopic to the constant loop at x. We simply form the *adjunction space* $X = A \cup_\phi c^2$ in which the closed disc $\mathbb{D}^2 \supset \mathbb{S}^1$ is attached to A via the continuous function $\phi : (\mathbb{S}^1, 1) \to (A, x)$ corresponding to the loop λ.

The adjunction space $X = A \cup_\phi c^2$ is formally defined in Section **10**.5 as the identification space obtained from the disjoint union $A \,\dot\cup\, \mathbb{D}^2$ by identifying each point $z \in \mathbb{S}^1 \subset \mathbb{D}^2$ with its image $\phi(z) \in A$. The identification map $p : A \,\dot\cup\, \mathbb{D}^2 \to A \cup_\phi c^2 = X$ embeds A as a closed subspace of X and it embeds the open disc $\mathbb{B}^2 = \mathbb{D}^2 - \mathbb{S}^1$ as the complementary subspace $c^2 = X - A$. The restriction $p\big|_{\mathbb{D}^2} : \mathbb{D}^2 \to X$ is a null-homotopy of $\phi = p\big|_{\mathbb{S}^1} : \mathbb{S}^1 \to X$. Thus, the loop $\lambda = \phi \circ e : \mathbb{I} \to A \subset X$ is path-homotopic in X to the constant loop at x.

More generally, let $\{\phi_\omega : (\mathbb{S}^1_\omega, 1) \to (A, x)\}$ be a collection of continuous functions indexed by $\omega \in \mathcal{W}$ and defined on the boundaries of a collection of disjoint closed discs $\{\mathbb{D}^2_\omega\}$. In the disjoint union $A \,\dot\cup\, (\dot\cup_\omega \mathbb{D}^2_\omega)$, we identify each point $z \in \mathbb{S}^1_\omega \subset \mathbb{D}^2_\omega$ with its image $\phi_\omega(z) \in A$, for each $\omega \in \mathcal{W}$.

The identification map $p : A \,\dot\cup\, (\dot\cup_\omega \mathbb{D}^2_\omega) \to X$ embeds A as a closed subspace of X and it embeds each open disc $\mathbb{B}^2_\omega = \mathbb{D}^2_\omega - \mathbb{S}^1_\omega$ as an open subspace c^2_ω of X, called a **2-cell** in X. The space X is denoted by $A \cup_\phi \{c^2_\omega\}$ and is called an ***adjunction space*** obtained by attaching the 2-cells $\{c^2_\omega\}$ to A. The continuous functions $\{\phi_\omega : \mathbb{S}^1_\omega \to A\}$ and the loops $\{\lambda_\omega = \phi_\omega \circ e : \mathbb{I} \to A\}$ are called the ***attaching maps*** and ***attaching loops*** for the 2-cells.

To picture the adjunction space $X = A \cup_\phi \{c^2_\omega\}$, view the attached 2-cells c^2_ω as open hemispheres with wrinkled equators attached to A:

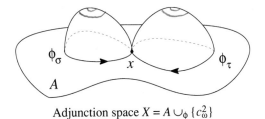

Adjunction space $X = A \cup_\phi \{c^2_\omega\}$

Example 3 The adjunction space $X = A \cup_k c^2$, in which a single 2-cell is adjoined via a constant attaching map $k : \mathbb{S}^1 \to A$, is homeomorphic to a one-point union $A \vee \mathbb{S}^2$ of A with a 2-sphere $\mathbb{S}^2 \cong \mathbb{D}^2 / \mathbb{S}^1$.

Example 4 The adjunction space $X = \mathbb{S}^1 \cup c^2_+ \cup c^2_-$, in which each of the 2-cells c^2_+ and c^2_- is adjoined to $\mathbb{S}^1$ via the identity attaching map $1_{\mathbb{S}^1} : \mathbb{S}^1 \to \mathbb{S}^1$, is homeomorphic to the 2-sphere $\mathbb{S}^2$.

When $\pi_1(A)$ is known, the fundamental group of the adjunction space $X = A \cup_\phi \{c_\omega^2\}$ is not hard to describe. The following heuristic argument is supported below by an application of the Seifert-Van Kampen Theorem.

Any loop in X may wander off A onto the attached 2-cells $\{c_\omega^2\}$. It is possible to preliminarily *homotop* the loop so that there is a free point in each attached 2-cell, and then *homotop* the loop away from the free points down into A. Thus, the homomorphism $j_\# : \pi_1(A) \to \pi_1(X)$, induced by the inclusion $j : A \subseteq X = A \cup_\phi \{c_\omega^2\}$, is an epimorphism.

Each attaching loop $\lambda_\omega = \phi_\omega \circ e : \mathbb{I} \to \mathbb{S}_\omega^1 \to A$ is homotopic over the attached 2-cell c_ω^2 in X to the constant loop x^*. So $\operatorname{Ker} j_\#$ contains the normal closure $\mathbb{N}(\{[\lambda_\omega]\})$ of their path-homotopy classes $[\lambda_\omega] \in \pi_1(A)$ ($\omega \in \mathcal{W}$).

Conversely, $\operatorname{Ker} j_\# \leq \mathbb{N}(\{[\lambda_\omega]\})$. If a loop f in A represents an element of $\operatorname{Ker} j_\#$, there exists a path-homotopy $H : f \simeq x^*$ in X. We can view the image of H as a blanket thrown over X with its edges resting on A and we can modify H without changing the position of the edges, so that the blanket rests entirely on A, save for finitely many small open discs in $\mathbb{I} \times \mathbb{I}$ that rest on attached 2-cells. By discarding these open discs in $\mathbb{I} \times \mathbb{I}$ and cutting slits in $\mathbb{I} \times \mathbb{I}$ to their boundary, we can convert the given homotopy $H : f \simeq x^*$ in X into a homotopy $K : f \simeq g$ in A between f and a product of conjugates,
$$g = g_1 \cdot \lambda_1^{\varepsilon_1} \cdot g_1^{-1} \cdot \ldots \cdot g_n \cdot \lambda_n^{\varepsilon_n} \cdot g_n^{-1},$$
of the attaching loops λ_ω (or inverses) of the 2-cells c_ω^2 on which H rested:

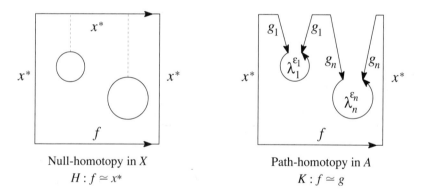

Null-homotopy in X
$H : f \simeq x^*$

Path-homotopy in A
$K : f \simeq g$

Thus, $\operatorname{Ker} j_\#$ equals the normal closure $\mathbb{N}(\{[\lambda_\omega]\})$ of the path-homotopy classes $[\lambda_\omega] \in \pi_1(A)$ ($\omega \in \mathcal{W}$) of the attaching loops. These unsubstantiated arguments suggest the following description of $\pi_1(X)$.

2.3 Theorem *The inclusion $j : A \subseteq X = A \cup_\phi \{c_\omega^2\}$ induces an epimorphism $j_\# : \pi_1(A) \to \pi_1(X)$ whose kernel is the normal closure of the set of path-homotopy classes $[\lambda_\omega] \in \pi_1(A)$ ($\omega \in \mathcal{W}$) of the attaching loops. Thus,*
$$\pi_1(X) \approx \frac{\pi_1(A)}{\mathbb{N}(\{[\lambda_\omega]\})}.$$

Proof: For each $\omega \in \mathcal{W}$, let O_ω denote the center point of the attached 2-cell c_ω^2. Removing the centers of all the open 2-cells, save the one indexed by $\sigma \in \mathcal{W}$, yields the open subset $U_\sigma = X - \cup \{O_\omega : \sigma \neq \omega \in \mathcal{W}\}$, which has the strong deformation retract $A_\sigma = A \cup_{\phi_\sigma} c_\sigma^2$. For any distinct pair of indices $\sigma \neq \tau$, the intersection $U_\sigma \cap U_\tau$ equals $X - \cup \{O_\omega : \omega \in \mathcal{W}\}$, which has A as a strong deformation retract. Since the open subsets $\{U_\sigma : \sigma \in \mathcal{W}\}$ are path-connected and have union X, the Seifert-Van Kampen Theorem (1.3) and the strong deformation retractions yield this amalgamated sum diagram:

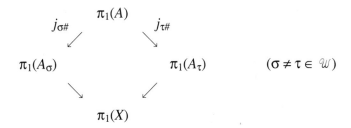

For each $\sigma \in \mathcal{W}$, $j_{\sigma\#} : \pi_1(A) \to \pi_1(A_\sigma = A \cup_{\phi_\sigma} c_\sigma^2)$ is an epimorphism whose kernel is the normal closure $\mathbb{N}([\lambda_\sigma])$. This results from an application of the Seifert-Van Kampen Theorem to the open subsets c_σ^2 and $A_\sigma - d_\sigma^2$, where d_σ^2 is the concentric closed disc in c_σ^2 with radius $\frac{1}{2}$. Their union is A_σ, their intersection is $c_\sigma^2 - d_\sigma^2$, and $\pi_1(c_\sigma^2) = \{1\}$. So $\pi_1(A_\sigma - d_\sigma^2) \to \pi_1(A_\sigma)$ is an epimorphism whose kernel is the normal closure of the subgroup

$$Im(\pi_1(c_\sigma^2 - d_\sigma^2) \to \pi_1(A_\sigma - d_\sigma^2)).$$

Moreover, 1_A extends to a homotopy equivalence $r : A_\sigma \to A_\sigma$ via the radial expansion of d_σ^2 onto c_σ^2 and a deformation retraction of $A_\sigma - d_\sigma^2$ onto A that sends the annulus $c_\sigma^2 - d_\sigma^2$ by a homotopy equivalence $c_\sigma^2 - d_\sigma^2 \simeq \mathbb{S}^1$ followed by the attaching map ϕ_σ. Therefore, it follows from the previous paragraph that $j_{\sigma\#} : \pi_1(A) \to \pi_1(A_\sigma)$ is an epimorphism whose kernel is the normal closure of the subgroup $\langle [\lambda_\sigma] \rangle = Im(\phi_{\sigma\#} : \pi_1(\mathbb{S}^1) \to \pi_1(A))$.

The above facts imply that $j_\# : \pi_1(A) \to \pi_1(X)$ is an epimorphism whose kernel is the normal closure of the path-homotopy classes $[\lambda_\omega] \in \pi_1(A)$ of the attaching loops of the 2-cells c_ω^2 ($\omega \in \mathcal{W}$). $\square$

2.4 Corollary *An adjunction space $X = (\vee_\alpha \mathbb{S}_\alpha^1) \cup_\phi \{c_\omega^2\}$, whose attaching loops $\lambda_\omega = \phi_\omega \circ e : \mathbb{I} \to \vee_\alpha \mathbb{S}_\alpha^1$ spell the words $r_\omega \in \mathbb{F}(\{x_\alpha\})$, has fundamental group $\pi_1(X) \approx \langle x_\alpha : r_\omega \rangle$.*

Proof: By hypothesis, the isomorphism $\pi_1(\vee_\alpha \mathbb{S}_\alpha^1) \approx \mathbb{F}(\{x_\alpha\})$ corresponds each path-homotopy class $[\lambda_\omega] \in \pi_1(\vee_\alpha \mathbb{S}_\alpha^1)$ with the word $r_\omega \in \mathbb{F}(\{x_\alpha\})$. Therefore, by Theorem 2.3, $\pi_1(X)$ is isomorphic to $\pi_1(\vee_\alpha \mathbb{S}_\alpha^1)/\mathbb{N}(\{[\lambda_\omega]\})$. This is the quotient group $\mathbb{F}(\{x_\alpha\})/\mathbb{N}(\{r_\omega\})$ presented by $\{x_\alpha : r_\omega\}$. $\square$

POLYGONAL COMPLEXES

Let $\mathcal{P} = \mathring{\cup}_\omega \mathbb{D}^2_\omega$ be the topological disjoint union of a family of polygonal versions of the closed disc $\mathbb{D}^2$. Let certain vertices of $\mathcal{P}$ be identified and let certain edges of $\mathcal{P}$ be identified according to linear homeomorphisms. The resulting identification space $X = \mathcal{P}/\sim$ is called a ***polygonal complex***. The images of the vertices, edges, and polygons of $\mathcal{P}$ are called the ***vertices***, ***edges***, and ***faces*** or ***0-cells***, ***1-cells***, and ***2-cells*** of the polygonal complex X. The union of the vertices and edges is called the ***1-skeleton*** of X.

When the polygonal complex X has a single vertex, we may view the 1-skeleton as a one-point union $\vee_\alpha \mathbb{S}^1_\alpha$ and X as an adjunction space of the form $(\vee_\alpha \mathbb{S}^1_\alpha) \cup_\phi \{c^2_\omega\}$. To describe each attaching map $\phi_\omega : \mathbb{S}^1_\omega \to \vee_\alpha \mathbb{S}^1_\alpha$, we label each edge of the polygon $\mathbb{D}^2_\omega$ with a symbol x_α for the 1-sphere $\mathbb{S}^1_\alpha$ that it becomes in the 1-skeleton $\vee_\alpha \mathbb{S}^1_\alpha$ and we direct that edge of $\mathbb{D}^2_\omega$ with an arrow corresponding to a counter-clockwise traverse of $\mathbb{S}^1_\alpha$. The ***boundary word*** $r_\beta = x_{\alpha_1}^{\pm 1} x_{\alpha_2}^{\pm 1} \ldots x_{\alpha_n}^{\pm 1}$ for the ω-face in X records the edge labels that are encountered on a counter-clockwise walk around the boundary of $\mathbb{D}^2_\omega$, with each label given an exponent +1 or −1 as the edge direction agrees or disagrees with the direction of the walk. Then the attaching loop,

$$\lambda_\omega = \phi_\omega \circ e : \mathbb{I} \to \mathbb{S}^1_\omega \to \vee_\alpha \mathbb{S}^1_\alpha,$$

spells the boundary word $r_\omega \in \mathbb{F}(\{x_\alpha\}) \approx \pi_1(\vee_\alpha \mathbb{S}^1_\alpha)$.

Therefore, by Corollary 2.4, we have this calculation:

2.5 Theorem *If a polygonal complex X has a single vertex, 1-skeleton $\vee_\alpha \mathbb{S}^1_\alpha$, and faces with boundary words $r_\omega \in \mathbb{F}(\{x_\alpha\})$, then $\pi_1(X) \approx \langle x_\alpha : r_\omega \rangle$.*

Here are three polygonal complexes to which this calculation applies:

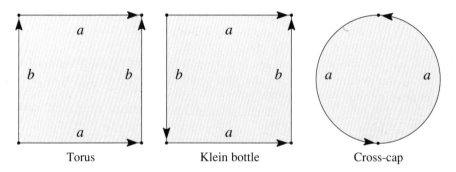

Torus　　　　　Klein bottle　　　　　Cross-cap

Example 5　The *torus* T results from a rectangular disc $\mathbb{D}^2$ by identifying opposite pairs of edges by translation. This produces T as a polygonal complex with a single vertex, 1-skeleton $\mathbb{S}^1_a \vee \mathbb{S}^1_b$, and a single face with boundary word $[a, b] = aba^{-1}b^{-1}$.

Therefore, $\pi_1(T) \approx \langle a, b : [a, b] \rangle$, a free abelian group of rank 2. This agrees with the calculation $\pi_1(T) \approx \pi_1(\mathbb{S}^1) \times \pi_1(\mathbb{S}^1)$ in Example **9.3.2**, which views the torus T as the topological product $\mathbb{S}^1 \times \mathbb{S}^1$.

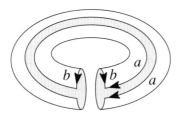
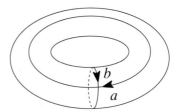

Edge identification   Loops $a, b : \mathbb{I} \to T$

Example 5: Torus

Example 6 The *Klein bottle* K results from a rectangular disc $\mathbb{D}^2$ by identifying one pair of opposite edges by translation and the other pair by inversion in the center of the rectangle. This produces K as a polygonal complex with a single vertex, 1-skeleton $\mathbb{S}_a^1 \vee \mathbb{S}_b^1$, and a single face with boundary word $aba^{-1}b$.

So $\pi_1(K) \approx \langle a, b : aba^{-1}b \rangle$. This is a non-abelian group with two generators and the non-commutativity law $ab = b^{-1}a$. This law says that a and a^{-1} commute with powers of b, at the expense of negating the exponent of the power. Therefore, each member of $\pi_1(K)$ can be uniquely represented as a two-syllable word $a^m b^n$, and multiplication of these representatives is given by $(a^m b^n)(a^r b^s) = a^{m+r} b^{(-1)^r n + s}$.

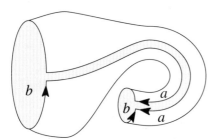
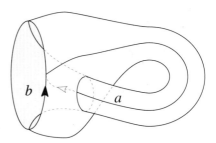

Edge identifications   Loops $a, b : \mathbb{I} \to T$

Example 6: Klein Bottle

Example 7 The *cross-cap* or *projective plane* $\mathbb{P}^2$ results from a circular disc $\mathbb{D}^2$ by identifying the pair of semi-circular edges via the antipodal map (top left of the four figures on p. 364). From this viewpoint, $\mathbb{P}^2$ is a polygonal complex with 1-skeleton $\mathbb{S}_a^1$ and a single face with boundary word aa. Thus, $\pi_1(\mathbb{P}^2) \approx \langle a : aa \rangle$, a cyclic group of order 2. So $\mathbb{P}^2$ has the curious feature that the exponential loop $a : \mathbb{I} \to \mathbb{S}_a^1$ is *essential* (not homotopic to a constant loop) in $\mathbb{P}^2$, but its square is *inessential* (homotopic to a constant loop) in $\mathbb{P}^2$.

$\mathbb{P}^2$ does not embed in $\mathbb{R}^3$; but, as is the case for the Klein bottle, some sense of this space can be obtained from a drawing of it having self-intersection. The second figure (top right) identifies the endpoints of the two semi-circular edges of $\mathbb{D}^2$ to produce a spherical *basket*. The third figure (bottom left) identifies the first halves of the two semi-circular edges into a *radius* of the spherical basket. $\mathbb{P}^2$ results when just the last halves of the two semi-circular edges are identified. This is not possible in $\mathbb{R}^3$. To get something in $\mathbb{R}^3$, the fourth figure identifies the last halves of the two semi-circular edges with the radius. Some sense of $\mathbb{P}^2$ is gained from the fourth figure (bottom right) when one keeps in mind these traveling instructions: there are two unrelated ways to cross the *radius*, traveling between the left front and the right rear and traveling between the right front and the left rear.

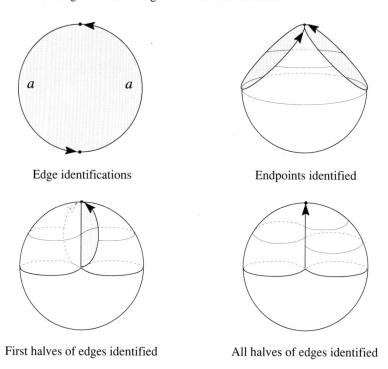

Example 7: Cross-cap or projective plane

3 Knot Theory

When the ends of a rope are allowed to be manipulated in space, it is possible, with enough patience, to tie or untie any knot in the length of rope. However, when the ends of the rope are permanently spliced together, or alternatively, when the rope is infinitely long and the ends are kept at infinity, it may not be possible to untie the knot. The latter situation is the knot theory that is explored mathematically in this section.

MATHEMATICAL KNOTS

Knot theory is a topological investigation of embeddings (placements) of one topological space in a larger one. The smaller space, representing the length of rope, is the real line $\mathbb{R}^1$; the larger space, representing the world about us, is real 3-space $\mathbb{R}^3$. As indicated above, it is necessary to restrict attention to embeddings $\mathbb{R}^1 \to \mathbb{R}^3$ that keep the ends of $\mathbb{R}^1$ at infinity. To mathematically express this requirement, we use the 1-point compactifications $\mathbb{S}^1 \cong \mathbb{R}^1 \cup \infty$ and $\mathbb{S}^3 \cong \mathbb{R}^3 \cup \infty$ of $\mathbb{R}^1$ and $\mathbb{R}^3$, as follows.

A **knot** is a *proper* embedding $\kappa : \mathbb{R}^1 \to \mathbb{R}^3$, that is, an embedding that extends to an embedding $\mathbb{S}^1 \to \mathbb{S}^3$ by the assignment $\infty \to \infty$. The image $K = \kappa(\mathbb{R}^1)$ is often used to denote the knot.

Intuitively, two knots are the same if over time we can manipulate the containing space until one knot is converted into the other. Thus, two knots J and K are **equivalent** if the identity $1_{\mathbb{S}^3}$ is homotopic through homeomorphisms of $\mathbb{S}^3$ to a homeomorphism $h : \mathbb{S}^3 \to \mathbb{S}^3$ such that $h(J) = K$. Otherwise, the knots J and K are called **inequivalent**.

Incidentally, a homotopy of homeomorphisms is called an **isotopy**. So the equivalence of knots requires a homeomorphism h that is isotopic to $1_{\mathbb{S}^3}$.

The **trivial knot** is the usual proper inclusion $\iota : \mathbb{R}^1 \subset \mathbb{R}^3$. A knot K is **unknotted** if it is equivalent to the trivial knot; otherwise, K is **knotted**.

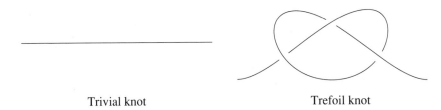

Trivial knot   Trefoil knot

A commonly held belief is that the trefoil knot is knotted, i.e., is not equivalent to the trivial knot. We are nowhere with our knot theory if we cannot prove this rigorously. But, the only homotopy invariant tool at our disposal is the fundamental group. It is useless to apply it to either the embedded space $\mathbb{R}^1$ or the containing space $\mathbb{R}^3$, as they are contractible spaces and so each one has a trivial fundamental group by Theorem **10**.3.9.

KNOT COMPLEMENTS

As we shall show, the complement $\mathbb{R}^3 - K$ of any knot K in $\mathbb{R}^3$ has a nontrivial fundamental group $\pi_1(\mathbb{R}^3 - K)$, called the **group of the knot K**. Moreover, for equivalent knots J and K, there is a homeomorphism h carrying J onto K and, hence, $\mathbb{R}^3 - J$ onto $\mathbb{R}^3 - K$. Therefore, the complements $\mathbb{R}^3 - J$ and $\mathbb{R}^3 - K$ of equivalent knots J and K have isomorphic fundamental groups. This gives the most basic method of distinguishing between knots:

3.1 Knot Theorem *Knots with non-isomorphic groups are inequivalent.*

To apply the Knot Theorem, we need to be able to calculate knot groups. Below, we give a general technique for determining a presentation for the group of a knot, once a suitable projection of it is drawn. But first we give a couple of basic examples.

TRIVIAL KNOT GROUP

We first consider the trivial knot $\iota : \mathbb{R}^1 \subset \mathbb{R}^3$. Its complement $\mathbb{R}^3 - \mathbb{R}^1$ strong deformation retracts to a 1-sphere $\mathbb{S}^1$ in two stages:

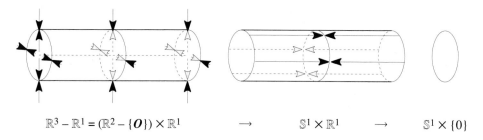

$\mathbb{R}^3 - \mathbb{R}^1 = (\mathbb{R}^2 - \{\boldsymbol{O}\}) \times \mathbb{R}^1 \qquad \rightarrow \qquad \mathbb{S}^1 \times \mathbb{R}^1 \qquad \rightarrow \qquad \mathbb{S}^1 \times \{0\}$

Therefore, the trivial knot group is isomorphic to $\pi_1(\mathbb{S}^1) = \mathbb{Z}$.

TREFOIL KNOT GROUP

For a second illustration of the general technique, we consider the embedding $\tau : \mathbb{R}^1 \rightarrow T \subset \mathbb{R}^3$ of the trefoil knot T (below left). In order to construct a convenient strong deformation retract of the knot complement $\mathbb{R}^3 - T$, we

(1) extend τ to an embedding $\tau : \mathbb{R}^1 \times \mathbb{D}^2 \rightarrow \mathbb{R}^3$ that loosely ties the solid cylinder $\mathbb{R}^1 \times \mathbb{D}^2$ into a *closed tubular neighborhood* $\mathbb{D}(T)$ of T in $\mathbb{R}^3$;

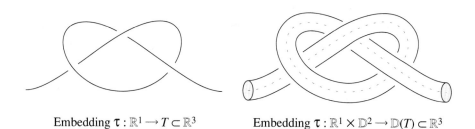

Embedding $\tau : \mathbb{R}^1 \rightarrow T \subset \mathbb{R}^3$ Embedding $\tau : \mathbb{R}^1 \times \mathbb{D}^2 \rightarrow \mathbb{D}(T) \subset \mathbb{R}^3$

(2) fully inflate the tubular neighborhood $\mathbb{D}(T)$ until its boundary contacts itself along a 1-point union of two circles, producing an *immersion* (i.e., local embedding) $\tau : \mathbb{R}^1 \times \mathbb{D}^2 \rightarrow \mathbb{R}^3$ that still embeds the open solid cylinder $\mathbb{R}^1 \times \mathbb{B}^2$ as an *open tubular neighborhood* $\mathbb{B}(T)$ of T in $\mathbb{R}^3$;

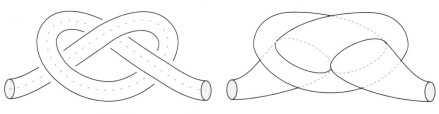

Embedding $\tau : \mathbb{R}^1 \times \mathbb{D}^2 \rightarrow \mathbb{D}(T) \subset \mathbb{R}^3$ Immersion $\tau : \mathbb{R}^1 \times \mathbb{D}^2 \rightarrow \mathbb{R}^3$

(3) strong deformation retract the knot complement $\mathbb{R}^3 - T$ to the immersed boundary cylinder $\tau(\mathbb{R}^1 \times \mathbb{S}^1)$, the *infinite trefoil shell*; and

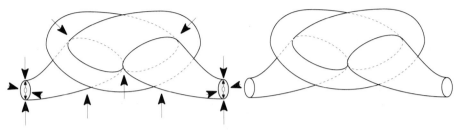

Retraction of $\mathbb{R}^3 - T$ Infinite trefoil shell $\tau(\mathbb{R}^1 \times \mathbb{S}^1)$

(4) discard the ends of the infinite trefoil shell $\tau(\mathbb{R}^1 \times \mathbb{S}^1)$, leaving a strong deformation retract $\mathbb{S}(T)$ of $\mathbb{R}^3 - T$, called the *trefoil shell*.

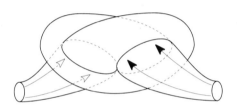
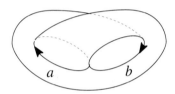

Retraction of infinite trefoil shell   Trefoil shell $\mathbb{S}(T)$

The trefoil shell $\mathbb{S}(T)$ is a polygonal complex with single vertex, 1-skeleton $\mathbb{S}_a^1 \vee \mathbb{S}_b^1$, and a single face with boundary word $abab^{-1}a^{-1}b^{-1}$. So, by Theorem 2.5, the trefoil knot group $\pi_1(\mathbb{R}^3 - T)$ has the presentation:

$$\langle a, b : abab^{-1}a^{-1}b^{-1} \rangle = \langle a, b : aba = bab \rangle.$$

It will follow from the Knot Theorem that the trefoil knot is knotted, once we show the trefoil knot group is not isomorphic to the trivial knot group $\mathbb{Z}$.

But its abelianization is $\mathbb{A}(a, b)/\langle ab^{-1} \rangle \approx \mathbb{Z}$. So the trefoil knot group and the trivial knot group $\mathbb{Z}$ are not distinguished by their abelianizations.

To get the job done, we have to analyze the relation, $aba = bab$, that is satisfied by the two generators of the trefoil knot group. It is also satisfied by the two transpositions (1 2) and (2 3) that generate the symmetric group $\mathbb{S}_3$ of all permutations of the three symbols $\{1, 2, 3\}$:

$$(1\ 2)(2\ 3)(1\ 2) = (1\ 3) = (2\ 3)(1\ 2)(2\ 3).$$

Therefore, there is an epimorphism $\langle a, b : abab^{-1}a^{-1}b^{-1} \rangle \to \mathbb{S}_3$ defined by the assignments $a \to (1\ 2)$ and $b \to (2\ 3)$. This property of the trefoil knot group $\pi_1(\mathbb{R}^3 - T)$ distinguishes it from the trivial knot group $\mathbb{Z}$; for there can be no epimorphism $\mathbb{Z} \to \mathbb{S}_3$ of the abelian (in fact, cyclic) group $\mathbb{Z}$ onto the non-abelian, non-cyclic group $\mathbb{S}_3$. *Thus, the trefoil knot is knotted.*

KNOT COMPLEMENT SPINES

The technique illustrated above with the trefoil knot applies to any knot K whose proper embedding $\kappa : \mathbb{R}^1 \times \mathbf{O} \to K \subset \mathbb{R}^3$ can be thickened to an embedding $\kappa : \mathbb{R}^1 \times \mathbb{D}^2 \to K \times \mathbb{D}^2 \subset \mathbb{R}^3$. Such knots are called **tame** knots. The tame knots are precisely the knots that avoid all *wild* behavior. For example, any knot consisting of finitely many line segments (and two infinite rays) is certainly a tame knot.

Our derivation of a presentation of the group of a tame knot K begins with a construction of a polygonal complex X that is a strong deformation retract of the knot complement $\mathbb{R}^3 - K$, a so-called **spine** of $\mathbb{R}^3 - K$.

The construction begins with a **central projection** of K, that is, the projected image of K on a plane separating the knot from an observer at a selected point of perspective $*$, as indicated below.

We put the knot in **regular position** in the sense that (1) there are at most double points in this projection, (2) the double points are finite in number, and (3) each double point represents a genuine crossing (i.e., an infinitesimal change of the perspective point $*$ cannot eliminate the crossing).

Each double point P is the projection of an **undercrossing point U** and an **overcrossing point O**, encountered in that order along a ray from the point of perspective $*$. The components of the complement, $K - \mathcal{O}$, of the set $\mathcal{O}$ of overcrossing points are called the **underpasses** of the knot K. We label the overcrossing points $O_1, \ldots, O_n$ and the underpasses $A_0, \ldots, A_n$, in the order that they are encountered along a traverse of the knot K in the positive direction. For convenience, we make all undercrossing points on any underpass A_i lie close to a selected point Q_i of A_i.

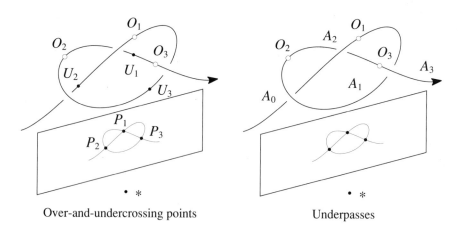

Over-and-undercrossing points   Underpasses

Notice that the overcrossing point O_i ($1 \leq i \leq n$) does not lie over the underpass A_i, but rather over an underpass A_{σ_i} for some $0 \leq \sigma_i \leq n$. In the example (above right), O_1 lies over A_2, O_2 lies over A_0, and O_3 lies over A_1. Also notice that an underpass, such as A_3, may not pass under anything.

We reparametrize the knot embedding $\kappa : \mathbb{R}^1 \times \boldsymbol{O} \to K \subset \mathbb{R}^3$ to satisfy

$$\kappa(0 \times \boldsymbol{O}) = Q_0 \in A_0, \ldots, \kappa(n \times \boldsymbol{O}) = Q_n \in A_n.$$

Then we select an extension $\kappa : \mathbb{R}^1 \times \mathbb{D}^2 \to K \times \mathbb{D}^2 \subset \mathbb{R}^3$ that loosely ties the solid cylinder $\mathbb{R}^1 \times \mathbb{D}^2$ into a *closed tubular neighborhood* $\mathbb{D}(K)$ of the knot K and makes the $n+1$ cross-sectional discs indexed by the integers $0 \leq k \leq n$,

$$D_0 = \kappa(\{0\} \times \mathbb{D}^2), \ldots, D_n = \kappa(\{n\} \times \mathbb{D}^2),$$

lie in planes containing the point of perspective $*$ and the points $Q_0, \ldots, Q_n$, respectively. Then we stretch these discs, drawing the boundary point 1 on each one to the point of perspective $*$, to form a 1-point union $\vee_i D_i$:

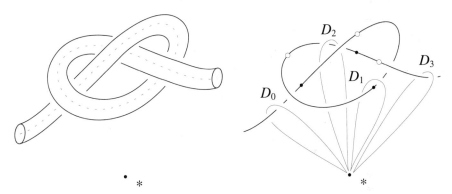

Closed tubular neighborhood　　　Stretched cross-sectional discs

Separating consecutive cross-sectional discs D_{i-1} and D_i is the overcrossing point O_i. The undercrossing point U_i below O_i lies on the underpass A_{σ_i}, and there is the cross-sectional disc D_{σ_i} strung on the underpass A_{σ_i} at Q_{σ_i}, a point that lies near the undercrossing point U_i and all other undercrossing points on A_{σ_i}. We inflate the image $\kappa([i-1, i] \times \mathbb{D}^2)$ of the solid cylinder section to make the arc $\beta_i = [i-1, i] \times \{1\}$ on the boundary cylinder ride the boundary circle $\mathbb{S}^1_{\sigma_i} \subset D_{\sigma_i}$ in the $\varepsilon_i = \pm 1$ direction, whichever is appropriate.

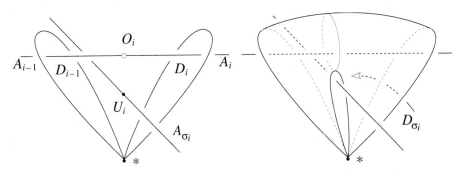

Overcrossing O_i　　　Modified section $\kappa([i-1, i] \times \mathbb{D}^2)$

Other undercrossing points U_k may lie on the same underpass $A_{\sigma_i} = A_{\sigma_k}$, so that other solid cylinder sections $\kappa([k-1, k] \times \mathbb{D}^2)$ may also ride along the same boundary circle $\mathbb{S}^1_{\sigma_i} = \mathbb{S}^1_{\sigma_k}$. These cylindrical sections are kept in the spatial ordering of the overcrossing points U_k above the underpass $A_{\sigma_i} = A_{\sigma_k}$.

This procedure produces an immersion $\kappa : \mathbb{R}^1 \times \mathbb{D}^2 \to \mathbb{R}^3$. It embeds the open solid cylinder $\mathbb{R}^1 \times \mathbb{B}^2$ as an open tubular neighborhood $\mathbb{B}(K)$ of K. Because $\mathbb{B}(K)$ is *fully inflated*, the immersed boundary cylinder $\kappa(\mathbb{R}^1 \times \mathbb{S}^1)$ is a strong deformation retract of the knot complement $\mathbb{R}^3 - K$. When the ends $\kappa((-\infty, 0) \times \mathbb{S}^1)$ and $\kappa((n, \infty) \times \mathbb{S}^1)$ of $\kappa(\mathbb{R}^1 \times \mathbb{S}^1)$ are discarded, the strong deformation retract $\mathbb{S}(K) = \kappa([0, n] \times \mathbb{S}^1)$ remains as a spine of $\mathbb{R}^3 - K$.

WIRTINGER PRESENTATIONS

There is no need to draw any of this procedure, beyond a central projection of the knot. The spine $\mathbb{S}(K)$ just produced has the structure of a polygonal complex with a single vertex $*$, one edge $\mathbb{S}^1_i = \kappa(\{i\} \times \mathbb{S}^1)$ for each of the underpasses A_i ($0 \le i \le n$), and one *cylindrical-like* face for each of the overcrossing points O_i ($1 \le i \le n$).

All edges in the 1-skeleton $\vee_i \mathbb{S}^1_i$ inherit from the immersion the counter-clockwise orientation about K. All faces attach similarly to $\vee_i \mathbb{S}^1_i$, but the boundary word of each one depends upon the direction of the traverse of the underpass beneath the overcrossing, as indicated below:

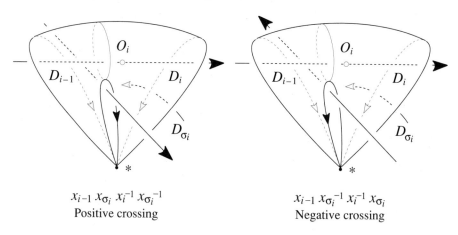

$x_{i-1} x_{\sigma_i} x_i^{-1} x_{\sigma_i}^{-1}$
Positive crossing

$x_{i-1} x_{\sigma_i}^{-1} x_i^{-1} x_{\sigma_i}$
Negative crossing

In each of the two cases displayed above, the boundary word for the face has the form

$$x_{i-1} x_{\sigma_i}^{\varepsilon_i} x_i^{-1} x_{\sigma_i}^{-\varepsilon_i}.$$

In the first case, called a ***positive crossing***, the underpass $A_{\sigma(i)}$ passes, from left to right, below O_i and the exponent ε_i is $+1$. In the second case, called a ***negative crossing***, the underpass $A_{\sigma(i)}$ passes, from right to left, below O_i and the exponent ε_i is -1.

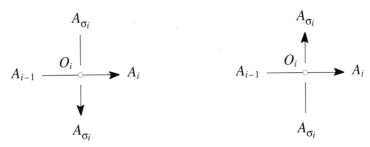

Positive crossing Negative crossing

Thus, by Theorem 2.5, the fundamental group of the spine $\mathbb{S}(K)$ of the knot complement $\mathbb{R}^3 - K$ has a presentation with one generator x_i for each of the underpasses A_i ($0 \leq i \leq n$) and one relator $x_{i-1} x_{\sigma_i}^{\varepsilon_i} x_i^{-1} x_{\sigma_i}^{-\varepsilon_i}$ for each of the overcrossing points O_i ($1 \leq i \leq n$):

$$\langle x_0, \ldots, x_n : x_1 x_{\sigma_1}^{\varepsilon_1} x_2^{-1} x_{\sigma_1}^{-\varepsilon_1}, \ldots, x_n x_{\sigma_n}^{\varepsilon_n} x_{n+1}^{-1} x_{\sigma_n}^{-\varepsilon_n} \rangle,$$

with $\varepsilon_i = \pm 1$ accordingly as the ith crossing is positive or negative.

This is called a **Wirtinger presentation** for the knot group $\pi_1(\mathbb{R}^3 - K)$ associated with the central projection of the knot K.

Example 1 A Wirtinger presentation for the trefoil knot is

$$\langle a, b, c, d : acb^{-1}c^{-1}, bac^{-1}a^{-1}, cbd^{-1}b^{-1} \rangle.$$

By use of the second and third relators, this simplifies to the presentation:

$$\langle a, c : acac^{-1}a^{-1}c^{-1} \rangle.$$

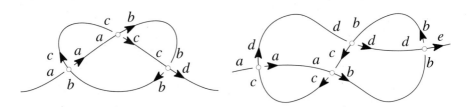

Example 1: Trefoil knot   Example 2: Figure-of-eight knot

Example 2 A Wirtinger presentation for the figure-of-eight knot is

$$\langle a, b, c, d, e : acb^{-1}c^{-1}, bd^{-1}c^{-1}d, cad^{-1}a^{-1}, db^{-1}e^{-1}b \rangle$$

or more simply,

$$\langle a, b, c, d : acb^{-1}c^{-1}, bd^{-1}c^{-1}d, cad^{-1}a^{-1} \rangle.$$

Notice that when there is no undercrossing point on the last (first) underpass, the last (first) generator appears only in the last (first, respectively) relator. That generator and relator may be deleted from the presentation.

Exercises

SECTION 1

1* Prove that a topological space X covered by a family of open subspaces $\{U_\alpha\}$ has the properties indicated in the opening paragraph of this section and is characterized by these properties.

2 Devise a bonding system whose amalgamated sum is $\mathbb{Z} \times \mathbb{Z}$.

3 Devise a bonding system whose amalgamated sum is $\langle x, y : x^3, y^3 = x^2 \rangle$.

4 Let $\mathbb{H}$ and $\mathbb{K}$ be normal subgroups of the group $\mathbb{G}$. Describe the amalgamated sum of the bonding system of inclusion homomorphisms $\{\mathbb{H} \geq \mathbb{H} \cap \mathbb{K} \leq \mathbb{K}\}$.

5 Let $\mathbb{K}$ be a subgroup of the group $\mathbb{G}$. Describe the amalgamated sum of the bonding system $\{\{1\} \leftarrow \mathbb{K} \leq \mathbb{G}\}$.

6 Let $\mathbb{G}$ and $\mathbb{H}$ be groups that share a subgroup $\mathbb{K}$. Describe the amalgamated sum of the system $\{\mathbb{H} \geq \mathbb{K} \leq \mathbb{G}\}$ of inclusion bonding homomorphisms.

7 Consider an indexed family of subgroups $\{\mathbb{G}_\alpha \leq \mathbb{G}, \alpha \in \mathscr{A}\}$ that contains the intersection of each pair of members $\mathbb{G}_\sigma \cap \mathbb{G}_\tau \leq \mathbb{G}$. Determine the amalgamated sum of the system of inclusion bonding homomorphisms $\{\mathbb{G}_\sigma \geq \mathbb{G}_\sigma \cap \mathbb{G}_\tau \leq \mathbb{G}_\tau\}$.

8 Let X be the union of an infinite nested family of open path-connected subspaces
$$U_1 \subseteq U_2 \subseteq \ldots \subseteq U_n \subseteq U_{n+1} \subseteq \ldots .$$
Prove: If each inclusion-induced homomorphism $\pi_1(U_n) \to \pi_1(U_{n+1})$ is trivial, then X is simply connected.

9 Apply the Seifert-Van Kampen Theorem to calculate the fundamental group of the 2-sphere $\mathbb{S}^2$, using an open covering by neighborhoods of the northern hemisphere, southern hemisphere, and equator.

10 Apply the Seifert-Van Kampen Theorem to calculate the fundamental group of the doubly-punctured real plane $\mathbb{R}^2 - \{<0,0>, <1,0>\}$.

11 Let X be a space that is the union of two open simply-connected subspaces U and V, whose intersection is path-connected and has fundamental group $\mathbb{G}$. Calculate the fundamental group of X.

12 Determine the fundamental group of the complement of the coordinate axes in $\mathbb{R}^3$.

SECTION 2

1 Prove that the n-times punctured plane has fundamental group $\mathbb{F}(x_1, \ldots, x_n)$, a free group with free rank n.

2 Consider the map $d_n : \mathbb{S}^1 \to \mathbb{S}^1$, $d_n(e^{i\theta}) = e^{in\theta}$, of degree n. The space obtained by adjoining a two-cell to $\mathbb{S}^1$ via d_n is denoted by P_n and is called the ***pseudo-projective plane of order n***. Determine the fundamental group of P_n.

3 Construct two- and three-dimensional spaces with fundamental group $\mathbb{Z} \times \mathbb{Z} \times \mathbb{Z}$.

SECTIONS 1-3

4 Construct the simplest spaces with the following fundamental groups.
   (a) $\mathbb{Z}_m \times \mathbb{Z}_n$ (direct product)   (b) $\mathbb{Z}_m * \mathbb{Z}_n$ (free product)

5 Construct polygonal complexes with the following fundamental groups.
   (a) $\langle a, b, c : acb^{-1}c^{-1}, bac^{-1}a^{-1} \rangle$ (*trefoil shell*)
   (b) $\langle a, b, c, d : ac^{-1}b^{-1}c, bd^{-1}c^{-1}d, cad^{-1}a^{-1} \rangle$ (*figure-of-eight shell*)
   Sketch embeddings of both these spaces in $\mathbb{R}^3$.

6 Prove that every group $\mathbb{G}$ arises as the fundamental group of some adjunction space $X = (\vee_\alpha \mathbb{S}^1_\alpha) \cup_\varphi \{c^2_\omega\}$ obtained by attaching 2-cells to a 1-point union of 1-spheres.

7 Let X be a polygonal complex with a single vertex and suppose that each face has some unshared edge, i.e., an edge whose label appears just once in all the boundary words. Prove that $\pi_1(X)$ is a free group.

8 Prove that the projective plane $\mathbb{P}^2$ is homeomorphic to the identification space obtained from the 2-sphere $\mathbb{S}^2$ by identifying antipodal points.

9 Calculate the fundamental group of the pinched torus $(\mathbb{S}^1 \times \mathbb{S}^1) / (\mathbb{S}^1 \times \{1\})$ and the funcamental group of the torus-with-membrane $(\mathbb{S}^1 \times \mathbb{S}^1) \cup (\mathbb{D}^2 \times \{1\})$.

A continuous function $f : X \to Y$ is an ***immersion*** of X into Y if locally it is an embedding, i.e., each $x \in X$ has a neighborhood $N(x)$ such that $f : N(x) \to f(N(x))$ is a homeomorphism. The drawing of the Klein bottle in Example 6 is an immersion into $\mathbb{R}^3$, but the drawing of the projective plane in Example 7 is not.

10 Construct an immersion of the projective plane into $\mathbb{R}^3$. (Werner Boy presented several such immersions in his thesis at the University of Göttingen in 1901.)

11 (a) Form an identification space X of $\mathbb{I}^3$ by identifying opposite faces of the cube via translation. Give a presentation of the fundamental group of X.
   (b) Form an identification space Y of $\mathbb{I}^3$ by identifying opposite faces of the cube via right-helix quarter turns. Give a presentation of the fundamental group of Y.

12 Identify the opposite edges of a regular hexagon via translations to produce an identification space H. Give a presentation of the fundamental group of H.

13 Determine the fundamental group of the complement of the coordinate axes in $\mathbb{R}^3$.

SECTION 3

1 Prove that the abelianization of every knot group is $\mathbb{Z}$.

2 Prove that the figure-of-eight knot is not equivalent to the trivial knot.

3 (a) Find presentations for the groups of the square and granny knots.
   (b) Prove that the square and granny knots are non-trivial knots.

square knot   granny knot

4 Prove that the square knot and the granny knot have isomorphic knot groups.

13

SURFACES

The 2-sphere, the torus, the Klein bottle, and the cross-cap are locally indistinguishable; all their points have open neighborhoods homeomorphic to $\mathbb{R}^2$. Any connected, compact, Hausdorff space in which each point has an open neighborhood homeomorphic to $\mathbb{R}^2$ is called a *surface*. The pin-cushion space in Section **1**.1 is not a surface; exactly one of its points fails to have open neighborhoods homeomorphic to $\mathbb{R}^2$. This chapter presents a classification of all surfaces, a result that is an early topological success story.

1   Polygonal Surfaces

An ***n-dimensional manifold*** or ***n-manifold*** is a second countable, Hausdorff space M in which each point has an open neighborhood U homeomorphic to real n-space $\mathbb{R}^n$. Then $(U, \phi : U \cong \mathbb{R}^n)$ is called a **neighborhood chart** in M.

The neighborhood charts in an n-manifold show that it is locally path-connected; hence, by Theorem **6**.3.11, its components and path-components coincide and they are open subsets. So any n-manifold is the topological disjoint union of its components by Theorem **5**.4.2. This reduces the study of n-manifolds to a consideration of connected n-manifolds.

There are just two connected 1-manifolds: the compact 1-manifold $\mathbb{S}^1$ and the non-compact 1-manifold $\mathbb{R}^1$ (Exercise 1).

The situation is richer in dimension 2. Compact, connected, 2-manifolds are called ***surfaces*** and they form two infinite families, as described below.

HANDLEBODY SURFACES

The 2-sphere and the torus are the simplest surfaces. We can view them as the boundary surfaces of solids with zero and one handle, respectively. From this viewpoint, they are the first two members of an infinite family of boundary surfaces of *handlebody solids* in real 3-space $\mathbb{R}^3$:

SECTION 1   POLYGONAL SURFACES

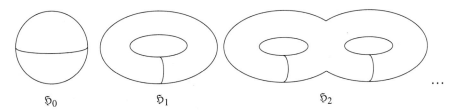

Family of handlebody surfaces

The gth member of the family $\mathfrak{H}_g$ ($g \geq 0$), called the **handlebody surface of genus g**, is a generalization of the torus that involves g handles. The pictures above show that small relative neighborhoods of the points of the subspace $\mathfrak{H}_g \subset \mathbb{R}^3$ are homeomorphic to $\mathbb{R}^2$. In other words, $\mathfrak{H}_g$ is a surface.

Each handlebody surface $\mathfrak{H}_g$ ($g \geq 0$) can be cut apart and flattened into a planar polygon, as in the following sketches:

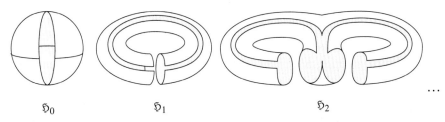

Cuts for the handlebody surfaces

Each cut made in the surface $\mathfrak{H}_g$ produces a pair of boundary edges of the polygon, which we give compatible directions and which we label with the same symbol in the following models:

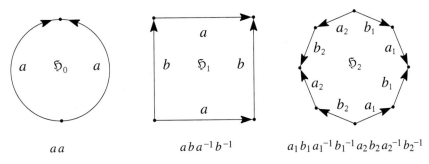

aa $aba^{-1}b^{-1}$ $a_1 b_1 a_1^{-1} b_1^{-1} a_2 b_2 a_2^{-1} b_2^{-1}$

Polygonal models and boundary words for the handlebody surfaces

The handlebody surface $\mathfrak{H}_g$ can be reconstructed from its polygonal model by identifying the paired directed edges and giving the result the identification topology. These polygonal models suggest the following general way to construct surfaces.

POLYGONAL SURFACES

Let $\mathcal{P}$ be the disjoint union of a finite family of polygonal versions of the closed disc $\mathbb{D}^2$, with the total number of edges even. Suppose that each edge of each polygon in $\mathcal{P}$ is paired with *exactly one other* edge of a polygon in $\mathcal{P}$ and that the paired edges are identified according to linear homeomorphisms. The resulting identification space $S = \mathcal{P}/\sim$ is compact, Hausdorff, and it acquires the structure of a polygonal complex (Section **12**.2). Its vertices, edges, and faces are the images of the vertices, edges, and polygons in $\mathcal{P}$.

1.1 Polygonal Gluing Theorem   *The identification space $S = \mathcal{P}/\sim$ defined by any pairing of the edges of a polygonal family $\mathcal{P}$ is a 2-manifold.*

Proof: Each interior point of a polygon P in $\mathcal{P}$ (i.e., a point not on an edge or at a corner vertex) has open disc neighborhoods in P that are unchanged in S. Each interior point p of an edge of a polygon P in $\mathcal{P}$ (i.e., not a corner vertex) is identified with exactly one other interior point on some other edge of some polygon P' in $\mathcal{P}$; and open half-disc neighborhoods of the two edge points in the polygons assemble into an open disc neighborhood of p in S.

Each corner vertex of a polygon P in $\mathcal{P}$ is shared by two edge ends of P. Since each edge end is identified with exactly one other edge end and since there are just finitely many edge ends, the corners of the polygons assemble into open disc neighborhoods of the vertices in S. Thus, S is a 2-manifold. $\square$

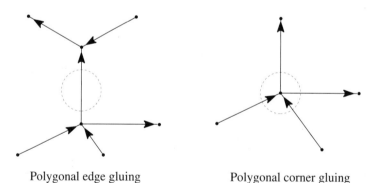

Polygonal edge gluing   Polygonal corner gluing

Working with a family of polygons instead of a single polygon presents only one complication. The 2-manifold S may not be connected, because the edge pairing may glue the polygons into separate component surfaces.

A compact, connected, 2-manifold S resulting from a pairing of the edges of a polygonal union $\mathcal{P}$ is called a ***polygonal surface***. The polygonal complex of vertices, edges, and polygonal faces on the surface S that arise from the corners, edges, and polygons of the family $\mathcal{P}$ is called a ***map*** on S and is denoted by Λ. The connectedness of S implies that it is possible to travel across shared edges from any face to any other face of the map Λ.

You may view the map Λ as a specific tiling of the topological surface S by polygonal faces. The surface S can support many different maps Λ, just as a polygon can be divided in many different ways into smaller polygons.

The edge identifications of the polygonal family $\mathcal{P}$ that produce the surface S are recorded in boundary labels, as follows. To begin, each edge of the map Λ is directed and labeled by a lower case letter, and then this direction and label is applied to the corresponding paired edges in $\mathcal{P}$. The **boundary label** $\mathscr{L}$ of a polygon P in $\mathcal{P}$ is the sequence of edge labels that are encountered on a walk along the boundary circuit of P. The label for an edge appears in $\mathscr{L}$ with an exponent $+1$ or -1, according to whether the edge direction agrees or disagrees with the direction of the walk. *When the map has a single vertex*, the boundary label $\mathscr{L}$ is the *boundary word* of the corresponding face in Λ, as described in Section **12.2**.

Example 1 The polygonal model provided (on p. 375) for the handlebody surface $\mathfrak{H}_g$ of genus $g \geq 1$ gives a map Λ_g with a single vertex, 1-skeleton $\vee_i (\mathbb{S}^1_{a_i} \vee \mathbb{S}^1_{b_i})$, and a single face with boundary label $\mathscr{L}(\Lambda_g) = a_1 b_1 a_1^{-1} b_1^{-1} \ldots a_g b_g a_g^{-1} b_g^{-1}$.

The polygonal model for the 2-sphere $\mathfrak{H}_0 = \mathbb{S}^2$ gives a map Λ_0 with two vertices, one edge, and one face with boundary label $\mathscr{L}(\Lambda_0) = a a^{-1}$.

For each genus $g \geq 0$, Λ_g is called the **standard map** on $\mathfrak{H}_g$.

CROSSCAP SURFACES

To invent a new surface and a map with a single face, we need merely devise a new boundary label in which each letter appears exactly twice. It determines a surface by the Polygonal Gluing Theorem (1.1), even though it may be impossible to assemble a rubber-sheet model in $\mathbb{R}^3$. This is the case for the surfaces $\{\mathcal{C}_g : g \geq 1\}$ with these polygonal models and boundary labels:

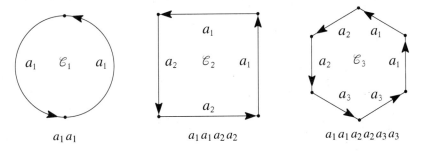

Polygonal models and boundary labels for the cross-cap surfaces

The first surface $\mathcal{C}_1$ in this family is the **cross-cap** or the **projective plane** $\mathbb{P}^2$. Despite the simplicity of its planar model, $\mathcal{C}_1$ does not embed as a subspace of $\mathbb{R}^3$. Performing the edge identification is like closing a coin purse whose zipper pull has been misinstalled to simultaneously run down the left side and up the right side of the zipper. It can't be done in $\mathbb{R}^3$.

The gth surface $\mathcal{C}_g$ ($g \geq 1$) of this family is called the ***cross-cap surface of genus g***. The polygonal model for the surface $\mathcal{C}_g$ ($g \geq 1$) gives a map Ψ_g with a single vertex, 1-skeleton $\vee_i \mathbb{S}^1_{a_i}$, and a single face with boundary label $\mathscr{L}(\Psi_g) = a_1 a_1 \ldots a_g a_g$. Ψ_g is called the ***standard map*** on $\mathcal{C}_g$.

ORIENTABILITY STATUS

When two opposite edges of the square $\mathbb{I} \times \mathbb{I}$ are identified via a linear homeomorphism, the result is either a cylinder or a Möbius strip (Examples **5.1.9** and **5.1.10**). The Möbius strip has the telling feature that a clockwise oriented disc, when pushed along the length of the strip, returns as a counterclockwise oriented disc:

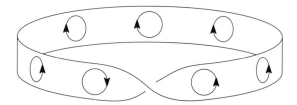

The non-orientability of the Möbius strip

Because of this feature, all surfaces that contain a copy of the Möbius strip are called ***non-orientable***. Surfaces that do not contain a copy of the Möbius strip are called ***orientable***.

1.2 Orientability Theorem
(a) The crosscap surfaces $\{\mathcal{C}_g : g \geq 1\}$ are non-orientable surfaces.
(b) The handlebody surfaces $\{\mathfrak{H}_g : g \geq 0\}$ are orientable surfaces.

Proof: (a) In the polygonal model for the cross-cap surface $\mathcal{C}_g$, all of the edges have the same direction. Therefore, a narrow strip placed in the polygon, with its ends resting on middle thirds of paired edges of the model, glues into a Möbius strip contained in $\mathcal{C}_g$.

(b) Each handlebody surface $\mathfrak{H}_g$ ($g \geq 0$) has been exhibited as a subspace of $\mathbb{R}^3$ that divides space into a bounded region and an unbounded region. So the painting experiment employed in Example **5.1.2** for the 2-sphere $\mathbb{S}^2 = \mathfrak{H}_0$ shows that there is no embedded Möbius strip in any $\mathfrak{H}_g$. □

DISTINGUISHING POLYGONAL SURFACES

The surfaces in the two infinite families, $\{\mathfrak{H}_g : g \geq 0\}$ and $\{\mathcal{C}_g : g \geq 1\}$, are distinguished from each other by their orientability status. For a rigorous proof that no surface belongs to both families and that all these surfaces are topologically distinct, we appeal to the fundamental group.

1.3 Surface Catalog *The surfaces $\{\mathfrak{H}_g\ (g \geq 0),\ \mathcal{C}_g\ (g \geq 1)\}$ are distinct.*

Proof: The orientable surface $\mathfrak{H}_0 = \mathbb{S}^2$ of genus $g = 0$ has $\pi_1(\mathfrak{H}_0) = \{1\}$.

The standard map Λ_g on the handlebody surface $\mathfrak{H}_g$ of genus $g \geq 1$ has a single vertex, 1-skeleton $\vee_i\ (\mathbb{S}^1_{a_i} \vee \mathbb{S}^1_{b_i})$, and single face with boundary word

$$\mathscr{L}(\Lambda_g) = a_1 b_1 a_1^{-1} b_1^{-1} \ldots a_g b_g a_g^{-1} b_g^{-1};$$

the standard map Ψ_g on the cross-cap surface $\mathcal{C}_g$ of genus $g \geq 1$ has a single vertex, 1-skeleton $\vee_i\ \mathbb{S}^1_{a_i}$, and single face with boundary word

$$\mathscr{L}(\Psi_g) = a_1 a_1 \ldots a_g a_g.$$

Thus, by Theorem **12.2.5**, $\pi_1(\mathfrak{H}_g)$ and $\pi_1(\mathcal{C}_g)$ have these presentations:

$$\mathbb{H}_g = \langle a_1, b_1, \ldots, a_g, b_g : a_1 b_1 a_1^{-1} b_1^{-1} \ldots a_g b_g a_g^{-1} b_g^{-1} \rangle \qquad (g \geq 1)$$

$$\mathbb{C}_g = \langle a_1, \ldots, a_g : a_1 a_1 \ldots a_g a_g \rangle \qquad (g \geq 1).$$

These fundamental groups $\{\mathbb{H}_g\ (g \geq 0),\ \mathbb{C}_g\ (g \geq 1)\}$ are non-isomorphic. According to Theorem **11.4.10**, they have distinct abelianizations:

$$Ab(\mathbb{H}_g) \approx (\mathbb{Z} \oplus \mathbb{Z}) \oplus \ldots \oplus (\mathbb{Z} \oplus \mathbb{Z}) \quad \text{and} \quad Ab(\mathbb{C}_g) \approx \mathbb{Z}_2 \oplus \mathbb{Z} \oplus \ldots \oplus \mathbb{Z}.$$

Therefore, by the topological invariance of the fundamental group, the catalogued surfaces are topologically distinct. □

Hereafter, $\mathfrak{H}_g\ (g \geq 0)$ is called the ***orientable surface of genus g*** and $\mathcal{C}_g\ (g \geq 1)$ is called the ***non-orientable surface of genus g***.

Although there are infinitely many other possible boundary labels for polygonal surfaces, we will prove in Section 2 that every polygonal surface is homeomorphic to one of the surfaces $\mathfrak{H}_g\ (g \geq 0)$ or $\mathcal{C}_g\ (g \geq 1)$.

Actually, Section 2 classifies all surfaces, since any surface arises as a polygonal surface. It was shown in 1925 by T. Radó that any (compact) surface can be *triangulated* or *tiled by triangles*, in the following sense.

A *triangulation* of a surface S is a polygonal complex for which each face has three distinct vertices and three distinct edges, and the intersection of distinct faces is either empty, a shared vertex, or a complete shared edge.

2 Classification of Surfaces

This section classifies all surfaces by proving that each one is homeomorphic to a unique surface in the catalog $\{\mathfrak{H}_g\ (g \geq 0),\ \mathcal{C}_g\ (g \geq 1)\}$. The method is to convert any map on a polygonal surface into one of the standard maps $\{\Lambda_g, \Psi_g\}$ via a sequence of modifications that preserve the underlying topological surface. Since the map determines the underlying surface, this conversion identifies the given polygonal surface.

MODIFICATIONS OF MAPS

Here are two inverse pairs of *modifications* of maps on surfaces.

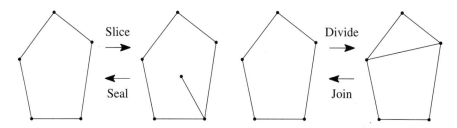

- To **slice** a face, insert a new edge connecting a vertex in its boundary to a new vertex in its interior. To **seal** a face, remove a dead-end edge and vertex that penetrates the face.

- To **divide** a face into two faces, insert a new edge across the face to connect the vertices at two of its corners. To **join** two distinct faces, remove a shared edge.

Two maps are called *equivalent* if they are related by a sequence of the slice, seal, divide, and join modifications, in any order. Equivalent maps reside on the same surface, because the modifications of a map do not alter the underlying topological surface, but rather simply resurvey it.

Here are two fundamental lemmas.

2.1 Lemma *Any map is equivalent to one with a single face.*

Proof: Since the surface is assumed to be connected, it is possible to travel across shared edges from any face to any other face of the map. If there is more than one face, then there must be some edge that is shared by two distinct faces. Removal of the shared edge joins the distinct faces and produces a new map with one less face. The lemma follows from an iteration of such modifications. □

2.2 Lemma *A map with a single face is equivalent to Λ_0 or to a map with a single face and a single vertex.*

Proof: The standard map Λ_0 on the sphere $\mathbb{S}^2$ has just two vertices and one edge. Although that edge in Λ_0 is a dead-end edge, it cannot be removed by a sealing modification, because it defines, rather than penetrates, the face. Any dead-end edge and vertex in a map (distinct from Λ_0) can be removed by a sealing modification. Given any map with a single face, an iteration of this process produces an equivalent map Λ with a single face, which is either the standard map Λ_0 or is free of dead-end edges and vertices.

If $\Lambda \neq \Lambda_0$ and Λ has distinct vertices, then there are some dividing and joining modifications that convert any vertex v into a dead-end vertex, as follows. Some edge e joins v to a vertex $v' \neq v$, and e and a second edge f form a corner of the single face at v. Insertion of a new edge g across this corner, connecting the vertices at the other ends of e and f, divides the single face. Then the removal of the edge f joins the two new faces. The result is a map with a single face and the same number of edges and vertices as the original map, but one less edge-end at v, even if both vertices of f were v.

Because these procedures eliminate one edge at the vertex v, they apply until the vertex v is a dead-end vertex, which can then be removed by a sealing modification. An iteration of this process converts the map Λ into either the standard map Λ_0 or one with a single vertex and a single face. □

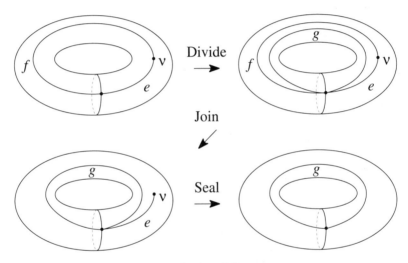

Lemma 2.2

We now consider any map Λ with a *single face* and a *single vertex*. Because there is a single face, Λ arises from edge identifications of a single polygon P according to some boundary label $\mathscr{L}(\Lambda)$. Our goal is to modify Λ until its boundary label $\mathscr{L}(\Lambda)$ is one of the standard boundary labels.

Paired edges of the polygon P that produces the map Λ have either the same or opposite direction and are called **coherent** or **incoherent**, accordingly. Paired edges with label e appear in the boundary label as $U e V e W$ or $U e^{-1} V e^{-1} W$ if they are coherent and as $U e V e^{-1} W$ or $U e^{-1} V e W$ if they are incoherent. (*Capital letters are used to abbreviate portions of the boundary label.*) The boundary label $\mathscr{L}(\Lambda)$ is free of **adjacent** incoherent pairs, such as in $U a^{-1} a V$ or $U a a^{-1} V$. The reason is that adjacent incoherent paired edges in the polygon would identify into a dead-end edge between distinct vertices, but the map Λ has a single vertex by hypothesis. There may be a foursome of **interlocked** and **adjacent** incoherent pairs as in the label $U a b a^{-1} b^{-1} V$

and there may be an *adjacent* coherent pair as in $UaaV$.

Example 1 The boundary label $\mathscr{L}(\Lambda_g) = a_1 b_1 a_1^{-1} b_1^{-1} \ldots a_g b_g a_g^{-1} b_g^{-1}$ of the standard map Λ_g on $\mathfrak{H}_g$ consists of interlocked, adjacent, incoherent pairs $a_i b_i a_i^{-1} b_i^{-1}$. The boundary label $\mathscr{L}(\Psi_g) = a_1 a_1 \ldots a_g a_g$ of the standard map Ψ_g on $\mathcal{E}_g$ consists of adjacent coherent pairs $a_i a_i$.

2.3 Boundary Label Algebra *The divide-join modifications of maps with a single face and a single vertex give these boundary label moves:*

M1 *A coherent pair can be made adjacent:*

$$UeVeW \longrightarrow UV^{-1}ffW.$$

M2 *An adjacent coherent pair can be separated:*

$$UccV^{-1}W \longrightarrow UeVeW.$$

M3 *Interlocked incoherent pairs can be made adjacent:*

$$UaVbWa^{-1}Xb^{-1}Y \longrightarrow UXcdc^{-1}d^{-1}WVY.$$

M4 *In the presence of an adjacent coherent pair cc, adjacent interlocked incoherent pairs $aba^{-1}b^{-1}$ are convertible into two adjacent coherent pairs:*

$$Xccaba^{-1}b^{-1}Y \longrightarrow XeeffggY.$$

Proof: For **M1**, we insert a new edge f that connects the terminal vertices of the two appearances of the edge e, and then remove the edge e:

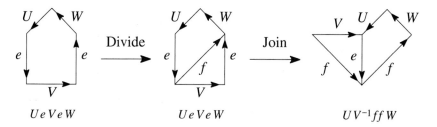

For **M2**, we insert a new edge connecting the terminal vertex of the first appearance of the edge c to the initial vertex of the segment W, and then remove the edge c:

For **M3**, we insert a new edge d that connects the terminal vertices of the two appearances of edge b, and then remove the edge a. Secondly, we insert a new edge c that connects the terminal vertices of the two appearances of the edge d, and then remove the edge b:

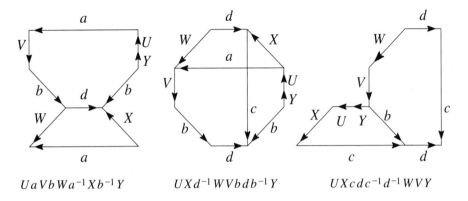

$UaVbWa^{-1}Xb^{-1}Y$ $\qquad$ $UXd^{-1}WVbdb^{-1}Y$ $\qquad$ $UXcdc^{-1}d^{-1}WVY$

For **M4**, apply the following sequence of boundary label moves:

$$\begin{aligned}
Xccaba^{-1}b^{-1}Y &= X\underline{cc}(ab^{-1}a^{-1})^{-1}(b^{-1}Y) & \text{(Group for \textbf{M2})} \\
&\rightarrow Xe(ab^{-1}a^{-1})e(b^{-1}Y) & \text{(Apply \textbf{M2})} \\
&= (Xea)\underline{b}^{-1}(a^{-1}e)\underline{b}^{-1}(Y) & \text{(Regroup for \textbf{M1})} \\
&\rightarrow (Xea)(a^{-1}e)^{-1}ggY & \text{(Apply \textbf{M1})} \\
&= (Xe)\underline{a}(e^{-1})\underline{a}(ggY) & \text{(Regroup for \textbf{M1})} \\
&\rightarrow (Xe)(e^{-1})^{-1}ff(ggY) & \text{(Apply \textbf{M1})} \\
&= XeeffggY. & \text{(Simplify)}
\end{aligned}$$

The underlined letters indicate where the moves are applied. $\qquad \square$

CLASSIFICATION OF SURFACES

The divide-join modifications do not change the vertices of a map. Therefore, any sequence of the four moves **M1**-**M4** converts a map with a single face and a single vertex into an equivalent map with identical features. It remains to show how these moves can put any map into a standard form.

2.4 Standardization Theorem *Any map on a surface is equivalent to one of the standard maps Λ_g or Ψ_g.*

Proof: By the Lemmas 2.1 and 2.2, it is sufficient to consider a map with a single face and a single vertex. By **M1**, all the coherent pairs of the boundary label can be made adjacent, as in $U_1 c_1 c_1 U_2 c_2 c_2 U_3 \ldots c_k c_k U_k$.

If this new boundary label involves no incoherent edge pairs, then it is $\mathscr{L}(\Psi_k) = c_1 c_1 c_2 c_2 \ldots c_k c_k$. If this new boundary label involves an incoherent edge pair as in $R a Z a^{-1} S$, then there must be another incoherent edge pair that is interlocked with the first, as in $U a V b W a^{-1} X b^{-1} Y$. The argument goes this way. Some edge in Z must be paired with some edge in R or S, else the terminal vertices of the a edges do not become identified with the initial vertices of the a edges. Because the coherent edge pairs are adjacent, it must be that some incoherent pair has one edge in Z and one edge in R or S.

By **M**3, the interlocked incoherent edge pairs can be made adjacent, without altering the adjacent coherent edge pairs. By induction, **M**3 produces a boundary label consisting entirely of adjacent coherent pairs cc and interlocked and adjacent incoherent pairs $a b a^{-1} b^{-1}$. If there is no coherent pair, this boundary label is $\mathscr{L}(\Lambda_g)$ for some g. If there is at least one coherent pair, then applications of **M**4 produce a boundary label consisting entirely of adjacent coherent pairs, namely $\mathscr{L}(\Psi_g)$ for some g.

Thus, each map is equivalent to one of the standard maps Λ_g or Ψ_g. □

2.5 Classification Theorem *Each surface is homeomorphic to a unique surface in the catalog* $\{\mathfrak{H}_g \, (g \geq 0), \mathcal{E}_g \, (g \geq 1)\}$.

Proof: By the concluding remarks of Section 1, any surface arises as a polygonal surface. Since the slice-seal and divide-join modifications of maps preserve the underlying surface, the Standardization Theorem (2.4) for maps implies that the surface is homeomorphic to at least one of the surfaces in the catalog. Because the surfaces in the catalog are topologically distinct by Theorem 1.3, any surface is homeomorphic to exactly one of them. □

The first published proof of this classification theorem for (compact) surfaces is due to H. R. Brahana and appeared in 1922.

It is usually quite tedious to apply all the steps of the Standardization Theorem to determine a surface from a bizarre map and boundary label for it. The following section shows how the orientability status and a simple arithmetic count of the vertices, edges, and faces of any map on the surface are sufficient to locate the surface in the catalog.

By the triangulability of 3-manifolds, established by E. Moise in 1952, any compact, connected 3-manifold arises by face identifications of a polyhedral version of the 3-ball $\mathbb{D}^3$ (see Exercise 3.20). But there is no analogous classification available for these manifolds.

SURFACES WITH BOUNDARY

A ***surface with boundary*** is a compact, connected, Hausdorff space S in which each point has an open neighborhood homeomorphic to either the open disc $\mathbb{B}^2$ or the half-disc $\mathbb{B}^2_+ = \{<x_1, x_2> \in \mathbb{B}^2 : 0 \leq x_2\}$. The ***manifold*** *interior* of S consists of its points that have an open disc neighborhood; the *manifold boundary* of S, denoted ∂S, is the complement of the interior.

For example, the Möbius strip is a surface with circular boundary.

In a surface with boundary S, the manifold interior is an open set. So the manifold boundary is a closed, hence, compact subspace. It follows that there are just finitely many components of ∂S and each one is a compact, connected, 1-manifold, i.e., a topological circle. Thus, a surface S^* without boundary results from adjoining a closed disc $\mathbb{D}^2$ to each component of ∂S. Conversely, a surface with boundary, $S - k\mathbb{B}^2$, results from a surface S by removing the interior $\mathbb{B}^2$ from each of $k \geq 1$ disjoint closed discs $\mathbb{D}^2$ in S.

The techniques of this section can also establish the following:

2.6 Classification Theorem   *Each surface with boundary is homeomorphic to a unique surface in this catalog:*

$$\{ \mathfrak{H}_g - k\mathbb{B}^2 \, (g \geq 0, k \geq 1), \ \mathcal{C}_g - k\mathbb{B}^2 \, (g \geq 1, k \geq 1) \}.$$

3 Euler Characteristic

Euler Characteristic of Surfaces

Consider a polygonal surface $S = \mathcal{P}/\!\sim$ and associated map Λ. Let V, E, and F denote the numbers of vertices, edges, and faces of Λ. The alternating sum $\chi(\Lambda) = V - E + F$ is called the ***Euler characteristic of the map*** Λ.

Each polygon P in the family $\mathcal{P}$ becomes a face of the map Λ and each pair of edges in $\mathcal{P}$ becomes an edge in the map Λ. So the numbers F and E are easy to determine from $\mathcal{P}$. The number V of vertices depends upon the pattern of edge identification. We must make all the edge identifications to discover which corner vertices of $\mathcal{P}$ coalesce into vertices of the map Λ on S, before we can count the distinct vertices of Λ that result.

> **Example 1**   The standard map Λ_0 on the orientable surface $\mathfrak{H}_0$ of genus $g = 0$ has Euler characteristic $\chi(\Lambda_0) = 2 - 1 + 1 = 2$.
>
> The standard map Λ_g on the orientable surface $\mathfrak{H}_g$ of genus $g \geq 1$ has Euler characteristic $\chi(\Lambda_g) = 1 - 2g + 1 = 2 - 2g$.
>
> The standard map Ψ_g on the non-orientable surface $\mathcal{C}_g$ of genus $g \geq 1$ has Euler characteristic $\chi(\Psi_g) = 1 - g + 1 = 2 - g$.

The Euler characteristic is useful in identifying polygonal surfaces. The following theorem shows that the Euler characteristic depends only upon the surface and not on the specific map being used to make the count.

3.1 Invariance Theorem   *Any two maps on the same surface have the same Euler characteristic.*

Proof: This invariance is a consequence of the following two facts:

(1) Equivalent maps have the same Euler characteristic because the slice and seal modifications preserve F and the difference $V - E$, and the divide and join modifications preserve V and the difference $F - E$.

(2) Any two maps on the same surface are equivalent because they are both equivalent to the standard map on that surface by the Standardization and Classification Theorems 2.4 and 2.5. □

This invariance and the calculations in Example 1 yield:

3.2 Euler's Formula
(a) Any map Λ on $\mathfrak{H}_g$ has Euler characteristic $\chi(\Lambda) = 2 - 2g$.
(b) Any map Ψ on $\mathcal{E}_g$ has Euler characteristic $\chi(\Psi) = 2 - g$.

So we may define the ***Euler characteristic*** $\chi(S)$ ***of a surface*** S to be the Euler characteristic $\chi(\Lambda)$ of any map Λ on S.

The Euler characteristics of the orientable surfaces,

$$\chi(\mathfrak{H}_0) = 2, \chi(\mathfrak{H}_1) = 0, \chi(\mathfrak{H}_2) = -2, \chi(\mathfrak{H}_3) = -4, \ldots, \chi(\mathfrak{H}_g) = 2 - 2g, \ldots,$$

and the Euler characteristics of the non-orientable surfaces,

$$\chi(\mathcal{E}_1) = 1, \chi(\mathcal{E}_2) = 0, \chi(\mathcal{E}_3) = -1, \chi(\mathcal{E}_4) = -2, \ldots, \chi(\mathcal{E}_g) = 2 - g, \ldots,$$

are not entirely distinct. So the Euler characteristic, alone, does not classify the surfaces. But as there is no duplication in either display, the Euler characteristic and the orientability status, together, do determine the surface.

3.3 Corollary *A surface is determined by any one of the following:*
(a) *its fundamental group,*
(b) *the rank and torsion coefficients of its abelianized fundamental group,*
(c) *its genus and orientability status, or*
(d) *its Euler characteristic and orientability status.*

In particular, two surfaces are homotopy equivalent if and only if they are homeomorphic.

Proof: The fundamental groups $\pi_1(\mathfrak{H}_g)$ and $\pi_1(\mathcal{E}_g)$ determine the abelianized fundamental groups

$$Ab(\pi_1(\mathfrak{H}_g)) = (\mathbb{Z} \oplus \mathbb{Z}) \oplus \ldots \oplus (\mathbb{Z} \oplus \mathbb{Z})$$

and

$$Ab(\pi_1(\mathcal{E}_g)) = \mathbb{Z}_2 \oplus (\mathbb{Z} \oplus \ldots \oplus \mathbb{Z}).$$

The rank and torsion coefficients of $Ab(\pi_1(\mathfrak{H}_g))$ and $Ab(\pi_1(\mathcal{E}_g))$ uniquely determine the genus and orientability status, which, in turn, uniquely determine the Euler characteristic and orientability status of the surface. □

It is fairly easy to determine the standard form of a polygonal surface obtained by edge-gluing a single polygon according to some boundary label. The orientability status is determined by the presence or absence of a coherent pair of edges, and the Euler characteristic is determined by a count of the edges and vertices that result from the edge pairings.

Example 2 The non-standard boundary label $\mathscr{L} = abcda^{-1}b^{-1}c^{-1}d^{-1}$ on an octagon determines an orientable surface S, because $\mathscr{L}$ contains no coherent pairs. The edge identifications produce a single vertex and four edges, so that S has Euler characteristic $\chi(S) = 1 - 4 + 1 = -2$. Thus, S is the orientable surface, $\mathfrak{H}_2$, of genus 2.

REGULAR POLYHEDRA

A *polyhedron* in $\mathbb{R}^3$ is a convex solid that is bounded by a finite collection of polygons that meet edge to edge. The polygons are called the *faces* of the polyhedron. A polyhedron is *regular* if its faces are congruent regular polygons and each vertex is shared by the same number of edges. For a regular polyhedron, let $p \geq 3$ denote the number of edges on each face and let $q \geq 3$ denote the number of edges at each vertex.

For example, the five **Platonic solids**, below, are regular polyhedra:

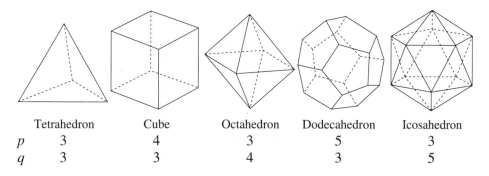

	Tetrahedron	Cube	Octahedron	Dodecahedron	Icosahedron
p	3	4	3	5	3
q	3	3	4	3	5

The mathematicians of ancient Greece used their geometric techniques for solids to prove that the Platonic solids are the only regular polyhedra. Their conclusion also follows from the combinatorial concept of maps on $\mathbb{S}^2$.

Any Platonic solid constructed using rubber faces would lose some of its geometric regularity if it were inflated and stretched. But its vertices, edges, and faces would always determine a map on $\mathbb{S}^2$ that is *regular* in the sense that there are p distinct edges around each face and q distinct edges at each vertex, for some fixed numbers $p \geq 3$ and $q \geq 3$. Surprisingly, there are no more regular maps on $\mathbb{S}^2$ than there are regular polyhedra. The Euler characteristic plays a regulatory role.

3.4 Theorem *The Platonic maps are the only regular maps on $\mathbb{S}^2$ that have $p \geq 3$ edges per face and $q \geq 3$ edges at each vertex.*

Proof: Consider any regular map on $\mathbb{S}^2$ with p edges per face and q edges per vertex. The count, Fp, of the edges, face by face, equals $2E$, since each edge is counted from two sides in the surface. The count, Vq, of the edges, vertex by vertex, equals $2E$, since each edge is counted from each of its two ends. The relations $Fp = 2E = Vq$ and Euler's Formula $V - E + F = 2$ yield the equation

$$\frac{2E}{q} - E + \frac{2E}{p} = 2,$$

which simplifies to

$$\frac{1}{q} + \frac{1}{p} = \frac{1}{2} + \frac{1}{E}.$$

Because $E > 0$, it follows that the integers $p, q \geq 3$ satisfy the relation $1/q + 1/p > \frac{1}{2}$. The only possible combinations for (p, q) are:

$(3, 3)$, $(3, 4)$, $(4, 3)$, $(5, 3)$, and $(3, 5)$.

For each of these five combinations for (p, q), the values of V, E, F are completely regulated by the three equations:

$$\frac{1}{q} - \frac{1}{2} + \frac{1}{p} = \frac{1}{E} \qquad F = \frac{2E}{p} \qquad V = \frac{2E}{q}.$$

The Platonic maps are the only possibilities, as indicated below. $\square$

Regular Maps	p	q	V	E	F
Tetrahedron	3	3	4	6	4
Cube	4	3	8	12	6
Octahedron	3	4	6	12	8
Dodecahedron	5	3	20	30	12
Icosahedron	3	5	12	30	20

When faces with just $p = 2$ edges are permitted, there is a regular map on $\mathbb{S}^2$ with q distinct edges at each vertex, for each $q \geq 2$. When just $q = 2$ edges are permitted at each vertex, there is a regular map on $\mathbb{S}^2$ with p distinct edges around each face, for each $p \geq 2$. Exercise 7 asks for drawings of these non-platonic regular maps.

PLANAR GRAPHS

A ***topological graph*** Γ is an identification space obtained from a disjoint union of line segments by identifying sets of their endpoints. The segments and endpoints determine the ***edges*** and ***vertices*** of the graph. So a topological graph is a one-dimensional analogue of a polygonal complex. In particular, the 1-skeleton of any polygonal complex is a topological graph.

Example 2 The *utilities graph* U_3 (below left) is a graph with six vertices and nine edges. Three of the vertices are the *houses* A, B, and C and three of the vertices are the *utilities* E, G, and W. The nine edges connect each one of the three *houses* to each one of the three *utilities*. The following projection of U_3 involves seven overcrossings of its nine edges.

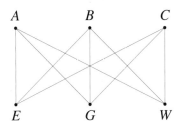

Example 2: Utilities graph U_3 Example 3: Complete 5-graph C_5

Example 3 A *complete* graph has a unique edge for each pair of distinct vertices. The complete graph with N vertices is called the ***complete N-graph*** and is denoted by C_N; it has $N(N-1)/2$ edges, one for each pair of distinct vertices.

A graph Γ of finitely many vertices and edges is called ***planar*** if it can be embedded in $\mathbb{R}^2$ so that each edge is one or more line segments of the plane. Or equivalently, Γ can be embedded in $\mathbb{S}^2$ as finitely many great arcs.

For example, the complete graphs C_2, C_3, and C_4 are planar graphs.

An embedding of a connected planar graph Γ in the 2-sphere $\mathbb{S}^2$ can be viewed as the 1-skeleton of a map Λ on $\mathbb{S}^2$, since the complementary regions in $\mathbb{S}^2$ are necessarily homeomorphic to open polygonal discs. Thus, by Euler's Formula (3.2), an embedded planar graph Γ always gives a map Λ on $\mathbb{S}^2$ with $F = 2 - V + E$ faces. This can be used to show that certain graphs, such as the complete 5-graph and utilities graph, are not planar, as follows.

3.5 Non-Planarity Theorem *The complete 5-graph C_5 and the utilities graph U_3 are not planar.*

Proof: There are $V = 5$ vertices and $E = 10$ edges in C_5. By Euler's Formula, the number of faces of a map Λ given by an embedding of C_5 in $\mathbb{S}^2$ would be

$$F = 2 - V + E = 2 - 5 + 10 = 7.$$

The border of each face of the map Λ would be a circuit in the graph C_5, and the shortest circuits in C_5 have length 3. Therefore, the count, M, of the edges of the map Λ, face by face, would satisfy the inequality $M \geq 3F = 21$.

But this count M would also equal $2E = 20$, since each edge would be counted from two sides. This would yield the contradiction $20 \geq 21$. Therefore, the graph C_5 cannot be planar.

The proof for U_3 is left as an exercise (Exercise 13). □

According to the following result, proved by C. Kuratowski in 1930, the non-planar graphs C_5 and U_3 constitute the heart of the planarity problem. For a proof, see Chapter 21 in Claude Berge, *The Theory of Graphs*.

3.6 Kuratowski's Theorem *A graph Γ is non-planar if and only if Γ contains a fractured version of either U_3 or C_5.*

To *fracture* a graph is to add a new vertex that subdivides an existing edge into two new edges that share the new vertex.

4 Poincaré Index Theorem

In this section, we prove Poincaré's index formula relating the Euler characteristic of a surface with the sum of indices of singularities of any tangent vector field or, more generally, any foliation on that surface.

FOLIATIONS ON SURFACES

Roughly speaking, a *non-singular foliation* of a 2-manifold S is a partition of S into one dimensional subspaces called the leaves of the foliation. The horizontal lines in $\mathbb{R}^2$ are leaves for a *model* non-singular foliation of $\mathbb{R}^2$. The rays emanating from the origin in $\mathbb{R}^2$ are leaves for a foliation of $\mathbb{R}^2$ that has the origin O as a *singular point*; no homeomorphism carries the radial leaves in a neighborhood of O to horizontal lines in $\mathbb{R}^2$. The circles centered on the origin are leaves for another foliation of $\mathbb{R}^2$ that has O as its sole singular point. The leaves of these three foliations are described by the complex variable conditions $Im\ z = c$, $Arg\ z = c$, and $|z| = c$, respectively.

More generally, for each integer $q \geq 1$, consider the function $z \mapsto z^{1/q}$ that carries each of the q sectors $k 2\pi/q \leq Arg\ z \leq (k+1) 2\pi/q$ ($0 \leq k < q$) of $\mathbb{R}^2$ onto upper or lower half space of $\mathbb{R}^2$. The family of horizontal lines in $\mathbb{R}^2$ pulls back to a foliation of $\mathbb{R}^2$ whose leaves are described by the condition $Im\ z^{1/q} = c$. These foliations are distinguished by the number, q, of leaves that limit at the origin; we call them the ***q-pronged foliations***:

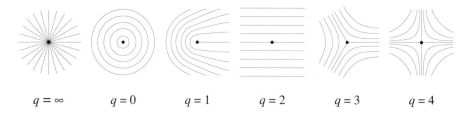

$q = \infty \qquad q = 0 \qquad q = 1 \qquad q = 2 \qquad q = 3 \qquad q = 4$

When $q \geq 2$, the q-pronged singularity is constructed from q hyperbolic sectors. We leave it to you to invent other singularities surrounded by any number of parabolic, elliptic, or hyperbolic sectors (see Section **9**.4).

SECTION 4   POINCARÉ INDEX THEOREM   391

A *foliation* $\mathcal{F}$ on a 2-manifold S is a partition of S into a finite set of exceptional points called *singular points* and a family of subsets called *leaves*, subject to these restrictions: for each *non-singular* or *regular point* p, there is a neighborhood chart $(U, \phi : U \to \mathbb{R}^2)$ sending the leaves of $\mathcal{F}$ in U to the horizontal lines in $\mathbb{R}^2$; for each singular point p, there is a neighborhood chart $(U, \phi : U \to \mathbb{R}^2)$ sending p to O and the leaves of $\mathcal{F}$ in U to the leaves of a q-pronged foliation ($q \neq 2$) or a generalization; finally, intersecting charts in S correspond under a *smooth* homeomorphism of open subsets of $\mathbb{R}^2$, i.e., one whose coordinate functions have continuous partial derivatives.

Example 1   Each smooth embedding of a handlebody surface $\mathfrak{H}_g$ in $\mathbb{R}^3$ determines a *contour foliation* whose leaves consist of points of equal height in $\mathbb{R}^3$. Any smooth mountainous terrain on a handlebody surface $\mathfrak{H}_g$ in $\mathbb{R}^3$ determines a *topographic foliation* whose leaves consist of points of equal elevation from the surface.

Example 1: Contour foliations

INDEX OF A SINGULARITY

When the integer q is odd, the leaves of the q-pronged foliation of $\mathbb{R}^2$ cannot be compatibly directed. Therefore, they cannot arise as the integral curves of a continuous vector field on $\mathbb{R}^2$. So foliations are generalizations of the phase portraits of vector fields on $\mathbb{R}^2$ discussed in Section **9.4**. But, because any choice of one of two opposite unit tangent vectors at a point of a leaf of a foliation extends continuously to unit tangent vectors of neighboring leaves, we can still define an index along any loop of regular points.

Consider any foliation $\mathcal{F}$ of an open subspace $S \subseteq \mathbb{R}^2$ and any loop $g : \mathbb{I} \to S$ through regular points in $S \subseteq \mathbb{R}^2$. Unit tangent vectors $v_g(t) \in \mathbb{S}^1$ to the leaves through the points $g(t)$ of the loop g, with their direction determined by continuous extension of some initial direction, define a path $v_g : \mathbb{I} \to \mathbb{S}^1$ that records the *angular variation* of the leaves traveled by g. Because g returns to its starting point, x, and the leaf through x has just two opposite unit tangent vectors at x as possibilities for $v_g(0)$ and $v_g(1)$, the path $v_g : \mathbb{I} \to \mathbb{S}^1$ is either a loop or a path joining antipodal points.

So the path-homotopy class $[v_g] \in \pi(\mathbb{S}^1)$ has *fractional winding number* $w(v_g) = L(v_g)/2\pi \in \frac{1}{2}\mathbb{Z}$, denoted by $Index(\mathcal{F}, g)$ and called the *index of $\mathcal{F}$ along g*. $Index(\mathcal{F}, g)$ is the fractional number $\frac{1}{2}k$ of twists in one's neck that result when one walks the loop g, keeping one's chest parallel to the x-axis and head continuously pointed along the leaves of $\mathcal{F}$ at each point along g.

Example 2 Consider the *supernova foliation* $\mathcal{S}$, depicted below, on $\mathbb{R}^2 - \mathbb{D}^2$. Along a curve g that circumscribes $\mathbb{D}^2$ the foliation $\mathcal{S}$ has $Index(\mathcal{S}, g) = +2$.

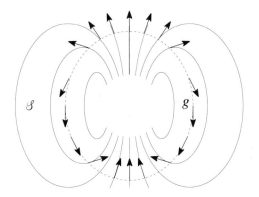

Example 2: Supernova foliation $\mathcal{S}$

The vector field Index Features (**9.4.1**) carry over to the foliation index.

Likewise, the ***index of a singular point*** $p \in S \subseteq \mathbb{R}^2$ of a foliation $\mathcal{F}$ is defined to be $Index(\mathcal{F}, g)$, for any loop g that circumscribes just the one singularity p. When $\mathcal{F}$ is a foliation on an arbitrary 2-manifold S, the index of each singular point is defined using a neighborhood chart to $\mathbb{R}^2$.

Then the following calculation holds on any surface (Exercise 1):

- *The index of a q-pronged singularity is* $\frac{1}{2}(2 - q)$.

INDEX FOR A FOLIATION

Let $\mathcal{F}$ be any foliation of a surface S. The ***index*** of $\mathcal{F}$, denoted by $Index(\mathcal{F})$, is defined to be the sum of the indices of the singularities that appear in $\mathcal{F}$.

Example 3 Three foliations of the 2-sphere $\mathbb{S}^2$ are depicted below. Each one has index $+2$, which curiously equals the Euler characteristic of $\mathbb{S}^2$:

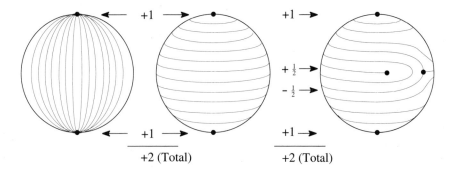

According to Poincaré's Index Theorem that follows, the curiosity in Example 3 is a result that holds for any foliation on any orientable surface.

SECTION 4   POINCARÉ INDEX THEOREM

4.1 Poincaré's Index Theorem   *For any foliation $\mathcal{F}$ on an orientable surface S, the Index of $\mathcal{F}$ equals the Euler characteristic of S:*

$$\text{Index}\,(\mathcal{F}) = \chi(S).$$

Proof: As a first case, let $\mathcal{F}_0$ be any foliation on the 2-sphere $\mathfrak{H}_0$. Rest $\mathfrak{H}_0$ on the origin O of a plane so that a non-singular point of the foliation $\mathcal{F}_0$ is on top. Then stereographic projection carries the foliation on the complement of the top point into a foliation of the plane $\mathbb{R}^2$. The projected foliation has all its singularities contained in a sufficiently large disc centered on O, and it has the shape of the supernova foliation of Example 2 on the complement of that disc. Since the singularities of $\mathcal{F}_0$ are undisturbed by the projection process, $\text{Index}(\mathcal{F}_0)$ is the sum of the indices of the projected singularities. By the Index Property (**9.4.2**) for foliations, that sum equals the index of the projected foliation along a loop g that surrounds its singularities. But, by the direct calculation made for the supernova in Example 2, this index along g equals +2. Therefore, $\text{Index}(\mathcal{F}_0) = +2$. This proves the theorem for $\mathfrak{H}_0$.

For the remaining cases, let $\mathcal{F}_g$ be a foliation of the orientable surface $S = \mathfrak{H}_g$ of genus $g \geq 1$. When $\mathfrak{H}_g$ is cut along a longitudinal circle α across one of its handles and two hemispherical discs $\mathbb{D}^2_+$ and $\mathbb{D}^2_-$ are attached to the cut ends, the result is the surface $\mathfrak{H}_{g-1}$ with one less handle:

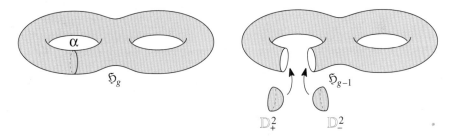

This *surgery* reduces the genus g by one; hence, it increases the Euler characteristic $\chi(\mathfrak{H}_g) = 2 - 2g$ by +2. Some longitudinal curve α consists of regular points of the foliation $\mathcal{F}_g$, so that $\mathcal{F}_g$ extends over the hemispherical discs $\mathbb{D}^2_+$ and $\mathbb{D}^2_-$ to a foliation $\mathcal{F}_{g-1}$ on $\mathfrak{H}_{g-1}$ that is free of singular points in a neighborhood of their boundary circles (Exercise 7). Because $\mathcal{F}_{g-1}$ is, in effect, the union of the foliation $\mathcal{F}_g$ on $\mathfrak{H}_g$ and a foliation $\mathcal{F}_0$ on the sphere $\mathbb{D}^2_+ \cup \mathbb{D}^2_-$, $\text{Index}(\mathcal{F}_{g-1})$ equals the sum of $\text{Index}(\mathcal{F}_g)$ and $\text{Index}(\mathcal{F}_0) = +2$.

Therefore, the surgery to eliminate one handle of $\mathfrak{H}_g$ has the same effect on the Euler characteristic of the surface $\mathfrak{H}_g$ as it has on the index of the foliation $\mathcal{F}_g$; it increases both by 2. If this operation is performed g times, the handles will be eliminated and the result will be a foliation $\mathcal{F}_0$ on the sphere $\mathfrak{H}_0$. Because the equality $\text{Index}(\mathcal{F}_0) = +2 = \chi(\mathfrak{H}_0)$ has been established for that simplest case, the surgery procedure proves the same equality, $\text{Index}(\mathcal{F}_g) = \chi(\mathfrak{H}_g)$, for any foliation $\mathcal{F}_g$ on any orientable surface $\mathfrak{H}_g$. □

Example 4 *Topology regulates topography.* Any mountainous terrain on a surface *S* gives a *topographic foliation.* Its index must equal the *topological invariant* $\chi(S)$.

Incidently, in the 1930's, M. Morse developed a theory, subsequently refined by S. Smale in the 1960's, that turns this theme around. Amazingly, *topography determines topology*: any smooth *n*-manifold *M* can be reconstructed from the knowledge of the singularities of a height function on *M*!

4.2 Corollary *The torus is the only orientable surface that admits a foliation or tangent vector field that is free of singularities.*

Proof: The contour foliation for a prone torus is non-singular. For every other orientable surface *S*, the Euler characteristic $\chi(S)$ is non-zero. Hence, by Poincaré's Index Theorem, any foliation $\mathcal{F}$ on *S* has $Index(\mathcal{F}) \neq 0$. Since $Index(\mathcal{F})$ equals the sum of the indices of the singularities in $\mathcal{F}$, there must be some singularity. See Exercise 5 for tangent vector fields. □

Example 5 *Satellite weather photographs never lie.* The foliation on the 2-sphere described by any air-stream pattern on the globe must have some singularity.

4.3 Corollary *On any foliated surface, an embedded disc whose boundary is a non-singular leaf must contain some singularity of the foliation.*

Proof: Otherwise, two copies of the foliated disc bounded by the closed leaf could be glued together to create a foliated sphere with no singularities. □

Example 6 *Any tornado on the globe swirls around some eye.* Similarly, *any closed contour curve in the mountainous terrain of any spherical planet must surround some mountain peak or valley bottom.*

Exercises

SECTION 1

1* Prove that $\mathbb{S}^1$ and $\mathbb{R}^1$ are the only connected 1-manifolds.

2 Sketch the location of the cuts on the orientable surface $\mathfrak{H}_3$ of genus 3 that will produce its standard polygonal model. Indicate in your drawing how the surface is opened into a polygon by these cuts. (Warning: The system of cuts for $\mathfrak{H}_3$ may not be the expected generalization of the cuts for $\mathfrak{H}_2$.)

3 (*a*) Prove that the corners of the standard polygonal model of the surface $\mathfrak{H}_g$ form an open disc neighborhood of a single vertex in $\mathfrak{H}_g$.
    (*b*) Prove that the corners of the standard polygonal model of the surface $\mathcal{E}_g$ form an open disc neighborhood of a single vertex in $\mathcal{E}_g$.

SECTION 1-4

4. For each of the following boundary labels, show how the corners of the polygon form open disc neighborhoods of the vertices in the resulting polygonal surface.
   (a) $abca^{-1}b^{-1}c^{-1}$
   (b) $abcdb^{-1}d^{-1}a^{-1}c^{-1}$

5. Prove that the Klein bottle (Example 12.2.6) is homeomorphic to the non-orientable surface $\mathcal{C}_2$ of genus 2. (Hint: Cut the model for the Klein bottle along one of its diagonals and reassemble it into a model for $\mathcal{C}_2$.)

6. Find a surface that contains three disjoint Möbius strips.

7. The polygonal complex D that results from a triangle with boundary label $aa^{-1}a$ is called the mathematical **dunce hat**. Sketch neighborhoods of the vertex, of an edge point, and of a face point. Is this space a surface? What is its fundamental group?

8. Prove that a projective plane results from the identification of the boundary of a disc with the boundary of a Möbius strip.

9. Prove that two Möbius strips glue edge-to-edge to form the Klein bottle.

10. Construct an economical triangulation for each of the following surfaces.
    (a) sphere $\mathfrak{H}_0$ (b) torus $\mathfrak{H}_1$
    (c) cross-cap $\mathcal{C}_1$ (d) Klein bottle $\mathcal{C}_2$

SECTION 2

1. For each of the following boundary labels, put the associated polygonal map Λ in standard form by some slice-seal and divide-join modifications.
   (a) $aba^{-1}b$ (b) $abca^{-1}b^{-1}c^{-1}$
   (c) $abacb^{-1}c^{-1}$ (d) $abcdb^{-1}d^{-1}a^{-1}c^{-1}$

2. Construct non-standard maps on the following surfaces:
   (a) $\mathfrak{H}_1$ and $\mathcal{C}_1$ (b) $\mathfrak{H}_2$ and $\mathcal{C}_2$
   For each of these maps, calculate the sum: #(vertices) − #(edges) + #(faces).

Let S_1 and S_2 be surfaces. The **connected sum** of S_1 and S_2 is the surface, $S_1 \# S_2$, that results when the interior of discs $D_1 \subset S_1$ and $D_2 \subset S_2$ are removed and their boundary circles are identified.

3. (a) Prove that $\mathfrak{H}_g$ is a connected sum, $\mathfrak{H}_1 \# \ldots \# \mathfrak{H}_1$, of g copies of the torus $\mathfrak{H}_1$.
   (b) Prove that $\mathcal{C}_g$ is a connected sum, $\mathcal{C}_1 \# \ldots \# \mathcal{C}_1$, of g copies of the cross cap $\mathcal{C}_1$.
   So $\mathfrak{H}_g$ is a **sphere with g handles** and $\mathcal{C}_g$ is a **sphere with g crosscaps**.

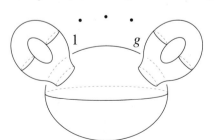
$\mathfrak{H}_g$: Connected sum of g tori

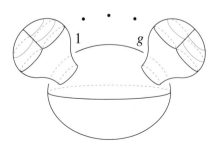
$\mathcal{C}_g$: Connected sum of g cross-caps

4   (a) Prove that $\mathbb{S}^2 \# S \cong S$ for each surface S.
    (b) Identify the connected sums $\mathfrak{H}_1 \# \mathcal{C}_1$ and (Klein bottle) $\# \mathcal{C}_1$.

5   (a) Let S be the identification space obtained by identifying in pairs some, but not all, of the edges of a polygon. Prove that S is a surface with boundary.
    (b) Using the fact that any surface is a polygonal surface, prove that any surface with boundary is a polygonal surface with boundary, as in (a).

6   Establish the Classification Theorem (2.6) for surfaces with boundary.

SECTION 3

1   Use the Euler characteristic and orientability status to determine the surfaces that support the polygonal maps associated with the following boundary labels.
    (a) $aba^{-1}b$ (b) $abca^{-1}b^{-1}c^{-1}$
    (c) $abacb^{-1}c^{-1}$ (d) $abcdb^{-1}d^{-1}a^{-1}c^{-1}$

2   Repeat Exercise 1 with the following boundary labels.
    (a) $c_1 c_2 \ldots c_{n-1} c_n c_1^{-1} c_2^{-1} \ldots c_{n-1}^{-1} c_n^{-1}$
    (b) $c_1 c_2 \ldots c_{n-1} c_n c_1^{-1} c_2^{-1} \ldots c_{n-1}^{-1} c_n$

3   (a) Express the Euler characteristic of a connected sum $S_1 \# S_2$ (see Exercise 2.3) in terms of the Euler characteristics of the two surfaces S_1 and S_2.
    (b) Identify the connected sum $\mathfrak{H}_g \# \mathcal{C}_{g'}$ for each genus $g \geq 0$ and $g' > 1$.

The **double** of a surface with boundary is the surface that results from identifying corresponding boundary points in two copies of the surface.

4   Classify the doubles of the surfaces with boundary in Exercises **4**.4.12 and **4**.4.17.

A **triangulation** of a surface S is a polygonal map Λ for which each face has exactly distinct three edges and three distinct vertices, and the intersection of distinct faces is either empty, a shared vertex, or a complete shared edge.

5   Prove that for any triangulation Λ of a surface S, with V vertices, E edges, and F faces, we have $3F = 2E$ and $2E \leq V(V-1)$. Deduce the following two statements.
    (a) $E = 3(V - \chi(S))$.   (b) $V \geq \frac{1}{2}(7 + \sqrt{49 - 24\chi(S)}\,)$.

6   Determine a minimum triangulation for each of the following spaces.
    (a) sphere $\mathfrak{H}_0$ (b) torus $\mathfrak{H}_1$
    (c) cross-cap $\mathcal{C}_1$ (c) Klein bottle $\mathcal{C}_2$

7* (a) Show that if faces with just $p = 2$ edges are permitted, there is a regular map on $\mathbb{S}^2$ with q edges at each vertex, for each $q \geq 2$.
    (b) Show that if just $q = 2$ edges are permitted at each vertex, there is a regular map on $\mathbb{S}^2$ with p edges around each face, for each $p \geq 2$.

8   (a) Determine how many pieces result when a 2-sphere is dissected by n great circles, no three of which share an intersection point.
    (b) Determine how many pieces result when a 2-sphere is dissected by n great circles, exactly three of which share each intersection point.

9   A contractible graph Γ is called a **tree**. Prove that in a finite, non-trivial, tree there exists a dead-end edge and vertex, hence, #(*vertices*) = #(*edges*) + 1.

SECTION 1-4   EXERCISES   397

10 Prove that the following are equivalent conditions for a map Λ on $\mathbb{S}^2$:
   (a) Λ has a single face.
   (b) #(vertices) = #(edges) + 1.
   (c) The edge graph Γ is a tree.

11 Find all maps on surfaces that have an equal number of vertices, edges, and faces.

12 For any map on a surface S, $p = 2E/V$ is the average number of edge ends per vertex and $q = 2E/F$ is the average number of edges per face. Prove that when no vertex is shared by just two edge ends, then $q \geq 3$ and $p \leq 6(1 - \chi(S)/F)$. Deduce:
   (a) Any such map on the sphere or cross-cap has some face with 5 or fewer edges.
   (b) Any such map on the torus has some face with six or fewer edges.

13* Use Euler's Formula to prove that the utilities graph U_3 is not planar.

14 Use Kuratowski's Theorem (3.6) to prove:
   (a) Any graph with eight or fewer edges is planar.
   (b) Any non-planar graph with nine edges is identical to the utilities graph U_3.
   (c) There are just five connected non-planar graphs with exactly ten edges.

15 (a) Use Euler's Formula to prove that the Petersen graph, below, is not planar.
   (b) Use Kuratowski's Theorem (3.6) to prove that the Petersen graph is not planar.

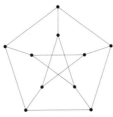

Petersen graph

16 (a) A *toroidal graph* is one that can be embedded on the torus surface T. Prove that the utilities graph U_3 and the complete 5-graph C_5 are toroidal graphs.
   (b) Test Euler's formula: $\chi(T) = V - E + F = 0$ using the toroidal graphs in (a).

17 Develop a theory of regular maps on non-spherical surfaces.

18 Determine whether the following two graphs are planar:

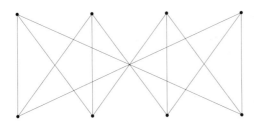 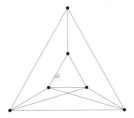

19 (a) Define the orientability and the Euler characteristic of surfaces with boundary.
   (b) Prove that surfaces with boundary are homeomorphic if and only if they have the same number of boundary components, the same orientability status, and the same Euler characteristic. Do homotopy and homeomorphism type coincide here?

20* Suppose that the faces of a map on the boundary $\mathbb{S}^2$ of $\mathbb{D}^3$ are identified in pairs. Prove that the identification space $\mathbb{D}^3/\sim$ is a 3-manifold if and only if $\mathbb{S}^2/\sim$ is a polygonal complex with Euler characteristic #(*vertices*) − #(*edges*) + #(*faces*) = 0.

SECTION 13.4

1* Prove that the index of the q-pronged singularity is $\frac{1}{2}(2 - q)$.

2 Construct a foliation of the sphere having a single singularity.

3 Deform some foliation of the orientable surface $\mathfrak{H}_2$ of genus 2 to bring its singular points into a single point. Can this be done for the other orientable surfaces?

4 Prove: Any singular foliation of the torus has some singularity with positive index.

Embed the handlebody surface $\mathfrak{H}_g$ into $\mathbb{R}^3$ so that there is a **tangent plane** $\mathbb{T}_x$ for each point $x \in \mathfrak{H}_g$, i.e., a 2-dimensional vector subspace $\mathbb{T}_x$ of $\mathbb{R}^3$ such that the hyperplane $\mathbb{T}_x + x$ is tangent to $\mathfrak{H}_g$ at x. A **continuous tangent vector field** V on $\mathfrak{H}_g$ is a continuous function $V : \mathfrak{H}_g \to \mathbb{R}^3$ such that $V(x) \in \mathbb{T}_x$ for each $x \in \mathfrak{H}_g$.

5* (*a*) Prove that the torus is the only orientable surface that admits a non-vanishing continuous tangent vector field.
 (*b*) Show that it is possible to barber and comb a hairy doughnut without introducing a bald spot, but that it is not possible to do the same for a hairy billiard ball.

6 Design a high-speed particle accelerator in the shape of some surface. Explain why the sphere would be a bad choice and why your choice is the only reasonable one.

7* Prove that any non-singular foliation $\mathcal{F}$ on a neighborhood of $\mathbb{S}^1$ extends to a foliation on $\mathbb{D}^2$.

8 (*a*) Give a topography on the sphere involving just three types of features: pits, peaks, and passes. Verify that the sum #(*pits*) − #(*passes*) + #(*peaks*) equals 2.
 (*b*) Give a topography on the torus involving just three types of features: pits, peaks, and passes. Verify that the sum #(*pits*) − #(*passes*) + #(*peaks*) equals 0.

9 (*a*) A foliation on $\mathfrak{H}_g$ ($g \geq 1$) has a single q-pronged singularity. What is q?
 (*b*) Use the standard polygonal model for $\mathfrak{H}_g$ ($g \geq 1$) to illustrate part (*a*).

10 Use the ∞-pronged foliation of $\mathbb{R}^2$ to construct a foliation for the crosscap $\mathcal{C}_1$.
 (*b*) Prove that any foliation $\mathcal{F}$ on the crosscap $\mathcal{C}_1$ has Index($\mathcal{F}$) = +1. (Hint: View the crosscap as the result of identifying antipodal points on $\mathbb{S}^2$ and construct a foliation on $\mathbb{S}^2$ that is carried to $\mathcal{F}$.)
 (*c*) Does the Klein bottle admit a non-singular foliation?

11 Use Exercise 10 and surgery to verify the Poincaré Index Theorem for the non-orientable surfaces.

12 (*a*) Let $\mathcal{F}$ be a foliation on a disc. Define the index of any singularity that appears on the boundary circle to be one-half the index of the singularity that results when a half-disc neighborhood of the singularity is glued along its boundary to a mirror image copy of itself. State and prove an index theorem for the foliation $\mathcal{F}$.
 (*b*) Devise a theory of foliations on surfaces with boundary.
 (*c*) Establish an index theory for foliations on surfaces with boundary.

14

COVERING SPACES

The exponential function $e : \mathbb{R}^1 \to \mathbb{S}^1$ appears in Chapter **9** in the calculation the fundamental group of the 1-sphere $\mathbb{S}^1$. This function *wraps* the real line $\mathbb{R}^1$ in such a way around $\mathbb{S}^1$ that loops in $\mathbb{S}^1$ can be unwound by lifting them to paths in $\mathbb{R}^1$, where they are more easily analyzed.

This chapter studies continuous functions $p : \hat{X} \to X$, called *covering projections*, which perform a similar wrapping of the *total space* $\hat{X}$ onto the *base space* X. The analysis of the path-homotopy classes in the base space lifts to a corresponding, but simpler, analysis in the total space. When the total space is completely unwrapped, i.e., simply connected, the covering projection provides a beautiful relationship between the fundamental group of the base space and the global topology of the total space.

1 Fundamentals

In this chapter, all spaces denoted by X or $\hat{X}$ are assumed to be connected and locally path-connected, hence, path-connected by Theorem **6**.3.11.

COVERING PROJECTIONS

Let $p : \hat{X} \to X$ be a continuous surjection. An open subset $U \subseteq X$ is ***evenly covered*** by p if its pre-image, $p^{-1}(U) \subseteq \hat{X}$, is the disjoint union of open subsets $\hat{U}$ of $\hat{X}$ for which $p|_{\hat{U}} : \hat{U} \to U$ is a homeomorphism. The continuous surjection $p : \hat{X} \to X$ is a ***covering projection*** if X has a covering $\{U\}$ by open subsets U that are evenly covered by p. Then the pair $(\hat{X}, p)$ consisting of p and its *total space* $\hat{X}$ is called a ***covering space*** of the *base space* X.

Let $p : \hat{X} \to X$ be a covering projection. Consider a path-connected open subset U of the base space X that is evenly covered by p. The open subsets $\hat{U} \subseteq p^{-1}(U) \subseteq \hat{X}$ that project under p homeomorphically onto U are the path-components of $p^{-1}(U)$. We argue this in two steps: because these subsets $\hat{U}$

project homeomorphically onto U, they are path-connected; because they are open in $p^{-1}(U)$, they are in separate components of $p^{-1}(U)$.

An open path-connected subset $U \subseteq X$ that is evenly covered by p is called an ***admissible*** subset of X. When U is an admissible subset and $\hat{U}$ is a path component of the pre-image $p^{-1}(U) \subseteq \hat{X}$, the inverse $s : U \to \hat{U}$ of the projection homeomorphism $p : \hat{U} \to U$ is called a ***section*** homeomorphism.

Since the base space X is locally path-connected and since an open subset of an evenly covered subset is also evenly covered by $p : \hat{X} \to X$, it follows that X has an open covering $\{U\}$ by admissible subsets U.

FIBERS OF A COVERING PROJECTION

Let $p : \hat{X} \to X$ be a covering projection. For each point $x \in X$, the pre-image $p^{-1}(x) \subset \hat{X}$ is called the ***fiber over*** x.

Let $U \subseteq X$ be an admissible subset and let $p^{-1}(U) = \overset{\circ}{\cup}_i \hat{U}_i$ be the partition of $p^{-1}(U)$ into its path-components (which necessarily project homeomorphically onto U). Then the fiber $p^{-1}(x)$ over a point $x \in U$ consists of one point from each of the (open) path-components $\hat{U}_i$. Therefore, the fiber $p^{-1}(x)$ over any point $x \in U$ is a discrete subspace of $\hat{X}$ and the fibers $p^{-1}(x)$ and $p^{-1}(z)$ over any two points $x, z \in U$ are in bijective correspondence.

Since X has a covering $\{U\}$ by admissible subsets U, the fiber over each point of X is discrete. Moreover, all the fibers in X are set equivalent, as the admissible subsets show that points of X whose fibers are equivalent to a specific fiber form an open and closed subset of the connected space X. Thus, $(\hat{X}, p, X, F)$ is a fiber bundle (**10**.4), with discrete fiber $F = p^{-1}(x)$.

The cardinality, N, of a fiber is called the ***number of sheets*** of the covering space $(\hat{X}, p)$ and p is called an ***N-fold covering projection***.

Example 1 The exponential function $e : \mathbb{R}^1 \to \mathbb{S}^1$, $e(t) = e^{2\pi i t}$, is a covering projection with infinitely many sheets. For any integer $n \neq 0$, the function $d_n : \mathbb{S}^1 \to \mathbb{S}^1$, $d_n(e^{it}) = e^{nit}$, is an $|n|$-fold covering projection.

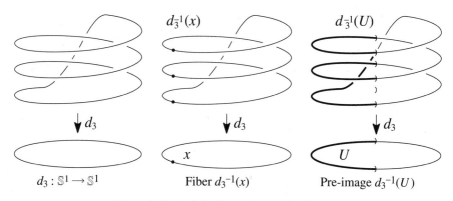

Example 1: A 3-fold covering projection

ASSEMBLING COVERING SPACES

The covering projection $e : \mathbb{R} \to \mathbb{S}^1$ can be assembled from infinitely many copies of the identification map $q : \mathbb{I} \to \mathbb{I}/\{0, 1\} \cong \mathbb{S}^1$. A large variety of covering projections can be similarly assembled from copies of an identification map that produces the base space. Here is one general procedure:

Let a space Y contain closed homeomorphic subspaces $A, B \subseteq Y$ that have disjoint open neighborhoods $U, V \subseteq Y$. Let $q : Y \to X$ be the quotient map that identifies A and B via a homeomorphism $F : A \to B$, as indicated below:

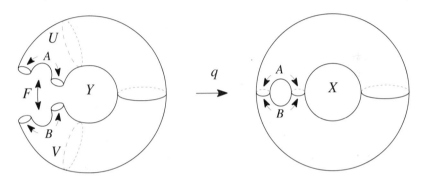

Identification map $q : Y \to X$

Let $\{Y_i\}$ be a collection of copies of Y indexed by $i \in \mathbb{Z}$ (or $\mathbb{Z}_n$), and let $Q : \dot\bigcup_i Y_i \to \hat{X}$ be the quotient map that identifies $A_i \subseteq Y_i$ with $B_{i+1} \subseteq Y_{i+1}$ via $F : A_i \to B_{i+1}$ for all $i \in \mathbb{Z}$ (or $\mathbb{Z}_n$). By the Transgression Theorem (5.1.9), copies $\{q_i : Y_i \to X\}$ of the quotient map q define a continuous function $p : \hat{X} \to X$ such that $p \circ Q|_{Y_i} = q_i$ for all $i \in \mathbb{Z}$ (or $\mathbb{Z}_n$), as indicated below:

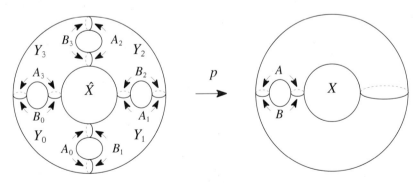

Covering projection $p : \hat{X} \to X$

1.1 Theorem *This construction gives a covering projection $p : \hat{X} \to X$.*

Proof: The open subset $X - q(A \cup B) \cong Y - (A \cup B)$ is evenly covered; its pre-image, $p^{-1}(X - q(A \cup B))$, is $\dot\bigcup \{Q(Y_i - (A_i \cup B_i)) \subseteq \hat{X} : i \in \mathbb{Z}$ (or $\mathbb{Z}_n)\}$.

The open subset $q(U\cup V)$ is evenly covered; its pre-image, $p^{-1}(q(U\cup V))$, is $\overset{\circ}{\cup}\,\{Q(U_i\cup V_{i+1})\subseteq \hat{X}: i\in \mathbb{Z}\text{ (or }\mathbb{Z}_n)\}$. So p is a covering projection. □

Example 2 In the previous displays, Y is the four-times punctured 2-sphere. Its boundary is expressed as two pairs $A = \mathbb{S}^1_{a'}\cup \mathbb{S}^1_{a''}$ and $B = \mathbb{S}^1_{b'}\cup \mathbb{S}^1_{b''}$ of two circles each. The identification of the boundary circles according to a homeomorphism $F: A \to B$ produces the orientable surface $\mathfrak{H}_2$ of genus 2. Using copies of the identification map $q: Y \to \mathfrak{H}_2$ indexed by $\mathbb{Z}_4$, the construction yields a covering projection $p: \mathfrak{H}_5 \to \mathfrak{H}_2$.

COVERING SPACES OF PRODUCTS AND SUBSPACES

Product maps provide another way to construct covering projections.

1.2 Theorem *The product $p \times q: \hat{X}\times \hat{Y} \to X\times Y$ of two covering projections $p: \hat{X}\to X$ and $q: \hat{Y}\to Y$ is a covering projection.*

Proof: Let $U \subseteq X$ and $V \subseteq Y$ be admissible for p and q, respectively. Then $U\times V \subseteq X\times Y$ is admissible for $p\times q$. For, if $p^{-1}(U)$ and $q^{-1}(V)$ are disjoint unions of path-components $\hat{U}_i$ and $\hat{V}_k$ that p and q project homeomorphically onto U and V, then $(p\times q)^{-1}(U\times V)$ is the disjoint union of path components $\hat{U}_i\times \hat{V}_k$ that $p\times q$ projects homeomorphically onto $U\times V$. □

Example 3 There is the covering projection $e\times e: \mathbb{R}^1\times \mathbb{R}^1 \to \mathbb{S}^1\times \mathbb{S}^1$ of the plane onto the torus. It factors into the covering projections $(e\times 1): \mathbb{R}^1\times \mathbb{R}^1 \to \mathbb{S}^1\times \mathbb{R}^1$ and $(1\times e): \mathbb{S}^1\times \mathbb{R}^1 \to \mathbb{S}^1\times \mathbb{S}^1$ of the plane onto the cylinder onto the torus:

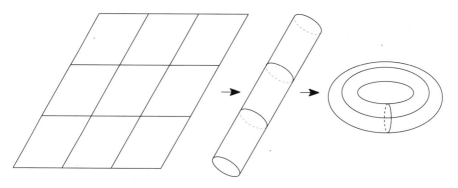

Example 3: $e\times e: \mathbb{R}^1\times \mathbb{R}^1 \to \mathbb{R}^1\times \mathbb{S}^1 \to \mathbb{S}^1\times \mathbb{S}^1$

Example 4 Another covering space of the torus $\mathbb{S}^1\times \mathbb{S}^1$ arises from the product $d_n\times d_m: \mathbb{S}^1\times \mathbb{S}^1 \to \mathbb{S}^1\times \mathbb{S}^1$ of two covering projections of $\mathbb{S}^1$ onto itself. The covering projection $d_n\times d_m$ wraps the torus around itself n times in the meridional axis and m times in the longitudinal axis.

Restricting a covering projection to subspaces of the base space and total space can yield a covering projection.

1.3 Theorem *Let $p: \hat{X} \to X$ be a covering projection and let $A \subseteq X$ be a path-connected and locally path-connected subspace. Then the restriction $p: \hat{A} \to A$ of p to a path-component $\hat{A}$ of the pre-image $p^{-1}(A) \subseteq \hat{X}$ is a covering projection.*

Proof: If $U \subseteq X$ is evenly covered by $p: \hat{X} \to X$, then $U \cap A \subseteq A$ is evenly covered by $p: \hat{A} \to A$. □

Example 5 The pre-image of the one-point union $\mathbb{S}^1 \vee \mathbb{S}^1 \subset \mathbb{S}^1 \times \mathbb{S}^1$ under the covering projection $e \times e: \mathbb{R}^1 \times \mathbb{R}^1 \to \mathbb{S}^1 \times \mathbb{S}^1$ is the lattice $L = (\mathbb{R} \times \mathbb{Z}) \cup (\mathbb{Z} \times \mathbb{R})$ of all horizontal and vertical lines in $\mathbb{R}^1 \times \mathbb{R}^1$ with integer valued intercepts. By Theorem 1.3, the restriction $e \times e: L \to \mathbb{S}^1 \vee \mathbb{S}^1$ is a covering projection of L onto $\mathbb{S}^1 \vee \mathbb{S}^1$.

Example 6 There is an enormous variety of other covering spaces of the one-point union $\mathbb{S}^1 \vee \mathbb{S}^1$. Here is a sampler:

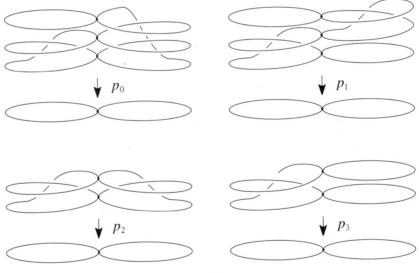

Example 6: $p_i: \hat{X}_i \to \mathbb{S}^1 \vee \mathbb{S}^1$

2 Liftings of Paths and Homotopies

A covering space offers an unwrapped version of the base space into which we can lift various investigations.

Let $p: \hat{X} \to X$ be a covering projection; let Y be an arbitrary space; and let $F: Y \to X$ be a continuous function into the base space X.

A continuous function $\hat{F}: Y \to \hat{X}$ to the total space $\hat{X}$ *covers*, or is a ***lifting*** of, $F: Y \to X$ provided that $F = p \circ \hat{F}: Y \to \hat{X} \to X$, as in the following commutative diagram:

$$\begin{array}{ccc} & & \hat{X} \\ & \hat{F}\nearrow & \downarrow p \\ Y & \xrightarrow{F} & X \end{array}$$

Lifting $\hat{F}$ of F

UNIQUENESS OF LIFTINGS

There are two basic problems involving liftings: existence and uniqueness.

2.1 Uniqueness of Liftings *Let $p : \hat{X} \to X$ be a covering projection; let $F : Y \to X$ be a continuous function to the base space X. If Y is connected, two liftings $\hat{F}_1, \hat{F}_2 : Y \to \hat{X}$ of F that agree at some point of Y are equal.*

Proof: By the connectedness of Y, it suffices to show that the set of coincidences, $B = \{y \in Y : \hat{F}_1(y) = \hat{F}_2(y)\}$, is open and closed in Y.

If $y \in B$, we use any admissible neighborhood $U \subseteq X$ of $x = F(y) \in X$ and the path-component $\hat{U} \subseteq \hat{X}$ of the pre-image $p^{-1}(U)$ containing the image point $\hat{F}_1(y) = \hat{F}_2(y) \in \hat{X}$. $\hat{U}$ is an open neighborhood of $\hat{F}_1(y) = \hat{F}_2(y)$ in $\hat{X}$; hence, $N(y) = \hat{F}_1^{-1}(\hat{U}) \cap \hat{F}_2^{-1}(\hat{U})$ is an open neighborhood of y in Y. $N(y)$ is contained in B, because the image points $\hat{F}_1(z), \hat{F}_2(z) \in \hat{U}$ of any $z \in N(y)$ have the same value, $p(\hat{F}_1(z)) = F(z) = p(\hat{F}_2(z))$, under the projection *homeomorphism* $p|_{\hat{U}} : \hat{U} \to U$ and thus must coincide in $\hat{U} \subseteq \hat{X}$. So B is open in Y.

If $y \notin B$, we use an admissible neighborhood $U \subseteq X$ of $x = F(y) \in \hat{X}$. The distinct image points $\hat{F}_1(y) \neq \hat{F}_2(y) \in \hat{X}$ must belong to path-components $\hat{U}_1, \hat{U}_2 \subseteq \hat{X}$ of the pre-image $p^{-1}(U)$. Now $\hat{U}_1 \cap \hat{U}_2 = \emptyset$; otherwise, $\hat{U}_1$ and $\hat{U}_2$ would be the same component $\hat{U}$. And the projection *homeomorphism* $p|_{\hat{U}} : \hat{U} \to U$ would map the distinct points $\hat{F}_1(y) \neq \hat{F}_2(y)$ to the same point:

$$p(\hat{F}_1(y)) = F(y) = p(\hat{F}_2(y)).$$

The open neighborhoods $\hat{U}_1$ and $\hat{U}_2$ in $\hat{X}$ of $\hat{F}_1(y)$ and $\hat{F}_2(y)$ yield an open neighborhood $N(y) = \hat{F}_1^{-1}(\hat{U}_1) \cap \hat{F}_2^{-1}(\hat{U}_2)$ of y in Y. $N(y)$ is contained in the complement B^c, because the images $\hat{F}_1(z)$ and $\hat{F}_2(z)$ in $\hat{X}$ of any $z \in N(y)$ lie in the disjoint path components $\hat{U}_1$ and $\hat{U}_2$, respectively. Because B^c contains an open neighborhood $N(y)$ of each of its points, B is closed in Y. □

EXISTENCE OF PATH LIFTINGS

The existence of liftings is easily verified for paths in the base space. We adapt the argument in Section **9**.2 for the exponential covering projection.

2.2 Path Lifting Property *Let $p : \hat{X} \to X$ be a covering projection; let $x_0 \in X$ and $\hat{x}_0 \in p^{-1}(x_0) \subseteq \hat{X}$ be selected basepoints. Any path $f : \mathbb{I} \to X$ with initial point x_0 has a unique lifting to a path $\hat{f} : \mathbb{I} \to \hat{X}$ with initial point $\hat{x}_0$.*

Proof: By the Lebesgue Covering Lemma (**7.3.8**), there is a partition of $\mathbb{I}$,

$$\mathcal{P} : 0 = t_0 < t_1 < \ldots < t_{n-1} < t_n = 1,$$

such that each partial path $f : [t_i, t_{i+1}] \to X$ takes values in some admissible subset $U_i \subseteq X$.

The crux is that each partial path $f : [t_i, t_{i+1}] \to U_i \subseteq X$ admits a lifting $\hat{f} : [t_i, t_{i+1}] \to \hat{X}$ beginning at any selected point $\hat{f}(t_i) \in p^{-1}(f(t_i)) \subseteq p^{-1}(U_i)$. If $\hat{U}_i$ is the path-component of $p^{-1}(U_i)$ containing the selected point $\hat{f}(t_i)$, then the composite of $f : [t_i, t_{i+1}] \to U_i$ with the section homeomorphism $s : U_i \to \hat{U}_i \subseteq \hat{X}$ is a lifting $\hat{f} : [t_i, t_{i+1}] \to \hat{X}$ with initial point $\hat{f}(t_i)$.

Then the full path lifting $\hat{f}$ is constructed inductively on consecutive subintervals of $\mathcal{P}$, as follows. The preliminary argument provides a lifting $\hat{f} : ([0, t_1], 0) \to (\hat{X}, \hat{x}_0)$ of the partial path $f : [0, t_1] \to U_1 \subseteq X$, and also shows how to extend a lifting $\hat{f} : ([0, t_i], 0) \to (\hat{X}, \hat{x}_0)$ of the partial path $f : [0, t_i] \to X$ to a lifting $\hat{f} : ([0, t_{i+1}], 0) \to (\hat{X}, \hat{x}_0)$ of the longer path $f : [0, t_{i+1}] \to X$. By induction, there results a path lifting $\hat{f}$ at $\hat{x}_0$ of the given path f. Its uniqueness follows from the Uniqueness of Liftings (**2.1**). □

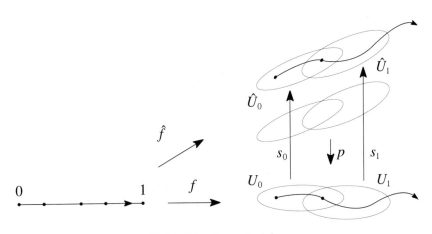

Path Lifting Property 2.2

Example 1 The covering projection $p : L \to \mathbb{S}^1_a \vee \mathbb{S}^1_b$ in Example 1.5 wraps the lattice L in $\mathbb{R}^2$ into the one-point union $\mathbb{S}^1_a \vee \mathbb{S}^1_b$. The loop $f_{ab} : \mathbb{I} \to \mathbb{S}^1_a \vee \mathbb{S}^1_b$ spelling the word $ab \in \mathbb{F}(a, b)$ lifts to a path $\hat{f}_{ab}$ in L that runs horizontally and then vertically, one unit each. The loop $f_{ba} : \mathbb{I} \to \mathbb{S}^1_a \vee \mathbb{S}^1_b$ spelling the word $ba \in \mathbb{F}(a, b)$ lifts to a path $\hat{f}_{ba}$ in L that runs vertically and then horizontally, one unit each.

LIFTING OF HOMOTOPIES

As in Section **9**.2 for the exponential covering projection, the lifting of a loop in the base space is rarely a loop in the total space, and liftings with the same initial point of paths with common endpoints rarely have the same terminal point. The crucial feature of path liftings is that their terminal points are intimately related with the homotopy class of the paths that they cover. Here is a result that describes aspects of this relationship.

2.3 Homotopy Lifting Property *Let $p : \hat{X} \to X$ be a covering projection. If paths $f, g : \mathbb{I} \to X$ have the same endpoints and are path-homotopic in X, then any two liftings $\hat{f}, \hat{g} : \mathbb{I} \to \hat{X}$ with the same initial point have the same terminal point and are path-homotopic in $\hat{X}$.*

Proof: We first show that any path-homotopy $H : \mathbb{I} \times \mathbb{I} \to X$ lifts to a path-homotopy $\hat{H} : \mathbb{I} \times \mathbb{I} \to \hat{X}$ with any prescribed value $\hat{H}(0, 0)$ in the fiber over $H(0, 0)$. By an application of the Lebesgue Covering Lemma (**7.3.8**), there exist partitions of the factors of the closed unit square $\mathbb{I} \times \mathbb{I}$,

$$\mathcal{P} : 0 = t_0 < t_1 < \ldots < t_n < t_{n+1} = 1 \quad \text{and} \quad \mathcal{R} : 0 = r_0 < r_1 < \ldots < r_m < r_{m+1} = 1,$$

such that each sub-rectangle $R_{i,k} = [t_i, t_{i+1}] \times [r_k, r_{k+1}]$ of $\mathbb{I} \times \mathbb{I}$ is carried by the homotopy H into some admissible subset $U_{i,k}$ of the base space X.

Each partial homotopy $H_{i,k} = H|_{R_{i,k}} : R_{i,k} \to U_{i,k} \subseteq X$ has a lifting $\hat{H}_{i,k} : R_{i,k} \to \hat{X}$ with any prescribed value $\hat{H}_{i,k}(t_i, r_k)$ in the fiber over the value, $H(t_i, r_k)$, of H at the corner point $<t_i, r_k> \in R_{i,k}$. This lifting $\hat{H}_{i,k}$ is obtained by composing the partial homotopy $H_{i,k} : R_{i,k} \to U_{i,k} \subseteq X$ with the appropriate section homeomorphism $s : U_{i,k} \to \hat{U}_{i,k} \subseteq \hat{X}$.

The desired lifting $\hat{H}$ is constructed inductively over the rectangles $R_{i,k}$ (ordered lexicographically), as follows. By the preliminary argument, the partial homotopy $H_{0,0} : R_{0,0} \to U_{0,0} \subseteq X$ lifts to a homotopy $\hat{H}_{0,0} : R_{0,0} \to \hat{X}$ with any prescribed value $\hat{H}(0, 0) = \hat{x}_0$ in the fiber over $H(0, 0) = x_0$.

Suppose that the lifting $\hat{H}$ has been defined on the union of the rectangles:

$$R_{0,0}, \ldots, R_{n,0};\ R_{0,1}, \ldots, R_{n,1};\ \ldots;\ R_{0,k}, \ldots, R_{i-1,k}.$$

Then, in particular, the image $\hat{H}(t_i, r_k)$ of the corner point $<t_i, r_k> \in R_{i,k}$ is defined. By the uniqueness of liftings, the partial lifting $\hat{H}_{i,k} : R_{i,k} \to \hat{X}$, with the prescribed value $\hat{H}(t_i, r_k)$, agrees with the given lifting $\hat{H}$ on the adjacent rectangles $R_{i-1,k}$ and $R_{i,k-1}$, since the intersections $R_{i,k} \cap R_{i-1,k}$ and $R_{i,k} \cap R_{i,k-1}$ are connected intervals that contain the corner point $<t_i, r_k>$ at which $\hat{H}_{i,k}$ and $\hat{H}$ agree. By the Gluing Theorem (**4.4.3**), this inductive process yields a continuous lifting $\hat{H} : \mathbb{I} \times \mathbb{I} \to \hat{X}$ of the original homotopy H.

Suppose now that H is a path-homotopy $f \simeq g : (\mathbb{I}, 0, 1) \to (X, x_0, x_1)$. Since the restrictions of H to the sides $\{0\} \times \mathbb{I}$ and $\{1\} \times \mathbb{I}$ are constant paths at x_0 and x_1, then the restrictions of the lifting $\hat{H}$ to the sides $\{0\} \times \mathbb{I}$ and $\{1\} \times \mathbb{I}$ are constant paths at $\hat{x}_0 \in p^{-1}(x_0)$ and $\hat{x}_1 \in p^{-1}(x_1)$, by the uniqueness of path liftings. Therefore, the lifting $\hat{H}$ is a path-homotopy between the unique liftings $\hat{f}, \hat{g} : (\mathbb{I}, 0, 1) \to (\hat{X}, \hat{x}_0, \hat{x}_1)$ of the paths f and g at the initial point $\hat{x}_0$. In particular, these liftings have the same terminal point. □

CONSEQUENCES

Let $p : \hat{X} \to X$ be a covering projection. The Homotopy Lifting Property (**2.3**) implies that a loop in $\hat{X}$ that is null-homotopic when projected into X is necessarily null-homotopic in $\hat{X}$. This proves the following:

2.4 Monotonicity Property *A covering projection $p : \hat{X} \to X$ induces a monomorphism, $p_\# : \pi_1(\hat{X}, \hat{x}) \to \pi_1(X, x)$, for all choices of basepoints $x \in X$ and $\hat{x} \in p^{-1}(x) \subseteq \hat{X}$.*

The subgroup $\pi_1(\hat{X}, \hat{x}) \approx p_\#(\pi_1(\hat{X}, \hat{x})) \leq \pi_1(X, x)$ captures the loops in X based at x that lift to loops in $\hat{X}$ based at $\hat{x} \in p^{-1}(x)$. Another choice of the fiber element $\hat{x} \in p^{-1}(x)$ can lead to a different subgroup of $\pi_1(X, x)$.

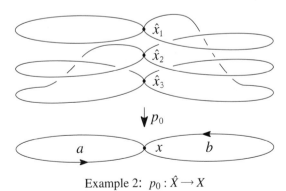

Example 2: $p_0 : \hat{X} \to X$

Example 2 Consider the 3-fold covering projection $p_0 : \hat{X} \to X$, depicted above. The fundamental group $\pi_1(X, x)$ of $X = \mathbb{S}_a^1 \vee \mathbb{S}_b^1$ is the free group with free basis $\{a = [e_a], b = [e_b]\}$, where $e_a, e_b : \mathbb{I} \to \mathbb{S}_a^1 \vee \mathbb{S}_b^1$ are the exponential loops on the left- and right-hand summands, respectively.

Of the subgroups $p_\#(\pi_1(\hat{X}, \hat{x}_i)) \leq \pi_1(X, x)$ ($i = 1, 2, 3$), only the first contains a, only the second contains ab^{-1}, and only the third contains ab.

By the Cofibration Theorem (10.5.5), a collapse of two of the edges that lie above $\mathbb{S}^1_b$ gives a homotopy equivalence between $\hat{X}$ and a one-point union of four copies of $\mathbb{S}^1$. Thus, $\pi_1(\hat{X}, \hat{x}_1)$, for example, is a free group of rank 4; its image $p_\#(\pi_1(\hat{X}, \hat{x}_1))$ in $\pi_1(X, x)$ is $\mathbb{H} = <a, b^3, bab, ba^{-1}b> \leq \mathbb{F}(a, b)$.

The subgroups $p_\#(\pi_1(\hat{X}, \hat{x})) \leq \pi_1(X, x)$, $\hat{x} \in p^{-1}(x) \subseteq \hat{X}$, are closely related. For any path $\hat{g}$ in $\hat{X}$ from $\hat{x}$ to $\hat{z} \in p^{-1}(x)$, let $p \circ \hat{g} = g$ be its projection to a *loop* in X based at x. The change of basepoint isomorphism

$$\omega_{[\hat{g}]} : \pi_1(\hat{X}, \hat{x}) \longrightarrow \pi_1(\hat{X}, \hat{z}), \quad \omega_{[\hat{g}]}([\hat{f}]) = [\hat{g}]^{-1} \cdot [\hat{f}] \cdot [\hat{g}],$$

corresponds to conjugation in $\pi_1(X, x)$ by $[g] \in \pi_1(X, x)$, as indicated below:

$$\begin{array}{ccc} \pi_1(\hat{X}, \hat{x}) & \xrightarrow{\omega_{[\hat{g}]}} & \pi_1(\hat{X}, \hat{z}) \\ p_\# \downarrow & & \downarrow p_\# \\ \pi_1(X, x) & \xrightarrow{[g]^{-1} \cdot (-) \cdot [g]} & \pi_1(X, x) \end{array}$$

Therefore, $p_\#(\pi_1(\hat{X}, \hat{z}))$ is the conjugate subgroup $[g]^{-1} \cdot p_\#(\pi_1(\hat{X}, \hat{x})) \cdot [g]$ to $p_\#(\pi_1(\hat{X}, \hat{x})) \leq \pi_1(X, x)$.

Conversely, any conjugate $[g]^{-1} \cdot p_\#(\pi_1(\hat{X}, \hat{x})) \cdot [g]$ of the subgroup $p_\#(\pi_1(\hat{X}, \hat{x})) \leq \pi_1(X, x)$ equals $p_\#(\pi_1(\hat{X}, \hat{z}))$ for some fiber element $\hat{z} \in p^{-1}(x)$. Namely, $\hat{z}$ is the terminal point of the lifting $\hat{g}$ of g at $\hat{x}$.

We summarize the previous observations, as follows.

2.5 Conjugacy Property Let $p : \hat{X} \to X$ be a covering projection and let $x \in X$ be a selected basepoint. Then the subgroups

$$p_\#(\pi_1(\hat{X}, \hat{x})) \leq \pi_1(X, x) \qquad (\hat{x} \in p^{-1}(x) \subseteq \hat{X})$$

constitute a complete conjugacy class of subgroups of $\pi_1(X, x)$.

Viewpoint: The conjugacy class captures the *loops* in X based at x that admit some lifting to a *loop* in $\hat{X}$.

Example 3 For the covering projection $p_0 : \hat{X} \to X$ in Example 2, the conjugacy class consists of the three conjugates $\mathbb{H}$, $b\,\mathbb{H}\,b^{-1}$, and $b^2\,\mathbb{H}\,b^{-2}$ of the subgroup $\mathbb{H} = <a, b^3, bab, ba^{-1}b> \leq \mathbb{F}(a, b)$.

Example 4 For the covering projection $e \times e : L \to \mathbb{S}^1_a \vee \mathbb{S}^1_b$ in Example 1.5, the conjugacy class consists of the single normal subgroup $\mathbb{N}(aba^{-1}b^{-1}) \leq \mathbb{F}(a, b)$.

Example 5 For the covering projection $p_3 : \hat{X} \to \mathbb{S}^1_a \vee \mathbb{S}^1_b$ in Example 1.6, the conjugacy class consists of the single normal subgroup $\mathbb{N}(a^2, b) \leq \mathbb{F}(a, b)$.

2.6 Action Property   Let $p : \hat{X} \to X$ be a covering projection and let $x \in X$ be a selected basepoint. There is a well-defined action (function),

$$\cdot : p^{-1}(x) \times \pi_1(X, x) \to p^{-1}(x),$$

that associates with each fiber element $\hat{x} \in p^{-1}(x)$ and group element $[f] \in \pi_1(X, x)$ a fiber element $\hat{x} \cdot [f] \in p^{-1}(x)$. The action also satisfies:
(a) $(\hat{x} \cdot [f]) \cdot [g] = \hat{x} \cdot ([f] \cdot [g])$;
(b) $\hat{x} \cdot [x^*] = \hat{x}$;
(c) for $\hat{x}, \hat{z} \in p^{-1}(x)$, there exists $[f] \in \pi_1(X, x)$ such that $\hat{x} \cdot [f] = \hat{z}$; and
(d) $\hat{x} \cdot [f] = \hat{x}$ if and only if $[f] \in p_\#(\pi_1(\hat{X}, \hat{x}))$.

Proof: The action of the path-class $[f] \in \pi_1(X, x)$ on the fiber element $\hat{x} \in p^{-1}(x)$ is defined to be the terminal point $\hat{f}_{\hat{x}}(1)$ of the lifting $\hat{f}_{\hat{x}}$ at $\hat{x}$ of any loop f representing the class $[f]$. By the Path and Homotopy Lifting Properties 2.2 and 2.3, the point $\hat{x} \cdot [f] = \hat{f}_{\hat{x}}(1)$ is well-defined. Moreover, this definition guarantees the conditions (a-d) listed above.   □

3 General Liftings

In this section we characterize those continuous functions into the base of a covering projection that admit a lifting to the total space. Applications of this characterization include a classification of the covering spaces of a given base space, and the universality of simply connected covering spaces.

CONDITIONS FOR LIFTING

Let $p : \hat{X} \to X$ be a covering projection and let Y be a connected, locally path-connected space. A lifting of a continuous function $F : Y \to X$ into the base space X is a continuous function $\hat{F} : Y \to \hat{X}$ into the total space $\hat{X}$ such that $F = p \circ \hat{F} : Y \to \hat{X} \to X$. If, in addition, $F(y) = x$ and $\hat{F}(y) = \hat{x}$, then $\hat{F}$ is called a lifting of F at $\hat{x} \in p^{-1}(x) \subset \hat{X}$. Then,

$$F_\# = p_\# \circ \hat{F}_\# : \pi_1(Y, y) \to \pi_1(\hat{X}, \hat{x}) \to \pi_1(X, x).$$

So the image $F_\#(\pi_1(Y, y))$ is a subgroup of the image $p_\#(\pi_1(\hat{X}, \hat{x})) \leq \pi_1(X, x)$.

The containment relationship $F_\#(\pi_1(Y, y)) \leq p_\#(\pi_1(\hat{X}, \hat{x}))$ in $\pi_1(X, x)$ does not involve the assumed lifting $\hat{F}$ and therefore it serves as a necessary condition for the existence of such a lifting of F at $\hat{x}$.

This necessary condition is also sufficient.

3.1 Lifting Criterion   A continuous function $F : (Y, y) \to (X, x)$ has a lifting $\hat{F} : (Y, y) \to (\hat{X}, \hat{x})$ at $\hat{x} \in p^{-1}(x) \subset \hat{X}$ if and only if one has:

$$F_\#(\pi_1(Y, y)) \leq p_\#(\pi_1(\hat{X}, \hat{x})) \leq \pi_1(X, x).$$

Proof: Were $\hat{F}$ the desired lifting, its composite $\hat{F} \circ g$ with a path $g : \mathbb{I} \rightarrow Y$ from y to z would be a path in $\hat{X}$ from $\hat{x} = \hat{F}(y)$ to $\hat{F}(z)$; moreover, $\hat{F} \circ g$ would cover the path $F \circ g : \mathbb{I} \rightarrow X$ from $x = F(y)$ to $F(z)$. Thus, $\hat{F}(z)$ would be the terminal endpoint $\widehat{(F \circ g)}(1)$ of the path lifting $\hat{F} \circ g = \widehat{(F \circ g)} : \mathbb{I} \rightarrow \hat{X}$ at $\hat{x}$ of the path $F \circ g : \mathbb{I} \rightarrow X$.

Accordingly, we define the value of $\hat{F} : (Y, y) \rightarrow (\hat{X}, \hat{x})$ at any point $z \in Y$ to be the terminal endpoint $\widehat{(F \circ g)}(1)$ of the path lifting $\widehat{(F \circ g)} : \mathbb{I} \rightarrow \hat{X}$ at $\hat{x}$ of the path $F \circ g : \mathbb{I} \rightarrow X$, where $g : \mathbb{I} \rightarrow Y$ is any path joining y to z in the path-connected space Y. We now show that this is independent of the choice of g.

For another choice of a path $h : \mathbb{I} \rightarrow Y$ joining y to z, the path product $g \cdot \bar{h} : \mathbb{I} \rightarrow Y$ is a loop in Y based at y. Because $F_{\#}(\pi_1(Y, y)) \leq p_{\#}(\pi_1(\hat{X}, \hat{x}))$, the loop $F \circ (g \cdot \bar{h}) : \mathbb{I} \rightarrow X$ at x in the base space X lifts to a loop at $\hat{x}$ in the total space $\hat{X}$. This implies that the two liftings $\widehat{(F \circ g)}$ and $\widehat{(F \circ h)}$ in $\hat{X}$ at $\hat{x}$ have the same terminal point $\widehat{(F \circ g)}(1) = \widehat{(F \circ h)}(1)$. So the defined assignment $\hat{F}(z) = \widehat{(F \circ g)}(1)$ is independent of the choice of the path g.

It follows from this definition that $\hat{F}(y) = \hat{x}$, $p \circ \hat{F} = F$, and $\hat{F}$ carries each path $f : \mathbb{I} \rightarrow Y$ into a path lifting $\widehat{(F \circ f)} : \mathbb{I} \rightarrow \hat{X}$ of the path $F \circ f : \mathbb{I} \rightarrow X$.

To show that $\hat{F}$ is continuous at a point $z \in Y$, let U be an admissible neighborhood of $F(z)$, let $W \subseteq F^{-1}(U) \subseteq Y$ be a path-connected neighborhood of z, and let $\hat{U}$ be the path-component of $p^{-1}(U)$ containing $\hat{F}(z)$. Both the restriction $\hat{F} : W \subseteq Y \rightarrow \hat{X}$ and the composite of $F|_W : W \rightarrow U$ with the section homeomorphism $s : U \rightarrow \hat{U} \subseteq \hat{X}$ send z to $\hat{F}(z)$. Both also carry each path f in W into the unique path lifting in $\hat{X}$ of the path $F \circ f$ in U. Thus, $\hat{F}$ equals the continuous function $s \circ F|_W$ on the path-connected neighborhood W. Since continuity is a local notion, this proves that $\hat{F}$ is continuous. $\square$

The basepoints $x \in X$ and $\hat{x} \in p^{-1}(x) \subset \hat{X}$ play a crucial role in the lifting criterion. Here is a revealing example.

Example 1 Consider the 3-fold covering projection $p_1 : \hat{X} \rightarrow X$, below. The inclusion $F : \mathbb{S}_a^1 \subset \mathbb{S}_a^1 \vee \mathbb{S}_b^1 = X$ admits a lifting at $\hat{x}_1$ but neither at $\hat{x}_2$ nor at $\hat{x}_3$. The image $F_{\#}(\pi_1(\mathbb{S}_a^1, 1)) = \langle\, [e_a]\, \rangle$ is contained in the subgroup $p_{1\#}(\pi_1(\hat{X}, \hat{x}_i))$ only for $i = 1$.

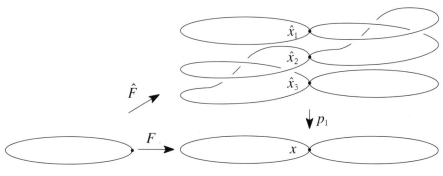

Example 1: $p_1 : \hat{X} \rightarrow X$

Example 2 The identity function $1 : \mathbb{P}^2 \to \mathbb{P}^2$ does not lift through the double covering $p : \mathbb{S}^2 \to \mathbb{P}^2$ obtained by identifying antipodal points; for any choice of basepoints, the lifting criterion takes the form of the impossible relation $\mathbb{Z}_2 \leq 0 \leq \mathbb{Z}_2$.

Example 3 The m-fold covering projection $d_m : \mathbb{S}^1 \to \mathbb{S}^1$ lifts through the n-fold covering projection $d_n : \mathbb{S}^1 \to \mathbb{S}^1$ if and only if $m\mathbb{Z} \leq n\mathbb{Z} \leq \mathbb{Z}$, i.e., n divides m.

3.2 Corollary   *If there are only trivial homomorphisms from $\pi_1(Y, y)$ to $\pi_1(X, x)$, then every continuous function $F : (Y, y) \to (X, x)$ admits a lifting $\hat{F} : (Y, y) \to (\hat{X}, \hat{x})$ at each fiber point $\hat{x} \in p^{-1}(x) \subseteq \hat{X}$.*
In particular, this is the case when Y is simply connected.

Example 4 Any continuous function $F : \mathbb{P}^2 \to \mathbb{S}^1$ admits a lifting through every covering of $\mathbb{S}^1$, since any homomorphism from $\pi_1(\mathbb{P}^2) = \mathbb{Z}_2$ to $\pi_1(\mathbb{S}^1) = \mathbb{Z}$ is trivial.

Example 5 When $n \geq 2$, $\mathbb{S}^n$ is simply connected by Theorem **9.2.7**. Then, any continuous function $\mathbb{S}^n \to \mathbb{S}^1$ admits a lifting through $e : \mathbb{R}^1 \to \mathbb{S}^1$ and so is inessential because $\mathbb{R}^1$ is contractible. Thus $[\mathbb{S}^n, \mathbb{S}^1]$ is trivial for all $n > 1$.
But Hopf's map $p : \mathbb{S}^3 \to \mathbb{S}^2$ is essential (Example **10.5.7**), so that $[\mathbb{S}^n, \mathbb{S}^m]$ is not trivial for all $n > m \geq 2$.

MORPHISMS BETWEEN COVERING SPACES

A **morphism** $\Phi : (\hat{X}_1, p_1) \to (\hat{X}_2, p_2)$ between covering spaces $(\hat{X}_1, p_1)$ and $(\hat{X}_2, p_2)$ of X is a continuous function $\Phi : \hat{X}_1 \to \hat{X}_2$ such that $p_2 \circ \Phi = p_1$:

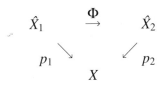

Covering spaces over X and their morphisms constitute a category $\mathcal{C}_X$. If $(\hat{X}, p)$ is a covering space of X, the identity map $1_{\hat{X}} : \hat{X} \to \hat{X}$ is a morphism $1 : (\hat{X}, p) \to (\hat{X}, p)$. If $\Phi : (\hat{X}_1, p_1) \to (\hat{X}_2, p_2)$ and $\Theta : (\hat{X}_2, p_2) \to (\hat{X}_3, p_3)$ are morphisms, their composite is a morphism $\Theta \circ \Phi : (\hat{X}_1, p_1) \to (\hat{X}_3, p_3)$.

A morphism $\Phi : (\hat{X}_1, p_1) \to (\hat{X}_2, p_2)$ is an **isomorphism** if it is an equivalence in $\mathcal{C}_X$, i.e., if there exists a morphism $\Theta : (\hat{X}_2, p_2) \to (\hat{X}_1, p_1)$ such that $\Theta \circ \Phi$ and $\Phi \circ \Theta$ are the identity morphisms. Two covering spaces are called **isomorphic** when they are equivalent in the category $\mathcal{C}_X$.

A morphism $\Phi : (\hat{X}_1, p_1) \to (\hat{X}_2, p_2)$ can be considered as a lifting of $p_1 : \hat{X}_1 \to X$ through the covering projection $p_2 : \hat{X}_2 \to X$. Thus, by the Uniqueness of Liftings (2.1), two morphisms $(\hat{X}_1, p_1) \to (\hat{X}_2, p_2)$ that agree at some point of $\hat{X}_1$ are equal, and two morphisms $(\hat{X}_1, p_1) \to (\hat{X}_2, p_2)$ and $(\hat{X}_2, p_2) \to (\hat{X}_1, p_1)$ that exchange a pair of points of $\hat{X}_1$ and $\hat{X}_2$ are inverse isomorphisms. So the Lifting Criterion (3.1) has these consequences:

3.3 Morphism Criterion Let $(\hat{X}_1, p_1)$ and $(\hat{X}_2, p_2)$ be covering spaces of X; let $x \in X$, $\hat{x}_1 \in p_1^{-1}(x) \subseteq \hat{X}_1$, and $\hat{x}_2 \in p_2^{-1}(x) \subseteq \hat{X}_2$ be selected basepoints.
(a) There exists a morphism $\Phi : (\hat{X}_1, p_1) \longrightarrow (\hat{X}_2, p_2)$ such that $\Phi(\hat{x}_1) = \hat{x}_2$ if and only if $p_{1\#}(\pi_1(\hat{X}_1, \hat{x}_1)) \leq p_{2\#}(\pi_1(\hat{X}_2, \hat{x}_2))$ in $\pi_1(X, x)$.
(b) There exists an isomorphism $\Phi : (\hat{X}_1, p_1) \longrightarrow (\hat{X}_2, p_2)$ such that $\Phi(\hat{x}_1) = \hat{x}_2$ if and only if $p_{1\#}(\pi_1(\hat{X}_1, \hat{x}_1)) = p_{2\#}(\pi_1(\hat{X}_2, \hat{x}_2))$ in $\pi_1(X, x)$.

3.4 Classification Theorem Two covering spaces $(\hat{X}_1, p_1)$ and $(\hat{X}_2, p_2)$ of X are isomorphic if and only if their conjugacy classes of subgroups,

$$\{ p_{1\#}(\pi_1(\hat{X}_1, \hat{x}_1)) : \hat{x}_1 \in p_1^{-1}(x) \subseteq \hat{X}_1 \}$$

and

$$\{ p_{2\#}(\pi_1(\hat{X}_2, \hat{x}_2)) : \hat{x}_2 \in p_2^{-1}(x) \subseteq \hat{X}_2 \},$$

coincide.

Proof: Two conjugacy classes of subgroups of $\pi_1(X, x)$ coincide if and only if they have some member in common. Therefore, the Classification Theorem follows from part (b) of the Morphism Criterion (3.3). □

In an abelian group $\mathbb{G}$, conjugation by an element $g \in \mathbb{G}$ is the identity automorphism $1_{\mathbb{G}} = g^{-1} \cdot (-) \cdot g : \mathbb{G} \longrightarrow \mathbb{G}$, and, therefore, each conjugacy class of subgroups of $\mathbb{G}$ consists of a single subgroup. It follows that the covering spaces of a space X whose fundamental group is abelian are classified up to isomorphism by the *subgroups* of $\pi_1(X, x)$.

Example 6 The covering projections $e : \mathbb{R}^1 \longrightarrow \mathbb{S}^1$ and $d_n : \mathbb{S}^1 \longrightarrow \mathbb{S}^1$ ($n \geq 1$) give all possible non-isomorphic covering spaces of the 1-sphere $\mathbb{S}^1$, since the only subgroups of $\pi_1(\mathbb{S}^1)$ are $\{0\}, \mathbb{Z}, 2\mathbb{Z}, \ldots, n\mathbb{Z}, \ldots$.
Similarly, up to isomorphism, the only covering spaces of the projective plane $\mathbb{P}^2$ are the double covering $p : \mathbb{S}^2 \longrightarrow \mathbb{P}^2$ and the identity covering $1 : \mathbb{P}^2 \longrightarrow \mathbb{P}^2$, since the only subgroups of $\pi_1(\mathbb{P}^2)$ are $\{0\}$ and $\mathbb{Z}_2$.

Example 7 The subgroups $m\mathbb{Z} \times n\mathbb{Z}$ ($m \geq 0, n \geq 0$) of the fundamental group $\mathbb{Z} \times \mathbb{Z}$ of the torus $\mathbb{S}^1 \times \mathbb{S}^1$ correspond to the following covering projections:

$$\begin{array}{lll} e \times e : & \mathbb{R}^1 \times \mathbb{R}^1 \longrightarrow \mathbb{S}^1 \times \mathbb{S}^1 & (m = 0 = n) \\ d_m \times e : & \mathbb{S}^1 \times \mathbb{R}^1 \longrightarrow \mathbb{S}^1 \times \mathbb{S}^1 & (m > 0 = n) \\ e \times d_n : & \mathbb{R}^1 \times \mathbb{S}^1 \longrightarrow \mathbb{S}^1 \times \mathbb{S}^1 & (m = 0 < n) \\ d_m \times d_n : & \mathbb{S}^1 \times \mathbb{S}^1 \longrightarrow \mathbb{S}^1 \times \mathbb{S}^1 & (m > 0 < n). \end{array}$$

Any other subgroup of $\pi_1(\mathbb{S}^1 \times \mathbb{S}^1)$ is isomorphic to one of the listed subgroups under an automorphism induced on π_1 by a homeomorphism $h : \mathbb{S}^1 \times \mathbb{S}^1 \longrightarrow \mathbb{S}^1 \times \mathbb{S}^1$ of the type introduced in Exercise 9.3.8. Thus, every covering space of $\mathbb{S}^1 \times \mathbb{S}^1$ is isomorphic to one of the displayed covering spaces, up to such a homeomorphism h.

In the previous two examples, there is a bijective correspondence between isomorphism classes of covering spaces of X and conjugacy classes of subgroups of $\pi_1(X, x)$. We show in Section 5 that this is always the case for a space X that admits a simply connected covering space.

UNIVERSAL COVERING SPACES

A covering space $(\hat{X}, p)$ of a space X is **universal** if there exists a morphism $\Phi : (\hat{X}, p) \to (\hat{X}', p')$ to any other covering space $(\hat{X}', p')$ of X.

Since a morphism between covering spaces is itself a covering projection (Exercise 3), it follows that a universal covering space $(\hat{X}, p)$ of X *covers* any other covering space of X. By the Classification Theorem (3.4), any two universal covering spaces of X are isomorphic (Exercise 4).

3.5 Theorem *A simply connected covering space is universal.*

Proof: This is a corollary of the Morphism Criterion (3.3). □

4 Fundamental Theorem

This section develops a description of the fundamental group of a space in terms of automorphisms of a simply connected covering space of it.

COVERING SPACE AUTOMORPHISMS

Let $(\hat{X}, p)$ be a covering space of X.

An *automorphism* of $(\hat{X}, p)$ is an isomorphism $\Phi : (\hat{X}, p) \to (\hat{X}, p)$.

There is always the identity automorphism $1_{\hat{X}}$. Moreover, the composite of two automorphisms of $(\hat{X}, p)$ is another one, and the inverse of any automorphism is another automorphism. Therefore, the set of all automorphisms of $(\hat{X}, p)$ is a group under the operation of composition.

The *automorphism group* of the covering space $(\hat{X}, p)$ is denoted by $A(p)$, or by $A(\hat{X} \to X)$ when the covering projection is unambiguous.

Example 1 The 3-fold covering $p_0 : \hat{X} \to X$ of the one-point union $X = \mathbb{S}^1 \vee \mathbb{S}^1$ in Example 1.6 and Example 3.1 admits only the identity automorphism $1_{\hat{X}}$ and so has the trivial automorphism group $A(p_0) = \{1_{\hat{X}}\}$. The reason is that an automorphism preserves the fiber $p_0^{-1}(x) = \{\hat{x}_1, \hat{x}_2, \hat{x}_3\}$ and is determined by its value at any one point. The possibilities $\Phi(\hat{x}_1) = \hat{x}_2$ and $\Phi(\hat{x}_1) = \hat{x}_3$ are ruled out by the Morphism Criterion (3.3).

Example 2 The automorphism group $A(e)$ of the exponential covering projection $e : \mathbb{R}^1 \to \mathbb{S}^1$ is the group of translations, $t_n : \mathbb{R}^1 \to \mathbb{R}^1$, of the real line by the integers $n \in \mathbb{Z}$. The translations are automorphisms and they achieve all possible assignments of the fiber element $0 \in e^{-1}(e(0)) = \mathbb{Z}$. So they constitute $A(e)$, by the Uniqueness of Liftings (2.1). The composite of two translations t_n and t_m is the translation t_{n+m}. So $A(e)$ is isomorphic to the additive group of integers $\mathbb{Z}$.

Similarly, the product covering projection $e \times e : \mathbb{R}^1 \times \mathbb{R}^1 \to \mathbb{S}^1 \times \mathbb{S}^1$ has automorphism group $A(e \times e)$ isomorphic to $\mathbb{Z} \times \mathbb{Z}$.

Example 3 The double covering $p : \mathbb{S}^2 \to \mathbb{P}^2$ of the projective plane has just two automorphisms: the identity homeomorphism $1_{\mathbb{S}^2} : \mathbb{S}^2 \to \mathbb{S}^2$ and the antipodal homeomorphism $\alpha : \mathbb{S}^2 \to \mathbb{S}^2$. As $\alpha \circ \alpha = 1_{\mathbb{S}^2}$, the automorphism group $A(p)$ is isomorphic to $\mathbb{Z}_2$.

THE FUNDAMENTAL CORRESPONDENCE

Let $p : \hat{X} \to X$ be a covering projection and let $x \in X$ be a selected basepoint. Recall from the Action Property (2.6), the well-defined action

$$\cdot : p^{-1}(x) \times \pi_1(X, x) \to p^{-1}(x).$$

This action associates to each fiber element $\hat{x} \in p^{-1}(x)$ and each group element $[f] \in \pi_1(X, x)$ a fiber element $\hat{x} \cdot [f] \in p^{-1}(x)$, namely, the terminal point $\hat{f}_{\hat{x}}(1)$ of the lifting $\hat{f}_{\hat{x}}$ at $\hat{x}$ of any representative f of the class $[f]$.

Similarly, the evaluation of automorphisms provides an action

$$\cdot : p^{-1}(x) \times A(p) \to p^{-1}(x).$$

This action associates to each fiber element $\hat{x} \in p^{-1}(x)$ and automorphism $\Phi \in A(p)$ the new fiber element $\Phi(\hat{x}) \in p^{-1}(x)$.

Let $\hat{x} \in p^{-1}(x)$ be a selected fiber element. We say the fundamental group element $[f] \in \pi_1(X, x)$ **corresponds** to the covering automorphism $\Phi \in A(p)$ if they have the same effect on $\hat{x}$, that is,

$$\hat{x} \cdot [f] = \hat{f}_{\hat{x}}(1) = \Phi(\hat{x}).$$

According to the Morphism Criterion (3.3) and the Conjugacy Property (2.5), $[f] \in \pi_1(X, x)$ corresponds to some automorphism if and only if

$$p_\#(\pi_1(\hat{X}, \hat{x})) = p_\#(\pi_1(\hat{X}, \hat{f}_{\hat{x}}(1))) = [f]^{-1} \cdot p_\#(\pi_1(\hat{X}, \hat{x})) \cdot [f].$$

This condition is an instance of the following algebraic situation.

Let G be a group and let $H \leq G$ be a subgroup. The **normalizer** $N_G(H)$ of H in G is the subgroup of G consisting of all elements $g \in G$ for which $H = g^{-1} \cdot H \cdot g$. So $N_G(H)$ is the largest subgroup of G in which H is normal.

Thus, $[f] \in \pi_1(X, x)$ corresponds to some (necessarily unique) automorphism if and only if $[f] \in \pi_1(X, x)$ belongs to the normalizer,

$$N_{\pi_1(X, x)}(p_\#(\pi_1(\hat{X}, \hat{x}))),$$

of the subgroup $p_\#(\pi_1(\hat{X}, \hat{x}))$ in $\pi_1(X, x)$. It follows that there is a function

$$\mu = \mu(\hat{x}) : N_{\pi_1(X, x)}(p_\#(\pi_1(\hat{X}, \hat{x}))) \to A(p)$$

given by $\mu([f]) = \Phi$ if and only if $\hat{x} \cdot [f] = \hat{f}_{\hat{x}}(1) = \Phi(\hat{x})$. This function μ is well-defined, since an automorphism is uniquely determined by its value on $\hat{x}$; it is surjective, since the action $\hat{x} \cdot : \pi_1(X, x) \to p^{-1}(x)$ is surjective.

4.1 Fundamental Theorem for Covering Spaces

Let $p : \hat{X} \to X$ be a covering projection; let $x \in X$ and $\hat{x} \in p^{-1}(x) \subseteq \hat{X}$ be selected basepoints. Then the function

$$\mu = \mu(\hat{x}) : \mathbb{N}_{\pi_1(X,\,x)}(p_\#(\pi_1(\hat{X}, \hat{x}))) \to A(p)$$

is a surjective homomorphism with kernel $p_\#(\pi_1(\hat{X}, \hat{x})) \leq \pi_1(X, x)$.

Hence, there is the induced group isomorphism

$$\bar{\mu}(\hat{x}) : \frac{\mathbb{N}_{\pi_1(X,\,x)}(p_\#(\pi_1(\hat{X}, \hat{x})))}{p_\#(\pi_1(\hat{X}, \hat{x}))} \approx A(p).$$

Proof: To show that μ is a homomorphism, let $\mu([f]) = \Phi$ and $\mu([g]) = \Psi$. That is, let the loops f and g at x have liftings $\hat{f}$ and $\hat{g}$ at $\hat{x}$ that terminate at $\hat{f}(1) = \Phi(\hat{x})$ and $\hat{g}(1) = \Psi(\hat{x})$, as in the figure, below. Then the lifting $\widehat{(f \cdot g)}$ at $\hat{x}$ of the loop $f \cdot g$ equals the path-product $\hat{f} \cdot (\Phi \circ \hat{g})$, and the terminal point $\widehat{(f \cdot g)}(1)$ is given by $\Phi(\hat{g}(1)) = \Phi(\Psi(\hat{x}))$. Thus, $\Phi \circ \Psi$ is the unique automorphism sending $\hat{x}$ to $\widehat{(f \cdot g)}(1)$. This implies the homomorphism property:

$$\mu([f] \cdot [g]) = \Phi \circ \Psi = \mu([f]) \circ \mu([g]).$$

The identity element of $A(p)$ is the identity automorphism $1_{\hat{X}}$, which sends $\hat{x}$ to $\hat{x}$. So, $[f] \in Ker\, \mu$ if and only if $\hat{x} \cdot [f] = \hat{f}(1) = \hat{x}$; equivalently, $[f] = p_\#([\hat{f}])$ belongs to $p_\#(\pi_1(\hat{X}, \hat{x}))$. Thus, $Ker\, \mu = p_\#(\pi_1(\hat{X}, \hat{x}))$.

Therefore, by Corollary **11.**4.6, μ induces a group isomorphism $\bar{\mu}(\hat{x})$, as indicated in the statement of the theorem. $\square$

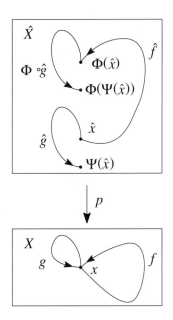

Fundamental Theorem 4.1

NORMAL COVERING SPACES

A covering space $(\hat{X}, p)$ is called **normal** if the conjugacy class of subgroups,
$$\{p_\#(\pi_1(\hat{X}, \hat{x})) : \hat{x} \in p^{-1}(x) \subseteq \hat{X}\},$$
consists of a single, hence, normal subgroup of $\pi_1(X, x)$.

This condition is independent of the choice of the basepoint $x \in X$, because any change of basepoint isomorphism $\omega : \pi_1(X, x_0) \to \pi_1(X, x_1)$ (Theorem 9.3.4) carries the conjugacy class of subgroups in $\pi_1(X, x_0)$ for $(\hat{X}, p)$ to the conjugacy class of subgroups in $\pi_1(X, x_1)$ for $(\hat{X}, p)$.

For normal covering spaces, the Fundamental Theorem (4.1) becomes:

4.2 Corollary *If $(\hat{X}, p)$ is a normal covering space of X, then*
$$\mu(\hat{x}) : \frac{\pi_1(X, x)}{p_\#(\pi_1(\hat{X}, \hat{x}))} \approx A(p)$$
for any choice of basepoints $x \in X$ and $\hat{x} \in p^{-1}(x) \subseteq \hat{X}$.

Simply connected covering spaces are normal and Corollary 4.2 applies:

4.3 Corollary *If $(\hat{X}, p)$ is a simply connected covering space of X, then*
$$\mu(\hat{x}) : \pi_1(X, x) \approx A(p)$$
for any choice of basepoints $x \in X$ and $\hat{x} \in p^{-1}(x) \subseteq \hat{X}$.

Corollary 4.3 provides an amazing correspondence between the group of homotopy classes of loops in the base space X and the group of automorphisms of the simply connected covering space $(\hat{X}, p)$. The latter group of homeomorphisms of $(\hat{X}, p)$ that respect the covering projection $p : \hat{X} \to X$ is a geometric representation of the fundamental group $\pi_1(X, x)$.

Example 4 Consider the three simply connected covering spaces
$$e : \mathbb{R}^1 \to \mathbb{S}^1, \quad e \times e : \mathbb{R}^1 \times \mathbb{R}^1 \to \mathbb{S}^1 \times \mathbb{S}^1, \quad \text{and} \quad p : \mathbb{S}^2 \to \mathbb{P}^2.$$
They have automorphism groups $A(e) \approx \pi_1(\mathbb{S}^1) = \mathbb{Z}$, $A(e \times e) \approx \pi_1(\mathbb{S}^1 \times \mathbb{S}^1) = \mathbb{Z} \times \mathbb{Z}$, and $A(p) \approx \pi_1(\mathbb{P}^2) = \mathbb{Z}_2$, respectively, as indicated in Examples 2 and 3.

THE PSEUDO-PROJECTIVE PLANE

The **pseudo-projective plane** of order $n \geq 1$ is the quotient space P_n of the closed unit disc $\mathbb{D}^2$ that results when each point of the boundary 1-sphere $\mathbb{S}^1$ is identified with its image under the rotation $\rho_n : \mathbb{S}^1 \to \mathbb{S}^1$ through $2\pi/n$ radians. The identification map $q : \mathbb{D}^2 \to P_n$ wraps the boundary 1-sphere $\mathbb{S}^1 \subset \mathbb{D}^2$ n-times around its image $q(\mathbb{S}^1)$, which is a topological circle that we call the **base** of P_n.

Here is an exploded view of this identification map in the case $n = 3$:

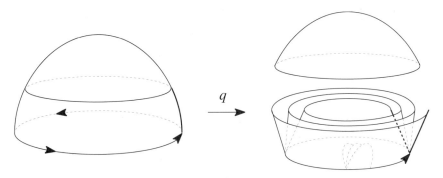

Identification map $q : \mathbb{D}^2 \to P_n$

A simply connected covering space of P_n arises from the collection $\{q_\sigma : \mathbb{D}^2_\sigma \to P_n \ (\sigma \in \mathbb{Z}_n)\}$ of n copies of the identification map q, as follows.

Let $\hat{P}_n$ be the identification space obtained from the disjoint union $\mathring{\cup}_\sigma \mathbb{D}^2_\sigma$ by identifying each boundary point $z \in \mathbb{S}^1_\sigma \subset \mathbb{D}^2_\sigma$ of one disc with its rotated image $\rho_n(z) \in \mathbb{S}^1_\sigma \subset \mathbb{D}^2_{\sigma+1}$ in the boundary of the next disc. This makes $\hat{P}_n$ a stack of the n discs $\mathbb{D}^2_\sigma$ ($\sigma \in \mathbb{Z}_n$) that share a common boundary 1-sphere, with each disc being rotated $2\pi/n$ radians more than the disc above it.

The identification maps $\{q_\sigma : \mathbb{D}^2_\sigma \to P_n \ (\sigma \in \mathbb{Z}_n)\}$ assemble into a continuous function $p : \hat{P}_n \to P_n$, below, that wraps each of the discs into P_n.

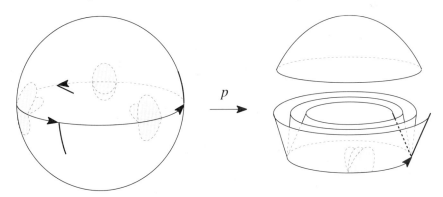

Covering projection $p : \hat{P}_n \to P_n$

An easy inspection shows that $p : \hat{P}_n \to P_n$ is a covering projection. Each point in the *base* of P_n has an admissible neighborhood that looks like a *book with n pages* bound to the base circle; the pre-image in $\hat{P}_n$ of that neighborhood is the disjoint union of n such books, their bindings evenly spaced around the boundary 1-sphere in $\hat{P}_n$. Also, each point *above the base* of $\hat{P}_n$ has an admissible open disc neighborhood whose pre-image in $\hat{P}_n$ is a disjoint union of n open discs that spiral up the stack in $\hat{P}_n$.

The total space $\hat{P}_n$ is simply connected and there is the fundamental isomorphism $\mu : \pi_1(\hat{P}_n) \approx A(p)$ of Corollary 4.3. The loop around the base of the pseudo-projective plane P_n generates the finite cyclic fundamental group $\pi_1(\hat{P}_n) = \mathbb{Z}_n$. The corresponding automorphism $\Phi \in A(p)$ is the homeomorphism, $\Phi : \hat{P}_n \to \hat{P}_n$, that rotates the discs in the stack through $2\pi/n$ radians, and carries the disc $\mathbb{D}_\sigma^2$ to the disc $\mathbb{D}_{\sigma+1}^2$ for each $\sigma \in \mathbb{Z}_n$. This automorphism generates $A(p)$ as a cyclic group of order n.

5 Intermediate Covering Spaces

EXISTENCE OF SIMPLY CONNECTED COVERING SPACES

Suppose that the space X has a simply connected covering space $(\hat{X}, p)$. Then, for any point $x \in X$ and admissible neighborhood U of x, the inclusion induced homomorphism $\pi_1(U, x) \to \pi_1(X, x)$ is trivial, because the inclusion $U \subseteq X$ factors through the simply connected covering space $\hat{X}$, via any section homeomorphism $U \to \hat{U}$, as $U \to \hat{U} \subseteq \hat{X} \to X$. This fact provides a necessary condition for the existence of a simply connected covering space:

A space X is **semilocally simply connected** if each point $x \in X$ has a path-connected open neighborhood U whose inclusion into X induces the trivial homomorphism $\pi_1(U, x) \to \pi_1(X, x)$. Such a neighborhood U is called **relatively simply connected.**

> **Example 1** The Hawaiian earrings space (Example 7.1.8) is not semilocally simply connected; it cannot have a simply connected covering space. The origin is the unique point of this space that has no relatively simply connected neighborhood.
>
> **Example 2** Every surface is semilocally simply connected.

This necessary condition for the existence of a simply connected covering space is also sufficient, as we shall show shortly. To guide our constructive proof, let's examine any covering space $(\hat{X}, p)$ of a space X.

There is an extension of the action of the fundamental group of the base space on a fiber of a covering space. Let $\pi(X)_x \subseteq \pi(X)$ denote those path classes with initial point x. There is the action $\cdot : p^{-1}(x) \times \pi(X)_x \to \hat{X}$ that associates to each fiber element $\hat{x} \in p^{-1}(x)$ and path class $[f] \in \pi(X)_x$ the element $\hat{x} \cdot [f] = \hat{f}_{\hat{x}}(1) \in p^{-1}(f(1))$, where $\hat{f}_{\hat{x}}$ is the lifting of f at $\hat{x}$.

5.1 Lemma $(\hat{X}, p)$ *is a simply connected covering space of X if and only if, for some (hence, any) choice of basepoints $x \in X$ and $\hat{x} \in p^{-1}(x) \subseteq \hat{X}$, the extended action $\hat{x} \cdot : \pi(X)_x \to \hat{X}$ is a bijection.*

Proof: The action $\hat{x} \cdot$ is always surjective, because any point $\hat{z} \in \hat{X}$ is the terminal point of some path $\hat{f} : \mathbb{I} \to \hat{X}$ that initiates at $\hat{x}$. The injectivity of the action $\hat{x} \cdot$ has this characterization: paths f and g in X with the initial point x are path-homotopic if their liftings $\hat{f}, \hat{g}$ in $\hat{X}$ at $\hat{x}$ have the same terminal point. The simple connectedness of $\hat{X}$ has this characterization: paths $\hat{f}, \hat{g}$ in $\hat{X}$ with common initial point $\hat{x}$ are path-homotopic if they have the same terminal point. Thus, the injectivity of the action $\hat{x} \cdot$ is equivalent to the simple connectedness of $\hat{X}$, because, by the Homotopy Lifting Property (2.3), paths f and g in X with the initial point x are path-homotopic if and only if their liftings $\hat{f}, \hat{g}$ in $\hat{X}$ at $\hat{x}$ are path-homotopic. □

Lemma 5.1 says that when $(\hat{X}, p)$ is a simply connected covering space of X, the covering space $\hat{X}$ is just a topologized version of the set $\pi(X)_x$ of path-classes with initial point x. And the covering projection $p : \hat{X} \to X$ is simply the endpoint assignment $\rho : \pi(X)_x \to X$, $\rho([f]) = f(1)$.

Now let X be any space. Given any $[f] \in \pi(X)_x$ and any path-connected neighborhood U of $\rho([f]) = f(1) = z$, let $([f], U)$ denote the set

$$\{[f] \cdot [g] \in \pi(X)_x : [g] \in Im(\pi(U)_z \to \pi(X)_z)\}.$$

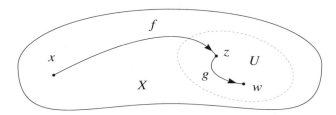

5.2 Lemma *Let U be a relatively simply connected open subset of any point $z \in X$, and consider the endpoint assignment $\rho : \pi(X)_x \to X$. Then,*
(a) *the restriction, $\rho : ([f], U) \to U$, is a bijection for each $[f] \in \pi(X)_x$ with endpoint $\rho([f]) = z$; and*
(b) *$U \subseteq X$ has pre-image $\rho^{-1}(U) = \overset{\circ}{\cup}\{([f], U) : [f] \in \pi(X)_x, \rho([f]) = z\}.$*

Proof: (a) $\rho : ([f], U) \to U$ is surjective, because each point of the path-connected U is an endpoint for some path-class $[g] \in Im(\pi(U)_z \to \pi(X)_z)$, where $z = \rho([f]) \in U$. ρ is injective on $([f], U)$, since paths g_1 and g_2 in U with initial point z and the same terminal point w define the same path-class $[g_1] = [g_2]$ in $Im(\pi(U)_z \to \pi(X)_z)$, by the triviality of $\pi_1(U, z) \to \pi_1(X, z)$.

(b) The sets $\{([f], U) : [f] \in \pi(X)_x, \rho([f]) = z\}$ are pairwise disjoint, because $[f_1] \cdot [g_1] = [f_2] \cdot [g_2]$ implies $g_1(1) = g_2(1)$, which implies $[g_1] = [g_2]$ in $Im(\pi(U)_z \to \pi(X)_z)$. The latter, in turn, implies $[f_1] = [f_2]$ in $\pi(X)_x$. Furthermore, $\rho^{-1}(U)$ is the union of these sets. □

5.3 Theorem *Any connected, locally path-connected, and semilocally simply connected space X has a simply connected covering space $(\hat{X}, p)$.*

Proof: The space X has a base $\{U\}$ of relatively simply connected neighborhoods. So the sets $([f], U) \subseteq \pi(X)_x$, where $[f] \in \pi(X)_x$ and U is a relatively simply connected neighborhood of the endpoint $\rho([f])$, serve as a base for a topologized version $\hat{X}$ of $\pi(X)_x$ (Exercise 1). Then $\rho : \hat{X} = \pi(X)_x \to X$ is continuous by Lemma 5.2(b), and it matches each member of the base $\{([f], U)\}$ for $\hat{X}$ with a member of the base $\{U\}$ for X by Lemma 5.2(a). Thus, each bijection $\rho : ([f], U) \to U$ is a homeomorphism of subspaces of $\hat{X}$ and X. So Lemma 5.2(b) provides the partition of the pre-image $\rho^{-1}(U)$ into its path-components $([f], U)$. This proves that $\rho : \hat{X} \to X$ is a covering projection with admissible neighborhoods $\{U\}$.

Any path $f : (\mathbb{I}, 0) \to (X, x)$ determines a parametrized family of paths $f_t : (\mathbb{I}, 0) \to (X, x)$, given by $f_t(s) = f(ts)$. The function $F : (\mathbb{I}, 0) \to (\hat{X}, [x^*])$ given by $F(t) = [f_t]$ covers f as $\rho([f_t]) = f_t(1) = f(t)$ for all $0 \leq t \leq 1$.

Moreover, F is continuous. Indeed, for any $t \in \mathbb{I}$ and any admissible neighborhood U of $f(t)$, F carries the component of $f^{-1}(U)$ containing t into $([f_t], U)$, and its composition with the homeomorphism $\rho : ([f_t], U) \to U$ is the continuous path f.

Since F joins $[x^*]$ and $[f]$, it follows that $\hat{X}$ is path-connected. Since F is the path lifting $\hat{f}_{\hat{x}}$ of f at $\hat{x} = [x^*]$, it also follows that the extended action $\hat{x} \cdot : \pi(X)_x \to \hat{X}$ is the identity function: $\hat{x} \cdot [f] = \hat{f}_{\hat{x}}(1) = [f]$. So $\hat{X}$ is simply connected, by Lemma 5.1. □

TRANSITIVE ACTION

Let $(\hat{X}, p)$ be a covering space of X. We say that the group of automorphisms $A(p)$ **acts transitively** on the fiber $p^{-1}(x)$ if, for each pair of fiber elements $\hat{x}, \hat{z} \in p^{-1}(x)$, there exists $\Phi \in A(p)$ such that $\Phi(\hat{x}) = \hat{z}$.

5.4 Transitivity Theorem *Let $(\hat{X}, p)$ be a covering space of X. The group of automorphisms $A(p)$ acts transitively on some (hence, each) fiber $p^{-1}(x)$ in $\hat{X}$ if and only if $(\hat{X}, p)$ is a normal covering space.*

Proof: By the Morphism Criterion (3.2), $A(p)$ acts transitively on the fiber $p^{-1}(x)$ if and only if the conjugacy class of subgroups,

$$\{p_\#(\pi_1(\hat{X}, \hat{x})) : \hat{x} \in p^{-1}(x) \subseteq \hat{X}\},$$

consists of a single, hence, normal subgroup of $\pi_1(X, x)$. □

INTERMEDIATE COVERING SPACES

Let $(\hat{X}, p)$ be a normal covering space of X. An intermediate covering space of $(\hat{X}, p)$ is a triple $(q, \overline{X}, h)$ associated with a factorization of p,

$$p = h \circ q : \hat{X} \to \overline{X} \to X,$$

into two covering projections $q : \hat{X} \to \overline{X}$ and $h : \overline{X} \to X$.

Because $(\hat{X}, p)$ is normal, then $q : \hat{X} \to \overline{X}$ is a normal covering projection, as we now prove. For each choice of basepoints $x \in X$ and $\hat{x} \in p^{-1}(x) \subseteq \hat{X}$,

$$p_\#(\pi_1(\hat{X}, \hat{x})) = h_\#(q_\#(\pi_1(\hat{X}, \hat{x}))$$

is a normal subgroup of $\pi_1(X, x)$. Hence, it is also a normal subgroup of $h_\#(\pi_1(\overline{X}, q(\hat{x})))$, in view of the containment relations

$$h_\#(q_\#(\pi_1(\hat{X}, \hat{x})) \leq h_\#(\pi_1(\overline{X}, q(\hat{x})) \leq \pi_1(X, x).$$

Since $h_\#$ is a monomorphism, this proves that $q_\#(\pi_1(\hat{X}, \hat{x}))$ is a normal subgroup of $\pi_1(\overline{X}, q(\hat{x}))$, so that $q : \hat{X} \to \overline{X}$ is a normal covering projection.

But even when $(\hat{X}, p)$ is a simply connected covering space of X, $(\overline{X}, h)$ need not be a normal covering space of X (Exercise 4(a)).

An intermediate covering space $(q, \overline{X}, h)$ of $(\hat{X}, p)$ is called **normal** provided that the two covering projections $q : \hat{X} \to \overline{X}$ and $h : \overline{X} \to X$ are normal.

Example 3   The triple $(e \times 1_{\mathbb{R}^1}, \mathbb{S}^1 \times \mathbb{R}^1, 1_{\mathbb{S}^1} \times e)$ is a normal intermediate covering space of $(\mathbb{R}^1 \times \mathbb{R}^1, e \times e : \mathbb{S}^1 \times \mathbb{S}^1 \to \mathbb{S}^1 \times \mathbb{S}^1)$, and the triple $(d_n, \mathbb{S}^1 d_m)$ is a normal intermediate covering space of $(\mathbb{S}^1, d_{mn} : \mathbb{S}^1 \to \mathbb{S}^1)$.

CONSTRUCTION OF INTERMEDIATE COVERING SPACES

There is an easy construction for intermediate covering spaces of a normal covering space $(\hat{X}, p)$ of X. Since a covering projection is an identification map, the base space X results from identifying each fiber $p^{-1}(x)$ in the total space $\hat{X}$ to a single point x. Since the automorphism group $A(p)$ acts transitively on the fibers of the normal covering projection p, two points $\hat{x}$ and $\hat{z}$ in $\hat{X}$ are identified by p if and only if there is an automorphism $\Theta \in A(p)$ sending $\hat{x}$ to $\hat{z}$. We can summarize this situation by saying that the normal covering projection $p : \hat{X} \to X$ results when we **mod out by the action** of the automorphism group $A(p)$ on $\hat{X}$.

In general, an intermediate covering space $(q_\mathbb{H}, \hat{X}_\mathbb{H}, p_\mathbb{H})$ of $(\hat{X}, p)$ results when we **mod out by the action of a subgroup** $\mathbb{H} \leq A(p)$. Here are the details. Given the subgroup $\mathbb{H}$, we introduce this relation on $\hat{X}$:

$\hat{x} \equiv_\mathbb{H} \hat{z}$ if and only if there exists $\Theta \in \mathbb{H}$ such that $\Theta(\hat{x}) = \hat{z}$.

The group axioms for $\mathbb{H}$ imply that $\equiv_\mathbb{H}$ is an equivalence relation on $\hat{X}$. The quotient space of $\hat{X}$ modulo $\equiv_\mathbb{H}$ is denoted by $\hat{X}_\mathbb{H}$ and is called the **orbit space** of $\mathbb{H}$. The associated quotient map $q_\mathbb{H} : \hat{X} \to \hat{X}_\mathbb{H}$ makes some of the identifications that the covering projection p does, and the Transgression Theorem (5.1.9) provides the factorization $p = h_\mathbb{H} \circ q_\mathbb{H} : \hat{X} \to \hat{X}_\mathbb{H} \to X$.

5.5   Theorem   *For each subgroup $\mathbb{H} \leq A(p)$ of the automorphism group of a normal covering space $(\hat{X}, p)$, the triple $(q_\mathbb{H}, \hat{X}_\mathbb{H}, h_\mathbb{H})$ is an intermediate covering space of $(\hat{X}, p)$ such that $A(q_\mathbb{H}) = \mathbb{H}$.*

Proof: Given $x \in X$, let $U(x)$ be an admissible neighborhood of x. Given the fiber element $\hat{x} \in p^{-1}(x) \subseteq \hat{X}$, let $\hat{U}(\hat{x})$ be the path component of $p^{-1}(U(x))$ containing $\hat{x}$. Then there is the decomposition

$$p^{-1}(U(x)) = \overset{\circ}{\cup} \{\hat{U}(\Phi(\hat{x})) : \Phi \in A(p)\},$$

since the points $\Phi(\hat{x})$, $\Phi \in A(p)$, constitute the fiber $p^{-1}(x)$.

Consider these projection and section homeomorphisms:

$$p(\hat{x}) : \hat{U}(\hat{x}) \cong U(x) \quad \text{and} \quad s(\hat{x}) : U(x) \cong \hat{U}(\hat{x}).$$

If $\Phi : (\hat{X}, p) \to (\hat{X}, p)$ is any automorphism, then, by the uniqueness of liftings, the restriction $\Phi|_{\hat{U}(\hat{x})}$ coincides with the composite function

$$s(\Phi(\hat{x})) \circ p(\hat{x}) : \hat{U}(\hat{x}) \cong U(x) \cong \hat{U}(\Phi(\hat{x})).$$

Thus, the quotient map $q_\mathbb{H} : \hat{X} \to \hat{X}_\mathbb{H}$ that identifies the point $\hat{x}$ with the points $\Theta(\hat{x})$ ($\Theta \in \mathbb{H}$) actually identifies the neighborhood $\hat{U}(\hat{x})$ with the neighborhoods $\hat{U}(\Theta(\hat{x}))$ ($\Theta \in \mathbb{H}$) via the homeomorphisms $s(\Theta(\hat{x})) \circ p(\hat{x})$. Moreover, no other identifications are made involving points of $\hat{U}(\hat{x})$. So,

$$q_\mathbb{H}^{-1}(q_\mathbb{H}(\hat{U}(\hat{x}))) = \overset{\circ}{\cup} \{\hat{U}(\Theta(\hat{x})) : \Theta \in \mathbb{H}\}$$

and each path-component $\hat{U}(\Theta(\hat{x}))$ ($\Theta \in \mathbb{H}$) projects homeomorphically under $q_\mathbb{H}$ to $q_\mathbb{H}(\hat{U}(\hat{x}))$. This shows that $q_\mathbb{H}(\hat{U}(\hat{x}))$ is an admissible neighborhood of $\hat{X}_\mathbb{H}$ and that $q_\mathbb{H} : \hat{X} \to \hat{X}_\mathbb{H}$ is a covering projection.

Since $q_\mathbb{H}$ mods out by the action of $\mathbb{H} \leq A(p)$, each $\Theta \in \mathbb{H}$ is an automorphism of the covering space $(\hat{X}, q_\mathbb{H})$. Since the images of $\hat{x} \in \hat{X}$ under the automorphisms $\Theta \in \mathbb{H}$ constitute the fiber $q_\mathbb{H}^{-1}(q_\mathbb{H}(\hat{x}))$, then $\mathbb{H}$ is the entire automorphism group $A(q_\mathbb{H})$ of the covering space $(\hat{X}, q_\mathbb{H})$.

For two automorphisms Φ and Ψ of $(\hat{X}, p)$ that determine the same coset $\mathbb{H}\Phi = \mathbb{H}\Psi$ of $A(p)$ modulo $\mathbb{H}$, the path-components $\hat{U}(\Phi(\hat{x}))$ and $\hat{U}(\Psi(\hat{x}))$ have the same image under $q_\mathbb{H}$ because $\Phi = \Theta \circ \Psi$ for some $\Theta \in \mathbb{H}$. For each coset $\mathbb{H}\Phi \in \mathbb{H} \setminus A(p)$, there is a well-defined subset $q_\mathbb{H}(\hat{U}(\Phi(\hat{x}))) \subseteq \hat{X}_\mathbb{H}$. Also, $U(x)$ is an admissible neighborhood for the projection $h_\mathbb{H} : \hat{X}_\mathbb{H} \to X$. Its pre-image, $h_\mathbb{H}^{-1}(U(x)) = q_\mathbb{H}(p^{-1}(U(x)))$, is the disjoint union

$$\overset{\circ}{\cup} \{q_\mathbb{H}(\hat{U}(\Phi(\hat{x}))) : \mathbb{H}\Phi \in \mathbb{H} \setminus A(p)\}.$$

And each path component $q_\mathbb{H}(\hat{U}(\Phi(\hat{x})))$ projects homeomorphically under $h_\mathbb{H}$ to $U(x)$. Thus, $h_\mathbb{H} : \hat{X}_\mathbb{H} \to X$ is a covering projection. □

THE CO-GALOIS CORRESPONDENCE

Let $(\hat{X}, p)$ be a normal covering space of X and let $(q, \overline{X}, h)$ be an intermediate covering space of $(\hat{X}, p)$. Any automorphism Φ of $(\hat{X}, q)$ is also an automorphism of $(\hat{X}, p)$, because the relation $q \circ \Phi = q$ implies the relations $p \circ \Phi = h \circ q \circ \Phi = h \circ q = p$. So we may view $A(q)$ as a subgroup of $A(p)$.

Two intermediate covering spaces $(q_1, \overline{X}_1, h_1)$ and $(q_2, \overline{X}_2, h_2)$ of $(\hat{X}, p)$

are called *isomorphic* if there is a homeomorphism $\Phi : \overline{X}_1 \to \overline{X}_2$ such that
$$\Phi \circ q_1 = q_2 \quad \text{and} \quad h_2 \circ \Phi = h_1,$$
as indicated below:

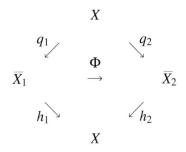

It follows that isomorphic intermediate covering spaces $(q_1, \overline{X}_1, h_1)$ and $(q_2, \overline{X}_2, h_2)$ of $(\hat{X}, p)$ determine the same subgroup, $A(q_1) = A(q_2)$, of $A(p)$.

5.6 Co-Galois Correspondence
Let $(\hat{X}, p)$ be a normal covering space of X.
(a) There is a bijective correspondence between the isomorphism classes of intermediate covering spaces of $(\hat{X}, p)$ and subgroups of the automorphism group $A(p)$, associating $(q, \overline{X}, h)$ with $A(q) \leq A(p)$.
(b) The intermediate covering space $(q, \overline{X}, h)$ is normal if and only if the subgroup $A(q) \leq A(p)$ is normal.
(c) When the intermediate covering space $(q, \overline{X}, h)$ is normal, the group of automorphisms $A(h)$ is isomorphic to the quotient group $A(p)/A(q)$.

Proof: (a) To each isomorphism class of an intermediate covering space $(q, \overline{X}, h)$ of $(\hat{X}, p)$, we associate the subgroup $A(q) \leq A(p)$ of automorphisms of $(\hat{X}, p)$ that are automorphisms of $(\hat{X}, q)$. Conversely, to each subgroup $\mathbb{H} \leq A(p)$ of the group of automorphisms of $(\hat{X}, p)$, we associate the intermediate covering space $(q_{\mathbb{H}}, \hat{X}_{\mathbb{H}}, h_{\mathbb{H}})$.

These correspondences are inverses. Given a subgroup $\mathbb{H} \leq A(p)$, we can form the intermediate covering space $(q_{\mathbb{H}}, \hat{X}_{\mathbb{H}}, h_{\mathbb{H}})$, and then form the automorphism group $A(q_{\mathbb{H}}) \leq A(p)$. By construction, $A(q_{\mathbb{H}}) = \mathbb{H}$.

Conversely, given an intermediate covering space $(q, \overline{X}, h)$, we can form the automorphism group $\mathbb{H} = A(q) \leq A(p)$, and then we can form the intermediate covering space $(q_{\mathbb{H}}, \hat{X}_{\mathbb{H}}, h_{\mathbb{H}})$. Since $(\hat{X}, q)$ is necessarily a normal covering space, we recover the projection $q : \hat{X} \to \overline{X}$ when we mod out by the action of $\mathbb{H} = A(q)$ in forming $q_{\mathbb{H}} : \hat{X} \to \hat{X}_{\mathbb{H}}$. Thus, there is a homeomorphism $\overline{X} \to \hat{X}_{\mathbb{H}}$ relating the two intermediate covering spaces $(q, \overline{X}, h)$ and $(q_{\mathbb{H}}, \hat{X}_{\mathbb{H}}, h_{\mathbb{H}})$. This proves that the two correspondences are inverses.

(b) By the correspondence of (a), it suffices to prove that $\mathbb{N} \leq A(p)$ is a normal subgroup if and only if $h_{\mathbb{N}} : \hat{X}_{\mathbb{N}} \to X$ is a normal covering projection.

When $\mathbb{N}$ is a normal subgroup of $A(p)$, each automorphism $\Phi \in A(p)$ respects the relation $\equiv_\mathbb{N}$. The reason is that the relation $\Theta(\hat{x}) = \hat{z}$ for some $\Theta \in \mathbb{N}$ implies that $\Phi \circ \Theta \circ \Phi^{-1}(\Phi(\hat{x})) = \Phi(\hat{z})$, where $\Phi \circ \Theta \circ \Phi^{-1} \in \mathbb{N}$ by the normality of the subgroup $\mathbb{N} \leq A(p)$. Therefore by the Transgression Theorem (5.1.9), each automorphism $\Phi \in A(p)$ induces a homeomorphism $\Phi_\mathbb{N} : \hat{X}_\mathbb{N} \to \hat{X}_\mathbb{N}$, which is necessarily an automorphism of $(\hat{X}_\mathbb{N}, h_\mathbb{N})$. Since $A(p)$ acts transitively on the fibers of $(\hat{X}, p)$, then $A(h_\mathbb{N})$ acts transitively on the fibers of $(\hat{X}_\mathbb{N}, h_\mathbb{N})$. This proves that $(\hat{X}_\mathbb{N}, h_\mathbb{N})$ is a normal covering space of X and $(q_\mathbb{H}, \hat{X}_\mathbb{H}, h_\mathbb{H})$ is a normal intermediate covering space of $(\hat{X}, p)$.

The fundamental homomorphisms for the normal covering spaces $(\hat{X}, p)$ and $(\hat{X}, q_\mathbb{H})$ give this commutative diagram:

$$\begin{array}{ccc} \pi_1(\hat{X}_\mathbb{H}, q_\mathbb{H}(\hat{x})) & \stackrel{h_{\mathbb{H}\#}}{\longrightarrow} & \pi_1(X, x) \\ \mu(\hat{x}) \downarrow & & \downarrow \mu(\hat{x}) \\ \mathbb{H} = A(q_\mathbb{H}) & \leq & A(p) \end{array}$$

When $h_\mathbb{N} : \hat{X}_\mathbb{N} \to X$ is a normal covering projection, the image subgroup $h_{\mathbb{H}\#}(\pi_1(\hat{X}_\mathbb{H}, q_\mathbb{H}(\hat{x})))$ of $\pi_1(X, x)$ is normal, for each choice of basepoints $x \in X$ and $\hat{x} \in p^{-1}(x) \subseteq \hat{X}$. Since the fundamental homomorphisms $\mu(\hat{x})$ are surjective and since $h_{\mathbb{H}\#}$ is a monomorphism whose image is a normal subgroup of $\pi_1(X, x)$, then $\mathbb{H}$ is a normal subgroup of $A(p)$.

(c) Let $\mathbb{N}$ be a normal subgroup of $A(p)$. The assignment $\Theta \to \Theta_\mathbb{N}$ introduced in (b) is a group homomorphism $A(p) \to A(h_\mathbb{N})$. This homomorphism is surjective: an automorphism $\Phi \in A(h_\mathbb{N})$ is uniquely determined by its effect on a fiber element, and, in view of the transitivity of $A(p)$, there is an induced automorphism $\Theta_\mathbb{N} \in A(h_\mathbb{N})$ duplicating this effect.

The kernel of the homomorphism $A(p) \to A(h_\mathbb{N})$ is exactly the subgroup $\mathbb{N} = A(q_\mathbb{N})$, because $\Theta_\mathbb{N} = 1 : \hat{X}_\mathbb{N} \to \hat{X}_\mathbb{N}$ if and only if $q_\mathbb{N} \circ \Phi = q_\mathbb{N}$. Thus, $A(h_\mathbb{N})$ is isomorphic to the quotient group $A(p)/\mathbb{N}$. □

The Co-Galois Correspondence (5.6), the Fundamental Theorem (4.1) of covering spaces, and the Classification Theorem (3.4) yield the following:

5.7 Classification Theorem
Let $(\hat{X}, p)$ be a simply connected covering space of X.
(a) There is a bijective correspondence between the isomorphism classes of (normal) intermediate covering spaces of $(\hat{X}, p)$ and (normal) subgroups of the fundamental group $\pi_1(X, x)$.
(b) There is a bijective correspondence between the isomorphism classes of covering spaces of X and conjugacy classes of subgroups of the fundamental group $\pi_1(X, x)$.

Proof: As $\hat{X}$ is simply connected, the fundamental homomorphisms in the previous diagram are isomorphisms. If we identify $\pi_1(X, x)$ and $A(p)$ via the fundamental isomorphism, then an intermediate covering space $(q_\mathbb{H}, \hat{X}_\mathbb{H}, h_\mathbb{H})$ corresponds uniquely to the subgroup $\mathbb{H} \leq \pi_1(X, x) \approx A(p)$. In particular, every subgroup $\mathbb{H} \leq \pi_1(X, x)$ gives rise to a covering space $(\hat{X}_\mathbb{H}, h_\mathbb{H})$ of X with $h_{\mathbb{H}\#}(\pi_1(\hat{X}_\mathbb{H}, q_\mathbb{H}(\hat{x}))) = \mathbb{H} \leq \pi_1(X, x)$. So the classification of covering spaces of X by their conjugacy classes of subgroups of $\pi_1(X, x)$ yields (b). □

6   Group Actions and Geometries

We consider group actions on a space for which the orbit construction produces a covering projection. When the group consists of isometries for some geometry on the space, the orbit space acquires a geometry of the same type. In this way, the surfaces classified in Chapter 13 acquire a geometry from the spherical, Euclidean, or hyperbolic plane accordingly as their Euler characteristic is positive, zero, or negative.

Orbit Maps

Under the operation of composition, the homeomorphisms of any space $\hat{X}$ constitute a group, denoted by *Homeo*$(\hat{X})$. Any subgroup $\mathbb{G} \leq$ *Homeo*$(\hat{X})$ is called a **group of homeomorphisms** of the space $\hat{X}$.

Given any group $\mathbb{G}$ of homeomorphisms, we introduce this relation on $\hat{X}$:

$\hat{x} \equiv_\mathbb{G} \hat{z}$ *if and only if there exists* $\Theta \in \mathbb{G}$ *such that* $\Theta(\hat{x}) = \hat{z}$.

The group axioms for $\mathbb{G}$ imply that $\equiv_\mathbb{G}$ is an equivalence relation. The equivalence class containing $\hat{x} \in \hat{X}$ is called the **orbit** of $\hat{x}$ and is denoted by $\mathbb{G}\hat{x} = \{\Theta(\hat{x}) : \Theta \in \mathbb{G}\}$. The set of these orbits is the quotient space of $\hat{X}$ modulo the relation $\equiv_\mathbb{G}$. It is denoted by $X = \hat{X}/\mathbb{G}$ and is called the **orbit space** of $\mathbb{G}$; the associated quotient map is denoted by $q_\mathbb{G} : \hat{X} \to \hat{X}/\mathbb{G}$ and is called the **orbit map**.

The following condition, which appeared in the construction of intermediate covering spaces (Theorem 5.5), characterizes those groups of homeomorphisms whose orbit map is a covering projection.

We call $\mathbb{G}$ a **properly discontinuous group** if every $\hat{x} \in \hat{X}$ has an open neighborhood $\hat{U}(\hat{x})$ whose images $\Theta(\hat{U})$ under $\Theta \in \mathbb{G}$ are pairwise disjoint.

One consequence of *proper discontinuity* is *discreteness*.

We call $\mathbb{G}$ a **discrete group** if each orbit is a discrete subspace of $\hat{X}$.

6.1   Orbit Projection Theorem   *Let $\hat{X}$ be a connected, locally path-connected topological space and let $\mathbb{G}$ be a properly discontinuous group of homeomorphisms of $\hat{X}$. Then the orbit map $q_\mathbb{G} : \hat{X} \to \hat{X}/\mathbb{G}$ is a normal covering projection with automorphism group $A(q_\mathbb{G}) = \mathbb{G}$.*

Proof: Since $\hat{X}$ is locally path-connected and $\mathbb{G}$ is properly discontinuous, every point $\hat{x} \in \hat{X}$ has an open path-connected neighborhood $\hat{U}(\hat{x})$ such that

$$q_{\mathbb{G}}^{-1}(q_{\mathbb{G}}(\hat{U})) = \overset{\circ}{\cup} \{\Theta(\hat{U}) : \Theta \in \mathbb{G}\},$$

and the pairwise disjoint images $\Phi(\hat{U})$ ($\Phi \in \mathbb{G}$) project homeomorphically to $q_{\mathbb{G}}(\hat{U})$ under $q_{\mathbb{G}}$. This shows that $q_{\mathbb{G}}(\hat{U})$ is an admissible neighborhood of $\hat{X}/\mathbb{G}$ and that $q_{\mathbb{G}} : \hat{X} \to \hat{X}/\mathbb{G}$ is a covering projection.

By construction, $\mathbb{G} \leq A(q_{\mathbb{G}})$. $\mathbb{G}$ acts transitively on the orbits of $\mathbb{G}$, which are the fibers of $q_{\mathbb{G}}$; thus, $A(q_{\mathbb{G}}) = \mathbb{G}$ by the Uniqueness of Liftings (2.1). So $(\hat{X}, q_{\mathbb{G}})$ is a normal covering space by the Transitivity Theorem (5.4). □

Example 1 The group $\mathbb{Z}^n$ of homeomorphisms of Euclidean n-space $\mathbb{E}^n$ generated by the n translations along the standard unit basis vectors is properly discontinuous. For instance, each open n-ball $B(x, \frac{1}{2})$ has pairwise disjoint images $\Theta(B(x, \frac{1}{2}))$ under the translations $\Theta \in \mathbb{Z}^n$. Thus, the orbit map $q : \mathbb{E}^n \to \mathbb{E}^n/\mathbb{Z}^n$ is a covering projection. The orbit space $\mathbb{E}^n/\mathbb{Z}^n$ is the n-cube $\mathbb{I}^n$ with opposite faces identified by the unit translations and, hence, is homeomorphic to the n-torus $T^n = \mathbb{S}^1 \times \ldots \times \mathbb{S}^1$.

A finite group $\mathbb{G}$ of homeomorphisms of a Hausdorff space $\hat{X}$ is properly discontinuous if and only if each $1_{\hat{X}} \neq \Theta \in \mathbb{G}$ is fixed point free (Exercise 1). So Theorem 6.1 applies if $\hat{X}$ is connected and locally path-connected.

Example 2 The antipodal map $\alpha : \mathbb{S}^n \to \mathbb{S}^n$, $\alpha(x) = -x$, generates a cyclic group $\mathbb{Z}_2 = \{1_{\mathbb{S}^n}, \alpha\}$ of order 2. $\mathbb{Z}_2$ is a properly discontinuous group of homeomorphisms whose orbit map is a double covering $q : \mathbb{S}^n \to \mathbb{P}^n$. The orbit space $\mathbb{P}^n$ is called *projective n-space*. By the Fundamental Theorem (4.1), $\pi_1(\mathbb{P}^n) \approx \mathbb{Z}_2$.

When $\mathbb{G}$ is a properly discontinuous group of isometries of some Riemannian metric on an n-manifold $\hat{X}$, the orbit space $X = \hat{X}/\mathbb{G}$ acquires the topological and geometric structure of the Riemannian n-manifold $\hat{X}$. The orbit manifold X acquires a Riemannian metric whose *geodesics*, i.e., curves of locally minimal arc length, are the images of geodesics in $\hat{X}$.

Example 3 *The cross-cap surface or projective plane $\mathbb{P}^2$ admits a spherical geometry, which is a geometry with constant positive curvature.* The identity and antipodal map $1_{\mathbb{S}^2}, \alpha : \mathbb{S}^2 \to \mathbb{S}^2$ are isometries of the spherical geometry of $\mathbb{S}^2$. Thus, the orbit surface $\mathbb{P}^2 = \mathbb{S}^2/\mathbb{Z}_2$ of Example 2 acquires the *spherical geometry* of $\mathbb{S}^2$.

The remainder of this section shows that each other orientable surface acquires a geometric structure by virtue of being an orbit surface of a group of isometries of some geometric plane.

GROUPS OF EUCLIDEAN ISOMETRIES

Homeomorphisms of $\mathbb{R}^2$ that preserve the Euclidean metric on $\mathbb{R}^2$ are called ***Euclidean isometries***. They form a subgroup, $Isom(\mathbb{E}^2)$, of the group

Homeo($\mathbb{R}^2$) of all homeomorphisms under composition.

Isom($\mathbb{E}^2$) includes the **translations** T_v by *vectors* $v \in \mathbb{R}^2$ and the **rotations** $R_{c,\alpha}$ about c by *angles* $\alpha \in \mathbb{R}/2\pi\mathbb{Z}$. These isometries $T_v, R_{c,\alpha} : \mathbb{R}^2 \to \mathbb{R}^2$ are expressed using the complex variable $z \in \mathbb{C} = \mathbb{R}^2$ by $T_v(z) = z + v$ and $R_{c,\alpha}(z) = e^{i\alpha}(z - c) + c$. Translations and rotations preserve the orientation of $\mathbb{R}^2$ and the subgroup of *Isom*($\mathbb{E}^2$) that they generate is the group, *Isom*$_+$($\mathbb{E}^2$), of all **orientation preserving** Euclidean isometries. *Isom*($\mathbb{E}^2$) also includes two types of orientation reversing isometries: the **reflections** R_l flipping $\mathbb{R}^2$ across the *lines* $l \subset \mathbb{R}^2$ and the **glide reflections** G_v translating $\mathbb{R}^2$ along the *vectors* $v \in \mathbb{R}^2$ and reflecting across them.

Any isometry carries each triangle of three non-collinear points into a congruent triangle and any two isometries that have the same effect on a triangle of non-collinear points coincide (Exercise 5(a)). Any isometry equals a product of one, two, or three reflections, because its effect on a triangle can be duplicated by such a product (Exercise 5(b)). Thus, each Euclidean isometry is either the identity, a reflection, a rotation (a product of two reflections in intersecting lines), a translation (a product of two reflections in parallel lines), or a glide reflection (Exercise 5(c)).

When $\mathbb{G} \leq Isom(\mathbb{E}^2)$ is a properly discontinuous group of Euclidean isometries of $\mathbb{R}^2$, the orbit map $q_{\mathbb{G}} : \mathbb{R}^2 \to \mathbb{R}^2/\mathbb{G}$ is a universal covering projection. The orbit space $S = \mathbb{R}^2/\mathbb{G}$ inherits a topological 2-manifold structure from $\mathbb{R}^2$ and also a geometry from the Euclidean plane $\mathbb{E}^2$. This Euclidean geometry is one of constant zero curvature, a *flat geometry*. Lines of $\mathbb{R}^2$ are carried to *geodesics* for the induced metric on the orbit manifold S.

A discrete group $\mathbb{G} \leq Isom(\mathbb{E}^2)$ is called a **bounded group**, a *frieze group*, or a **wallpaper group** depending upon whether its subgroup of translations is minimally generated by zero, one, or two translations. There are two types of bounded groups: the *cyclic group* $\mathbb{Z}_n$ generated by a rotation through the angle $2\pi/n$ and the *dihedral group* $\mathbb{D}_{2n}$ generated by the reflections in a pair of lines that make an angle π/n. There are seven types of frieze groups; each one is generated a single translation T_v or glide reflection G_v, and possibly a reflection in a line perpendicular or parallel to T_v or G_v, or a rotation of order 2, or both. Finally, there are 17 types of wallpaper groups; each is generated by two translations T_v and T_σ, possibly a rotation of order 1, 2, 3, 4, or 6 that carries v into σ, and possibly an orientation-reversing isometry.

Only four of these types of discrete groups of Euclidean isometries act properly discontinuously on the entire Euclidean plane $\mathbb{R}^2$:

Example 4 *The cylinder $\mathbb{S}^1 \times \mathbb{R}$, the infinite Möbius strip, the torus T^2, and the Klein bottle K admit a Euclidean geometry. Any group $\mathbb{G} \leq Isom(\mathbb{E}^2)$ generated by either a translation or glide reflection, or two non-parallel translations, or a translation and a perpendicular glide reflection is properly discontinuous. Its orbit space $\mathbb{R}^2/\mathbb{G}$ is the cylinder $\mathbb{S}^1 \times \mathbb{R}$, the infinite Möbius strip, the torus T^2, or the Klein bottle K, accordingly. In this way, these orbit 2-manifolds acquire the flat geometry of $\mathbb{E}^2$.*

GROUPS OF HYPERBOLIC ISOMETRIES

The **Poincaré model** for the hyperbolic plane $\mathbb{H}^2$ uses the open unit ball $\mathbb{B}^2$ and the Riemannian metric $ds^2/(1 - r^2)$, where ds is Euclidean arc length and r is Euclidean distance from the origin $O \in \mathbb{B}^2$. This means that the **hyperbolic arc length** of a curve $g(t) = \langle x(t), y(t) \rangle$ $(a \leq t \leq b)$ in $\mathbb{B}^2$ is

$$\int_a^b \frac{\sqrt{x'(t)^2 + y'(t)^2}\ dt}{1 - (x(t)^2 + y(t)^2)}.$$

This model is conformally accurate in that hyperbolic angles agree with Euclidean angles. The unit 1-sphere $\mathbb{S}^1$ bounds the hyperbolic plane but is not part of it. In fact, the hyperbolic distance from O to $x \in \mathbb{R}$,

$$\int_0^r \frac{dx}{1-x^2} = \tanh^{-1}(r),$$

tends to infinity as x approaches $\mathbb{S}^1$. The boundary sphere $\mathbb{S}^1$ is called the **sphere at infinity** and is denoted by $\mathbb{S}^1_\infty$. The points of the open unit ball $\mathbb{B}^2$ are called **ordinary points**; the points of $\mathbb{S}^1_\infty$ are called **ideal points**.

An **isometry** of $\mathbb{H}^2$ is any function that preserves the hyperbolic length of curves in $\mathbb{H}^2$. They form a group that we denote by $Isom(\mathbb{H}^2)$. We shall not examine the hyperbolic metric to determine the isometries of $\mathbb{H}^2$. It is sufficient for our purposes to know that $Isom(\mathbb{H}^2)$ is generated by the *Euclidean inversions* of the extended Euclidean plane $\mathbb{E}^2_\infty = \mathbb{E}^2 \cup \{\infty\}$ in the *Euclidean circles* that are orthogonal to $\mathbb{S}^1$. The definitions follow.

Consider a Euclidean circle C in $\mathbb{E}^2_\infty$ with center $A \in \mathbb{E}^2_\infty$ and radius r. **Inversion in C** is the function of $\mathbb{E}^2_\infty$ that trades A with ∞ and sends each other point P to the unique point P' on the ray from A through P such that $|AP||AP'| = r^2$. Thus, all the points of the circle C are fixed and the points *inside* the circle C are traded with those *outside* C. Any Euclidean line is considered as a Euclidean circle centered at ∞, and Euclidean inversion in l is Euclidean reflection across l. All Euclidean inversions preserve angles, reverse orientation of $\mathbb{E}^2_\infty$, and carry circles to circles (Exercise 6).

A **hyperbolic line** in $\mathbb{H}^2$ is defined to be $h = C \cap \mathbb{B}^2$, where C is a Euclidean circle in $\mathbb{R}^2$ meeting $\mathbb{S}^1$ orthogonally. Since C is orthogonal to $\mathbb{S}^1$, Euclidean inversion in C restricts to a homeomorphism of $\mathbb{B}^2$. This restriction is called **reflection** R_h of the hyperbolic plane $\mathbb{H}^2$ in the hyperbolic line $h = C \cap \mathbb{B}^2$. These reflections are hyperbolic isometries and the hyperbolic lines are the **geodesics** in $\mathbb{H}^2$, i.e., the curves that locally minimize hyperbolic distance (Exercise 7).

As in $\mathbb{E}^2$, each isometry of $\mathbb{H}^2$ is determined by its effect on three noncollinear points; hence, each one is a product of one, two, or three hyperbolic reflections. So any *orientation-preserving* hyperbolic isometry equals a product, $R_l \circ R_k$, of two hyperbolic reflections. The three types of pairs $\{k, l\}$ of hyperbolic lines give three types of orientation-preserving isometries of $\mathbb{H}^2$:

When k and l intersect at an ordinary point $P \in \mathbb{B}^2$, the isometry $R_l \circ R_k$ is called ***elliptic***; it *rotates* $\mathbb{H}^2$ about P leaving setwise invariant the curves of points equidistant from P. When k and l share an ideal endpoint $Q \in \mathbb{S}^1_\infty$, $R_l \circ R_k$ is called ***parabolic***; it leaves setwise invariant the horospheres that are *centered on Q* and mutually tangent to k and l. All other hyperbolic lines k and l have a unique mutually orthogonal hyperbolic line m. Then $R_l \circ R_k$ is called ***hyperbolic***; it translates $\mathbb{H}^2$ along m and leaves setwise invariant each curve of points equidistant from m.

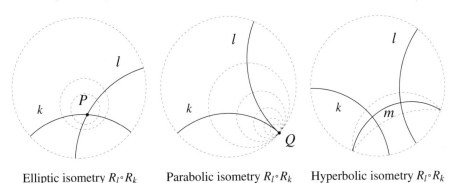

Elliptic isometry $R_l \circ R_k$ Parabolic isometry $R_l \circ R_k$ Hyperbolic isometry $R_l \circ R_k$
Invariant curves   Invariant horospheres   Invariant curves

Any pair of hyperbolic lines k and l has a *bisecting* hyperbolic line b such that the elliptic, parabolic, or hyperbolic isometry $R_b \circ R_k$ carries k into l:

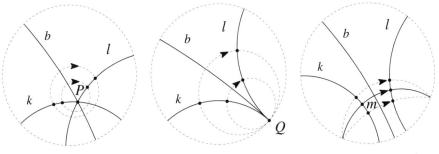

Elliptic isometry $R_b \circ R_k$ Parabolic isometry $R_b \circ R_k$ Hyperbolic isometry $R_b \circ R_k$
rotates k to l swings k to l translates k to l

The hyperbolic plane $\mathbb{H}^2$ is much more accommodating than the Euclidean plane $\mathbb{E}^2$. The angle sum for any Euclidean triangle is π, while the angle sum for any hyperbolic triangle is $\pi - Area$. Thus, in $\mathbb{H}^2$ there are two extremes for a triangle: a tiny triangle is nearly Euclidean, with angle sum close to π; and one with vertices at infinity has angle sum 0 and maximal area π. Also, an n-sided polygon in $\mathbb{H}^2$ has angle sum $(n-2)\pi - Area$. For example, as a regular hyperbolic octagon expands from area 0 to 6π, its angle sum shrinks from 6π to zero. The octagon on p. 430 has angle sum 2π.

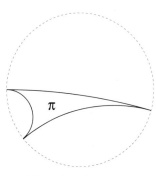

Ideal triangle, Area π Regular octagon, Area 4π

The existence of regular polygons in $\mathbb{H}^2$ with angle sum 2π makes it possible to give each handlebody surface $\mathfrak{H}_g$, with genus $g \geq 2$, the geometry of the hyperbolic plane $\mathbb{H}^2$:

6.2 Geometrization Theorem *When $g \geq 2$, there is a subgroup $\mathbb{H}_g$ of Isom($\mathbb{H}^2$) that acts properly discontinuously with orbit surface $\mathfrak{H}_g$. Thus, the surface $\mathfrak{H}_g$ acquires a hyperbolic geometry.*

Proof: Consider the regular $4g$-sided polygons concentric with $\mathbb{H}^2$. Somewhere between a tiny one with nearly Euclidean angle sum $2\pi(2g-1)$ and one with vertices at infinity and angle sum 0 there is one with angle sum 2π. When its edges are identified in the manner that produces the handlebody surface $\mathfrak{H}_g$, the eight $\pi/4$ corner angles coalesce into a full 2π angle around the vertex. This is necessary in order that $\mathfrak{H}_g$ inherit a geometry from $\mathbb{H}^2$ at the vertex. But it is also sufficient. For each edge pair of the regular $4g$-gon, there is a hyperbolic isometry that translates one edge to the other along the unique hyperbolic line orthogonal to both of them. These $2g$ hyperbolic isometries (translations) generate a subgroup of *Isom*($\mathbb{H}^2$) that acts properly discontinuously to tile the entire hyperbolic plane with non-overlapping copies of the regular $4g$-gon. The orbit surface is $\mathfrak{H}_g$.

This produces a covering projection $p : \mathbb{H}^2 \to \mathfrak{H}_g$. Its automorphism group is generated by the $2g$ hyperbolic isometries constructed above; it is isomorphic to $\pi_1(\mathfrak{H}_g) = \mathbb{H}_g$, according to the Fundamental Theorem 4.1. □

Exercises

SECTION 1

1. Prove:
   (a) A covering projection is a ***local homeomorphism***: each point of the total space has an open neighborhood that projects homeomorphically onto an open subset of the base space.

(b) A continuous local homeomorphism is an open function.
(c) A continuous open surjection is an identification map.
(d) Every covering projection is an identification map.

2 Prove that none of the implications in Exercise 1 (a)-(d) has a valid converse, by verifying the following assertions:
(a) The exponential function $e : (0, 2) \to S^1$ is a continuous, surjective, local homeomorphism that is not a covering projection.
(b) The squaring function $\mathbb{R}^1 \to [0, \infty)$, $x \to x^2$, is a continuous, open, surjection that is not a local homeomorphism.
(c) The exponential function $e : [0, 1] \to S^1$ is an identification map that is not an open function.
(d) The projection $p : \mathbb{R}^2 \to \mathbb{R}^1$, $p(<x, y>) = x$, is an identification map that is not a covering projection.

3 Find a continuous surjective function $p : \hat{X} \to X$ whose fibers are discrete and set equivalent to each other, but that is not a covering projection.

4 Prove that the quotient map $p : S^2 \to \mathbb{P}^2$ (Exercise 5.1.16) that produces the projective plane by identifying antipodal points of the 2-sphere is a covering projection.

5 (a) Construct at least three different 4-fold covering spaces of $S^1 \vee S^1$.
(b) Construct a 2-fold covering of the torus onto the Klein bottle.

6 Show that the topological product of denumerably many copies of the exponential covering projection $e : \mathbb{R}^1 \to S^1$ is not a covering projection.

The *pseudo-projective plane* P_n of degree $n > 0$ is the quotient space of the closed unit disc $\mathbb{D}^2$ that results when each point of the boundary 1-sphere S^1 is identified with its image under the rotation $\rho_n : S^1 \to S^1$ through $2\pi/n$ radians.

7 Construct a simply connected, n-sheeted covering space of the pseudo-projective plane P_n. Make a drawing of your construction for the case $n = 3$.

8 Tile the entire plane $\mathbb{R}^2$ with copies of the polygonal model of the Klein Bottle K. Show that this describes a covering projection $p : \mathbb{R}^2 \to K$.

9 Restrict the covering projection $e \times 1_{S^1} : \mathbb{R}^1 \times S^1 \to S^1 \times S^1$ to the subspace $S^1 \vee S^1$.

10 (a) Let $m, n \geq 1$. Prove that the composition of an m-fold and an n-fold covering projection is an mn-fold covering projection.
(b) Prove that the composition of two covering projections need not be another.

SECTION 2

1 Let $(\hat{X}, p)$ be a simply connected covering space of X. Prove:
(a) Two paths in X with the same endpoints are path-homotopic if and only if their liftings in $\hat{X}$ with the same initial point have the same terminal point.
(b) For any basepoints $x \in X$ and $\hat{x} \in p^{-1}(x) \subseteq \hat{X}$, the action $\hat{x} \cdot : \pi_1(X, x) \to p^{-1}(x)$ is a bijection.

2 Apply Exercise 1 to the following covering projections:
(a) $e : \mathbb{R}^1 \to S^1$, the exponential covering of the 1-sphere (Example 1.1),
(b) $e \times e : \mathbb{R}^1 \times \mathbb{R}^1 \to S^1 \times S^1$, the product covering of the torus (Example 1.3),

(c) $p : \mathbb{S}^2 \to \mathbb{P}^2$, the double covering of the projective plane (Exercise 1.4), and
(d) $p : \mathbb{R}^2 \to K$, the covering of the Klein bottle by the plane (Exercise 1.8).

3   (a) Construct a double covering projection $p : T \to K$ of the torus onto the Klein bottle and describe the monomorphism $p_\# : \pi_1(T) \to \pi_1(K)$.
    (b) Prove that there is no double covering of the Klein bottle onto the torus.

4   (a) Construct a double covering projection $p : \mathfrak{H}_3 \to \mathfrak{H}_2$ and describe the monomorphism $p_\# : \pi_1(\mathfrak{H}_3) \to \pi_1(\mathfrak{H}_2)$.
    (b) Construct a covering projection $p : \mathfrak{H}_g \to \mathfrak{H}_2$ ($g \geq 1$) with $g - 1$ sheets and describe the monomorphism $p_\# : \pi_1(\mathfrak{H}_g) \to \pi_1(\mathfrak{H}_2)$.

5   (a) Construct a double covering projection $p : \mathcal{C}_{2g} \to \mathcal{C}_{g+1}$ ($g \geq 1$) and describe the monomorphism $p_\# : \pi_1(\mathcal{C}_{2g}) \to \pi_1(\mathcal{C}_{g+1})$.
    (b) Construct a covering projection $p : \mathcal{C}_{2g+1} \to \mathcal{C}_3$ ($g \geq 1$) with $2g - 1$ sheets and describe the monomorphism $p_\# : \pi_1(\mathcal{C}_{2g+1}) \to \pi_1(\mathcal{C}_3)$.
    (c) Show how suitable compositions of the covering projections in (a) and (b) give covering projections $p : \mathcal{C}_g \to \mathcal{C}_3$ for all $g \geq 2$.

6   Determine the conjugacy class of subgroups of $\pi_1(\mathbb{S}^1_a \vee \mathbb{S}^1_b) = \mathbb{F}(a, b)$ for each of the covering projections p_i ($i = 0, 1, 2, 3$) in Example 1.6.

SECTION 3

1   Construct representatives of all isomorphism classes of covering spaces of the pseudo-projective plane P_n of order n.

2   Prove that any covering projection with simply connected base space is a homeomorphism.

3* Let $(\hat{X}_1, p_1)$ and $(\hat{X}_2, p_2)$ be covering spaces of X, and let $\Phi : (\hat{X}_1, p_1) \to (\hat{X}_2, p_2)$ be a morphism. Prove that $\Phi : \hat{X}_1 \to \hat{X}_2$ is a covering projection.

4* Prove that any two universal covering spaces of the same space are isomorphic.

5   Prove that the composition of two covering projections is a covering projection if the base space has a universal covering space.

6   Show that a simply connected covering space $(\hat{X}, p)$ of any space X covers every covering space $(\overline{X}, q)$ of X; prove that there exists a covering projection $r : \hat{X} \to \overline{X}$ such that $p = q \circ r : \hat{X} \to \overline{X} \to X$.

7   Construct a covering projection $p : \hat{X} \to X$ and a continuous function $F : X \to X$ such that F has no lifting through p, but $F^2 = F \circ F$ does.

8   Show that every map of the 2-sphere $\mathbb{S}^2$ into the torus $\mathbb{S}^1 \times \mathbb{S}^1$ is inessential.

9   Must a (self-)morphism $\Phi : (\hat{X}, p) \to (\hat{X}, p)$ of a covering space be an isomorphism?

A *topological group* $\mathbb{G}$ is a space with a group structure for which multiplication $\mathbb{G} \times \mathbb{G} \to \mathbb{G}$, $<g, h> \to g \cdot h$, and inversion $\mathbb{G} \to \mathbb{G}$, $g \to g^{-1}$, are continuous.

10 Prove that if $(\hat{X}, p)$ is covering space of a connected, locally path-connected, topological group X, then $\hat{X}$ has the structure of a topological group such that the covering projection p is a homomorphism.

SECTION 4

1. Determine the automorphism group $A(d_n)$ of the n-fold covering $d_n : \mathbb{S}^1 \to \mathbb{S}^1$ of the 1-sphere onto itself. Describe the fundamental isomorphism for this covering space.

2. Determine the automorphism group $A(e \times e)$ and describe the fundamental isomorphism for the covering projection $e \times e : \mathbb{R}^1 \times \mathbb{R}^1 \to \mathbb{S}^1 \times \mathbb{S}^1$.

3. Restrict the covering projection $e \times 1 : \mathbb{R}^1 \times \mathbb{S}^1 \to \mathbb{S}^1 \times \mathbb{S}^1$ to the subspace $\mathbb{S}^1 \vee \mathbb{S}^1$ of $\mathbb{S}^1 \times \mathbb{S}^1$. Determine the automorphism group and describe the fundamental isomorphism for this covering space of $\mathbb{S}^1 \vee \mathbb{S}^1$.

4. Determine the automorphism group $A(p)$ of the covering projection $p : L \to \mathbb{S}^1 \vee \mathbb{S}^1$ of Example 1.5 and describe the fundamental isomorphism for this covering space.

5. (a) Determine the identification space H that results when opposite faces of a hexagon are identified under translation. Use hexagons to construct a simply connected covering space $\hat{H}$ of H.
   (b) Determine the automorphism group $A(p)$ of the covering projection $p : \hat{H} \to H$ in (a) and describe the fundamental isomorphism for this covering space.

6. (a) Determine the identification space X that results when opposite faces of the cube $\mathbb{I}^3$ are identified under translation. Use cubes to construct a simply connected covering space $\hat{X}$ of X.
   (b) Determine the automorphism group $A(p)$ of the covering projection $p : \hat{X} \to X$ and describe the fundamental isomorphism for this covering space.

7. (a) Construct a simply connected covering space $\hat{X}$ of the torus-with-membrane:
$$X = (\mathbb{S}^1 \times \mathbb{S}^1) \cup (\mathbb{D}^2 \times \{1\}) \subset \mathbb{D}^2 \times \mathbb{S}^1.$$
   (b) Determine the automorphism group $A(p)$ of the covering projection $p : \hat{X} \to X$ and describe the fundamental isomorphism for this covering space.

   The *torus with pseudo-projective membranes* $Z_{m,n}$ is the polygonal complex modeled on the presentation $\langle a, b : a^m, b^n, aba^{-1}b^{-1} \rangle$. $Z_{m,n}$ has a single vertex, 1-skeleton $\mathbb{S}^1_a \vee \mathbb{S}^1_b$, and three faces with boundary words a^m, b^n, and $aba^{-1}b^{-1}$.

8. (a) Construct a simply connected covering space $\hat{Z}$ of $Z = Z_{m,n}$.
   (b) Determine the automorphism group $A(p)$ and the fundamental isomorphism for the covering projection $p : \hat{Z} \to Z$ in (a).

9. Construct a simply connected covering space of each of these one-point unions:
   (a) $\mathbb{S}^1 \vee \mathbb{S}^1$, and
   (b) $P_n \vee P_m$, where P_k denotes the pseudo-projective plane of order k.
   Describe the fundamental isomorphism for each of these covering spaces.

SECTION 5

1*. Let X be semilocally simply connected. Prove that the sets $([f], U) \subseteq \pi(X)_x$, where $[f] \in \pi(X)_x$ and U is a relatively simply connected neighborhood of the endpoint $\rho([f])$, serve as a base for a topology on $\pi(X)_x$.

2. Let $\mathbb{Z}_{2n}$ be the automorphism group of the simply connected covering space of the

pseudo-projective plane of degree $2n$. Build the intermediate covering space associated with the subgroup $\mathbb{Z}_2 \leq \mathbb{Z}_{2n}$.

3 Build the intermediate covering space associated with the subgroup $n\mathbb{Z} \leq \mathbb{Z}$ of the automorphism group of the covering projection $e \times 1 : \mathbb{R}^1 \times \mathbb{S}^1 \to \mathbb{S}^1 \times \mathbb{S}^1$.

4* (a) Construct an intermediate covering space $(q, \overline{X}, h)$ of some normal covering space $(\hat{X}, p)$ of some space X for which $(\overline{X}, h)$ is a non-normal covering space of X.
(b) Construct non-isomorphic intermediate covering spaces $(q_i, \overline{X}_i, h_i)$ $(i = 1, 2)$ of a covering space $(\hat{X}, p)$ of some space X for which $(\overline{X}_i, h_i)$ $(i = 1, 2)$ are isomorphic covering spaces of X.

SECTION 6

1* Prove that a finite group of homeomorphisms of a Hausdorff space is properly discontinuous if and only if the non-identity homeomorphisms have no fixed points.

2 Prove: If $(\hat{X}, p)$ is a normal covering space of X, then the automorphism group $\mathbb{G} = A(p)$ is properly discontinuous and the orbit space $\hat{X}_\mathbb{G}$ is homeomorphic to X.

3 Let $\mathbb{G}$ be the group of homeomorphisms of $\mathbb{R}^2$ generated by two clockwise rotations about distinct points through (a) 60° or (b) 90°. Find a maximal subspace $\hat{X} \subset \mathbb{R}^2$ that has $\mathbb{G}$ as a properly discontinuous group of homeomorphisms. Determine the covering projection $p : \hat{X} \to \hat{X}_\mathbb{G}$ described in Theorem 6.1.

4 Prove: Each group $\mathbb{G}$ of isometries of $\mathbb{S}^2$, $\mathbb{E}^2$, or $\mathbb{H}^2$ is generated by its orientation preserving subgroup $\mathbb{G}_+$ and any single orientation reversing isometry $\Theta \in \mathbb{G}$.

5* Prove that in the Euclidean plane:
(a) Any isometry carries each triangle of three non-collinear points into a congruent triangle and any two isometries that have the same effect on a triple of non-collinear points coincide.
(b) Any isometry equals a product of one, two, or three reflections, because its effect on a triangle can be duplicated by such a product.
(c) Each Euclidean isometry is either the identity, a rotation, a reflection, a translation, or a glide reflection.

6* Prove that Euclidean inversions preserve angles, reverse orientation of the extended Euclidean plane $\mathbb{E}^2_\infty$, and carry Euclidean circles to Euclidean circles.

7* Prove that hyperbolic reflections preserve the hyperbolic metric and that the hyperbolic lines locally minimize hyperbolic distance.

8 Use Brouwer's fixed point theorem (**9.2.5**) and some geometry to classify the orientation-preserving isometries of $\mathbb{H}^2$. Accept that each one has a continuous extension to the sphere at infinity $\mathbb{S}^1_\infty$ and then prove that there are just these three possibilities for the fixed points of $\Theta \in \mathit{Isom}_+(\mathbb{H}^2)$ in the closed unit disc $\mathbb{D}^2$:
(a) Θ has a unique fixed point in $\mathbb{B}^2$ (elliptic type);
(b) Θ has a unique fixed point on $\mathbb{S}^1_\infty$ (parabolic type); or
(c) Θ has two fixed points on $\mathbb{S}^1_\infty$ (hyperbolic type).

9 Give each non-orientable surface of genus $g \geq 2$ a hyperbolic geometry.

15

CW COMPLEXES

The polygonal complexes, the knot complement spines, the polygonal maps on surfaces, and the topological graphs in Chapters **12** and **13** are examples of *cell complexes*, i.e., spaces partitioned into *cells* homeomorphic to open balls. *CW complexes* are cell complexes that satisfy two topological restrictions that make them especially convenient for investigations in homotopy theory. They were formulated by J. H. C. Whitehead in 1949.

1   Cell Complexes

This section presents CW complexes and their basic properties.

CELLULAR DECOMPOSITIONS

Let $\mathbb{B}^n \subset \mathbb{D}^n$ be the open and closed unit n-balls in Euclidean space $\mathbb{E}^n$, with boundary $(n-1)$-sphere $\mathbb{S}^{n-1} = \mathbb{D}^n - \mathbb{B}^n$ ($n \geq 1$). Let X be a Hausdorff space.

A *characteristic map* $\psi : \mathbb{D}^n \to X$ is a continuous function such that $\psi : \mathbb{B}^n \to \psi(\mathbb{B}^n)$ is a homeomorphism and $\psi(\mathbb{S}^{n-1}) = \psi(\mathbb{D}^n) - \psi(\mathbb{B}^n)$. Then $\bar{c}^n = \psi(\mathbb{D}^n)$ is called a *closed n-cell* in X; $c^n = \psi(\mathbb{B}^n)$ is called an *open n-cell* in X; and $\dot{c}^n = \psi(\mathbb{S}^{n-1})$ is called the *boundary* of these n-cells.

For simpler notation, we often denote the cell by c, its dimension by $|c|$, and the characteristic map by $\psi : \mathbb{D}^{|c|} \to X$.

We extend the terminology to dimension $n = 0$ by setting $\mathbb{D}^0 = \{O\} = \mathbb{B}^0$. Then any constant function $\psi : \mathbb{D}^0 \to X$ serves as a characteristic map for a singleton *0-cell* $c^0 = \psi(\mathbb{B}^0) = \psi(\mathbb{D}^0) = \bar{c}^0$ with *boundary* $\dot{c}^0 = \varnothing$.

Since $\mathbb{D}^n$ is compact and X is Hausdorff, then any characteristic map $\psi : \mathbb{D}^n \to X$ is a closed continuous surjection, hence, an identification map. So a closed n-cell $\bar{c}^n$ is a crumpled version of the closed n-ball $\mathbb{D}^n$ in which only points of the boundary sphere $\mathbb{S}^{n-1}$ are identified among themselves.

This terminology warrants the following clarifications (Exercise 1):

435

1.1 Theorem *In a Hausdorff space X,*
(a) *a closed n-cell $\bar{c}^n$ is the topological closure in X of the open n-cell c^n;*
(b) *the boundary $\dot{c}^n$ is the topological boundary of the open n-cell c^n; but*
(c) *the open n-cell c^n need not be an open set in X.*

Example 1 The continuous functions $\psi_\pm : \mathbb{D}^n \to \mathbb{S}^n$ given by
$$\psi_\pm(x_1, \ldots, x_n) = (x_1, \ldots, x_n, \pm\sqrt{1 - \Sigma_i x_i^2})$$
are characteristic maps for the **hemispherical n-cells** $c_\pm^n = \{x \in \mathbb{S}^n : \pm x_{n+1} > 0\}$, both of which have equatorial boundary $\dot{c}_\pm^n = \mathbb{S}^{n-1}$.

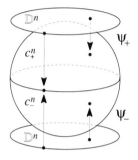
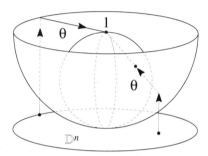

Example 1: $\psi_\pm : \mathbb{D}^n \to \mathbb{S}^n$ Example 2: $\theta : \mathbb{D}^n \to \mathbb{S}^n$

Example 2 The continuous function $\theta : \mathbb{D}^n \to \mathbb{S}^n$, given by
$$\theta(x_1, \ldots, x_n) = (2(\Sigma_i x_i^2) - 1, 2x_1\sqrt{1 - \Sigma_i x_i^2}, \ldots, 2x_n\sqrt{1 - \Sigma_i x_i^2}),$$
is a characteristic map for the spherical n-cell $c^n = \mathbb{S}^n - \{1\}$ with boundary $\dot{c}^n = \{1\}$.

A **cellular decomposition** of a Hausdorff space X is a partition $\mathcal{E} = \{c\}$ of X into pairwise disjoint open cells c of various dimensions $|c| \geq 0$ such that the boundary $\dot{c}$ of each open cell $c \in \mathcal{E}$ meets only open cells $c' \in \mathcal{E}$ of lower dimension: for any cells $c, c' \in \mathcal{E}$ such that $\dot{c} \cap c' \neq \emptyset$, we have $|c'| < |c|$.

Example 3 $\mathcal{E} = \{c_\pm^0 = \{\pm 1\}, c^1 = (-1, +1)\}$ is a cellular decomposition of $\mathbb{D}^1$.

Example 4 $\mathcal{E} = \{c^0 = 1, c^n = \mathbb{S}^n - \{1\}\}$ is a minimal cellular decomposition of $\mathbb{S}^n$.

Example 5 The torus $X = \mathbb{S}^1 \times \mathbb{S}^1$ has a cellular decomposition into the four cells:
$$c^0 = \{1\} \times \{1\} \qquad\qquad c^2 = (\mathbb{S}^1 - \{1\}) \times (\mathbb{S}^1 - \{1\})$$
$$c_a^1 = (\mathbb{S}^1 - \{1\}) \times \{1\} \qquad\qquad c_b^1 = \{1\} \times (\mathbb{S}^1 - \{1\}).$$

Example 6 Let $\mathcal{E}$ be a cellular decomposition of X. For each integer $n \geq 0$, the collection $\mathcal{E}^n = \{c^k \in \mathcal{E} : k \leq n\}$ is a cellular decomposition of the subspace $X^n = \cup \{c^k \in \mathcal{E} : k \leq n\}$. $\mathcal{E}^n$ is called the **n-skeleton** of $\mathcal{E}$.

Example 7 The $2n + 2$ hemispherical cells $\{c_\pm^0, c_\pm^1, \ldots, c_\pm^n\}$ in the equatorial spheres $\mathbb{S}^0 \subset \mathbb{S}^1 \subset \ldots \subset \mathbb{S}^n$ (Example 1) constitute a cellular decomposition of $\mathbb{S}^n$.

COMPLEXES AND SUBCOMPLEXES

A *cell complex* $(X, \mathcal{E})$ consists of a Hausdorff space X and a cellular decomposition $\mathcal{E}$ of X. A *subcomplex* $(A, \mathcal{D})$ of a cell complex $(X, \mathcal{E})$ is a cell complex for which A is a subspace of X and $\mathcal{D}$ is a subfamily of $\mathcal{E}$.

Example 8   The n-skeleton $(X^n, \mathcal{E}^n)$ of Example 6 is a subcomplex of $(X, \mathcal{E})$.

1.2   Theorem   *Let $(A, \mathcal{D})$ be a subcomplex of a cell complex $(X, \mathcal{E})$. Then $c \in \mathcal{E}$ and $c \cap A \neq \emptyset$ imply that $c \in \mathcal{D}$ and $\bar{c} \subseteq A$.*

Proof: Since $\mathcal{E}$ is a partition of X and since $\mathcal{D} \subseteq \mathcal{E}$ is a partition of $A \subseteq X$, then $c \in \mathcal{E}$ and $c \cap A \neq \emptyset$ imply $c \in \mathcal{D}$. Since $\mathcal{D}$ is a cellular decomposition of A, the cell $c \in \mathcal{D}$ has a characteristic map $\psi : \mathbb{D}^{|c|} \to A \subseteq X$. Then, by Theorem 1.1, the image $\psi(\mathbb{D}^{|c|}) \subseteq A$ is the closure $\bar{c}$ of c in A and in X. □

The subcomplex property described in Theorem 1.2 suggests the following definition.

Let $(X, \mathcal{E})$ be a cell complex. A subset $A \subseteq X$ is *cellularly closed* (relative to $\mathcal{E}$) if $c \in \mathcal{E}$ and $c \cap A \neq \emptyset$ imply $\bar{c} \subseteq A$.

When $A \subseteq X$ is cellularly closed, the characteristic map $\psi : \mathbb{D}^n \to X$ for any cell $c \in \mathcal{E}$ such that $c \cap A \neq \emptyset$ can be viewed as a characteristic map $\psi : \mathbb{D}^{|c|} \to A$ for c in A, as $\psi(\mathbb{D}^{|c|}) = \bar{c} \subseteq A$. Thus, the collection

$$\mathcal{E}|_A = \{c \in \mathcal{E} : c \cap A \neq \emptyset\}$$

is a cellular decomposition of the cellularly closed subspace A.

Notice that arbitrary unions and arbitrary intersections of cellularly closed subsets are cellularly closed. Thus, the intersection $C(B)$ of all cellularly closed subsets of X containing a given subset $B \subseteq X$ is the smallest cellularly closed subset of X containing B. $C(B)$ is called the *cellular closure* of B.

Example 9   For any cell complex $(X, \mathcal{E})$ and point-cell pair $x \in c \in \mathcal{E}$, there are these cellular closures:

$$C(\{x\}) = C(c) = C(\bar{c}) = c \cup C(\overset{\bullet}{c}).$$

Example 10   For a subcomplex $(A, \mathcal{D})$ of $(X, \mathcal{E})$ and subset $B \subseteq X$, one has that

$$A \cap C(B) = \cup \{\bar{c} : c \subseteq C(B), c \in \mathcal{D}\}.$$

A cell complex $(X, \mathcal{E})$ is *finite* if the cellular decomposition $\mathcal{E}$ consists of a finite number of cells.

A cell complex $(X, \mathcal{E})$ has *finite type* if the cellular decomposition $\mathcal{E}$ consists of a finite number of cells in each dimension.

A cell complex $(X, \mathcal{E})$ is *n-dimensional* if has some n-dimensional cells, but none of any higher dimension; equivalently, $X^{n-1} \neq X^n = X$.

CLOSURE FINITENESS AND COHERENCE

The unrestricted concept of cell complex offered so far permits such pathological examples as these:

- Any space X partitioned into its individual points as 0-cells.
- $\mathbb{D}^n$ partitioned into the points of $\mathbb{S}^{n-1}$ as 0-cells and $\mathbb{B}^n$ as an n-cell.

Here is the first of two restrictions that will rule out such examples.

A cellular decomposition $\mathcal{E}$ of a space X is **closure finite** if the boundary $\dot{c}$ of each cell $c \in \mathcal{E}$ meets only finitely many open cells $c' \in \mathcal{E}$.

Any finite cellular decomposition $\mathcal{E}$ of any space X is closure finite.

1.3 Theorem *A cellular decomposition is closure finite if and only if the cellular closure of each cell is finite.*

Proof: For any cell $c \in \mathcal{E}$, let $A \subseteq X$ be the union of all cells $e \in \mathcal{E}$ that initiate a sequence of cells in $\mathcal{E}$,

$$e = c_1, \ldots, c_i, c_{i+1}, \ldots, c_k = c,$$

such that $c_i \cap \dot{c}_{i+1} \neq \emptyset$ for all $1 \leq i \leq k-1$. The union A contains c, is cellularly closed ($e \cap A \neq \emptyset \Rightarrow e \subset A \Rightarrow \bar{e} = e \cup \dot{e} \subset A$), and is contained in any cellularly closed set containing c. This implies that $A = C(c)$. Thus, $\mathcal{E}$ is closure finite if and only if the cellular closure $C(c)$ of each cell $c \in \mathcal{E}$ involves only finitely many cells. $\square$

The second of two restrictions to be imposed on cell complexes involves a weak topology (Section **5**.4) determined by the closed cells.

A cellular decomposition $\mathcal{E}$ of X is **coherent** if X has the weak topology relative to the family of closed cells $\bar{\mathcal{E}} = \{\bar{c} : c \in \mathcal{E}\}$; then, $F \subseteq X$ is closed in X if and only if $F \cap \bar{c}$ is closed in the subspace topology on $\bar{c}$, for all $c \in \mathcal{E}$.

A finite cellular decomposition $\mathcal{E}$ of any space X is always coherent.

Example 11 The Hawaiian earring $HE = \cup_i C_i$ (Example **7**.1.8) partitions into a 0-cell $c^0 = \{O\}$ and circular 1-cells $c_i^1 = C_i - \{O\}$ with boundary $\dot{c}_i^1 = c^0$, for $i \geq 1$. This cellular decomposition is not coherent with the subspace topology on $HE \subset \mathbb{R}^2$; e.g., $HE \cap (0, 2]$ is closed in the weak topology but not in the subspace topology.

Let $(X, \mathcal{E})$ be a cell complex. For each cell $c \in \mathcal{E}$, let $\psi_c : \mathbb{D}^{|c|} \to X$ be a characteristic map. Define the continuous function

$$\Psi : \mathring{\cup} \{\mathbb{D}^{|c|} : c \in \mathcal{E}\} \to X$$

by $\Psi|_{\mathbb{D}^{|c|}} = \psi_c$ for all $c \in \mathcal{E}$. Then, by Theorem **5**.4.3, we have:

1.4 Coherence Theorem *A cellular decomposition $\mathcal{E}$ of X is coherent if and only if $\Psi : \mathring{\cup} \{\mathbb{D}^{|c|} : c \in \mathcal{E}\} \to X$ is an identification map.*

Here are two advantages that the weak topology offers.

1.5 Continuity Properties *Let X have the weak topology relative to a family $\{X_\alpha : \alpha \in \mathcal{A}\}$ of closed subspaces. Then:*
(a) $f: X \to Y$ is continuous $\Leftrightarrow$ $f|_{X_\alpha}$ is continuous for all $\alpha \in \mathcal{A}$.
(b) $H: X \times \mathbb{I} \to Y$ is continuous $\Leftrightarrow$ $H|_{X_\alpha \times \mathbb{I}}$ is continuous for all $\alpha \in \mathcal{A}$.

Proof: (a) This is Theorem **5.4.4**.
(b) Since X has the weak topology relative to the family of closed subsets $\{X_\alpha : \alpha \in \mathcal{A}\}$, then $X \times \mathbb{I}$ has the weak topology relative to the family of closed subsets $\{X_\alpha \times \mathbb{I} : \alpha \in \mathcal{A}\}$ (Exercise 4). Thus, (a) applies.   □

Coherence, with closure finiteness, has this characterization (Exercise 7):

1.6 Coherence Characterization *For a closure finite cellular decomposition $\mathcal{E}$ of X, coherence means that $F \subseteq X$ is closed in X if and only if $F \cap D$ is closed in D, for each finite subcomplex $(D, \mathcal{D})$ of $(X, \mathcal{E})$.*

CW COMPLEXES

The partition of a space X into its individual points taken as 0-cells is a closure finite cellular decomposition that is not coherent with the topology on X, unless X is discrete. The partition of $\mathbb{D}^n$ into the individual points of $\mathbb{S}^{n-1}$ as 0-cells and $\mathbb{B}^n$ as an n-cell is a cellular decomposition that is coherent with the usual topology on $\mathbb{D}^n$, but is not closure finite.

A **CW complex** $(X, \mathcal{E})$ consists of a Hausdorff space X, together with a coherent, closure finite cellular decomposition $\mathcal{E}$ of X. In this notation, C stands for the closure finite property; W for the weak topology.

This concept was formulated by J. H. C. Whitehead in the late 1940's as the generalization of simplicial complexes appropriate for homotopy theory.

Example 12   Any finite cell complex is a CW complex.

Example 13   Any *polygonal complex* (Section **12**.2) is a 2-dimensional CW complex with special attaching maps for the 2-cells. Any topological *graph* (Section **13**.3) is a 1-dimensional CW complex, and conversely.

Example 14   The integers and the intervening open intervals constitute a cellular decomposition $\mathcal{E}$ of $\mathbb{R}$ into 0-cells and 1-cells. $\mathcal{E}$ is closure finite and coherent with the usual topology on $\mathbb{R}$. This provides $\mathbb{R}$ with the structure of a CW complex.

Example 15   The union $\mathbb{S}^\infty$ of the nested equators, $\mathbb{S}^0 \subset \mathbb{S}^1 \subset \ldots \subset \mathbb{S}^n \subset \ldots$, given the weak topology, and the hemispherical cellular decomposition

$$\mathcal{E}^\infty = \{c_\pm^0, c_\pm^1, \ldots, c_\pm^n, \ldots\}$$

constitute an infinite CW complex $(\mathbb{S}^\infty, \mathcal{E}^\infty)$.

The second of the pathological examples discussed above shows that coherence is not inherited by subcomplexes of coherent cell complexes. But, in the presence of closure finiteness, coherence is inherited:

1.7 Subcomplex Theorem *If $(A, \mathcal{D})$ is a subcomplex of a CW complex $(X, \mathcal{E})$, then A is a closed subspace of X and $(A, \mathcal{D})$ is a CW complex.*

Proof: Since $\mathcal{E}$ is closure finite, so is the subfamily $\mathcal{D} \subseteq \mathcal{E}$. By the Coherence Characterization (1.6), coherence is equivalent, in the presence of closure finiteness, to the fact that closed sets are those sets with closed intersections with all finite subcomplexes.

For each finite subcomplex D of X, th intersection $A \cap D$ is a finite subcomplex of X; hence, $A \cap D$ is closed in X because it is the finite union of the closed cells that it contains. This proves that A is a closed subspace of X. Then any closed subset $F \subseteq A$ of the subspace A is closed in X; hence, $F \cap D$ is closed for all finite subcomplexes D of A (equivalently, of X), and conversely. This shows that $\mathcal{D}$ is coherent with the subspace topology on A. □

Example 16 Each skeleton of a CW complex is a CW complex.

The Subcomplex Theorem aids the following analysis of compact subspaces in a CW complex.

1.8 Compactness Theorem *Let $(X, \mathcal{E})$ be a CW complex. A closed subspace $A \subseteq X$ is compact if and only if A intersects only finitely many open cells $c \in \mathcal{E}$.*

Proof: ($\Rightarrow$) A subset $E \subseteq A$, consisting of a single point from each open cell $c \in \mathcal{E}$ with $A \cap c \neq \emptyset$, is a discrete subspace of X. Indeed, each subset $F \subseteq E$ is closed because $F \cap D$ is finite, hence, closed for each finite subcomplex $D \subseteq X$. But the discrete subspace $E \subseteq A$ is necessarily finite since A is compact. Thus, $A \cap c \neq \emptyset$ for only finitely many open cells $c \in \mathcal{E}$.

($\Leftarrow$) If $A \cap c \neq \emptyset$ for just $c = c_1, \ldots, c_k \in \mathcal{E}$, then A can be expressed as the finite union of the subspaces $A \cap \bar{c}_i$. The latter are compact subspaces as A is closed and $\bar{c}_i = \psi(\mathbb{D}^{|c_i|})$ is compact. Thus, A is compact. □

1.9 Corollary *In a CW complex, the cellular closure of a compact subspace is a finite subcomplex.*

1.10 Corollary *A CW complex $(X, \mathcal{E})$ is compact if and only if $\mathcal{E}$ is finite.*

1.11 Closure Finiteness Theorem *A coherent cellular decomposition $\mathcal{E}$ of a Hausdorff space X is closure finite if and only if, for every $n \geq 0$, $\mathcal{E}^n$ is a coherent cellular decomposition of the subspace X^n.*

Proof: ($\Rightarrow$) By the Subcomplex Theorem (1.7), each skeleton $(X^n, \mathcal{E}^n)$ of a CW complex $(X, \mathcal{E})$ is a CW complex. In particular, $\mathcal{E}^n$ is a coherent cellular decomposition of X^n.

($\Leftarrow$) Suppose that, for every $n \geq 0$, $\mathcal{E}^n$ is a coherent cellular decomposition of X^n. Then $\mathcal{E}^n$ is closure finite for every $n \geq 0$, by the following inductive argument. $\mathcal{E}^0$ is closure finite since 0-cells have empty boundary. If $\mathcal{E}^n$ is closure finite, then $(X^n, \mathcal{E}^n)$ is a CW complex. Because the boundary $\overset{\bullet}{c}{}^{n+1} = \psi(\mathbb{S}^n) \subseteq X^n$ of each $(n+1)$-cell c^{n+1} is compact, it meets at most finitely many open cells $c^k \in \mathcal{E}$ ($k \leq n$) by the Compactness Theorem (1.8) applied to the CW complex $(X^n, \mathcal{E}^n)$. This proves that $\mathcal{E}^{n+1}$ is closure finite.

This induction shows that $\mathcal{E}$ is closure finite. $\square$

CONSTRUCTION OF CW COMPLEXES

Given any map $\Phi : \mathring{\cup}_i \mathbb{S}_i^{n-1} \to A$ into a space A, there is an adjunction space X obtained from the disjoint union $A \mathring{\cup} (\mathring{\cup}_i \mathbb{D}_i^n)$ by identifying each point $x \in \mathbb{S}_i^{n-1} \subset \mathbb{D}_i^n$ with its image $\Phi(x) \in A$, for each index i. The resulting identification map is denoted by

$$\Psi : A \mathring{\cup} (\mathring{\cup}_i \mathbb{D}_i^n) \to A \cup_\Phi (\mathring{\cup}_i \mathbb{D}_i^n) = X.$$

Then each restriction,

$$\psi_i = \Psi|_{\mathbb{D}_i^n} : \mathbb{D}_i^n \to X,$$

is a characteristic map for the n-cell $c_i^n = \Psi(\mathbb{B}_i^n)$ of X, and $\Psi|_A : A \to X$ is an embedding of A as a closed subspace of X, by Theorem **5.4.5**. The identification topology on the space X is the same as the weak topology with respect to the closed subsets $\{A, \overline{c_i^n}\}$, by Theorem **5.4.3**.

The space X is called the ***adjunction space*** obtained by ***attaching n-cells*** $\{c_i^n\}$ via the ***attaching maps*** $\{\phi_i = \Phi|_{\mathbb{S}_i^{n-1}} : \mathbb{S}_i^{n-1} \to A\}$.

1.12 Construction Theorem *Let X be the union of closed subsets*

$$X_0 \subseteq X_1 \subseteq \ldots \subseteq X_{n-1} \subseteq X_n \subseteq \ldots .$$

Suppose that
(a) X_0 is a discrete subspace of X with points $\{c^0\}$;
(b) X_n is obtained from X_{n-1} by attaching n-cells $\{c^n\}$; and
(c) X has the weak topology with respect to the closed sets $\{X_n : n \geq 0\}$.
Then X is Hausdorff, $\mathcal{E} = \{c^n : n \geq 0\}$ is a cellular decomposition of X, and $(X, \mathcal{E})$ is a CW complex.

Proof: The Hausdorff property is left to Exercise 10. The characteristic maps $\{\psi_{c^n} : \mathbb{D}^n \to X_n\}$ provided by (b) serve as characteristic maps for the n-cells $\{c^n\}$ in X. Thus, the partition $\mathcal{E} = \{c^n : n \geq 0\}$ of X by the cells in (a) and (b) is a cellular decomposition of X. This partition $\mathcal{E}$ is coherent:

F is closed in X $\Leftrightarrow$ $\forall\, n \geq 0$, $F \cap X_n$ is closed in X_n

$\Leftrightarrow$ $\forall\, n \geq 0$, $F \cap X_n \cap X_{n-1}$ is closed in X_{n-1} and $\forall\, c^n$, $F \cap X_n \cap \bar{c}^n$ is closed in $\bar{c}^n$

$\Leftrightarrow$ $\forall\, n \geq 0$ and $\forall\, c^n$, $F \cap \bar{c}^n$ is closed in $\bar{c}^n$

$\Leftrightarrow$ F is closed in the topology coherent with $\mathcal{E}$.

By the same argument, $\mathcal{E}^k$ is a coherent cellular decomposition of $X^k = X_k$ for all $k \geq 0$. So, by the Coherence Theorem (1.4), $\mathcal{E}$ is also closure finite. □

The Construction Theorem has the following converse (Exercise 11).

1.13 Reconstruction Theorem *Let $(X, \mathcal{E})$ be a CW complex. Then X is the union of its skeletons $\{X^n : n \geq 0\}$ and they have these properties:*
(a) *X^0 is a discrete space;*
(b) *X^n is obtained from X^{n-1} by attaching n-cells $\{c^n \in \mathcal{E}\}$; and*
(c) *X has the weak topology with respect to the skeletons $\{X^n : n \geq 0\}$.*

More generally, a CW complex can be reconstructed beginning with any one of its subcomplexes, as follows:

1.14 Relative Reconstruction Theorem *Let $(A, \mathcal{D})$ be a subcomplex of the CW complex $(X, \mathcal{E})$. Then X is the union of the **relative skeletons***

$$A = A \cup X^{-1} \subseteq A \cup X^0 \subseteq A \cup X^1 \subseteq \ldots \subseteq A \cup X^{n-1} \subseteq A \cup X^n \subseteq \ldots,$$

which have these properties:
(a) *$A \cup X^0$ is a topological disjoint union of A and a discrete space;*
(b) *$A \cup X^n$ is obtained from $A \cup X^{n-1}$ by attaching n-cells $\{c^n \in \mathcal{E} - \mathcal{D}\}$; and*
(c) *X has the weak topology with respect to the skeletons $\{A \cup X^n : n \geq 0\}$.*

COFIBRATIONS AND CW COMPLEXES

In Section 10.4, a closed subspace inclusion $j : A \subseteq X$ is called a *cofibration* if j has the homotopy extension property for every space Y. This means that every partial homotopy $H : (A \times \mathbb{I}) \cup (X \times \{0\}) \to Y$ of every continuous function $h : X \times \{0\} \to Y$ extends to a homotopy $K : X \times \mathbb{I} \to Y$.

This is an important homotopy property. We shall show that the inclusion of any subcomplex into a CW complex is a cofibration.

1.15 Lemma *If X is obtained from A by attaching n-cells in a dimension $n \geq 0$, then the inclusion $j : A \subseteq X$ is a cofibration.*

Proof: When $n = 0$, the hypothesis means that X is the topological disjoint union of A and a discrete space $\{c^0\}$ whose points are called 0-cells. Then

$j : A \subseteq X$ has the homotopy extension property, because the product $X \times \mathbb{I}$ is the topological disjoint union of $A \times \mathbb{I}$ and copies, $c^0 \times \mathbb{I}$, of the interval.

When $n \geq 1$, the hypothesis means that there is a continuous function $\Phi : \mathaccent"7017{\cup}_i \mathbb{S}_i^{n-1} \to A$ and X is the adjunction space $A \cup_\Phi (\mathaccent"7017{\cup}_i \mathbb{D}_i^n)$ obtained from the disjoint union $A \mathaccent"7017{\cup} (\mathaccent"7017{\cup}_i \mathbb{D}_i^n)$ by identifying each point $x \in \mathbb{S}_i^{n-1} \subset \mathbb{D}_i^n$ with its image $\Phi(x) \in A$, for each index i. The inclusion $j : A \subseteq X = A \cup_\Phi (\mathaccent"7017{\cup}_i \mathbb{D}_i^n)$ is a cofibration by the Induced Cofibration Theorem (**10.4.3**), because the inclusion $\mathaccent"7017{\cup}_i \mathbb{S}_i^{n-1} \subset \mathaccent"7017{\cup}_i \mathbb{D}_i^n$ is a cofibration by Example **10.4.8**. □

1.16 Lemma *If X has the weak topology relative the closed subspaces*

$$A = X_{-1} \subseteq X_0 \subseteq X_1 \subseteq \ldots \subseteq X_{n-1} \subseteq X_n \subseteq \ldots \subseteq \cup_n X_n = X$$

and each inclusion $X_{n-1} \subseteq X_n$ is a cofibration, then the inclusion $j : A \subseteq X$ is a cofibration.

Proof: Since the inclusions are cofibrations, any partial homotopy,

$$H : (A \times \mathbb{I}) \cup (X \times \{0\}) \to Y,$$

of a continuous function $h : X \times \{0\} \to Y$ (i.e., $H|_{A \times \{0\}} = h|_{A \times \{0\}}$) extends inductively to partial homotopies,

$$K_n : (X_n \times \mathbb{I}) \cup (X \times \{0\}) \to Y, \qquad (n \geq 0)$$

of h. Since by induction $K_n|_{X_{n-1} \times \mathbb{I}} = K_{n-1}|_{X_{n-1} \times \mathbb{I}}$ for $n \geq 0$, the function $K : X \times \mathbb{I} \to Y$ given by $K|_{X_n \times \mathbb{I}} = K_n$ for $n \geq 0$ is well-defined. The function K is continuous by the Continuity Properties (1.5), because X has the weak topology with respect to the closed subspaces $\{X_n : n \geq -1\}$. Finally, K is a homotopy of h since K extends the partial homotopy H of h. □

1.17 Subcomplex Cofibration Theorem *If $(A, \mathcal{D})$ is a subcomplex of a CW complex $(X, \mathcal{E})$, then the inclusion $j : A \subset X$ is a cofibration.*

Proof: Apply Lemmas 1.15-16 and the Relative Reconstruction Theorem. □

1.18 Corollary *Any quotient map collapsing a contractible subcomplex of a CW complex is a homotopy equivalence.*

Proof: Apply 1.17 and the Cofibration Theorem (**10.5.5**). □

Example 17 The lattice $X = (\mathbb{R} \times \mathbb{Z}) \cup (\mathbb{Z} \times \mathbb{R})$ of all horizontal and vertical lines in $\mathbb{R}^2$ with integer intercepts is a 1-dimensional CW complex with 0-cells $\{m\} \times \{n\}$ and 1-cells $(m, m+1) \times \{n\}$ and $\{m\} \times (n, n+1)$, where $m, n \in \mathbb{Z}$. The subcomplex $A = (\mathbb{R} \times \mathbb{Z}) \cup (\{0\} \times \mathbb{R})$ of all horizontal lines and a single vertical line is contractible. Thus, $X \simeq X/A$. The latter space is a one-point union of denumerably many circles, corresponding to the vertical 1-cells $\{m\} \times (n, n+1)$ ($m \neq 0, n \in \mathbb{Z}$) in $X - A$.

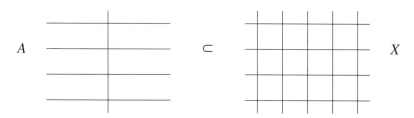

Example 17: $A \subset X$

2   Fundamental Group of a CW Complex

Calculations of the fundamental group of CW complexes are summarized in the following three theorems:

2.1   Theorem   *The fundamental group of a connected graph is a free group with free basis in bijective correspondence with its 1-cells that lie outside a maximal contractible subcomplex.*

2.2   Theorem   *The fundamental group of a connected 2-dimensional CW complex is the quotient of the free fundamental group of its 1-skeleton modulo a normal subgroup associated with the attaching loops of its 2-cells.*

2.3   Theorem   *The fundamental group of a connected CW complex is isomorphic to the fundamental group of its 2-skeleton.*

The proofs of these three theorems follow.

FUNDAMENTAL GROUP OF GRAPHS

The proof of Theorem 2.1 begins with the following lemma.

2.4   Lemma   *In any connected CW complex, there exists a contractible 1-dimensional subcomplex that contains the 0-skeleton.*

Proof: If v_0 is a selected 0-cell in a connected CW complex X, every other 0-cell v_k can be ***typed*** by the minimum number, k, of 1-cells e_i ($1 \leq i \leq k$) in a connected 1-dimensional subcomplex joining v_k to v_0. By this definition, we can select, for each 0-cell v_k of type $k \geq 1$, a specific 1-cell $e_k = \lambda(v_k)$ that joins v_k to a 0-cell v_{k-1} of type $k-1$. And then we can form the subcomplex T of all the 0-cells v_k of X, together with the selected 1-cells $e_k = \lambda(v_k)$. Then each 0-cell v_k of type k is contained in a unique minimal connected subcomplex $T(v_k)$ of T joining v_k to v_0. Namely, $T(v_k)$ is the arc,

$$T(v_k) = v_k \cup e_k \cup v_{k-1} \cup \ldots \cup v_i \cup e_i \cup v_{i-1} \cup \ldots \cup v_1 \cup e_1 \cup v_0,$$

of k distinct 1-cells e_i and $k+1$ distinct 0-cells v_i constructed inductively by forming $e_i = \lambda(v_i)$ and $v_{i-1} = \dot{e}_i - v_i$ for decreasing indices $1 \leq i \leq k$.

We define a contraction $H : T \times \mathbb{I} \to T$, as follows. First of all, we make $H|_{T \times \{0\}} = 1_T$ and $H(T \times \{1\}) = \{v_0\}$. Then we define H on $T^0 \times \mathbb{I}$ by sending the interval $\{v_k\} \times \mathbb{I}$ above any 0-cell v_k homeomorphically onto the arc $T(v_k)$, so that $H(v_k, 0) = v_k$ and $H(v_k, 1) = v_0$. At this point, H carries the four boundary edges of the square $\bar{e}_k \times \mathbb{I}$ above any 1-cell $e_k = \lambda(v_k) \in T$ into the subcomplex $T(v_k)$. Since $T(v_k)$ is contractible, being homeomorphic to an interval, H extends continuously over the square $\bar{e}_k \times \mathbb{I}$, by Exercise **10**.2.5. By the Continuity Property 1.5(*b*), these extensions define a continuous contraction $H : T \times \mathbb{I} \to T$. □

A contractible graph (1-dimensional CW complex) is called a *tree*. A *maximal tree* in the 1-skeleton of a CW complex is a tree that is contained in no larger tree in the 1-skeleton. A tree that contains all the 0-cells in a CW complex is maximal (Exercise 1). Thus, Lemma 2.4 shows that the 1-skeleton of every connected CW complex contains a maximal tree.

Proof of Theorem 2.1: Given a connected graph X, let T be a maximal tree in X. By the Cofibration Theorem (**10**.5.5), the quotient map $q : X \to X/T$, collapsing the contractible subcomplex T to a point, is a homotopy equivalence. The quotient space X/T is homeomorphic with a 1-point union $\vee \mathbb{S}^1$ of copies of the 1-sphere $\mathbb{S}^1$ in bijective correspondence with the 1-cells of X that lie outside T. Therefore, the fundamental group $\pi_1(X) \approx \pi_1(X/T)$ is a free group, by Corollary **12**.2.2. For a free basis in $\pi_1(X)$, we take path-homotopy classes of loops $f(e) : \mathbb{I} \to X$, each of which remains in T, save for a one-way excursion across one of the 1-cells $e \subset X - T$. More formally, for each 1-cell $e \in X - T$, say with boundary $\dot{e} = v_- \cup v_+$, we form the loop

$$f(e) = \bar{g}(v_-) \cdot \psi(e) \cdot g(v_+) : \mathbb{I} \to X,$$

where $\psi(e) : \mathbb{I} \to X$ is a characteristic map for e and $g(v_-), g(v_+) : \mathbb{I} \to X$ are the arcs in T joining the boundary 0-cells v_- and v_+ to the base 0-cell v_0. □

2.5 Corollary *A graph is a tree if and only if it is simply connected.*

Proof: A tree is contractible and so is simply connected. Conversely, a simply connected 1-dimensional CW complex has trivial fundamental group; hence, by Theorem 2.1, it coincides with any maximal tree that it contains. □

More generally, we have:

2.6 Corollary *Two connected graphs are homotopy equivalent if and only if they have isomorphic fundamental groups.*

Proof: A collapse of maximal trees replaces the connected graphs by homotopy equivalent one-point unions $\vee_\sigma \mathbb{S}^1_\sigma$ and $\vee_\tau \mathbb{S}^1_\tau$. Any fundamental group homomorphism $h : \pi_1(\vee_\sigma \mathbb{S}^1_\sigma) \to \pi_1(\vee_\tau \mathbb{S}^1_\tau)$ is induced by a based continuous function $f : (\vee_\sigma \mathbb{S}^1_\sigma, 1) \to (\vee_\tau \mathbb{S}^1_\tau, 1)$ (Exercise 2(*a*)); and any two such continuous functions inducing the same homomorphism on π_1 are homotopic relative the union point 1 (Exercise 2(*b*)). It follows that inverse group homomorphisms $h : \pi_1(\vee_\sigma \mathbb{S}^1_\sigma) \leftrightarrow \pi_1(\vee_\tau \mathbb{S}^1_\tau) : k$ are induced by homotopy inverses $f : \vee_\sigma \mathbb{S}^1_\sigma \leftrightarrow \vee_\tau \mathbb{S}^1_\tau : g$ (Exercise 2(*c*)); and conversely, by the homotopy invariance of π_1 (Theorem **10**.3.9). □

For a finite connected 1-dimensional CW complex X, Theorem 2.1 and its corollaries are expressible using this numerical count: the alternating sum

$$\chi(X) = \#\,(0\text{-cells } v) - \#\,(1\text{-cells } e)$$

is defined and is called the ***Euler characteristic*** of X.

Construct a maximal tree T in X as in the lemma for Theorem 2.1. Except for the base 0-cell v_0, the cells of T arise in pairs: each 0-cell v_k of type k pairs with a selected 1-cell $e_k = \lambda(v_k)$. Thus, $\chi(T) = \chi(v_0) = 1$. The collapse of T in X to a 0-cell v produces the homotopy equivalent complex X/T. On one hand, $\chi(X/T) = \chi(X)$ since T and its replacement v have $\chi(T) = 1 = \chi(v)$; on the other hand, $\chi(X/T) = 1 - \#\,(1\text{-cells in } X - T)$ by a direct count. Therefore, we have the calculation: $\#\,(1\text{-cells in } X - T) = 1 - \chi(X)$.

This calculation yields the following specializations of Theorem 2.1 and its two corollaries.

2.1 Theorem (*finite case*) *The fundamental group of a finite connected graph X is a free group with free rank $1 - \chi(X)$.*

2.5 Corollary (*finite case*) *A finite connected graph is a tree if and only if it has Euler characteristic $\chi(X) = 1$.*

2.6 Corollary (*finite case*) *Two finite connected graphs are homotopy equivalent if and only if they have the same Euler characteristic.*

HIGHER-DIMENSIONAL CW COMPLEXES

Theorem 2.2 describes the fundamental group of any connected 2-dimensional CW complex. This is only a slight generalization of the calculations made in Section **12**.2; its proof utilizes the collapse of a maximal tree and the Cell Sliding Theorem (**10**.5.7), as follows.

Proof of Theorem 2.2: When the connected 2-dimensional CW complex X has a single 0-cell, the skeletal pair (X^1, X^0) is homeomorphic to a 1-point union $(\vee_\alpha \mathbb{S}^1_\alpha, 1)$. When, in addition, the attaching maps of the 2-cells c^2_ω of

X are based continuous functions $\phi_\omega : (\mathbb{S}^1, 1) \to (\vee_\alpha \mathbb{S}^1_\alpha, 1)$, the complex X is an adjunction space $X = (\vee_\alpha \mathbb{S}^1_\alpha) \cup_\phi \{c^2_\omega\}$ to which Corollary **12**.2.4 applies directly. The fundamental group $\pi_1(X)$ is the quotient of the free group generated by path-homotopy classes of the exponential loops on the 1-skeleton summands $\mathbb{S}^1_\alpha$, modulo the normal closure of the path-homotopy classes of the attaching loops $\lambda_\omega = \phi_\omega \circ e : \mathbb{I} \to \vee_\alpha \mathbb{S}^1_\alpha$ of the 2-cells c^2_ω.

The proof is complete, because any connected 2-dimensional CW complex is homotopy equivalent to an adjunction space of the special form above. First, a collapse of a maximal tree in the 1-skeleton, as in the proof of Theorem 2.1, produces a homotopy equivalent 2-dimensional CW complex with a single 0-cell. The attaching map for each 2-cell is homotopic in the connected 1-skeleton to a continuous function based at the single 0-cell. Then an application of the Cell Sliding Theorem (**10**.5.7) produces a homotopy equivalent adjunction space to which Corollary **12**.2.4 applies. □

As for graphs and polygonal maps on surfaces, the ***Euler characteristic*** of a finite 2-dimensional CW complex X is defined as the alternating sum:

$$\chi(X) = \#(\text{0-cells}) - \#(\text{1-cells}) + \#(\text{2-cells}).$$

Like the situation for finite graphs (Corollary 2.6) and surfaces (Section **13**.3), the Euler characteristic is an invariant of the homotopy type of a finite 2-dimensional CW complex. This fact is too difficult to verify with the tools presented in this text. Unlike the case for finite graphs and surfaces, neither the fundamental group π_1, nor the Euler characteristic χ, nor a combination of them, completely determines the homotopy type of a finite, connected, 2-dimensional CW complex.

Example 1 The first counter-examples were constructed by W. Metzler in 1976. The polygonal complexes modeled on the presentations

$$\{\, a, b, c : a^5, b^5, c^5, [a, b], [a, c], [b, c] \,\}$$

and

$$\{\, a, b, c : a^5, b^5, c^5, [a, b^2], [a, c], [b, c] \,\}$$

of $\mathbb{Z}_5 \times \mathbb{Z}_5 \times \mathbb{Z}_5$ (as in Corollary **12**.2.4) are the simplest of his examples. They are homotopy inequivalent, finite, 2-dimensional CW complexes that have isomorphic fundamental groups and equal Euler characteristics.

Theorem 2.3 analyzes the fundamental group of any connected CW complex; its proof begins with this lemma.

2.7 Lemma *If X is obtained from A by attaching cells of dimension $n > 2$, the inclusion $i : A \subseteq X$ induces an isomorphism $i_\# : \pi_1(A) \to \pi_1(X)$.*

Proof: This is minor revision of the proof of Theorem **12**.2.3, with n-cells appearing instead of 2-cells. As before, the Seifert-Van Kampen Theorem

(12.1.3) implies that $\pi_1(X)$ is an amalgamated sum:

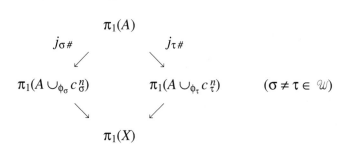

Because $n > 2$, $j_{\sigma\#} : \pi_1(A) \approx \pi_1(A_\sigma = A \cup_{\phi_\sigma} c_\sigma^n)$ for each $\sigma \in \mathcal{W}$. This results from an application of the Seifert-Van Kampen Theorem to the open subsets c_σ^n and $A_\sigma - d_\sigma^n$, with union A_σ and intersection $c_\sigma^n - d_\sigma^n$, where d_σ^n is the concentric closed n-ball in c_σ^n with radius $\frac{1}{2}$. Both $\pi_1(c_\sigma^n)$ and $\pi_1(c_\sigma^n - d_\sigma^n)$ are trivial, since $c_\sigma^n \simeq *$ and $c_\sigma^n - d_\sigma^n \simeq \mathbb{S}^{n-1}$ ($n > 2$). This application shows that $\pi_1(A_\sigma - d_\sigma^n) \to \pi_1(A_\sigma)$ is an isomorphism. Since A is a strong deformation retract of $A_\sigma - d_\sigma^n$, then $j_{\sigma\#} : \pi_1(A) \approx \pi_1(A_\sigma)$ also.

Taken together, the above facts imply that $j_\# : \pi_1(A) \approx \pi_1(X)$. $\square$

Proof of Theorem 2.3: By Lemma 2.7, the inclusions of the skeletons induce fundamental group isomorphisms:

$$\pi_1(X^2) \to \pi_1(X^3) \to \dots \to \pi_1(X^n) \to \pi_1(X^{n+1}) \to \dots .$$

These isomorphisms and the Compactness Theorem (1.8) imply that the homomorphism $i_\# : \pi_1(X^2) \to \pi_1(X)$ induced by the inclusion $i : X^2 \subseteq X$ is an isomorphism, as follows.

Any loop $f : \mathbb{I} \to X$ takes values in a finite subcomplex of X, hence, in some skeleton X^n. By the isomorphism $\pi_1(X^2) \approx \pi_1(X^n)$, the loop f is path-homotopic to one in X^2. So $i_\#$ is surjective.

By the Compactness Theorem (1.8), any null-homotopy $H : \mathbb{I} \times \mathbb{I} \to X$ in X of a loop $f : \mathbb{I} \to X^2$ in X^2 takes values in a finite subcomplex of X and so in some skeleton X^n. Thus, f is null-homotopic in X^n, hence, also in X^2, by the isomorphism $\pi_1(X^2) \approx \pi_1(X^n)$. So $i_\#$ is injective. $\square$

Example 2 For relatively prime integers p and q, let $\rho_{p,q} : \mathbb{D}^2 \to \mathbb{D}^2$ be the rotation through $2\pi(q/p)$ radians. In $\mathbb{S}^2 \subset \mathbb{D}^3$, identify each point $(x, y, -z)$ of the southern hemisphere with the point $(\rho_{p,q}(x, y), z)$ of the northern hemisphere. The resulting quotient space $L_{p,q}$ is called the **lens space** for the relatively prime integers p and q. We view $\mathbb{D}^3$ as a CW complex with p evenly spaced 0-cells and 1-cells on the equator $\mathbb{S}^1$, a southern and a northern hemispherical 2-cell, and single 3-cell $\mathbb{B}^3$. Then, $L_{p,q}$ becomes a CW complex with the pseudo-projective plane P_p as 2-skeleton and a single 3-cell. By Theorem 2.3, $L_{p,q}$ has cyclic fundamental group $\mathbb{Z}_p$ for all q.

Unlike the situation for surfaces, there is no homotopy classification of compact, connected, 3-dimensional manifolds by their fundamental groups, and their homotopy and topological classifications differ. Both of these facts are illustrated by the lens spaces introduced in Example 2.

J. H. C. Whitehead (1941) gave this *homotopy classification of the lens spaces*: $L_{p,q} \simeq L_{r,s}$ if and only if $p = r$ and $qs \equiv \pm m^2$ (*mod p*) for some m. A combinatorial characterization by K. Reidemeister (1935) and a triangulation result by E. Moise (1951) give this *topological classification for the lens spaces*: $L_{p,q} \cong L_{r,s}$ if and only if $p = r$ and $q \equiv \pm s^{\pm 1}$ (*mod p*).

Example 3 The lens spaces provide the simplest compact connected manifolds not classified by their fundamental group. For example, $L_{5,1}$ and $L_{5,2}$ have fundamental group $\mathbb{Z}_5$, yet $L_{5,1} \not\simeq L_{5,2}$. The lens spaces also provide the simplest compact connected 3-manifolds that distinguish between homotopy and homeomorphism type. For example, $L_{7,1} \simeq L_{7,2}$, yet $L_{7,1} \not\cong L_{7,2}$.

CONTRACTIBILITY OF CW COMPLEXES

By Corollary 2.5, a path-connected 1-dimensional CW complex is contractible if and only if it is simply connected. On the other hand, the 2-sphere $\mathbb{S}^2$ is simply connected, but it is not contractible (though this fact is not established in this text). What is required to assure the contractibility of higher dimensional CW complexes is higher order *connectedness*.

Path-connectedness of a space Y may be expressed as the property that every continuous function $\mathbb{S}^0 \to Y$ extends to a continuous function $\mathbb{D}^1 \to Y$. Similarly, the simple connectedness of a path-connected space Y may be expressed as the property that every continuous function $\mathbb{S}^1 \to Y$ extends to a continuous function $\mathbb{D}^2 \to Y$. This suggests the following definitions.

Let $0 \le N \le \infty$. A space Y is *N-connected* if, for $0 \le n < N+1$, every continuous function $\mathbb{S}^n \to Y$ extends to a continuous function $\mathbb{D}^{n+1} \to Y$.

If Y is a contractible space, then every continuous function into Y is inessential (Theorem **10**.3.3). And every inessential continuous function $\mathbb{S}^n \to Y$ extends to a continuous function $\mathbb{D}^{n+1} \to Y$ (Exercise **10**.2.5). Thus, a contractible space is ∞-connected.

One of the virtues of CW complexes in homotopy theory is that the converse is valid for them: *an ∞-connected CW complex is contractible*. This result and several related ones are proved next.

2.8 Extension Lemma *Let A be a subcomplex of the CW complex X; and let Y be any space. Suppose that, whenever $X - A$ has $(n+1)$-cells, each continuous function $\mathbb{S}^n \to Y$ has a continuous extension $\mathbb{D}^{n+1} \to Y$. Then each continuous function $A \to Y$ has a continuous extension $X \to Y$.*

Proof: Consider a sequence of continuous functions $f_n : A \cup X^n \to Y$ with $f = f_{-1} : A \to Y$ and $f_n = f_{n+1}|_{A \cup X^n}$ for all $-1 \le n$. By the Continuity Properties (1.5) and the Relative Reconstruction Theorem (1.14), a continuous extension $f_\infty : X \to Y$ of f is defined by $f_\infty|_{A \cup X^n} = f_n$ for all $-1 \leq n$.

We construct such a sequence $\{f_n : -1 \leq n\}$ inductively, as follows. We define the first extension $f_0 : A \cup X^0 \to Y$ arbitrarily on each 0-cell of X outside A. Then we suppose that an extension $f_n : A \cup X^n \to Y$ is defined for some $-1 \leq n$. The $(n+1)$st relative skeleton $A \cup X^{n+1}$ is the adjunction space

$$(A \cup X^n) \cup_\Phi (\mathring{\cup}_i \mathbb{D}_i^{n+1})$$

for some attaching map $\Phi : \mathring{\cup}_i \mathbb{S}_i^n \to A \cup X^n$. By hypothesis on Y,

$$f_n \circ \Phi : \mathring{\cup}_i \mathbb{S}_i^n \to A \cup X^n \to Y$$

has a continuous extension $F : \mathring{\cup}_i \mathbb{D}_i^{n+1} \to Y$. Then, by the Adjunction Theorem (**5.4.6**), f_n and F determine a continuous extension

$$<f_n, F> : A \cup X^{n+1} = (A \cup X^n) \cup_\Phi (\mathring{\cup}_i \mathbb{D}_i^{n+1}) \to Y$$

of $f_n : A \cup X^n \to Y$. This completes the proof. □

2.9 Contraction Theorem *Let $0 \leq N \leq \infty$. An N-dimensional CW complex is contractible if and only if it is N-connected.*

Proof: Let Y be an N-dimensional CW complex. Then $Y \times \mathbb{I}$ is a CW complex with cells $c^n \times \{0\}$, $c^n \times \{1\}$, and $c^n \times (0, 1)$, where the c^n are cells of Y of dimension $0 \leq n < N+1$. So $Y \times \mathbb{I}$ is obtained from its subcomplex $Y \times \{0, 1\}$ by attaching $(n+1)$-cells $c^n \times (0, 1)$ in dimensions $1 \leq n+1$, where $n < N+1$.

If Y is N-connected, the Extension Lemma (2.8) provides a continuous function $H : Y \times \mathbb{I} \to Y$ with $H|_{Y \times \{0\}} = 1_Y$ and $H|_{Y \times \{1\}} = c_y$, i.e. a contraction of Y. Thus, an N-connected, N-dimensional CW complex Y is contractible.

The converse was verified before the Extension Lemma. □

Example 4 *The infinite dimensional sphere $\mathbb{S}^\infty$ (Example 1.15) is ∞-connected, hence, contractible.* The Compactness Theorem (1.8) shows that every continuous function $\mathbb{S}^n \to \mathbb{S}^\infty$ takes values in some skeleton $\mathbb{S}^k$ and so is inessential in either hemisphere of the next skeleton $\mathbb{S}^{k+1}$, which has $\mathbb{S}^k$ as its equator.

Example 5 *The doubled harmonic comb DHC (Exercise **3.4.12**) is ∞-connected* (Exercise 7), *but is not contractible* (Example **10.3.7**). Therefore, by the Contraction Theorem (2.9), *DHC* is not homotopy equivalent to a CW complex.

HIGHER HOMOTOPY GROUPS

According to the Contraction Theorem (2.9), non-contractibility of a CW complex Y is captured by the essential continuous functions $\mathbb{S}^n \to Y$, or, more formally, by the sets of homotopy classes $[\mathbb{S}^n, 1; Y, y]$ for all dimensions $0 \leq n$. The first two of these sets are old acquaintances in disguise.

$[\mathbb{S}^0, 1; Y, y]$ *is equivalent to the set* $\pi_0(Y, y)$ *of path-components of Y*, with preferred one $P(y)$. The homotopy class $[f]$ of each continuous function $f: (\mathbb{S}^0, 1) \to (Y, y)$ corresponds to the path-component $P(f(-1))$.

$[\mathbb{S}^1, 1; Y, y]$ *is equivalent to the fundamental group* $\pi_1(Y, y)$ *of* Y. The homotopy class $[f]$ of each continuous function $f: (\mathbb{S}^1, 1) \to (Y, y)$ corresponds to the path-homotopy class of the associated loop $f \circ e : \mathbb{I} \to Y$ at y. The group operation in $\pi_1(Y, y)$ can be expressed in $[\mathbb{S}^1, 1; Y, y]$ as

$$[f] \cdot [g] = [<f, g> \circ \mu_1 : \mathbb{S}^1 \to \mathbb{S}^1 \vee \mathbb{S}^1 \to \mathbb{S}^1],$$

where $\mu_1 : \mathbb{S}^1 \to \mathbb{S}^1 \vee \mathbb{S}^1$ wraps the northern and southern hemisphere of $\mathbb{S}^1$ around the left- and right-hand summands of $\mathbb{S}^1 \vee \mathbb{S}^1$, resp. (Exercise **10**.2.9).

In general, $[\mathbb{S}^n, 1; Y, y]$ is denoted by $\pi_n(Y, y)$ for all dimensions $0 \le n$. When $1 \le n$, $\pi_n(Y, y)$ is a group under the operation

$$[f] \cdot [g] = [<f, g> \circ \mu_n : \mathbb{S}^n \to \mathbb{S}^n \vee \mathbb{S}^n \to \mathbb{S}^n],$$

where $\mu_n : \mathbb{S}^n \to \mathbb{S}^n \vee \mathbb{S}^n$ is a continuous function that wraps the two hemisphere of $\mathbb{S}^n$ around the two summands of $\mathbb{S}^n \vee \mathbb{S}^n$ (Exercise 9). Below, μ_n is the radial projection map of $\mathbb{S}^n$ onto the union of two half-scale copies s_+^n and s_-^n of $\mathbb{S}^n$ that are specially placed in the $(n+1)$-ball $\mathbb{D}^{n+1}$ that $\mathbb{S}^n$ bounds.

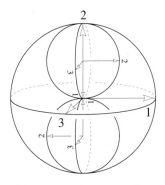
Placement of $s_+^n \vee s_-^n \subset \mathbb{D}^n$

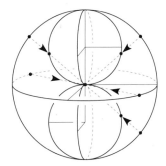
Radial projection $\mu_n : \mathbb{S}^n \to \mathbb{S}^n \vee \mathbb{S}^n$

These ***higher dimensional homotopy groups*** π_n, $2 \le n$, were introduced in a series of four papers, 1934-1936, by W. Hurewicz (1904-1956). They are studied in texts on algebraic topology.

We are content to use them to express the Contraction Theorem this way: *A CW complex is contractible if and only if all its homotopy groups vanish.*

A stronger result relating CW complexes and homotopy groups is this next one due to J. H. C. Whitehead. We neither use nor prove this result.

2.10 Equivalence Theorem *A continuous function* $f: X \to Y$ *between CW complexes is a homotopy equivalence if and only if the induced functions*

$$f_\# : \pi_n(X, x) \to \pi_n(Y, f(x))$$

are bijections for all $x \in X$ *and all* $0 \le n$.

ASPHERICAL CW COMPLEXES

With the Equivalence Theorem in mind, it is surprising that the fundamental group alone classifies all surfaces. Corollary 13.3.3 shows that two surfaces are homotopy equivalent (even homeomorphic) if and only if they have isomorphic fundamental groups. One explanation for this homotopy classification via π_1 is the asphericity of the surfaces with Euler characteristic $\chi \leq 0$.

A space Y is **aspherical** if it is path-connected and, for $2 \leq n$, each continuous function $\mathbb{S}^n \to Y$ has a continuous extension $\mathbb{D}^{n+1} \to Y$.

In other words, an aspherical space Y is a path-connected space whose homotopy groups $\pi_n(Y, y)$ in dimensions $2 \leq n$ are all trivial. It has only its fundamental group $\pi_1(Y, y)$ to express its non-contractibility.

The next result relates asphericity to higher order connectedness.

2.11 Theorem *Let Y have a simply connected covering space $\hat{Y}$. Then Y is aspherical if and only if $\hat{Y}$ is ∞-connected (contractible if a CW complex).*

Proof: When $n \geq 2$, any continuous function $F : \mathbb{S}^n \to Y$ ($2 \leq n$) lifts through the covering projection $p : \hat{Y} \to Y$ to a continuous function $\hat{F} : \mathbb{S}^n \to \hat{Y}$, by the Lifting Criterion (14.3.1). If $\hat{Y}$ is ∞-connected, the lifting $\hat{F}$ has an extension $\hat{F} : \mathbb{D}^{n+1} \to \hat{Y}$. Then $p \circ \hat{F} : \mathbb{D}^{n+1} \to \hat{Y} \to Y$ extends $F : \mathbb{S}^n \to Y$. This proves that Y is aspherical when $\hat{Y}$ is ∞-connected.

Conversely, every continuous function $\hat{F} : \mathbb{S}^n \to \hat{Y}$ ($2 \leq n$) projects to a continuous function $F = p \circ \hat{F} : \mathbb{S}^n \to Y$. If Y is aspherical, the projection $F = p \circ \hat{F}$ has an extension $F : \mathbb{D}^{n+1} \to Y$. By the Lifting Criterion (14.3.1), the extension has a lifting $\hat{F} : \mathbb{D}^{n+1} \to \hat{Y}$ that extends $\hat{F} : \mathbb{S}^n \to \hat{Y}$. This proves that the simply connected $\hat{Y}$ is ∞-connected when Y is aspherical. $\square$

Example 6 The 2-sphere $\mathbb{S}^2 = \mathfrak{H}_0$ is the double cover of the projective plane $\mathbb{P}^2 = \mathcal{C}_1$. So neither is aspherical, since $\mathbb{S}^2$ is not contractible.

Example 7 The torus $T = \mathfrak{H}_1$ and the Klein bottle $K = \mathcal{C}_2$ are aspherical. Being orbit spaces of properly discontinuous groups of isometries of the Euclidean plane (see Example 14.6.3), they have $\mathbb{R}^2$ as a (contractible) covering space.

Example 8 All the other surfaces, $\mathfrak{H}_g$ ($2 \leq g$) and $\mathcal{C}_g$ ($3 \leq g$), are aspherical. The Geometrization Theorem (14.6.2) and its extension to the non-orientable surfaces show that these surfaces are orbits spaces of properly discontinuous groups of isometries of the hyperbolic plane, and so have $\mathbb{B}^2$ as a (contractible) covering space.

2.12 Asphericity Theorem *When X and Y are aspherical CW complexes with fundamental groups $\Pi = \pi_1(X, x)$ and $\Xi = \pi_1(Y, y)$, the assignment*

$$\eta : [X, x; Y, y] \to \mathrm{Hom}\,(\Pi, \Xi), \quad \eta([f]) = f_\# : \pi_1(X, x) \to \pi_1(Y, y),$$

is a bijection.

Proof: We may assume that the aspherical CW complexes X and Y have 1-skeletons $X^1 = \vee_\sigma \mathbb{S}_\sigma^1$ and $Y^1 = \vee_\tau \mathbb{S}_\tau^1$, with union points x and y, respectively. For we can collapse maximal trees in X and Y to produce homotopy equivalent CW complexes with trivial 0-skeletons, as in Theorem 2.1.

Suppose now that continuous functions $f, g : (X, x) \to (Y, y)$ induce the same homomorphism $h : \Pi \to \Xi$. Then, in $\pi_1(Y, y) = [\mathbb{S}^1, 1; Y, y]$,

$$[f \circ i_\sigma] = f_\#([i_\sigma]) = h([i_\sigma]) = g_\#([i_\sigma]) = [g \circ i_\sigma]$$

for each inclusion $i_\sigma : \mathbb{S}_\sigma^1 \subset X^1$ into the 1-point union. This means there is a homotopy $f|_{\mathbb{S}_\sigma^1} \simeq g|_{\mathbb{S}_\sigma^1} : (\mathbb{S}_\sigma^1, 1) \to (Y, y)$ for each index σ. So the continuous function $H : X \times \{0, 1\} \to Y$, given by $H|_{X \times \{0\}} = f$ and $H|_{X \times \{1\}} = g$, extends by a homotopy $H : (X^1 \times \mathbb{I}, \{x\} \times \mathbb{I}) \to (Y, y)$ over the 1-skeleton. Now, $X \times \mathbb{I}$ is obtained from its subcomplex $(X \times \{0, 1\}) \cup (X^1 \times \mathbb{I})$ by attaching $(n+1)$-cells $c^n \times (0, 1)$ in dimensions $3 \leq n+1$. Since Y is aspherical, the Extension Lemma (2.8) implies that the continuous function

$$H : (X \times \{0, 1\}) \cup (X^1 \times \mathbb{I}) \to Y$$

extends to a homotopy $H : (X \times \mathbb{I}, \{x\} \times \mathbb{I}) \to (Y, y)$ from f to g. This proves that the assignment η is injective.

Now let $h : \Pi \to \Xi$ be any homomorphism. By Theorems 2.2 and 2.3, the inclusions of the skeletons give these epimorphisms and isomorphisms:

$$\begin{array}{ccccc} \pi_1(X^1, x) & \to & \pi_1(X^2, x) & \approx & \pi_1(X, x) = \Pi \\ \downarrow f_{1\#} = h_1 & & \downarrow f_{2\#} = h & & \downarrow f_\# = h \\ \pi_1(Y^1, y) & \to & \pi_1(Y^2, y) & \approx & \pi_1(Y, y) = \Xi \end{array}$$

Now h lifts through the epimorphisms to a homomorphism h_1 on the free fundamental groups of the 1-skeletons $X^1 = \vee_\sigma \mathbb{S}_\sigma^1$ and $Y^1 = \vee_\tau \mathbb{S}_\tau^1$. By Exercise 2(b), h_1 is induced by a continuous function $f_1 : (X^1, x) \to (Y^1, y)$.

The 2-skeleton X^2 is the adjunction space $X^1 \cup_\Phi (\mathring{\cup}_\omega D_\omega^2)$ for some attaching map $\Phi : \mathring{\cup}_\omega \mathbb{S}_\omega^1 \to X^1$. The composite function

$$f_1 \circ \Phi : \mathring{\cup}_\omega \mathbb{S}_\omega^1 \to X^1 \to Y^1 \subset Y^2$$

has a continuous extension $F : \mathring{\cup}_\omega D_\omega^2 \to Y^2$, as the homomorphism $f_{1\#} = h_1$ carries $[\Phi|_{\mathbb{S}_\omega^1}] \in Ker\,(\pi_1(X^1, x) \to \pi_1(X^2, x))$ into $Ker\,(\pi_1(Y^1, y) \to \pi_1(Y^2, y))$. By the Adjunction Theorem (5.4.6), there is the continuous extension

$$f_2 = <f_1, F> : X^2 = X^1 \cup_\Phi (\mathring{\cup}_\omega D_\omega^2) \to Y^2$$

of $f_1 : X^1 \to Y^1$. As Y is aspherical and X is obtained from its subcomplex X^2 by attaching $(n+1)$-cells c^{n+1} in dimensions $3 \leq n+1$, the Extension Lemma (2.8) provides an extension $f : X \to Y$ of f_2. Then $f_\# = h : \Pi \to \Xi$, since h_1 induces h on these quotient groups. This proves that η is surjective. □

The Asphericity Theorem shows that the homotopy theory of aspherical CW complexes is identical to the group theory of their fundamental groups.

2.13 Corollary *Aspherical CW complexes are homotopy equivalent if and only if they have isomorphic fundamental groups.*

Proof: By the Asphericity Theorem, inverse group homomorphisms of the fundamental groups induce homotopy inverse continuous functions between the aspherical CW complexes. □

You may wonder what groups arise as the fundamental group of an aspherical CW complex. Actually, when infinite dimensional CW complexes are allowed, there is no restriction on the group (Exercise 11). But otherwise, any bound on the dimension makes restrictions on the group.

A group Π has ***geometric dimension*** $\leq n$ if there is an aspherical n-dimensional CW complex X with $\pi_1(X) \approx \Pi$, and the ***geometric dimension*** of Π is the least such dimension n.

For example, no finite group has finite geometric dimension, though we can't prove this fact here. By a result of Papakyriakopolous in 1962, the complement of a knot in $\mathbb{S}^3$ is aspherical. So the knot complement spines in Section **12**.3 show that all knot groups have geometric dimension ≤ 2.

3 Covering Spaces of CW Complexes

A fundamental fact is that a CW complex on the base space of a covering projection lifts to a CW complex on the total space. We shall prove this, plus a little more involving the following terminology.

COVERING COMPLEXES

Let X be a Hausdorff space. If $\psi : \mathbb{D}^{|c|} \to X$ is a characteristic map for a cell c in X, the pair (c, ψ) is called an ***oriented cell*** in X. A complete family

$$\Psi = \{\psi_c : \mathbb{D}^{|c|} \to X : c \in \mathcal{C}\}$$

of characteristic maps for a CW complex $(X, \mathcal{C})$ is called an ***orientation***. Then the triple $(X, \mathcal{C}, \Psi)$ is called an ***oriented CW complex***.

Let $p : \hat{X} \to X$ be a covering projection. Each lifting through p of a characteristic map $\psi : \mathbb{D}^{|c|} \to X$ is a characteristic map $\hat{\psi} : \mathbb{D}^{|c|} \to \hat{X}$, because

$$\hat{\psi} : \mathbb{B}^{|c|} \leftrightarrow \hat{\psi}(\mathbb{B}^{|c|}) : \psi^{-1} \circ p$$

are inverse homeomorphisms and

$$\hat{\psi}(\mathbb{D}^{|c|} - \mathbb{B}^{|c|}) = \hat{\psi}(\mathbb{D}^{|c|}) - \hat{\psi}(\mathbb{B}^{|c|}).$$

So each lifting $\hat{\psi}$ gives an oriented cell $(\hat{c}, \hat{\psi})$ in $\hat{X}$ that *lies over* the oriented cell (c, ψ) in X. Since $\psi = p \circ \hat{\psi}$, then $c = p(\hat{c})$, $\bar{c} = p(\bar{\hat{c}})$, and $\dot{c} = p(\dot{\hat{c}})$. So by the uniqueness and existence of liftings (Sections **14**.2 and **14**.3), we have:

3.1 Lemma *If $p : \hat{X} \to X$ is a covering projection, then the oriented cells $(\hat{c}, \hat{\psi})$ in $\hat{X}$ that lie over an oriented cell (c, ψ) in X exhaust the inverse image $p^{-1}(c) \subset \hat{X}$ and they are precisely its path-components.*

3.2 Covering Complex Theorem *Let $p : \hat{X} \to X$ be a covering projection. An oriented CW complex $(X, \mathcal{E}, \Psi)$ on the base space X determines an oriented CW complex $(\hat{X}, \hat{\mathcal{E}}, \hat{\Psi})$ on the total space $\hat{X}$ whose oriented cells are preserved by p and by all covering automorphisms $\Phi \in A(p)$.*

Proof: Since $\mathcal{E}$ is a partition of X into cells, Lemma 3.1 implies that the cells $\hat{c}$ in $\hat{X}$ that lie over all the cells $c \in \mathcal{E}$ constitute a partition $\hat{\mathcal{E}}$ of $\hat{X}$ into cells. If the boundary of $\hat{c}$ meets $\hat{c}'$, then the boundary of c meets c'. Since $(X, \mathcal{E})$ is a cell complex, then $|\hat{c}'| = |c'| < |c| = |\hat{c}|$. Thus, $(\hat{X}, \hat{\mathcal{E}})$ is a cell complex.

We now establish the coherence of $\hat{\mathcal{E}}$. By the Coherence Theorem (1.4), the orientations Ψ and $\hat{\Psi}$ define continuous functions,

$$\Psi : \dot{\cup} \{\mathbb{D}^{|c|} : c \in \mathcal{E}\} \to X \quad \text{and} \quad \hat{\Psi} : \dot{\cup} \{\mathbb{D}^{|\hat{c}|} : \hat{c} \in \hat{\mathcal{E}}\} \to \hat{X},$$

that detect the coherence of $\mathcal{E}$ and $\hat{\mathcal{E}}$. The continuous function,

$$P : \dot{\cup}_{\hat{c}} \mathbb{D}^{|\hat{c}|} \to \dot{\cup}_c \mathbb{D}^{|c|},$$

sending $\mathbb{D}^{|\hat{c}|}$ identically onto $\mathbb{D}^{|c|}$ ($c = p(\hat{c})$) is open and gives $\Psi \circ P = p \circ \hat{\Psi}$.

Consider $W \subset \hat{X}$ such that $\hat{\Psi}^{-1}(W)$ is open in $\dot{\cup}_{\hat{c}} \mathbb{D}^{|\hat{c}|}$. For $\hat{x} \in W$, let $U(x)$ be an admissible neighborhood of $x = p(\hat{x})$, and let $\hat{U}(\hat{x})$ be the path component of $p^{-1}(U(x))$ containing $\hat{x}$. Then the pre-image

$$\hat{\Psi}^{-1}(W \cap \hat{U}(\hat{x})) = \hat{\Psi}^{-1}(W) \cap \hat{\Psi}^{-1}(\hat{U}(\hat{x}))$$

is open in $\dot{\cup}_{\hat{c}} \mathbb{D}^{|\hat{c}|}$, and its image under the open function P is the open subset

$$P(\hat{\Psi}^{-1}(W \cap \hat{U}(\hat{x}))) = \Psi^{-1}(p(W \cap \hat{U}(\hat{x})))$$

in $\dot{\cup}_c \mathbb{D}^{|c|}$. Because $\mathcal{E}$ is coherent on X, Ψ is an identification map by (1.4). So $p(W \cap \hat{U}(\hat{x}))$ is open in X. Because $p : \hat{U}(\hat{x}) \to U(x)$ is a homeomorphism between open subsets of $\hat{X}$ and X, $W \cap \hat{U}(\hat{x})$ is open in $\hat{X}$. We conclude that W contains a neighborhood of each of its points $\hat{x}$; hence, W is open in $\hat{X}$. This proves that $\hat{\Psi}$ is an identification map, so $\hat{\mathcal{E}}$ is coherent on $\hat{X}$ by (1.4).

In fact, $\hat{\mathcal{E}}$ is an extremely coherent cellular decomposition of $\hat{X}$ in that, for each $n \geq 0$, $\hat{\mathcal{E}}^n$ is coherent on the n-skeleton $\hat{X}^n$. Clearly, $\hat{\mathcal{E}}^0$ is coherent on the discrete 0-skeleton $\hat{X}^0$. For $k > 0$, the restriction $p : \hat{X}^n \to X^n$ is a covering projection by Theorem **14**.1.3. Therefore, $\hat{\mathcal{E}}^n$ is coherent on $\hat{X}^n$, by the previous analysis of $\hat{\mathcal{E}}$ on $\hat{X}$. It follows from the Closure Finiteness Theorem (1.11) that $\hat{\mathcal{E}}$ is closure finite. Thus, $(\hat{X}, \hat{\mathcal{E}})$ is a CW complex. $\square$

$(\hat{X}, \hat{\mathcal{C}}, \hat{\Psi})$ is called the *oriented covering complex* of $(X, \mathcal{C}, \Psi)$.

UNIVERSAL COVERING COMPLEXES

Any path-connected CW complex is locally path-connected and semilocally simply connected (Exercise 1). Therefore, it has a simply connected covering space, by Theorem **14.5.3**. By the Covering Complex Theorem (3.2), this simply connected covering space is a CW complex.

There is a much more direct way to construct a simply connected covering complex of any given path-connected CW complex. To develop guidelines for this construction, we investigate a typical simply connected covering complex.

Let $(\hat{X}, \hat{\mathcal{C}})$ be the simply connected covering complex of the CW complex $(X, \mathcal{C})$, with covering projection $p : \hat{X} \to X$. We select 0-cells $x \in X$ and $\hat{x} \in p^{-1}(x) \subseteq \hat{X}$, and we use the fundamental isomorphism (Theorem **14.4.1**),

$$\mu(\hat{x}) : \pi_1(X, x) \to p^{-1}(x) \leftarrow A(p),$$

to identify the fundamental group and the automorphism group, denoting both of them by Π.

If $(\hat{c}, \hat{\psi})$ is any selected oriented cell in $\hat{X}$ that lies over an oriented cell (c, ψ), then any other one over (c, ψ) is a transported version, $\pi(\hat{c}, \hat{\psi}) = (\pi(\hat{c}), \pi \circ \hat{\psi})$, of $(\hat{c}, \hat{\psi})$ by a non-identity automorphism $\pi \in \Pi \equiv A(p)$.

Any oriented 1-cell $(c_\alpha^1, \psi_\alpha)$ that is a loop at the basepoint x represents an element $\pi_\alpha = [\psi_\alpha] \in \Pi \equiv \pi_1(X, x)$. By the definition of the fundamental isomorphism $\mu(\hat{x})$, the lifting $\hat{\psi}_\alpha$ at $\hat{x}$ of the loop $\psi_\alpha : \mathbb{D}^1 \to X$ terminates at $\pi_\alpha(\hat{x})$. So the oriented 1-cell $(\hat{c}_\alpha^1, \hat{\psi}_\alpha)$ has initial 0-cell $\hat{x}$ and terminal 0-cell $\pi_\alpha(\hat{x})$. Any other oriented 1-cell $\pi(\hat{c}_\alpha^1, \hat{\psi}_\alpha) = (\pi(\hat{c}_\alpha^1), \pi \circ \hat{\psi}_\alpha)$ that lies over $(c_\alpha^1, \psi_\alpha)$ has initial 0-cell $\pi(\hat{x})$ and terminal 0-cell $\pi \pi_\alpha(\hat{x})$.

UNIVERSAL COVERING COMPLEX CONSTRUCTION

The preceding analysis provides sufficient guidelines for the construction of a simply connected covering CW complex of any given CW complex.

We first indicate the construction in case X has a single 0-cell x, which is taken as its basepoint for the fundamental group $\Pi \equiv \pi_1(X, x)$.

Let $\hat{X}^0 = \{\hat{x}_\pi : \pi \in \Pi\}$ denote a 0-dimensional CW complex whose 0-cells are in bijective correspondence with Π. We let Π act as a group of homeomorphisms of $\hat{X}^0$, permuting its 0-cells by $\pi'(\hat{x}_\pi) = \hat{x}_{\pi'\pi}$ for $\pi, \pi' \in \Pi$. If $\hat{x}_1$ is denoted by $\hat{x}$, then $\hat{x}_\pi$ is $\pi(\hat{x})$.

Each oriented 1-cell $(c_\alpha^1, \psi_\alpha)$ in X is a loop at the basepoint x, and so represents an element $\pi_\alpha = [\psi_\alpha] \in \Pi \equiv \pi_1(X, x)$. For each oriented 1-cell $(c_\alpha^1, \psi_\alpha)$ and for each $\pi \in \Pi$, we attach an oriented 1-cell $(\hat{c}_{\pi,\alpha}^1, \hat{\psi}_{\pi,\alpha})$, with initial 0-cell $\pi(\hat{x})$ and terminal 0-cell $\pi \pi_\alpha(\hat{x})$. The result is a graph $\hat{X}^1$. By this construction, Π extends to a group of homeomorphisms of $\hat{X}^1$, permuting its oriented 1-cells by $\pi'(\hat{c}_{\pi,\alpha}^1, \hat{\psi}_{\pi,\alpha}) = (\hat{c}_{\pi'\pi,\alpha}^1, \hat{\psi}_{\pi'\pi,\alpha})$ for $\pi, \pi' \in \Pi$. If

SECTION 3   COVERING SPACES OF CW COMPLEXES   457

$(\hat{c}^1_{1,\alpha}, \hat{\psi}_{1,\alpha})$ is denoted by $(\hat{c}^1_\alpha, \hat{\psi}_\alpha)$, then $(\hat{c}^1_{\pi,\alpha}, \hat{\psi}_{\pi,\alpha})$ is $\pi(\hat{c}^1_\alpha, \hat{\psi}_\alpha)$.

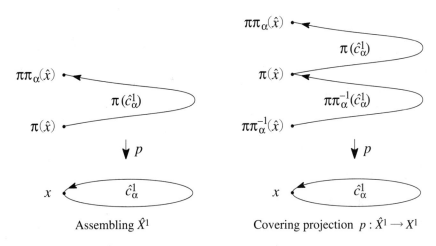

Assembling $\hat{X}^1$ Covering projection $p : \hat{X}^1 \to X^1$

$\hat{X}^1$ is path-connected because the fundamental group $\Pi \equiv \pi_1(X, x)$ is generated by the the path-homotopy classes $\pi_\alpha = [\psi_\alpha] \in \Pi$ of the loops in the one-point union $X^1 = \{x\} \cup \{c^1_\alpha\}$ (Exercise 2(a)).

The constant function $\hat{X}^0 \to x$ extends uniquely to a covering projection $p : \hat{X}^1 \to X^1$ by the projection of each oriented 1-cell $(\hat{c}^1_{\pi,\alpha}, \hat{\psi}_{\pi,\alpha})$ of $\hat{X}^1$ to the oriented 1-cell $(c^1_\alpha, \psi_\alpha)$ of X^1, i.e., $p \circ \hat{\psi}_{\pi,\alpha} = \psi_\alpha$ for all α and $\pi \in \Pi$. This is a covering projection since $\hat{X}^1$ has the correct local structure at each 0-cell $\pi(\hat{x})$: over each oriented 1-cell $(c^1_\alpha, \psi_\alpha)$, there is an oriented 1-cell $\pi(\hat{c}^1_\alpha, \hat{\psi}_\alpha)$ that initiates at $\pi(\hat{x})$ and there is an oriented 1-cell $\pi\pi_\alpha^{-1}(\hat{c}^1_\alpha, \hat{\psi}_\alpha)$ that terminates at $\pi\pi_\alpha^{-1}\pi_\alpha(\hat{x}) = \pi(\hat{x})$.

Moreover, Π is the automorphism group $A(p)$ of this covering projection. Since Π acts transitively on the fibers, $(\hat{X}^1, p)$ is a normal covering space.

3.3 Theorem *Any loop* $f : \mathbb{I} \to X^1$ *at x lifts through the covering projection* $p : \hat{X}^1 \to X^1$ *at each fiber point* $\pi(\hat{x}) \subset \hat{X}^0$ *to a path* $\hat{f} : \mathbb{I} \to \hat{X}^1$ *that terminates at* $\pi\pi'(\hat{x})$, *where* $\pi' = [f] \in \Pi = \pi_1(X, x)$.

Proof: By construction, the statement is true for each characteristic loop $\psi_\alpha : \mathbb{D}^1 \to X^1$ and its inverse $\psi_\alpha^{-1} : X^1 \to \mathbb{D}^1$. Any other loop f at x is path-homotopic to some *cellular loop*

$$\psi_{\alpha_1}^{\varepsilon_1} \cdot \ldots \cdot \psi_{\alpha_n}^{\varepsilon_n} : \mathbb{D}^1 \to X,$$

where $\varepsilon_i = \pm 1$ $(1 \le i \le n)$ and $\pi_{\alpha_1}^{\varepsilon_1} \ldots \pi_{\alpha_n}^{\varepsilon_n} = \pi'$. This cellular loop lifts at $\pi(\hat{x})$ to a cellular path passing successively through these 0-cells:

$$\pi(\hat{x}), \ \pi\pi_{\alpha_1}^{\varepsilon_1}(\hat{x}), \ \ldots, \ \pi\pi_{\alpha_1}^{\varepsilon_1} \ldots \pi_{\alpha_n}^{\varepsilon_n}(\hat{x}) = \pi\pi'(\hat{x}).$$

Then the Path Lifting Property (**14**.2.2) implies that the lifting of f at $\pi(\hat{x})$ also terminates at $\pi\pi'(\hat{x})$, and the proof is complete. $\square$

3.4 Corollary *Any loop $f : \mathbb{I} \to X^1$ at x lifts through the covering projection $p : \hat{X}^1 \to X^1$ to a loop $\hat{f} : \mathbb{I} \to \hat{X}^1$ at $\hat{x}$ if and only if it has path-homotopy class $[f] = 1 \in \Pi = \pi_1(X, x)$.*

When X has more than the single 0-cell x, the construction of the covering projection $p : \hat{X}^1 \to X^1$ proceeds similarly, using a maximal tree $T \subset X$ as a substitute for x. The 0-skeleton $\hat{X}^0 = \{\pi(\hat{c}^0) : \pi \in \Pi, c^0 \in X^0\}$ has cells in bijective correspondence with $\Pi \times X^0$. Each oriented 1-cell $(c_\alpha^1, \psi_\alpha)$ in X has an initial 0-cell $c_{-,\alpha}^0$ and a terminal 0-cell $c_{+,\alpha}^0$ in T. So there is a loop f_α based at x that involves the one-way excursion, ψ_α, across the 1-cell c_α^1 and otherwise remains in T; it represents an element $\pi_\alpha = [f_\alpha] \in \Pi$. For each oriented 1-cell $(c_\alpha^1, \psi_\alpha)$ and for each $\pi \in \Pi$, we attach an oriented 1-cell $\pi(\hat{c}_\alpha^1, \hat{\psi}_\alpha)$ to $\hat{X}^0$ with initial 0-cell $\pi(c_{-,\alpha}^0)$ and terminal 0-cell $\pi\pi_\alpha(c_{+,\alpha}^0)$. The result is a graph $\hat{X}^1$. In $\hat{X}^1$, those cells that are indexed by a specific $\pi \in \Pi$ contain a copy $\hat{T}_\pi$ of T. The reason is that the loop f_α associated with any oriented 1-cell $(c_\alpha^1, \psi_\alpha)$ in T remains in T, so it represents $\pi_\alpha = 1 \in \Pi$.

As before, the projection function $\hat{X}^0 \to X^0$, $p(\pi(\hat{c}^0)) = c^0$, extends uniquely to a covering projection $p : \hat{X}^1 \to X^1$ by the projection of each oriented 1-cell $\pi(\hat{c}_\alpha^1, \hat{\psi}_\alpha)$ of $\hat{X}^1$ to the oriented 1-cell $(c_\alpha^1, \psi_\alpha)$ of X^1. And, the automorphism group, $A(p) = \Pi$, acts as a group of homeomorphisms of $\hat{X}^1$, permuting its 0-cells and 1-cells. Finally, Theorem 3.3 and Corollary 3.4 remain valid in this general case.

3.5 Construction Theorem *Any path-connected oriented CW complex $(X, \mathcal{E}, \Psi)$ has a simply connected oriented CW covering complex $(\hat{X}, \hat{\mathcal{E}}, \hat{\Psi})$.*

Proof: The construction of $p : \hat{X} \to X$ proceeds inductively over the skeletons of X, beginning with the normal covering projection $p : \hat{X}^1 \to X^1$ in Theorem 3.3. Corollary 3.4 implies that the attaching map $\psi_\omega|_{\mathbb{S}^1} : \mathbb{S}^1 \to X^1$ for each 2-cell c_ω^2 in X lifts through $p : \hat{X}^1 \to X^1$ to continuous functions $\hat{\psi}_{\pi,\omega}|_{\mathbb{S}^1} : \mathbb{S}^1 \to \hat{X}^1$ (Exercise 3). We use these liftings, which are permuted by $\Pi = A(p)$, to attach oriented 2-cells $\pi(\hat{c}_\omega^2, \hat{\psi}_\omega)$ to $\hat{X}^1$, forming $\hat{X}^2$. The covering projection $p : \hat{X}^1 \to X^1$ extends to a covering projection $p : \hat{X}^2 \to X^2$ by the projection of each oriented 2-cell $\pi(\hat{c}_\omega^2, \hat{\psi}_\omega)$ of $\hat{X}^2$ to the oriented 2-cell $(c_\omega^2, \psi_\omega)$ of X^2 (Exercise 4(b)).

The space $\hat{X}^2$ is simply connected. For this, we consider the commutative diagram with vertical *monomorphisms* induced by the covering projections and the horizontal *epimorphisms* induced by the inclusions of the skeletons:

$$\begin{array}{ccc} \pi_1(\hat{X}^1, \hat{x}) & \xrightarrow{j_\#} & \pi_1(\hat{X}^2, \hat{x}) \\ p_\# \downarrow & & \downarrow p_\# \\ \pi_1(X^1, x) & \xrightarrow{i_\#} & \pi_1(X^2, x) \end{array}$$

By Corollary 3.4,

$$Im \ (p_\# : \pi_1(\hat{X}^1, \hat{x}) \to \pi_1(X^1, x)) = Ker \ (i_\# : \pi_1(X^1, x) \to \pi_1(X^2, x)).$$

Then a simple diagram chase shows that $\pi_1(\hat{X}^2, \hat{x})$ must be the trivial group. Thus, $\hat{X}^2$ is simply connected covering space of X^2.

The attaching maps of higher dimensional cells c^{n+1} ($n \geq 2$) have simply connected domain $\mathbb{S}^n$; so they lift to any covering space of the n-skeleton. Thus, the remainder of the inductive construction of the simply connected covering CW complex $\hat{X}$ proceeds easily (Exercises 4 and 5). □

Example 1 The minimal cellular decomposition of the *pseudo-projective plane* P_n of order $n > 0$ is $P_n = c^0 \cup c^1 \cup_n c^2$. The 2-cell is attached by an n-fold wrap around the circular 1-skeleton $P^1 = c^0 \cup c^1$. Thus, the fundamental group $\Pi = \pi_1(P_n, c^0)$ is the cyclic group $\mathbb{Z}_n = \langle a : a^n \rangle$.

The simply connected covering complex $\hat{P}_n$ consists of the cells:

$$a^k(\hat{c}^0) \qquad\qquad a^k(\hat{c}^1) \qquad\qquad a^k(\hat{c}^2). \qquad (a^k \in \mathbb{Z}_n)$$

As the oriented 1-cell (c^1, ψ^1) represents an infinite cyclic generator $a \in \pi_1(P^1, c^0)$, the oriented 1-cell $a^k(\hat{c}^1, \hat{\psi}^1)$ has initial 0-cell $a^k(\hat{c}^0)$ and terminal 0-cell $a(a^k(\hat{c}^0)) = a^{k+1}(\hat{c}^0)$. Thus, the 1-skeleton $\hat{P}^1$ is a circle complex, and $p : \hat{P}^1 \to P^1$ is the n-fold circular wrap. The generator a of the automorphism group $\Pi = A(p)$ is a rotation $a : \hat{P}^1 \to \hat{P}^1$ carrying $a^k(\hat{c}^i)$ to $a^{k+1}(\hat{c}^i)$ for all $i = 0, 1$ and $0 \leq k < n$.

The attaching map of the suitable oriented 2-cell (c^2, ψ^2) is a path product of n copies of the characteristic map ψ^1. So it lifts at $\hat{c}^0$ as a path product of the characteristic maps $a^k \circ \hat{\psi}^1$ of the circle of 1-cells $a^k(\hat{c}^1)$ $(a^k \in \mathbb{Z})$ that form $\hat{P}^1$. Each 2-cell $a^k(\hat{c}^2)$ $(a^k \in \mathbb{Z})$ is attached to $\hat{P}^1$ via this lifting, rotated by a^k. The n-fold wrap of $\hat{P}^1$ onto P^1 extends to a covering projection $p : \hat{P}_n \to P_n$ by the projection of each oriented 2-cell $a^k(\hat{c}^2, \hat{\psi}^2)$ onto the oriented 2-cell (c^2, ψ^2).

The generator $a \in \Pi = A(p)$ rotates the 1-skeleton and cyclically permutes the 2-cells up the stack in $\hat{P}_n$. All this is described in Section **14.4**.

Example 2 The *torus-with-membranes* is the 2-dimensional CW complex,

$$T = T_{m,n} = c^0 \cup c_a^1 \cup c_b^1 \cup c^2 \cup c_a^2 \cup c_b^2,$$

modeled on the presentation $\{a, b : aba^{-1}b^{-1}, a^m, b^n\}$. Its fundamental group $\Pi = \pi_1(T, c^0)$ is the finite abelian group

$$\mathbb{Z}_m \times \mathbb{Z}_n = \langle a, b : aba^{-1}b^{-1}, a^m, b^n \rangle.$$

We orient T so that its oriented 1-cells (c_a^1, ψ_a^1) and (c_b^1, ψ_b^1) represent a and b in the free group $\mathbb{F}(a, b) = \pi_1(T^1, c^0)$ and so that the attaching maps of its oriented 2-cells:

$$(c^2, \psi^2) \qquad\qquad (c_a^2, \psi_a^2) \qquad\qquad (c_b^2, \psi_b^2)$$

represent $aba^{-1}b^{-1}$, a^m, and b^n in $\mathbb{F}(a, b) = \pi_1(T^1, c^0)$, respectively.

The simply connected covering complex $\hat{T} = \hat{T}_{m,n}$ consists of cells

$$\hat{c}^0 \qquad \hat{c}_a^1 \qquad \hat{c}_b^1 \qquad \hat{c}^2 \qquad \hat{c}_a^2 \qquad \hat{c}_b^2$$

and their translates under $a^i b^k \in \Pi = \mathbb{Z}_m \times \mathbb{Z}_n$.

Each oriented 1-cell $a^i b^k(\hat{c}_a^1, \hat{\psi}_a^1)$ has initial 0-cell $a^i b^k(\hat{c}^0)$ and terminal 0-cell $a^{i+1} b^k(\hat{c}^0)$; each oriented 1-cell $a^i b^k(\hat{c}_b^1, \hat{\psi}_b^1)$ has initial 0-cell $a^i b^k(\hat{c}^0)$ and terminal 0-cell $a^i b^{k+1}(\hat{c}^0)$. The 1-skeleton $\hat{T}^1$ can be drawn as a lattice of n meridional circles and m longitudinal circles on a large torus, partitioning it into mn curved rectangular regions. And $p : \hat{T}^1 \to T^1$ is an m-fold circular wrap of each meridional circle onto $\mathbb{S}_a^1 = c^0 \cup c_a^1$ and an n-fold circular wrap of each longitudinal circle onto $\mathbb{S}_b^1 = c^0 \cup c_b^1$. The generators a and b of the automorphism group $\Pi = A(p)$ are meridional and longitudinal rotations $a, b : \hat{T}^1 \to \hat{T}^1$.

The attaching map of the oriented 2-cell (c^2, ψ^2) lifts at $\hat{c}^0$ to a traverse of the oriented 1-cells

$$(\hat{c}_a^1, \hat{\psi}_a^1) \quad \text{and} \quad a(\hat{c}_b^1, \hat{\psi}_b^1),$$

in the positive direction, and then the oriented 1-cells

$$aba^{-1}(\hat{c}_a^1, \hat{\psi}_a^1) \quad \text{and} \quad aba^{-1}b^{-1}(\hat{c}_b^1, \hat{\psi}_b^1) = (\hat{c}_b^1, \hat{\psi}_b^1),$$

in the negative direction. So the 2-cell $\hat{c}^2$ fills in one rectangle of the lattice $\hat{T}^1$, and its translates $a^i b^k (\hat{c}^2)$ complete the process, forming a large torus.

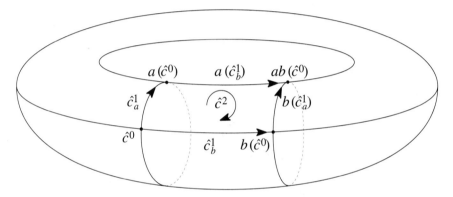

Simply connected covering complex $\hat{T} = \hat{T}_{m,n}$

The cells

$$\{a^i(\hat{c}^0), \ a^i(\hat{c}_a^1), \ a^i(\hat{c}_a^2) \ : a^i \in \mathbb{Z}_m\}$$

form a simply connected covering complex $\hat{P}_m$ of the pseudo-projective plane $P_m = c^0 \cup c_a^1 \cup_m c_a^2$ in T; each translate $b^k(\hat{P}_m)$ ($b^k \in \mathbb{Z}_n$) is a copy of $\hat{P}_m$ in $\hat{T}$.

The cells

$$\{b^k(\hat{c}^0), \ b^k(\hat{c}_b^1), \ b^k(\hat{c}_b^2) \ : b^k \in \mathbb{Z}_n\}$$

form a simply connected covering complex $\hat{P}_n$ of the pseudo-projective plane $P_n = c^0 \cup c_b^1 \cup_n c_b^2$ in T; and each translate $a^i(\hat{P}_n)$ ($a^i \in \mathbb{Z}_m$) is a copy of $\hat{P}_n$ in $\hat{T}$.

The wrap of $\hat{T}^1$ onto T^1 extends to a covering projection $p : \hat{T} \to T$ by the projection of each oriented 2-cell in $\hat{T}$ onto the corresponding oriented 2-cell in T. The generators $a, b \in \Pi = A(p)$ are meridional and longitudinal rotations $a, b : \hat{T} \to \hat{T}$.

APPLICATIONS

By the direct construction just presented, every CW complex X has a simply connected covering complex $(\hat{X}, p)$. Thus, the Classification Theorem **14.5.7** applies to covering complexes of every CW complex.

To the conjugacy class of a subgroup $\mathbb{H}$ of the fundamental group $\Pi = \pi_1(X, x)$ of a CW complex X, there corresponds a covering complex of X, and it arises as the orbit space of the simply connected covering complex $(\hat{X}, p)$ modulo a subgroup of the automorphism group $\Pi = A(p)$.

As applications of these covering complexes of CW complexes, we establish and illustrate some purely group-theoretical results. These applications are based on the facts that each group $\mathbb{G}$ can be realized as the fundamental group of some CW complex X, and that each subgroup $\mathbb{H} \leq \mathbb{G}$ can be realized by some covering complex of X.

3.6 Nielsen-Schreier Subgroup Theorem
Any subgroup of a free group is a free group.

Proof: Let $\mathbb{E}$ be a subgroup of a free group $\mathbb{F}$. There is a 1-dimensional CW complex X and covering complex $(\hat{X}_\mathbb{E}, p_\mathbb{E})$ such that the pair of groups $p_{\mathbb{E}\#}(\pi_1(\hat{X}_\mathbb{E})) \leq \pi_1(X)$ is isomorphic to the pair of groups $\mathbb{E} \leq \mathbb{F}$.

By the Covering Complex Theorem (3.2), $\hat{X}_\mathbb{E}$ is a 1-dimensional CW complex. By Theorem 2.1,

$$\mathbb{E} = p_{\mathbb{E}\#}(\pi_1(\hat{X}_\mathbb{E})) \approx \pi_1(\hat{X}_\mathbb{E})$$

is a free group. □

More can be said by analyzing the covering complex $\hat{X}_\mathbb{E}$.

Example 3 Any subgroup of a finitely generated *abelian* group is finitely generated. *But a subgroup of a finitely generated free group need not be finitely generated.* The covering projection $(e \times e)|_L : L \to \mathbb{S}^1 \vee \mathbb{S}^1$ (Example **14**.1.5) corresponds to the commutator subgroup $[\mathbb{F}, \mathbb{F}] \leq \mathbb{F}$ of the free group $\mathbb{F} = \mathbb{F}(a, b)$. The fundamental group $\pi_1(L) = [\mathbb{F}, \mathbb{F}]$ of the lattice L is the free group on denumerably many generators corresponding to the 1-cells outside a maximal tree in L (see Example 1.17).

Here is another peculiar feature of free groups not shared by abelian groups. The smaller the subgroup $\mathbb{E}$ of a finitely generated free group $\mathbb{F}$, the larger the index of $\mathbb{E}$ in $\mathbb{F}$ and the larger the free rank of $\mathbb{E}$:

3.7 Corollary
If $\mathbb{F}$ is a free group on m generators and $\mathbb{E}$ is a subgroup of $\mathbb{F}$ of index n, then $\mathbb{E}$ is a free group on $1 - m + nm$ generators.

Proof: Choose X and $\hat{X}_\mathbb{E}$ in the proof of the preceding theorem, so that X has a single 0-cell and m 1-cells. Since the index, $[\mathbb{F}, \mathbb{E}]$, equals n, then the covering projection $p : \hat{X}_\mathbb{E} \to X$ has n sheets. Therefore, $\hat{X}_\mathbb{E}$ has n 0-cells and nm 1-cells. By the finite case of Theorem 2.1, the fundamental group $\mathbb{E} = p_{\mathbb{E}\#}(\pi_1(\hat{X}_\mathbb{E})) \approx \pi_1(\hat{X}_\mathbb{E})$ is free with rank $1 - \chi(\hat{X}_\mathbb{E}) = 1 - m + nm$. □

Exercises

SECTION 1

1* Use the characteristic map for a cell in a Hausdorff space to prove Theorem 1.1.

2 (a) Construct a cell complex $(X, \mathcal{E})$ consisting of a single cell in each of the dimensions $0 < 1 < 2 < 3$ such that the cellular closure of each cell is the union of all the cells of lower dimension.
(b) Find a topological space that doesn't admit a cellular decomposition having an odd number of cells. Find a space that doesn't admit a cellular decomposition having an even number of cells.
(c) Find a minimal cellular decomposition of the Möbius strip.

3 (a) Show that a CW complex is locally path-connected, so that its components and path-components coincide.
(b) Show that a CW complex has the weak topology with respect to the family of its components.

4* Prove: If X has the weak topology relative to the closed subsets $\{A_\alpha : \alpha \in \mathcal{A}\}$, then $X \times \mathbb{I}$ has the weak topology relative to the closed subsets $\{A_\alpha \times \mathbb{I} : \alpha \in \mathcal{A}\}$.
(Hint: Use the Theorem 5.4.3 and Whitehead's Theorem (7.4.4).)

5 Let $(X, \mathcal{E})$ be a CW complex and let $(Y, \mathcal{D})$ be a closure finite cell complex. Prove that if $p : X \to Y$ is an identification map and if the cellular closure in $(Y, \mathcal{D})$ of the image $p(c)$ of each open cell $c \in \mathcal{E}$ is finite, then $(Y, \mathcal{D})$ is a CW complex.

A space is ***cellular*** if it has some closure finite, coherent, cellular decomposition.

6 Prove:
(a) Real space $\mathbb{R}^n$ is a cellular space.
(b) The Hawaiian earring (Example 7.1.8) is not a cellular space.
(c) A non-cellular space can be the union of disjoint cellular subspaces.
(d) The solid torus $\mathbb{S}^1 \times \mathbb{D}^2$ is a cellular space.

7* Establish the Coherence Characterization (1.6).

8 Let A be an open or closed subspace of a CW complex $(X, \mathcal{E})$ and let Y be an arbitrary topological space. Prove:
(a) $f : A \to Y$ is continuous $\Leftrightarrow f|_{A \cap \bar{c}}$ is continuous for all $c \in \mathcal{E}$.
(b) $H : A \times \mathbb{I} \to Y$ is continuous $\Leftrightarrow H|_{(A \cap \bar{c}) \times \mathbb{I}}$ is continuous for all cells $c \in \mathcal{E}$.

9 Prove Corollaries 1.9 and 1.10 that follow the Compactness Theorem (1.8).

10* Prove that the space X in the Construction Theorem (1.12) is Hausdorff.

11* Prove the Reconstruction and Relative Reconstruction Theorems (1.13) and (1.14).

A ***relative CW complex*** (X, A) is a pair, together with a sequence of subspaces:
$$A = (X, A)^{-1} \subseteq (X, A)^0 \subseteq \ldots \subseteq (X, A)^{n-1} \subseteq (X, A)^n \subseteq \ldots$$
with union X such that
(a) $(X, A)^n$ is obtained from $(X, A)^{n-1}$ by attaching n-cells $\{c^n\}$ $(n \geq 0)$.
(b) X has the weak topology with respect to the family $\{(X, A)^n : n \geq -1\}$.

SECTIONS 1-3
EXERCISES 463

12 Let C be Cantor's subspace of the unit interval $\mathbb{I}$ with its Euclidean topology. Determine whether or not the pair $(\mathbb{I}, C)$ is a relative CW complex.

13 Prove:
(a) If $(A, \mathcal{D})$ is a subcomplex of a CW complex $(X, \mathcal{E})$, then (X, A) is a relative CW complex.
(b) If (X, A) is a relative CW complex and $(A, \mathcal{D})$ is a CW complex, then $(X, \mathcal{E})$ is a CW complex whose cells are those in $(A, \mathcal{D})$ and (X, A).
(c) If (D, S) is a relative CW complex and $\phi: S \to A$ is any continuous function, then the adjunction space pair $(X = A \cup_\phi D, A)$ is a relative CW complex.
(d) If (X, A) is a relative CW complex, then X/A is a cellular space.

14 (a) Prove that if (X, A) is a relative CW complex, then $(X \times \mathbb{I}, A \times \mathbb{I})$ is a relative CW complex with n-cells of the form $c^n \times \{0\}$, $c^n \times \{1\}$, and $c^{n-1} \times (0, 1)$, where c^n and c^{n-1} are cells of (X, A).
(b) Prove that if X is a CW complex, then $(X \times \mathbb{I}, X \times \{0, 1\})$ is a relative CW complex with n-cells of the form $c^{n-1} \times (0, 1)$, where c^{n-1} are $(n-1)$-cells of X.

15 Prove:
(a) Any connected CW complex contains a contractible 1-dimensional subcomplex that contains the 0-skeleton.
(b) Any connected CW complex is homotopy equivalent to a CW complex that has a single 0-cell.

SECTION 2

1* Prove that a tree $T \subseteq X^1$ that contains all the 0-cells in a CW complex X is maximal. (Hint: Analyze the quotient homotopy equivalence $T' \to T''/T$ for any larger tree $T \subseteq T' \subseteq X^1$.)

2* Consider one-point unions $\vee_\sigma S^1_\sigma$ and $\vee_\tau S^1_\tau$ of 1-spheres, with union point 1. Prove:
(a) Any fundamental group homomorphism $h : \pi_1(\vee_\sigma S^1_\sigma) \to \pi_1(\vee_\tau S^1_\tau)$ is induced by a based continuous function $f : (\vee_\sigma S^1_\sigma, 1) \to (\vee_\tau S^1_\tau, 1)$.
(b) Any two based continuous functions $f, g : (\vee_\sigma S^1_\sigma, 1) \to (\vee_\tau S^1_\tau, 1)$ inducing the same homomorphism $h : \pi_1(\vee_\sigma S^1_\sigma) \to \pi_1(\vee_\tau S^1_\tau)$ are homotopic relative 1.
(c) Inverse group homomorphisms $h : \pi_1(\vee_\sigma S^1_\sigma) \leftrightarrow \pi_1(\vee_\tau S^1_\tau) : k$ are induced by homotopy inverses $f : \vee_\sigma S^1_\sigma \leftrightarrow \vee_\tau S^1_\tau : g$.

3 Prove that the fundamental group of the one-point union of CW complexes is the free product of their individual fundamental groups.

4 Establish a Seifert-Van Kampen theorem for CW complexes:
Let A and B be connected subcomplexes of a CW complex X such that $A \cap B$ and $A \cup B$ are connected. Then:
(a) Construct a maximal tree T for $A \cup B$ such that $A \cap T$ is a maximal tree for A, $B \cap T$ is a maximal tree for B, and $A \cap B \cap T$ is a maximal tree for $A \cap B$; and
(b) Show that the fundamental group $\pi_1(A \cup B)$ is the amalgamated sum of the inclusion induced bonding system $\pi_1(A) \leftarrow \pi_1(A \cup B) \to \pi_1(B)$.

5 Let A and B be connected subcomplexes of a CW complex X such that $A \cap B$ and $A \cup B$ are connected. Can $A \cap B$ and $A \cup B$ be simply connected while A or B is not?

6 (a) Determine the fundamental group of the quotient space obtained from the torus by identifying two of its points.
(b) Repeat the problem in (a) for the Klein bottle.

7* Prove that the doubled harmonic comb (Example 3.4.12) is ∞-connected.

8 Use Example 1.17 to determine a free basis for the commutator subgroup of the free group $\mathbb{F}(a, b)$.

9* Let $\mu_n : \mathbb{S}^n \to \mathbb{S}^n \vee \mathbb{S}^n$ be the radial projection map onto the specially placed one-point union $s_+^n \vee s_-^n \subset \mathbb{D}^{n+1}$ (see p. 451) and let $\lambda_n : \mathbb{S}^n \to \mathbb{S}^n$ be negation of the $(n+1)^{\text{st}}$ coordinate. Prove:
(a) There are the following homotopies relative $1 \in \mathbb{S}^n$:
$$\langle 1_{\mathbb{S}^n}, *\rangle \circ \mu_n \simeq 1_{\mathbb{S}^n} \simeq \langle *, 1_{\mathbb{S}^n}\rangle \circ \mu_n : \mathbb{S}^n \to \mathbb{S}^n \vee \mathbb{S}^n \to \mathbb{S}^n;$$
$$(1_{\mathbb{S}^n} \vee \mu_n) \circ \mu_n \simeq (\mu_n \vee 1_{\mathbb{S}^n}) \circ \mu_n : \mathbb{S}^n \to \mathbb{S}^n \vee \mathbb{S}^n \to \mathbb{S}^n \vee \mathbb{S}^n \vee \mathbb{S}^n,$$
$$\langle 1_{\mathbb{S}^n}, \lambda_n\rangle \circ \mu_n \simeq * \simeq \langle \mu_n, 1_{\mathbb{S}^n}\rangle \circ \mu_n : \mathbb{S}^n \to \mathbb{S}^n \vee \mathbb{S}^n \to \mathbb{S}^n.$$
(b) $\pi_n(Y, y) = [\mathbb{S}^n, 1; Y, y]$ is a group under the operation $[f] \cdot [g] = [<f, g> \circ \mu_n]$.
(c) When $n \geq 2$, $\mu_n \simeq \tau \circ \mu : \mathbb{S}^n \to \mathbb{S}^n \vee \mathbb{S}^n \to \mathbb{S}^n \vee \mathbb{S}^n$, where τ interchanges the summands of $\mathbb{S}^n \vee \mathbb{S}^n$. (Hint: When $n \geq 2$, rotation of $\mathbb{R}^{n+1}$ through π radians about the first axis followed by rotation through π radians about the second axis interchanges the summands of $s_+^n \vee s_-^n$ within $\mathbb{D}^{n+1}$.)
(d) When $n \geq 2$, the nth homotopy group $\pi_n(Y, y) = [\mathbb{S}^n, 1; Y, y]$ is abelian.

10 Prove that any homomorphism between the fundamental groups of 2-dimensional CW complexes is induced by some continuous function between them.

11* When X is obtained from A by attaching cells of dimensions $> n$, then the inclusion induced function $[\mathbb{S}^k, 1; A, x] \to [\mathbb{S}^k, 1; X, x]$ is a bijection for all $k < n$. Use this fact to attach cells of all dimensions > 2 to any 2-dimensional CW complex to produce an aspherical CW complex with the same fundamental group Π.

12 Construct aspherical CW complexes with fundamental groups $\mathbb{Z}$ and $\mathbb{Z}_2$.

13 Let (T, t) be a torus with basepoint and let (K, k) be a Klein bottle with basepoint. Determine the following homotopy sets:
(a) $[T, t; \mathbb{S}^1, 1]$,
(b) $[T, t \ T, t]$, and
(c) $[T, t; K, k]$.

SECTION 3

1* Prove that any path-connected CW complex is locally path-connected and semilocally simply connected.

The **Cayley graph** $\Gamma(\mathbb{G}, S)$ for a group $\mathbb{G}$ with respect to a subset $S \subseteq \mathbb{G}$ has vertex set $\mathbb{G}$ and edge set $\mathbb{G} \times S$, with the edge (g, s) joining vertices g and $s \cdot g$ for each $g \in \mathbb{G}$ and each $s \in S$.

2* (a) Prove that the Cayley graph $\Gamma(\mathbb{G}, S)$ is connected if and only if S is a set of generators of $\mathbb{G}$.
(b) Let X be a CW complex with a single 0-cell and fundamental group Π. Show

that the 1-skeleton $\hat{X}$ of the simply connected covering complex $\hat{X}$ of X is a Cayley graph $\Gamma(\Pi, S)$. What is the set of generators S?

3* Prove: Any continuous function $\psi : \mathbb{S}^1 \to X^1$ that is inessential in X^2 lifts through the covering projection $p : \hat{X}^1 \to X^1$ constructed for Theorem 3.5.

4* Let X be obtained from A by attaching n-cells ($n \geq 3$). Prove:
   (a) Any (normal) covering space $(\hat{A}, p)$ of A extends to a (normal) covering space $(\hat{X}, p)$ of X with this property: Each admissible neighborhood $U \subseteq A$ for $(\hat{A}, p)$ extends to an admissible neighborhood $V \subseteq X$ for $(\hat{X}, p)$.
   (b) The result in (a) holds when $n = 2$, provided that the attaching map for each 2-cell of X lifts into $\hat{A}$ at all fiber points.

5* Let (X, A) be a relative CW complex with 2-skeleton $(X, A)^2 = A$ (see Exercise 1.12). Prove that every (normal) covering space $(\hat{A}, p)$ of A extends to a (normal) covering space $(\hat{X}, p)$ of X such that $(\hat{X}, \hat{A})$ is a relative CW complex with 2-skeleton $(\hat{X}, \hat{A})^2 = \hat{A}$.

6 Draw the simply connected covering complex $\hat{T}_{m,n}$ of the torus-with-membranes $T_{m,n}$ in the case $m = 2$ and $n = 3$. Describe generators of its automorphism group.

7 Construct the simply connected covering complexes of minimal CW complexes on the torus, the projective plane, and the Klein bottle.

8 Construct the simply connected covering complex of the polygonal complexes modeled on these group presentations:
   (a) $\langle a, b : a^2, b^2, (ab)^4 \rangle$, for the dihedral group of order 8, and
   (b) $\langle a, b : a^2 = b^2, b^2 = (ab)^2 \rangle$, for the quaternionic group.

9 Construct the simply connected covering complex of the minimal CW complex on $\mathbb{S}^1 \vee \mathbb{S}^1$, the one-point union of two copies of the 1-sphere $\mathbb{S}^1$.

10 Construct the simply connected covering complex of the minimal CW complex on the lens space $L_{p,q}$.

11 Identify antipodal points in the infinite dimensional sphere $\mathbb{S}^\infty$ to produce the infinite dimensional *projective space* $\mathbb{P}^\infty$. Prove:
   (a) The identification map $p : \mathbb{S}^\infty \to \mathbb{P}^\infty$ is a covering projection.
   (b) $\mathbb{P}^\infty$ is an aspherical CW complex with fundamental group $\mathbb{Z}_2$.

12 Construct an aspherical CW complex with fundamental group $\mathbb{Z}_3$.

BIBLIOGRAPHY

BARNSLEY, M. F., *Fractals Everywhere*, Academic Press, New York, 1988.

BENDIXSON, I., "Sur les courbes définies par des équations différentielles," *Acta Math.* Vol. 24, 1-88, 1901.

BERGE, CLAUDE, *The Theory of Graphs*, John Wiley and Sons Inc., New York, 1962.

BOY, WERNER, *Math. Annalen*, Vol. 57, 1903, 173-184 (Doctoral Thesis, University of Göttingen, 1901).

BRAHANA, H. R., "Systems of circuits on two-dimensional manifolds," *Ann. Math.* Vol. 23, 144-168, 1922.

BROUWER, L. E. J., "Über die natürlichen Dimensionsbegriff," *Journal. f. Math.*, Vol. 142, 146-152, 1913.

BROUWER, L. E. J., "Über Abildung von Mannigfaltigkeiten," *Math. Ann.*, Vol. 71, 97-115, 1911.

CAYLEY, A., "Contour and Slope Lines," *Philosophical Magazine*, London, Ser. 4, Vol. XVIII (1859), 264-268; Collected Works, Vol. IV, 198-211.

CECH, E., "On bicompact spaces," *Ann. of Math.*, (2) 38, 823-844, 1937.

CHINN, W. G., AND STEENROD, N. E., *First Concepts of Topology*, Random House, Inc., New York, 1966.

DIEUDONNÉ, J., "Une généralization des espaces compacts," *J. Math. Pures Appl.*, Vol. 23, 65-76, 1944.

DOLD, A., "Partitions of unity in the theory of fibrations," *Annals of Mathematics*, Vol. 78, 223-255, 1963.

DOWKER, C. H., "An embedding theorem for paracompact metric spaces," *Duke Math. J.*, Vol. 14, 639-645, 1947.

DUGUNDJI, J., *Topology,* Allyn and Bacon, Boston, 1965.

HILBERT, D., AND COHN-VOSSEN, S., *Geometry and the Imagination*, Chelsea, New York, 1956.

HUEBSCH, W., "On the covering homotopy theorem," *Annals of Math.*, Vol. 71, 555-567, 1955.

HUREWICZ, W. AND WALLMAN, H., *Dimension Theory*, Princeton University Press,

Princeton, N. J., 1948.

HUREWICZ, W., "On the concept of fibre space," *Proc. Nat. Acad. Sci, U. S. A.*, Vol. 41, 956-961, 1955.

KELLEY, J. L., *General Topology*, Springer-Verlag, New York, 1975.

KLINE, M., *Mathematical Thought from Ancient to Modern Times*, Chapter 50, "The Beginnings of Topology," Oxford University Press, New York, 1972.

KURATOWSKI, G., "Sur le problème des courbes gauches en topologie," *Fund. Math.*, Vol. 15-16, 271, 1930.

MASSEY, W., *Algebraic Topology: An Introduction*, Harcourt, Brace and World, Inc., New York, 1967.

MAXWELL, J. CLERK, "Hills and Dales," *Philosophical Magazine*, ser. 4, Vol. XL, 421-427, 1870; Collected Works, Vol. II, 233-240.

MENDELSON, B., *Introduction to Topology*, Allyn and Bacon, Inc., Boston, 1975.

METZLER, W., "Über den Homotopietyp zweidimensionaler CW-Komplexe und Elementartransformationen bei Darstellungen von Gruppen durch Erzeugende und definierende Relationen," *J. Riene und Ang. Math.*, Vol. 285, 7-23, 1976.

MICHAEL, E., "A note on paracompact spaces," *Proc. Amer. Math. Soc.*, Vol. 4, 831-838, 1953.

MOISE, E., "Affine structures in 3-manifolds, V: The triangulation theorem and Hauptvermutung," *Ann. Math.*, Vol. 56, 96-144, 1952.

MUNKRES, J. R., *Topology: A First Course*, Prentice-Hall, Englewood Cliffs, NJ, 1975.

NAGATA, J., "On a necessary and sufficient condition of metrizability," *J. Inst. Polytech. Osaka City Univ.*, Vol. 1, 93-100, 1950.

PEANO, G., *Selected Works*, Ed. H. C. Kennedy, Toronto University Press, 1973.

PEANO, G., "Sur une courbe, qui remplit une aire plane," *Math. Annalen*, Vol. 36, 157-160, 1890.

POINCARÉ, H.: "Mémoire sur les courbes définies par une équation différentielle," Journal de Math. 7. 375-422 (1881); 8, 251-296 (1882); 1, 167-244 (1855); 2, 151-217 (1886); *Oeuvres de Henri Poincaré*, Vol.b 1, 3-84, 90-161, 167-221.

RADO, T.,"Über den Begriff der Riemannschen Fläsche." *Acta. Litt. Sci. Szeged.*, 2, 101-121, 1925.

RICHARDS, I., "On the classification of non-compact surfaces," *Trans. Amer. Math. Soc.*, Vol. 106, 259-269, 1963.

SEIFERT, H., AND THRELFALL, W., *Lehrbuch der Topologie*, Chelsea, New York, 1947.

SINGER, I. M., AND THORPE, J. A., *Lecture Notes on Elementary Topology and Geometry*, Scott, Foresman, Greenview, Il., 1967.

SMIRNOV, YU. M., "A necessary and sufficient condition for metrizability of a topological space," *Doklady Akad. Nauk S. S. S. R. N. S.*, 77, 197-200, 1951.

SPANIER, E. H., *Algebraic Topology,* McGraw Hill, New York, 1966.

STONE, A. H., "Paracompactness and product spaces," *Bull. Amer. Math. Soc.*, Vol. 54, 977-982, 1948.

STONE, M. H., "Applications of the theory of Boolean rings to general topology," *Trans. Amer. Math. Soc.*, Vol. 41, 375-481, 1937.

INDEX

A

α-tuple 158
Abel, N. 331
abelian
    group 331
    sum 348
abelianization of group 345
absolute retract 241
accumulation point 43, 51, 57, 140
Accumulation Theorem 89
Action Property 409
adjacent edge pair 381
adjunction space 165, 303, 359, 441
Adjunction Theorem 167
admissible partition 272, 354
admissible subset 400
affine function 83
Alexandroff's Theorem 217
alphabet 337
amalgamated
    sum 351
    sum diagram 352
    union 167, 350
amalgamating system 351
analytic 286
and 8
angular variation 286
annulus 5
antipodal-preserving 278
antisymmetric 30
apex of cone 309
arbitrarily near 62
arc 218
Arc Characterization 225
aspherical space 452
asphericity of surfaces 452
Asphericity Theorem 452

attached cell 321
attaching
    loop 359
    map 321, 359
attractor 98
automorphism
    of covering space 413
    group of covering space 413
Axiom of Choice 32

B

B-adic system 18, 20
    expansion 19, 31
Baire Category Theorem 72
Baire space 213
Baire, R. 72
Banach, S. 72, 73
Base Characterization 153
base 9 number system 105, 106
base space 312
base for a topology 153
basepoint 280
BDY set boundary 61, 127
Bendixson's Index Formula 292, 295
Bernstein 26
Bernstein-Schröder Theorem 26
bijective 21
binary
    number system 224
    operation 331
    relation 28
Bolzano, B. 25
Bolzano-Weierstrass
    property 90, 207
    Theorem 90
bonding system 351
Borsuk-Ulam Theorem 87, 279

468

boundary 127, 435
    label 377, 381
    word 362, 375
Boundary Label Algebra 382
bounded 79
    above 16, 46
    below 16, 46
    subset 69
box in $\mathbb{R}^n$ 91
box topology 174
Brahana, H. R. 384
Brouwer, L. E. J. 108, 277

C

C1-**C**3 closed set properties 118
Cantor's
    metropolis 194
    Power Theorem 28
    shuffle 27
    space 21, 54, 67, 147, 162, 177
    Teepee 193
    Theorem 199
    uniform metropolis 194
Cantor, G. 17, 25
category 296
    covering spaces 411
    groups 334
    homotopy 306
    k-spaces 298
Cauchy, A. (1789-1857) 47
Cauchy Characterization 69
Cauchy-Schwarz inequality 76
Cauchy sequence 47, 68
Cayley graph 464
Cech, E. 245
Cell Sliding Theorem 322
cell 359, 362
    complex 437
cellular
    closure 437
    decomposition 436
    space 462
cellularly closed 437
central projection of knot 368
chain 30
change of base-point isomorphism 281
Characteristic Feature
    topological disjoint union 157
    topological product 160
characteristic function 80
characteristic map 435
characterization
    CLS(A) 128
    INT(A) 127
Circle Characterization 226

Classification Theorems
    covering spaces 412, 424
    surfaces 384
    surfaces with boundary 385
Closed Function Theorem 95
closed
    ball in metric space 55
    function 136
    integral curve 290
    n-ball $D(x, r)$ 50
    n-cell 435
    set in metric space 59
    set in $\mathbb{R}^n$ 53
    sets in topological space 118
    tubular nbd $\mathbb{D}(K)$ 369
    unit n-ball $\mathbb{D}^n$ 51
closure 61, 127
closure finite cellular decomposition 438
Closure Finiteness Theorem 440
CLS set closure 61, 127
co-compact 227
co-countable topology 117
co-finite
    set 117
    topology 117
Co-Galois Correspondence 423
codomain of function 21
cofibration 315
Cofibration Theorem 320
Cohen, P. 32
Coherence Characterization 439
Coherence Theorem 438
coherent cellular decomposition 438
coherent edge pair 381
Coincidence Theorem 133
coinduced topology 169, 171
commutator
    group element 342
    subgroup 345
Compact Cubes Theorem 92, 200
Compact Image Theorem 95, 202
compact space 198, 207
compactification 216, 244
    Alexandroff 217
    Stone-Cech 245
compactly generated topology 215
Compactness Characterization 94
compactness of $\mathbb{I}$ 197
Compactness Theorem 440
compactum 219
complement 11
complete graph C_N 389
complete metric space 69
completely
    Hausdorff space 259
    normal space 124
    regular space 241
completion of a metric space 72, 83

components of a space 181
composition
    category 296
    functions 24
    homomorphisms 334
    telescope 26
Composition Theorem 68
concatenation 337
conclusion 9
Cone Contraction Theorem 309
cone over space 309
congruence modulo n 29
Conjugacy Property 408
conjugate group elements 282, 341
conjunction 8
connected
    metric space 85
    topological space 175
connected sum 395
connectivity relation 181
constant
    function 80
    path 188
Construction Theorem 441
    simply connected covering 458
Continuity Characterizations 65, 131
Continuity Properties 439
continuity 63
    at a point 81, 131
continuous
    family of paths 266
    function 132
    tangent vector field 398
    vector field 285
continuum 231
contour foliation 391
contractible space 307
contraction 73, 81
contraction (homotopy) 307
Contraction Mapping Principle 73, 88
Contraction Theorem 450
contractivity scalar 73
Contrapositive Rule 10
contravariant functor 300
convergence
    metric space 58
    real line 45
    topological space 121
Convergence Characterization 47
convex set 89, 302
coordinate chart 312
coordinates 16
corresponds 414
Coset Partition Lemma 340
cosets 340
countable complement topology 117
countable set 27
countably compact space 207

covariant functor 300
cover 29
covering 197, 254
    projection 399
    space 313, 399
Covering Complex Theorem 455
covers 274, 403
cross-cap surface 363, 377, 378
cube 91
cube bounding space 243
cubical spiral 114
curvilinear path 293
cut point 179, 194, 219
    order 194
cut set 179
CW complex 439
cyclic group 333
cylinder 5, 15

D

d-bounded set 69, 79
D-expansion 225
D-intervals 225
D1-**D**3 metric properties 49
De Morgan laws 9, 10, 12
decreasing sequence 46
deformable into subspace 317
deformation 317
    retract 317
    retraction 317
Deformation Theorem 318
degree 284
deleted Tychonoff plank 236
dense subset 62, 75, 129
denumerable set 27
derived set 44, 51, 57, 137
Detachment Rule 9, 10
diagonal or identity relation 28
diameter 69, 79, 209
Dieudonné, J. 251
dimension theory 108
direct
    limit 323
    product 299, 335
    sum 300, 348
    system 323
directed distance in $\mathbb{S}^1$ 272
directed set 120, 121
disconnected
    metric space 85
    topological space 175, 219
discrete group action 425
discrete
    metric δ 55
    topology 117

disjoint subsets 10
disjoint union of sets 156
disjoint union topology 157
disjunction 8
dispersion point 179
distance 55
    point to set 62, 67
distinguishes points 249
distinguishes points and closed sets 249
diverge 45, 58
divide-join modifications 380
doubled
    harmonic broom 194
    harmonic comb 113, 308
    Hawaiian earrings 202
    topologist's sine curve 60
DU1-DU2 disjoint union properties 156
dual
    category 297
    operations 37
dunce hat 166, 395

E

edges 362
    in graph 388
elementary replacements of words 337
elements 10
elliptic isometry of $\mathbb{H}^2$ 429
elliptic sector 292
embedded space 5
embedding 144
Embedding Theorem 243, 249
empty set 11
empty word 337
enumeration of a set 27
epimorphism 334
equal functions 21
equal sets 10
equivalence in category 297
equivalence class 29
Equivalence Class Properties 29
equivalence relation 29
Equivalence rule 7
Equivalence Theorem 451
equivalent
    compactifications 246
    knots 365
    polygonal maps 380
    sets 24
    words 337
essential function 301
essential length 273
Euclidean
    distance 42, 49
    isometry 426

n-space 50
norm 49
Euler characteristic
    CW complex 447
    graph 446
    polygonal maps Λ 385
    surface S 386
Euler's Formula 386
evaluation function 249
even
    covering 261
    decimal expansion 20
    Peano space 196
    ternary expansion 20
evenly covered subset 399
excluded point topology 117
Existence Theorem
    abelianizations 345
    cut points 220
    free groups 339
    products of groups 336
    sums of groups 339
existential quantifier 10
expanding earrings 201
exponential
    covering projection 400
    function 146, 274
extension of function 239, 314
Extension Lemma 449
Extreme Value Theorem 96

F

F-bundle 312
F_σ set 214
faces of a map 362
faces of a polyhedron 387
factor spaces 159
factors of product set 14
family of sets 13
fiber 274, 312, 400
fiber bundle 312
fibered product 314
fibration 312
Fibration Theorem 319
figure-of-eight knot 373
filterbase 121
finite
    cell complex 437
    complement topology 117
    intersection property 198
    intervals 11
    set 27
finite type cell complex 437
FIP Theorem 198
first category set 130, 214

first countable space 151
first infinite ordinal ω 35
first uncountable ordinal Ω 36
fixed point 73, 88, 276
Fixed Point Theorem 88, 277
flattening of sphere 279
foliation 390, 391
Fort topology 126
fractional winding number 391
Fréchet 122
free
    basis 339
    group 339
    product 300, 337
free-abelian
    basis 348
    group 346, 348
Free Point Lemma 278
full subcategory 298
function 21
    affine 83
    characteristic 80
    closed 133
    composite 24
    constant 80
    continuous 132
    distance 55
    equality 21
    evaluation 249
    exponential 146, 274
    fixed point free 276
    identity 22
    inclusion 22
    injective 22
    inverse 22
    one-to-one 22
    onto 22
    open 133
    product 22
    projection 22
    quotient 29, 149
    surjective 22
    Urysohn 38
functorial properties 272
fundamental
    group 280, 324
    groupoid 270
    set 268
Fundamental Theorem
    covering spaces 415
    finitely generated abelian groups 336

G

G_δ set 214
generators 343

geodesic 426, 428
geometric dimension of a group 454
Geometrization Theorem 430
glide reflections 427
Gluing Theorem 133
Gödel, K. 32
granny knot 373
graph of a function 39
graph 388
    complete 389
    planar 389
    topological 388
    utilities 389
greatest lower bound 17
group 331
    axioms 328
    homomorphism 333
    identity element 328
    presentation 343
groupoid homomorphism 271

H

hairy
    billiard ball 398
    doughnut 398
half-disc topology 137
Ham Sandwich Theorem 279, 293
handlebody surface of genus g 375
harmonic
    broom 180
    comb 113, 308
    filter 113
    hayrick 60
    maze 113
    rake 180
    target 186
Hausdorff 122, 123
    dimension 75, 84
    distance in hyperspace 97
    p-dimensional measure 84
    property 60
Hausdorff, F. 71
Hausdorff's metric completion 83
Hawaiian earrings 201, 230
Heine-Borel Theorem 93
Heine-Borel-Lebesgue Theorem 201
Hilbert space $\mathbb{H}$ 79, 83, 230
homeomorphic spaces 101, 134
homeomorphism 101, 134
    group 425
    types 135
Homeomorphism Theorem 104, 203
homomorphism 332
    epimorphism 334
    monomorphism 334

Homomorphism Theorem 342
homotopic functions 301
homotopy 301
    category 306
    classes 305
    equivalences 306
    equivalent spaces 306
    extension property 314
    groups 277, 451
    invariance of π_1 311
    inverse 306
    lifting property 312
    for pairs 324
    property 310
    relative subspace 304
    types 306
Homotopy Adjunction Theorem 304
Homotopy Lifting Property 406
Homotopy Transgression Theorem 303
Hopf's
    bundles 313
    map 168, 313
house with two rooms 327
Hurewicz, W. 313, 451
Hurewicz's Fibration Theorem 313
hyperbolic
    arc length 428
    isometry of $\mathbb{H}^2$ 429
    line 428
    plane $\mathbb{H}^2$ 428
    sector 292
hyperspace of compact subspaces 99
Hyperspace Contraction Theorem 98
hypothesis 9

I

ideal points of hyperbolic plane 428
identification
    map 146
    space 146
    topology 146
Identification Space Construction 145
identity
    function 22
    morphism 297
image 334
    function 23
    point 21
    subset 23
Image and Pre-image Properties 23
immediate successor 35
immersion 373
implication 9
implies 9
inadequacy of sequences 120

inclusion function 22
incoherent edge pair 381
increasing sequence 46
index
    foliation 391, 392
    singular point 287, 392
    subgroup 368
    vector field 286
Index Features 287
Index Formula 289
indiscrete topology 117
indiscriminant relation 28
induced
    fibration 314
    function 268, 271
    homomorphism 271, 281
Induced Cofibration Theorem 316
Induced Fibration Theorem 314
induced topology 168, 171
inequivalent knots 365
inessential 301
infimum of topologies 152
infinite intervals 11
infinite projective space 465
initial object in category 297
initial point of path 188
injections for sum 299, 336
injective function 22
integers modulo n 29, 331
integral curve 285
Integral Curve Index 290
interior 61, 127
interlocked incoherent pairs 381
intermediate covering spaces 420
Intermediate Value Theorem 87
intersection 11
Interval Theorem 86
Invariance Theorem 385
inverse
    functions 24
    limit 323
    morphisms 297
    system 323
inversion in Euclidean circle 428
Inversion Rule and Theorem 24
invertible morphism 297
isolated point 45, 52
    singularity 287
isometry of metric space 67
isometry of $\mathbb{H}^2$ 428
isomorphic
    covering spaces 411
    groups 334
    intermediate covering spaces 422
isomorphism
    covering spaces 411
    groups 334
isotopy 365

J

join 380
juxtaposition 337

K

K-Factory 216
k-space 215
kernel 334
Key Feature
    identification topology 148
    subspace topology 144
Klein bottle 363
    asphericity 452
knot complements 365
    asphericity 454
    spines 368
knot 365
    group 365
Knot Theorem 365
knotted 365
Koch's curve 115
Kolmogorov 122
Kuratowski, G. 390
Kuratowski's
    closure properties 139
    Theorem 390

L

$\mathscr{L}$ boundary label 377
Lagrange's Theorem 341
Laws of Set Theory 12, 13
least upper bound 16
leaves of foliation 391
Lebesgue Covering Lemma 210
Lebesgue measure 83
Lebesgue number 210
left coset 340
left homotopy inverse 325
lens space 448
letter in alphabet 337
lexicographic order relation 31
lifting 312, 403
Lifting Criterion 409
Lifting Lemma 275
limit 45, 58, 120, 121
limit ordinals 36
Limit-Accumulation Properties 48, 58
Lindelöf space 207

line with two origins 137
Lipschitz condition 74
local homeomorphism 430
Local Index Formula 292
locally
    compact 213
    connected 184
    finite 174, 253, 254
    path-connected 191
logic 8
logically equivalent propositions 9
long line 130
longitudinal loop 285, 301
loop in a space 280
lower bound 16
lower semicontinuous 81

M

M1-**M**4 boundary moves 382
manifold 374
    boundary 385
    interior 384
map 376
mapping
    cone 327
    cylinder 318
    torus 326
    wedge 327
max metric 57
maximal
    element 30
    tree 445
maximum 17
Maxwell, J. C. 295
members of a set 10
Menger, S. 108
meridional loop 285, 301
mesh of partition 272
metric 55
    completion 83
    Euclidean 49
    Hausdorff 97
    max 57
    radar screen 57
    taxicab 57
    space 55
    subspace 56
    topology 117
    uniform 55
metrizable
    space 117, 247
    topological product 247
Michael's Theorem 256
minimum 17
Minkowski's inequality 76

INDEX

Möbius strip 5, 6, 144
    non-orientability 378
mod out action 421
modifications of maps 380
Moise, E. 384, 449
monomorphism 334
Monotonic Limits Theorem 46
monotonicity conditions 46
Monotonicity Property 407
Moore-Smith convergence 120
morphism
    category 296
    covering spaces 411
Morphism Criterion 412
Morse, M. 394

N

n-cell 321, 435
N-connectedness 449
n-dimensional cell complex 437
n-dimensional manifold 191
n-dimensional torus 205
N-fold covering projection 400
n-point compactification 216
n-skeleton 436, 437
n-tuples 16
Nagata, J. 247
Nagata-Smirnov Metrization Thm 253
natural comb 112
Naturality Property 268
necessary condition 9
negation of proposition 8
negative group element 331
negative knot crossing 370
neighborhood 43, 50, 55, 118, 122
    base 151
    chart 374
Neighborhood Property 50
nested closed intervals 17
Nested
    Boxes Theorem 91
    Intervals Theorem 17
    Set Theorem 71, 83
net 121
non-accumulation point 43, 52, 57
non-cut point 179
non-degenerate basepoint 358
non-limit ordinal 36
non-orientable surface of genus g 379
non-orientable surfaces 378
Non-Planarity Theorem 389
non-singular point 391
Normal Form Lemma 338
normal $(T_4 + T_1)$ space 241
    closure in group 341

covering space 416
    intermediate covering space 421
    space $(T_4 + T_1)$ 120, 236
    subgroup 341
Normality Property 206
    metric spaces 60
normalizer 414
not 8
nowhere dense subset 63, 129
null-homotopic
    function 301
    loop 282
null-homotopy 301
numerically equivalent metrics 56

O

objects in a category 296
one-to-one function 22
one-point union 357
onto function 22
open
    ball in metric space 55
    covering 197, 254
    function 133
    hemispheres 272
    interval 42
    n-ball 50
    n-cell 435
    set in metric space 58
    set in $\mathbb{R}^n$ 53
    set in topological space 118
    tubular knot nbd $\mathbb{B}(K)$ 366
    unit n-ball $\mathbb{B}^n$ 51
or 9
orbit 425
    map 425
    space 421, 425
Orbit Projection Theorem 425
Order Completeness Property 16
order
    cut point 194
    group 331
    group element 333
    preserving function 31
    relations on $\mathbb{R}$ 16
    topology 120, 152
    type 31
ordered pair 14
ordinal 33, 34
    order relation 34
Ordinal Theorem 34
Orientability Theorem 378
orientable surface of genus g 379
orientable surfaces 378
orientation of cell complex 454

orientation preserving isometry 427
oriented
    cell 454
    covering complex 456
    CW complex 454

P

P1-**P**2 product set properties 158
pairwise disjoint sets 29
Pancake Theorem 88, 109
Papakyriakopolous 454
parabolic
    isometry of $\mathbb{H}^2$ 429
    sector in vector field 292
paracompact space 254
partial homotopy 314
partial order 30
partially ordered set 30
particular point topology 118
partition of a set 30
partition of unity 257
path-component 189
path-connected space 16
path-connectivity relation 189
path-homotopy 266
    classes 268
Path-Homotopy Classification 275
Path-Homotopy Invariance 273
path lifting 274
Path Lifting Property 404
path in space 186
Peano, G. (1858-1932) 105
Peano coordinate system 106
Peano's Curve 105, 107
perfect set 54, 76
phase portrait 286
Picard's Theorem 75
pin-cushion space 6
pits, peaks, passes 295, 398
planar graph 389
Platonic solids 387
Poincaré, H. 105
Poincaré
    Index Formula 289
    Index Theorem 393
    model for $\mathbb{H}^2$ 428
polygonal
    complex 362
    model 375
    surface 376
Polygonal Gluing Theorem 376
polyhedron 387
positive knot crossing 370
powers of group element 333
pre-image 23

precise refinement 254
presentation of group 344
prime ideal topology 154
prime space 193
product
    function 22
    path 188, 269
    set 10, 158
    topology 118, 158
product in category 298
product of groups 335
Product Metrization Theorem 248
projection functions 22, 158, 335
projective
    plane 169, 363, 377
    space 426, 465
proper 298
    embedding 365
    subset 10
properly discontinuous 425
propositions 8
pseudo-metric 80, 81
pseudo-projective plane 372, 416, 431
    covering complex 459
punctured torus 7

Q

q-pronged foliations 390
quotient
    function 29, 149
    map 149
    set 29
    space 149, 150

R

radar screen metric 57
rank of abelian group 336
Rational Density Property 16
rays 11
real
    line 16
    n-space 16, 49, 117
Reconstruction Theorem 442
reduced word 338
reducible space 193
refinement 253, 256
reflection
    Euclidean 427
    hyperbolic 428
reflexive relation 29
regular
    polyhedron 387

INDEX 477

regular (continued)
    position 369
    space 123, 233
    spherical map 387
regular point
    for foliation 391
    for vector field 286
Reidemeister, K. 449
Relation by Partition 30
relative
    CW complex 462
    homotopy 304
    neighborhood 119
    skeletons 442
    topology 118
Relative Reconstruction Theorem 442
relatively closed 119
relatively open 119
relatively simply connected nbd 418
relators 343
restriction 68, 239
retract 276, 317
retraction 276, 317
reverse path 188
right coset 340
right homotopy inverse 325
right interval topology 153, 209
rotations 427
Rules of Containment 10
Rules of Inference 9

S

σ-locally finite 251, 255
sausage-link space 6
saw space 104
scalar product 49
Schröder 26
Schröder-Bernstein Theorem 26, 183
seal modification of map 380
second category set 130, 214
second countable space 154
section homeomorphism 400
section of $\prod_\alpha X_\alpha$ 159
seed for attractor 100
Seifert-Van Kampen Theorem 353
semilocally simply connected space 418
separable space 130
separated 122, 175, 219
separation order 222
    topology 223
separation properties 122
separation of space 85, 175, 219
separation subintervals 223
sequence 22, 120
sequentially compact 91, 207

set 10
    complement 11
    intersection 11
    product 14
    union 11
Shrinkability Lemma 258
Sierpinski 117
Sierpinski's
    dust 102
    Theorem 231
    triangle 115
similarity of $\mathbb{R}^n$ 73
simple closed curve 232
simply connected space 266
singular point
    foliation 391
    vector field 286
skeleton 362
slice 14, 161
slice modification of map 380
slice-seal modifications 380
Smale, S. 394
Smirnov, Y. 249
smooth homeomorphism 391
space
    adjunction 165, 301, 359, 441
    Baire space 213
    base of covering projection 312
    Cantor's 54, 67, 147, 177
    Cantor's Teepee 193
    cellular 462
    compact 198
    complete metric 69
    completely normal 124
    completely regular 243
    connected 85, 175
    contractible 307
    covering 313, 399
    disconnected 85, 175
    discrete 55, 147
    Euclidean 50
    even Peano 196
    factor 159
    Hawaiian earrings 230
    Hilbert 79, 83, 230
    homeomorphic 101, 134
    homotopy equivalent 306
    hyperspace 97
    identification 146
    intermediate covering 420
    lens 448
    Lindelöf space 207
    line with two origins 137
    metric 54
    metrizable 117
    normal ($T_4 + T_1$) 123, 234
    orbit 425
    paracompact 254

space (continued)
    path-connected 186
    pin-cushion 6
    prime 193
    projective 465
    quotient 149
    real 16
    reducible 193
    regular ($T_3 + T_0$) 123, 233
    sausage-link 6
    saw 104
    second countable 154
    semilocally simply connected 418
    separable 130
    simply connected 266
    sphere 5, 50
    sphere with g crosscaps 395
    sphere with g handles 395
    tangent-disc 138
    tapered saw 104
    torus 5, 15, 362
    torus with membranes 433, 459
    total 312
    totally bounded metric 94
    totally disconnected 176
    universal covering 413
    wire eight and handcuff 6
sphere 5, 50
sphere at infinity 428
sphere with g crosscaps 395
sphere with g handles 395
spine of knot complement 368, 370
square knot 373
standard map
    Λ_g on $\mathfrak{D}_g$ 377
    Ψ_g on $\mathcal{E}_g$ 378
Standardization Theorem 383
star-like subset 89
stereographic projection 135
Stone-Čech compactification 245
Stone, A. H. 251
Stone, M. H. 245
Stone's Theorem 251
straight-line homotopy 302
strong deformation
    retract 317
    retraction 317
Subbase Theorem 152
subbase for a topology 152
subcategory 298
subcomplex 437
Subcomplex Cofibration Theorem 443
Subcomplex Theorem 440
subcovering 198
subgroup 332
subsequence 22
subset 10
subspace 118, 144

Subspace Construction 143
sufficient condition 9
sum in category 299
sum of groups 336
supernova foliation 392
support 80, 257
supremum of topologies 152
surface 374
    with boundary 384
Surface Catalog 379
surjective function 22
syllogism rule 9
symmetric relation 29

T

T1-**T**3 topology axioms 116
tail of a sequence 22, 45, 120, 121
tame knot 368
tangent-disc space 138
tangent plane 398
tapered saw space 104
tautology rule 9
taxicab metric 57
terminal object in a category 297
terminal point of path 186
Tietze, H. 122, 237
Tietze's Extension Theorem 239
topographic foliation 391
topological
    cube 206
    disjoint union 157
    graph 388
    group 432
    invariant 135
    pair 324
    product 118, 159
    property 103
    space 116
    subbase 152
    subspace 118
Topological Invariance 103
topologically equivalent
    metrics 56
    spaces 101, 117
topologist's sine curve 113
topology 116
    base 153
    box 174
    co-countable 117
    co-finite 117
    coinduced 169, 171
    compactly generated 215
    countable complement 117
    discrete 117
    disjoint union 157

topology (continued)
    excluded point 117
    finite complement 117
    Fort 126
    half-disc 137
    identification 146
    indiscrete 117
    induced 168, 171
    metric 117
    order 120, 152
    prime ideal 154
    product 118, 158
    quotient 149
    relative 118
    right interval 153, 209
    separation order 223
    subspace 118, 144
    tangent-disc 137
    weak 164
toroidal graph 397
torsion coefficients 336
torus 5, 15, 362
    asphericity 452
torus with membranes 433
    covering complex 459
total space 312
totally
    bounded metric space 94
    disconnected space 176
    ordered set 30
    path-disconnected 189
Transfinite Induction 32
Transgression Theorem 148
Transitive Action 420
transitive relation 29, 121
Transitivity Property
    identification spaces 148
    subspaces 144
Transitivity Theorem 420
translations 425
tree 396, 445
trefoil
    knot 5, 365
    shell 367
triangle
    equality 272
    inequality 49
triangulation 379, 396
trivial knot 365
Tychonoff's Lemma 236
Tychonoff's Theorem 205
typed 0-cell 444

U

uncountable set 27

undercrossing point 368
underpasses 368
uniform
    continuity 210
    metric 55
union 11
Uniqueness Theorems
    abelianizations 344
    free groups 339
    liftings 404
unity 257
universal
    covering complex construction 456
    covering space 413
    quantifier $\forall$ 10
universe 11
unknotted 365
upper bound 16, 30
upper semicontinuous 81
Urysohn
    function 238
    Metrization Theorem 250
    Theorem 238
Urysohn, P. 108, 237
utilities graph U_3 389

V

van der Waerden 338
vector sum 49
vertices 362, 388
volume of rectangle 83

W

weak topology 164
well-chained 227
well-ordered set 30
Well-Ordering Principle 32
Whitehead, J. H. C. 449
Whitehead's Theorem 212
winding number 283
    fractional winding number 391
wire eight and handcuff 6
Wirtinger presentations 370
word in alphabet 337

Z

zero 331
Zorn's Lemma 32

SYMBOLS AND NOTATION

Homotopy

$\pi_0(X)$	path-component set 189
$x^* : \mathbb{I} \to X$	constant path 188
$\bar{f} : \mathbb{I} \to X$	reverse path 188
$f \cdot g : \mathbb{I} \to X$	product path 188
$f \simeq g$	path-homotopy 266
$L(f)$	essential length 273
$w(f)$	winding number 283
$P(X)$	set of paths 267
$\pi(X)$	fundamental set 268
$\pi_1(X, x)$	fundamental group 280
$F_\# : \pi_1(X, x) \to \pi_1(Y, y)$ homo. 281	
$f \simeq g$	homotopy relation 301
$X \simeq Y$	homotopy equivalence 306
$[f]$	homotopy class 305
$[X, Y]$	homotopy set 305
$f \simeq g$ rel A	relative homotopy 304
$[X, A; Y, B]$	rel. homotopy set 305
π_n	homotopy groups 277, 471

Category Theory

$\mathcal{C}$	category 296
$(P, \{p_\alpha\})$	product in category 298
$(S, \{i_\alpha\})$	sum in category 299
$\mathcal{G}$	category of groups 334
$\mathcal{A}b$	category of abelian groups 348
$\mathcal{S}$	category of sets 297
$\mathcal{T}$	topological category 297
$\mathcal{T}_k$	category of k-spaces 300
$\mathfrak{H}$	homotopy category 306

Group Theory

$(\mathbb{G}, \cdot)$	group 324
$(\mathbb{Z}, +)$	group of integers 331
$(\mathbb{Z}_n, +_n)$	integers modulo n 331
$\mathbb{Z}_n^*$	invertible integers mod n 331
$\varphi : \mathbb{G} \to \mathbb{H}$	homomorphism 332
$Ker\ \varphi$	kernel subgroup 334
$Im\ \varphi$	image subgroup 334
$\mathbb{G} \approx \mathbb{H}$	group isomorphism 334
$K \leq G$	subgroup 332
$x \cdot K$	left coset 340
$K \cdot x$	right coset 340
$\mathbb{G}/\mathbb{K}$	set of left cosets 341
$[\mathbb{G} : \mathbb{K}]$	subgroup index 341
$\langle S \rangle$	subgroup generated by S 332
$\{X, R\}$	group presentation 343
$[\mathbb{G}, \mathbb{G}]$	commutator subgroup 345
$Ab(\mathbb{G})$	abelianization 345
$\mathbb{N}(S)$	normal closure 341
$\mathbb{N}_\mathbb{G}(\mathbb{H})$	normalizer of $\mathbb{H}$ in $\mathbb{G}$ 414